Alina Kabata-Pendias

Arun B. Mukherjee

Trace Elements from Soil to Human

Alina Kabata-Pendias
Arun B. Mukherjee

Trace Elements from Soil to Human

With 26 Figures and 209 Tables

Springer

Authors

Prof. Dr. hab. Alina Kabata-Pendias

IUNG-PIB
ul. Czartoryskich 8
24-100 Puławy
Poland
Phone: +48 81 886 34 21
Fax: +48 81 886 45 47

Dr. Arun B. Mukherjee

Environmental Sciences
Department of Biological and Environmental Sciences
P.O. Box 27
FI-00014 University of Helsinki
Finland
Phone: +358 9 191 58539
Fax: +358 9 191 58462
E-mail: arun.mukherjee@helsinki.fi

Library of Congress Control Number: 2007920909

ISBN-10 3-540-32713-4 **Springer Berlin Heidelberg New York**
ISBN-13 978-3-540-32713-4 **Springer Berlin Heidelberg New York**

Springer is a part of Springer Science+Business Media
springeronline.com

Cover design: WMXDesign, Heidelberg
Typesetting: Stasch · Bayreuth (stasch@stasch.com)
Production: Agata Oelschläger

Printed on acid-free paper 30/2132/AO – 5 4 3 2 1 0

Preface

The understanding of fundamental principles and phenomena that control the transfer of trace elements in soil-plant-human chain can contribute to the protection of the environment and human health. Many books, articles, and reports have already described both fundamental and/or detail problems related to these topics.

The intention of the authors is to provide up-to-date and selected interdisciplinary data for the concise presentation of existing knowledge on trace element transfer in the food chain, from soil to human. To accomplish this, the inclusion of appropriate data has been necessary. This book inevitably leaves publications of many investigators' uncited. The authors regret that such approach was necessary.

This volume is composed of two parts. Part I – *Biogeochemistry of the Human Environment* – presents fundamental information on biogeochemical properties of environmental compartments (soil, water, air, plants, humans) concerning trace elements. Part II – *Biogeochemistry of Trace Elements* – provides detailed data of the behavior and the occurrence of trace elements in the environment. There is a close relationship in the biogeochemical behavior between elements and their position in the Periodic System. Therefore, the format of this book follows the elemental sequences of the contemporary Periodic Table. The book provides data on the production usage, and on the occurrence of trace elements in soils, waters, air, plants, and humans (animals). Environmental stress, and biological functions of these elements are widely discussed.

The authors hope that information, presented in this book, will encourage young scientists to undertake further studies for better understanding of all factors that influence cycling of trace elements in a given ecosystem, and to develop the most effective methods for the effective remediation of contaminated sites. Finally, this knowledge will improve the assessment of health and ecological risk. The authors will be satisfied if this book fires the imagination of some readers and encourages them to study the biogeochemistry of trace elements in greater depth.

Alina Kabata-Pendias, Arun B. Mukherjee
January 2007

Acknowledgments

Sincere thanks of the authors should be given to a number of colleagues and to some organizations in Europe and the USA. It is impossible, however, to mention all of them. Nevertheless, some individuals will be recognized.

The authors express special thanks to Professor Albert Page and Professor Joseph Rule for their considerable contribution as reviewers of all the texts. Their contributions are of real importance, and their help and guidance is greatly appreciated.

We also wish to thank: Dr. Brian J. Alloway, Professor Anthony Dexter, Dr. Michala Harthill, Dr. Pekka Kauppi, Dr. Pekka Nuorteva, Dr. Grzegorz Siebielec, Professor Barbara Szteke, and Mr. Joseph W. Gorsuch for their encouragement and help with the editorial work.

Special acknowledgments are due to Dr. Wiesław Sadurski, who was willing to discuss and verify the correct chemical formulas. We received editorial support from Dr. Alina Bochniarz. Mr. Darek Wysocki, M.Sc., and Mr. Esa Tulisalo were always ready and available to assist with our computer work.

The easy access to references enabled us to use a versatile database with the assistance of the library staff at: The Central Agricultural Library (CBR) in Puławy, The Library of the Polish Geological Institute in Warsaw, and The Viikki Science Library of the Helsinki University in Helsinki.

The support and help of Ms. Agata Oelschläger and Mr. Armin Stasch from the Springer Publishing House is greatly appreciated.

Alina Kabata-Pendias, Arun B. Mukherjee
January 2007

Past to Future

Inscription on the Sybilla's Temple
erected 1801 in Puławy, Poland

Contents

Presentation of Data

Basic units of the International System (SI Unit) are used in this book. Mean contents refer to arithmetic mean value, unless otherwise specified.

The concentrations of a trace element in soil, plant, animal, and human samples are based on the total content by weight of the element in air-dried or oven-dried (at 40 °C) material. Otherwise it is indicated as follows:

AW – ash weight
FW – fresh weight or wet weight
BW – body weight

All data are given for topsoils, unless otherwise indicated. If not identified, a content of an element in environmental samples is given as, so called "total", i.e., measured in totally digested sample or measured directly in a sample.

Units

Mt – million metric tons (10^3 kt)
kt – thousand metric tons (10^3 t)
t – metric ton (10^3 kg)
kg – kilogram (10^3 g)
g – gram (10^{-3} kg)
mg – milligram (10^{-3} g)
µg – microgram (10^{-3} mg)
ng – nanogram (10^{-3} µg)
pg – picogram (10^{-3} ng)
fg – femtogram (10^{-3} pg)
pm – picometer (10^{-12} m); 1 Å = 100 pm
ha – hectare (10 000 m^2)
l – liter (1 dm^3)
m^3 – cubic meter (10^3 dm^3)
Bq – becquerel
mBq – millibecquerel
Ci – curie; 1 Ci = 3.7×10^{10} Bq
mCi – millicurie
nCi – nanocurie
pCi – picocurie

Symbols

°C – temperature in Celsius degrees
cmol(+) kg^{-1} – centimoles of positively charged ions (cations) in 1 kg of sample (formerly meq / 100 g; 1 cmol(+) kg^{-1} = 1 meq / 100 g)
Eh – redox potential (volts, millivolts)
pH – negative logarithm, base 10, of hydrogen ion concentration (activity)
M – molar concentration of solution of an element (compound)
$t_{1/2}$ – half life-time of an radionuclide
K_d – dissociation constant: product of the activities of cations and anions divided by the activity of the unionized electrolyte

The Periodic Table of Elements

1	2	3	4	5	6	7	8	9	10	11	12	13	14	15	16	17	18
1 H																	2 He
3 Li	4 Be											5 B	6 C	7 N	8 O	9 F	10 Ne
11 Na	12 Mg											13 Al	14 Si	15 P	16 S	17 Cl	18 Ar
19 K	20 Ca	21 Sc	22 Ti	23 V	24 Cr	25 Mn	26 Fe	27 Co	28 Ni	29 Cu	30 Zn	31 Ga	32 Ge	33 As	34 Se	35 Br	36 Kr
37 Rb	38 Sr	39 Y	40 Zr	41 Nb	42 Mo	43 Tc	44 Ru	45 Rh	46 Pd	47 Ag	48 Cd	49 In	50 Sn	51 Sb	52 Te	53 I	54 Xe
55 Cs	56 Ba	57 * La	72 Hf	73 Ta	74 W	75 Re	76 Os	77 Ir	78 Pt	79 Au	80 Hg	81 Tl	82 Pb	83 Bi	84 Po	85 At	86 Rn
87 Fr	88 Ra	89 ** Ac	104 Rf	105 Db	106 Sg	107 Bh	108 Hs	109 Mt									

* Lanthanide series	58 Ce	59 Pr	60 Nd	61 Pm	62 Sm	63 Eu	64 Gd	65 Tb	66 Dy	67 Ho	68 Er	69 Tm	70 Yb	71 Lu
** Actinide series	90 Th	91 Pa	92 U	93 Np	94 Pu	95 Am	96 Cm	97 Bk	98 Cf	99 Es	100 Fm	101 Md	102 No	103 Lr

The symbols given in thin letters indicate the elements not occurring naturally in the environment.

Glossary (Abbreviations, Acronyms)

AAAc	ammonium acetic acid
AAAcEDTA	mixture of ammonium acetic acid with Na_2EDTA
AAP	acid available particulate
AD	Alzheimer disease
ADI	acceptable daily intake
AMAP	Arctic Monitoring and Assessment Programme
A/N	ratio of anthropogenic to natural sources
AROMIS	Assessment and Reduction of Heavy Metal Inputs into Agro-Ecosystems
ASB	alkaline-stabilize biosolids
ASM	artisanal and small-scale gold mining
ATSDR	Agency for Toxic Substances and Disease Registry
AHM	Asian herbal medicines
BAC	biological absorption coefficient
BAT	biological tolerance values at workplace
BC	Before Christ
BCF	biological concentration factor
BIM	biologically induced minerals
B-x	blood level of a given element
Cc	continental crust concentration of element
CEC	cation exchangeable capacity
CLPP	community level physiological profile
Clarke	Clarke's data for mean values for chemical elements in a given geological material
COPR	chromite-ore processing residue
Cw	dissolved concentration of element in water
Cw/Cc	ratio of element concentration in water to its content in the continental crust
DBT	dibutyltin
DDI	daily dietary intake
DM-X	dimethylated element
DMT	dimethyltin
DNA	de(s)oxyribonucleic acid, carrier of genetic information

DOC	dissolved organic carbon
DOM	dissolved organic matter
DTPA	diethylenetriaminepentaacetic acid
DU	depleted uranium
EC	European Commission
EDTA	ethylenediaminetetraacetic acid
EF_c	enrichment factor-ratio of element concentration in air to its content in the Earth's crust, normalized to the reference element (Al)
EPA	Environmental Protection Agency
ESADDI	estimated safe and adequate daily dietary intake
FA	fly ash
FAO	Food and Agriculture Organization of the United Nation
FDA	Food and Drug Administration
GIS	geographical information system
GEMS	Global Environment Monitoring System
GERM	Geochemical Environmental Reference Methods
HA	humic acids
HELCOM	Helsinki Commission
HI	hazardous index
HDL	high density lipoproteins
HRE	heavy rare earth
IHC	interactive health communication
IAEA	International Atomic Energy Agency
IARC	International Agency for Research on Cancer
IBA	index of bioaccumulation
ICRP	International Commission on Radiological Protection
IDD	Iodine Deficiency Disorders
IHC	Interactive Health Communication,
IHR	International Health Regulation
IPCS	International Program on Chemical Safety
IUPAC	International Union of Pure and Applied Chemistry
IUR	International Union of Radioecology
JECFA	Joint FAO/WHO Expert Committee on Food Additives
KD	Keshan disease
KBD	Kashin-Beck disease
LDL_{50}	lethal dose, the simple dose of an element which causes the death of 50% of a population in a specific period of time
LDL_0	the lowest lethal dose

LDH	low density lipoproteins
LOAEL	lowest-observed-adverse-effect level
LRE	light rare earth
LTI	lowest threshold intake
MAC	maximum allowable concentration
MAK	maximum concentration of a chemical substance on air at the workplace (after German)
MBT	monobutyltin
MCL	maximum concentration level
MCRA	Monte Carlo Risk Assessment
Me-X	methylated metal
MF	modifying factors
ML	maximum level
MPL	maximum permissible limit
MMT	methylocyclopentadienyl manganese tricarbonyl
MM-X	monomethylated metal
MND	motor neuron disease
MPC	maximum permissible concentration
MSW	municipal solid waste
MTD	maximum tolerable dose
NAS	National Academy of Sciences
NIOSH	National Institute for Occupational Safety and Health
NOAEL	no-observed-adverse-effect level
NOEC	no-observed-effect concentration
OEL	occupational exposure limit
OECD	Organization for Economic Cooperation and Development
OTC	organo-tin compound
OSHA	Occupational Safety and Health Administration (in USA)
PbB	lead in blood
PCBs	polychlorinated biphenyls
PGM	platinum group metal
PM	particulate matter (e.g., PM10 = particle 10 μm in diameter)
PT_{50}	phytotoxicity threshold corresponding to 50% growth retardation
PTDI	provisional tolerable daily intake
PTWI	provisional tolerable weekly intake
PMTD	provisional maximum tolerable daily intake
RDI	recommended daily intake
RDA	recommended dietary allowance
REE	rare earth element
RfD	reference dose
RNA	ribonucleic acid, structural element of the cytoplasm and cell nucleus

SeCys	selenocysteine
SEFcrust	soil enrichment factor in relation to crust mean concentrations (ratio of mean soil content to crustal content)
SeGSH	selenoglutathione
SeMC	selenomethylocysteine
SeMet	selenomethionine
SOD	superoxide dismutase
SETAC	Society of Environmental and Toxicological Chemistry
SOM	soil organic matter
SPTF	soil-plant transfer factor
TAV	trigger action value
TBT	tributyltin
TF	transfer factor (ratio of an element content in plant to its concentration in soil)
TDI	total dietary intake
TDS	Total Diet Study
TI	tolerable intake
TIC	trace inorganic contaminants
TLV	threshold limit value
TUIL	tolerable upper intake level
TWA	time weighted average
UF	uncertainty factors
UL	tolerable upper intake level
UNEP	United Nations Environmental Program
UNESCO	United Nations Educational, Scientific and Cultural Organization
UPAC	International Union of Pure and Applied Chemistry
USEPA	US Environmental Protection Agency
USGS	United States Geological Survey
WHO	World Health Organization
W/S	worms/soil concentration ratio

Introduction

The quality of human life depends on the chemical composition of food and of the surroundings. Recent improvements and new methods in analytical chemistry and increasing fields of environmental investigations have added substantially to our knowledge of the biogeochemistry of trace elements. In the last three decades there has been a real "explosion" of research data and various publications on occurrence and behavior of almost all trace elements including both elements of known and unknown physiological functions in organisms. In order to realize the vast significance of the biogeochemistry of trace elements, it is essential to gather the knowledge acquired over this period into one comprehensive compilation.

Soil is not homogenous, and the microscale heterogeneity creates a real problem in representative sampling. Also the variability in sampling procedure from plants and other organisms is a subject of concern and this has made the assessment and evaluation of some data almost impossible. Therefore, quantitative comparisons of analytical data for soils, plants and human/animal tissues have often been difficult.

Different chemical preparations of samples (e.g., HNO_3 microwave decomposition, ashing with aqua regia, total digestion) as well as different instrumental methods used for the determination of elements (e.g., ICP-MS, ICP-OES, F-AAS, ETA-AAS) have an influence on final results. Luckily, analytical quality assurance and the use of reference materials have decreased uncertainties of analytical data. Therefore, each measurement of trace elements builds up a database and contributes to a better understanding of their overall distribution and behavior in given media and in the total environment.

A better understanding of the biogeochemical processes that control trace element cycling and comprehensive dataset on the abundance of trace elements in abiotic and biotic environmental compartments may be a key to better management of trace elements in the environment that is prerequisite to sustainable land use and, presumably, to diminish health risks due to trace inorganic pollutants.

The term "trace elements" has never been defined precisely. It has been used in geochemistry for chemical elements that occur in the Earth's crust in amounts less than 0.1% (1 000 mg kg^{-1}) and also in biological sciences, for elements at similar concentrations. Therefore some elements that are "trace" in biological materials are not "trace" in terrestrial ones (i.e., iron). The term "trace elements" is related to their abundance and includes elements of various chemical properties: metals and metalloids.

Common trace cations descriptors are "trace metals" and/or "heavy metals". The trace metalloids are simple "trace elements". The other terms: "micronutrients", "essential elements", and "toxic elements" are related to their physiological functions and are rather confusing since their effects on organisms and health depend upon concentrations. All these terms are inadequate, and a great deal of confusion has occurred in the literature where authors have been imprecise in their use of these terms. Especially the term "heavy metals" has recently become a subject of a broad discussion that emphasizes its non-precise definition. Duffus (2002) has written: "Over the past two decades, the term 'heavy metals' has been widely used... and related to chemical hazards. It is often used as a group name for metals and semimetals (metalloids) that have been associated with contamination and potential toxicity or ecotoxicity." This term is based on various criteria (i.e., atomic weight, atomic number, density, chemical properties etc.). Thus, the inconsistent use of the term "heavy metals" reflects inconsistency in the scientific literature. This term has never been defined by any authoritative body, such as IUPAC.

Other terms that need to be defined are related to chemical speciation and fractionation of elements. Thanks to new developments in analytical instrumentation and methodology, the identification and measurement of element species in a particular system is possible. In an attempt to end the confusion regarding the usage of the term speciation, three IUPAC Divisions collaborated to consider the issue (Templeton et al. 2000). Their definitions for the recommended use of term speciation are following:

- *Chemical species.* Chemical elements: specific form of an element defined as to isotopic composition, electronic or oxidation state, and/or complex or molecular structure
- *Speciation analysis.* Analytical chemistry: analytical activities of identifying and/or measuring the quantities of one or more individual chemical species in a sample
- *Speciation of an element, speciation.* Distribution of an element among defined chemical species in a system
- *Fractionation.* Process of classification of an analyte or a group of analytes from a certain sample according to physical or chemical properties

Part I Biogeochemistry of the Human Environment

The Biosphere

The biosphere, also called the ecosphere, is the natural environment of living organisms and is the complex biological epidermis of the Earth whose dimensions are not precisely defined. It consists of the surficial part of the lithosphere, a lower part of the atmosphere, and the hydrosphere. Several ecosystems have been developed within the biosphere. Each ecosystem is a fundamental division of the total environment consisting of living organisms in a given area and having a *balanced* cycling of chemical elements and energy flow.

Among the principal resources of which man disposes, are terrestrial ecosystems consisting of soil and water, and associated animal and plant life. Ecosystems are functional environmental units, having *balanced* cycles of chemical elements, organic materials and energy flow. There is a homeostatic interrelationship between the non-living media (abiotic compartments) and the living organisms (biotic compartments). However, a significant part of the ecosystems has already been considerably modified by humans, and these processes will continue.

The energy for life is derived from the radiant energy of the sun, which drives the chemical reaction of photosynthesis. The other sources of energy, e.g., geothermal, gravitation, and electrical, are of negligible importance in the total energy flow, but may determine specific conditions of some ecosystems.

Organisms have adjusted during the course of evolution and life to the chemistry of their environment and have developed their biochemistry in close connection to the composition of the natural environment. These phenomena have been easily observed, mainly in microorganisms and plant populations that have evolved tolerance to high concentrations of trace elements either in natural geochemical provinces, or under man-induced conditions.

Most of the chemical elements for life on the land are supplied mainly from the soil overlying the surficial lithosphere (Fig. I-1.1). Although mechanisms of biological selection of chemical elements allow plants to control, to a certain extent, their chemical composition, this barrier is somewhat limited in respect to trace elements. Therefore concentrations of trace elements in plants are often positively correlated with the abundance of these elements in growth media. This creates several problems for plants, animals and humans associated either with deficiency or with excess. Thus, questions of how and how much of an element is taken up by organisms have been hot topics of research in recent decades. Usually the quantitative differences between essential amounts and biological excesses of trace elements are very small. A proper balance between trace and major elements plays a significant role in biochemical processes.

Fig. I-1.1.
The transfer of chemical elements in schematic terrestrial trophic chain

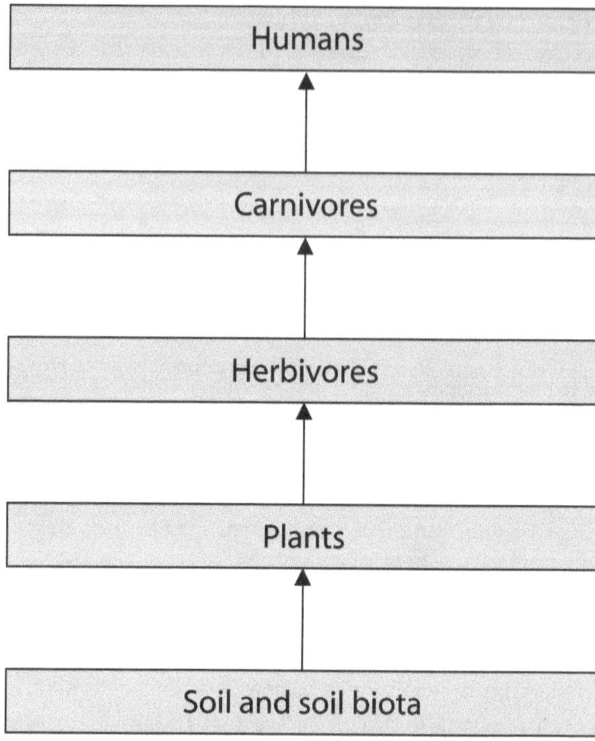

The bioavailability of these elements is variable and is controlled by specific properties of abiotic and biotic media as well as by physical and chemical properties of a given element.

The biochemical functions of essential trace elements are already known. A great number of trace elements are known to have a biological role, often as cofactors or part of cofactor in enzymes and as structural elements in proteins. Some of them also are used in several processes of electron transfer. Non-essential elements seem to be involved in vital processes but their biochemical functions are not yet understood. The essentiality of other trace elements, possible at very minor concentrations, may be discovered in the future. Most of trace elements that are essential to humans are also essential to plants. Unfortunately, contents of most elements that may be harmful to humans and animals are not toxic to plants. This has created an increased transfer of some elements in the food chain.

The survival of mankind is a story of food. Both, lack of food and bad quality of food have created throughout the centuries serious problems for people. Nowadays it is calculated that over 3 billion people worldwide suffer from either deficiency or toxicity of some trace elements.

Here is a place to remind Paracelsus' (1538) statement:

> All substances are poisonous, there are none which is not a poison; the right dose is what differentiates a poison from a remedy.

The anthroposphere. Many ecosystems have been considerably modified by humans and therefore it has become necessary to distinguish the anthroposphere – the sphere of man's settlement and activity. The anthroposphere does not represent a separate sphere, but may be applied to any part of the biosphere that has been changed under an influence of technical civilization.

While geological, geochemical and biological alterations of the lithosphere have been very slow, changes introduced or stimulated by humans have been accumulated extremely quickly in recent decades of the past century. Anthropogenic changes, associated mainly with chemical pollution, lead most often to a degradation of the natural human environment. Among all chemical pollutants, trace elements are of a special ecological, biological and health significance.

The production of energy and the consumption of natural resources are the main source of trace elements as contaminants. However, agricultural activities and especially application of sewage sludge, manure, mineral fertilizers (NPK), and pesticides also contribute significantly to the trace metal status of agroecosystems.

Bowen (1979) has suggested that when the rate of mining of a given element exceeds the natural rate of its cycling by a factor of ten or more, the element should be considered a potential pollutant. Thus, the potentially most hazardous trace metals to the biosphere may be: Ag, Au, Cd, Hg, Pb, Sb, Sn, Te, W. Also those elements that are essential to plants and humans, such as: Cr, Cu, Mn, and Zn, may be released, in some regions, in excessive amounts.

Soils

Soil is not only a part of the ecosystem but also occupies a basic role for humans, because the survival of man is tied to the maintenance of its productivity. Soil functions as a filtering, buffering, storage, and transformation system protect against the effects of trace element pollution. Soil is effective in these functions only as long as its capacity for cation exchange and its biological activity are preserved. The frequent association of trace element pollution with acid deposition (mainly S, NO_x, and HF) greatly complicates the overall effects in the environments.

Soil is the main source of trace elements for plants both as micronutrients and as pollutants. It is also a direct source of these elements to humans due to soil ingestion affected by "pica-soil", geophagia, dust inhalation, and absorption through skin.

The soil-plant transfer of trace elements is a part of chemical element cycling in nature. It is a very complex process governed by several factors, both natural and affected by humans. Thus, the prediction of trace element uptake by plants from a given growth medium should be based on several biotic and abiotic parameters that control their behavior in soil.

Soils contain trace elements of various origins: *(i) lithogenic* – inherited from the lithosphere (parent material), *(ii) pedogenic* – from lithogenic sources but forms changed due to soil-forming processes, and *(iii) anthropogenic* – elements deposited onto and/or into soils as results of human's activities. Soil processes and anthropogenic factors control the behavior of all these elements (Table I-2.1). It has been assumed that the behavior of trace elements in soils and in consequence their phytoavailability differ as to their origin. Several recent reports have indicated that regardless of the forms of the anthropogenic trace metals, their availability to plants is significantly higher than those of natural origin (Kabata-Pendias and Pendias 2001).

Soils of several regions of the world (and especially of Europe) have been and will be in the future subjected to mineral fertilization, pesticide application, waste disposal, and industrial pollution. All these human activities affect both chemical and physical soil properties, and will lead to changes in the behavior of trace elements in soils. The impact of soil acidification, alkalization, salinity and losses of SOM on the uptake of trace elements by vegetation, particularly by crop plants, have already become serious issues for the environment and for human health.

Weathering. Weathering, the basic soil-forming process is a complex set of interactions between the lithosphere, atmosphere, and hydrosphere that occur in the biosphere and are powered by solar energy. It can be described chemically as the processes of

Table I-2.1. Influence of origin of trace elements on their behavior in soil

Origin	Association	Phase	Form[a]	Bioavailablity
Lithogenic	Minerals or bound to minerals	Solid	Residual	Very slight
Pedogenic	CM, SOM, and OX	Solid	Fixed by CM, SOM, and OX	Slight
	Simple and complex ions	Aqueous	Free ions and nonionic forms	Easy
Anthropogenic	Minerals, SOM, and PS	Solid	Mainly exchangeable and chelated	Moderate and easy
Pedogenic and anthropogenic	Simple and complexed ions	Aqueous	Free ions and nonionic forms	Easy

Symbols: *CM* – clay fraction, *SOM* – soil organic matter, *OX* – oxides and hydroxides, *PS* – particle surface.
[a] Operational fractions according to the sequential extraction (Tessier et al. 1979).

Table I-2.2. Geochemical classification of trace elements (based on data after Rankama and Sahama *vide* Fairbridge 1972)

Siderophile		Chalcophile		Lithophile	
6	Mo	**5**	(Ta)	**1**	Li, Rb, Cs
7	Re	**6**	(Mo)	**2**	Be, Sr, Ba, Ra
8	Fe, Ru, Os	**8**	Fe, Ru	**3**	Sc, Y, Lanthanides, Actinides
9	Co, R, Ir	**9**	(Co)	**4**	Ti, Zr, Hf
10	Ni, Pd, Pt	**10**	(Ni), (Pd), (Pt)	**5**	V, Nb, Ta
11	Au	**11**	Cu, Ag	**6**	Cr, Mo, W
14	Ge, Sn, (Pb)	**12**	Zn, Cd, Hg	**7**	Mn
15	As	**13**	Ga, In, Tl	**8**	(Fe)
		14	(Ge), (Sn), Pb	**9**	(Co)
		15	As, Sb, Bi	**10**	(Ni)
				12	(Zn), (Cd)
				13	B
				14	(Ge), (Sn), (Pb)
				15	(As)
				17	F, Cl, Br, I

Bold numbers are the Periodic Table's groups. *Symbols in parenthesis* are given for secondary geochemical tendencies of an element.

dissolution, hydration, hydrolysis, oxidation, reduction, and carbonization. All these processes are guided by the rules of enthalpy and entropy, and they lead to the formation of mineral and chemical components that are relatively stable and equilibrated in the particular soil environment (Pédro 1979). The behavior of elements during weathering and pedogenic processes are highly associated with their geochemical properties, which are base for the geochemical classification (Table I-2.2). The majority of elements reveal lithophilic character, which indicates a tendency to form oxygen compounds, as well as silicates, carbonates, phosphates, and sulfates.

Clay minerals, the principal products of weathering and soil formation, are due to water-rock interaction processes. Two types of compounds released by organic matter or organisms are believed to be particularly involved in weathering processes: carbonic acid, formed from the CO_2 released during decay of organic matter, and organic chelates.

Processes of weathering (degradation) and neoformation of minerals as well as formation of amorphous minerals and organo-mineral substances greatly affect the forms and adsorption of trace elements in soil. The affinity of trace elements for soil constituents is strongly influenced by their electrochemical properties and is closely related to the specific surface area and cation exchange capacity (CEC) of the minerals. Some clay minerals, such as montmorillonite, imogolite, vermiculite, and amorphous allophanes reveal the highest sorption capacity (Table I-2.3).

Table I-2.3. Surface area and sorption capacity (CEC) of some soil minerals[a]

Mineral	Total or specific surface ($m^2 g^{-1}$)	Cation exchange capacity (cmol(+)/kg)
Allophane	100 – 880	5 – 350
Kaolinite	7 – 30	3 – 22
Halloysite	10 – 45	3 – 57
Montmorillonite	280 – 800	80 – 150
Illite	65 – 100	20 – 50
Chlorite	25 – 150	10 – 40
Vermiculites[b]	50 – 800	10 – 200
Gibbsite	25 – 58	50[c]
Goethite	41 – 81 (305–412)[d]	160[d]
Mn-oxides	32 – 300	150 – 230
Imogolite	900 – 1 500	30 – 135
Zeolite	720	350 – 615
Palygorskite	5 – 30	–
Sepiolite	20 – 45	21 – 24
Muscovite	60 – 100	10 – 40
Biotite	40 – 100	10 – 40
Quartz	2 – 3	7

[a] Data sources: Baldock and Nelson 2000, Bolt 1979, Brümmer 1986, Kabata-Pendias and Pendias 2001, Tan 1998, Sparks 1995.
[b] Dioctahedral and trioctahedral vermiculites.
[c] Amorphous Al-oxides.
[d] Amorphous Fe-oxides.

Table I-2.4. Behavior of trace elements in various weathering conditions (after Kabata-Pendias and Pendias 2001)

Degree of mobility	Condition	Trace elements
High	Oxidizing and acid	B, Br, and I
	Neutral or alkaline	B, Br, F, I, Li, Mo, Re, Se, U, V, W, and Zn
Medium	Reducing	B, Br, and I
	Oxidized and acid	Li, Cs, Mo, Ra, Rb, Se, Sr, F, Cd, Hg, Cu, Ag, and Zn
Medium	Mainly acid	Ag, Au, Cd, Co, Cu, Hg, and Ni
	Reducing, with variable potential	As, Ba, Cd, Co, Cr, F, Fe, Ge, Li, Mn, Nb, Sb, Sn, Sr, Tl, U, and V
Low	Oxidizing and acid	Ba, Be, Bi, Cs, Fe, Ga, Ge, La, Li, Rb, Si, Th, Ti, and Y
	Neutral or alkaline	Ba, Be, Bi, Co, Cu, Ge, Hf, Mn, Ni, Pb, Si, Ta, Te, and Zr
Very low	Oxidized and acid	Al, Au, Cr, Fe, Ga, Os, Pt, Rh, Ru, Sc, Sn, Ta, Te, Th, Ti, Y, and Zr
	Neutral or alkaline	Ag, Al, Au, Cu, Co, Fe, Ga, Ni, Th, Ti, Y, and Zr
	Reducing	Ag, As, Au, B, Ba, Be, Bi, Cd, Co, Cu, Cs, Ge, Hg, Li, Mo, Ni, Pb, Re, Se, Te, Th, Ti, U, V, Y, Zn, and Zr

Factors controlling the complex set of chemical weathering processes of minerals in rocks and soils are: *(i)* oxidation, *(ii)* reduction, *(iii)* hydrolysis, *(iv)* hydration, *(v)* solution, and *(vi)* chelation. All of these reactions are controlled by chemical equilibria of the particular environment that can be illustrated by the Eh-pH diagrams for the given geochemical reactions (Garrels and Christ 1965). In various weathering environments, the behavior of trace elements differs (Table I-2.4). The electrochemical properties of elements and the stability of host minerals controls the mobility of trace elements. In an arid environment, physical weathering generally occurs, chemical transformation being rather poor due to the lack of water. In a tropical climate with high temperature and humidity, chemical weathering is favored, resulting in the formation of more clay minerals.

Due to the variable charge of both clays and organic particles, different organo-mineral complexes are formed that may greatly influence element mobility in soils (Tan 1998). This author has emphasized that various bindings between microorganisms, like bacteria and fungi, and soil clays and organic matter, may also occur and may have serious environmental importance.

Pedogenic processes. Pedogenic processes cannot be easily distinguished from weathering processes as they take place simultaneously and at the same sites; most often they are closely interrelated. The principal types of these processes include: *(i)* podzolization, *(ii)* alkalization, *(iii)* aluminization, *(iv)* laterization, *(v)* sialization, and *(vi)* hydromorphic processes. All these processes control the distribution and behavior of trace elements in distinct layers of soil profiles that are related to sorption and desorption and to the formation of various species of elements.

The main soil parameters governing these processes are: *(i)* pH and Eh values, *(ii)* amount and mineral composition of the fine granulometric fraction (<0.2 mm), *(iii)* amount and kind of organic matter, *(iv)* oxides and hydroxides of Fe, Mn and Al, and *(v)* microorganisms.

In most soil conditions the effects of pH on the solubility of trace elements, especially of microcations, is more significant than that of redox potential (Chuang et al. 1996). In some soils, however, redox potentials have a crucial impact on the behavior of some metals, in particular of Co, Cr, Fe and Mn (Bartlett 1999). The diversity of ionic species of trace elements and their various affinities to complex inorganic and organic ligands make possible the dissolution of each element over a relatively wide range of pH and Eh.

Smith and Huyck (1999) described metal mobility under different environmental conditions. Although it is rather difficult to predict trace element mobility in soils and other terrestrial compartments, the authors referred to the capacity of an element to move with fluids after dissolution in surficial environments. The following conditions and behavior of trace elements were distinguished:

- Oxidizing and acid, pH <3: *(a)* very mobile – Cd, Co, Cu, Ni, and Zn, *(b)* mobile – Hg, Mn, Re, and V, and *(c)* somewhat mobile and scarcely mobile – all other metals
- Oxidizing in the absence of abundant Fe-rich particles, pH >5: *(a)* very mobile – Cd and Zn, *(b)* mobile – Mo, Re, Se, Sr, Te, and V, and *(c)* somewhat mobile and scarcely mobile – all other metals
- Oxidizing with abundant Fe-rich particulates, pH >5: *(a)* very mobile – none, *(b)* mobile – Cd and Zn, and *(c)* somewhat mobile and scarcely mobile – all other metals
- Reducing in the absence of hydrogen sulfide, pH >5: *(a)* very mobile – none, *(b)* mobile – Cd, Cu, Fe, Mn, Pb, Sr, and Zn, and *(c)* somewhat mobile and scarcely mobile – all other metals
- Reducing with hydrogen sulfide, pH >5: *(a)* very mobile – none, *(b)* mobile – Mn and Sr, and *(c)* scarcely mobile to immobile – all other metals

It is evident that Fe/Mn-rich particulates and hydrogen sulfide are most significant among the abiotic factors in controlling trace metal behavior in the terrestrial environment (Sparks 1995; Bartlett 1999). The relationship between Fe contents and some trace metals indicates an impact of soil Fe on the distribution of some trace metals. However, this relationship is significantly high only for Ni (Fig. I-2.1).

All metals that are especially susceptible to redox reactions, e.g., Co, Cr, Fe, and Mn play a significant role in dissolution/precipitation reactions in soil. This is best illustrated by the behavior of Mn:

- The reduction of Mn^{3+} to Mn^{2+} by: Fe^{2+}, Cr^{3+}, Co^{2+}, reduced S, phosphate ligands (e.g., $P_2O_7^{4-}$), phenols, and other easily oxidizing organic compounds
- The oxidation of Mn^{2+} to Mn^{3+} or Mn^{4+} by: any hydroxyl free radicals or atmospheric oxygen
- The Mn^{3+} is an extremely reactive redox species that quickly disappears, either by accepting or by donating an electron

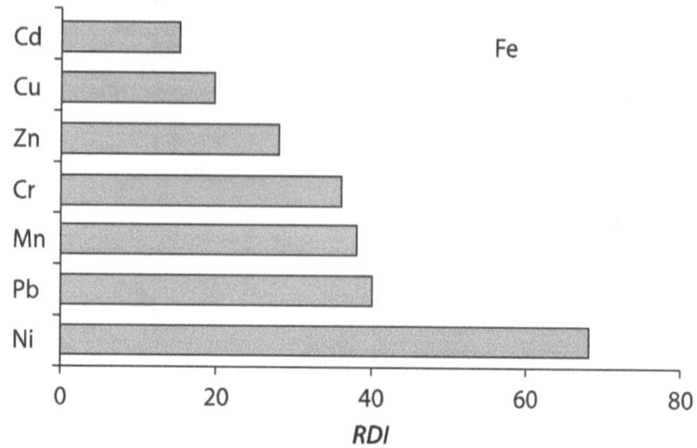

Fig. I-2.1. Relative distribution index (RDI) of statistically-significant relationships between trace metals and Fe content of mineral soils, at the 99% confidence level (after Kabata-Pendias and Pendias 2001)

Other soil minerals such as, carbonates, phosphates, sulfides, sulfates and chlorides may have important influences on the distribution and behavior of trace elements in soils developed under specific geological and climatic conditions (Kabata-Pendias and Sadurski 2004).

Soil biota. Living organisms in soils composed of fauna and flora of various dimensions (macro-, meso-, and micro-) are abundant in soils, especially in surface layers. Functions of living organisms are reflected in biological activity of soils and greatly contribute to bio-physico-chemical processes in soils and soil fertility.

Some species of soil mesofauna (invertebrates) may easily reflect chemical changes in soil and are used as sensitive indicators of the trace metal status of soils (Ma 1987). According to this author the concentration factor (ratio of worms to soil) varies from 0.1 to 190, being the lowest for Pb and the highest for Cd. Depending upon soil properties, this factor is also variable for a given element. Also Edwards et al. (1998) have concluded, based on an earthworm bioassay, that many factors hampered quantitative determination of the degree of metal bioavailability in soil. However, there is always observed a general rule, that Cd is the most available to earthworms (Worms/Soil ratio: 1–37) whereas Cu and Pb are the least available (W/S ratio < 1) (Kabata-Pendias and Pendias 2001).

Trace element contamination causes changes in biochemical parameters of soils such as biomass and species diversity, and microbial activity. Most of these parameters have been used as indices of trace element status of soils, however, there is not always a clear relationship. The impact of soil microorganisms and enzymes on all redox and dissolution/precipitation processes is crucial (Burns and Dick 2002; Naidu et al. 2001; Kostyuk and Bunnenberg 1999). The physicochemical relationship between bacteria and mineral surfaces leads to diverse effects of dissolution and secondary precipitation of trace elements, including changes in their valence and/or conversion into organomineral compounds. The water-extractable F in soils is also positively correlated with the microbial biomass (Tschenko and Kandeler 1997). The microbial activities are associated with following processes:

- uptake or release of an element from a cell
- charge alteration of an ion
- complexation of an element by various produced compounds
- immobilization of an element
- mobilization of an element, often due to the methylation

Biological methylation of some elements, such as As, Hg, Se, Te, Tl, Pb, and In may greatly influence their behavior in soils. Microorganisms reveal a variable and often high sorption capacity to trace elements. Cumulative pools of trace elements in the mass of microbiota has been calculated as follows (g ha^{-1}): Ni, 350; Cu, 310; Zn, 250; Co, 150; Mo, 148; and Pb 8.4, which corresponds to 0.002 to 0.21% of their total abundance in the 20-cm topsoil level (Kovalskiy et al. 1980). Microbioaccumulation of trace elements may be of great importance both in the cycling of elements in soil and in their phytoavailability. Increased levels of trace elements can disturb soil microbial processes. On the other hand, bacterial impact on trace metal behavior may be significant, especially in soil amended with sewage sludge, where bacterial leaching of metals have a practical application (Olson 1986). Apparently, soil microorganisms are more sensitive than other terrestrial organisms to increased levels of metals, thus the assessment of their toxicity has recently become the subject of investigations (Brooks et al. 2005). These authors reported that N fixation in clover is the most sensitive to an excess of Cd and Zn. However, the adverse concentration effects on the potential nitrification rate differs greatly between soils and varied for Cd from 7 to 55 mg kg^{-1} and for Zn from 107 to 1764 mg kg^{-1}. Increased levels of F (up to 189 mg F kg^{-1}) is reported to decrease microbial activity to 5–20% of those in the uncontaminated soil (Tschenko and Kandeler 1997).

The microbial activity is highly governed by soil properties. As microorganisms are involved in SOM transformations, their activities are associated with organic matter content. Czaban and Wróblewska (2005) observed that microbial transformation of Cd species from the solid phase to species soluble in DTPA was lower in fine-texture soil with low SOM content (1.1% C) than in sandy soil with higher SOM content (4.2% C). Smolders et al. (2004) have suggested that the most significant factor controlling different reactions of microbial processes to increased Zn concentrations is the variable community tolerance. Community level of physiological profiles (CLPPs) of soil microorganisms have not changed under different levels of Cd, Zn, and Pb that indicate their developed tolerance to increased contents of metals (Niklińska et al. 2004).

There are a large number of biologically induced minerals (BIM) especially of Fe and Mn. The major metabolic processes that cause deposition and/or dissolution of Fe and Mn minerals are oxidation and reduction of metal, metal sulfate oxidation, and metal sulfide reduction (Frankel and Bazylinski 2003).

Mycorrhizal fungi play very specific and important role in the uptake of trace elements by plants. There is evidence that mycorrhiza play a role in the micronutrient supply to plants (Turnau et al. 1996). The presence of mycorrhizal symbiosis has also been demonstrated in plants, especially hyperaccumulating plants which are being used in the phytoextraction of metals (Skinner et al. 2005; Turnau et al. 2005). Arbuscular mycorrhizal fungi (AMF) exhibit various mechanisms in controlling metal uptake by plants (Biró at al. 2006).

Soil solution. Transfer of trace elements between soil phases should be considered as the main process controlling their behavior and bioavailability. The aqueous phase of the soil (soil solution) is composed of water with a colloidal suspension, free and/or complexed, and dissolved substances of various compounds, including bio-inorganic

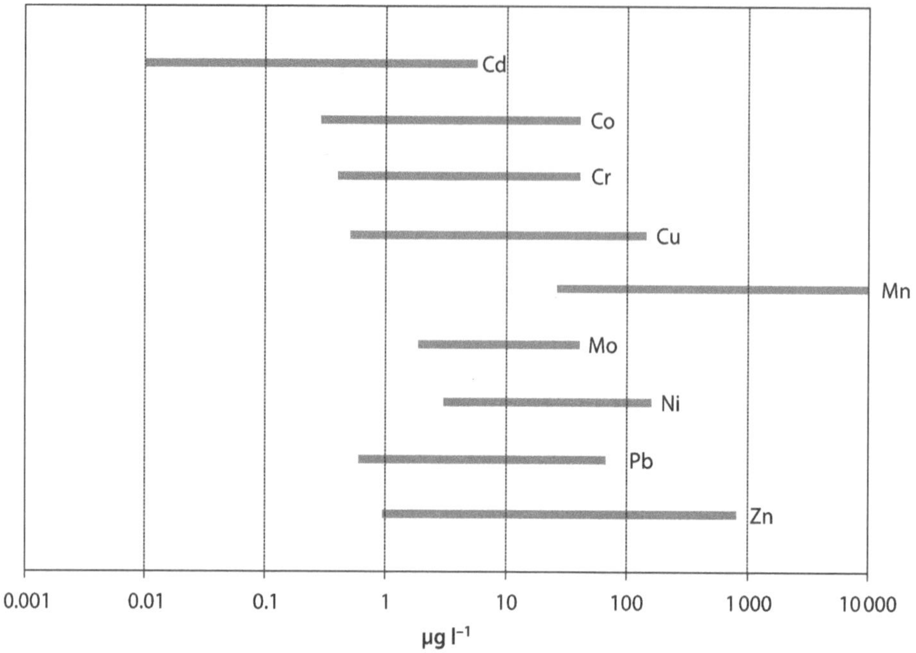

Fig. I-2.2. Common ranges of trace metals in the solution of uncontaminated soils

complexes. Concentrations of trace elements in soil solution are closely correlated with their mobility and availability. However, the soil solution is constantly and rapidly changing in amount and chemical composition of species due to contact with highly diverse soil solid phase and by the uptake of water and ions by plant roots. Data on trace element concentrations in soil solution can be useful for the prediction of their phytoavailability and efficiency/toxicity to crops. Chemistry of the soil solution provides useful information on soil processes that are important to both agricultural and environmental sciences. Mineral particles of biological origin associated with some protists are also found in soil solution (Keller 1997).

Methods used for obtaining soil solution differ widely, and therefore, it is difficult to determine adequate mean contents of trace elements in soil solutions. As Wolt (1994) stated "No one approach to obtain soil solution is appropriate to all applications".

The concentrations of some elements measured in the solution obtained by various techniques from uncontaminated soils are (after Kabata-Pendias and Pendias 2001; Wolt 1994), as follows (in $\mu g\, l^{-1}$): As, 0.01; B, 5–800; Be, 0.1; Cd, 0.08–5; Co, 0.08–29; Cr, 0.01–29; Cu, 0.5–135; Mn, 25–8 000; Mo, 0.0004–30; Ni, 0.2–150; Pb, 0.005–63; Se, 0.06; Sn, 0.2; and Zn, 0.1–750. Most commonly reported concentrations of some metals (Co, Cr, Cu, Mo, Ni, and Pb) in soil solution are in the range between 0.05 and 100 $\mu g\, l^{-1}$ (Fig. I-2.2).

In general, the total contents of trace elements in solutions of uncontaminated mineral soils range from 1–100 $\mu g\, l^{-1}$, while in contaminated soils these values can be much higher. In both kinds of soils, however, these are negligible portions of the total soil metals. The transfer factor, calculated as ratios of metals in solution of contaminated soils to their contents in solution of the control soils usually decreases in the following order: Cd > Ni > Zn > Cu > Pb > Cr.

Common ion pairs are mainly hydroxides, however several other complexes are also likely to occur in solutions of various soils (Table I-2.5). Ion pairs, specific forms of oppositely charged ions that are thermodynamically stable can occur in solutions as manifold and of variable compositions. Complex compounds of a relatively high stability constant occurring in soils solution are: BeF_2^0, $CdCl_2^0$, $Cd(OH)_2^0$, $CdSO_4^0$, $Co(OH)_2^0$, $Cr(OH)_3^0$, $Cu(OH)_2^0$, $FeCl_2^0$, $FeCl_3^0$, $Hg(OH)_2^0$, $MnSO_4^0$, $Ni(OH)_2^0$, $PbSO_4^0$, $Pb(OH)_2^0$, $VO(OH)_3^0$, $Zn(OH)_2^0$.

Chemistry of the soil solution provides useful information on soil processes that are important to agricultural and environmental sciences. Data on concentrations of trace elements in a "real" soil solution are useful for the prediction of their availability, toxic effects on crops and on biological activities of soils. McBride et al. (1997) have formed an equation for the metal solubility as a function of some soil properties (e.g., pH, SOM, and total metal content). The SOM content was often, but not always, a statistically significant variable in predicting metal solubility, especially Cu. These authors stated that "Although total metal content alone is not generally a good predictor of metal solubility or activity, it assumed great importance when comparing metal solubility in soils having similar pH and organic matter content".

Table I-2.5. Inorganic ionic species of trace elements and iron in soil solutions (after Kabata-Pendias and Sadurski 2004)

Element	Cations	Anions
Ag	Ag^+	$AgCl_2^-$, $AgCl_3^{2-}$, $Ag(SO_4)_2^{3-}$
As	**As^{3+}**	AsO_2^-, $HAsO_4^{2-}$, $H_2AsO_3^-$
B		**BO_3^{3-}**, $H_2BO_3^-$, HBO_3^{2-}
Be	Be^{2+}, $BeOH^+$, $Be_5(OH)_7^{3+}$	BeO_2^{2-}, $Be(OH)_3^-$, $Be(CO_3)_2^{2-}$, $Be(OH)_4^{2-}$
Cd	$CdCl^+$, $CdOH^+$, $CdHCO_3^+$, **$CdHS^+$**	$CdCl_3^-$, $Cd(OH)_3^-$, **$Cd(OH)_4^{2-}$**, **$Cd(HS)_4^{2-}$**
Co	Co^{2+}, Co^{3+}, $CoOH^+$	$Co(OH)_3^-$
Cr	Cr^{3+}, $CrOH^{2+}$	$HCrO_3^{2-}$, CrO_4^{2-}, $Cr(OH)_4^-$, $Cr(CO_3)_3^{3-}$, $HCrO_4^-$
Cu	Cu^{2+}, $CuOH^+$, $Cu_2(OH)_2^{2+}$	$Cu(OH)_3^-$, **$Cu(OH)_4^{2-}$**, $Cu(CO_3)_2^{2-}$
F	AlF^{2+}, AlF_2^+	F^-, AlF_4^-
Fe	Fe^{2+}, $FeCl^+$, $Fe(OH)_2^+$, $FeH_2PO_4^+$	$Fe(OH)_3^-$, **$Fe(OH)_4^{2-}$**, $Fe(SO_4)_2^-$
Hg	Hg_2^{2+}, $HgCl^+$, $HgCH_3^+$	$HgCl_3^-$, **HgS_2^{2-}**
I		I^-, I_3^-, IO_3^-, **$H_4IO_6^-$**
Mn	Mn^{2+}, $MnOH^+$, $MnCl^+$, $MnHCO_3^+$	MnO_4^-, $HMnO_2^-$, $Mn(OH)_3^-$, **$Mn(OH)_4^{2-}$**
Mo		MoO_4^{2-}, $HMoO_4^-$
Ni	Ni^{2+}, $NiOH^+$, $NiHCO_3^+$	$HNiO_2^-$, $Ni(OH)_3^-$
Pb	Pb^{2+}, $PbCl^+$, $PbOH^+$	$PbCl_3^-$, $Pb(CO_3)_2^{2-}$
Se		**SeO_3^{2-}**, SeO_4^{2-}, **HSe^-**, $HSeO_3^-$
V	VO^{2+}	$H_2VO_4^-$, HVO_4^{2-}, VO_3^-
Zn	Zn^{2+}, $ZnCl^+$, $ZnOH^+$, $ZnHCO_3^+$	ZnO_2^{2-}, $Zn(OH)_3^-$, $ZnCl_3^-$

The symbols given in bold letters indicate the ions occurring only in extreme pH and Eh regimes.

The partitioning of trace elements between the soil and soil solution determines their mobility and bioavailability. However, the prediction of properties of soil solution is rather difficult, and sophisticated techniques must be applied for the extraction of undiluted soil solution as well as special conditions for the storage of soil/sediment samples.

The concentrations of free metal species in soil solution are controlled by several factors of which the most significant are thermodynamic/kinetic parameters. Mathematical approaches to modeling soil solution – solid phase equilibria are broadly described in numerous publications (Sposito et al. 1984; Waite 1991; Wolt 1994; Sparks 1995; Suarez 1999), and several models for calculating activity coefficients for trace metals have been overviewed and discussed. Waite (1991) concluded that: "mathematical modeling clearly has a place in extending the information that can be obtained on trace element species distributed by other methods and will be of practical use in systems for which determination of concentrations of all species of interest is impossible because of sensitivity constraint or other analytical difficulties".

Trace element speciation. As soils consist of heterogeneous mixtures of different organic and organic-mineral substances, clay minerals, oxides and hydroxides of Fe, Mn, and Al, and other solid components as well as a variety of soluble substances, the binding mechanisms for trace elements and forms of their occurrence in soils are manifold and vary with the composition and physical properties of soils.

Great confusion exists in the use of the term "speciation" in environmental sciences. According to Ure et al. (1993), the definition of speciation in the context of soils, sediments and sewage sludge refers to processes of identification and quantification of the different defined species, forms and phases in which an element occurs in investigated materials. However, most often "speciation" also means the description of the amounts and kinds of existing forms of trace elements. Whichever approach is taken, the species, forms or phases are defined: *(i)* functionally, *(ii)* operationally, and/or *(iii)* as specific chemical compounds or oxidation states.

Considerable controversy has developed over selective extraction methods to determine the amounts of trace elements associated with various soil phases and compounds. Determination of various fractions of these elements is broadly used, especially for two purposes: to link with the potential bioavailability, and to predict the mobility affecting their transport within the soil profile and to ground waters. The speciation of trace metals in soils is not stable and relatively easy transformation of their forms in soils is observed.

Various concentrations of different reagents and variable soil/solution ratios over broad pH ranges are used for the selective extraction procedures. In consequence, each method gives operational groups of metals that are not comparable. However, each method gives an approximate fractionation of metals occurring in soils, from easily mobile to strongly sorbed/fixed forms. All these methods are used for the determination of commonly distinguished metal species which are, in general: *(i)* easily exchangeable or water soluble, *(ii)* specifically sorbed, e.g., by carbonates, or phosphates, *(iii)* organically bound, *(iv)* occluded by Fe/Mn oxides and hydroxides, and *(v)* structurally bound in minerals (residual).

Metal species resulting from the partitioning of the total metal content that is associated with various solid soil fractions are usually estimated using specific operational extraction procedures. Comprehensive reviews of methods applied for sequential extraction have been given in several publications (Tessier et al. 1979; Salomons and Förstner 1984; Brümmer 1986, 1986; Kersten and Förstner 1991; Sauvé 2001; Mortvedt et al. 1991; Ure and Davidson 1995).

Some authors distinguish also water-soluble fraction of metals, which actually corresponds to a dilute salt solution extract (Boust and Saas 1981; Mathur and Levésque 1988). Solid phase fractionation methods allow separate additional fractions of metals, such as silicate clays, sulfides, and specifically sorbed or precipitated forms.

Various methods used to evaluate the pool of soluble (phytoavailable) trace elements in soils are based mainly on extraction by various solutions at different concentrations (as described by Houba et al. 1999; McLaughlin 2001):

- Acids: mineral acids at various concentrations
- Chelating agents: e.g., EDTA, DTPA
- Buffered salts: e.g., AAAc
- Neutral salts: $CaCl_2$, $MgCl_2$, NH_4NO_3, $Sr(NO_3)_2$
- Other extractants, like *Coca Cola*, which has been proposed for use in routine soil testing

Several other techniques like electrodialysis, diffusion through membranes, diffusion gradients in thin films as well as bioindicators have also been proposed (Kabata-Pendias and Sadurski 2004).

According to Houba et al. (1999), solutions of a neutral salt, such as $CaCl_2$, are the most suitable for the extraction of bioavailable pools of metals. Karczewska (2002) has reported the highest proportions of metals (up to above 90% of total contents of Cu, Pb, and Mn; and up to above 60% of Zn) to be extracted from surface horizons of polluted soils by the following solutions: 0.02 M EDTA-Na, 1 M HCl, and 0.1 M HCl.

Depending upon the variability in physico-chemical characteristics of metals, their affinity to soil components governs their speciation. Rule (1999) broadly reviewed recent literature on the phase distribution of trace metals in soils, and concluded that the highest proportions of most metals are found either in the residual or in Fe-Mn oxide fractions in both natural and contaminated soils. A high association with the exchangeable fraction was observed only for Cd in a few soils. Cadmium and Zn were also abundant in the carbonate form in some soils. The association of metals with organic matter was generally low for Cd and Ni, and high for Cu. Herbert (1997) found that Cu, Pb, Ni and Zn are bound primarily to cation exchange sites and to organic matter in the reference soil, while in the polluted soils, metal partitioning is dominated by Fe oxide fractions, despite the high organic matter content. However, the proportion of different species is highly variable and depends on elements, soils, and seasons (e.g., climatic conditions).

Kernsten and Förstner (1991) presented the affinity of metals for different minerals:

- Sulfides of Fe and Pb are likely to sorb most metals
- Carbonates fix predominantly Cd, Cu, Pb, and Mn
- Fe/Mn nodules (deep-sea) sorb mainly Fe, Mn, Cu, Zn, Ni, and Cr
- Fe crystalline oxides hold, as reducible fractions, Fe, Mn, Cd, and Zn
- Chlorite reveals variable capacity to fix metals, predominantly as reducible and residual fractions

It is necessary to emphasize, however, that during analytical procedures several artifacts can occur and create unreal results.

The speciation of trace metals in soils is related to their biogeochemical reactivity and to several physicochemical soil conditions. The determinations of various fractions of these elements are broadly used especially for two purposes, to link with the potential bioavailability, and to predict the mobility affecting their transport to groundwater. The speciation of metals in soils is not stable, and the transformation of their forms in soil is observed. An example can be the relatively rapid transformation of metals (Cd, Zn, Pb) applied to soils as oxides into exchangeable and carbonate species (Kabata-Pendias and Pendias 2001).

Trace element speciation analysis provides a basis for the description of forms and compounds in which they occur in soils. Other definitions are also used to define the properties of trace elements in soils. The general term "availability" has been recently described as: geoavailability, geochemical availability, biogeochemical availability and bioavailability (Wolt 1994; Smith and Huyck 1999). "Geoavailability" is defined as that portion of the total content of a chemical element or a compound in an earth material that can be liberated to the surficial (soil) environment. "Geochemical availability" is related to that fraction of a chemical element that participates in transport processes affecting the spatial distribution and changes over time. "Biogeochemical availability" is defined as the quantity-intensity factor calculated as a ratio of total content to the concentration of a trace element in soil solution. "Bioavailability" refers to the fraction of a trace element that is or has been available to an organism.

In spite of intensive investigations (sequential extractions and specific extractants) on the prediction of the availability of trace metals, good estimations of general bioavailability of a given trace element are needed. However, based on results of long-term experiments, some specific extractants, as well as weak neutral salt solutions seem to be adequate to assess the impact of trace elements on plants and soil bioactivity.

Trace element interactions. Interactions between chemical elements both micro and macro may have an antagonistic and/or synergistic character. Although most of these interactions are associated with physiological processes in plants some reactions are also related to pedochemical processes. These are mainly impacts of major elements on the distributions and forms of some trace elements in soils. The main antagonistic interactions (or relationships) in soils, often associated with microbiota, are reported by Kabata-Pendias and Pendias (2001) for:

- Ca and B, Ba, Cd, Co, Cr, Cs, Li, Mn, Ni, Pb, Sr, and Zn
- P and As, Cr, Hg, Mo, Mn, Ni, Pb, Rb, Se, and Zn
- Mg and Cr, Mn, Zn, Ni, Co, and Cu
- Fe and Co, Ni, Mn, Mo, Cr, and Zn
- Mn and As, Cr, Cu, Mo, Ni, V, and Zn
- Cu and Mo
- Cd and Zn

Interactions between elements in soils are of multivariant character but may be mainly attributed to the sorption sites of soils particles and also to the surface root cells and to root exudates.

Trace element occurrence. Contents of trace elements in soil materials from natural as well as from contaminated sites show great variability in both the horizontal and vertical dimensions. The heterogeneity of soils, especially at the microscale, creates real problems in representative sampling that have serious impacts on the reproducibility and comparability of analytical data. This is of great concern in the assessment of background (pristine) contents of trace elements in soils. Although a real pristine status of trace elements in soils does not exist nowadays, even in remote regions, some values of trace elements in "uncontaminated" soils are prerequisites as reference values for the evaluation of soil pollution. Several methods were developed to calculate background contents of trace elements in soils. The best-known methods are, as presented (from various sources) by Kabata-Pendias and Pendias (2001):

- GB: Geochemical Baseline
- GD: Index of Geochemical Distribution
- IGL: Index of Geochemical Load
- IPD: Index of Pedogenic Distribution
- PEF: Pedochemical Enrichment Factor
- FSP: Factor of Soil Parameters
- SCV: Spatial Concentration Variability
- GIS and MA: Geographical Information System and Multivariate Analyses

The fine granulometric soil fraction (<0.02 mm) is composed mainly of clay minerals and therefore exhibits a high ability to bind trace ions. Thus, in most soils there is observed a positive correlation between the content of the clay fraction and amounts of trace elements, in particular cations. The highest relationship, more than 50% of relative distribution index (RDI) has been documented for Zn, Fe, Ni and Cr (Fig. I-2.3). This is clearly demonstrated in "background" contents of trace elements in sandy (low content of clay fraction) and loamy (high contents of clay fraction) soils (Table I-2.6). Clay minerals and other aluminosilicates represent several groups of minerals of variable structure, and can be both primary (inherited from parent material) and secondary (formed in soils).

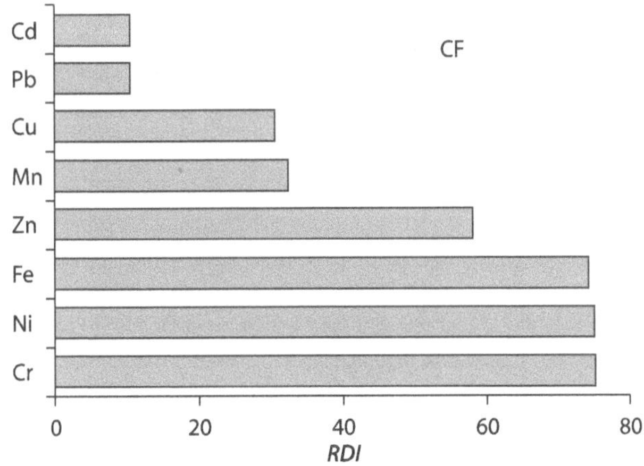

Fig. I-2.3. Relative distribution index (RDI) of statistically-significant relationships between trace metals and content of clay fraction, <0.02 mm (CF) in mineral soils, at the 99% confidence level

Table I-2.6. Mean (arithmetic) and maximum values (mg kg^{-1}) of the background contents of some trace elements[a] in two common soil kinds on the worldwide scale

Element	Podzol, sandy soils (<20% fraction <0.02 mm)		Cambisols, loamy soils (>20% fraction <0.02 mm)	
	Mean	Maximum	Mean	Maximum
As	4.4	30	8.4	27
B	22	134	40	128
Ba	330	1500	520	1500
Cd	0.37	2.7	0.45	1.61
Co	5.5	65	10	58
Cr	47	530	51	1100
Cu	13	70	23	100
F	130	1100	385	800
Hg	0.05	0.7	0.1	1.1
I	2.3	10	1.7	8.3
Li	22	72	46	130
Mn	270	2000	525	9200
Mo	1.3	3.7	2.8	7.2
Ni	13	110	26	110
Pb	22	70	28	70
Sc	5	30	8	20
Se	0.25	1.32	0.34	1.9
Sr	87	1000	210	1000
V	67	260	76	330
Zn	45	220	60	362

[a] Related to aqua regia soluble elements or other forms defined as total contents.

Organic matter also plays a significant role in sorption of trace elements. Most kinds of SOM have a relatively high specific surface area and common range of CEC (at pH 8) is calculated to be in the range from 60 to 300 cmol(+)kg^{-1} (Baldock and Nelson 2000). The CEC values of SOM are governed to a great extent by acidity and increase from 36 cmol(+)kg^{-1} at pH 2.5 to 215 cmol(+)kg^{-1} at pH 8 (Sparks 1995). The contribution of SOM to the CEC is significant, and varies from 25 to 90%, depending on soil categories.

In several soils or in some layers of soil horizons, oxides and hydroxides of Fe and Mn play very important roles in the distribution and behavior of trace elements, and may fix great amounts of some trace metals, especially of Co (up to 0.3%) and Ni (up to 0.5%) (see Table II-7.3). The Fe and Mn hydroxides occur in soils as concretions and/or nodules, as well as coatings on soil particles, and as fillings in cracks and veins.

The large database of recent surveys of trace elements in soils allows general estimations to be made of the concentration means in soils of various countries (Table I-2.7). The background populations of trace elements in soils, both uncontaminated and contaminated, seem to be log-normally distributed, though with much higher extremes in contaminated ones.

Many elements are likely to be accumulated in the surface soil layer (SEFcrust > 1). However, some elements, e.g., Te, Se, Pd, Pt, Cd, Ge, Cl, As, Sb, Br, and Hf are significantly enriched in soils as compared to their crustal abundance (SEFcrust from >2 to 16).

The surface soil layer, enriched in organic matter (mainly humus) is an important sink for trace element deposition from the atmosphere. The investigation of trace metals in organic surface soils in Fennoscandia, sampled in 1995, has shown a general impact of long-range transport of pollutants (Steinnes and Rühling 2002). There was observed a much smaller influence of regional differences in geochemistry of the underlying bedrocks than of aerial pollution. The ranges of trace metals were following (in mg kg^{-1}, expressed in range of < and >): As, 0.4–2; Cd, 0.2–1; Cu, 4–12; Fe, 1 000–5 000; Hg, 0.1–0.5; Ni, 2–10; Pb, 10–60; V, 3–15; and Zn, 30–110.

Soil pollution. Both terms, contamination and pollution, are used synonymously. According to the definition given by Knox et al. (1999) trace element contaminated soils are not considered to be polluted unless a threshold concentration exists that begins to affect biochemical and biological processes.

Soil pollution is as old as man's ability to smelt and process ores, and goes back as far as the Bronze Age (2500 BC). Ernst (1998b) comprehensively reviewed ancient metal contamination from the Bronze Age to Roman times. Almost all human activities have resulted in increased levels of trace elements in soils. The trace elements from anthropogenic sources exist mainly on the surfaces of soils as reactive forms and may occur as: water-soluble, exchangeable, associated with organic matter, carbonate, oxides of Fe, Al and Mn and silicates (Ma and Uren 1998; Ramos et al. 1994). Sources of trace inorganic contaminants (TIC) are:

- Industrial activities
  - Mining and smelting of metalliferous ores
  - Brick and pipe manufacture
  - Cement manufacture
  - Others, e.g., ceramic, glass, chemical plants
- Power generation
  - Burning of fossil fuel
  - Nuclear reactors
  - Incineration of municipal wastes
- Agricultural practices
  - Soil amendment with sewage sludge
  - Application of manure
  - Mineral fertilizers
  - Pesticides, fumigation
- Transport and urban-derived pollution
- Long-range transport of pollutants

Energy and mineral consumption by humans is the main cause of contamination of the biosphere. The ratio of mined trace elements to their deposition in contemporary geological sediments may indicate their potential threat to the environment. Based on such calculations, Bowen (1979) has found that the most dangerous to the biosphere are following elements: Ag, Au, Cd, Cr, Cu, Hg, Pb, Sb, Tl, and Zn. Among the

Table I-2.7. Commonly reported mean background contents of trace elements in continental crust and surface soils (mg kg^{-1})

Element	Crustal average[a]	A	B	C	D	E	SEFcrust
Antimony, Sb	0.2	0.62	0.25	0.78	–	0.66	2.89
Arsenic, As	1.8	4.7	3.8	–	–	7.2	2.91
Barium, Ba	400	362	608	350	–	580	1.19
Beryllium, Be	3	1.9	1.3	1.4	–	0.92	0.46
Bismuth, Bi	0.2	0.7	0.16	0.33	–	–	1.98
Boron, B	15	–	5.1	–	–	33	1.27
Bromine, Br	2	–	–	–	10.5	0.85	2.84
Cadmium, Cd	0.1	1.1	0.17	0.33	0.18	<0.01–41	4.40
Cerium, Ce	60	49	60	52	89	75	1.08
Cesium, Cs	3	8	1.7	5.4	4.6	–	1.64
Chlorine, Cl	130	380	–	–	–	–	2.92
Chromium, Cr	100	42	22	58	86	54	0.52
Cobalt, Co	10	6.9	7.1	18	17	9.1	1.16
Copper, Cu	55	14	17	48	109	25	0.77
Dysprosium, Dy	3	0.7	4.1	3.9	5.6	–	1.19
Erbium,, Er	2.8	1.6	2.2	2.2	3.1	–	0.81
Europium, Eu	1.2	1.2	0.79	1.2	1.5	–	0.98
Fluorine, F	625	264	–	–	269	430	0.51
Gadolinium, Gd	5.4	2.2	3.4	4.2	5.5	–	0.71
Gallium, Ga	15	1.2	8.9	20	31	17	1.04
Germanium, Ge	1.5	1.2	19	–	1.9	1.2	3.88
Gold, Au	0.004	0.002	<0.005	–	0.002	–	0.75
Hafnium, Hf	3	3	7.6	2.5	12.7	–	2.15
Holmium, Ho	0.8	1.1	0.87	0.73	1	–	1.16
Indium, In	0.06	–	<0.04	0.09	0.11	–	1.33
Iodine, I	0.5	2.4	–	–	13	1.2	11.1
Iridium, Ir	0.0001	–	<0.04	–	–	–	–
Lanthanum, La	30	26	33	23	34	37	1.02
Lead, Pb	14	25	18	24	22	19	1.54
Lithium, Li	20	28	17	13	24	24	1.06
Lutetium, Lu	0.3	0.34	0.39	0.31	0.52	–	1.30
Manganese, Mn	900	418	411	–	535	550	0.53
Mercury, Hg	0.07	0.1	0.043	–	0.053	0.09	1.02
Molybdenum, Mo	1.5	1.8	0.58	1.3	1.6	0.97	0.83
Neodymium, Nd	28	19	29	22	32	46	1.06

Table I-2.7. *Continued*

Element	Crustal average[a]	A	B	C	D	E	SEFcrust
Nickel, Ni	20	18	13	26	25	19	1.01
Niobium, Nb	20	12	12	10	25	11	0.70
Osmium, Os	0.0005	–	–	–	–	–	–
Palladium, Pd	0.004	–	0.04	–	0.003	–	5.38
Platinum, Pt	0.004	–	<0.04	–	0.002	–	5.25
Praseodymium, Pr	8.2	7.6	7.7	5.3	8.4	–	0.88
Rhenium, Re	0.001	–	<0.04	–	–	–	–
Rhodium, Rh	0.001	–	<0.04	–	–	–	–
Rubidium, Rb	90	50	116	70	18	67	0.71
Ruthenium, Ru	0.001	–	<0.04	–	–	–	–
Samarium, Sm	4.7	3.1	4.5	4.4	6.7	–	0.99
Scandium, Sc	11	9.5	10	21	–	8.9	1.12
Selenium, Se	0.05	0.7	0.23	–	0.47	0.39	8.95
Silver, Ag	0.06	0.1	0.11	0.1	0.05	–	1.50
Strontium, Sr	375	147	163	190	–	240	0.49
Tantalum, Ta	2	1.1	1.1	1.7	2.3	–	0.78
Tellurium, Te	0.005	–	<0.08	–	–	–	16.0
Terbium, Tb	0.6	0.4	0.48	0.74	0.9	–	1.05
Thallium, Tl	0.5	0.6	0.23	0.49	0.36	–	0.84
Thorium, Th	7.2	8.2	8.1	9	11	9.4	1.27
Thulium, Tm	0.5	0.46	0.32	0.3	0.5	–	0.79
Tin, Sn	2.5	–	1.8	2.4	–	1.3	0.73
Titanium, Ti	4400	–	3700	–	15480	2900	1.67
Tungsten, W	1.5	1.2	1.3	1.3	1.4	<0.16–0.17	0.71
Uranium, U	2	3.7	4.4	1.9	2.9	2.7	1.56
Vanadium, V	135	60	69	180	320	80	1.05
Ytterbium, Yb	2.2	2.1	2.9	2.1	3.2	3.1	1.22
Yttrium, Y	33	12	27	21	27	25	0.68
Zinc, Zn	70	62	65	89	73	60	1.00
Zirconium, Zr	165	300	308	92	421	230	1.64

[a] Values are compiled from Mason and Moore (*vide* Hedrick 1995) and Reimann and Caritat (1998). Given are mean values for various soils in different countries: *A* – worldwide data after Kabata-Pendias and Pendias (1999, 2001); *B* – agricultural soils of Sweden after Eriksson (2001a); *C* – agricultural soils of Japan, after Takeda (2004); *D* – medians for soils of Parana State, Brazil, after Licht (2005); *E* – data for the U.S. soils, after Burt et al. (2003) and Shacklette and Boerngen (1984); *SEFcrust* – soil enrichment factor in relation to crust mean concentrations (ratio of "grand" mean soil content to upper continental crust content).

Table I-2.8. World mining of the elements in 2000 and 2003 (t, kt, Mt yr^{-1})

Element	2000	2003	Element	2000	2003
Aluminum, Al	139 Mt	148 Mt	Lithium, Li	138 kt	–
Antimony, Sb	130 kt	155 kt	Manganese, Mn	21.38 Mt	24.35 Mt
Arsenic, As	27.9 kt	38.8 kt	Mercury, Hg	2.2 kt[a]	–
Barium, Ba	266 t	–	Molybdenum, Mo	–	127.4 kt
Beryllium, Be,	2.26 t	2.55 t[b]	Nickel, Ni	1.14 Mt	1.2 Mt
Bismuth, Bi	4.28 kt	4.47	Niobium, Nb	–	27 kt
Boron, B	4.2 Mt	4.4 Mt	Palladium, Pd	140 t	230 t
Bromine, Br	540 kt	–	Platinum, Pt,	365 t	–
Cadmium, Cd	19.21 kt	16.87 kt	Rhenium, Re	43 t	39.3 t
Cesium, Cs	5.9 Mt	–	Selenium, Se	–	2.31[d] kt
Chromium, Cr	13.91 Mt	15.83 Mt	Silver, Ag	17.75 kt	18.21 kt
Cobalt, Co	35.6 kt	43.03 kt	Strontium, Sr	320 kt	–
Copper, Cu	13.23 Mt	13.68 Mt	Tantalum, Ta	1.22 kt	–
Fluorine, F	4.4 kt	4.5 kt[c]	Thallium, Tl	15 t	–
Gallium, Ga	70 t	203 t	Tellurium, Te	313 t	243 t
Germanium, Ge	2.4 kt	–	Tin, Sn	247.3 kt	257.7 kt
Gold, Au	–	2.35 kt	Titanium, Ti	4.31 Mt	4.22 Mt
Hafnium, Hf	145 t	335 t	Tungsten, W	35.74 kt	50.0 kt
Indium, In	335 t	–	Uranium, U	34.86 kt	35.37 Kt
Iodine, I	18 kt	–	Vanadium, V	43 kt	–
Iron, Fe	571 Mt	–	Zinc, Zn	8.73 Mt	9.17 Mt
Lead, Pb	3.1 Mt	6.8 Mt	Zirconium, Zr	750 kt	1.1 Mt

t – Metric ton, *kt* – thousand tons, *Mt* – million tons. Sources: WMSY (2004), and USGS (2004).
[a] Data for 1999.
[b] Given as Be-ores.
[c] Given as fluorspar.
[d] Data for 1995, after Reimann and Caritat (1998).

least dangerous elements are: Ga, La, Nb, Sr, Ta, and Zr. Data for the world mining of trace elements differ between 2000 and 2003, and also depend on the source of data (Table I-2.8). The residence time of trace metals in surface soils depends on several soil properties of which pH and Eh parameters are the most significant. As Bowen (1979) has reported, the maximum residence time (in years) may be: *(i)* 1 000 for Cd, Hg, and Be, *(ii)* 3 000 for Zn and As, and *(iii)* 6 000 for Pb.

The natural sources of trace elements are associated mainly with continental and volcanic dust, although some elements easily released from the ocean surface (e.g., I and Se) contribute to their elevated contents in soils of some ecosystems. According to Korzh (1991) the deposition of trace elements from oceans *via* the atmosphere onto soils are at the rate (in t yr^{-1}):

- 10^4–10^6: Li, Rb, Mo, I, Ba, B, Br, F, and Sr
- 10^2–10^3: Cr, Mn, Fe, Cu, Ga, Se, Y, Zr, Cd, W, TI, Ti, V, Ni, Zn, As, Sb, Cs and U
- 1 –10: Be, In, Eu, Pr, Nd, Sm, Gd, Dy, Ho, Er, Yb, Hf, Ta, Re, Au, Hg, and Pb

Cosmic particles that are deposited on the Earth at rates of 100–1 000 t d^{-1} may also be, though mainly unrecognized, a source of some trace elements (Dodd 1981). However, as indicated by data presented by Kjølholt (1997), the atmospheric deposition of trace metals from natural sources is negligible compared with their loads with products used for soil amendment (Table I-2.9). The highest amounts of trace metals are added to agricultural soils with the application of sewage sludge, compost with biosolids, and fly ash (Table I-2.10). This may change, however, depending on the soil system and on variable factors that affect the trace element status of soils (Fig. I-2.4). Inputs of trace metals to soil on both animal and crop farms may be quite significant, however

Table I-2.9. Loads of trace elements (g ha^{-1} yr^{-1}) from various products applied on Danish agricultural soils and from aerial deposition (after Kjølholt 1997)

Source	Cd	Cr	Cu	Hg	Ni	Pb	Zn
Sewage sludge	2	65	390	2.4	50	96	1 900
Fertilizers (P_2O_5)	0.25	3.3	67	≪0.01	0.61	0.26	4.3
Manure, cattle + pigs	1.9	11	630	<0.4	21	3.8	820
Compost, all	3.2	53	320	0.66	89	140	970
Compost, MSW[a]	3.8	98	320	1.2	150	350	1 200
Liming	0.6	<1	3	0.003	4	<3	13
Irrigation	0.04	0.6	1.2	0.008	3.3	1	21
Emissions from tractors	0.009	0.009	0.043	0.0004	2.7	0.13	0.11
Atmospheric deposition	0.15	10	7	0.07	7	12	4

[a] *MSW* – Municipal Solid Waste.

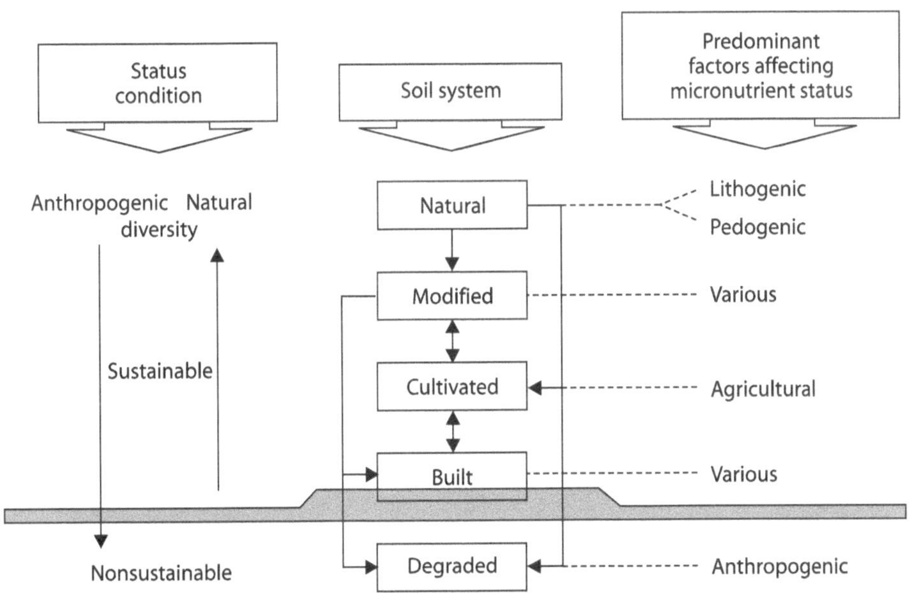

Fig. I-2.4. Factors affecting the trace element status of various soil systems

Table I-2.10. Mean contents of trace elements in fertilizers, sewage sludge, and fly ash (adopted from Eriksson 2001a, unless otherwise stated)

Element	NPK-S[a]	P20[b]	SS[c]	SM[d]	Fly ash[e]	Element	NPK-S[a]	P20[b]	SS[c]	SM[d]	Fly ash[e]
Antimony, Sb	0.031	0.20	2.4	0.16	3.8	Neodymium, Nd	13	594	11.3	2.9	–
Arsenic, As	0.59	9.1	4.7	1.3	60	Nickel, Ni	6.2	11	20	6.1	87.9
Barium, Ba	3.3	123	310	72	398	Niobium, Nb	1.6	0.83	4.5	0.34	–
Beryllium, Be	0.066	0.14	<0.6	0.13	11	Palladium, Pd	<0.001	<0.001	0.16	0.01	–
Bismuth, Bi	0.003	0.06	0.73	0.013	–	Platinum, Pt	0.006	0.003	<0.04	<0.001	–
Boron, B	24	7.6	61	28	509	Praseodymium, Pr	–	136	2.8	0.8	–
Cadmium, Cd	0.05	0.11	1.2	0.25	1.3	Rhenium, Re	0.0002	0.0007	<0.04	0.0008	–
Cerium, Ce	27	1027	24	7.1	112	Rhodium, Rh	<0.001	0.002	0.04	<0.001	–
Cesium, Cs	0.073	0.32	0.63	0.14	16.1	Rubidium, Rb	<0.001	6.8	15	224	106
Chromium, Cr	8.9	12	33	11	118	Ruthenium, Ru	<0.001	0.003	0.08	<0.001	–
Cobalt, Co	0.95	5	6.2	1.1	29.2	Samarium, Sm	2.6	113	1.8	0.49	–
Copper, Cu	1.5	109	390	113	71.8	Scandium, Sc	0.41	3.5	3.2	0.27	–
Dysprosium, Dy	2.1	35	1.7	0.33	9.5	Selenium, Se	0.041	0.032	1.3	0.89	1.6
Erbium, Er	1	9.1	1	0.18	–	Silver, Ag	<0.0005	<0.0005	7.4	0.018	–
Europium, Eu	0.87	25	0.3	0.11	–	Strontium, Sr	53	1720	170	76	720
Gadolinium, Gd	3.1	87	2	0.77	–	Tantalum, Ta	0.071	0.011	0.94	0.0026	1.5
Gallium, Ga	0.25	0.69	3.5	0.45	10.8	Tellurium, Te	<0.0005	<0.0005	0.16	0.006	–
Germanium, Ge	0.019	0.16	4.3	0.05	6.6	Terbium, Tb	0.42	9.8	0.34	0.066	1.6
Gold, Au	0.001	0.003	0.79	0.002	–	Thallium, Tl	0.026	0.11	0.15	0.026	1.4
Hafnium, Hf	0.1	0.48	1.3	0.45	–	Thorium, Th	1.9	48	2.4	0.35	22.1
Holmium, Ho	0.38	0.024	0.4	0.07	–	Thulium, Tm	0.13	089	0.21	0.025	–
Indium, In	<0.005	<0.005	0.15	<0.01	–	Tin, Sn	0.075	0.34	22	1.3	8.7
Iridium, Ir	0.0002	0.001	<0.4	<0.0001	–	Titanium, Ti	78	259	1800	57	–
Lanthanum, La	12	422	16	4.1	56.5	Tungsten, W	0.04	0.023	7.9	0.11	4.5
Lead, Pb	0.53	9.1	33	2	52	Uranium, U	0.36	9.6	10	2.2	22.9
Lithium, Li	0.66	3.6	4.3	1.5	235	Vanadium, V	6.7	17	18	6.1	208
Lutetium, Lu	0.096	0.56	0.19	0.02	0.7	Ytterbium, Yb	0.75	5	1.1	0.16	4.8
Manganese, Mn	323	307	280	250	325	Yttrium, Y	14	114	11	2.1	43.8
Mercury, Hg	0.012	0.024	1.1	0.02	52	Zinc, Zn	15	54	550	680	221
Molybdenum, Mo	0.13	0.56	6.7	6.8	14.6	Zirconium, Zr	3.3	21	53	2.6	194

[a] Fertilizers manufactured by Hydro Agri, Sweden. [b] Phosphorus fertilizer manufactured in Sweden. [c] Sewage sludge produced in Sweden. [d] Solid pig manure. [e] After Llorens et al. (2000).

they vary depending on the farm, the country, and the metal (Table I-2.11). The highest annual net gain has been calculated, among European countries, for Cu and Zn in dairy farms in France. The balance for the same elements was negative in dairy farms of Norway. Crop farms may generate higher inputs, than animal farms, of trace metals to soils. However, there is also observed a great variation between farm types and this depends mostly on crop and soil types. Balances for trace metals in crop-farm soils are presented in the AROMIS database (Eckel et al. 2005) as follows (in g ha^{-1} yr^{-1}):

- Cd: from −0.3 in Denmark to 4.4 in France
- Cr: from 14 in Denmark to 100 in France
- Cu: from −35 in France to 347 in Germany
- Ni: from −3.3 in Denmark to 33 in France
- Pb: from 1.3 in Norway to 139 in France
- Zn: from −115 in Germany to 838 in France

Table I-2.11. Trace metal balance (g ha^{-1} yr^{-1}) for livestock farms in some EU countries, selected data from the AROMIS data base (after Eckel et al. 2005)

Country	Farm	Cd	Cr	Cu	Ni	Pb	Zn
Czech Republic							
Input/output	Pigs	2.1/0.9	19/8.5	87/57	25/21	24/6.5	609/389
Balance		1.2	11	30	4	17	220
Finland							
Input/output	Dairy	0.5/0.04	15/0.4	123/6.0	23/2.9	13/0.8	629/36
Balance		0.5	15	117	20	12	593
France							
Input/output	Pigs-1	7.5/0.7	60/7.6	1556/632	17/21	13/4.3	3279/1495
Balance		6.7	53	924	−3.9	8.8	1784
Input/output	Dairy	6.1/0.4	–	1066/29	–	17/3.7	2746/152
Balance		5.7	–	1037	–	13	2594
Germany							
Input/output	Pigs	2.0/1.5	43/2.6	408/37	36/15	62/4.4	1482/206
Balance		0.5	41	371	21	58	1276
Input/output	Dairy	2.5/0.8	34/1.9	939/31	38/16	40/3.6	959/161
Balance		1.8	32	908	21	36	798
Norway							
Input/output	Dairy	1.5/1.4	250/36	63/201	100/67	32/31	404/535
Balance		0.1	215	−138	32	1.3	−131
United Kingdom							
Input/output	Dairy	3.1/0.1	27/7.9	109/8.6	22/16	59/3.4	393/40
Balance		3.0	19	100	6	55	354

Budgets of trace metals in soils of forest ecosystems are also variable, however, they are more often negative than in agricultural soils. The highest negative balances were calculated (in the 1970s and 1980s) for spruce forest soils (in Sweden) and for Fe and Mn (Tyler, Bergkvist *vide* Kabata-Pendias and Pendias 2001).

Trace element budgets have been calculated for various ecosystems, and the ratio of input/output, show that for the majority of elements, the accumulation rate in surface soils is positive. However, in some European forest soils with a high rate of mobility and leaching of trace metals, a decrease of these elements is observed, even with relatively high inputs to soils. The highest losses, under such conditions, have been reported mainly for Mn (up to 6 100 g ha^{-1} yr^{-1}) and for Fe (up to 11 000 g ha^{-1} yr^{-1}) (Kabata-Pendias and Pendias 2001). It has been estimated, that in most European ecosystems, and in particular in agricultural soils, the input of most metals is higher than their losses. This is especially pronounced for Zn, Pb and Cd. The calculation of Cd budgets for two soils, in Germany and Poland, has indicated that the point source of pollution and long-range transport is about equal to nonpoint source of pollution, which is mainly from agricultural practices. The net gain balances for Cd range (in g ha^{-1} yr^{-1}) between 4.5–7.5 and 1.3–40 for Poland and Germany, respectively (Kabata-Pendias and Brümmer 1991).

The transport of dissolved trace elements may take place through the soil solution (diffusion) and with the moving soil solution (mass flow, leaching). In soils formed under a humid and cool climate, the leaching of elements downward through the profiles is greater than their accumulation, unless there is a higher input of these elements into the soils. In warm and dry climates, and to some extent in humid hot climates, upward translocation of trace elements in the soil profiles is relatively common. Several soil properties control leaching processes, of which the most important may be considered soil texture and the Eh-pH system.

Radionuclide contamination. Nuclear power generation has been utilized since the mid-1950s. As Fuge (2005) has described, accidental leakage and permitted effluence have released some radionuclides, mainly to the immediate environment of the nuclear installations. The explosion at the Chernobyl nuclear station (presently in Ukraine) in April 1986 caused widespread contamination that seriously affected some European countries and several other parts of the world.

More important radioactive isotopes released by the Chernobyl accident, according to Becker-Heidmann and Scharpensel (1990), are as follows:

- Short half-life time (days): ^{141}Ce, ^{132}Te, ^{133}Xe, ^{131}I, ^{99}Mo, ^{95}Zr, and ^{89}Sr
- Longer half-life time (years): ^{134}Cs, ^{85}Kr, ^{137}Cs, ^{103}Ru, and ^{90}Sr
- Long half-life time (hundreds of years): $^{238-240}Pu$, and ^{241}Am

The artificial radioactivity of soils is caused by settling of the released radionuclides. The fallout of Chernobyl's isotopes varied greatly and the nonhomogenous distribution is partly associated with the wet precipitation (rainfall) during that period.

The behavior of isotopes in soil depends on several soil properties, as well as on the specific behavior of each radionuclide. The distribution, fate, and phytoavailability of ^{137}Cs (half-life 30 years) have been investigated most frequently (Sheppard 2003).

The highest radioactive contamination of surface soils of Poland, after the Chernobyl explosion was observed in the southwestern part of the country. The concentrations of ^{137}Cs in soils of those regions were at the range 5–13 kBq m^{-2}. During the period 1988–2001 the country average concentration of this isotope diminished from about 4.6 to 3.2 kBq m^{-2} (Biernacka et al. 2004).

Urban soils. Contamination of urban soils has recently become the subject of several studies (Kasimov 1995; de Miquel et al. 1997; Kabata-Pendias and Krakowiak 1998; Umfitseva and Terekhina 2005). Recently reported levels of some trace metals in various soils, e.g., in China may be a real hazard to the environment and health (Table I-2.12).

There are several sources of trace element pollution in the urban environment, among which roadside dust may be a significant one, due to their increased contents in fuels and road materials (Table I-2.13).

Based on investigations of soil pollution in Oslo and Madrid, some trace metals have been distinguished as so-called "urban" contaminants: Ba, Cd, Pb, Sb, Ti, and Zn (de Miquel et al. 1997). However, the pollution of cities is specific and varies with local conditions. The trace elements increased in soils of some cities as compared with surrounding soils are reported as follows:

- Berlin: Cd, Cu, Hg, Ni, Pb, Sn, Th, Tl (Birke and Rauch 1994)
- Minsk: Cd, Pb, and Zn (Pietukhova et al. 1996)
- Novosibirsk: As, Br, Cd, Cu, Hg, Mo, Pb, Zn, and Zr (Ilyin 1997)
- Sankt-Petersburg: Ba, Cr, Cu, Pb, V, W, and Zn (Umfitseva and Terekhina 2005)
- Toronto, increase in house dust against street dust: Cd, Cu, Hg, Pb, Sn (Rasmussen et al. 2001)
- Warsaw: Ba, Cd, Cr, Cu, Li, Ni, Pb, Sr, and Zn (Kabata-Pendias and Krakowiak 1998)

Table I-2.12. Concentrations (the highest reported contents) of some trace elements in soils contaminated from industrial sources in selected countries (mg kg^{-1}) (various sources, as compiled by Kabata-Pendias and Pendias 2001, unless otherwise indicated)

Element	Canada	China[a]	Japan	Poland	UK	USA
As	2000	2500	2470	2000	900	2000
Cd	36	800	88	270	468	1500
Co	127	300	–	–	67	85
Cu	3700	>2000	206	620	2000	–
Hg	5.7	100	–	–	3.4	40
Pb	12123	>10000	2100	8000	4563	13000
Mo	–	100	–	–	–	–
Ni	26000	8000	–	–	1150	–
Zn	1397	>10000	5400	13800	4500	80000

[a] Chinese soils polluted either from parent material or from industrial sources; after He et al. (2005).

Table I-2.13. Trace metals in samples contributing to roadside dust (after Ozaki et al. 2004). Samples collected during the period 2001–2002 in Japan

Sample	Cd	Hg	Cu	As	Sb	Pb	V	Cr	Ni	Zn
	μg kg^{-1}		mg kg^{-1}							
Exhausted soot (bus)	49.5	13	29.7	7.5	1.1	10.5	346	32.5	5.3	1450
Road marking (white)	510	4.3	2.2	2.5	0.6	7.4	5.9	2.7	1.2	20.8
Road marking (yellow)	413	7.2	1.5	0.9	2.1	12000	26	820	0.5	8.4
Road marking (red)	365	20	1.9	6.7	10.2	5290	53.6	561	1.6	13.6
Asphalt	50	11.3	9.8	3.1	0.5	6.9	38.4	38.4	27.5	4
Auto tire	4040	ND	0.8	0.7	2.2	2.2	10.4	0.9	0.4	12700
	μg kg^{-1}						mg kg^{-1}			
Regular gasoline	240	15	193	54	43	295	ND	ND	3.5	13
Premium gasoline	347	47	246	ND	5.4	173	ND	ND	4.6	13.6
Gas oil	266	0.7	179	11	16	13	ND	ND	2.9	4.4
Kerosene	171	2.7	18.6	11	18	7	ND	ND	1.2	2.5

ND – Not detected.

As reported by Umfitseva and Terekhina (2005), most of other trace elements also are increased in urban soils as compared with soils of surrounding areas (Table I-2.14). Increased levels of some trace metals in house dust as compared with both street dust and garden soils may be a significant source of metallic pollutants, and may pose a risk especially to children.

Soil remediation. The remediation of soils refers to practices of either removing contaminants or converting them into to less mobile species – that is, into less bioavailable forms. Remediation methods are based on physical, chemical and biological approaches. The selection of a method is generally based on the nature of the contaminants, the soil type, the characteristics of the contaminated site (e.g., chemical and physical soil properties, and the length of the soil contamination period). Also remediation costs and regulations of the country are involved in remediation projects (Alloway 1995; Mukherjee 2001).

Methods for remediating metal-polluted soils have been recently widely investigated and discussed (Sparks 1995; Adriano et al. 1998; Cunningham and Berti 2000; Iskandar 2001; Knox et al. 2000). The remediation of soils and sites contaminated with radionuclides is of special concern. Several techniques, *in situ* and non-*in situ* are applied to clean up soils contaminated with trace metals. Due to the complexity of soils and the presence of multiple contaminants, only a few of these techniques have been applied successfully in practice and some of them are quite costly. Most commonly used remediation technologies (in-situ and non-in-situ methods) are:

Table I-2.14.
Trace elements and iron in soils of central Sankt-Petersburg and of the surrounding region ($mg\,kg^{-1}$) (after Umfitseva and Terekhina 2005)

Element	Surrounding region		Down town	
	Mean	Maximum	Mean	Maximum
Ag	<0.1	0.1	0.4	30
As	2.6	6	5.7	–
Ba	202	300	533	50000
Cd	0.2	0.6	1.5	190
Co	4.1	7	11.5	500
Cr	12.5	30	92.2	30000
Cu	18	30	120	15000
Fe[a]	13.9	50	38	200
Hg	<0.1	0.1	0.8	53.5
Mo	1.1	1.5	3.2	500
Mn	118	300	323	7000
Ni	15.3	30	47.7	5000
Pb	19.1	44	223	29448
Sn	1.4	3	15	2000
Sr	162	283	117	3000
Ti	1522	5000	3011	30000
V	16.2	30	24.4	5000
W	1	1	9	1500
Y	17	50	20	70
Zn	43	70	774	X[b]
Zr	140	300	155	1500

[a] Fe content in $g\,kg^{-1}$.
[b] Maximum Zn concentration not established due to extremely high values.

- Leaching/washing/flushing – soil with water or surfactant (a surface active substance), not very commonly practiced, limited due to the need for large quantities of water, results depend strongly on physical soil parameters
- Solidification/stabilization/immobilization – soil amendments with materials having a high capacity to bind metals in possible slightly mobile fractions and/or immobilize by keeping a neutral soil pH, for effectiveness required site-specific data, relatively low cost
- Biodegradation – substances containing trace metals are decomposed due to microbial degradation, trace pollutants are mobilized and washed-out, long-term effects, practiced in specific sites
- Vitrification – pollutants are immobilized with an electric current, seldom used, very costly
- Isolation/containment – minimizing downward migration of pollutants by installing subsurface barriers: e.g., clay layer, organo-clay layer, plastic liner
- Encapsulation – covering small sites with a layer of material of a low permeability (e.g., clay) to prevent water percolation and wind-blown dust

- Immobilization of trace metals, e.g., by steel shots, containing mainly Fe oxides, which adsorb or occlude several trace metals, by phosphoric compounds which decrease the mobility of most metals
- Phytoremediation – the phytoextraction techniques provide an adequate method under reasonable yield of plants that hyperaccumulate metals, promising in practice, needs development of technology, relatively low cost
- Removal – the contaminated soil is exposed to chemical extraction and/or thermal treatment to remove volatile elements or compounds, and to other leaching or immobilizing processes, high costs
- Excavation – the contaminated soil is removed and disposed elsewhere (e.g., in prepared landfills), high cost and possible problem with groundwater contamination

The problem of concern in soil remediation actions is the cost. Phytoremediation techniques are likely to be less costly than those based on conventional technologies. At present, an urgent requirement for phytoextraction is to increase the yield of plants that hyperaccumulate metals from soils, and to develop adequate technologies for the utilization of the plant materials.

Agricultural practices have also been applied to soil remediation. Most commonly, the uptake of metals by plants is diminished by keeping a neutral soil pH and by amendments with materials having a high capacity to bind metals in possibly slightly

Table I-2.15. Mean (arithmetic) contents of trace elements in street dust, garden soils and house dust of Ottawa, Canada (mg kg^{-1}) (after Rasmussen et al. 2001; samples collected in 1993)

Element	Street dust	Garden soil	House dust
Ag	0.22	0.31	2.5
As	1.3	3.0	7.3
Ba	576	766	492
Bi	0.05	0.08	1.67
Cd	0.37	0.3	6.46
Co	8.31	8.36	8.92
Cr	43.3	44.8.	86.7
Cu	65.84	13.19	206.08
Hg	0.03	0.11	3.63
Li	7.4	11.3	6.3
Mn	431.5	525.3	269.3
Mo	1.39	0.64	3.16
Ni	15.2	16.3	62.9
Rb	37.2	52.0	25.3
Se	0.5	0.7	1.2
Sn	3.02	1.65	54.84
Sr	459	360	255
Te	0.08	0.04	0.08
Tl	0.21	0.29	0.14
U	0.82	1.17	0.58
V	34.0	46.8	24.8
Zn	112.5	113.7	716.9

mobile fractions. Various materials are used for soil amendments and remediation. These are mainly: lime, phosphate fertilizers, zeolites, montmorillonite clays, humic (organic) matter, and biosolids. The beneficial effects of these treatments are broadly discussed by some authors (Adriano et al. 1995; Cunningham and Berti 2000; Iskandar 2001; Pierzynski et al. 2000).

In general, the phytotoxicity of trace metals are severe at low pH, thus liming is a very common practice to prevent their phytoavailability. However, site effects of liming are a decreased mobility of other metal micronutrients. This has been clearly demonstrated by Siebielec and Chaney (2006) who found Mn deficiency in plants grown in soil limed to remediate Ni-phytotoxicity.

According to Adriano et al. (2004), using agricultural chemicals (e.g., lime, phosphate compounds, organic compounds) metals may be immobilized in soils. Such a natural attenuation has been reported by several authors to be very effective in reducing of the bioavailability of metals (Knox et al. 2000a,b; Basta et al. 2001; Naidu et al. 2006).

The Maximum Allowable Concentrations (MAC) for trace metals, differ greatly between countries (Table I-2.16). Generally, lower concentrations are for light sandy soils and for low range of soil pH 5–6. In most countries, the permissible value for soil Pb is 100 mg kg^{-1}, whereas in the UK, it is 300 mg kg^{-1}. In the Netherlands, the permissible value of soil Cd is reported to be 0.8 mg kg^{-1}, but in other European countries this value is 3 mg kg^{-1} (McGrath 1992; Gzyl 1995).

Table I-2.16.
Ranges of Maximum Allowable Concentrations (MAC) and Trigger Action Value (TAV) for trace metals in agricultural soils (mg kg^{-1})

Metal	MAC[a]	TAV[b]
Ag	–	2 – 40
As	15 – 20	10 – 65
Ba	–	400 – 600
Be	10	10 – 300
Cd	1 – 5	2 – 10
Co	20 – 50	30 – 100
Cr_{total}	50 – 200	50 – 450
Cr^{6+}	–	3 – 25
Cu	60 – 150	60 – 500
Hg	0.5 – 5	1.5 – 10
Mo	4 – 10	5 – 20
Ni	20 – 60	75 – 150
Pb	20 – 300	50 – 300
Sb	10	10 – 20
Se	–	3 – 10
Sn	–	35 – 50
V	150	100 – 340
Zn	100 – 300	200 – 1 500

[a] Values reported most commonly in the literature, compiled from Kabata-Pendias and Sadurski (2004) and Chen (1999).

[b] Values proposed in some European countries, compiled from various reports, documents and internet data.

Phytoremediation. Phytoremediation is an environmental cleanup strategy in which plants and their root-bound microbial community are used to absorb and remove trace elements in the contaminated soil and to translocate them to plant tissues (Robinson et al. 2003).

Field trails of accumulation of Ni and Zn in wild plants grown in metal contaminated sludge confirmed the feasibility of phytoextraction. In this process, plants act as bio-pumps to remove contaminants and water from growth media. These processes can be divided into three following sections:

- Phytostabilization – binding of trace metals (metalloids) in an aerobic environment in the root zone
- Phytoextraction – increased uptake of trace metals, mainly by hyperaccumulating plants, process that should be continued for several seasons to obtain an effect
- Phytofiltration/rhizofiltration – absorption of metals (metalloids) in root tissues due to chelation and surface absorption from both soil and groundwater

Recently, several papers have been published on this topic (Baker et al. 1994; Cunningham et al. 1995; McGrath et al. 2001; Fitz and Wenzel 2002; Wenzel et al. 2003; Prasad 2004b, 2006; Chaney et al. 1997).

Hyperaccumulators are plants and/or genotypes that accumulate metals much above common concentrations. At present, there are about 400 species that are known as hyperaccumulator plants (Robinson et al. 2003). As Greger (1999) showed, based on a literature review, hyperaccumulators may contain trace metals in leaves above the following levels (in mg kg^{-1}):

- 100 – Cd
- 1 000 – Co, Cu, Ni, and Pb
- 10 000 – Mn and Zn

Various wild plants species are known as hyperaccumulators, depending on growth conditions and contaminant. Most commonly *Alyssum*, *Thlaspi*, and *Astragalus* species are proposed for the phytoextraction of Ni, Zn, and Se, respectively. These plants, however, usually give a low yield and this limits their effectiveness. Thus, a number of cereal crops, such as wheat, rice, rye, oats, barley, corn, and sorghum, have been recently proposed to be used for phytoremediation, since they can tolerate relatively high concentrations of metals (Huang and Cunningham 1996; Kashem and Singh 2001; Schmidt 2003; Shtangeeva et al. 2001).

The application of phytoremediation procedures for the clean up of highly contaminated soils has been widely discussed (Shtangeeva 2006). Calculated amounts of trace elements that are removed with plant yield are relatively small. Even accumulator plants withdraw, mainly from topsoils, about 10% of Cd, the most readily available metal (Table I-2.17). The utilization of crops enriched in metals is a serious problem that has to be always included in the application of any phytoremediation methods. Such biomass could be used for the production of ethanol and straw (of corn and cereals) and can be burned for the production of heat and electricity. Seeds of sunflower, which is quite effective in the accumulation of Cu and Pb from contaminated soils, can be

Table I-2.17. Trace element removal with plants from soils (after Kabata-Pendias 1994)

Element	Content of soil ($kg\ ha^{-1}$)	Output with plant yield			
		Reference plant		Accumulator plant	
		($g\ ha^{-1}$)	(%)[a]	($g\ ha^{-1}$)	(%)[a]
Mn	810	1 000	0.1	5 000	0.6
Cr	150	50	0.03	500	0.3
Zn	135	100	0.3	1 500	1
B	90	100	0.1	2 500	2
Cu	45	100	0.2	500	1
Ni	39	50	0.1	100	0.3
Mo	6	30	0.5	250	4
Cd	1.5	1	0.06	100	10

[a] Percentage of the total content of topsoil.

used for the production of bio-fuels (Kayser et al. 2000). Any yield of biomass that contains elevated amounts of metals is a hazardous waste which needs specific technologies to be utilized for the production of heat and energy or other secondary products. The ash from biomass power plants containing metals (e.g., Zn and Cd) can be recycled to recover these metals (MacDougall et al. 1997).

Some plants reveal a special capability to accumulate a specific element. For example, ladder brake (*Pteris vittata* L.), a terrestrial fern, can contain large amounts of As, up to 23 000 $mg\ kg^{-1}$ (Ma et al. 2001). The unique properties of Chinese brake fern has great importance in the clearing of As contaminated soils by the phytoremediation process. The laboratory study conducted by Tu et al. (2002) indicated that this plant might take up to 7 230 mg As kg^{-1} within 20 weeks. Tu and Ma (2003) reported that the addition of phosphates is an important strategy in improving the removal of As from contaminated soils by use of Chinese brake.

The phytovolatilization method can be applied for elements which are highly volatile, such as Se, As and Hg. It is not clear whether or not terrestrial plants can volatilize As in significant quantities. However, volatilization of As may occur by the rhizospheric bacteria of terrestrial plants (Salt et al. 1998). It is also reported that transgenic plants (*Arabidopsis*) are very effective for volatilization of Hg (Watanabe 1997). Rugh et al. (1996) suggested a molecular genetic approach to the transfer of Hg resistance genes to plant species as a potential means of phytovolatilization of Hg^0. According to these authors this technique may be useful to remove Hg from contaminated soils. In addition, it is possible to manipulate the transfer of Hg^0 in plant shoot tissue to Hg^{2+} that may be another option for the Hg phytoremediation. Some plants may convert Se into dimethyl selenium that is nontoxic and may volatilized from plant tissues.

Phytovolatilization mechanisms are not yet fully understood. There are observations that some commercial vegetables (e.g., broccoli, cabbage) and rice are quite effective in the phytovolatilization. These processes, however, are controlled by soil factors, such as sulfates and salinity of soils.

Economic aspects are very significant in phytoremediation practices (Thewys 2006). There is not much information in this field, as well as on farmers' approaches to this problem. Lewandowski et al. (2005) have recently studied this topic in Germany and widely discussed the income expected from cleaned up areas and the length of period the area can be used for the production of food crops. This information is quite valuable for farmers and landowners. The authors have distinguished important parameters, such as: *(i)* length of time needed for remediation and *(ii)* time for production of crops after remediation.

The cost of and socio-economic aspects of phytoremediation has been recently discussed and two alternative crops have been considered: rape and willow (Thewys 2006). However, using the biomass as a source of energy, especially rape oil for biodiesel creates a risk of uncontrolled emission of trace inorganic pollutants.

According to Chaney et al. (1999b), hyperaccumulator plants offer a new sustainable economic system based on the use of biomass from the phytoremediation for power generation factories. There is a belief that phytoremediation processes may be economically acceptable in about 10 years time. However, possible risks of food and forage chain contamination and the reaction of plants when overloaded with metallic pollutants, have become open questions (Keller 2006).

The effectiveness of some plants in the extraction of metals from constructed wetland purification systems have recently become widely investigated in both laboratory and field conditions (Ofelder et al. 2006; Marmiroli et al. 2006). Some plants (e.g., *Phragmites*, *Salix*, and *Populus*) may remove more than 50% of metals (Ba, Cd, Cu, Ni, Pb, Sr) from the growth media (Samecka-Cymerman et al. 2004).

Waters

Freshwater. Water plays fundamental functions in processes both geochemical and biochemical. It is also a main carrier for all chemical elements; its amount and chemical composition control element cycling in water-air-soil systems. Thus, water is probably the most studied medium that governs the forms of trace elements of which investigated Cr, Se, Cu, As, Pb, Cd, and Hg have been studied the most frequently (Das et al. 2001).

All surface waters (oceans, seas, rivers, lakes, ponds) form the so-called hydrosphere of mass of which has been estimated at about 1.35×10^{18} t. Waters of oceans and seas contain approximately 98% of the hydrosphere mass. All waters contain various ions at different concentrations. The main ions dissolved in waters are: Na^+, K^+, Mg^+, Ca^+, Cl^-, SO_4^{2-}, HCO_3^-, and these also occur as different species adsorbed by inorganic and organic colloidal particles. Concentrations of these elements differ greatly being usually much higher in sea than in river waters. So-called secondary elements (C, N, P, S, and Si) as well as trace elements occur in all water systems, in highly variable concentrations, depending on several factors of which pollution plays a significant role.

Water pollution by trace elements is an important factor in both geochemical cycling of trace elements and in environmental health. The hydrocycle of trace elements plays a significant role in each aquatic and terrestrial ecosystem. Especially cycling of trace metals in the oceans and their role in the photosynthetic fixation of carbon by phytoplankton is of great importance. Ecological consequences of trace element pollution of waters are difficult to predict and to assess.

Most trace elements, especially trace metals, do not remain in soluble forms in waters, for a longer period. They are present mainly as suspended colloids or are fixed by organic and mineral substances. On the other hand, easily volatile elements such as Br and I can reach higher concentrations in surface waters, from which they can easily vaporize under favorable climatic conditions. In addition, microbial alkylation of several elements, e.g., Hg, Se, As, Te, and Sn may greatly affect their volatilization from surface waters. Trace element speciation in water control their behavior and toxicological risk. The bioavailability of trace elements to both unicellular and higher organisms is the result of complex reactions between the ligands present in the aqueous medium and those of living cells. Practically, all fractions of trace elements, truly dissolved and associated with suspended particulate matter, may be bioavailable to aquatic organisms.

Several elements have been identified in water as simple cations (e.g., Li^+, Co^{2+}, Ni^{2+}, Rb^+, Cs^+). Some metals (e.g., Cd, Cu, Hg, Pb, Zn), however, are known to occur in complex forms. It may be illustrated on the example of Pb that can be present in water as cation: Pb^{2+}, $PbCl^+$, $Pb_3(OH)_4^{2+}$, $Pb_4(OH)_4^{4+}$, and as anion: $PbCl_3^-$, $Pb(OH)^-$, $Pb(OH)_3^-$.

Vignati (2004) has distinguished four groups of elements based on their fractionations in freshwater:

- Terrigenous metals: Al, Ti, and Pb, associated mainly with the particulate phase (often >90% of the total) and with the high molecular weight colloids in the filterable phase
- Trace cations: Co, Cu, Fe, Mn, Ni, and Zn, variably distributed between particulate, colloidal and truly dissolved phases depending on master environmental variables. Mn-Fe oxides and organic matter are significant carrier phases
- Oxyanions: As, Mo, Sb, U, slightly associated with suspended particulate matter and colloids, are preferentially transported in the truly dissolved fraction
- Metals of variable characters: Cr and V, may occur in various forms and phases depending on variations in water parameters

The main sources of trace elements in ocean and sea basins are riverine fluxes, however, calculations of these values differ depending upon authors (Table I-3.1). The contribution of anthropogenic pollution to the trace element pool in waters is very significant. As Matschullat (1997) calculated, the fluxes of anthropogenic trace metals to the Baltic Sea, with both river flow and atmospheric deposition amounts to over 90% of total input of Cd, Hg, and Pb, and about 80% of Cu and Zn. Although it varies for each water basin and season, the total contribution of anthropogenic sources should be regarded as very important.

Trace element concentrations of seawaters seem to be relatively stable, whereas in river waters their contents might be very variably between both rivers and seasons. Nevertheless some average values have been calculated for various water systems and show distinct differences among sea and river waters (Table I-3.2).

Table I-3.1. World average riverine fluxes of trace elements to oceans and seas (kt yr^{-1})

Element	After Kitano (1992)	After Gaillardet et al. (2003)
As	100	23
Ba	–	860
B	–	380
Be		0.33
Co	–	5.5
Cd	0.7– 3	3
Cr	35	5.5
Cu	50	55
Fe	7000	2470
Hg	<0.1	–
Li	–	69
Mn	400 – 1000	1270
Mo	20	16
Ni	10	30
Pb	40	3
V	–	27
W	–	3.7
Zn	200	23

Table I-3.2. Commonly reported mean concentrations (or ranges) of trace elements in natural sea and river waters, as compiled by different authors ($\mu g\,l^{-1}$)

Element	Seawater[a]	Seawater[b]	North Pacific Ocean[c]	River[a]	River[d]
Ag	0.04	0.003	0.002	0.03	0.004
Al	2	–	0.03	300	2.6 – 1080
As	7	–	1.2	2	0.11 – 2.71
B	4440	4400	4500	15	1.5 – 150
Ba	13	11.7	15	10	3 – 80
Be	0.006	0.0002	0.0002	0.3	0.0006 – 0.61
Bi	0.02	–	0.0003	0.02	–
Br	67000	67000	67000	14	–
Cd	0.1	0.07	0.07	0.2	0.0006 – 0.42
Co	0.01	0.003	0.001	1	0.02 – 0.43
Cr	0.3	0.4	0.21	1	0.29 – 11.46
Cs	0.3	0.3	0.31	0.02	0.0006 – 0.02
Cu	0.2	0.12	0.15	2	0.23 – 2.59
F	1300	–	1300	100	–
Fe	1	0.04	0.03	300	31 – 739
Ga	0.03	0.02	0.001	0.09	0.003 – 0.12
Ge	0.05	0.005	0.006	–	0.001 – 0.08
Hg	0.02	0.006	0.0001	0.07	–
I	60	60	58	2	–
Li	180	178	180	2	0.16 – 4.5
Mn	0.2	0.01	0.02	15	1.3 – 50.7
Mo	7	11	10	1.7	0.04 – 2.69
Nb	0.01	0.001	<0.005	–	0.002 – 0.01
Ne	0.1	–	0.003	–	–
Ni	0.5	0.48	0.48	1.6	0.35 – 5.06
Pb	0.03	0.001	0.003	1.5	0.007 – 3.8
Rb	120	124	120	1	0.3 – 4.04
Sb	0.2	–	0.2	0.2	0.005 – 0.27
Se	0.2	0.17	0.1	0.2	0.02 – 0.23
Sc	0.001	0.0006	0.0007	0.7	0.05 – 1.58
Si	2200	–	2800	7000	–
Sn	0.004	0.0005	0.0005	0.009	–
Sr	7900	7800	7800	70	2.9 – 237.8
Ti	1	0.023	0.007	5	0.23 – 5.8
Tl	0.02	0.012	0.013	0.4	0.008 – 0.04
U	3.2	3.2	3.2	0.4	0.004 – 1.4
V	2.5	1	2	0.5	0.009 – 1.77
W	0.1	1	0.01	0.03	0.1 – 180
Y	0.01	0.013	0.02	–	0.02 – 1.4
Zn	2	0.39	0.35	18	0.27 – 27
Zr	0.03	–	0.002	0.8	0.004 – 0.71

[a] After Kabata-Pendias and Pendias (1999). [b] According to GERM (www.earthref/org).
[c] After Nozaki (2002). [d] After Gaillardet et al. (2003).

Concentrations of trace elements in rainwater are highly variable depending on local conditions and in particular on anthropogenic pollutants. Their concentrations in samples collected in Sweden during 1999 (Eriksson 2001a) are in most cases much lower than in river waters, with the exception of Pb and Zn with mean contents of 1.5 and 11 $\mu g\ l^{-1}$, respectively, which are fairly similar to those reported in river waters.

Increased concentrations of trace metals in surface run-offs contribute to their higher levels in river and water reservoirs. Several factors control their levels, however, in mountain areas the land slope seems to be very important. Wiśniowska-Kielan and Klima (2005) reported that contents of some trace metals (e.g., Cr) in run-offs are similar to those in rainwater, whereas others, e.g., Cd (max. 9.9 $\mu g\ l^{-1}$) and Ni (max. 24 $\mu g\ l^{-1}$) are significantly higher than in rainwater.

Bottom sediments. In the most cases trace element concentrations in bottom sediments and/or in plankton are adequate indications of water pollution. Soluble fractions of trace elements are, in most aquatic environments, rapidly absorbed either by clay or organic compounds and deposited in sediments or they are caught by plankton and root tissues of aquatic plants. Generally, the speciation of trace metals in sediments is very similar to that in soil. Both phytoplankton and vascular water plants are known to selectively concentrate trace elements of both geochemical and anthropogenic origins. Higher accumulation by mesoplankton has been observed for Cd, Cu, and Zn, while seaweeds take up Cd, Zn, Mn, and Pb more readily (Szefer 1998). Thus, concentrations of trace elements in selected samples of aquatic compartment reflect either chemical composition of bedrocks or anthropogenic influence.

Trace elements which are absorbed and thus immobilized in bottom sediments constitute a potential hazard to water quality and aquatic life as they may be released as a result of physico-chemical changes. These changes are most commonly stimulated by a change in the redox conditions and by microbial activity.

The anthropogenic sources of trace elements in waters and sediments are associated mainly with mining of coal and mineral ores and with manufacturing and municipal waste waters. The most frequently described example of industrial metal pollution is that of Minamata, Japan. Sediment samples taken about 1 and 1.5 km away from the discharge point contained 22 and 12 mg Hg kg^{-1}, respectively (Förstner 1979).

Several regulations of acceptable trace element pool in various waters and discharged wastewaters have been set up at national level. The most crucial are health-related limits for potable waters (Table I-3.3). However, limits for mine-discharge waters are also of great importance for total "geochem" balance in the environment (Pluta 2005).

Ground waters. About 99% of the world's fresh, available water is groundwater that is a basic source of the domestic, industrial, and agricultural water supply (Bhattacharya and Mukherjee 2002). Therefore the chemical composition of groundwater and its contamination has become recently of great concern. All contaminants, both inorganic and organic are likely to pass from soil to groundwater. The transfer of trace elements depends on several properties of the soil media as well as on the geochemical properties of an element. Thus, elements that predominate in soils in easily mobile forms are of higher hazard than those that are relatively stable in prevalent soil conditions.

Table I-3.3.
Threshold limits of trace elements in drinking water, irrigation water and coal mine water ($\mu g\ l^{-1}$)

Element	Drinking water[a]	Drinking water[b]	Irrigation water[c]	Coal mine water[d]
Al	n.e.	10	5 000	–
As	10	0.5	100	100
B	500	10	<1 000–2 000[e]	1 000
Ba	700	–	–	2 000
Be	n.e.	–	100	400
Cd	3	0.03	10	500
Cr (+6)	50	1	100	500
Cu	2 000	3	200	500
F	1 500[f]	–	1 000	25 000
Fe	n.e.	500	5 000	10 000
Hg	1	0.07	–	60
Li	n.e.	10	2 500	–
Mn	400	10	200	–
Mo	70	0.5	10	–
Ni	20	0.5	200	2 000
Pb	10	0.1	5 000	500
Sb	20	0.2	–	–
Se	10	0.2	20	–
U-234	15	–	–	–
V	n.e.	0.5	100	–
Zn	n.e.	15	2 000	500

n.e. – Not established due to uncertainty of data and/or not enough data for health based guideline values or no health concern at concentrations normally observed in drinking water.
[a] Health based limit value or maximum concentration limits (WHO 2004).
[b] Commonly found concentrations in drinking water in Poland. (*http://eko.ch.pw.edu.pl/kons/aimzs.doc*).
[c] Recommended maximum concentrations (NAS 1972, Pratt 1972).
[d] Highest allowed concentrations in coal mine effluent (Pluta 2005).
[e] 0.2 mg B l^{-1} is essential, but at 1–2 mg l^{-1} is toxic.
[f] Target is 800–1 200 µg F l^{-1}, to maximize benefits and minimize harmful effects.

The chemical quality of groundwater is of special importance, as it is a source of both drinking and irrigation waters and therefore has a significant impact on trace element transfer into the food chain. Groundwater pollution by leachate from landfills is considered to be the most important. Not measured and not controlled leachate entrance into the groundwater may contribute a source of contamination long after dumping has ceased (Bjerg et al. 2003). According to these authors, the highest concentration in landfill leachate is observed for Zn, up to 1 000 mg l^{-1}.

Ground waters also may contain elevated concentrations of elements resulting from weathering processes of rocks. Common pollutants of natural origin are As and F that in some regions have significant ecological and health effects.

Arsenic contamination of groundwater is relatively well known and has been widely investigated in several regions: *(i)* alluvial plains and deltas (e.g., the Bengal Delta Plain), *(ii)* inland basins in arid and semi-arid areas (e.g., Mexico, Argentina), *(iii)* sulfide mining-related regions, and *(iv)* geothermal waters (Naidu et al. 2006; Smidley and Kinniburgh 2005).

Natural high-F groundwater is associated with several parameters among which geology and climatic conditions seem to be the most significant. Increased F level in waters occurs mainly in arid climatic regions and in areas where F-rich minerals are abundant. In several countries, defluoridation water treatments had to be applied (Edmunds and Smedley 2005).

Drinking waters. Concentration of trace elements in drinking water is of special concern in relation to health. Several ecological studies have indicated an influence of the quality of drinking water on some diseases. Liu et al. (2000) have reported a significantly increased levels of Ti, V, Fe, Cu and Sr in drinking water collected from areas with high incidence of gastric carcinoma. Increased levels of As in waters pose very great threats to human health in many parts of the world. Also high-F waters are known to have detrimental effects on health in several regions.

The quality of drinking water is very important since it is consumed every day during the whole lifetime. In recent years, carcinogenic and non-carcinogenic risks of ground waters are of a special interest. Momot and Synzynys (2005) have calculated daily intake of several metals by inhabitants of Obninsk (Middle Russia) and assessed that it may be a risk of oncological diseases to 4 of 100 persons and of non-oncological diseases to 1 of 1 000 persons. The daily doses of metals to this population are as follows (in $\mu g\ kg^{-1}$ BW): Ag, <1.0; Al, 4.01; As, 1.28; B, 2.57; Be, 0.04; Cd, 1.03; Cr, 2.1; Hg, 0.17; Pb, 0.43; and Zn, 0.17.

Aquatic biota. Aquatic plants are known for a high ability to accumulate increased amounts of trace elements and for the resistance to high concentrations of some metals. Samecka-Cymerman and Kempers (2004) reported that macrophyte (*Myriophyllium spicatum*) survived in surface polluted water, at tissue contents (in $mg\ kg^{-1}$) up to: 1 040 Cu, 6 660 Mn, and 57 Co.

The worldwide literature on trace elements in phytoplankton and seaweeds has been presented by Szefer (2002). Aquatic plants can be considered suitable biomonitors for trace elements and radionuclides. The special trends in the distribution of some metals in aquatic plants (algae and weeds) indicate variable pollution of both water and sediments. They also are sensitive indicators of temporal changes. After the Chernobyl accident (1986), elevated levels of several radionuclides have been reported in aquatic plants of the Baltic Sea. Recently, the main source of radionuclides in this sea is the Chernobyl fallout being transported by riverine runoff entering the sea (Szefer 2002b).

The biomonitoring of the pollution of river waters (from the Odra catchment, Poland) indicated a great accumulation of trace metals by *Elodea canadensis* (Cymerman and Kempers 2003). This macrophyte seems to be a suitable bioindicator of metal pollution. Enrichment ratios of metals between this macrophyte and water varied as follows:

- Cd: 6 670 – 255 000
- Co: 84 – 3 130
- Cr: 2 830 – 117 200
- Cu: 906 – 23 600
- Fe: 19 – 176
- Ni: 1 690 – 97 700
- Pb: 31 700 – 823 000
- Sr: 29 – 159
- Zn: 145 – 4 450

Ciesielski et al. (2006) studied the hepatic distribution of trace elements in the Baikal seals and have concluded that their content is, in principle, affected by physiological factors and not by anthropogenic (pollution) ones. On the other hand, aquatic biota are a good indicator of Hg levels in water. Fish captured in rivers from goldmining area in Brazil contained enhanced level of Hg; higher levels (410–550 $\mu g\ kg^{-1}$) were found in larger, older and carnivorous fishes than in detrivore and herbivore ones (130–170 $\mu g\ kg^{-1}$) (Moreira 1996). Some associations among trace metals in tissues of Antarctic seals may result from pollution (Szefer et al. 1994).

Air

Air pollution has arisen from both natural (meteoric, terrestrial, marine, volcanic, erosion and surface winds, forest fires, biogenic) and anthropogenic (coal and fuel combustion, industry, automobile, agriculture) sources. The steady global increase of trace element concentrations in the atmosphere has been observed and monitored in some countries for over 30 years. The majority of trace element emitters have been located in the northern hemisphere (mainly between 40–55° N). Buat-Ménard (1984) calculated that emissions of trace elements in the Northern Hemisphere are several times higher than in the Southern Hemisphere and are about 80% and 30%, respectively of anthropogenic origin. However, at the global scale, the natural emissions of trace elements cannot be neglected because large amounts of dusts containing trace elements come from natural sources. The estimated principal trace elements emissions for natural sources are as follow: 50% of Cr, Mn, and V, and >20% of Cu, Mo, Ni, Pb, Sb and Zn. Volcanic activities may contribute over 20% of the atmospheric Cd, Hg, As, Cr, Cu, Ni, Pb, and Sb. Sea salt aerosols may also contribute about 10% of total trace element emissions to the atmosphere (Allen et al. 2001). Differentiating natural and anthropogenic sources of metals is not easy and some methods for monitoring various sources of metal pollution have been discussed (Dias and Edwards 2003).

Retrospective studies of trace element enrichment in ice layers, ombrotrophic peatlands and tree-rings show a steady increase of several trace metal contents due to anthropogenic emissions. However, recent observations have shown downward trends for several metals, e.g., Fe, Pb, Cr, Cd, Hg, As, and Be. A significant reduction of the Pb content in airborne particulates has recently been reported for some urban sites (Puxbaum and Limbeck 2004).

Investigations of ice cores have indicated that anthropogenic emissions of Pb and Cd have dramatically increased over the last 200 years. However, volcanic eruptions also have contributed greatly to increased levels of Pb, Bi, Tl, and Cd, especially during the pre-industrial Holocene period.

Metals in the atmosphere, whether anthropogenic or natural in origin, are associated mainly with particulate matter, except for Hg, which occurs mostly in gaseous form. These elements are adsorbed by particulates, both inorganic and organic, mainly of diameter between 0.01 μm to 100 μm. Particles of larger diameter have a tendency to precipitate quickly near the discharge sources, but particle sizes less than 10 μm and low density exhibit long term existence, from days to months, and may be transported to distant regions. There is evidence that trace elements of anthropogenic sources are widely distributed in the atmosphere and occur at high latitudes in mountainous regions and in the Arctic, causing "Arctic Haze" which creates poor visibility, and affects the atmospheric radiation budget (Heidum 1986).

The calculated EF_c values near unity in air dust, normalized to the ratio of the given element in the crust to the reference element (Al) indicates that the primary source of most elements in the atmosphere is terrestrial dust (Puxbaum and Limbeck 2004). There are also various studies on the origin of Arctic air pollution and it is generally agreed that there are three main sources: anthropogenic, soil dust, and sea-salt. The major contributor of anthropogenic pollutants in the Arctic air is Eurasia (Maenhaut et al. 1989; Pacyna 1995). In summer, aerosol concentrations are lower than in other seasons. The fate of Arctic pollutants is not well understood, but it is known that they are removed somehow and end up in the ocean mainly in the North Atlantic, Norwegian Sea and the Bering Sea (Pacyna and Oehme 1988). These phenomena have significant effects on trace metal levels in aquatic biota and affect the quality of marine food (Dehn et al. 2006).

In the last three decades, large amounts of data on trace elements in air have been presented. However, the calculated global atmospheric inventory differs depending on the sources used (Table I-4.1 and I-4.2). Nevertheless, there are general similarities concerning relative ranges and sources. These may be summarized, based on the highest emissions, as follows:

Table I-4.1. Global atmospheric emission inventory for trace metals (kt yr^{-1})

Element	Anthropogenic[a]	Natural[a]	Total[b]	A/N[c] ratio
As	5	12	18.82	0.43
Cd	3	1.3	7.57	2.3
Cr	14.7	44	30.48	0.33
Cu	25.9	28	35.37	0.93
Hg	2.2	2.5	3.56	0.88
Mn	11	317	38.27	0.03
Mo	2.6	3	–	0.87
Ni	95.3	30	55.65	3.2
Pb	119.3	12	332.35	9.9
Sb	1.6	2.4	3.51	0.67
Se	4.6	9.3	6.32	0.49
V	240	28	86	8.6
Zn	57	45	131.88	1.3

[a] After Pacyna and Pacyna (2001).
[b] After Livett (1992).
[c] *A/N* – ratio of total anthropogenic to natural emission, after Pacyna and Pacyna (2001).

- Total (>50 kt yr^{-1}): Pb > Zn > V > Ni
- Anthropogenic (>50 kt yr^{-1}): V > Pb > Ni > Zn
- Natural (>10 kt yr^{-1}): Mn > Zn > Ni > Cr > V > Cu
- Combustion (>10 kt yr^{-1}): V > Ni > Pb > Cr, Al
- Nonferrous metal (>10 kt yr^{-1}): Zn > Cd > Cu > Pb
- A/N emission ratio (>1): Pb > V > Ni > Cd > Zn

Anthropogenic sources exceed natural ones for some metals, in the order as follows: Pb > V > Ni > Cd > Zn.

Several other trace elements, not presented in Table I-4.1 and I-4.2, are also reported to be significantly increased in air. These elements are, most often, from coal combustion that is clearly indicated by their high increased levels in fly ash (Fig. I-4.1). Concentrations of trace elements in FA are different from different geological coal basins and with the technology applied for cleaning gas from the power plant.

Table I-4.2. Worldwide emissions of trace metals from major anthropogenic sources to the atmosphere (kt yr^{-1}) (after Pacyna and Pacyna 2007)

Element	SFFC[a]	Industry			Waste disposal
		Nonferrous metals	Iron and steel	Cement	
As	0.81	3.46	0.35	0.12	0.12
Cd	0.69	22.17	0.06	0.02	0.04
Cr	10.15	–	2.83	1.34	0.43
Cu	7.08	18.07	0.14	–	0.62
Hg	1.48	0.16	0.03	0.13	011
In	–	0.05	–	–	–
Mn	9.42	0.06	1.06	–	0.51
Mo	2.64	–	–	–	–
Ni	86.11	8.88	0.04	0.13	0.13
Pb	11.69	14.82	2.93	0.27	0.82
Sb	0.73	0.55	0.007	–	0.27
Se	4.10	0.47	0.007	0.003	0.02
Sn	3.52	0.32	–	–	0.12
Tl	1.82	–	–	–	–
V	240.08	0.008	0.07	–	0.02
Zn	9.42	40.87	2.12	2.67	1.93

[a] *SFFC* – stationary fossil fuel combustion.

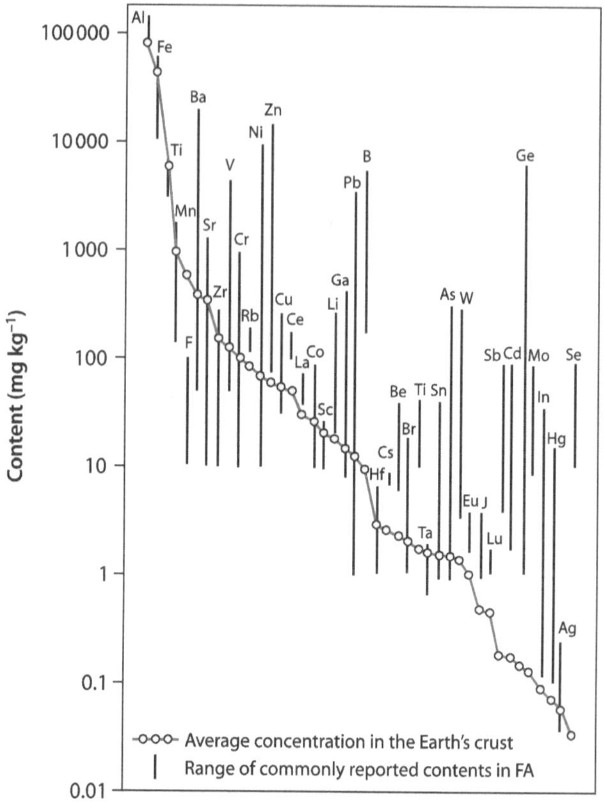

Fig. I-4.1.
Commonly reported ranges of trace elements in fly ash as compared to their abundance in the Earth's crust

There is no estimation of world production and consumption of fly ash (FA) as a byproduct from coal combustion, but Smith (2005) calculated the production of FA for 30 countries at 400 Mt yr^{-1}, between 2001 and 2002. China and India have dominated the production of FA, followed by the USA and Europe. Due to better gas cleaning equipment, FA may contain less trace elements. However, some countries, e.g., India, due to high ash content (>32%) in coal, emits over 100 Mt yr^{-1} of FA enriched in trace elements.

The atmospheric deposition of trace elements, both wet and dry, contributes to contamination of all environmental compartments (e.g., soil, water, and vegetation). Trace metal deposition has been reported even for relatively remote sites, e.g., rural areas across New Zealand (Gray et al. 2003). The ranges of mean values of the atmospheric deposition in this country, over a period of two years, were as follows (in g $ha^{-1} yr^{-1}$): Cd, 0.09–0.36; Cr, 7–50; Cu, 18–55; Ni, 6–16; Pb, 10–50; and Zn, 87–1 347.

Moss analysis, including the moss-bag method, has been used as a good indicator for aerial pollution. Results of studies have indicated a steady decrease in aerial deposition of almost all trace elements during the last 25 years. Especially noticeable decreases are observed for Cd, Cr and Pb (Table I-4.3). Significant variation in average trace metal contents of moss has been reported for European countries during the period 1990–1992 (Rühling 1994). The ranges of the metals determined in mosses, *Pleurozium*

Table I-4.3. Mean concentrations of trace elements in mosses growing in the Scandinavian countries, collected in different period ($mg\,kg^{-1}$)

Element	Sampling period			Element	Sampling period		
	1975[a]	1990–1995[b]	2000[a]		1975[a]	1990–1995[b]	2000[a]
Antimony, Sb	0.41	0.15	0.11	Molybdenum, Mo	0.26	0.23	0.11
Arsenic, As	0.64	0.23	0.12	Neodymium, Nd	0.79	0.49	0.28
Barium, Ba	21.9	31	15.2	Nickel, Ni	1.85	3.6	0.65
Beryllium, Be	0.033	0.023	0.011	Niobium, Nb	0.2	0.11	0.08
Bismuth, Bi	0.113	0.033	0.016	Praseodymium, Pr	0.112	0.13	0.04
Cadmium, Cd	0.63	0.18	0.18	Rubidium, Rb	36.6	13.0	30.9
Cerium, Ce	1.0	1.10	0.36	Samarium, Sm	0.071	0.1	0.025
Cesium, Cs	0.69	0.26	0.51	Scandium, Sc	0.33	–	0.17
Chromium, Cr	1.65	2.6	0.79	Selenium, Se	0.48	0.1	0.22
Cobalt, Co	0.25	0.37	0.12	Silver, Ag	0.103	0.037	0.02
Copper, Cu	7.5	7.0	4.5	Strontium, Sr	9.8	15	9.9
Dysprosium, Dy	0.053	0.07	0.019	Tellurium, Te	–	0.005	–
Erbium, Er	0.029	0.037	0.011	Terbium, Tb	0.01	0.013	0.004
Europium, Eu	0.02	0.02	0.01	Thallium, Tl	0.152	0.12	0.066
Gadolinium, Gd	0.077	0.095	0.028	Thorium, Th	0.11	0.07	0.039
Gallium, Ga,	0.30	1.1	0.09	Thulium, Tm	0.004	0.005	0.001
Germanium, Ge	0.091	1.1	0.015	Tin, Sn	0.40	0.25	0.09
Hafnium, Hf	0.013	0.008	0.005	Titanium, Ti	23.3	53	8.5
Holmium, Ho	0.01	0.013	0.004	Tungsten, W	0.075	0.093	0.023
Indium, In	0.011	–	0.001	Uranium, U	0.057	0.05	0.015
Lanthanum, La	0.51	0.55	0.19	Vanadium, V	6.6	2.8	1.2
Lead, Pb	47.7	8.6	4.2	Thallium, Tl	0.152	0.12	0.066
Lithium, Li	0.34	0.22	0.11	Ytterbium, Yb	0.025	0.032	0.009
Lutetium, Lu	0.004	0.004	0.001	Yttrium, Y	0.27	0.31	0.1
Manganese, Mn	290	310	289	Zinc, Zn	58.4	45	32.2
Mercury, Hg	0.086	0.075	0.051	Zirconium, Zr	0.51	0.35	0.2

[a] *Pleurozium schreberi* collected from south Sweden (Rühling and Tyler 2004). [b] *Hylocomium splendens* collected from Norway (Berg and Steinnes 1997).

schreberi, *Hylocomium splendens* and *Hypnum cupressiforme*, from various countries, are as follows (in mg kg^{-1}):

- Cd: Portugal 0.22 – Slovenia 1.38
- Cr: UK 0.76 – Spain 5.36
- Cu: Swiss 4.8 – Romania 26.3
- Fe: UK 255 – Romania 7856
- Ni: Denmark 1.5 – Romania 11.2
- Pb: Lithuania 9.5 – Slovakia 60.5
- V: UK 1.7 – Romania 15.0
- Zn: Estonia 31.6 – Slovakia 173.0

Moss monitoring carried out in Poland, in 1995, indicated regional variation in metal contents of *Pleurozium schreberi*. Grodzińska et al. (2003) reported the following ranges of mean metal concentrations (in mg kg^{-1}) in moss sampled in Poland, from a rural-forested region (Baltic region) and from an industrialized region (south region), as follows (respectively):

- Cd: 0.34 – 1.07
- Cr: 1.4 – 2.6
- Cu: 1 – 9.7
- Fe: 299 – 763
- Hg: 0.34 – 0.22
- Ni: 1.73 – 1.65
- Pb: 11 – 38.3
- V: 5 – 4.8
- Zn: 39 – 69

Aerial particles and aerosols contain variable amounts of various species of trace elements. Manecki (1984) reported that urban dust of Kraków (Poland) contained the following amounts of metals (in weight-%): Fe, 33–38; Mn, 0.3–0.5; Pb, 0.1–0.4; Zn, 0.05–0.06; Cd, 0.002–0.01; and Cr, 0.008. In gaseous emission of a municipal incinerator in Vienna the major constituent among trace metals are: $HgCl_2$, HgCl, HgO, and Hg. However, halides of Pb, Zn, and Cd were also identified in gaseous states, as well as small quantities of CdO (Puxbaum and Limbeck 2004).

Dmuchowski (2005) has applied the moss-bag method for monitoring of air pollution in Warsaw. The results indicated a relatively small variation in time, with an exception of Pb that has decreased by about half, during 10 years. The ranges of mean concentrations of metals in moss-bags exposed in Warsaw during the period 1992–2004 varied, respectively, as follows (in mg kg^{-1}):

- Cd: 0.4 – 0.35
- Cr: 2.13 – 2.0
- Cu: 5.9 – 5.9
- Fe: 638 – 552
- Pb: 13.5 – 7.9
- Zn: 52 – 53

Trace elements in the atmosphere occur in various chemical forms, of which organometallic compounds are of special concern. Also the size of aerial particles, especially PM10 (diameter 10 μm) that are easily inhaled, are of a great significance. These compounds are of both anthropogenic and natural origin. The natural processes, mainly the methylation, have a significant influence on forms and fate of aerial pollutants. Volatile methylated compounds of several trace elements have been detected in air, particularly in highly polluted environments. Biogenic methyl and methyl/hydride species of metals (As, Hg, Pb. Sb, Sn, and Tl) and of metalloids (Br, Cl, I, Se, and Te) have been reported to occur in air (Brinckman et al. 1982). These compounds are relatively easily soluble and thus bioavailable and are of a greater toxicity than inorganic forms (Puxbaum and Limbeck 2004). The chemical composition of rainwater is closely associated with air pollution and greatly contributes to the deposition of pollutants. Eriksson (2001a) reported ranges of trace element concentrations in rainwaters, collected in Sweden, during 1999, as follows (in $\mu g\, l^{-1}$):

- Above 10: Al, Zn
- Range 0.1 – 1: As, B, Ba, Cr, Cu, Mn, Ni, Pb, Rb, Se, Sn, Sr, Ti, V
- Range 0.01 – 0.1: Ag, Cd, Ce, Co, Ge, Fe, La, Li, Mo, Nd, Pt, Sb, W, Y, Zr
- Range 0.001 – 0.01: Au, Be, Bi, Dy, Ga, Gd, Hf, Hg, Nb, Pr, Sc, Sm, Te, Th, Tl, U
- Below 0.001: Er, Eu, Lu, Pd, Rh, Ru, Ta, Tb, Tm, Yb

Seasonal as well as regional differences in trace element distribution are evident. Many anthropogenically-derived elements are elevated in winter season and in urban areas. Very high and variable concentrations of trace metals have been measured in urban air (Table I-4.4). Several factors influence their atmospheric abundance, but most commonly they are industrial technologies and fuel combustion. The highest concentrations have been reported for Cu, Zn, and Fe in Bursa, Turkey, and for Pb in Taejon, Korea. A close relationship between the Pb concentration and the contents of other trace

Table I-4.4. Concentrations of trace metals in air of worldwide cities (ng m^{-3}) (after Tasdemir et al. 2005)

Element	Bursa, Turkey (2003)	La Plata, Argentina (1993)	Gothenburg, Sweden (2000)	Seville, Spain (1996)	Sao Paulo, Brazil (1998)	Chicago, USA (1993–'95)	Taejon, Korea (1997–'99)
Cd	0.7	0.4	0.8	0.2	–	–	3.2
Co	1.1	–	–	0.3	–	–	1.5
Cr	10.7	4.3	55	–	4.2	4.2	25.1
Cu	396.6	29.5	404	9.57	10.2	18.7	41.1
Fe	2165.4	1183	20690	260.8	674	–	1633
Mn	51.1	25.5	261	7.4	15.2	31.3	50.3
Ni	5.1	3.2	–	1.4	5.1	–	37.9
Pb	77.2	64.5	113	43.7	16	48.4	243
Zn	250.5	273	759	–	94	135.8	240

Table I-4.5.
The revised guideline values, for concentrations and time exposure of selected trace metals for Europe (after WHO 2005) and in workplace air

Element	Guideline value ($\mu g\ m^{-3}$)	Average time	Workplace air (TWA, $\mu g\ m^{-3}$)[a]
As	0.0015	UR[b]/lifetime	200
Cd	5000	Yearly	2
Cr^{6+}	0.04	UR[c]/lifetime	10
Hg	1.0	Yearly	50
Mn	0.15	Yearly	–
Ni	0.0004	UR[b]/lifetime	–
Pb	0.5	Yearly	150

[a] Guideline of UNEP (2004): *TWA* – time-weighted average airborne concentration when calculated over a normal eight-hour working day for five-day working week.
[b] *UR* is the excess risk of dying from cancer following lifetime exposure.

metals in the urban atmosphere has been observed. Sakata (1998) has proposed a method based on Pb isotope ratios for the estimation of pollution sources in urban air.

Extremely high Hg concentrations in air result from roasting of Au/Hg amalgam and from the purification processes. Atmospheric Hg concentration in area with goldmining in Brazil reached 292 000 ng m^{-3}, whereas its contents in rural and urban areas varied between 0.1–10 and 0.5–50 ng m^{-3}, respectively (Moreira 1996).

Trace pollutants in air have a significant impact on human health. They may be absorbed by aerial parts of vegetables and thus included into the food chain. All compounds of trace elements present in air can be inhaled both in the open air and indoors. Especially occupational exposure is of great concern and should be monitored because clean air is considered to be a basic requirement for humans. The WHO (2005) has updated the air quality guidelines for Europe. It is clearly mentioned that the "Guidelines values" are not just numbers, but they incorporate recommendations that will help to reduce harmful effects of organic and inorganic particulates to humans and vegetables. However, there is limited information for guideline values for inorganic elements (Table I-4.5).

Household dust. House dusts may be a significant source of several trace elements. Fergusson and Kim (1991) reported that concentrations of Pb, Zn, Cu, Cd, As, Sb, Co, and Cl in house dusts exceed their abundance in soil-borne dusts. Especially four metals, Pb, Cd, Cu, and Zn are increased in house dusts of worldwide towns and cities (Table I-4.6). A great proportion of the metals is associated with two phases, carbonate and Fe-Mn hydroxides. The bioavailability of house-dust metals is not well known, however, some elements are fairly soluble in stomach fluids but less in lung fluids (Rasmussen et al. 2001).

The mode of cooking and heating can be a dominant source of indoor trace elements. It has been estimated that two thirds of all households in the developing countries still depend on unprocessed wood, crop residues and animal dung burned in poorly

Table I-4.6.
Concentrations of trace metals in common household dusts of worldwide towns and cities ($mg\,kg^{-1}$) (as compiled by Rasmussen 2004 from various sources)

Element	Concentration	
	Range of GM[a]	The highest
Cd	0.86 - 7.7	8040
Cu	121 - 311	48800
Pb	80 - 1007	36900
Zn	475 - 1324	114800

[a] Geometric mean.

functioning wood stoves, for their daily cooking and heating requirements (World Resources Institute 1998 *vide* ITDG Project 2002). These materials as well as coal-heating generate airborne particles and smoke are enriched in some trace elements. In improperly ventilated buildings, metal-bearing smoke may be of a great health risk. Dramatic examples of indoor intoxications in China, due to the combustion of coal enriched in As or F, have been described (Finkelman et al. 2003).

Plants

Trace element concentrations in plants reflect, in most cases, their abundance in growth media (soil, nutrient solution, water) and in ambient air. The metabolic fate and role of each element in plants can be characterized in relation to some basic processes such as: *(i)* uptake (absorption), *(ii)* transport within plants, concentration, and speciation, *(iii)* metabolic processes, *(iv)* deficiency and toxicity, and *(v)* ionic competition and interaction. Some trace elements, particularly trace metals such as Cu, Fe, Mn, Mo, and Zn play a key role in plant metabolisms and are constituents of several enzymes.

Plants exhibit various tendencies in the uptake of trace elements, and these depend on several factors both internal and external. During the evolution of plants, several mechanisms have developed that have resulted in adaptation to and tolerance of new or chemically imbalanced environments. Therefore, plant responses to trace elements in the environment should always be investigated for a particular system. Nevertheless, some generalization for trace element contents in plants and their physiological aspects can be made (Table I-5.1).

Absorption. Plants exhibit a variable, and sometimes specific, ability to absorb trace elements from soil. Plants absorb trace elements by roots from soil (or other growth media) and by above-ground parts from aerial deposition.

Foliar uptake is governed by several factors, mainly by surface properties of leaves and by aerial deposition of elements (contaminants). Foliar absorption usually consists of two phases – nonmetabolic cuticular penetration and metabolic transport across cell membranes.

Foliar absorption is significant in both aerial contamination of plants and foliar applications of fertilizers. Elements absorbed by leaves are partly translocated to roots and are also leached from plant foliage, especially by acid rain. Under specific conditions, and especially with atmospheric pollution, a high proportion of trace metals may enter plant tissues. Dalenberg and van Driel (1990) have calculated that 73 to 95% of the total Pb content of field crops are from aerial deposition on leaf surfaces.

The root absorption can be both active (metabolic) and passive (nonmetabolic). In each case, however, the rate of trace element uptake is positively correlated with their available (mobile) pool at the root surface. The plant uptake of elements from growth media is evaluated by a ratio of element concentration in plants to element concentration in soils and is called: Biological Absorption Coefficient (BAC), Index of Bioaccumulation (IBA), and/or Transfer Factor (TF). Values of these indexes differ

Table I-5.1. Approximate common concentrations of trace elements in mature leaf tissues generalized for various plant species (mg kg^{-1}) (after Kabata-Pendias and Pendias 2001)

Element	Sufficient or normal	Excessive or toxic[a]	Tolerable in crop plants
Ag	0.5	5 – 10	–
As	1 – 1.5	5 – 20	0.2[b]
B	10 – 100	50 – 200	100
Be	<1 – 7	10 – 50	–
Cd	0.01 – 0.2	5 – 30	0.05 – 0.5
Co	0.02 – 1	15 – 50	5
Cr	0.1 – 0.5	5 – 30	2
Cu	5 – 30	20 – 100	5 – 20
F	5 – 30	50 – 500	–
Hg	–	1 – 3	0.2[b]
Li	3 – 5	5 – 50	–
Mn	30 – 300	400 – 1000	300
Mo	0.2 – 5	10 – 50	–
Ni	0.1 – 5	10 – 100	1 – 10
Pb	5 – 10	30 – 300	0.5 – 10
Se	0.01 – 2	5 – 30	–
Sb	7 – 50	150	–
V	0.2 – 1.5	5 – 10	–
Zn	25 – 150	100 – 400	50 – 100

[a] Values are not given for very sensitive or highly tolerant plant species.
[b] FW basis.

highly for both plants and elements, however, the general tendency for the phytoavailability of elements can be summarized for terrestrial plants, at a given values of TF, as follows:

- 10^1: Cd
- 1: B, Br, Cs, Rb
- 10^{-1}: Ag, As, Co, Cu, Ge, Hg, Mo, Pb, Sr, Te, Zn
- 10^{-2}: Be, Cr, Li, F, I, Mn, Ni, Sb
- 10^{-3}: Ba, Bi, Ga, Fe, Sc, Se, V, Tl, Zr

Sheppard (2003) has widely discussed the great variability in plant/soil concentration ratios and emphasized the significance of generic data. Transfer processes are controlled by several highly changeably parameters and thus on-site data can significantly differ from the generic data. For example, TF values for Cs in *Vaccinium* vary from 0.7 to 2.7, calculated in Poland and Canada, respectively. This value for vascular plants is recommended at 1.0 and according to the IUR/IAEA estimations it is 0.05 (Sheppard 2003). Generally, trace element concentrations in plants are highly associated with the chemical composition of growth media. Based on this relationship, there is a possibility to identify the origin of food plants that occurred in markets (Szefer and Nriagu 2006).

Various fungi are well known to accumulate a high proportion of easily available and/or easily volatile elements such as Hg, Se, Cd, Cu, Zn, as well lanthanides and actinides.

One of the most important factors that determines phytoavailability of a trace element is its binding to soil constituents. In general, plants readily take up all species of elements that are in the soil solution. However, plant roots reveal a great capability to take up also slightly mobile forms of elements due to various root exudates that change the pH of the root ambient solution and chelate elements. Roots of plants grown in metal-contaminated sites can develop mechanisms to protect the plant from excessive uptake of a metal (Santa-Maria and Cogliatti 1998). However, root exudates from some hyperaccumulating plants do not enhance metal mobilization (Zhao et al. 2001).

In agricultural practices, several extractants have been used for quite a long time to estimate the phytoavailable pool of elements, and with various results. However, since a number of soil and plant parameters, as well as climatic factors, have a significant impact on the absorption of trace elements by roots any applied method must be related to a given soil and plant conditions. Methods used for the evaluation of soluble (bioavailable) trace elements are based on extractions by various solutions: *(i)* mineral acids, *(ii)* chelating agents, *(iii)* buffered salts, *(iv)* neutral salts, and *(v)* other solutions, such as *Coca Cola* (containing phosphoric acid). There are also proposed some other techniques, such as electrodialysis, diffusion through membrane, and diffusive gradient in thin-film (Kabata-Pendias and Sadurski 2004). However, neutral salt solutions (mainly $CaCl_2$) and chelating agents (EDTA and DTPA) have been used most commonly and broadly.

Deficiency and Toxicity. In spite of great progress that has been made in understanding the mechanisms and external factors that control the uptake of trace elements by plants, there is still not a simple model to predict their phytoavailability, especially in contaminated environments. Several models have been used to predict the bioavailability of trace metals, and in particular of Cd, Zn, Cu, and Pb (McLaughlin 2001). These models, however, are limited to a given plant and specific growth conditions, and thus the application to crop plants and field condition is still uncertain.

Based on a huge database of observations it has been possible to characterize soil properties that can affect deficiency of some elements to crop plants (Table I-5.2). The most common symptoms of micronutrient deficiency in sensitive plants are: *(i)* chlorosis and necrosis mainly of young leaves, *(ii)* wilting, *(iii)* melanism: brown, violet, red, *(iv)* stunted growth, and *(v)* leaf deformation. Schematic plant response to changes in concentrations of essential and nonessential elements differs at the ranges of low contents (Fig. I-5.1).

It is much more difficult to define soil properties and concentrations of trace element that are related to toxic effects of element excess. An assessment of toxic contents

Table I-5.2. Soil factors contributing to micronutrient deficiency

Element	Soil	Factors			
		pH	SOM	Water regime	Others
B	Podzols, gleysols, rendzinas, ferrasols	Acid, neutral	Very high or low	Flooded soil	$CaCO_3$, light texture
Co	Podzols, histosols, rendzinas, solonetz	Alkaline, acid	High	High moisture	$CaCO_3$, high Mn and Fe
Cu	Histosols, podzols, rendzinas, solonetz	Various	Low or high	High moisture	High N, P, Zn, light texture
Fe	Rendzinas, ferrasols, solonetz	Alakline	High or low in acid soil	Poor drainage	High P, Mn, $CaCO_3$
Mn	Podzols, rendzinas, histosols	Strongly acid, alkaline	Very high	Moisture extreme	High Fe, $CaCO_3$
Mo	Podzols, ferrasols	Acid	–	Good drainage	High Fe, Al, S
Se	Podzols, histosols, ferrasols	Acid	High	Waterlogging	High Fe, S
Zn	Podzols, rendzinas, solonetz	Strongly acid, alkaline	Low	–	High P, N, $CaCO_3$

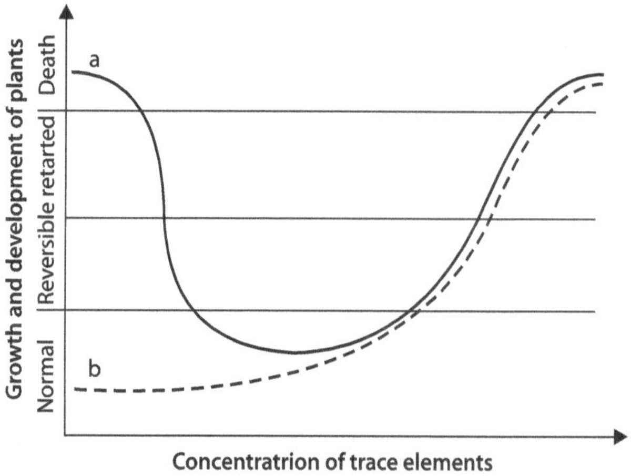

Fig. I-5.1.
Schematic diagram of plant response to deficient and toxic concentrations of trace elements in growth media: **a** essential trace elements; **b** nonessential trace elements

and effects on plants is very complex since this depends on many factors both external and internal. Plants reveal a great behavioral plasticity under chemical stress (Fig. I-5.2).

General effects of toxicity on common cultivars are: *(i)* stunting and damaged roots, *(ii)* chlorotic, necrotic, brown, and dark green spots, mainly on old leaves, *(iii)* wilting and dying-off of older leaves. Krämer et al. (1999) have supported a concept that the excess of trace metals induce biochemical changes in plants similar to responses elicited by pathogen attack. Nevertheless, the damage to photosynthesis apparatus involved in several metabolic processes is the most significant toxic effect of the excess of trace

Fig. I-5.2.
Effect of increasing concentrations of trace elements on behavior of plants: **a** tolerant species, no behavioral changes; **b** development of behavioral tolerance; and **c** nontolerant species

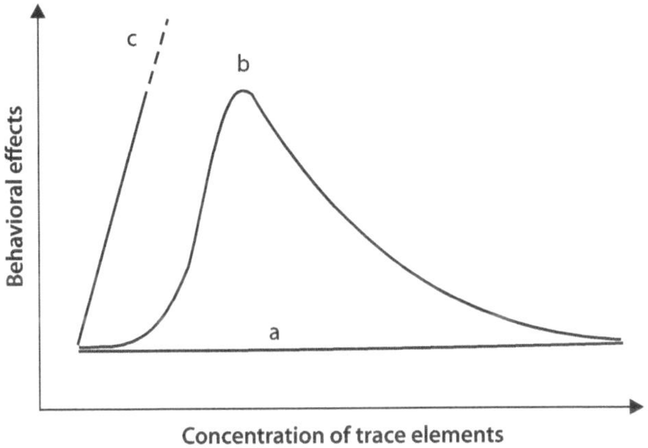

metals. Prasad and Hagemeyer (1999) summarized the main plant reactions to the toxicity of element excesses, as follows:

- Changes in permeability of cell membranes – Ag, Au, Br, Cd, Cu, F, Hg, I, Pb, and U
- Reactions of thiol group with cations – Ag, Hg, Pb
- Competition for sites with essential metabolites – As, Sb, Se, Te W, F
- Affinity for reactions with phosphate, ADP, and ATP groups – Al, Be, Sc, Y, Zr, lanthanides, and possible several other trace metals
- Replacement of essential ions (mainly major cations) – Cs, Li, Rb, Se, and Sr
- Occupation of site for essential groups such as phosphate and nitrate – compounds of As, B, Br, F, Se, Te, and W

A number of studies have been carried out on trace element toxicity to plants and the results vary with each experiment and each plant. However, it can be generalized that the most toxic trace metals for both higher plants and certain microorganisms are: Hg, Cu, Ni, Pb, Co, Cd, Ag, Be, and Sn. Several plants may adapt relatively readily to the excess of trace elements, but they may be very sensitive to a particular element. In many higher plants mechanisms of adaptation to, or avoidance of the stress factor have been developed. These mechanisms depend on plant species, specific element and environmental conditions. The most important processes are:

- Complexing and chelating of ions outside roots, in root tissues, and in plant cells
- Selective uptake of ions
- Binding to cell walls
- Immobilization in various organs in form of different immobile compounds, including minerals
- Restricted influx through the plasma membrane
- Release by leaching from foliage
- Excretion from leaf tips in the form of salts
- Volatilization of toxic compounds
- Efflux of element excess from roots
- Removal with falling leaves/needles

Several other mechanisms of adaptation include also the alteration of metabolic patterns that either increase the enzyme system that is inhibited or reduce the metabolic pathway associated with toxic ions. Tyler et al. (1989) concluded that: "The action of the tolerance mechanism often imparts a demand for metabolic energy, which would make tolerant forms less competitive under normal conditions than their nontolerant outerparts". Therefore, the growth of plants under the impact of elevated contents of trace elements is highly limited. In most cases, the morphology of those plants is altered, and salt excretion can be observed at the tips of leaves in some plants grown in soil polluted with trace metals. Forest declines have been often observed in contaminated regions.

Speciation. Trace metals in plants apparently occur mainly in organic complexes, however, data on forms of their occurrence are rather scant. Most metals form various complexes with both small and macromolecular organic compounds that are present in xylem and phloem of plants (Łobiński and Potin-Gautier 1998; Schaumlöffel et al. 2003. Common chelating agents are organic acids (i.e., carboxylic, amino, mercaptic, mugineic, phytic), proteins, polysaccharides, lingnins, pectins, DNA, metallothioneins, phytoferritins, etc. Inorganic ligands like Cl, F, and sulfate are also important complexants in plants (Prasad and Hagemeyer 1999). Plants are able to synthesize various metallo-complexes such as methylmercury, organoarsenic, and organoselenium (Schaumlöffel et al. 2006). Relatively often selenomethionine has been found in wheat, soybean, and Se-enriched yeast (Shibita et al. 1992; Dauchy et al. 1994).

Organo-Pb species were of a special concern before the abatement Pb additives, into gasoline as an anti-knock agent. Several Pb species, such as Me_3Pb^+, Me_2Pb^{2+}, Et_3Pb^+, Et_2MePb^+, have been reported to be absorbed by plants through roots or by foliage from the atmosphere (van Cleuvenbergen and Adams 1991).

Various organoarsenic species (different methylated compounds) have been found in aquatic plants and apparently this causes a higher As concentration in marine plants than in terrestrial ones (Kuehnelt et al. 2000). Microbial methylation has resulted in methylate-species of several elements, including trace metals.

The phytochelation of Cd and Cu has been demonstrated in cell cultures of rice, soybeans, and maize (Vacchina et al. 2000; Chassaigne et al. 2001). Pavlikova et al. (2005) separated, using organic solvents and water, species of As, Cd, and Zn from spinach biomass. The highest proportion of As was extracted by water and identified as phytin, phytic acid, pectin, and metallothioneins. Zinc was extracted mainly by methanol and acidified (HCl) methanol; isolated fractions were: compounds with organic acids, (including mugineic and phytic acids) flavone glucuronides, oligopeptides, proteins, lignins, and pectin. Cadmium was extracted predominantly by acidified methanol as fractions of: proteins, phytic acid, lignins and pectin. However, the largest amount of Cd remained as non-extractable residues, mainly as polysaccharides.

Interactions. Interactions of chemical elements may greatly contribute to a chemical balance of plants. Both, antagonistic and synergistic relations either between trace elements or between major and trace elements are variably and may occur inside the cells, within the membrane surface, and also surrounding plant roots. Interrelationships in different soils and/or plant species are of multivariant character and quite complex, being at times both antagonistic and synergistic (Table I-5.3).

Table I-5.3. Interactions between major and trace elements in plants

Major element	Antagonistic element	Synergistic element
Ca	Al, B, Ba, Be, C d, Co, Cr, Cs, Cu, F, Fe, Li, Mn, Ni, Pb, Sr, and Zn	Cu, Mn, and Zn
P	Al, As, B, Be, Cd, Cr, Cu, F, Fe, Hg, Mn, Mo,Ni, Pb, Rb, Se, Si, Sr, and Zn	Al, B, Cu, F, Fe, Mn, Mo, and Zn
Mg	Al, Ba, Be, Cr, Mn, F, Ni[a], Co[a], and Fe[a]	Al and Zn
K	Al, B, Cd, Cr, F, Hg, Mn, Mo, and Rb	–
S	As, Ba, Fe, Mo, Pb, Se, and Zn	F and Fe
N	B, Cu, F, and Mn	B, Cu, Fe, and Mo
Cl	Br and I	–
Na	Mn	–
Si	B and Mn	–

[a] Reported for microorganisms.

The main major antagonistic elements are Ca, P, and Mg that interfere the absorption and metabolism of several trace elements. Some synergistic effects, however, have also been observed for antagonistic pairs of elements, depending on the specific reaction of plants or specific properties of growth media. Most important for practical application are the antagonistic interactions between Ca and P and some trace cations and anions.

The greatest number of antagonistic reactions has been observed for Fe, Mn, Cu and Zn with several other elements, including As, Cr, Cr, Fe, Mo, Ni, Pb, Rb, and Se (Kabata-Pendias and Pendias 2001). Deficiency of one trace element essential to a plant can facilitate uptake of other micronutrients. The compensatory absorption of trace elements under deficiency stress is reported for several elements and most commonly for Mn and Cu, for Fe and Cu, Zn, and Mn (Rengel et al. 1998). Broadly studied Cd/Zn, Cd/Cu and Cu/Zn relationships are very complex, and as Sharma et al. (1999) have described, when only one element of these binary ratios exceeds the toxic level, the joint effect is either synergistic or additive. There is a controversy regarding Cd/Zn interactions, however, Zn seems to play an important role in controlling the Cd uptake in Cd-contaminated soils, whenever the ratio Zn/Cd is higher than 100 (Kabata-Pendias and Terelak 2004).

Synergistic interactions are not commonly observed. Some reported interactions between for Cd and other metals such as Fe, Ni, Pb may be artifacts resulting from the destruction of physiological barriers under the stress of elevated metal contents.

Bioindication. Indicator species are organisms that serve as a measure of chemical status of a given site or area. These may be various animal and plants organisms and their organs, or products (i.e., honey). Using plant chemical status for geochemical prospecting is a very old practice for the exploration of metal ores. The response of plants to the chemistry of the environment is controlled by several external and biochemical factors. Nevertheless, the trace element contents of plants is a promising indicator for the study of chemical properties and changes of the terrestrial environment. Basic characteristics of a good bioindicator are summarized by Witting (1993) as follows:

- Suitable accumulation rate of several or selected elements
- Toxitolerance and no sensitivity to the accumulated element
- Present in large amounts in the ecosystem under investigation
- Wide distribution in various environments
- No seasonal differences in availability and applicability
- Existence of a correlation between accumulation and input to the ecosystem

Based on these characteristics, mosses, lichens and mushrooms are especially suitable phytoindicators. However, plants that occur more commonly and are easy to sample are relatively often used as indicators. The Coefficient of Specific Relative Accumulation (CSRA) calculated as a ratio of the content of an element in a given plant to its average content in all other plants grown on the same site indicates that dandelion is especially useful bioindicator (Table I-5.4).

The biota reveals a potential to develop a resistance to anthropogenic loads of trace elements to ecosystems. However, the biodiversity and biological activity of polluted ecosystems is very poor, as observed in vicinities of most industrial plants. Sensitive phytoindicators such as mosses, lichens and mushrooms are among the first plants that disappear in such regions. Therefore, dandelion that easily adapt to contaminated ecosystems is suitable for biomonitoring. Metal concentrations in dandelion leaves sampled from different regions of Poland clearly indicate metal pollution in the industrial region (Table I-5.5). The RDB value identifies a very high relative deviation from background values of all metals in plants from industrial region. The order of relative accumulation of metals, as expressed by RDB factor, is the following: Ni > Cd > Zn > Pb > Mn > Cu.

The suitability of the dandelion as versatile phytomonitor has been confirmed by Rule (1994) who found the worldwide background concentrations of metals fairly similar in plants grown in various countries (Bulgaria, Poland, and USA). These vary (in mg kg^{-1}) for leaves: Cd, 0.2–1; Cr, 0.2–0.5; Cu, 5–20; Mn, 15–170; Ni, 0.5–6.2; Pb, 0.8–6.5; and Zn, 20–110; and for roots: Cd, 0.1–0.8; Cr, 0.1–2.8; Cu, 5–25; Mn, 5–50; Ni, 0.2–6.6; Pb, 0.2–5; and Zn, 10–60.

Table I-5.4. Transfer factor (TF) for trace metals in plants (leaves) grown under the same conditions (after Kabata-Pendias and Krakowiak 1998)

Plant	Cd	Cu	Fe	Mn	Pb	Zn
Dandelion, *Taraxacum officinale* Web	3.42	2.6	2.72	8.43	4.04	6.8
Plantain, *Plantago major* L.	0.93	0.89	0.85	0.23	0.57	0.54
Bromgrass, *Bromus unioloides* L.	0.11	0.52	0.49	0.06	0.26	0.09
Horse bean, *Ficia faba minor* L.	0.23	1.49	0.71	0.82	0.27	0.45
False acacia, *Robinia pseudoacacia* L.	1.49	1.02	0.76	0.49	1.33	0.25

Table I-5.5. Metals in leaves of dandelion (*Taraxacum officinale*) grown in different region of Poland (after Kabata-Pendias and Pendias 2001)

Metal	W Whole country		Industrial, SW region		Rural, NE region	
	GM	*RDB*	*GM*	*RDB*	*GM*	*RDB*
Cd	0.5	−20	1.2	50	0.4	−50
Cr	0.8	25	0.7	14	0.4	−50
Cu	9.4	−27	13.4	10	7	−71
Mn	60	0	74	18	42	−18
Ni	3.4	50	4.2	52	1.9	−5
Pb	1.1	−81	3	33	0.5	−300
Zn	45	0	72	37	40	−12

GM – Geometric mean contents, in mg kg^{-1}. *RDB* – Relative deviation to background values = ([AM – RF] / RF) × 100; *AM* – Arithmetic mean of measured contents, *RF* – Reference values (mg kg^{-1}): Cd, 0.6; Cr, 0.6; Cu, 12; Mn, 60; Ni, 2; Pb, 2; Zn, 45.

Some other organisms such as, e.g., roots of aquatic plants, soil microorganisms, are also considered good bioindicators of contamination of a given environment. Also enzymes that are inhibited by the excess of trace metals, particularly by Cd, Co, Cu and Zn, are known as good indicators (Nowak et al.1999).

Humans

Organisms have developed their internal biochemistry in close connection to the composition of the natural environment. Humans, as well as all mammals, unlike prokaryotes and other lower organisms, are not able to adapt easily to any change in the chemical composition of their surroundings. Changes in trace element concentrations are of especially vital importance. The homeostatic balance of chemical elements in an organism is the basic requirement of good health. Ionic relationships within any organism are very fragile and governed by several factors. Their balance is controlled by factors such as bioavailability of an element, capability of tissues or organs to accumulate and excrete an element and by interactions among elements that might vary from antagonistic to synergistic depending mainly on their quantitative ratio.

Trace elements, both those of vital physiological functions, and those not essential play fundamental roles in the normal development and health of organisms. Functions of most of the essential trace elements in metabolic and growth processes have been relatively well recognized. Their crucial roles are attributed commonly to functions of metallo-enzymes. Especially essential for the activity of physiologically important enzymes that function as catalysts are Cu, Zn, and Fe (Kleczkowski et al. 2004).

Geochemical anomalies of the bedrock, variable soil parameters, agricultural practices, and anthropogenic pollution influence the trace element contents of food crops and other plants resulting in a dietary intake of trace elements. Diseases and/or impaired metabolism, attributable to trace element deficiency or excess, are some times not easy to assess, especially at early stages of development. Moreover, often effects of their imbalanced supply may be quite variable (Table I-6.1 and I-6.2).

Trace elements that occur and/or may occur in human body are as follows:

- Essential: As, B, Br, Co, Cl, Cu, Cr, F, Fe, I, Li, Mn, Mo, Se, Si, V, and Zn
- Possible essential: Al, Ba, Ge, Ni, Rb, Sn, Sr, Ti
- Non-essential: Ag, Au, **Be**, **Bi**, **Cd**, Cs, Hf, **Hg**, In, Ir, **Pb**, Sb, REEs, Ta, Te, **Tl**, U, Y, Zr. (In bold are elements of high toxicity to humans)

Most common deficiencies for the worldwide population are reported for Fe and I. There are estimations that about 50% of the worldwide population suffer from Fe deficiency, especially in Africa and Asia, which is associated mainly with low Fe contents in food plants. Iodine deficiency occurs commonly in regions with light sandy soils developed from young geologic formations. It is not likely to occur in close-to-sea areas.

Table I-6.1. Consequences of deficiency and excess of some essential trace elements (adopted from various sources, mainly from Abernathy et al. 1993 and Plumlee and Ziegler 2003)

Element	Deficiency	Excess
Co	Anemia, anorexia	Cardiomyophaty, marrow defects, excess of red blood cells
Cu	Anemia, and tissue defects	Necrotic hepatitis, hemolysis, hyperglycemia
Cr	Cr^{3+}-defective glucose metabolism, hyperlipidemia	Lesions in skin, intetinal mucosa, pulmonary edema, lung cancer
F	Tooth decay, growth retardation	Fluorosis – variable effects, mottling of tooth enamel
Fe	Anemia	Siderosis, hemachromamtosis, cardiac failure
I	Goiter, impaired neurological function	Hyperthyroidism
Li	Depression	Impaired central nervous system, cardiovascular and renal effets
Mo	Defects in keratosis, growth retardation	Molybdenosis, defect in Cu metabolism, diarrhea
Mn	Skeletal and cartilage deformitis	Manganism, neurological disorders, liver cirrhosis
Se	Cardiac myopathy, osteoarthropathy, membrane malfunction	Selenosis, liver and kidney damage, fetal toxicity, cancer
V	Tooth defects	Nervous disturbances
Zn	Anorexia, anemia, impaired keratosis, teratogenic effects	Anemia, tissue lesions

Table I-6.2. Consequences of excess of some non-essential trace elements (adopted from various sources, mainly from Anke 2004a and Plumlee and Ziegler 2003)

Element	Toxicity symptom
Al	Osteomalacia, neurotoxicity, dementia, possible Alzheimer's disease agent
As	Variable disorders including nervous system, liver and kidney failure, intestine tract distress, anemia, skin cancer
Be	Berylliosis, pneumonitis, cancer
Cd	Cardiomyopathy, liver and kidney damage, gastroenteritis, pneumonitis, osteomalcia, cancer
Hg	Nervous and gastric systems disorders, kidney and pulmonary damage, a potent teratogen
Pb	Nervous system disorders, hematologic effects, kidney disease, intestine tract distress, hypertension
Ni	Gastric, liver, and kidney defects, neurological effects, emphysema and lung cancer

Contents of trace elements in the human body vary greatly, and are controlled by several external and internal factors. The highest variation in reported concentrations of most trace elements (with an exception of Cu and Zn) in human serum, generally by an order of three magnitudes, is either a real natural differentiation or artifact affected by sampling and analytical methods (Dybczyński 1998). Nevertheless, calculations of average concentrations of these elements in human tissues have been accepted (Table I-6.3).

There are not many data on the elemental speciation in biological fluids/tissues. According to Cornelis and De Kimpe (1994) trace elements may occur in a complex

Table I-6.3. Trace element concentrations commonly reported for some human tissues (adapted from the database http://eko.ch.pw.edu.pl/kons/aimzs.doc)

Element	Mean content (mg kg^{-1})		Mean concentration (µg l^{-1})		Reference man[a] (mg kg^{-1})
	Kidneys	Liver	Serum	Milk	
Al	5	5	1	15	2.6
As	0.01	0.03	1	0.5	0.26[b]
B	0.5	1	1	5	0.3
Br	5	0.2	500	1 520	2.9
Cd	1.5	0.8	0.1	1	0.71
Co	0.1	0.1	0.1	0.5	0.02
Cr	0.1	0.1	0.1	1.5	0.09
Cs	0.01	0.1	0.7	0.5	0.02
Cu	14	20	1 000	280	1
Fe	320	600	1 090	500	60
I	0.2	0.2	50	80	0.19
Li	0.01	0.01	0.8	2	–
Mn	1	5	0.4	5	0.17
Mo	0.3	1	0.5	1	<0.13
Ni	0.5	0.1	0.3	12	0.14
Pb	3	5	3	10	1.7
Rb	5	5	160	100	9.7
Sb	0.05	0.05	0.3	1	0.03
Se	0.5	2	90	18	0.11
Sr	0.05	0.05	5	10	4.6
Ti	0.5	0.5	5	10	–
V	0.2	0.1	0.05	0.8	–
Zn	150	250	900	1 500[b]	33

[a] Concentrations in whole adult body, after Li (2000).
[b] Questionable data.

matrix, such as biological fluids, in various forms, as: *(i)* simple anions (e.g., Br^-, F^-), *(ii)* chelated compounds (e.g., Fe-citrate), and *(iii)* build into simple metabolites (e.g., As in monometylarsonic acid, etc.). These authors studied the relative distribution of injected ^{74}As in the body fluids and found the highest concentration of As in urine (53%) and in the red blood cells (28%). In other tissues, the ^{74}As relative concentrations varied from 0.1% in pancreas to 2.7% in liver.

Concentrations of trace elements in brain tissue have been investigated and authors have suggested that further studies are needed with special regards to the correlation with health and disease (Andrási et al. 1999). According to these authors, the distribution of some trace metals in brain of normal humans varies (in mg kg^{-1}), as follows:

- Cd: from 0.19 in substantio nigra, right side to 0.42 in putamen, left side
- Cu: from 14 in genu corporis callosi to 38 in putamen
- Fe: from 124 in genu corporis callosi to 819 in putamen
- Mn: from 1 in genu corporis callosi to 3.1 in putamen
- Pb: from 0.16 in vermis cerebelli to 14.3 in caput nuclei caudate, right side
- Zn: from 37 in genu corporis callosi to 85 in putamen

Of the greatest nutritional concern are concentrations of Cd, Hg, and Pb in human milk. These values are highly variable upon several factors. Ursinova and Masanova (2005) have summarized data for trace metals in human milk in worldwide countries as follows (range of means, μg l^{-1}):

- Cd: Canada, 0.08 – Germany, 24.6
- Hg: Austria, 0.23 – Germany, 5.5
- Pb: Sweden, 0.5 – Mexico, 61.8

Environmental pollution has a significant influence on trace metal level in human tissues that is clearly illustrated by higher metal contents in bones of men from an industrial region than of men from a city (Table I-6.4). Skeletons of ancient inhabitants of the region of Cu mining activity in southern Jordan, especially during the Bronze

Table I-6.4. Geometric means of trace elements in surface part of the head of thighbones of adult men from two different locations in Poland (mg kg^{-1}) (after Kwapuliński et al. 2004)

Metal	City of Łódź	Industrial region, Silesia
Ni	0.13	0.20
Cd	0.14	0.79
Mn	1.01	2.86
Pb	2.40	2.54
Cu	2.68	4.28
Cr	2.94	4.69
Fe	29.39	42.20
Zn	44.07	49.59

Age and Byzantine Age, contain markedly enhanced amounts of Cu (103 mg kg^{-1}) and Pb (35 mg kg^{-1}). However, these metals were not distributed in bones uniformly and chemical fingerprints of skeletons differed between these two Ages (Pyatt et al. 2005). Contemporarily, a serious problem of pollution stress to humans is in the Arctic due to a combined effect of long distance transport of inorganic trace pollutants and biomagnification in the ocean food chains. This results in a human exposure to trace elements at a level which is often in excess of international accepted limits for intake (Hansen 1996).

Behavioral habits as well as some diseases can also affect contents of some trace elements in human organs. For example, smokers and persons with lung cancer have in lung tissues higher amounts, especially of As, Cd, and Ni than control persons, non-smokers (Table I-6.5). Wąsowicz et al. (2003) reported lower concentrations of Se (49.7 µg l^{-1}) and Zn (0.86 µg l^{-1}) in plasma of patients with pulmonary cancer than in persons of the reference group (Se: 54.3 and Zn: 0.93 µg l^{-1}), whereas the Cu level was higher in cancer patients than in healthy ones, 1.53 and 1.17 µg l^{-1}, respectively. However, these results do not confirm the role of these trace elements in lung cancer development. The decrease of Cu concentrations in human organs with age has been observed. Livers of the babies aged 1 day to 12 months contain, on average, 119 mg Cu kg^{-1}. With increasing age, the Cu contents of livers decrease to 18–20 mg kg^{-1} and remain at these concentrations through the life (Müller 2006).

There is not much information on forms of trace elements in human body. The most comprehensive data have been presented by Szpunar et al. (2003). According to these authors, predominating complexations of trace elements in biological systems are by organic compounds:

- Proteins and their constituents (*via* sulfur atom): Cu, Zn, Cd, Hg, and Ag (e.g., methionine, cysteine, phytochelatins, metallothioneins, albumine)
- Proteins and their constituents (*via* oxygen atom): Mo, Mn, Fe, Co, Ni, Cu, and Zn (e.g., histidine, tyrosine, glutamic acid, aspargine, metalloenzymes)
- Nucleic acids and their constituents: Cr, Ni, Pt, and Ru
- Metabolites with metalloid-carbon covalent bond: As, Se, and I (e.g., methyloarsonic acid, arsenobetaine, arsenosugar, selenoaminoacids, selenopeptides, selenoproteines)
- Complexes with polysaccharides: Sr, Ba, La, Pb (e.g., rhamnogalacturoran-II)
- Tetrapyrrol ligands: V, Fe, Co, Ni (e.g., porphyrins, cobalamins)
- Small organic ligands: Al, Ni, and Fe (e.g., citric acid, nicotianamine)

Table I-6.5. Mean concentrations of selected trace elements in human lung tissues (mg kg^{-1}) (after Chen and Wang 1990)

Person group	As	Cd	Ni	Se	Zn
Smokers with lung cancer	0.168	0.733	1.028	0.718	56.3
Non-smokers with lung cancer	0.141	0.370	0.528	0.639	57.6
Control-smokers	0.142	0.318	0.428	0.531	54.1
Control-non-smokers	0.078	0.404	0.340	0.480	60.7

Increased accumulation of some elements in certain organs may be an affect of both their physiological functions or immobilization and exclusion from metabolic processes. Increased contents of trace metals are observed in some biominerals formed in the human body. Many elements, such as Fe, Cu, Co, Sr, Ba, Zn, Cr, Pb, Te, Nb, Mo, Mn, Ti, Th, Ag, Cs, Sc, and Au (given in an occurrence frequency order) have been identified in kidney stones (Tynaliev and Usupbaev *vide* Kabata-Pendias and Pendias 1999). Some metals (Zn and Mg) are reported to increase in cancer tissues with simultaneously a decrease in the serum of those persons (Pasternak 1997). This may be associated with increased glutathione levels in carcinoma cells (Chen et al. 1997). Hasumi et al. (2003) have reported the association of Zn and Cd in prostate glands and suppressing effects of the Zn treatment on the growth of cancer cells. Floriańczyk and Grzybowska (2006) have reported a higher Zn level (16.6 mg kg^{-1}) in breast tumors than in tissues of benign breast lesions (7.6 mg kg^{-1}).

Suitable biomarkers for assessing human exposure to trace elements have been subjects of several studies for quite some time. Most commonly, hard dental tissues (including enamel), scalp hair and other ectoderm derivates have been found to be good indicators to assess human exposure (also prenatal exposure) to trace elements (Ericson 2001; Pereira et al. 2004; Rakovič et al. 1996).

Release of excess of trace metals from the human body follows normal physiological processes. However, there has been proposed, for more than 40 years, chelation therapy to remove excess trace metals from the human body. Ibrahim et al. (2006) widely described chelation therapies of patients intoxicated by some metals. The therapy with EDTA has been approved by the FDA as a treatment for lead poisoning. EDTA and other chelating drugs, used for intravenous infusions, can lower high blood levels of metals such as Cd, Cr, Hg, Mn, Pb, Zn and several others, by attaching to the metal molecules, which helps the body to remove them through the urine. However, chelation therapy is not a solution for the long-term toxicity of trace metals in human body.

Sata et al. (1998) investigated the behavior of eight metals in urine and blood following CaEDTA injection. The authors have concluded that only Cr and Mn absorbed in human tissues might be mobilized by this therapy. Other metals, e.g., Pb and Hg, have also been shown to be decreased in human tissues following the chelation therapy (Liu et al. 2002; Risher and Amler 2005). However, a number of studies have shown no clear relief from DMSA (trade name, drug composition not given) for humans, on the other hand the DMSA treatments have poisoned the patients (Englund et al. *vide* Risher and Amler 2005).

A few recent studies have indicated that the chelating agent may redistribute a metal from hard tissue to the brain causing fatal injury (Hoover and Aposhian *vide* Flora et al. 2005). Thus, there is still a controversy concerning chelating therapy. These procedures have not been yet sufficiently supported by the controlled studies (Chisolom 2001; Liu et al. 2002).

Different types of chelators have been used against a widespread As exposure to humans. The chelator, BAL (British Anti Lewisite) has been selected for the treatment of As poisoning patient in the USA (Klaassen 1985). This therapy, however, after intramuscular injection, resulted in pain and discomfort.

Exposure sources. The general population is exposed to trace elements mainly by ingestion of drinking water and food and by inhalation of air. Adults inhale daily approximately 20 m^3 of air and intake about (above) 2 l of water.

Humans intake trace elements from various sources by three absorption pathways:

- Gastrointestinal tract, from: food, water, drugs, soil, and aerosols
- Respiratory tract, from: aerosols, and gases
- Skin, from: soil, water, aerosols, gases, and others

All exposure pathways finally result in the absorption across the body's membrane by passive or active diffusion and can be deposited in specific tissues. Soft tissue organs, especially the liver and kidneys are likely to accumulate most trace elements. Easily bioavailable forms of elements are those which compounds can be dissolved by body fluids (i.e., gastrointestinal and respiratory tracts).

The food chain is considered the main tract for transfer of trace elements to humans. There are several possible routes of their (TE) transfer, *via* food chain, from soil to human:

- TE → soil → plant → animal→ human
- TE → soil → plant → human
- TE → soil → animal → human
- TE → soil → microbiota, mezobiota → human
- TE → soil → airborne dust → animal, human
- TE → soil-geophagia → human
- TE → soil → groundwater → drinking water → animal, human
- TE → soil/sediment → surface water → aquatic biota → human

The oral intake of trace elements has became to be of the greatest concern and several official regulations have been established for their safety intake. Some attempt has been made by the UNEP (2004) and European Communities to accept provisional toxic concentrations of metals as potential hazards to humans (Table I-6.6 and I-6.7).

A special source of some trace metals is associated with various orthopedic and stomatological metallic implants and prosthesis. Various alloys are used for orthopedic surgery. The most commonly they are alloys of Co, Cr, Mo, Ni and W. Several precious and semi-precious metals are also used. Recently Ti-base alloys are increasingly used for surgical implants. Metals are released into body tissues, due to the corrosion (in some cases due to the abrasion) of alloys. There are variable responses of cells on increased levels of metals, from tolerance to allergy and inflammatory. Most of the metals released from metallic implants can be recovered in the human organism in implant-surrounding tissues, in human fluids, and in ectoderm derivates (e.g., nails, hair).

Trace element intake due to soil ingestion both involuntary and deliberate may be significant especially for children. There was observed a higher content of Pb in blood in children present pica for soils than in no-pica soil. Intaken mean amounts of some metals by children, 1–4 years of age, were as follows (in mg d^{-1}): Al, 136; Ti, 208; V, 148; Zr, 113; and Y, 97 (Calabrese and Stanek *vide* Abrahams 2005).

Table I-6.6. Trace element oral intake of toxic hazards to humans (adopted from UNEP 2004)

Element	Compound	LDL_{50} (mg kg^{-1} BW)
As	–	763
Cd	–	225
Cr (+6)	$Na_2Cr_2O_7$	50
Cu	$CuSO_4$	300
Hg	$HgCl_2$	1
Pb	–	450 (LDL_0)[a]
Sb	–	7000
Se	–	6700
Te	–	83
Tl		6 (LDL_0)[a]
Zn	–	3000

[a] The lowest toxic dose (*http://www.ehs.ucsb.edu/units/labsfty/labrsc/lstoxicology.htm*).

Table I-6.7. Setting maximum levels for Pb and Cd in foodstuffs by European Communities (mg kg^{-1} FW; in parenthesis data for kidneys) (Commision Regulation 2001)

Product	Pb	Cd
Cows' milk and products	0.02	–
Cereals, including buckwheat, legumes, rice and pulses	0.2	0.1
Meat of cattle, sheep, pig and poultry	0.1	0.05
Offal cattle, sheep, pig and poultry	0.5	0.5 (1.0)
Muscle meat of fish	0.2 – 0.4	0.05 – 0.1
Crustaceous excluding crabs	0.5	0.5
Bivalve mollusks	1.0	1.0
Vegetables, including herbs, fungi and potatoes	0.1	0.05 – 0.1
Leafy vegetables, including brassica and cultivated fungi	0.3	0.2
Fruits, including juices	0.05 – 0.1	0.05
Fats and oils, including milk fat	0.1	–

Significantly, however, the most often unrecognized sources of ingested metals are those released from kitchen utensils. Marzec et al. (2006) have reported that cookware, especially enamel and stainless steel, may be a source of Cr, Ni, Zn, and Fe in popular soups, however, their concentrations are significantly lower than accepted limits for standard acid solution.

In addition to the food-chain transfer, the exposure to trace elements from other sources may be important. Among them, the ingestion from some herbal medicines has been recently of concern. Especially notable, contents of trace metals, in Asian herbal medicines (AHMs) have not been controlled until the present. Ernst (2002) has described elevated contents, mainly of Hg, Pb, and As, in some AHMs, possible due to: *(i)* intentional addition for alleged medicinal properties, *(ii)* contamination during manufacture, and *(iii)* increased metal levels in plants grown on polluted soil. This author has concluded that trace metals in AHMs represent a potentially serious problem that puts consumers at risk. Beneficial and/or toxic effects of trace elements in medicinal plants have been discussed by Szentmihalayi et al. (2006).

Contents of Pb and Cd in foodstuffs are presently of the greatest concern, and several regulations have been enacted to established allowable maximum levels of these metals in various food products. The ranges of maximum concentrations in food products, as have been set by the EC, vary (in mg kg^{-1} FW) for Pb from 0.02 in milk to 1 in bivalve mollusks, and for Cd from 0.05 in meat and vegetables to 1 in bivalve mollusks (Table I-6.7). All types of these mollusks are of a great health risk due to trace metal concentrations.

Maximum level for Hg in foodstuffs is set by the Commission Regulation (2001) for fishery products only. For most fish it is 0.5 mg kg^{-1} FW, excluding several fish like halibut, catfish, pike, sharks etc. for which this value is 1.0 mg kg^{-1} FW.

The intake of trace elements by various groups of population has recently been widely studied. Estimated daily intake of metals of the greatest concern vary significantly between European countries, which may have resulted from different methods used for the estimation of dietary exposure (Table I-6.8).

Table I-6.8. Mean or range intake of Cd, Hg, and Pb by adults with total diet in some European countries (after Winter-Sorkina et al. 2003) and in Nigeria (after Ojo et al. 2006) (μg $person^{-1}$ d^{-1})

Country	Cd	Hg	Pb
Belgium	45 – 50	14[c]	293
Denmark	20	3.5 – 7	42
Finland	9[c]	–	–
France[a]	30	9	34
Germany	21	31	246
Italy	35 – 64[d]	8[f]	108
Poland[b]	35/28	5.7/3.2	100/79
Sweden	8.5 – 11[e]	–	–
Swiss	9[a]	<5[d]	25[d]
Netherlands	14	0.6 – 1.4	55[e]
Nigeria	310/510[g]	–	171/214[g]

[a] Data for France (duplicate meals) after Noël et al. (2003).
[b] After Marzec and Schlegel-Zawadzka (2004), mean intake by male in 1990 and 2002, respectively.
[c] Hospital lunches.
[d] Canteen meals.
[e] Duplicate diet.
[f] Sea food consumer.
[g] Intale by children and adults, respectively.

Table I-6.9. Daily intake of trace metals by some population groups of USA, based on the TDS studies 1991–1996 (μg $person^{-1} d^{-1}$, and % PTDI) (adapted from Egan et al. 2002)

Element	Children, 6 years		Adults, 60–65 years			
	μg $pers^{-1} d^{-1}$	% PTDI	Women		Men	
			μg $pers^{-1} d^{-1}$	% PTDI	μg $pers^{-1} d^{-1}$	% PTDI
As	20.2	44	75	53	95.5	56
Cd	8 – 9	34 – 43	9 – 12	14 – 17	12 – 15	15 – 19
Hg	0.8	5	2	4	2.6	4
Pb	6 – 13	4 – 17	7 – 16	1 – 7	5 – 20	2 – 7

Diet, including drinking water, is the main source of trace element intake by humans. Extension studies for monitoring chemical contaminants and radionuclides in food have been carried out since 1961 under the program of the Total Diet Studies (TDS) in USA (Egan et al. 2002). These studies have identified four trace metals of concern, As, Cd, Hg, and Pb that have been controlled by JECFA regulation (WHO 1999). Fourteen groups of population have been distinguished in this program, from infants to adults (>70 years). Daily intakes of these elements were the highest among adult males (60–65 years), however, the highest relative intakes, expressed in % of the PTDI, were by infant (2 years) and children (6 years) (Table I-6.9). Results of TDS carried out in the UK in 1994 have indicated that, similarly as in other countries, the dietary exposure for most of the elements are decreasing and are below the relevant PTWIs and PMTDs (Gillian et al. 1999).

The total diet studies conducted in the UK, in 1994, for 30 elements (Table I-6.10) have shown that the dietary exposure (in μg $person^{-1} d^{-1}$) is the highest (>1 000) for: Al, B, Cu, Fe, Mn, Sn, Sr, and Zn, medium (>100) for: Ba, Cr, Mo, and Ni, low (>10) for: As, Cd, Co, Li, Pb, and Se, and very low (>1 to <1) for: Au, Bi, Ge, Ir, Hg, Pd, Pt, Rh, and Tl.

The estimated ingestion of trace elements does not provide information on their intake by body. These processes are control by several factors, of which bioavailability of a given element is of great importance. The bioavailability of Cd and Pb from cocoa powders was determined by their solubility in water and in simulated physiological liquids. Both, metal contents and the extractability were quite variable, depending mainly upon their origin (Table I-6.11).

Decreasing intake of trace elements by populations has been observed during recent decades in some countries. Larsen et al. (2002) have reported a decrease of intake by adults in Denmark, during the periods 1983–1987 and 1993–1997, as follows (in μg $person^{-1} d^{-1}$): Cd, 20 and 16; Hg, 7 and 3.5; Ni, 7 and 3.5; Pb, 42 and 18; and Se, 51 and 48. Similar trends in trace metal intake by adults are observed in Poland (Table I-6.8).

Slow decrease in concentrations of trace metals, particularly of Pb, in food plants has been reported for European countries (Winter-Sorkina 2003; Szteke 2006). This trend is clearly illustrated by results of monitoring trace elements in edible plants. The decrease in Pb contents in cereal grains from 0.3 to 0.15 mg kg^{-1}, during the period 1995–2003, has been observed (Fig. I-6.1), whereas Cd contents are relatively stable, and average at about 0.04 mg kg^{-1} in rye and about 0.08 mg kg^{-1} in wheat (Fig. I-6.2).

Table I-6.10. Daily intake by the population (μg person^{-1} d^{-1}) and mean contents of trace elements in some foodstuff (μg kg^{-1} FW), as estimated by the UK, TDS in 1994 (after Gillian et al. 1999)

Element	Daily intake	Miscellaneous cereals	Green vegetables	Potatoes	Meat products	Milk
Al	11000	78000	1800	2200	3200	<270
As	63	10	3	5	4	2
Au	1	2	<0.4	0.4	1	0.4
B	1500	900	2000	1400	400	<400
Ba	580	790	380	160	270	70
Bi	0.4	3	0.7	0.1	0.4	0.1
Cd	14	20	6	30	7	1
Co	12	10	9	20	8	2
Cr	340	100	200	100	200	300
Cu	1200	1800	840	1300	1500	50
Ge	4	4	<2	<2	2	<2
Ir	2	2	<1	<1	<1	1
Fe	14000	32000	11000	20	23000	4100
Hg	4	4	2	3	3	0.7
Li	16	20	10	10	10	3
Mn	4900	6800	2000	1900	1400	30
Mo	110	230	150	90	120	30
Ni	130	170	110	100	60	<20
Pb	24	20	10	10	10	<1
Pd	1	0.9	0.6	0.5	0.6	<0.3
Pt	0.2	0.1	0.1	<0.1	<0.1	<0.1
Rh	0.3	0.1	<0.1	<0.1	<0.1	0.3
Ru	4	<0.2	<2	<2	<2	2
Sb	3	4	<1	1	4	<1
Se	43	30	10	8	120	20
Sn	2400	20	20	<20	310	<20
Sr	1300	1300	1600	810	640	290
Tl	2	1	3	2	1	<1
Zn	8400	8600	11000	4500	25000	3500

Table I-6.11. Solubility of Cd and Pb in cocoa powders originated from eight countries (after Mouncou et al. 2003)

Fraction	Cd	Pb
Total content (μg kg^{-1})	125 – 1 833	133 – 229
Extractable by (% of total content)		
Water	7 – 37	<0.01 – 10
Various enzymes	<0.01 – 50	<0.01 – 29
Intestinal juice[a]	7 – 48	<0.01 – 8
Gastric juice[a]	7 – 55	<0.01 – 7

[a] Simulated intestinal and gastric solutions.

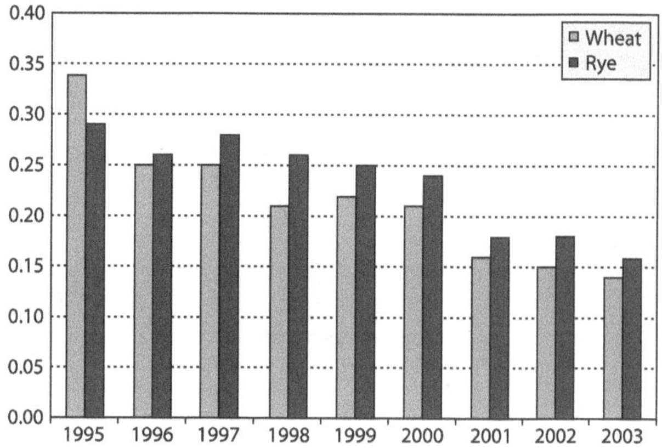

Fig. I-6.1. Lead contents (mg kg^{-1}) in cereals (wheat and rye) of Poland, as a function of time (based on data from Szteke 2006)

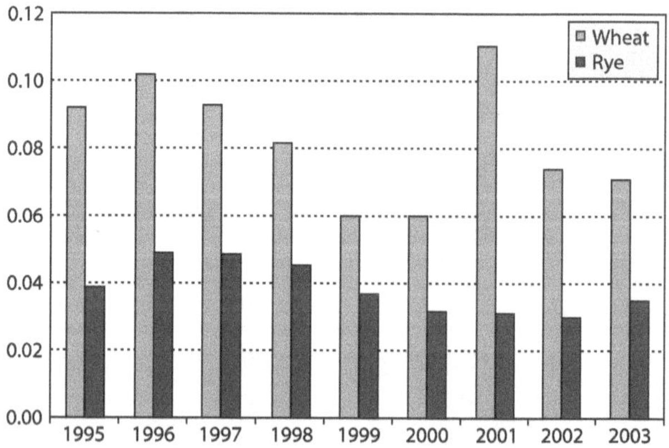

Fig. I-6.2. Cadmium contents (mg kg^{-1}) in cereals (wheat and rye) of Poland, as a function of time (based on data from Szteke 2006)

Data presented by Szefer and Nriagu (2006) have indicated a great variation in the distribution of trace elements in staple food of worldwide countries (Table I-6.12). Also concentrations of some trace elements in wine differ upon origin and apparently also due to various technologies (Table I-6.13).

Table I-6.12. Ranges of concentrations of some trace elements in staple animal and plant food of worldwide countries (adapted from Szefer and Nriagu 2007) (in FW samples)

Commodity	Cd (µg kg^{-1})	Cu (mg kg^{-1})	Hg (µg kg^{-1})	Pb (µg kg^{-1})	Se (µg kg^{-1})	Zn (mg kg^{-1})
Cattle meat	1 – 20[b]	0.7 – 3	0.4[a]	20 – 100[b]	–	0.05 – 67
Pork meat	10[c]	1.1 – 7	2[c]	20[c]	–	22 – 220[b]
Poultry meat	2 – 5[b]	0.5 – 0.9[b]	4 – 20[b]	<10 – 20[b]	150 – 300	5.7 – 16
Fish[d]	3 – 20	<1 – 2.3	40 – 100	20 – 40	200 – 400	4.3 – 16.2
Shrimp	10 – 100[b]	2.4 – 27	20 – 100	4 – 50	300[a]	1.3 – 12.5
Wheat flour	20 – 30	1.5 – 1.6	–	10 – 40	400[g]	6.5 – 9.2
Carrot	10 – 100	0.1 – 0.7[b]	–	10 – 30	0	
Lettuce	40 – 50	0.05 – 0.2[b]	–	10 – 600	<1	1 – 2.2
Onion, bulb	10 – 20	0.3 – 0.9[b]	–	10 – 100	<40 – 100	1.5 – 3.4
Potato	10 – 10	0.2 – 1.4[b]	–	30 – 60	10 – 30	2.8 – 4.5
Tomato	1 – 50	0.3 – 0.9[b]	–	1 – 200	–	0.8 – 1.3
Mushrooms	1 500 – 2 500[e]	2.2[f]	1 000 – 1 300[e]	1 100 – 1 400[e]	100[f]	3.6[f]
Apple	<10	<0.9	–	<30	<40	<1
Plum	<4	0.5	–	<20	–	1.2
Strawberry	<0.3 – 20	0.6	–	<20	–	1.5

[a] Mean value for the most countries.
[b] Range of mean values.
[c] One mean value.
[d] Various fish of UK and USA.
[e] *Agaricus* sps.
[f] Unidentified sps from USA.
[g] Buckwheat flour.

Table I-6.13. Range of trace elements in different wines produced in various countries

Element	Wines of various origin[a] ($\mu g\ kg^{-1}$)	Element	Italian wines[b] ($\mu g\ l^{-1}$)
Cr	100	Ag	5 – 20
Co	0.04 – 0.2	As	3 – 30
Cu	1 – <500	Br	10 – 700
Fe	900 – 5 200	Cd	0.01 – 1
Mn	400 – 2 600	Hg	<0.1
Mo	0.01	I	100 – 600
Ni	10 – 100	F	50 – 500
Se	<20	Li	10 – 200
Zn	100 – 700	Pb	10 – 300
V[c]	6.6 – 90		

[a] After Galani-Nikolakaki and Kallithrakas-Kontos (2007).
[b] After Aceto et al. (2002).
[c] After Teissèdre et al. (1998) for wines of France and California.

Table I-6.14. Mean contents ($\mu g\ kg^{-1}$ FW) in food plants and mean intakes ($ng\ d^{-1}$) of Cd and Pb by the whole population of Denmark (after Winter-Sorkina et al. 2003)

Plant	Cd		Pb	
	Content	Intake	Content	Intake
Wheat	43.4	4 779.6	24.5	2 698.2
Head lettuce	33.3	112.9	7	23.7
Broccoli	3.7	10.8	–	–
Beetroot	17.1	60	11.1	38.9
Onion	8.2	113.1	17	64.8
Potato	22.2	3 075.4	2	277.1
Carrot	27.9	338.1	17.1	207.3
Celery	151.5	36.4	30	7.2
Leek	34.3	274.7	11.7	93.7
Tomato	1.3	33.2	–	–
Rasberry	80	91.2	78	88.9

Highest amounts of Cd and Pb in the diet of the whole population of Denmark are from food plants (wheat and potatoes), and from meat (beef and pork) (Table I-6.14 and 7.15). Some fish, herring and salmon, are a significant source of Hg in the Danish diet, apparently not because of high concentrations but due to high consumption (Table I-6.15). It is recommended in Denmark to measure Cd concentrations in beverages (coffee), cereals, and beef and pork meat (Winter-Sorkina et al. 2003).

Table I-6.15. Mean contents (µg kg^{-1} FW) in animal food and mean intakes (ng d^{-1}) of Cd, Hg, and Pb by the whole population of Denmark (after Winter-Sorkina et al. 2003)

Commodity	Cd		Hg		Pb	
	Content	Intake	Content	Intake	Content	Intake
Beef	4.8	209	0.8	34.8	6.2	269.9
Pork	1.9	110	0.4	23.2	0.7	40.6
Poultry	0.8	16.6	–	–	0.3	6.2
Kidney/cow	649	–	4.7	–	124	–
Liver/chicken	26.1	1.3	–	–	3	0.2
Herring	13.2	23.9	34.2	61.9	22.5	40.7
Sardine	40.5	2.4	23.3	1.4	72.4	4.3
Mackerel	10.6	3.2	32.8	9.8	27.1	1.9
Salmon	2.1	2.6	27	32.9	15.7	19.2
Trout	5	0.3	10	0.5	7.5	0.4
Crab	84.3	1.7	51.3	1	30	0.6
Shrimp	19.2	8.1	26.7	11.2	68.3	28.7
Oyster	148	–	37.5	1	239	–

Reference intake. Risk assessment and characterization of the health hazard associated with the exposure to excess of trace elements, especially *via* the food chain, have been recently a hot topic of several international and non-government organizations. The risk assessment for trace elements has been proven to be very difficult. Smolders (2005) discussed the risk assessment for Cd, as a case study, and emphasized importance of soil as a critical pathway to the food chain.

Ibrahim et al. (2006) have emphasized the importance of exposure characterization in term of routes, intensity, exposure media, spatial scales and time scales (i.e., duration, frequency, timing). The framework for the integration of health and ecological risk assessment has been developed by some groups, mainly of WHO, EC, USEPA, JECFA, OECD, and IPCS. Also UNEP and UNESCO have been collaborating in several GEMS programs that are collective efforts to protect human health and preserve natural resources.

According to the EC Commission Regulation (2001), maximum levels (ML) for Cd, Pb, and Hg in foodstuffs have to be controlled. MLs for Hg have been given for various fish only and is set up at 1 mg kg^{-1} FW. These values for Cd in main food products vary (in mg g^{-1} FW) from 0.05 in meat and vegetables to 1.0 in kidney and mollusks, and for Pb vary from 0.1 to 1.0, respectively (Table I-6.16). In general, four main groups of foodstuffs can be distinguished as a source of trace metals: *(i)* mollusks and crustaceous – high exposure, *(ii)* offal and fish – medium exposure, *(iii)* meat (muscle), cereals and vegetables – low exposure, and *(iv)* fruits and milk – very low exposure.

Table I-6.16.
Maximum levels for Cd, and Pb ($mg\ kg^{-1}$ FW) in some foodstuffs set by EC Commission Regulation (2001)

Product	Cadmium	Lead
Meat (cattle, sheep, pig, poultry)	0.05	0.1
Liver (cattle, sheep, pig, poultry)	0.5	0.5[a]
Kidney (cattle, sheep, pig, poultry)	1.0	0.5[a]
Sea fish, various	0.1	0.4
Crustaceans (excluding crabs)	0.5	0.5
Bivalve molluscs	1.0	1.0
Cereals (excluding bran and germ)	0.1	0.2
Soybeans	0.2	–
Fruits, some vegetables	0.05	0.1
Leafy vegetables, fresh herbs	0.2	0.3
Potatoes, fungi	0.1	0.1

[a] MLs are given for edible offal of animals.

The evaluation of default safety factors in health risk assessment has been recently a subject of many workshops and discussions. The problem of cancer and non-cancer risk has been widely investigated. Kimmel and Vu (2001) have concluded that the field of risk assessment is changing and will continue to do so. Contemporary activities contribute to develop the framework for human health risk assessment.

Present literature strongly implies that there are links between health conditions, particularly cancer, and the geographical sites all over the world. This has indicated that the human health is related to permanent chemical characteristics of the environment. Among chemical elements that may be involved in this relationship, some trace elements of both natural and anthropogenic origin are of special importance. The human exposure to Cd, Hg, and Pb has been presently considered to be of the greatest health risk (Commission Regulation (EC) No 466/2001, OJ L77, 16.3.2001, p. 1). However, in some regions elevated As in food chain, and especially in waters, is also a real health hazard.

Various approaches and strategies have been suggested for integrating and harmonizing health risk assessment and for characterizing safe food. The common strategies are estimations of dietary exposure. Thus, in many countries monitoring trace elements (often together with major elements) in foodstuffs has been carried out. Recently the method of Monte Carlo Risk Assessment (MCRA) has been applied for the estimation of dietary intake (including geophagia) of metals and of health risk (Stanek et al. 2001; Winter-Sorkina et al. 2003). MCRA, in collaboration with other programs, based on consumption and chemical concentration data, provide following options: *(i)* acute and chronic risk assessment, *(ii)* modeling of residue levels and processing effects, *(iii)* assess the uncertainty of percentiles, and *(iv)* comparison with deterministic estimates.

In spite of several approximations, some provisional data for allowable intake of trace metals and nutritional and toxicological references have been established (Table I-6.17 and I-6.18).

All consumers are concern about technology-based and chemical-based information on foodstuffs. Communication on food safety is very important issue. Patric et al. (1999) have discussed needs for relevant policy associated with the evaluation of interactive health communication (IHC).

Abernathy et al. (1993) discussed several aspects related to reference doses and recommended dietary allowance for essential trace elements. They pointed out several problems associate with uncertainty (UF) and modifying (MF) factors. The variable bioavailability of different chemical forms and from various exposure sources needs to have special consideration when RDA and RfD values are proposed. These values cited after the EPA by Abernathy et al. (1993) differ significantly for some element (Table I-6.19).

Table I-6.17. Reference intakes for contaminant elements (μg kg^{-1} BW) (source: WHO 1999, cited by Egan et al. 2002)

Element	PTWI	PTDI
As[a]	15	2.1
Cd	7	1
Hg[b]	5	0.71
Pb	25	3.6

[a] For inorganic As.
[b] For methylmercury.

Table I-6.18. Nutritional and toxicological daily references values established for adults by various organizations (adopted from data summarized by Leblanc et al. 2005)

Element (unit)	Nutritional value, LTI[a]	Toxicological value, UL[b]
Cu (mg)	0.6	10 – >35
Mn (mg)	0.75	–
Mo (μg)	17.5	600
Ni (μg)	–	600
Se (μg)	20	150 – 300
Zn (mg)	4–5	25 – 60

[a] *LTI* – Lowest threshold intake.
[b] *UL* – Upper level.

Table I-6.19. Recommended dietary allowance (RDA), estimated safe and adequate daily dietary intakes (ESADDI) and reference dose (RfD) for some trace elements for adults (μg kg^{-1} BW d^{-1}) (after Abernathy et al. 1993)

Element	RDA (A) and ESADDI (B)	RfD
Cr	0.6 – 3 (B)	1000
Mn	25 – 60 (B)	5/140[a]
Se	0.9 (A)	5
Zn	190 (A)	90 – 300

[a] For water and food, respectively.

Part II Biogeochemistry of Trace Elements

Trace Elements of Group 1 (Previously Group Ia)

The trace alkali elements of Group 1 are: lithium (Li), rubidium (Rb), and cesium (Cs). Their common characteristic is a single electron in the outermost energy level and highly reactive chemical behavior. These cations do not usually form complex ionic species but can be bound in some chelates and organometallic compounds. Although their properties differ (Table II-1.1), they are quite similar, especially Rb and Cs, in their behavior in crystallochemical and geochemical processes. Lithophilic elements, Li, Rb, and Cs, are closely associated with the major crustal components and are likely to enter silicate minerals.

II-1.1 Lithium (Li)

II-1.1.1 Introduction

Lithium is the lightest metal widely and relatively uniformly distributed through the Earth's crust, but is likely to be concentrated in acidic igneous rocks and in argillaceous sedimentary rocks. Its abundance in upper continental crust averages at a value of above 20 mg kg^{-1} (Table II-1.2).

As a highly reactive cation of relatively small ionic radii (Table II-1.1), Li easily replaces cations having sufficiently similar atomic radii in geochemical processes: Mg^{2+}, Fe^{2+}, Al^{3+}, Ti^{4+}. It is associated with minerals of the late pegmatitic stage, e.g., micas, some pyroxenes, and tourmaline, as well as with some sedimentary aluminosilicates,

Table II-1.1. Selected properties of trace elements of Group 1

Element	Atomic number	Atomic mass	Atomic radius[a] (pm)	Density (20 °C, g cm^{-3})	Valence	Melting point (°C)
Li, lithium	3	6.94	205	0.53	+1	180.5
Rb, rubidium	37	85.47	298	1.52	+1[b]	38.9
Cs, cesium	55	132.91	334	1.87	+1	28.4

[a] Approximately average values.
[b] The oxidation states can also vary from +2 to +6.

Li

Table II-1.2. Abundance[a] of trace elements of Group 1 in the environment

Environmental compartment	Li	Rb	Cs
Earth's crust (mg kg^{-1})	20 – 25	90 – 110	<1 – 6
Igneous rocks (mg kg^{-1})			
Mafic	0.5 – 10	2 – 45	0.05 – 1.5
Acid	15 – 40	100 – 200	2 – 5
Sedimentary rocks (mg kg^{-1})			
Argillaceous	50 – 75	120 – 200	5 – 10
Sandstones	10 – 40	10 – 45	0.5 – 2
Calcareous	2 – 20	5 – 30	0.5 – 2
Raw energy material (mg kg^{-1})			
Coal, hard[b]	16 – 74	<1 – 140	<1 – 15
Fly ash	235	106	–
Crude oil	0.02 – <3	–	–
Petrol	0.001 – <0.6	–	–
Soils (mg kg^{-1})	32	50	7.5
Light sandy	5 – 70 (22)[c]	30 – 50	0.8 – 30 (5)
Medium loamy	2 – 130 (46)	100 – 120	2.5 – 20 (8)
Heavy loamy	9 – 175 (53)	55 – 140	5 – 20 (10)
Calcareous	6 – 105 (56)	–	5 – 15 (8)
Organic	0.01 – 3 (1.3)	10 – 20	–
Waters (µg l^{-1})			
Snow[e]	0.07 – 0.15	0.05 – 0.2	–
Rain[e]	0.04 – 0.12	0.08 – 0.19	0.01 – 0.5
Ocean, sea	170 – 200	100 – 200	0.3 – 0.6
River	0.2 – 3.5[d]	0.2 – 6.2	0.003 – 0.02
Air (ng m^{-3})			
Urban[e]	2 – 6.6	2	14 – 18[f]
South Pole	–	–	0.06 – 0.20[f]
Greenland	0.2 – <1.0[e]	<1	0.02 – 0.04[f]

[a] Presented are common, possible background, values from various sources as compiled by Kabata-Pendias and Pendias (1999, 2001), unless otherwise indicated.
[b] Data after Finkelman (1999) and Llorens et al. (2000).
[c] Values in parenthesis are given arithmetic means.
[d] Data after Gaillardet et al. (2003) and Anderson et al. (1988).
[e] After Reiman and Caritat (1998), for remote and polluted regions.
[f] Values for Cs are in pg m^{-3}.

and phosphates. However, the Li content in phosphorites is reported to be rather low, 1–10 mg kg^{-1}. Common Li minerals, all associated with pegmatite bodies, are: petalite, $(Li,Na)AlSi_4O_{10}$; spodumene, $LiAlSi_2O_6$; lepidolite, $K_2Li_3Al_4Si_7O_{21}(OH,F)_3$; and eucryplite, $LiAlSiO_4$.

During weathering, Li is released easily from the primary minerals in oxidizing and acid environment, and then is incorporated in clay minerals and Fe-Mn hydroxides, and is also easily absorbed by organic matter. Its concentrations in coal differ widely depending upon the condition of coal formation. In the US coal, Li contents vary from 0.17 to 370 mg kg^{-1}, and in coal ash range from 50 to 1 064 mg kg^{-1} (Finkelman 1999) Globally, mean Li content (mg kg^{-1}) in coals is 65, while the range in crude oil is 0.02 to <3, and in benzene 0.001 to <0.6 (Table II-1.2).

The environmental distribution index (EDI) for Li is as follows: soil/sandstone: 0.7; plant/soil: 0.1; animal tissue/plant: 0.008; animal tissue/sandstone: 0.0003. Natural cycling of Li indicates its decrease in biotic components as compared with geological material.

II-1.1.2
Production and Use

Lithium is obtained mainly from spodumene (also from petalite and lepidolite) and from subsurface brine. The global annual production in 1995 was estimated as 9.6 kt and in 2000 exceeded 183 kt (USGS 2004).

Lithium is used in organic synthesis, plastic production, special alloys, and in ceramic and glass (mainly optic) manufacturing. It is also added to some rocket fuels, lubricants and as cooling agent in nuclear reactors. Aircraft manufactures use Al-Li alloys to reduce the weight of aircraft. In recent years, extensive uses of Li have been in batteries for video cameras, computers, telephones, and cordless tools. It is also used in the pharmaceutical industry.

II-1.1.3
Soils

The abundance of Li in surface soils seems to be highly correlated with the fine granulometric fraction (<0.02 mm). Its average contents increase from 4.2 mg kg^{-1} in light sandy soil to 14.8 mg kg^{-1} in heavy loamy soils (Kabata-Pendias and Krakowiak 1995). Also positive relationship between Li (ranging from 4.9 to 8.7 mg kg^{-1}) and organic matter content was reported for surface soils (0–25 cm depth) from various regions of Poland (Motowicka-Terelak et al. 1997). However, the texture of mineral soils is the most significant factor controlling the Li status of soil, while all other soil parameters, such as SOM, CEC, and pH are of a much less importance. Mean Li content in soils, at the global scale, also indicates the lowest value for sandy soils, and higher for heavy loams and calcareous soils (Table II-1.2). Median Li concentrations in soils of Japan vary from 21 to 45 mg kg^{-1}, in Andosols (volcanic soils) and Cambisols, respectively (Takeda et al. 2004). Surface soils of the Piedmont region of the southeastern US contain Li from 11.5 to 33.3 mg kg^{-1} (four soil samples), and from the Coastal Plain, from 3.7 to 5.8 mg kg^{-1} (two soil samples). Higher contents of Li are found in deeper soil horizons, with the greatest amounts found in the parent rocks, up to 59.9 mg kg^{-1}. Contents of soluble Li in these soils (extracted with 1 M NH_4Cl) are very low and do not exceed 3–5% of the total Li content in surface layers (Anderson et al. 1988). The means of extractable Li (0.1 N ammonium acetate) in soils of Jordan range from 13.1 to 25.4 mg kg^{-1} (Haddadin et al. 2002). Swedish arable soils contain Li in the range from 1 to 36 mg kg^{-1}, at average value of 17 mg kg^{-1} (Eriksson 2001a).

Lithium is very soluble and mobile during hypergenic processes, as well as in the initial stage of soil formation. Later, it may become more stable due to its firm binding with clay minerals, hydrous oxides of Fe and Mn, and organic matter. However it may be relatively easily leached down the soil profile, especially in humid temperate climatic conditions. In the arid climatic zones, it follows the upward movement along with other soluble salts of chlorites, sulfates and borates. These processes explain the relatively higher Li contents of different soil units that have developed under hot and dry climate, e.g., solonchaks, kastanozems, and prairie soils.

II-1.1.4
Waters

Lithium occurs in all kinds of waters, most often sorbed by colloids in the form of hydroxides LiOH, but seldom as the simple ion Li^+. Its contents in the dissolved load of worldwide rivers vary between 0.16 and 2.46 µg l^{-1} and average 1.84 µg l^{-1}. Riverive flux of Li is estimated as 69 kt yr^{-1} (Gaillardet et al. 2003). Concentrations of Li in ocean waters are not very differentiated (Table II-1.2) and range from 170 to 200 µg l^{-1}. The global average Li concentration for world ocean water is 180 µg l^{-1} (Reimann and Caritat 1998). Surface sediments from the Baltic Sea contain Li in the range of <10.0–84.0 mg kg^{-1} (Szefer 2002a).

In ground waters, Li varies from <0.05 to 150 µg l^{-1} (Matschullat 1997). Lower contents of Li are found in ground waters in the temperate humid zones than in the dry, hot regions, where Li in well waters is reported to reach up to 15 000 µg l^{-1}. In waters used for irrigation Li contents in the range 5–100 µg l^{-1} can be harmful to crop plants.

Rainwater collected in Sweden during 1999 contained Li from 0.03 to 0.11 µg l^{-1} at an average value of 0.06 µg l^{-1} (Eriksson 2001a).

Lithium concentrations in drinking water of Germany are highly variable depending on the source rocks of the water resources. Average concentrations range from 4.2 to 60 µg l^{-1}, the lowest being for waters derived from new red sandstone and the highest for those from shell limestone and marshy lowlands (Anke et al. 1995a).

II-1.1.5
Air

There is a paucity of data on Li occurrence in air. Its content in atmospheric dust is probably related to its concentrations in fine soil granulometric fractions. The maximum Li concentrations at various regions are given as 2.3 and 8.9 ng m^{-3} (Reimann and Caritat 1998).

The maximum concentration of LiOH in the air at the work place was established in Germany as 25 µg m^{-3} (Schäfer 2004a).

II-1.1.6 Plants

Usually Li is easily available to plants. The plant content of this element is believed to be a good indicator of the Li status in soils. There are, however, considerable differences in plants' ability to take up this element and to tolerate its increased concentrations. Plants of the *Rosaceae* (mean Li content 2.9 μg kg^{-1}) and *Solanaceae* (mean Li content 1.9 μg kg^{-1}) families are known to have the highest tolerance to Li and also to accumulate high amounts of this element, especially when grown on Li enriched soils like e.g., solonchak and solonetz.

Lithium shares the K^+ transport carrier and therefore is easily transported to above-ground part of plants, being concentrated mainly in leaf tissues (Table II-1.3). Relatively high contents of Li are reported for lettuce (average 15 mg kg^{-1}) grown in Hungary (Szentmihalyi et al. 1985). Some foodstuff plants accumulate a higher proportion of Li in above-ground parts than in storage roots or bulbs. However, the ratio of root to shoot for Li in other plants, especially grasses and clovers differs which shows that some plants are likely to concentrate more Li in roots than in tops. This can also be an effect of the source of Li, for example, in industrial regions dandelion accumulates significantly higher amount of this metal in leaves than in roots, while in the rural areas plant roots contain only a bit less Li than tops. This phenomenon clearly indicates an aerial source of Li in plants in industrial areas (Kabata-Pendias 1995). A relative accumulation of Li by leaves of dandelion (*Taraxacum officinale*) grown in

Table II-1.3. Lithium in vegetables and fruits (mg kg^{-1}) (data from Anke et al. 1995a and Kabata-Pendias and Pendias 2001, unless otherwise indicated)

Plant	Average content
Tomato, fruits	6.7
Celery, leaves	6.6
Beat, leaves	6.2
Cucumber, fruits	5.0
Asparagus	2.2
Potato, tubers	1.6
Apple, fruits	1.5
Cabbage, leaves	0.5
Lettuce, leaves	0.3
Carrot, roots	0.2
Onion, bulbs	0.06
Citrus mitis, fruits[a]	51.3
Citrus limon, fruits[a]	49.4
Citrus reliculata, fruits[a]	93.0

[a] Fruits from Jordan, after Haddadin et al. (2002).

Warsaw is much higher than the accumulation of other trace elements and indicates anthropogenic sources of this element (Kabata-Pendias and Krakowiak 1997).

Some interactions of Li with other elements should also be considered. There have been reported antagonistic effects from Rb and possibly Zn, whereas synergistic effects from Fe and Mn. Also Ca inhibits Li uptake by plants, thus the addition of lime to high-Li soil may reduce toxic effects of this element.

Citrus fruits of Jordan, grown on soils with a relatively high level of extractable Li contain much higher amounts of this metal than other fruits and vegetables (Table II-1.3). Leaves of citrus trees contain Li at above 100 mg kg^{-1}, and its concentrations in leaves of red pepper (*Capsicum annuum*) range from 149 to 319 mg kg^{-1}, and average 256 mg kg^{-1} (Haddadin et al. 2002).

Contents of Li in moss (*Hylocomium splendens*) growing in Norway in the period 1990–1995 ranged from 0.0027 to 2.6 mg kg^{-1} (average 0.22) and have not changed as compared with Li contents in moss sampled earlier (Berg and Steinnes 1997).

Although Li is not known yet as an essential plant nutrient, there is some evidence that Li can affect plant growth and may play some metabolic function in halophytes. However, the observed stimulation may also be related to the influence of some other factors, including secondary effects of anions associated with Li salt used for experiments.

Most of data demonstrate the detrimental effects of Li to plants, of which *Citrus* trees are the most susceptible to an excess of this element. Several data have indicated that at low concentrations Li stimulates the growth and development of crop plants. Lithium has been used in fertilizers to increase starch levels and biomass of potatoes (Schäfer 1997). Based on several experimental results, it can be stated that Li affects various plant systems at the membrane level. However, it still has not been proven that Li is an essential nutrient for plants because its function seems to be non-specific and Li substitutes for other monovalent cations.

Increased Li contents in soil can be toxic to some plants. Citrus trees are probably the most susceptible to an excess of Li, and their growth in salt-enriched soils can be significantly reduced due to high Li contents. In high-Li soils, damage to root tips, injured root growth and chlorotic and necrotic spots on leaves have been observed for corn.

Although several microorganisms are relatively sensitive to increased levels of Li, some fungi, *Penicillium* and *Aspergillus*, are known to adapt easily to such growth conditions.

II-1.1.7
Humans and Animals

Humans. Lithium occurs in mammalian tissues within the very narrow range from <0.02 to 0.08 mg kg^{-1}, being the most concentrated in skin (Jørgensen 2000). The average Li content is estimated for total soft tissues of humans as 0.006 mg kg^{-1} (Li 2000). It is likely to be concentrated in tooth enamel where probably plays a protective function against decay processes. It is also slightly concentrated in kidneys and pancreas.

Lithium does not appear to serve a normal cofactor function in any enzyme or transport system, but is considered an essential element for man and animals. It is known to have an affinity for enzymes activated by Ca and Mg. Its deficiency apparently disturbs protein metabolism and may have an effect on several biological systems, such as endocrine, cardiovascular, neuromuscular, renal, and dermatological (Anke et al. 1995a). There are some observations that Li influences several hormonal, metabolic, neurological and immunological processes (Duda and Pasternak 2003). It is reported to stimulate body's natural defense against viral infection and is proposed as a supporting drug in the treatment of HIV (Gallichio 2003).

Lithium salts have been used in the psychiatry for over 50 years. It is now known that Li may stabilize mood swings by altering neurotransmission at the synaptic level in the brain. The cation Li^+ can substitute for Na^+ and K^+ in compounds that regulate the ratio of these elements in cells and may have a positive impact on several cell processes and on the immune system (Skotnicki 1990). Lithium salts have been used to control several nervous manifestations, however, a long term therapy is not without risk, and therefore those patients should be monitored for serum Li levels, which should not exceed 0.5–1.5 mMol l^{-1} (Hameed and Vohora 1990). People taking Li compounds to control their moods swings may have problems with being thirsty because of alteration of the action of protein (aquaporin) that controls water transport within the body (Chaccour et al. 1977).

The most important source of Li for humans is drinking water, although the highest Li contents are in animal foodstuffs, from 1.7 mg kg^{-1} in herrings to 7.4 mg kg^{-1} in eggs (Anke et al. 1995a). The mean Li intake from all sources (food, water, and air), by humans in the USA has been calculated at 2 mg d^{-1} per person (Schäfer 2004). In Germany, the average intake of Li in diet and water is calculated as 0.18–0.35 mg d^{-1} for women, and 0.18–0.55 mg d^{-1} for men. Intake of Li by man is very variable due to different consumption of beverage, which is a significant source of Li. Studies on the human exposure to the Li in drinking water have not indicated neither positive nor negative effects.

Animals. Animals, as investigated in goats, are sensitive to both deficiency and excess of Li in feed (Anke 2004b). At low Li ration a lower activity of several serum enzymes was observed. The supplementation of more than 100 mg Li kg^{-1} ration decreased feed consumption by experimental animals, and 50 mg Li kg^{-1} of feed badly affected test animals (Haddadin et al. 2002; Schäfer 2004a).

II-1.2 Rubidium (Rb)

II-1.2.1 Introduction

The upper continental crust contains Rb at around 110 mg kg^{-1} (range 90–310). Rubidium behaves similarly to Li in the lithosphere and thus it is likely to be concentrated rather in acidic igneous rocks and sedimentary argillaceous rocks (Table II-1.2). Although Rb is closely linked with K, its binding forces in silicates are stronger than those of K, and therefore the K: Rb ratio continually decreases during soil-forming processes.

Rubidium is considered to be a dispersed element and does not form minerals of its own. It is associated mainly with some feldspars (e.g., pollucite) and micas (e.g., lepidolite), and is likely to replace K in some silicates. Micas seem to be the most important Rb carriers, and contain up to 0.1% of this element.

Average global Rb content in coal is reported at 25 mg kg^{-1}, and in US coal at 21 mg kg^{-1} (Finkelman 1999; Llorens et al. 2000).

II-1.2.2 Production and Use

World mining of Rb was estimated at <1 t in 1988, and its capacity of production in 1997 was established as 5 t (USGS 2004).

Elemental Rb is highly reactive and can ignite spontaneously in air. It is used in electronics, special glass and in the production of semi-conductors and photocells.

II-1.2.3 Soils

The Rb content in soils is largely inherited from the parent rocks, as is indicated by the highest mean Rb content, 100 to 120 mg kg^{-1}, in soils over granites and gneisses. Also soils of heavy texture contain relatively high amounts of Rb (Table II-1.2) that is the result of easy adsorption onto clay minerals, especially on micaceous clays. A significant positive correlation between the Rb contents and clay fraction in loess deposit was established (Vaganov and Nizharadze 1981). Due to the small diameter of hydrated ion (Table II-1.1) Rb is likely to be relatively active in soil processes.

The lowest concentrations of Rb (30–50 mg kg^{-1}) are found in sandy and organic soils, as well as in peat bogs, 1–10 mg kg^{-1} (Markert and Lieth 1987). Relatively high Rb contents were reported by Shacklette and Boerngen (1984) for US soils, for which mean values range from 30 to 120 mg kg^{-1}, being the lowest for organic soils and the highest for soils developed from granites and gneisses. Results presented by Govindaraju (1994) for Rb in reference soil samples give a bit higher Rb contents in soils, in the range from 69 to 1 141 mg kg^{-1} in soils from the USA, and from 15 to 140 mg kg^{-1} for soils from China. Japanese soils contain Rb within the range from 36 to 100 mg kg^{-1}, being the lowest in Andosols and the highest in Gleysols (Takeda et al. 2004). Swedish arable soils contain Rb in the range from 70 to 190 mg kg^{-1}, at average value of 116 mg kg^{-1} (Eriksson 2001a).

The content of Rb in manure is very low, and reported as 0.06 mg kg^{-1}, whereas phosphate fertilizers contain Rb at an average value of 5 mg kg^{-1}, and sewage sludges – up to about 100 mg kg^{-1}. Its content in sewage sludges collected in Sweden varied between 2.5 and 53 mg kg^{-1} (Eriksson 2001a).

II-1.2.4 Waters

The Rb concentration in ocean and seawaters is fairly stable and ranges from 100–200 µg l^{-1} (Table II-1.2). Its contents in the dissolved load of worldwide rivers water vary between 0.2 and 6.5 µg l^{-1} and average 1.6 µg l^{-1}. The global riverine flux of Rb is calculated to be 61 kt yr^{-1} (Gaillardet et al. 2003).

Rubidium content was reported by Suszkin (*vide* Szefer 2002a) to range from 0.69 to 6.7 μg l^{-1} in rainwater sampled in 1976 around Baltic Sea region, whereas median concentrations in unfiltered rain-water in the Northern Europe range from 0.08 to 0.15 μg l^{-1} (Reimann and Caritat 1998), and in Sweden from 0.06 to 0.25 μg l^{-1} (Eriksson 2001a). Ground waters contain Rb within the range of 0.05–2.5 μg l^{-1} (Kabata-Pendias and Pendias 1999).

As relatively strongly adsorbed element, it is likely to concentrated from around 20 to over 100 mg kg^{-1} in surface sediments of the Baltic Sea (Szefer 2002a).

Aquatic biota. Soft tissues of mussels from the Baltic Sea contain Rb within the range 3.2–14.0 mg kg^{-1}, whereas in muscles of fish it ranges from 0.51 to 5.7 mg kg^{-1} FW (Szefer 2002a).

II-1.2.5
Air

There are very little data on the occurrence of Rb in the atmosphere. Cawse (1987) reported that the concentration of Rb in air above the UK was from <1 to <6 ng m^{-3}, and its deposition in inhabited regions was from <10 to <40 g ha^{-1}, during the period 1972–1981. Rubidium concentrations in air at the global scale are calculated at <1–2 ng m^{-3}, for remote and polluted regions, respectively (Reimann and Caritat 1998).

II-1.2.6
Plants

Rubidium, as other monovalent cations, is easily taken up by plants. The radius of Rb (298 pm) is similar to radius of K (277 pm) and therefore Rb can substitute at K sites in plants, but cannot substitute for K in metabolic processes. High Rb contents in plants might become toxic. Symptoms of its toxicity are dark green leaves, wilting and stunted foliage, and short brown roots; however, in field conditions they are practically unknown.

Despite the chemical similarity of Rb to K there is a significant difference in uptake of these metals. The availability of Rb to several plants increases with soil acidity (Tyler 1983), while K is not easily taken up from acid soils. Drobner and Tyler (1998) reported that a shortage or low availability of K increases Rb uptake from acid soils, and suggested that leaf Rb content may be used as a method to assess early stage deficiency of K in plants grown on acid soils.

Most of the higher plants species contain Rb in the range of 20–70 mg kg^{-1}. However, Rb contents in plant foodstuffs vary highly, from 1 to 20 mg kg^{-1} and are the lowest in grains and the highest in soybean seeds (Table II-1.4.). Fodder plants sampled in Poland in 1995 contained Rb in the range of about 33 and 5 mg kg^{-1}. In most cases, the highest contents were in plants from industrial regions (Kośla et al. 2001).

The distribution pattern of Rb in leaves of rice plants was similar to that of K and decreased from above 4 to around 1 mg kg^{-1}, from young to old leaves respectively (Tsukada et al. 2002b). The mean concentration of Rb in wheat and barley grains grown in Sweden are 2.6 and 4.7 mg kg^{-1}, respectively (Eriksson 2001a).

Rb

Table II-1.4.
Rubidium in vegetables and fruits (mg kg^{-1}) (data compiled by Kabata-Pendias and Pendias 2001)

Plant	Mean content
Wheat, grains[a]	2.6
Barley, grains[a]	4.7
Cereal, grains	4
Corn, grains	3
Cabbage, leaves	12
Lettuce, leaves	14
Avocado, fruits	20
Apple, fruits	50
Beans, pods	51
Soybean, seeds	220

[a] After Eriksson (2001a).

Mosses from the Nordic countries, collected in 1985, contained Rb at levels from 14 to 37 mg kg^{-1} (Rühling et al. 1987), and a fairly similar range for Rb was reported in moss sampled in Norway during the period 1990–1995 (Berg and Steinnes 1997).

Some fungi and also bacteria are known to accumulate increased amounts of Rb.

II-1.2.7
Humans and Animals

Humans. Rubidium is present in all tissues of mammals within the range from 8 to 30 mg kg^{-1}, with the lowest value for skin and the highest for liver (Jørgensen 2000). Other sources reported the Rb contents in soft tissues of human's organs, liver, kidney and brain, to range from 1 to 15 mg kg^{-1} (Kabata-Pendias and Pendias 1999). Total average soft tissue content of Rb in "reference man" has been calculated at 9.7 mg kg^{-1} (Li 2000). The Rb concentrations in human blood range from 0.9 to 6.8 mg l^{-1} (mean 2.8), in serum from 0.06 to 0.40 mg l^{-1}, and in urine from 450 to 840 µg l^{-1} (Reimann and Caritat 1998; Ward and Abou-Shakra 1993). The distribution of Rb in dairy cows' tissues shows a regular pattern of the highest concentration in liver, from >25 to <40 mg kg^{-1} and the lowest in rib, about 5 mg kg^{-1} (Kośla et al. 2001).

Rubidium is associated with the K cycle in organisms but does not substitute for it. There is some evidence that Rb is involved in brain functions, but specific roles have not yet been identified (Hameed and Vohora 1990). This element is antagonistic to Li, and its increased level in food can be harmful to some metabolic processes related to Li.

Amounts of Rb in food above 200 mg kg^{-1} might be harmful and its concentrations above 1 000 mg kg^{-1} in human diet are of a health risk. Most of Rb compounds can irritate the skin, by contact or when breathed. The most toxic Rb compounds are fluoride, hydroxide, and cyanide.

Daily intake of Rb by adults with a mixed diet is estimated at around 1.7 mg. In Germany, during the period of 1988–1992, it varied from 1.3 to 1.7 mg d^{-1} per person, being a bit higher for men than for women (Anke and Angelov 2004).

Animals. Experiments carried out by Anke (2004b) have indicated that small amounts of Rb might be essential for animals, and the normative requirement was estimated as 0.3–0.4 mg kg^{-1} feed. However, its increased level in the diet can cause neuromuscular hyperirritability spasms.

II-1.3 Cesium (Cs)

II-1.3.1 Introduction

Various sources list the Cs abundance in the continental crust within the range from <1 to 6 mg kg^{-1} (Table II-1.2). It is a relatively rare element of the naturally occurring alkali metals in the terrestrial environment. It has lithophilic properties and behaves in the geochemical environments similarly to Rb, but appears to have a greater affinity to aluminosilicates. Thus, it is concentrated mainly in acidic igneous rocks and in argillaceous sediments where its contents range from 5 to 10 mg kg^{-1} (Table II-1.2). It is a component of three minerals, of which only pollucite, $H_4Cs_4Si_9O_{27}$, containing between 5 and 32% Cs_2O, is of commercial importance. Most Cs, however, is dispersed in some common minerals, like micas and feldspars. Mean Cs content in coal is about 1 mg kg^{-1}.

II-1.3.2 Production and Use

Annual global production of Cs is estimated as 5.9 Mt (USGS 2004). There are relatively few commercial uses of metallic Cs. It is used in: vacuum tubes, rocket propellant, photovoltaic cells, and accurate atomic clocks. Compounds of Cs have also been employed as catalyst in organic synthesis.

Radioactive ^{137}Cs has been approved for the sterilization of some foods (e.g., wheat flour, potatoes), sewage, and surgical equipments, thus reducing the numbers of disease-causing bacteria. It may also be used in cancer treatment (to be placed within the body).

II-1.3.3 Soils

There is little information on the Cs status of soils. The common range for different soils, based on various data can be established as <1–30 mg kg^{-1} (Table II-1.2). In the reference soil samples, Cs contents range from 0.41 to 21.4 mg kg^{-1}, being higher in soils from China than from the USA (Govindaraju 1994). In agricultural soils of Canada, Cs is reported to range between 0.2 and 7.1 mg kg^{-1} (Reimann and Caritat 1998). Experimental surface soils of Japan have Cs at the level of around 3.3 mg kg^{-1}, whereas its contents in agricultural soils are at levels of 1–11 mg kg^{-1}, and in pine forest litter vary from 0.29 to 0.98 mg kg^{-1} (Tsukada et al. 1998, 2002a; Tsukada and Nakamura 1999). Swedish arable soils contain Cs in the range from 0.3 to 4.5 mg kg^{-1}, at average value of 1.7 mg kg^{-1} (Eriksson 2001a).

Cs

Dust samples from roadside in the vicinity of a metals factory in India contain Cs from 1.53 to 3.63 mg kg^{-1}. Bottom ash of municipal solid waste incinerations contains Cs at concentrations of 0.44–2.01 mg kg^{-1} (ATSDR 2002a).

Based on the available data on Cs content in soils, it is rather difficult to relate its occurrence to soil textural groups. Studies with radiocesium confirmed an increased capacity of fine granulometric fraction, clay minerals, apatite, and other minerals to bind this cation (Paasikallio 1999). Although Cs and most of its compounds are generally easily water soluble, they are very likely to be strongly adsorbed by clay soil fractions. Clay minerals that strongly influence the Cs absorption are mainly illites (Nakao et al. 2003).

II-1.3.4 Waters

Cesium abundance in waters is low, but significantly higher in oceans and seas (0.03–0.4 μg l^{-1}) than in rivers and streams (0.0004–0.02 μg l^{-1}). Japanese streams contain Cs up to 0.093 μg l^{-1} (Tanizaki et al. *vide* ATSDR 2002a). The world average Cs concentration in river waters is estimated at a value of 0.011 μg l^{-1} and riverine flux at 0.4 kt yr^{-1} (Gaillardet et al. 2003).

Cesium cations in water are rapidly transformed to CsOH and easily adsorbed by suspended solids and sediments where it forms insoluble, immobile complexes. Thus, Cs is almost absent in waters, particularly in ground waters. Ground and drinking waters of Nigeria contain Cs at averages values of 0.61 and 0.35 μg l^{-1}, respectively (Asubiojo et al. *vide* ATSDR 2002a).

Stream sediments of some rivers are likely to concentrate Cs, up to 52 mg kg^{-1}. Its contents in surface sediments of the Baltic Sea range from 3.1 to 10 mg kg^{-1} (Szefer 2002a). The concentrations of Cs in rainwater sampled in 1976 around the Baltic Sea region were reported by Suszkin (*vide* Szefer 2002a) to range from 0.06 to 0.53 μg l^{-1}. Rainwater collected in Sweden during 1999 contained Cs from 0.004 to 0.011 μg l^{-1}, with an average value of 0.007 μg l^{-1} (Eriksson 2001a).

Aquatic biota. Szefer (2002a) reported, after various authors, Cs contents in soft tissue of mussels from the Baltic Sea to range from 0.001 to 0.05 mg kg^{-1}. It is relatively easily taken up and held by algae, and ranges in plankton of the Baltic Sea between 0.03 and 0.23 mg kg^{-1} (Brzezińska et al. 1984).

II-1.3.5 Air

There is little information on the occurrence of stable Cs in air. The background Cs content in air might be 0.02–0.04 ng m^{-3}, reported for Norway and Shetland Islands, respectively (Bowen 1979). High Cs contents, up to 14–18 ng m^{-3}, are reported for air above cities, respectively in Germany and Japan, whereas the highest content of Cs in air from Northern America was found to be 3 ng m^{-3}. In 1972–1981, Cs concentrations in air from Great Britain ranged from 0.1 to 0.34 ng m^{-3}, and its deposition in England was calculated as 0.6–1.2 g ha^{-1} (Cawse 1987). At the global scale, median Cs concentrations are reported to range from 0.045 to 0.4 ng m^{-3}, for the remote and polluted regions, respectively (Reimann and Caritat 1998).

II-1.3.6 Plants

Cesium is not essential to plants and thus its distribution in plant samples has not been studied intensively. However, this element is relative active in soils and easily taken up by plants, especially from sandy acid soils. Based on few data, the common range of Cs in various plants is estimated at <0.1–3 mg kg^{-1} (Kabata-Pendias and Pendias data 2001). Usually lower concentrations of Cs are found in green leaves and are lower in plants from arid regions than from other areas. Roots of higher plants contain about twice as much Cs (0.32 mg kg^{-1}) as old leaves (0.16 mg kg^{-1}), whereas much smaller amounts of Cs are in young leaves – 0.07 mg kg^{-1} (Ozoliniya and Kiunke 1978). Tsukada et al. (2002a) has observed increasing Cs levels from below 0.01 to 0.02 mg kg^{-1} in older leaves of rice. According to these authors, over 90% of the total Cs content was in non-edible parts of rice plants. Also up to 77% of Cs was accumulated in non-edible leaves of cabbage (Tsukada and Hasewaga 2002). A very broad range of Cs was noticed for mushrooms sampled from pine forests in Japan, from 0.005 to 1.6 mg kg^{-1} (Tsukada et al. 1998).

Cesium contents in tea-leaves vary from 0.01 to 0.31 mg kg^{-1}, and are the highest in jasmine tea of China (Inadira et al. 1984). Rühling et al. (1987) found Cs to range from 0.3 to 0.48 mg kg^{-1} in moss samples from Nordic countries in 1985, and mosses collected in Norway in the period 1990–1995 had a mean Cs content at 0.26 mg kg^{-1}, with a range of 0.019–3.1 mg kg^{-1} (Berg and Steinnes 1997). The highest content of Cs in mosses from Germany is reported to be 9.8 mg kg^{-1} (Reimann and Caritat 1998).

White et al. (2003) observed that, although Cs and K are chemically similar, different transport proteins control their uptake and delivery to the xylem. Thus, the Cs/K concentration ratio differs among plant species.

II-1.3.7 Humans

Cesium occurs in tissues of all terrestrial and aquatic animals, and is distributed fairly uniformly in all types of tissues within the common range 0.01–0.05 mg kg^{-1}. The highest content is found in liver and heart tissues (Jørgensen 2000). The mean concentration of Cs in soft tissues of "reference man" is given as 0.021 mg kg^{-1} (Li 2000). In human fluids Cs averages (in µg l^{-1}) in: blood, 3; serum, 1.5; and urine, 7 (Reimann and Caritat 1998; Li 2000). When its intake by organisms is increased, then it is likely to be accumulated in soft tissues.

Cesium exhibits an antagonistic reaction against potassium with respect to the biological availability.

The daily dose of Cs with a regular adult's diet is calculated as 10 µg and its content in an average diet is 9.4 mg kg^{-1}.

Cs

II-1.3.8 Cesium-137

Radioactive isotopes, ^{134}Cs and ^{137}Cs, are produced by nuclear fission in nuclear power plants and occur also in fallout from nuclear weapons. Environmental concern with radiocesium in soils and food chain is largely a consequence of two events: *(i)* the worldwide fallout in the period 1952–1958, and *(ii)* the fallout from Chernobyl in 1986.

One of many Cs isotopes, ^{137}Cs is of special concern because it is a by-product of atomic energy production and nuclear weapons testing. Anthropogenic sources of this radionuclide include the mining and processing of contaminated ores. Also ash from coal burning and municipal waste incinerations may be a source of ^{137}Cs.

The geochemical characteristics of this long-lived radioisotope are fairly similar to those of the stable Cs. Thus, when it is deposited on soils it becomes strongly adsorbed by clay fractions, clay minerals and organic matter, and appears to migrate quite slowly in soil (Meriwether et al. 1988). The fate of ^{137}Cs was widely studied after the fallout following the Chernobyl accident. At this accident, ^{134}Cs was also released, but this radionuclide is of a much less concern because it was released in a significantly smaller quantity and has a relatively short half-life ($t_{1/2}$ is 2.1 years) compared with ^{137}Cs ($t_{1/2}$ is 30 years). The deposition of both radionuclides was not only large in the zone surrounding the reactor, but also in several European countries, as well as on other continents.

Soils. Results of investigations about a decade after the Chernobyl accident indicated that up to 90% of the inventory of radiocesium was still contained in the top 3 to 7 cm soil layer and ranged from 103 to 1 500 kBq m^{-2} (Kagan and Kadatsky 1996). Increased contents in clay fraction, organic matter and Fe and Mn sesquioxides decrease the mobility, and thus the migration rate of this nuclide (Arapis et al. 1997; Grin et al. 1999). Soil organic matter, especially of higher molecular weight, is likely to accumulate a large proportion of the ^{137}Cs and therefore influences its relatively low mobility (Agapkina et al. 1995). Horizontal distribution of Cs radionuclides in soils seems to be correlated with seasonal microbiota activity (Kostyuk and Bunnenberg 1999).

Reported concentrations of ^{137}Cs in other soils vary from 1.7–28.2 Bq kg^{-1} in paddy field soils, and from 1 to 37 Bq kg^{-1} in agricultural soils of Japan (Tagami and Ushida 1998; Tsukada and Nakamura 1999). Soils of the of Louisiana, USA, were reported to contain ^{137}Cs at the surface within the range from 1 to 92 Bq kg^{-1} (Meriwether 1988). Soils of lowland of Poland, analyzed in the period 1992–1993, contained both Cs radionuclides (137 and 134) up to 8 kBq m^{-2} (Strzelecki et al. 1993), while the surface soils of the Tatra Mountain in Poland contain ^{137}Cs within the range from 160 to 650 Bq kg^{-1}, at about the same time (Górski and Uziak 1991). In Russia (Bryansk region), soils after the Chernobyl accident, accumulated ^{137}Cs in the range of 703–899 Bq m^{-2}, mainly in the surface layer. Its fraction insoluble in water varied from 40 to 93% of the total content (Korobova et al. 1998). Soils of Kazakhstan sampled in 1992 contained ^{137}Cs in the range of average values from 82 to 120 mCi km^{-2} (Panin 2004). Soils of wetland in South Carolina, USA, contain ^{137}Cs in the range of 0.2 to 10.1 Bq kg^{-1}, with the highest values for moisture-regime location. In fresh alluvial sediments, most of the Cs was held by either amorphous Fe-oxyhydroxides or organic matter (Kaplan et al. 2005).

Soil samples collected during 1999–2001 in France had the activity of ^{137}Cs above 10 kBq m^{-2} in the surface layer (Renaud et al. 2003). These authors demonstrated a close relationship between ^{137}Cs deposition and the rainfall of the first week of May 1986, right after the Chernobyl accident. The impact of industrial pollution on the levels of ^{137}Cs has been reported by Outola et al. (2003) for forest soils surrounding the Cu-Ni smelter in Finland. Soil samples taken in 2000 had the total deposition of ^{137}Cs in the range 15–25 kBq m^{-2} being the highest at the distance of 2 km from the smelter.

Waters. The background ^{137}Cs concentration in surface waters is ordinarily between 0.01 and 1.2 µg l^{-1}. After the accident in Chernobyl, increased levels of radiocesium in various water basins have been observed. The concentration of ^{137}Cs in the Black Sea was between 2.7 and 8.1 pCi l^{-1}, for the period of 1991–1996, whereas in the Mediterranean Sea was 0.13 pCi l^{-1} at about that same period (Strezov et al. and Molero et al. vide ATSDR 2002a).

The maximum concentrations of ^{134}Cs and ^{137}Cs in drinking water have been estimated as 900 and 1 000 pCi l^{-1}, respectively, under the condition that there was no other intake of radioactivity (ATSDR 2002a).

Bottom sediments of pond around Chernobyl, sampled in 1992 contained radiocesium at an average value of around 1 Bq g^{-1}, but some samples had up to 11 Bq g^{-1} (Jagoe 1997). Surface sediments of the Baltic Sea (Gdansk Basin) collected before 1990 accumulated ^{137}Cs in the range between 100 and >200 mBq g^{-1}. Below 20 cm depth this radionuclide was undetectable (Pempkowiak 1991).

Aquatic Biota. Freshwater fish can absorb increased amounts of ^{137}Cs. For example, pike from lakes of Finland contained this radionuclide at concentrations of 66 297 and 18 567 pCi kg^{-1} in the period 1988 and 1992, respectively (ATSDR 2002a). The concentrations of radiocesium in fish (*Carassius carassius*) from ponds in the vicinity of Chernobyl varied greatly, but did not exceed the level of 8 Bq kg^{-1} (Jagoe et al. 1997). Stingray (*Dasyatis akajei*) fished off Japanese Sea in 2001 contained ^{137}Cs within the range from 0.15 to 0.31 Bq kg^{-1} (Isoyama 2003). However, as reported by Malek et al. (2003), radiocesium may be removed to a considerable extent from fish by dressing and cooking.

Air. There are rather limited data on ^{137}Cs in the atmosphere. This radionuclide was reported in the range of 0.072 to 0.14 pg m^{-3} in the South Pole during 1974–1975. Its concentration in the Arctic air ranged from 10 to 60 pg m^{-3} during 1990–1991 (ATSDR 2002a). The greatest ^{137}Cs concentration following the accident at the Chernobyl nuclear power station was around the reactor location with values of 1 756–8 919 pCi m^{-3} (Buzulukov and Dobrynin *vide* ATSDR 2002a). The maximum concentration of ^{137}Cs at that time in New York City was reported as 0.26 pCi m^{-3} (Freely et al. *vide* ATSDR 2002a).

Plants. Radioactive Cs, similarly to the stable Cs, is easily absorbed by higher plants and microorganisms. However, some data indicated that radioactive Cs is more easily taken up by rice than the stable Cs, apparently due to a higher biological activity (Tsukada and Nakamura 1999). Schimmack et al. (1989) reported that the rates of vertical migration of ^{134}Cs and ^{137}Cs deposited on grassland soil by fallout from the reactor accident in Chernobyl was considerably faster (0.2 and 0.3 cm h^{-1}) than of those from the long-term migration of ^{137}Cs from the global fallout of weapon-testing.

The phytoavailability of radiocesium from peat soil to grass is relatively high. Addition of some minerals can inhibit its uptake. Paasikallio (1999) reported that the most effective agent for the reduction in the uptake of ^{134}Cs, especially in the long-term, was biotite and a biotite-zeolite mixture.

Plants take up more ^{137}Cs from light sandy soils than from loamy soils and soils rich in organic matter (Table II-1.5). Some plants, such as *Amaranthus* and *Helianthus* species, exhibit a relatively great accumulation of this radionuclide and are suggested for the phytoremediation of polluted sites (Dushenkov and Sorochinsky 1999). White et al. (2003) stated that the specific mechanism by which Cs is taken up by root cells can be adapted to develop plant species for the phytoextraction of radiocesium from soils. Jegadeeswari et al. (2005) reported a much higher accumulation of radiocesium by sorghum roots (mean activity 696 Bq pot^{-1}) than by whole plant (mean 8.52 Bq pot^{-1}).

Knox et al. (2001) investigated the ^{137}Cs distribution and bioavailability in both aquatic and terrestrial sites of wetlands on Savannah River Sites, USA. Almost in all types of sediments the largest proportion of this radionuclide was absorbed by amorphous Fe oxyhydroxides. The concentration of ^{137}Cs in plants of aquatic sites varied from 0.2 to 8.1 Bq g^{-1} and in plants of terrestrial sites from 0.03 to 5.9 Bq g^{-1}.

Levels of ^{137}Cs in food plants of the USA were non-detectable, but in honey it was 181.1 pCi kg^{-1} (6.7 Bq kg^{-1}), which is about 200 times lower than the regulatory level. Green vegetables and fruits sampled in the UK after the Chernobyl accident contained radiocesium at an average value of 2703 pCi kg^{-1}. The average concentration of ^{137}Cs in onions from Denmark was 0.21 μg kg^{-1}, and in wheat flour from Pakistan was between 6.7 and 11.2 μg kg^{-1} (ATSDR 2002a).

The transfer of radiocesium in the food chain has been recently investigated. Tsukada et al. (2003) calculated that the geometric mean of the soil-to-grass transfer factor (TF) for ^{137}Cs was 0.13, which was 6 times higher than that for the stable Cs (TF 0.022). The values of both factors decreased with increasing K content in the soil. The geometric mean of TF value for radiocesium in the chain of grass-to-milk was 0.0027, which was fairly similar to that for stable Cs. The authors concluded that data on the food-chain transfer of stable Cs might help to understand the fate of radiocesium. The fraction of

Table II-1.5.
Concentration of ^{137}Cs (nCi g^{-1}) in plants grown on various soils with addition of 0.03 nCi kg^{-1} (data compiled by Kabata-Pendias and Pendias 2001)

Plant and parts	Sandy soil	Loamy soil	Chernozem
Cabbage, leaves	34.9	6.3	8.3
Carrot, roots	6.5	–	1.5
Beet, roots	5.1	2.2	1.6
Potato, tubers	5.9	2.3	2.3
Cucumber, fruits	8.1	2.5	2.4
Tomato, fruits	2.5	0.6	1.2

transferable ^{137}Cs seems to be larger than that of stable Cs in most of the soil-plant chain, which has been confirmed by the relatively high soil-to-potato TF, which ranged between 0.004 and 0.16 (Tsukada and Nakamura 1999). Soil-to-potato transfer of ^{137}Cs is of a real concern in Belarus (Knatko et al. 2000).

A high accumulation of radiocesium in mushroom pericarp is of a special concern due to its transfer to the food chain. Therefore, the health-related limit for ^{137}Cs in edible mushrooms has been established in Europe at 600 Bq kg^{-1} FW (*vide* Kabata-Pendias and Pendias 1999). The median content of radionuclides in mushrooms (65 species) collected in 1990 in Japan was 60 Bq kg^{-1} for ^{137}Cs, and <7 Bq kg^{-1} for ^{134}Cs (Yoshida and Muramatsu 1994). The soil-mushroom transfer factor was calculated to range from 5.5 to 13. Sugiyama et al. (1993) reported clear correlations between wild mushroom-to substrate ratios of ^{137}Cs concentrations and those of stable Cs. Concentrations of ^{137}Cs in wild mushrooms from Japan range from 0 to 1 070 Bq kg^{-1} FW, with an average value of 87.5 Bq kg^{-1}. Mushrooms sampled in pine forests in Japan in the autumn of 1992 contained ^{137}Cs in the range of <4 to 570 Bq kg^{-1} FW, whereas substratum had this radionuclide with the range from 53 to 261 Bq kg^{-1} FW. There was no observed relationship between ^{137}Cs in mushrooms and substratum (Tsukada et al. 1998). Mushrooms collected in 1998 in forests of Belarus exhibited a good correlation between ^{137}Cs and stable Cs (Yoshida et al. 2003).

The behavior of radioactive Cs isotopes in forest ecosystems is of a special concern because forest litter, the most surficial layer, has a great capacity for their accumulation. Also a moorland ecosystem can fix and hold a large amount of ^{137}Cs. The mycorrhiza of heathers prevents the transfer of ^{137}Cs from roots to top parts of the plant. Thus a relatively smaller content of this radionuclide, up to 52 Bq kg^{-1}, was noticed in heather (*Erica*) than in moss (*Sphagnum*), which had up to 1 992 Bq kg^{-1} (Horrill and Clint 1994).

Lichens from Greece contained radiocesium at levels of 2.6×10^4–3.3×10^4 pCi kg^{-1} in 1986, and had fallen to 3 324–7 892 pCi kg^{-1} in 1996 (ATSDR 2002a). The concentrations of both isotopes ^{137}Cs and ^{134}Cs in the lichen (*Stereocaulon vesuvianum*), sampled in Italy in autumn 1986, indicated an increase with elevation of Mt. Vesuvius, respectively from 460 to 1 020 and from 1 330 to 2 500 Bq kg^{-1}. In 1999 only ^{137}Cs was found within the range from 270 to 570 Bq kg^{-1}, from low and high elevation sites, respectively (Adamo et al. 2004). Isomura et al. (1993) observed a significant decrease of ^{137}Cs in moss (*Hylocomium splendens*) sampled in Japan, during the period of 1984–1991, however, a negligible radioactivity derived from the Chernobyl accident was noticed only in the period 1986–1988. The concentrations of ^{137}Cs varied within the ranges 0.9–34.3 and 38.0–359 Bq kg^{-1}, respectively for Lowland and Mt. Fuji. The contents of ^{134}Cs were much smaller and did not exceed 9.7 Bq kg^{-1} in mosses from Mt. Fuji.

Humans. The mean concentrations of ^{137}Cs in tissues of persons not exposed to this radionuclide vary from 0.106 to 2.440 pCi kg^{-1} FW, being the lowest in bones and the highest in gonads (ATSDR 2002a).

Radiocesium is most effectively taken up by cancer cells and thus it might be used for cancer testing and therapy (Brewer 1984).

Animals. Radiocesium was relatively easily transferred to animal tissues right after the Chernobyl accident. The content of ^{137}Cs in deer muscles (in Lithuanian forest) decreased after about a decade from 200 to 10 Bq kg^{-1} (Drebickas et al. 1996). In 1995, caribou from Canada (Saskatchewan) contained ^{137}Cs in tissues at levels 58, 228, and 553 Bq kg^{-1}, respectively in bone, liver, and kidney (ATSDR 2002a). The average ^{137}Cs contents in livers of albatross sampled from Indian Ocean during 1994, were at values of 0.014 to 0.029 ng kg^{-1} (ATSDR 2002a).

Trace Elements of Group 2 (Previously Group IIa)

The trace elements of Group 2, beryllium (Be), strontium (Sr), and barium (Ba), belong to the alkaline earths and behave similarly to Ca and Mg. Their physical properties, especially sizes of their ionic radii are fairly similar to those of Ca, and they may substitute for each other, however, the small ionic radius of Be prevents its replacement by other cations (Table II-2.1). All the alkaline earths are associated with the carbon cycle that strongly controls their behavior in the terrestrial environment. Radium (Ra), which occurs as several radionuclides, also belongs to this group, and is a product after the decay chain of U and Th.

Radionuclides of this group, and especially ^{90}Sr and ^{226}Ra, could present significant risk of environmental hazard.

II-2.1 Beryllium (Be)

II-2.1.1 Introduction

The abundance of Be in the Earth's upper crust averages 3 mg kg^{-1}, but the range of 4–6 mg kg^{-1} has been cited quite frequently (Grew 2002; Rossman 2004). Beryllium is the lightest of the alkaline earths and although widely distributed, exists in relatively small quantities and is likely to concentrate in acid igneous rocks and in argillaceous sediments, where its contents reach up to 6 mg kg^{-1}, (Table II-2.2). However, some alkaline rocks tend to be enriched in this element.

Table II-2.1. Selected properties of trace elements of Group 2

Element	Atomic number	Atomic mass	Atomic radius[a] (pm)	Density (20 °C, g cm^{-3})	Valence	Melting point (°C)
Be, beryllium	4	9.01	140	1.85	+2	1 283
Sr, strontium	38	87.63	245	2.46	+2	704
Ba, barium	56	137.33	278	3.5	+2	714
Ra, radium	88	226.03	–	5.0	+2	700

[a] Approximately average values.

Be

Table II-2.2. Abundance[a] of trace elements of Group 2 in the environment

Environmental compartment	Be	Sr	Ba
Earth's crust (mg kg^{-1})	4 – 6	260 – 370	250 – 584
Igneous rocks (mg kg^{-1})			
Mafic	0.3 – 1.0	140 – 460	250 – 400
Acid	2 – 6.5	60 – 400	400 – 1 200
Sedimentary rocks (mg kg^{-1})			
Argillaceous	2 – 6	300 – 450	500 – 800
Sandstones	0.2 – 1.0	20 – 140	100 – 320
Calcareous	0.2 – 2.0	460 – 600	50 – 200
Energy raw material (mg kg^{-1})			
Coal, hard	1 – 3	30 – 300	75 – 330
Fly ash	10	720	398
Crude oil	0.0005	–	0.3 – 5
Petrol	0.0004	0.1	0.1 – 0.2
Soils (mg kg^{-1})	1.9	147	362
Light sandy	0.1 – 1.7	5 – 1 000	85 – 780
Medium loamy and silty	0.73	20 – 1 000	200 – 1 500
Heavy loamy	2.5 – 4.9	20 – 3 100	200 – 1 500
Calcareous	2.5	5 – 1 000	150 – 1 500
Organic	–	70 – 500	10 – 700
Waters (µg l^{-1})			
Snow	–	0.01 – 0.08	–
Rain	<0.01 – 0.22[b]	0.7 – 383	
Ocean, seas	–	1 680 – 2 590[c]	2 – 15
River	<0.008 – 0.6[d]	300 – 5 000	32 – 56
Air (ng m^{-3})	–	20 – 50[f]	–
Urban	0.1 – 100[e]	–	<100
Rural (USA)	0.1 – 0.3	50	0.2 – 90[g]
South Pole	0.1 – 0.4	0.8	0.02 – 0.73[g]

[a] Presented are common, possible background, values from various sources as compiled by Kabata-Pendias and Pendias (1999 and 2001), and Reimann and Critat (1998), unless otherwise indicated.
[b] Data from Veselý et al. (2002).
[c] For the Baltic Sea, after Szefer (2002a).
[d] Data from Gaillardet et al. (2003).
[e] Data for Japanese cities after Kitagishi and Yamane (1981).
[f] Common range, after ATSDR (2002a).
[g] For remote and polluted areas, respectively (Reimann and Caritat 1998).

Organic matter reveal a sorption capacity for Be, and thus its concentrations in coal vary broadly fróm 11 to 330 mg kg^{-1} and average 2.5 mg kg^{-1} (Finkelman 1999; Llorens et al. 2000). Other data for Be in coal give the range 1.8–2.2 mg kg^{-1}, and in fly ash – up to 47 mg kg^{-1} (Stadnichenko et al. 1961). However, higher contents of Be, from 100–1 000 mg kg^{-1}, have also been reported, and the highest Be concentration, up to 2 000 mg kg^{-1}, was found in coals from the North Bohemian Basin in the Czech Republic (Veselý et al. 2002). Crude oil may content from 0.0005 to 0.5 mg Be kg^{-1}, depending upon local geochemical conditions.

Beryllium exhibits lithophilic affinity; it is similar to Al in geochemical behavior and can substitute for Si and Al in some minerals, and it can strongly bond, especially to montmorillonitic clays. Over 100 naturally occurring minerals contain essential amounts of Be. The most common its own mineral is beryl $Be_3Al_2Si_6O_{18}$, associated mainly with pegmatite bodies, where crystals of beryl may comprise a few percent of the total volume. Other minerals like bertrandite, $Be_4Si_2O_7 \cdot H_2O$; bromeline, BeO; and chrysoberyl, $BeAl_2O_3$; are relatively rare. Of these minerals, beryl and bertrandite are of commercial importance.

Beryllium occurs in several isotope species (8), of which ^{9}Be is stable, and ^{7}Be and ^{10}Be are both cosmogenic and produced by nuclear reaction induced by cosmic rays or accelerator mass spectrometry. Like Bourles (1992) described, both radionuclides exhibit a great affinity for either inorganic or organic particles of soil and ^{10}Be is especially likely to be concentrated in organic-rich sediments and soils.

II-2.1.2
Production and Use

Industrial use of Be has been highly increased in recent years. Its global mining is calculated as 2.55 kt in 2003 (WMSY 2004). Beryllium oxide is the most important commercial product and the Cu-Be alloy is the most commonly used in the metallurgy. Together with the use in electronic and electrical components (i.e., electrical insulators, submarine cables), these applications accounted for more than 80% of its consumption. It is also used in several precision instruments for aircraft/satellite structures, rocket propellants, atomic reactors, high-speed computers, high technology ceramics, as well as in other special facilities.

II-2.1.3
Soils

The abundance of Be in soils has not been investigated on a large scale. The mean background contents for worldwide soils range from 0.48 to 3.52 mg kg^{-1} (Kabata-Pendias and Pendias 2001). Geometric means for Be in various soils of Poland (range 0.03–2.8 mg kg^{-1}, $N = 671$) are 0.33 and 0.73 mg kg^{-1}, for light sandy and heavy loamy soils, respectively. A close positive relationship was reported for Be and clay fraction contents in soils (Kabata-Pendias and Motowicka-Terelak 1996), and it is likely to be more strongly bound by clay minerals than by organic matter. Results presented by Govindaraju (1994) for the Be in reference soil samples from different countries give the range from 1.4 to 4.4 mg kg^{-1}. The mean Be contents in soils of the USA is estimated to be 0.6 mg kg^{-1} (ATSDR 2002a). In Japanese soils it varies from 0.5 mg kg^{-1} in Andosols to 1.5 mg kg^{-1} in Gleysols (Takeda et al. 2004). Swedish arable soils contain Be in the range from <0.5 to 1.8 mg kg^{-1}, at an average value of 1.3 mg kg^{-1} (Eriksson 2001a). Average Be contents in Florida and California soils are given as 0.46 and 1.14 mg kg^{-1}, respectively (ATSDR 2002a).

Most of recent data for soil Be are related to contaminated soils. Other considerable concentrations were reported for some soils in Alaska, up to 300 mg kg^{-1} and for soils over Be-rich granite in Czech Republic, up to 15 mg kg^{-1} (Veselý et al. 2002). Soils from the North-Bohemian industrial region of the Czech Republic contain Be from 2.1 to 5.2 mg kg^{-1} (Podlešáková and Němeček 1994). The degree of soil pollution in this region reflects the isolines of fly ash deposition, as an effect of the increased Be content in coal. Soils from the San Joaquin Valley of central California, described as "uncontaminated" and "highly contaminated" contain Be at mean levels, 2.0 and 3.2 mg kg^{-1}, respectively (Wilson et al. 1994). Average Be contents in Florida and California soils are given as 0.46 and 1.14 mg kg^{-1}, respectively (ATSDR 2002a). Concentrations of Be in the range 0.46–4.7 mg kg^{-1} are reported for soils at the Nevada Nuclear Weapons Test Site, USA (Patton *vide* Veselý et al. 2002). High Be contents, up to 55 mg kg^{-1}, was reported for soils in the neighborhood of the Be factories in Japan (Asami 1988).

Beryllium occurs most often as divalent cation (Table II-2.1), and its complex ions also are mainly divalent: $(BeO_2)^{2-}$, $(Be_2O_3)^{2-}$, and $(Be_20)^{2+}$. Common anionic complexes in alkaline environments are $Be(OH)CO_3^-$, and $Be(CO_3)_2^{2-}$. Although Be appears to be rather immobile in soils, its salts, such as $BeCl_2$ and $BeSO_4$, are easily soluble and available, and therefore potentially toxic to plants. Its concentration in solutions of various soils ranges from 0.2 to 1.1 µg l^{-1}, mainly as simple cations, Be^{2+} and $BeOH^+$, and anions, BeO_2^{2-} and $Be(OH)_3^-$ (Kabata-Pendias and Sadurski 2004). As reported by Krám et al. (1998), Be is especially highly mobile under the condition of acidic (pH 3.4) forest soil, which can have serious environmental consequences in temperate climate regions. Most Be in soil does not dissolve easily in soil solution and remains in bound forms; its binding increases with increasing pH. Also soil organic compounds (mainly fulvic acid) reveal a greater sorption capacity for Be with increasing pH. There is observed a decrease of Be mobility with dissolved Al and dissolved organic carbon concentration (Veselý et al. 2002).

Some fertilizers, and especially phosphorus fertilizers, containing Be within the range <0.2–13.5 µg kg^{-1} might be a source of Be contamination. Also, the application of sewage sludge with elevated Be contents can be significant source of this element in soils.

II-2.1.4
Waters

Beryllium concentration in surficial waters ranges broadly from ng l^{-1} to µg l^{-1}. Its common contents in rivers are established for the range <0.008–0.6 µg l^{-1} (Table II-2.2). World average in dissolved load (<0.2 µm) for Be is calculated by Gaillardet et al. (2003) for 0.0089 µg l^{-1}, and world riverine flux – for 0.33 kt yr^{-1}. These authors have calculated that the chemical mobility of Be during erosion and transport is very low. However, pH of the terrestrial environment is the main factor controlling the Be solubility, and there is an inverse relationship between the Be concentration in waters and pH. The highest Be concentration in fresh water is at about pH 4. The median concentration of Be in water at pH 7 was about 0.03 µg l^{-1}, while at pH 4 was about 1 µg l^{-1}. The mean Be content of shallow ground water in the Czech Republic was 3.3 µg l^{-1} at pH 5.2 (Veselý et al. 2002). In polluted rivers of the industrial region of Poland Be ranges from 0.2–1.7 µg l^{-1} (Lis *vide* Kabata-Pendias and Pendias 1999).

The average value of Be in drinking water of the USA is 0.19 μg kg^{-1} (ATSDR 2002a), and the EPA US standard for drinking water is 4 μg Be l^{-1} (Rossman 2004). Vaessen and Szteke (2000) reviewed Be concentration in tap water and reported range for the mean values from 0.008 μg kg^{-1} (Germany) to <0.1 μg kg^{-1} (The Netherlands). However, the highest concentration of Be, 1.24 μg kg^{-1}, was found in tap water of Saudi Arabia. Mean Be concentrations in mineral (bottled) drinking water range from <0.1 to 0.17 μg kg^{-1}, being the highest for Poland.

Calculated mass balance of Be in watersheds indicates that the output in surface waters from acidified area, and especially from moorland, is higher than the atmospheric flux (Neal et al. *vide* Veselý et al. 2002). High concentrations of Be occur in deep ground water, where F content is high, apparently due to the formation of the mobile Be-F complex. So, mobility of Be is enhanced by formation of soluble F and organic complexes, but several other factors, e.g., Al and dissolved organic matter contents, can also influence the Be behavior in waters (Krám et al. 1998). Common species of this element in soft water are Be-F and Be-OH complexes.

In acidic conditions, Be may be easily mobilized from fly ash and slag, and its increased levels in aquatic environments around coal power plants might be a reason for real concern. The EPA guidelines established the threshold concentration of Be for the universal treatment of wastewater as 820 μg l^{-1} (ATSDR 2002a).

Beryllium in wet deposition (rain) ranges generally from <0.01 to 0.22 μg l^{-1} (Table II-2.2). Its average concentration of 0.36 μg l^{-1} was reported for mist in upland catchments in Wales (Neal et al. *vide* Veselý et al. 2002). The Be concentration in throughfall in different forests was investigated in the Bohemia-Lysina, Czech Republic, and fairly similar concentrations, within the range 0.017–0.032 μg l^{-1}, were reported for beech and spruce ecosystems. It has been suggested that leaching of Be from foliage is a possible source of this element in these waters (Veselý et al. 2002). Rain and snow melt-water from Kola region (Russia) contain Be in the range 0.1–0.3 μg l^{-1} (Reimann and Caritat 1998). Rainwater collected in Sweden during 1999 contained Be from 0.33 to 1.2 μg l^{-1} at an average value of 0.76 μg l^{-1} (Eriksson 2001a).

In water, Be as bivalent cation (Table II-2.1) is likely to occur in several, mainly hydroxyl compounds, i.e., in acid solution, Be_2OH^+, $Be_3(OH)_3^{3+}$, and in basic solution $Be_5(OH)_7^{3+}$. However, in typical range of pH in water (pH 6–8), slightly soluble $Be(OH)_2$ is the main species that controls the solubility of this element (Cotton and Wilkinson 1989).

Aquatic biota. The toxicity of Be to fresh water organisms decreased with water hardness (content of Ca + Mg), most probably because of the formation of non-toxic complexes between Be and HCO_3^- or other ligands, e.g., with SO_4 and CO_3 (Moore 1991). Laboratory tests indicate that the direct application of Be on the surface of fish gills is extremely toxic (Jagoe *vide* Veselý et al. 2002).

II-2.1.5
Air

Data for Be contents in air are rather scarce, although its concentration in urban air is relatively broadly studied. Its content in air varies from 0.1 to 100 ng m^{-3} and is the highest over Japanese cities (Table II-2.2). Beryllium range for urban regions of the

USA is reported for 0.1 to <0.5 ng m^{-3} ($N = 397$), but in urban Detroit, Michigan, it varied between 0.02 to 2.0 ng m^{-3}, in the period 1982–1992 (Del Delumyea et al. 1997). Average Be concentration in ambient air in the USA is reported to be 0.03 ng m^{-3}, but the median concentration in cities is 0.2 ng m^{-3} (ATSDR 2002a). Near the Be processing industry in Mumbai, India, Be level in air was measured ($N = 397$) at 0.42–0.48 ng m^{-3} (Thorat et al. 2001).

Concentration of Be in air is of a great environmental concern because inhaled rich-Be dust is acutely toxic to humans and animals and induce lung cancer. The concentration of Be 200 ng m^{-3} in air in the workplace has been adopted as an action level by the US Department of Energy (DOE) in 1999 (Rossman 2004). Much smaller concentration, 1.3 ng Be m^{-3} is given as hazardous ambient air in some states of the USA (ATSDR 2002a).

The highest Be emission is related to coal combustion. Annual emission of Be in Europe was calculated by Pacyna (1990) for 50 t, while total 450 t of Be bearing compounds (including Be) were released into atmosphere of the USA during the 1990s (www.epa.gov/triexplorer/chemicla.htm).

II-2.1.6
Plants

Mechanisms of the Be absorption by plants seem to be similar to those involved in the uptake of major divalent cations, Mg^{2+} and Ca^{2+}. However, these elements reveal antagonistic interactions, and Be is known to substitute for Mg, especially in some plants. Plants can easily take up Be when it occurs in mobile forms in soils.

Mean contents of Be in plants vary broadly from <1 to >100 μg kg^{-1}. Some plants of the *Leguminosae* and *Cruciferae* families reveal a great ability to accumulate Be, particularly in their root tissues. In some plant shoots increased levels of Be have been observed, e.g., lettuce leaves – 33 μg kg^{-1}, and tomato fruits – 240 μg kg^{-1} (Krampitz 1980). Vaessen and Szteke (2000) reviewed Be contents in food plants of several countries and gave range of mean values for Be from 0.07 to <25 μg kg^{-1} FW, with the lowest for beans, and the highest for corn (Table II-2.3).

Moss samples from Norway, collected in the period 1990–1995 contained Be within very broad range, from <0.4 to 370 μg kg^{-1} (Berg and Steinnes 1997). It has been already indicated that increased level of Be in plants, over that of baseline concentrations, can be a good indicator of pollution, mainly from industrial sources (Sarosiek et al. 1998). Beryllium in tissues of trees grown in the polluted region of Bohemia in the Czech Republic showed the highest concentrations in spruce needles (310 μg kg^{-1}) and in beech leaves (230 μg kg^{-1}), while its contents in stem wood and bark varied within the range 3–78 μg kg^{-1} (Skřivan et al. *vide* Veselý et al. 2002).

The toxicity of Be to plants has been frequently reported, and was observed at the concentration range form 10 to 50 mg kg^{-1}. However, according to the results of Hlušek (2000), above 2 mg Be kg^{-1} soil affected yield decrease of lettuce, up to 40%. The Be concentration in nutrient solution ranging from 2 to 16 mg l^{-1} is highly toxic to plants, reducing root length of collard and of wheat by 50%; concentrations >8 mg Be l^{-1} totally inhibited seed germination (Gettier and Adriano 1987). Increased levels of Be in plants reduce growth, inhibit the uptake of Ca, Mg, and partly of P, and degrade some proteins and enzymes.

Table II-2.3. Beryllium in food plants (μg kg^{-1} FW) (adapted from Vaessen and Szteke 2000, unless otherwise indicated)

Plant	Range	Mean
Cereal grain	–	<0.5
Beans	<0.001 – 0.07	0.07
Cabbage	–	0.2
Green pepper	–	42
Lettuce	–	16
Carrots[a]	<25	–
Mushrooms[b]	<3 – 36	9
Potatoes, tubers	0.4 – 59	–
Rice	3 – 5	4
Tomato	0.2 – 17	–
Banana, pulp	<DL – 18	4.2
Peanut, kernel[a]	–	0.3
Corn, field[a]	–	<25

DL – Detection limits.
[a] After ATSDR (2002a).
[b] European wild.

II-2.1.7 Humans

In mammals soft tissues, Be ranges from 0.9 μg kg^{-1} in liver to about 40 μg kg^{-1} in skin (Jørgensen 2000). Its abundance in a reference man is calculated to be 0.5 μg kg^{-1} (Li 2000). Concentrations of Be average in human fluids are following (in μg l^{-1}): about 1.0 in whole blood, 0.15 in serum, and 0.07–1.0 in urine. Data reported by ATSDR (2002a) gave background average levels of Be humans as follows (in μg kg^{-1}): lung, 20–210; brain, 80; kidneys and spleen, 70; liver and muscle, 40; heart, 30; and bones, 20.

The respiratory tract in humans and animals is the primary target of Be inhalation. Also skin exposures to Be may result in local immune or inflammation reactions. Oral ingestion of Be has not been associated with any clinical diseases, possible due to low rate of absorption estimated at 0.006% of ingested Be (Rossman 2004).

Beryllium is highly toxic to all organisms, especially when inhaled as aerosol dust of its compounds like BeF_2, BeO. Probably $Be(OH)^+$, as well as other Be species, can stimulate in cells enriched with Be the antigen-specific inflammatory response and inhibit the function of several enzymes, and particularly alkaline phosphatase. Ca-metabolism may also be disturbed under the increased Be levels. Although Be compounds are not biotransformed in the body, its soluble species are partly converted to less soluble forms in the lung.

Symptoms of chronic low concentration of Be toxicity include inflammation of the respiratory tract (berylliosus), however, Be is also considered as a carcinogenic agent. The risk of skin and respiratory track exposure to increase Be levels should be especially considered in a given environment. Some impaired health effects might be summarized as follows (ATSDR 2002a):

- Cardiovascular – artrial and ventrical hypertrophy
- Carcinogenic/mutagenic – granulomatous lensions, especially in the lung
- Endocrinologic – impact on several glands (i.e. testicular, prostate)
- Hematologic – a few differences in erythrocytes
- Hepatic – possible hepatocellular (carcinoma)
- Lungs – granulomatous inflammation, mainly in experimental animals

Chronic Be disease is observed mainly in workers of Be manufacturing facilities. The sensitivity to increased levels of Be may highly differ among the population. Human exposure to Be and its compounds occurs mainly in the workplace. The normal Be inhalation by adults in the USA is estimated to be <0.6 ng d^{-1} (at Be < 0.03 ng m^{-3}). It may highly increase at the work place where 8-hour time weighed average is allowed to be 2–5 µg m^{-3} (ATSDR 2002a). The exposure of adults from drinking water is calculated to be 1 µg d^{-1}, at Be concentration in water 0.5 µg l^{-1}. The daily average intake of Be via hospital diets in Japan was determined to be 84.4 µg $person^{-1}$ while other calculations have reported the total daily Be intake to range from 5 to 100 µg d^{-1} (Vaessen and Szteke 2000; Muto et al. *vide* ATSDR 2002a). The EPA guidelines applicable to Be is 20 µg kg^{-1} d^{-1} for food of adults (ATSDR 2002a).

II-2.2
Strontium (Sr)

II-2.2.1
Introduction

Strontium is a relatively common element in the Earth's crust and its contents range between 260 and 730 mg kg^{-1}. It is likely to concentrate in mafic igneous rocks and in carbonate sediments (Table II-2.2). Both geochemical and biochemical characteristics of Sr are similar to those of Ca. Strontium reveals lithophilic affinity and is associated with Ca, and in smaller extends with Mg. The Ca/Sr ratio, in particular environments, may reflect, to a certain extend, the abundance of Sr and its biogeochemical properties. Strontium occurs mainly as bivalent cation, but its chelated forms play an important role in the cycling.

Geological occurrence of Sr is associated mainly with calcareous rocks and sulfur deposits. The distribution of Sr in minerals is largely controlled by Ca, the ionic radius of which is fairly similar to that of Sr, thus it is likely to concentrate in calcium-rich minerals and to be incorporated in calcite and aragonite. Clay minerals have a large capacity to absorb Sr, and most argillaceous sediments are enriched in this element (Table II-2.2).

Commercially important minerals are strontianite, $SrCO_3$; and celestite, $SrSO_4$. Strontium occurs also in several other minerals, e.g., nordite or belovite, which are in association with Na and lanthanides. During weathering and transportation, Sr is incorporated in clay minerals and strongly fixed by organic matter. However, most Sr is likely to precipitate as biogenic carbonates, largely in the form of invertebrate shell material.

II-2.2.2
Production and Use

Strontium world mine production in 2000 was 320 kt (USGS 2004).

Strontium does not have direct applications as a pure metal. However, its various compounds are broadly used (e.g., above 80% of all Sr consumed in the USA) in the manufacturing of ceramics and glass, i.e., television tubes, pyrotechnic materials. All color televisions in the USA are required by law to contain Sr in the faceplate glass of the picture tube to block X-ray emissions (ATSDR 2002a). Other uses of Sr compounds are in various technologies of metallurgy, luminescent paint pigments, and some medicines.

II-2.2.3
Soils

Strontium content of soils is highly controlled by parent rocks and climate. Its mean amount in surface soils is approximately 250 mg kg^{-1}, but the content of 1 000 mg Sr kg^{-1} is often reported for various soils (Table II-2.2). Geometric means (for over 4 000 samples) of Sr contents in soils of Poland show a relation with textural soil groups, as follows (in mg kg^{-1}): sandy soils, 40; and loamy soils, 80 (Kabata-Pendias and Pendias 1999). However, in some calcareous soils and soils from industrial regions, Sr contents above 1 000 mg kg^{-1} have been reported (Lis and Pasieczna 1995). Soils from the San Joaquin Valley of the central California contain Sr, at mean levels from 236 to 246 mg kg^{-1} (Wilson et al. 1994). Median Sr concentrations in Japanese agricultural soils range from 32 to 130 mg kg^{-1}, in Acrisols and Gleysols, respectively (Takeda et al. 2004). Swedish arable soils contain Sr in the range from 112 to 258 mg kg^{-1}, at an average value of 163 mg kg^{-1} (Eriksson 2001a).

Strontium is moderately mobile in soils, and the predominate cation, Sr^{2+}, is likely to be sorbed in hydrated form by clay minerals and Fe oxides and hydroxides. In calcareous soils, Sr may precipitate as strontianite ($SrCO_3$) and became less mobile. Its distribution in soil profiles follows the general trends of soil solution circulation; thus, in acid soils is likely to be leached down the profile, while in calcareous soils and in SOM rich soil-layers, Sr can be concentrated in upper horizons.

Main source of Sr pollution are associated with coal combustions and sulfur mining. Strontium in phosphorites may be concentrated up to 2 000 mg kg^{-1} and thus it can be also a local source of soil contamination due to phosphate fertilizers. Some soil-amendment materials containing elevated amounts of Sr might be a source of pollution. Average contents of Sr in these materials are (in mg kg^{-1}): P-fertilizers, 610; limestone, 610; manure, 80; communal sludges, 75; and industrial sludges, 270.

II-2.2.4
Waters

Data on the Sr occurrence in waters are relatively scanty. Common Sr compounds (carbonates and sulfates) are slightly soluble. However, depending on local conditions, its concentration can highly differ. Gaillardet et al. (2003) gave a very broad range for

Sr in river waters, from 3 to 238 $\mu g\,l^{-1}$, the world average at a value of 60 $\mu g\,l^{-1}$, and world riverine flux at 2 240 kt yr^{-1}. In industrial regions and some areas with calcareous bedrocks, Sr concentration in waters can be highly increased. Some river waters of industrial region of Poland contain above 300 μg Sr l^{-1}, whereas in uncontaminated rivers Sr ranges from 10 to 35 $\mu g\,l^{-1}$ (Szefer 2002a).

The median Sr concentration in seawater is approximately 8 $mg\,l^{-1}$ (Reimann and Caritat 1998). Nozaki (2005) reported average Sr concentration as 7.8 $mg\,kg^{-1}$ in the North Pacific. However, a much lower Sr concentration, such as <0.5 $mg\,l^{-1}$, in the Pacific northwest of the USA, is also reported (ATSDR 2002a).

Rainwater collected in Sweden during 1999 contained Sr from 0.52 to 3.3 $\mu g\,l^{-1}$ at an average value of 1.2 $\mu g\,l^{-1}$ (Eriksson 2001a).

Aquatic biota. Most of aquatic organisms can accumulate large amounts of Sr and this process shows an inverse correlation to levels of Ca in water. Mussels from the Baltic Sea are likely to accumulate Sr up to 100 $mg\,kg^{-1}$ (Szefer 2002a).

II-2.2.5
Air

Distribution of stable Sr in air has not been investigated much. Its common range is given as 20–50 ng m^{-3}. In some industrial areas, especially close to coal power stations, Sr in air can be significantly enriched. Median Sr contents in air is reported for world-remote regions to be 0.81 ng m^{-3}, while in polluted areas it can reach up to 50 ng m^{-3} (Reimann and Caritat 1998).

Strontium is emitted during thermal processes as oxide, SrO, which in contact with water can be transferred to ionic form Sr^{2+} that can be easily trapped by both biotic and abiotic environmental components.

II-2.2.6
Plants

The content of Sr in plants is highly variable (Table II-2.4) and seems to be the lowest in grains (means 1.5–2.5 $mg\,kg^{-1}$) and the highest in vegetable leaves (means 45–74 $mg\,kg^{-1}$) and in tops of legume feed plants (means 219–662 $mg\,kg^{-1}$). Strontium in mosses sampled in Norway in the period 1990–1995 ranges from 2.8 to 51 $mg\,kg^{-1}$ and averages 15 $mg\,kg^{-1}$ and was somewhat lower than data reported earlier (Berg and Steinnes 1997). In the European forest ecosystem, Sr in blueberries (*Vaccinium* sp.) is reported to range 4.5–5.5 and 2.9–3.9 $mg\,kg^{-1}$, in Russia and Germany, respectively (Markert and Vtorova 1995). Although Sr seems to be not very readily transported from roots to shoots, most often its contents in plant tops are relatively high, which may be an effect of aerial deposition. In most cases, however, Sr is likely to be accumulated in roots (Fig. II-2.1).

Although Sr is apparently not a plant micronutrient, it is absorbed following the plant's metabolic requirements for Ca. Interactions between Sr and Ca are complex, and although they can compete with each other, Sr cannot replace Ca in biochemical functions. Increased level of Ca in growth media may both inhibit and stimulate Sr uptake by plants, depending of several soil and plant factors. Most often, however,

Table II-2.4.
Strontium in food and forage plants ($mg\ kg^{-1}$) (data compiled by Kabata-Pendias and Pendias 1999, 2001)

Plant	Range	Mean
Wheat, grains	0.5 – 2.3	1.5
Oats, grains	1.8 – 3.2	2.5
Corn, grains	0.06 – 0.4	–
Lettuce, leaves	–	74
Cabbage, leaves	1 – 150	45
Spinach, leaves	45 – 70	–
Onion, bulbs	10 – 88	50
Carrot, roots	1.5 – 131	25
Tomato, fruits	0.4 – 91	9
Apple, fruits[a]	0.5 – 1.7	0.9
Bean, pods	1.5 – 67	18
Clover, tops	95 – 850	219
Lucerne, tops	50 – 1 500	662
Grass, tops	6 – 37	24

[a] Fresh weight basis.

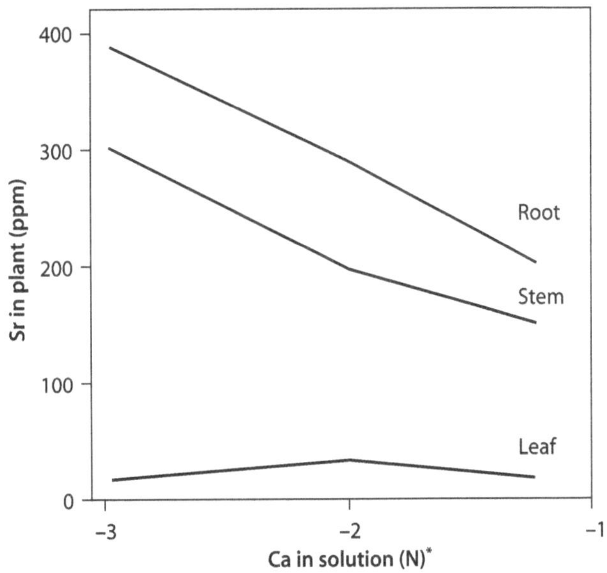

Fig. II-2.1.
Effects of Ca on Sr distribution in bush beans exposed 48 hr to 10^{-3} N Sr in solution culture. *Ca normality is given in powers of ten. These correspond to the Ca concentration (in $g\ l^{-1}$): 0.02, 0.2, and 2, respectively (adopted from Wallace and Romney 1971)

addition of Ca to soil increases its concentration in the soil solution and decreases Sr phytoavailability (Fig. II-2.1). Takeda et al. (2005) concluded that the exchangeable Sr/Ca ratio in the soil could be a good predictor of the Sr concentration in komatsuna (*Brassica rapa*). The highest uptake of Sr is from acid light sandy soils.

Both Sr deposited on the plant surface and taken up by roots can be easily absorbed and translocated to other parts of the plant. Soil-to-plant transfer ratios are relatively high and calculated to range 0.017–1.0 (ATSDR 2002a).

II-2.2.7 Humans

Strontium occurs in all mammalian tissues in the range from 0.09 to 0.24 mg kg^{-1}. Its highest concentration is in kidneys and lowest in brain (Jørgensen 2000). According to data presented by Li (2000), Sr contents range from 0.08 mg kg^{-1} in brain to 0.3 mg kg^{-1} in lymph. It is accumulated mainly in bones (138 mg kg^{-1}) and other hard tissues (e.g., hair 4.2 mg kg^{-1}). In human fluids, Sr concentrations vary as follows (in μg l^{-1}): serum, 40; blood, 27; saliva, 11; and milk, 20 (ATSDR 2002a).

Biochemical functions of Sr are not well known, but as reported D'Haese et al. (2002), its small quantities are needed for proper processes of the calcification of bones and teeth.

Daily intake of Sr by humans varies greatly, depending mainly on the locality. Its intake by adults (70 kg body weight) from food was calculated by Hamilton (1987) for 1.9 mg at the world scale and only for 0.9 mg for inhabitants of the Great Britain. Recent compilation by Seifert et al. (2002) has indicated that daily Sr intake by adults from different countries varies between 1.0 and 2.3 mg, and its availability from the nutrition is estimated to be about 20%. Fairly similar values have been given by Anke (2004a), being lower for women (1.8 mg d^{-1}) and higher for man (2.2 mg d^{-1}). The total exposure of adults to Sr in the USA is approximately estimated per adults as follows (in mg d^{-1}): from inhalation, 3.3; from drinking water, 2; and from the diet, 1.3. In the UK, combining air, water and diet exposure is estimated as 3.3 mg d^{-1} (ATSDR 2002a).

There is not much evidence of Sr toxicity to man, other than in the biogeochemical provinces where increased level of Sr in the food chain, and especially in water is affected by bedrocks of increased Sr levels (Kovalskiy 1974). Seifert (2004) has stated that Sr seems to be relatively low toxic to humans and animals. It has been even used medicinally against various diseases, and its effect on bone formation/mineralization in the treatment of osteoporosis and osteoarthritis has been discussed (D'Haese et al. 2002). However, these authors have described that effects of stable Sr on bone are dose-dependent, and while low doses may stimulate bone formation, high doses of the element may induce a mineralization defect. Young organisms are more sensitive to Sr excess than older organisms. Inadequate levels of Ca and vitamin D in the diet increases the severity of skeletal effects, such as a reduction in bone mineralization, and alteration in the chemical composition of organic bone matrix.

The excess of Sr in food and water inhibits the metabolic processes of Ca, P, and also Cu and Co. At the improper Ca/Sr ratio (e.g., below 8), deformation of bones and other symptoms of disordered metabolisms are likely to occur (Rikhanov 1997). However, there is no evidence that the stable Sr at levels present in the pristine environment has any deleterious effects on humans and animals. It has been suggested that an alimentary Sr intake by adults of 1 to 4 mg d^{-1} is normal and without any risk (Seifert et al. 2002).

II-2.2.8 Strotium-90

Strontium exists in the geological formations in several isotopic species, with atomic mass number from 79 to 102, of which ^{88}Sr is the most abundant stable isotope. The ^{87}Sr: ^{86}Sr ratio is used in age determinations, since ^{87}Rb decays to ^{86}Sr. Products of several nuclear processes are ^{89}Sr and ^{90}Sr, however the last radionuclide (^{90}Sr) is considered to be one of the most biologically hazardous for humans. The basis of the adverse effects of ^{90}Sr is related to ionizing radiation that is in the interaction of free radicals with cellular macromolecules, including DNA. Isotope ^{89}Sr is used as a cancer therapeutic to alleviate bone pain, while isotope ^{85}Sr has been used in various radiologic instruments (ATSDR 2002a).

Sources of ^{90}Sr are associated with nuclear reactions, and have been tracked since nuclear energy became used on the global and regional scales. During the period 1945–1980, radioactive Sr was released into the atmosphere mainly from aboveground detonations of nuclear weapons. Its continuous sources increased greatly after the Chernobyl accident in 1986, and its relatively long live ($t_{1/2}$: 29 years) may lead to steadily increasing concentrations in various environmental compartments.

Soils. Radionuclide ^{90}Sr is of special environmental concern. Its geochemical behavior is similar to stable Sr, and thus is highly mobile in acidic conditions. The distribution of this radionuclide has been recently monitored, particularly in the regions affected by the deposition of Chernobyl emissions.

About a decade after that accident (1986–1995), mineral soils of Belarus, in the most affected area, contained ^{90}Sr at the range from 1.64 to 109 kBq m^{-2}. Alluvial and organic-rich soils accumulated higher amounts of ^{90}Sr, up to 340 Bq kg^{-1}, than light mineral soils. This radionuclide is likely to occur in easily exchangeable forms and thus is characterized by relatively high migration ability (Kagan and Kadatsky 1996; Kuzniecow and Generalova 1996). Soils of Tomsk (industrial region of Russia) contained this radionuclide, in the period of the 1990s, in the range of 111–121 Bq kg^{-1} (Rikhanov 1997). Soils of Kazakhstan sampled in 1992 contained ^{90}Sr in the range of average values from 48 to 70 mCi m^{-2} (Panin 2004).

The speciation of this radionuclide in various soils of semi-natural landscapes indicates that easily soluble fractions make up to 90% the total content, and are likely to occur as organic fractions in the soil solution (Agapkina et al. 1995; Korobova et al. 1998). The migration of this radionuclide in soil profiles is relatively extensive, but depends largely upon soil properties, and particularly on both kind and content of SOM. The migration velocity for ^{90}Sr has been calculated as 0.7–1.5 cm yr^{-1}, but a great part of the residual ^{90}Sr still is located in the upper 10 cm soil layer (Arapis et al. 1997; Shagalova et al. 2000).

Waters. Sources of ^{90}Sr in waters are from both aerial deposition and industrial sludges. This radionuclide in aerial dusts is relatively easily soluble in water, and thus easily deposited on all terrestrial compartments.

Concentration of ^{90}Sr in surface of the Pacific Ocean has decreased steadily since the early 1960s to present days, approximately from 23 to 8.1 pCi m^{-3} in seawater (ATSDR

2002a). In 1969, water of the Northern Sea was assayed to contain ^{90}Sr at the level 30 mBq l^{-1}, and by the 1980s the level of this radionuclide increased up to 40 mBq l^{-1}. At that period, drinking water in Berlin had ^{90}Sr activity within 3.7–14.0 mBq l^{-1} (Becker-Heidemann and Scharpensel 1990). The median ^{90}Sr in drinking water of the USA in 1995, was 0.1 pCi l^{-1} (4 mBq l^{-1}) and ranged up to 0.5 pCi l^{-1}. Ground water sampled at that time had median and average concentrations of this nuclide at 1.9 and 1.4 nCi l^{-1}, respectively (ATSDR 2002a).

Air. A contribution of ^{90}Sr to global Sr concentration is estimated for about 2.4%. Some times its elevated presence in air can be of great risk to health. The concentration of ^{90}Sr in air of Silesia, the industrial region of Poland, was at levels 3–6.5 µBq m^{-3}, whereas after the Chernobyl accident its concentration in air of the central lowland of Poland was 47 mBq m^{-3}, but decreased, within period of two weeks, to 15 µBq m^{-3} (Grabowski et al. 1990).

Fall out of the ^{90}Sr with aerosols in the vicinity of Tomsk, Russia, was in the 1990s within the range 21–166 Bq m^{-2} (Rikhanov 1997). The average total annual wet deposition of ^{90}Sr in the USA during similar period was 5 pCi m^{-2} (0.2 Bq m^{-2}) (ATSDR 2002a).

Plants. Plants, easily absorb ^{90}Sr, similarly to stable Sr. Thus, some plants are proposed for use in phytoremediation; for example, alamo switchgrass (*Panicum virginatum* L.), can accumulate up to 44% of the total amount of this radionuclide from soils (Entry and Watrud 1998). Also *Amaranthus* species are reported to be effective in the phytoextraction of the radionuclide from soils, as has been investigated in the Chernobyl exclusion zone (Dushenkov and Sorochinsky 1999). The transfer factor (TF) for ^{90}Sr, expressed as its ratio in fresh weight plants to concentration in soils varies highly, and has been calculated as follows, 0.4 for potato tubers, 0.8 for vegetable leaves, and for 20 vegetable roots (Kabata-Pendias and Pendias 1999).

The uptake of ^{90}Sr by plants depends upon the content of its easily mobile fractions in soils and on the plants capability for uptake. Korobova et al. (1998) reported the following order of the radionuclide uptake by plants: legumes > herbs > grasses. Soils amendment with Ca, Mg, and K fertilizers inhibit its availability to plants.

Humans and Animals. The accumulation of ^{90}Sr in food and feedstuffs has been of a great concern. Milk and milk products can be one of the most serious sources of ^{90}Sr to humans in regions contaminated by this isotope.

The process of the discrimination of this radionuclide occurs during its transfer in the food chain, and the ratio of its concentrations in bones and soils is estimated to be relatively low, at value of 0.05. However, humans and animals are subjected to its additional sources from air dust and water. Bones of children are more susceptible to concentrate ^{90}Sr than skeletons of adults (Rikhanov 1997). Radioactive Sr is mainly taken up into bone and retains there for a long time. At very high exposure to ^{90}Sr in bone, bone marrow is damaged (at critical average level 8 pCi l^{-1}). The health of organisms exposure to higher doses of ^{90}Sr is harmfully affected. Toxic effects of the ^{90}Sr are following:

- Development of various tumors, especially in bones
- Impaired bone mineralization
- Blood cells (red and white) reduction
- Leukemia
- Loss of platelets resulting in abnormal bleeding
- General fatigue and susceptibility to infection disease

The impact of radioactive Sr on human reproduction has not been observed yet. However, these effects are known to have occurred in animals.

Similarly, as stable Sr, the radionuclide is relatively easily extracted from the organisms, especially by young, while in elders it is likely to be accumulated, especially in bones where Sr becomes relatively immobile.

II-2.3 Barium (Ba)

II-2.3.1 Introduction

Barium is common and quite ubiquitous element; its mean content in the Earth's crust amounts to 425 mg kg^{-1}, and ranges from 550 to 668 mg kg^{-1} in the upper continental crust. Barium has a lithophilic affinity and is likely to concentrate in acid igneous rocks and argillaceous sediments, ranging widely in various rocks from 250 to 1 200 mg kg^{-1} (Table II-2.2). Its ionic radius is fairly similar to the ionic radius of K, and thus usually follows the K fate in geochemical processes.

During weathering, Ba is not very mobile because is easily precipitated as sulfates and carbonates, and also strongly adsorbed by clays. Its great affinity for the concentration in Mn soil nodules (up to 0.2%) and Mn soil minerals (up to over 5%) is an effect of easy displacement of other sorbed alkaline earth metals. Those metals, on the other hand, can also readily substitute for Ba in aluminosilicates.

Contents of Ba in coals range from 75 to 330 mg kg^{-1}, but in some lignite coal it can be concentrated up to 1 420 mg kg^{-1} (Churey et al. 1979). Average Ba in coals of the USA is estimated as 170 mg kg^{-1}, but may reach the concentration up to 22 000 mg kg^{-1} (Finkelman 1999). In crude oil, Ba ranges from 0.3 to 5 mg kg^{-1}.

The most common naturally occurring minerals are barite, $BaSO_4$; and hollandite, $Ba_2Mn_8O_{16}$. The carbonate-Ba compound – whiterite, $BaCO_3$, occurs relatively often. Also, Ba occurs in some silicate minerals as impurities.

II-2.3.2 Production and Use

World mining of Ba, mainly from intrusions in granites and shale, was estimated as 2.9 Mt in 1995 (Reimann and Caritat 1998).

Industrial uses of Ba are wide and variable. It is widely used in the production of brick, tile, glass, and ceramics. It is applied in the synthetic rubber industry. The use of Ba is broad in the chemical industry, especially for the paint production.

The barite mineral ($BaSO_4$), due to its relatively heavy weight, is used extensively (as a drilling mud) in oil- and gas-drilling fluids. Also, different Ba compounds have several important medical uses. Because, some Ba compounds are normally not absorbed after oral intake, they are used for the X-ray examination. However, some uses of Ba have been abandoned due to their toxicity.

II-2.3.3
Soils

Barium contents in surface soils are fairly similar to its distribution in parent rocks (Table II-2.2). The reported Ba average range for soils on the world scale is from 84 to 960 mg kg^{-1}, being the lowest in organic soils and the highest in loamy and clay soils. Govindaraju (1994) reported the Ba content in reference soils from China to range within 18–1 210 mg kg^{-1}, and in soils of the USA, from 290 to 240 mg kg^{-1}. The high Ba content in Chinese soils is related to the Au ore field (Peters 2002). Monitoring on 218 surface-soil samples from throughout Poland gave the Ba range of 20–130 mg kg^{-1} and did not show any changes within the period 1995–2000 (Terelak et al. 2002). Swedish arable soils contain Ba in the range from 383 to 778 mg kg^{-1}, at an average value of 608 mg kg^{-1} (Eriksson 2001a). An interim soil quality criterion of 750 mg Ba kg^{-1} for agricultural soils has been established in Canada (Jaritz 2004).

Although Ba is easily adsorbed, especially by oxides and hydroxides, it is relatively mobile in most soils. Its concentration in soil solution shows considerable variation, from 43 µg l^{-1} in loamy soil to 307 µg l^{-1} in sandy soil (Turski et al. 1989). Easily soluble mineral, hollandite, is often responsible for upwards migration of Ba in aridic soils, where it is known to concentrate at surface varnish layer. However, in soils of temperate humid climate zones, Ba can be fixed by Fe and Mn hydrous oxides and become immobile.

Barium from aerial sources and from P-fertilizers can influence a steady increase of this metal in rural soils of some regions. Also, soils in surroundings of phosphate fertilizer plants have increased Ba levels in the surface layer.

II-2.3.4
Waters

Barium is a common component of all surface waters, but occurs at relatively low levels, unless waters are polluted. World river waters contain Ba within the range from 4 to 73 µg l^{-1}, and average 23 µg l^{-1} (Gaillardet et al. 2003). Background reported ranges for river and stream unfiltered waters are between 32 and 56 µg l^{-1} (Table II-2.2).

In surface fresh waters, Ba is reported to range between 7 and 15 000 µg l^{-1} (Jaritz 2004); in polluted rivers of Poland it has reached up to 136 µg l^{-1} (Lis *vide* Kabata-Pendias and Pendias 1999). In streams of Karpaty Piedmont, Ba concentrations range 25–47 µg l^{-1} (Strączyk and Somerlik 2003). In both, natural and treated waters, there are sufficient concentrations of sulfate anions to bond the Ba cation, so that its levels do not exceed usually the range of 1–1.5 mg l^{-1}.

Barium released from industries and mining is deposited mainly in waters and bottom sediments. Pollution sources are often from coal and oil combustion. Waters of some coal mines (i.e., in Upper Silesia, Poland) are enriched in Ba and affect its increased concentration in waters of rivers surrounding the settling mine ponds water. Its accumulation up to >1%, in river bottom sediments was observed (Pluta 2001). Coal-mine waters and stream waters of polymetallic ore deposits can contain high amounts of Ba, up to 2.5 g l^{-1}. However, when those waters are flowing through old coal mines, its concentration can be significantly reduced due to the precipitation of $BaSO_4$ (Pluta 2001).

Rainwater collected in Sweden during 1999 contained Ba from 0.33 to 1.2 µg l^{-1}, at an average value of 0.76 µg l^{-1} (Eriksson 2001a).

Aquatic bryophyte (*Scapania undulata*), used for the bioassay, can accumulated up to 5 500 mg Ba kg^{-1} (Samecka-Cymerman 1988).

II-2.3.5
Air

Range for the Ba concentrations in air of remote and polluted regions of the world are reported as 0.02–0.73 and 0.2–90 ng m^{-3}, respectively (Reimann and Caritat 1998). In urban air of the USA, the concentration of Ba is reported to be between 0.2 and 28 ng m^{-3}. The Ba distribution in air was observed in city vicinities at concentrations of up to 100 ng m^{-3} (ATSDR 2002a). Its source is mainly from oil and coal combustion.

The occupational limit for Ba is estimated as 500 µg m^{-3} (ATSRD 2002a).

II-2.3.6
Plants

Although Ba is reported to be commonly present in plants, it is apparently not an essential component of plant tissue. Barium mean contents in most plants range from 2 to 13 mg kg^{-1}, with an exception of blueberries in which highly elevated Ba levels are reported (Table II-2.5). Much higher concentrations of Ba are also found in different trees and shrubs grown in area of arid climate (Schaklette *vide* Kabata-Pendias and Pendias 1999). The highest contents of Ba are reported for Brazil nuts, depending on the location of growth, from up to 3 000 to 4 000 mg kg^{-1} (Shacklette et al. *vide* Kabata-Pendias and Pendias 2001, www.orau.org/ptp/collection).

Plants can take up Ba quite easily from acid soils, however there are only a few reports on toxic Ba concentrations in plants, i.e., the growth of barley and beans was inhibited in soil containing 2 000 mg Ba kg^{-1} (Chaudry et al. 1977). Possible toxicity of this element to plants may be reduced by the addition of Ca, Mg, and S salts to the growth media, as effects of antagonistic interactions between these elements, but most probably due to the formation of slightly soluble $BaCO_3$ ad $BaSO_4$.

II-2.3.7
Humans

Contents of Ba in mammalian tissues vary largely from <0.007 to 0.15 mg kg^{-1}, being the lowest in liver and the highest in skin (Jørgensen 2000). According to data presented by Li (2000), Ba in soft tissues of humans varies from 0.006 mg kg^{-1} in

Ba

Table II-2.5.
Barium in food and forage plants (in mg kg^{-1}) (data compiled by Kabata-Pendias and Pendias 2001, unless otherwise indicated)

Plant	Range	Mean
Wheat, grain[a]	–	3.2
Barley, grian[a]	–	5.1
Cereal, grains	4.2 – 6.6	5.5
Bean, seeds	1 – 15	8
Carrot, rrots	2 – 50	13
Onion, bulbs	3 – 75	12
Potato	1.3 – 35	5
Lettuce, leaves	9 – 11	–
Cabbage, leaves	–	4.8
Tomato, fruits	2.3	2.1
Apple, fruits	1.5 – 2	1.5
Orange, fruits	–	3.1
Bluberries-R, fruits[b]	8 – 60	–
Bluberries-G, fruits[b]	132 – 181	160
Meadow clover[a]	–	12
White clover[a]	–	14

[a] After Anke et al. (2002).
[b] Berries sampled in Russia – *R*, and Germany – *G*, after Markert and Vtorova (1995).

brain to 0.8 mg kg^{-1} in lymph. Its contents in all tissues of humans decrease with age and the highest is in babies below 1-year age (Anke et al. 2002). As Jaritz (2004) described, liver of adults contain 4–20 mg Ba kg^{-1}, whereas Ba amounts up to 144 mg kg^{-1} are reported for liver of children. The biomarkers used for Ba are: bone, blood, urine, and feces.

Barium is not metabolized in the body, but it may be metabolically transported or incorporated into complexes of tissues. Excessive Ba decreases K concentration at the cellular levels, apparently due to the similarity in their ion radii. Toxic effects are related to the disorder of digestion and respiratory trucks and also to the inhibition of the mineralization of bones. There is some evidence that the cardiovascular system may be one of primary targets of Ba toxicity (hypertension, abnormality in heart rhythm). Also carcinogenic effects of Ba compounds have been observed.

In most food products (e.g., farinaceous, cocoa, sugar, cheese, fish) Ba is detected at average values of 1–6 mg kg^{-1}. Also some plants can be a significant source of this element (Table II-2.5). However, the largest Ba concentrations are in black tea leaves, at an average value of 10.5 mg kg^{-1} (Anke et al. 2002), and in Brazil nuts. here are indications that Ba in Brazil nuts is present in a very easily assimilatable form, however, due a relatively low consumption of nuts, it does not appear to be hazardous to humans (Lisk et al. 1988; Stoewsand et al. 1988).

It has been calculated that above 50% of the total Ba intake by people comes from vegetables, while animal foods contribute about 30%, and beverages about 10%. As Anke et al. (2002) reported, the Ba intake by adults, with mixed diets, varies largely, depending on the location and gender, the lowest is for women in Ronneburg (0.37 mg d^{-1}) and the highest for men in Jena (2.27 mg d^{-1}). Average daily intake of Ba is estimated for around 0.5 mg, and the intake of 70 µg Ba kg^{-1} BW is considered harmless (Jaritz 2004). Average-sized adults exposed to 1.5 mg Ba every day for 70 years would not experience adverse health effects (www.hepn.com/tri/barium.htm).

Information on the absorption of Ba from gastrointestinal track is confusing, because its insoluble compounds can also be a significant source of this element. It seems to be rapidly transported in blood plasma. However, 20% of an ingested dose is excreted in the feces within 24 hours. The toxicity of Ba salts depends highly on their solubility in water. Compounds such as: $BaCl_2$, $Ba(NO_3)_2$, and $BaCO_3$ are very toxic and may affect hypertension and abnormality in heart rhythm.

The dietary intake of Ba by humans from drinking water may be a significant source. Acceptable Ba concentration in drinking water does not exceed 1 mg l^{-1} in most countries; the WHO (2004) regulation established the maximum Ba as 1.1 mg l^{-1}, and in Russia this limit is given as 4 mg l^{-1} (Kabata-Pendias and Pendias 1999).

II-2.4 Radium (Ra)

II-2.4.1 Introduction

Radium is a naturally occurring metal and is ubiquitous in all environmental compartments at very low concentrations. It exists mainly as divalent radioactive cation composed of several isotopes and has chemical properties that are similar to Ba, Sr, and Ca (Table II-2.1). It is a lithophilic element and occurs as impurities in several minerals of zeolite group, its typical mineral is radiobarite, $(Ba,Ra)SO_4$.

The most common and, so-called, stable isotopes are ^{226}Ra and its daughter ^{222}Ra. Most data on the environmental distribution are for the ^{226}Ra, and its half-live time is estimated at 1 599 or 1 622 years. All Ra radionuclides are products of the U and Th decay chain. Thus, the chemistry of U and Th controls in what minerals Ra is likely to occur.

The radionuclide ^{226}Ra is a natural component of the environment, and is also of the anthropogenic origin, mainly from phosphate and potassium fertilizer, coal combustion and cement factories. The mean concentration of ^{226}Ra in coal is on the order of 1 pCi g^{-1} (ATSDR 2002a).

Radium has been used as a radiation source for radiotherapy, for several nuclear facilities and in radiometric dating methods. Until the 1960s, Ra was a component of the luminous paints used for various instruments, and watch and clock dials (ATSDR 2002a).

II-2.4.2 Soils

Radium content in the continental crust is estimated as 0.9 ng kg^{-1}, its range in various rocks is between 0.1 and 1.1 ng kg^{-1}, and range in soils – between 0.03 and 1.6 ng kg^{-1}. In some soils, e.g., near coal-fired power plants, Ra level in surface soils was elevated up to 8 pCi g^{-1} (Jaworowski and Grzybowska *vide* ATSDR 2002a).

The cycling and behavior of ^{226}Ra has been recently a subject of various studies, often together with other radionuclides. It is accumulated in soil organic matter and is sparingly mobile in soil (exchangeable fraction amounts 0.5–10% of the total).

II-2.4.3 Waters

There is scarcity of information regarding the occurrence of Ra in waters, although the most interesting phase of Ra geochemistry is associated with the ocean. The radioactive equilibrium between ^{230}Th and ^{228}Ra is used to study large-scale-time oceanic mixing processes (Moore 1972).

Gaillardet et al. (2003) gave few data on the Ra abundance in river waters within the range between 0.002 and 0.09 pg l^{-1}, with the highest values for Asian rivers and the world average at 0.024 pg l^{-1}. Its occurrence in seawaters is calculated for around 1 pg l^{-1}.

The typical concentrations of ^{226}Ra in drinking water of the USA range from 0.9 to 10 pCi l^{-1}, and of ^{228}Ra, from 1.4 to 6.4 pCi l^{-1} (ATSDR 2002a). The MAC value for Ra in drinking water in Russia has been established for 0.1 pg l^{-1} (Reimann and Caritat 1998).

Aquatic biota. Aquatic organisms accumulate easily Ra from water. Bioconcentration factors for fish from lakes polluted with U-processing waste range from 1 to 60 for flesh portions, and from 40 to 1 800 for bone. The transfer chain for Ra are apparently bottom-feeding organisms consumed by predatory fish (ATSDR 2002a).

The most significant source of Ra is from leaching of U-mine tailings and from the release of ore processing effluents. It has been reported in ATSDR (2002a) that the untreated U-milling effluent contains ^{226}Ra up to 2.2 μCi l^{-1} (81 kBq l^{-1}). Concentrations of ^{226}Ra and ^{228}Ra found in surface and ground waters are generally low. However, there are specific geographic regions where elevated concentrations of Ra may occur.

Waters discharged from some coal mines are reported to be enriched in ^{232}Ra. After the reaction of polluted waters with sulfates during the flow through old coal mines, the concentration of this radionuclide can be reduced from 23.3 Bq l^{-1} to 0.3 Bq l^{-1} (Pluta 2001).

II-2.4.4 Air

The combustion of coal is the most common and important source of Ra in the atmosphere. After combustion, volatilized Ra condenses onto fly ash, where may be concentrated within the range of 1–3 pCi g^{-1}. Thus, Ra transport in the atmosphere is mainly by the movement of particulate matter.

Dust samples collected from the air of New York City contained ^{226}Ra at 8×10^{-5} pCi m^{-3} and ^{228}Ra at 1.5×10^{-4} pCi m^{-3} (ATSDR 2002a).

II-2.4.5 Plants

The content of Ra in plants is roughly estimated as 0.03–1.6 ng kg^{-1}. Plants grown on soil with total Ra at 8.48 Bq kg^{-1}, and exchangeable Ra at 0.62 Bq kg^{-1}, contained this isotope in the range from about 30 to above 1 000 mBq kg^{-1} (Kabata-Pendias and Pendias 1999).

Radium is relatively easily taken up by plants, mainly through the root system, but its phytoavailability varies depending on the specific plant species and soil conditions. Soil-plant transfer factor may range from 0.001 to 6.5, being lower for fruits than for grains (Rayno *vide* ATSDR 2002a). The order of the transfer factor from soil to plants was, as follows: leaf-vegetables > root-vegetables > grasses > corn stems > grains (Rosiak and Pietrzak-Flis 1998). The concentration of ^{226}Ra in vegetables collected in Poland during the period 1995–1996 varied from 8.66 (cucumber) to 59.4 (carrot) mBq kg^{-1}.

II-2.4.6 Humans

Jørgensen (2000) reported Ra contents in liver and kidneys of mammals, at values of 1–8 ng kg^{-1}. Radium enters the body by digestion and breathing. There is not evidence that it may be taken in through the skin.

Pietrzak-Flis et al. (2001) presented the average intake of ^{226}Ra by adults in different countries as follows (in mBq d^{-1}): Japan, 23; Poland, 42; Italy, 52; and USA, New York, 52. Its intake in 1995 by adults in Poland was calculated as 15.4 Bq yr^{-1}. Estimated intake of ^{226}Ra intake by American adults (in mBq d^{-1} kg^{-1} BW) is around 2×10^{-4} from water, 0.11×10^{-4} from food, and 1.9×10^{-4} from air (ATSDR 2002a).

Medium content of ^{226}Ra in food of the USA is estimated as 0.6 pCi kg^{-1}. The greatest sources of the radionuclide are cereals, potatoes and milk, including milk products (Pietrzak-Flis et al. 2001).

A potential Ra exposure for humans is related to the consumption of meet and milk from cattle that graze on forage grown in soils contaminated with Ra. Although the values of its transfer factors of forage to milk and of forage to flesh are very low, from 3.8 to 6.8×10^{13}, respectively, cattle food products are considered to be the most significant source of Ra radionuclides.

There is no evidence that the log-term exposure of humans to Ra at normal environmental contents may be of a risk health. However, exposure to higher levels over a long period may result in harmful effects, mainly such as: anemia, cataracts, and tumors, especially bone cancer. It also may be easily deposited in eyes. Increased oral exposure to Ra of humans and animals affects several processes in immunology, neurology, hematology and reproductivity.

Trace Elements of Group 3 (Previously Group IIIb)

Geochemical and biochemical properties of the elements of Group 3, as well as their abundance in the biosphere are highly divergent. Two elements, scandium (Sc) and yttrium (Y) are rather rare in the environment, and usually exhibit an affinity for oxygen and their oxidation state is mainly +3 (Table II-3.1). Together with two other elements of this group, lanthanum (La) and actinum (Ac), two subgroups of elements are distinguished as: lanthanides and actinides, of which many elements are either natural or artificial radionuclides.

II-3.1 Scandium (Sc)

II-3.1.1 Introduction

Scandium is likely to occur in higher amounts in mafic igneous rocks and in argillaceous sediments. In natural environments Sc^{3+} occurs mainly which can substitute for Al^{3+}, Fe^{3+}, and also Ti^{3+}, and it is mainly associated with ferromagnesian minerals and biotite; it is also reported to be enriched, up to 36 mg kg^{-1}, in some phosphorites. Its mineral, rarely occurring thorveitite, $Sc_2(Si_2O_7)$, is the primary source of Sc.

Some organic raw materials such as peats, coal, and crude oil can accumulate increased levels of Sc, and thus their combustion can be a source of environmental enrichment of this metal. Its mean concentration in US coal is 4.2 mg kg^{-1}, but may reach up to 100 mg kg^{-1} (Finkelman 1999).

II-3.1.2 Production and Use

Scandium is obtained as byproduct at uranium processing and its global annual production has been estimated at 537 kt (Reimann and Caritat 1998). It is used mainly for cathode-ray tubes, lasers, and lightening and fluorescent materials (e.g., high intensity lights, color television).

Table II-3.1. Selected properties of scandium and yttrium

Element	Atomic number	Atomic mass	Ionic radius (pm)	Density ($g\ cm^{-3}$)	Valence	Melting point (°C)
Sc, scandium	21	44.96	88.5 – 102	2.9	+3	1541
Y, yttrium	39	88.91	104 – 121.5	4.4	+3	1522

II-3.1.3 Soils

Average contents of Sc in worldwide soils range between 1.5 and 16.6 mg kg^{-1}, being the lowest in sandy and the highest in loamy soils (Table II-3.2). Govindaraju (1994) reported Sc to range from 5 to 28 mg kg^{-1} in the reference soils of China, and from 5 to 18 mg kg^{-1} in soils of the USA. Soils of San Joaquin Valley of central California contain Sc within the range of 9.2–12 mg kg^{-1} and its distribution did not show any relation to contamination of soils (Wilson et al. 1994). In Japanese agricultural soils, its median concentrations vary from 13 to 28 mg kg^{-1} in Acrisols and Andosols, respectively (Takeda et al. 2004). Swedish arable soils contain Sc in the range from 3.5 to 16 mg kg^{-1}, at an average value of 10 mg kg^{-1} (Eriksson 2001a). Average contents of Sc in soils of Poland range from 1.11 to 4.38 mg kg^{-1} and do not reveal any increase in soils of the industrial region (Buhl and Poledniok 2000). Sandy soils of Poland contain Sc in the range from <2 to 5 mg kg^{-1} (mean 2) and loamy soils from 3 to 10 mg kg^{-1}, at the mean value of 5.6 mg kg^{-1} (Dudka 1992). Vaganov and Nizharadze (1981) reported a higher level of Sc in silty deposits (mean 5 mg kg^{-1}) than in calcareous deposits (mean 0.5 mg kg^{-1}).

Scandium is likely to form complexes such as $Sc(H_2O)_6^{3+}$ and $Sc(H_2O)_5OH^{2+}$ in aqueous phases, and also shows an affinity for complexing with PO_4, SO_4, CO_3, as well as with some organic compounds. Phosphate fertilizers contain relatively high amounts of Sc, in the range of 7–36 mg kg^{-1}, and may elevate its concentration in surface agricultural soils (Kabata-Pendias and Pendias 2001).

II-3.1.4 Waters

Reimann and Caritat (1998) calculated the median Sc concentration in worldwide ocean waters at 0.6 ng l^{-1}, and Nozaki (2005) reported Sc in waters of the North Pacific to be at 0.7 ng kg^{-1}.

Concentrations of Sc differ widely in river waters, in African rivers its range is between 0.055 and 0.087 μg l^{-1}, whereas river waters of South America contain Sc between 0.530 and 1.770 μg l^{-1}. The world average of Sc in rivers is estimated as 1.2 μg l^{-1} (Gaillardet et al. 2003). Ground water of Norway contains Sc in the range of 0.5–9.6 μg l^{-1} (Reimann and Caritat 1998). Scandium concentrations in drinking water from remote regions of Poland range from 0.01 to 0.35 μg l^{-1}, with an average value of 0.07 μg l^{-1} (Macioszczyk *vide* Kabata-Pendias and Pendias 1999).

Table II-3.2. Abundance[a] of scandium and yttrium in the environment

Environmental compartmment	Sc	Y
Earth's crust (mg kg^{-1})	16 - 30	20 - 33
Igneous rocks (mg kg^{-1})	–	–
Mafic	5 - 35	0.5 - 20
Acid	3 - 15	20 - 33
Sedimentary rocks (mg kg^{-1})	–	–
Argillaceous	10 - 15	25 - 40
Sandstones	1 - 3	15 - 250
Calcareous	0.5 - 5	4 - 30
Soils (mg kg^{-1})	9.5	12
Light sandy	0.8 - 14	3 - 28
Medium loamy and silty	3 - 8	7 - 30
Heavy loamy	2.5 - 28	7 - 60
Calcareous	6 - 11	–
Organic	5 - 20[b]	–
Water (µg l^{-1})	–	–
Rain	0.73	0.003 - 0.05
Sea	0.001	0.008
River	1.2[c]	0.05 - 1.4
Air (pg m^{-3})	–	–
Urban	30 - 3000	6.0[d]
South Pole	0.06 - 0.21	–

[a] Presented are common, possible background, values from various sources as compiled by Kabata-Pendias and Pendias (1999 and 2001), unless otherwise indicated.
[b] Data for forest soils.
[c] World average, after Gaillardet et al. (2003).
[d] After Reimann and Caritat (1998).

Snow from the Kola Peninsula contained Sc in the range of 0.01–0.5 µg l^{-1}, from remote and polluted areas, respectively (Reimann and Caritat 1998). Rainwater collected in Sweden during 1999 contained Sc at concentrations from 2 to 12 ng l^{-1} and an average value of 4 ng l^{-1} (Eriksson 2001a).

II-3.1.5 Air

Air from the South Pole contains Sc in the range of 0.06–0.21 pg m^{-3}, air above Greenland contains Sc at around 40 pg m^{-3}, and in air from some urban regions of the USA, Sc concentration may be up to 3 000 pg m^{-3}. The world median values are cited by Reimann and Caritat (1998) to range from 19 to 30 pg m^{-3}. Cawse (1987) reported the atmospheric deposition of Sc in the UK to range between 0.3 and 2 g ha^{-1} yr^{-1}, and Sc concentrations in air to range from 30 to 90 pg m^{-3}. Wet deposition of Sc in Sweden is calculated to be 0.04 g ha^{-1} yr^{-1} (Eriksson 2001a).

Fly ash of lignite coal contains up to 5.2 mg Sc kg^{-1} (average 3.7) and might be a local source of increased levels of this metal in the environment.

II-3.1.6 Plants

There is a paucity of data on Sc contents in plants. The commonly reported range for Sc in various plants is 0.002–0.25 mg kg^{-1} (Kabata-Pendias and Pendias 2001). In Swedish cereal grains, Sc contents are reported to be <0.01 mg kg^{-1} (Eriksson 2001a).

Scandium is likely to increase in roots and old leaves. Relatively high levels of Sc have been found in tea leaves, in the range from 0.013 to 0.14 mg kg^{-1}, at an average value of 0.037 mg kg^{-1} (Inadira et al.1984). This supports the observation that the greatest amount of Sc is taken up by plants from sandy soils, which in the most cases are also acid soils. Tea plantations are usually on very acid soils, often of pH <4.

Shtangeeva et al. (2004) studied the effect of Sc addition to germinating wheat seeds. This addition decreased the uptake of K and Ca and increased contents of Sc from 0.03 to 0.18 mg kg^{-1} in control and experimental seedlings, respectively.

Contents of Sc in lichens and bryophytes range from 0.3 to 0.7 mg kg^{-1}, and in fungi from <0.02 to 0.3 mg kg^{-1} (Bowen 1979). Moss samples collected in Norway in the period 1990–1995 contained Sc in the range of 0.02–3.1 mg kg^{-1} (Berg and Steinnes 1997).

II-3.1.7 Humans

Scandium occurs in tissues of terrestrial and aquatic organisms at very low concentrations from <0.1 to 3 000 μg kg^{-1} and is accumulated more in aquatic animals than in terrestrial ones. Bones of man and animals exhibit a tendency to concentrate Sc. Jørgensen (2000) reported Sc to occur in mammalian tissues from 0.06 μg kg^{-1} in heart to 8.0 μg kg^{-1} in muscles.

Scandium is dangerous to human health, but mainly in the working environment. Thus, it must be treated and handled with extreme care. The excess of Sc, both inhaled and ingested, can cause lung embolisms, and damage to the liver and cell membranes.

II-3.2 Yttrium (Y)

II-3.2.1 Introduction

The occurrence of Y in the Earth's crust is relatively common and fairly uniform in different rocks, being somewhat less in mafic rocks (0.5–22 mg kg^{-1}) than in acid rocks (28–44 mg kg^{-1}) and sandstones (5–50 mg kg^{-1}).

Geochemical behavior of Y is very similar to the alkaline earths and other rare earth elements and, particularly to the lanthanides. Thus, its higher contents are associated with minerals that concentrate rare earth elements (REEs). This element exhibits lithophilic affinity, has the oxidation state +3 (Table II-3.1) and is incorporated, together with REEs, in several minerals, such as: oxides, carbonates, silicates and phosphates (e.g., xenotime, YPO_4; fergusonite, $YNbO_4$). Minerals of Y are associated mainly with monazite veins and pegmatites.

Contents of Y in coal range from 7 to 12 mg kg^{-1} and does not indicate its sorption by organic substances, although in some coal samples it may reach up to 170 mg kg^{-1} (Finkelman 1999). In fly and bottom ashes it is concentrated to about 44 mg kg^{-1} (Llorens et al. 2001).

II-3.2.2 Production and Use

Global production of Y (as Y_2O_3) was calculated in 1995 at 1.15 kt (Reimann and Caritat 1998) and in 2001 at 2.4 kt (Dobrowolski 2004). It has a relatively broad industrial application in ceramic and glass industries, high-temperature bricks, laser, color monitors, and various catalysts (e.g., in the production of plastics). Some Y compounds have luminescence properties and are used for various lasers.

II-3.2.3 Soils

The most comprehensive data are for Y in soils of the USA, and give the range of average contents from 16 to 33 mg kg^{-1}. The highest contents (up 150 mg kg^{-1}) have also been cited for various soil units (Shacklette and Boerngen 1984; Gough et al. 1988). Data of other authors, however, gave smaller amounts of Y in soils of various countries, within the range of <2–70 mg kg^{-1} (Kabata-Pendias and Pendias 2001). The average contents of Y in reference soils range from 22 to 24 mg kg^{-1}, for China and the USA, respectively (Govindaraju 1994). In Japanese agricultural soils median contents of Y varied from 8.8 to 24 mg kg^{-1}, in Acrisols and Andosols, respectively (Takeda et al. 2004). Soils of San Joaquin Valley of central California contain Y in the range between 17 and 27 mg kg^{-1}, being the highest in contaminated sites (Wilson et al. 1994). Swedish arable soils contain Y in the range of 9–41 mg kg^{-1}, at an average value of 27 mg kg^{-1} (Eriksson 2001a). Sandy soils of Poland contain Y in the range from <2 to 17 mg kg^{-1} (mean 4.8) and loamy soils from 7 to 19 mg kg^{-1} (mean 12.4) (Dudka 1992).

II-3.2.4 Waters

The mean concentration of Y in worldwide ocean waters is estimated at 0.013 µg l^{-1} (Reimann and Caritat 1998), and at 0.017 µg kg^{-1} in waters of the North Pacific (Nozaki 2005).

Gaillardet et al. (2003) reported data for Y in river waters only for Europe (range 0.05–1.40 µg l^{-1}) and for North America (range 0.028–0.217 µg l^{-1}), and have estimated world average at 0.04 µg l^{-1}. The concentrations of Y in European stream waters vary between 14 and 98 µg l^{-1}, being the highest for Romania (Reimann and Caritat 1998). These authors cited the maximum Y concentrations in snow from the Kola Peninsula to range from 0.01 to 0.7 μg l^{-1}, for remote and polluted regions, respectively. Rain waters collected in Sweden during 1999 contained Y from 0.003 to 0.033 μg l^{-1}, at an average value of 0.01 μg l^{-1} (Eriksson 2001a).

Bottom sediments of streams are reported to contain relatively high amounts of Y, with the maximum values of 731, 485, and 2005 mg kg^{-1} for Scotland, Germany and Austria, respectively (Reimann and Caritat 1998). However, the top layer of bottom sediments of the Swedish river of the Baltic catchment contains Y at the range of 13–23 mg kg^{-1} (Szefer 2002a).

II-3.2.5 Air

Atmospheric dust contains Y in the range of 0.2–2 mg kg^{-1}. The maximum content of 6.0 ng m^{-3} is reported for air in polluted areas (Reimann and Caritat 1998).

Wet deposition of Y in Sweden has been estimated at 1 g ha^{-1} yr^{-1} (Eriksson 2001a).

II-3.2.6 Plants

As Connor and Shacklette (1975) reported, Y was found at detectable levels in about 10% of the plants studied. The reference content of Y in plants is calculated at 0.02 mg kg^{-1} (Markert 1992). However, results of real analysis gave quite variable amounts for Y in plants, in the range of 0.01 to 3.5 mg kg^{-1}, depending on soil and climatic factors (Kabata-Pendias and Pendias 2001).

Relatively high contents of Y are reported for mosses and lichens. The data for Y in these plants, collected in 1970 in Scandinavia, indicate its contents range from 0.2 to 2.0 mg kg^{-1} (Erämetsä and Ylіroukanen 1971). Mosses collected during the period 1990–1995 contained fairly similar amounts of this element, in the range of 0.04–2.3 mg kg^{-1} (Berg and Steinnes 1997). Much smaller mean values are given by Eriksson (2001a) for Swedish wheat and barley grains: 0.79 and 1.3 μg kg^{-1}, respectively. Increased levels of Y in mosses and lichens, as compared with higher plants, clearly indicate that main source of this metal is atmospheric deposition.

II-3.2.7 Humans

Yttrium is reported to occur in all human and animal tissues at the average level of about 20 μg kg^{-1} and is likely to be concentrated in bones, up to 70 μg kg^{-1} (Li 2000). This author reported Y contents to range from 4 μg kg^{-1} in brain to 20 μg kg^{-1} in lung. Human liquids, blood and lymph, contain Y at averages values of 5 and 60 μg kg^{-1}, respectively.

The inhalation of aerial dust from combustion of Y-enriched materials may be a risk for human health. No adverse effects of Y have been reported on the health of mammals. However, an increased level of Y may reduce the activity of some enzymes and hormones, possibly due to its great affinity to phosphorus-containing ligands and some metal-binding proteins (Dobrowolski 2004).

II-3.3 Lanthanides

II-3.3.1 Introduction

Lanthanides, also called rare earth elements (REEs) comprise of a group of 15 elements (Table II-3.3), of which only one, promethium (Pm) does not occur naturally in the Earth's crust, while the other fourteen are relatively abundant in rocks and soils (Tables II-3.4 and II-3.5). The terrestrial distribution of the REEs shows a general peculiarity; their contents decrease with an increase in atomic weights, and, according to the Oddon-Harkins rule, the element with an even atomic number is more abundant than the next element with an odd atomic number. This rule also governs the distribution of all elements in the universe, but is much less pronounced than in the case of lanthanides. Although several factors influence the abundance of lanthanides in soils, this rule can also be observed in the average contents of lanthanides (Fig. II-3.1).

Geochemical properties of lanthanides are fairy similar; they occur mainly as +3 cations (Table II-3.3), exhibit an affinity for oxygen, and are likely to be concentrated in acid igneous rocks and in argillaceous sediments (Table II-3.4). Lanthanides are constituents of several different minerals and are also likely to be concentrated in phos-

Table II-3.3. Selected properties of the lanthanide elements (natural)

Element	Atomic number	Atomic mass	Ionic radius[a] (pm)	Density ($g\ cm^{-3}$)	Valence[b]	Melting point (°C)
La, lanthanum	57	138.9	117	6.15	+3	918
Ce, cerium	58	140.2	115	6.77	**+3**, +4	798
Pr, praseodymium	59	140.9	113	6.77	**+3**, +4	931
Nd, neodymium	60	144.2	143	7.00	+2, +3, +5	1021
Sm, samarium	62	150,4	135	7.52	**+3**, +2	1074
Eu, europium	63	151.9	131	5.24	**+3**, +2	822
Gd, gadolinium	64	157.3	108	5.90	**+3**, +2, +1	1313
Tb, terbium	65	158.9	106	8.23	**+3**, +4	1356
Dy, dysprosium	66	162.5	121	8.55	+2, +3, +4	1412
Ho, holmium	67	164.9	104	8.79	+3	1474
Er, erbium	68	167.3	103	9.06	+3	1529
Tm, thulium	69	168.9	117	9.32	**+3**, +2	1545
Yb, ytterbium	70	173.0	116	6.97	**+3**, +2	819
Lu, lutetium	71	174.9	100	9.84	+3	1663

[a] Given is one, the lowest value.
[b] Valence values in bold are for the main oxidation states.

Table II-3.4. Abundance of REEs in rocks (mg kg^{-1}) (compiled from Heiserman 1997, Reimann and Caritat 1998, and Kabata-Pendias and Pendias 2001)

Element	Earth's crust	Igneous rocks		Sedimentary rocks		
		Mafic	Acid	Argillaceous	Sandstone	Calcareous
La	16 – 39	2 – 70	30 – 150	30 – 90	17 – 40	4 – 10
Ce	33 – 67	4 – 60	80 – 250	3 – 90	25 – 80	7 – 20
Pr	3.9 – 9.2	1 – 15	6 – 30	6 – 10	4 – 9	1 – 2.5
Nd	16 – 42	2 – 30	18 – 80	18 – 35	16 – 48	5 – 9
Sm	3.5 – 7.1	0.1 – 1.7	6 – 11	5 – 7	4 – 10	1 – 2
Eu	1.1 – 2.0	0.01 – 4	1 – 2	1 – 2	0.7 – 2	0.2 – 0.4
Gd	3.3 – 6.2	0.1 – 8	4 – 10	5 – 7.5	3 – 10	1.3 – 2.7
Tb	0.6 – 1.2	0.1 – 1.2	1 – 2.5	0.9 – 1.1	1.6 – 2	0.2 – 0.4
Dy	3.7 – 5.2	0.05 – 7	5 – 8	4 – 6	2.6 – 7.2	0.8 – 2
Ho	0.8 – 1.3	0.1 – 1.5	1.3 – 2	1 – 1.8	0.05 – 2	0.2 – 0.3
Er	2.1 – 3.5	0.1 – 1	3.4 – 0.7	2.5 – 4	1.5 – 6	0.4 – 1.7
Tm	0.3 – 0.5	0.1 – 0.6	0.3 – 0.7	0.2 – 0.6	0.3	0.03 – 0.2
Yb	2.0 – 3.2	0.1 – 3.5	3 – 4.5	2.2 – 4	1.2 – 4.4	0.3 – 1.6
Lu	0.3 – 0.8	0.1 – 0.6	0.5 – 1.2	0.2 – 0.7	0.8 – 1.2	0.03 – 0.2

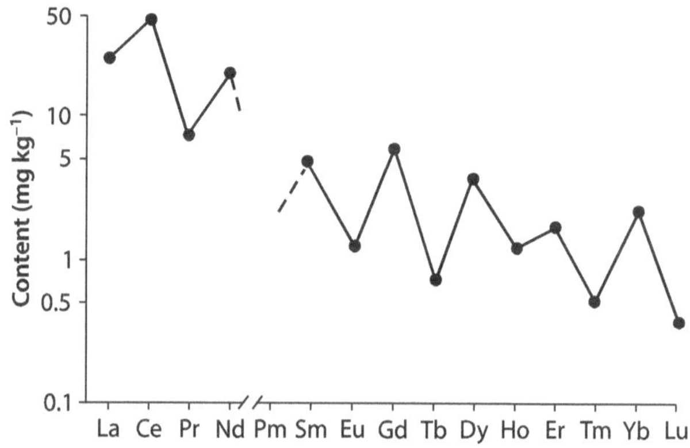

Fig. II-3.1. Average contents of lanthanides in surface soils (promethium, Pm, does not occur in the natural environment)

phorites being incorporated in relatively common minerals such as: monazite, $(CeLa)PO_4$; bastnaside, $(CeF)CO_3$; cheralite, (Ce, La, Y, Th) PO_4; and xenotime, YPO_4. The content of Ce is relatively high in phosphate fertilizers, at around 20 mg kg^{-1}. Christensen et al. (2004) studied the REEs in 70 natural calcite samples and found the variation in the distribution of light and heavy REEs depending upon the origin of calcite.

Organic compounds play a significant role in the distribution of lanthanides and affect their increased levels in humus horizons of soils and in forest-soil litter.

Table II-3.5. Average contents of REEs in soils (mg kg^{-1}) and in soil leachates (ng l^{-1})

Element	A	B	C	D	E	F
La	26.1	35.1	35.2	8.4 – 31.4	33	0.6
Ce	48.7	75.1	97.4	15.8 – 64.4	60	1.3
Pr	7.6	5.4	8.4	1.5 – 5.0	7.7	0.15
Nd	19	3.6	29.3	7.6 – 28.6	29	0.65
Sm	3.1	6.7	5.5	1.8 – 4.0	4.5	0.1
Eu	4.8	1.2	0.8	0.44 – 1.43	0.8	bdl[a]
Gd	1.23	3.2	4.8	1.77 – 4.54	3.4	0.1
Tb	6.03	0.5	0.6	0.27 – 0.83	0.5	bdl
Dy	0.71	4.3	2.9	1.68 – 4.68	4.1	0.15
Ho	1.08	1.0	0.5	0.36 – 0.95	0.9	bdl
Er	1.58	2.6	1.4	1.1 – 2.72	2.2	0.05
Tm	0.46	0.3	0.2	0.16 – 0.40	0.3	bdl
Yb	2.06	2.4	1.1	1.11 – 2.64	2.9	bdl
Lu	0.34	0.4	0.2	0.16 – 0.40	0.4	bdl

[a] *bdl* – Below detection limit.
References:
A – Data compiled by Kabata-Pendias and Pendias (2001) for various soils, mainly from Europe.
B – Data for reference soils, after Govindaraju (1994).
C – Data for 27 soils of China, after Zhu et al. (1997).
D – Range of means for 77 soils of Japan, after Yoshida et al. (1998).
E – Mean value for topsoil samples from Sweden (Eriksson 2001).
F – Dissolved (ng l^{-1}) in HCl (0.001 N) from surface soil (5 cm) of the Vosges Mts., France, in ng l^{-1}, after Aubert et al. (2002).

II-3.3.2
Production and Use

Annual production of lanthanides plus Sc and Y was estimated to be 52.9 kt in 1990; and 83.3 kt in 2000 (Di Francesco and Hedrick *vide* Goering 2004). The total world production of REEs (in oxide equivalent) was in 2003 estimated as 75.15 Mt (WMSY 2004).

Two major Ce sources for the commercial production are monazite, and bastnaside. Monazite is especially resistant to weathering of granitic rocks and is concentrated, by sedimentary processes, often in beach sands.

The use of lanthanides is relatively broad in various industries from glass production (mainly as colorants) to sophisticated electronic devices, as well as to catalytic converters, metallic alloys, rechargeable batteries, radars, etc.

Lant

II-3.3.3 Soils

There are scanty data on individual REEs in soils, however, the trend of their distribution in various soils is apparently related to the geological origin of parent rocks and their mineral composition.

The distribution of lanthanides in different soils is quite similar when comparing mean values after various authors. In most cases their lowest values are for sandy soils, and the highest for loamy soils (Table II-3.5). Acid soils contain usually less the REEs than alkaline soils, most probably due to the easy removal of soluble hydroxide complexes. There are data indicating a strong positive relation between several lanthanides (La, Ce, Pr, Nd, Sm) and fine granulometric fractions (<0.002 and <0.02 mm) in soils (Chojnicki 2000). Especially HREs are likely to concentrate in the fine granulometric soil fraction (<0.002 mm). In general, REE due to the affinity to SOM, are about ten times more abundant in peats than in surrounding mineral soils (Markert 1987).

Two subgroups of lanthanides can be distinguished: light metals (LREs), from La to Gd – more basic and more mobile than metals of the second group (HREs), from Tb to Lu (Table II-3.3). During weathering processes, lanthanides become fractionated, and metals of the LREs subgroup are more likely to be concentrated in weathered

Table II-3.6. Average contents of REEs in sewage sludges and farmyard manure (mg kg^{-1})

Element	Sewage sludge		Stable manure	
	USA[a]	Sweden[b]	USA[a]	Sweden[c]
Lu	2.7	16	24	110
Ce	41.9	24	55	180
Pr	4.3	2.8	11	22
Nd	2.5	11	2.5	78
Sm	3.5	1.8	5.2	14
Eu	3.7	0.3	0.7	3
Gd	6.8	2.0	1.5	13
Tb	1.4	0.3	0.3	2
Dy	0.7	1.7	1.0	10
Ho	0.3	0.4	0.4	2.1
Er	1.2	1.0	0.7	5.9
Tm	0.4	0.2	0.1	0.8
Yb	0.6	1.1	1.8	5.4
Lu	0.12	0.2	0.6	0.8

[a] After Furr et al. (*vide* Kabata-Pendias and Pendias 2001).
[b] Unweighted mean value, after Eriksson (2001a).
[c] Element's contents given in mg kg^{-1} of P contents in manure, after Eriksson (2001a).

material than the HREs. They also occur in higher amounts in soils than the metals of the HREs group (Table II-3.5). Zhu et al. (1997) reported that LREs range in soils of China from 76.2 to 447.7 mg kg^{-1}, whereas HREs range from 9.2 to 55.2 mg kg^{-1}. Some authors have also distinguished a medium weight group: MHs containing elements from Sm to Ho (Xu et al. 2002). Agricultural soils of Japan have the following median contents of HREs (in mg kg^{-1}): Tb, 0.73; Dy, 3.8; Ho, 0.73; Er, 2.1; Tm, 0.31; Yb, 2.0; and Lu, 0.31 (Takeda et al. 2004). Light sandy soils of Poland contain smaller amounts of some lanthanides than silty and loamy soils; geometric means are as follows (in mg kg^{-1}): La, 9.0–22.0; Ce, 16.9–42.8; Nd, 7.3–19.2; and Yb, 1–1.2 (Dudka 1992).

Lanthanides are commonly found in sewage sludges and manure at quite high amounts, fairly similar to their abundance in soils (Table II-3.6). Thus, these amendment materials as well as phosphate fertilizers might be considered as their source in cultivated soils.

All lanthanides are slightly mobile in soils. Their extractability with chelating agents (e.g., EDTA) is low and does not exceed 10% of the total contents (Zhu et al. 1997; Wyttenbach et al. 1998). The global mobility of REEs, calculated as the ratio of water-dissolved concentrations to those of the upper continental crust is relatively low; at about a 0.01 ratio value (Gaillardet et al. 2003). The solubility of soil REEs is very small; hundreds of times lower than their total contents. Nevertheless, their concentrations in soil solution are relatively high and exceed, by a hundred times those found in rainwater and throughfall (Table II-3.7).

Table II-3.7. Concentrations of REEs in different aquatic samples (ng l^{-1})

Element	Rain water[a]	Rain water[b]	Throughfall[b]	Soil solution[b,c]	Stream water[b]
La	21	1.68	20.1	188.4	26.6
Ce	36	1.38	29.8	114.5	73.1
Pr	4.4	0.17	3.90	12.8	14.3
Nd	15	0.76	16.9	58.0	84.9
Sm	3.3	0.63	4.00	17.6	38.7
Eu	0.6	0.19	0.96	3.9	7.0
Gd	2.6	0.86	4.15	19.8	50.0
Tb	0.4	0.22	0.69	3.6	8.4
Dy	1.8	1.48	3.62	19.0	46.8
Ho	0.4	0.30	0.74	3.2	8.1
ER	0.9	0.83	2.07	8.7	22.5
Tb	0.2	0.17	0.34	1.7	3.4
Yb	0.7	0.96	2.80	10.9	25.5
Lu	0.13	0.15	0.51	1.3	3.8

[a] Rainwater from Sweden collected during 1999 (Eriksson 2001a).
[b] Data for the Vosges Mts. region, adapted from Aubert et al. (2002).
[c] In HCl (0.001 N) soil leachates from surface soil (5 cm) of the Vosges Mts., France.

II-3.3.4 Waters

Concentrations of lanthanides in river waters vary highly, however, a general rule of their distribution can also be observed (Table II-3.8). The highest average concentration is for Ce, 0.263 μg l^{-1} and the lowest for Lu, 0.002 μg l^{-1}. Fairbridge (1972) gave 10.2 μg l^{-1} for the total REE concentration in Pacific Ocean water, with the highest value for La (2.9 μg l^{-1}) and the lowest for Lu (0.12 μg l^{-1}). As Gaillardet et al. (2003) described, pH and DOC of waters control the REE concentrations, as well as some associated elements, such as Th and Y.

Rain waters collected in the Vosges Mts., France, contain smaller amounts of some REEs than rainwater sampled in Sweden (Table II-3.7). Somewhat higher concentrations of REEs were found in snow than in rain waters (Aubert et al. 2002).

II-3.3.5 Air

There are very limited data on lanthanide abundance in air. Concentrations of some REEs in air from the South Pole and/or Greenland, as compiled by Kabata-Pendias and Pendias (2001), are as follows (in pg m^{-3}): Ce, 0.8–4.9; Eu, 0.004–0.02; La, 0.2–110; Sm, 0.03–12; and Tb, 1–5. Concentrations of these elements in air from urban or industrial regions are increased by factors of 10 to 10 000, which clearly indicate their anthropogenic enrichment.

Table II-3.8.
Range and world average concentrations of lanthanides in the dissolved load (<0.2 μm) of rivers (ng l^{-1}) (after Gaillardet et al. 2003) and in the North Pacific Ocean (ng kg^{-1}) (after Nozaki 2005)

Element	River, range	River, mean	Pacific, mean
La	1 – 538	120	5.6
Ce	7 – 1 177	263	0.7
Pr	1 – 179	40	0.7
Nd	5 – 5 347	152	3.3
Sm	1 – 184	36	0.6
Eu	0.3 – 29	10	0.2
Gd	0.7 – 134	40	0.9
Tb	0.3 – 29	6	0.2
Dy	1 – 86	30	1.1
Ho	0.1 – 8	7	0.4
Er	1 – 71	20	1.2
Tm	0.1 – 9	3	0.2
Yb	0.7 – 59	17	1.2
Lu	0.3 – 11	2	0.23

II-3.3.6
Plants

Although the distributions of lanthanides in plants have not been studied very intensively until the recent decade, Robinson et al. (1958), about 50 years ago, reported high concentrations (up to 2 300 mg kg^{-1}) of total REEs in hickory trees and pointed out a relationship between their content in trees and their exchangeable pool in the soil. Also Laul et al. (1979) have emphasized a close relationship between the REE content in plants and their occurrence in soils. Contents of REEs in plants vary widely, from <1 to >10 000 μg kg^{-1}, and their concentration orders, similarly as in terrestrial materials, decrease with increase of the atomic number, and follow somehow the Oddon-Harkins rule (Table II-3.9). The same rule can be observed in the distribution of lanthanides in edible mushrooms (Table II-3.10). Wheat grains from China contain higher amounts of all REEs than grains from Sweden (Table II-3.9). Soil dressing with these elements had variable effect on their content in wheat grains (Liang et al. 2005).

Table II-3.9. Lanthanide elements (average and range) in terrestrial plants (μg kg^{-1})

Element	Moss[a]	Vegetables[b]	Cheat-grass[c]	Wheat shoots[d]	Wheat grains[e]	Wheat grains[f]	Pine needles[g]
La	550	0.4 – 2 000	100	155	1.7	16.8	26 – 300
Ce	1 100	2 – 50	300	265	3.4	26.9	370
Pr	130	1 – 2	40	30	0.4	1.9	62 – 130
Nd	490	10	150	121	0.9	6.8	150 – 160
Sm	100	0.2 – 100	40	20	0.2	1.1	30 – 32
Eu	20	0.04 – 70	8	<10[h]	<0.1	0.3	4.9 – 5.3
Gd	95	<2	40	21	0.2	0.2	23 – 25
Tb	13	0.1 – 1	9	<8[h]	<0.1	0.1	12 – 22
Dy	70	–	-	15	0.1	0.7	20 – 22
Ho	13	0.06 – 0.1	<20	<8[h]	<0.1	0.1	3.9 – 5.1
Er	37	0.5 – 2	<500	8	<0.1	1.8	6 – 7
Tm	5	0.2 – 4	50	<2[h]	<0.1	1.8	1.1 – 1.7
Yb	32	0.08 – 20	20	8	<0.1	1.5	8.2 – 8.5
Lu	4	0.01 – 60	3	<2[h]	<0.1	1.5	1.9 – 2.0

[a] After Berg and Steinnes (1997).
[b] After Kabata-Pendias and Pendias (2001).
[c] After Laul et al. (1979).
[d] After Li et al. (2001).
[e] Grains from Sweden, after Eriksson (2001).
[f] Grains from China, after Liang et al. (2005).
[g] After Markert (1987).
[h] One determination only.

Lant

Table II-3.10.
Lanthanide elements (mean and range of means) in wild edible mushrooms sampled during 1997–1998 from northern Poland ($\mu g\ kg^{-1}$) (after Falandysz et al. 2002)

Element	King Boletus (*Boletus edulis*)	Other edible mushrooms
La	27	5.3 – 70
Ce	56	9.3 – 160
Pr	6.5	1.7 – 58
Nd	23	6.7 – 67
Sm	7.3	12 – 14
Eu	1.9	0.7 – 2.2
Gd	8.1	1.4 – 9.5
Tb	1.3	0.8 – 1.6
Dy	3.7	1.1 – 9.9
Ho	1.5	0.7 – 1.6
Er	2.9	0.7 – 4.5
Tm	0.9	0.7 – 0.9
Yb	2.2	0.7 – 3.4
Lu	–	0.7 – 1.2

Wyttenbach et al. (1998) observed variable absorption of REEs by plants from a forest ecosystem and have related this phenomena to organic complexes or changes in the oxidation state of individual REEs. Plant sampled in 1999 along highways in Germany contained (in $\mu g\ kg^{-1}$): La, 53–380; Ce, 110–820; and Nd 48–330. Always the highest concentrations were in dandelion (Djingova et al. 2003).

The sum of REE contents of food plants (edible parts) is reported by Liu (1988) to be in ($mg\ kg^{-1}$): tomato, 0.05; corn, 0.2; rice, 0.6; wheat, 1.0; and sesame, 3.6. Cabbages from the background area of Denmark contain, at average, La, Ce, Pr, Nd and Gd within the range of 0.15–2.6 $\mu g\ kg^{-1}$ FW, while concentrations of Sm, Tb, Dy, Ho, Er, Tm, Yb, and Lu in sprouts ranged from 0.005 to 0.06 $\mu g\ kg^{-1}$ FW (Bibak et al. *vide* Goering 2004).

Some authors (Liu 1988; Xijie et al. 1986) have reported good effects of the addition of REEs on plant growth and yield. These authors, however, have not specified the compounds of RREs used for the experiments, and have not discussed a possible function of these elements in plant metabolism. There is a report, that REEs are applied in Chinese agriculture to improve crop nutrition (Xu et al. 2002). The authors reported that the common dose of REEs at <0.23 $kg\ ha^{-1}\ yr^{-1}$ does not affect the quality of maize grains. The uptake of REEs by maize plants by both tops and roots increased with increasing doses, and was higher for the LREs than for the HREs. Xu et al. (2002) have not discussed any processes that might be related to the effects of REEs on the yield of crop plants. Li et al. (2001) reported that the availability of REEs to crop plants correlates well with their 0.1 M HCl soluble pool. Although there are some other reports on the stimulating effect of REEs on several processes in plants such as, for example, seed germination, root growth, nodulation, chlorophyll production, these elements have not yet been proved to be essential to plants. Shtangeeva and Ayrault (2006) reported that Eu added to a nutrient solution at the concentration of 0.01 mg l^{-1} disturbed the mineral balance in the plant tissues. The synergetic effect of Eu + Ca on the yield of wheat seedlings, observed by these authors, needs to be understood.

There is no evidence of toxic effects of REEs to plants, however, they have impairing effects on cell membranes of vascular plants and on the Ca metabolism in microorganisms.

II-3.3.7
Humans

Lanthanides occurring in oxide forms are apparently slightly available to both humans and animals. Other salts of REEs, however, might be easier absorbed by humans. Their danger is mainly from aerial dust inhalation by humans.

Levels of lanthanides, as compiled by Goering (2004), in human tissues are as follows (μg kg^{-1} AW): liver, 5; kidneys, 2; lung, 4; testes, 2; and bones, 200–1 000. Average concentrations of individual lanthanides in lungs of adults from urban environments varied from 4.2 to 30.5 μg kg^{-1} FW, being the lowest for Tb and Yb, and the highest for Ce. Thus, the distribution of lanthanides in human tissues also follows the general rule of lanthanide distribution in the terrestrial materials.

Chloride salts of lanthanides (Ce, Pr, Eu, Dy, Lu, and Y) administered by intravenous injection to rats were distributed after one day with over 78% of the dose to liver, bones, and spleen, and retained in organs for 45 days (Nakamura et al. *vide* Goering 2004). Lanthanides are likely to concentrate in lung, liver and kidney in individuals exposed to increased levels of these elements, mainly in smelter surroundings. Chronic exposure to the inhalation of dust increased in lanthanides has been reported to induce pulmonary fibrosis. However, due to the low toxicity of lanthanides, they are generally not considered to be hazardous to health.

The most important lanthanide feature is their affinity for Ca^{2+} sites. Thus, binding of lanthanide ions may alter a structure of some molecules, and change their biological functions. There are numerous reports on antitumor influence of some lanthanides, however, there is still a great paucity of defined experimental results. It is likely that lanthanides might affect biochemical pathways, and thus may alter physiological processes of humans and animals (Pałasz and Czekaj 2000).

II-3.4
Actinides

II-3.4.1
Introduction

Among the actinide series that obtains the elements from atomic number 89 (Ac) to 103 (Lr), only four elements: Ac, Th, Pa, and U occur naturally in the terrestrial environment (Table II-3.11), and only two elements exist in relatively large quantities as long-lived nuclides; two isotopes of U (^{238}U and ^{235}U) and one of Th (^{232}Th). Other actinide elements also have been found to occur as results of natural nuclear reactions of U, but in very negligible traces. These are mainly Ac, Pa, of which some isotopes have a relatively long half-life.

Table II-3.11. Selected properties of naturally occurring actinide elements

Element	Atomic number	Atomic mass	Ionic radius (pm)	Density ($g\ cm^{-3}$)	Valence[a]	Melting point (°C)
Ac, actinum	89	227	126	10.07	+3	1 050
Th, thorium	90	232	108 – 134[b]	11.72	+2, +3, **+4**	1 750
Pa, protactinum	91	231	92 – 109[b]	15.37	+3, +4, **+5**	1 600
U, uranium	92	238	59 – 100[b]	11.95	+2[c], **+6**	1 132
Pu, plutonium[d]	94	242[e]	–	19.84	+3, **+4**, +5	3 232

[a] Valence values in bold are for the main oxidation states.
[b] Value for the highest oxidation states.
[c] The valence values can vary from +2 to +6.
[d] Small amounts are generated in the natural reactors (Zhang et al. 2002).
[e] For the most stable isotope of Pu.

Table II-3.12. Thorium and uranium in rocks and soils[a] ($mg\ kg^{-1}$)

Rocks/soils	Thorium	Uranium	^{238}U ($pCi\ g^{-1}$)[b]
Earth's crust	3.6 – 9.6	0.9 – 1.7	0.2 – 0.3
Igneous rocks			
Mafic	1 – 14.0	0.3 – 3.0	0.2; 0.3[c]
Salic	10 – 23	2.5 – 6	1.3; 1.6[c]
Sedimentary rocks			
Argillaceous	10 – 12	3 – 4	1.0
Sandstones	2 – 4	0.5 – 0.6	<0.3 – 1.0
Limestones	2 – 3	2.2 – 2.5	0.7
Soils			
in Europe	3.4 – 10.5	0.8 – 11.0	–
in the USA	6.1 – 7.6	2.3 – 3.7	0.6

[a] Range of average values, as compiled by Kabata-Pendias and Pendias (2001), unless otherwise indicated.
[b] Adopted from ATSDR (2002a).
[c] The median and mean value, respectively.

Distribution and behavior of Th and U has received most attention, as the most abundant actinide elements in the terrestrial environments. Salic igneous rocks and argillaceous sediments contain usually more of these elements than do acid rocks and sandstones or limestones (Table II-3.12).

Geochemically, U and Th are coherent. They exist in the +4 and +6 oxidations states in most geological environments (Table II-3.11). They may substitute for REEs in some minerals and are associated with apatite and sphene. The common carriers of Th are monazite and cheralite – Ce minerals. Thorium has also own minerals, e.g., thorite, $ThSiO_4$ and thioriate, ThO_2, and can be associated with several other minerals. A major commercial ore of U is uranitite and pitchblende, of similar composition: $UO_2 + UO_3$.

The mobility of Th and U during weathering processes depends highly on the host minerals. When these elements become mobilized, they are very likely to be quickly precipitated as hydroxides or absorbed and form organic complexes and various stable compounds such as oxides, carbonates, phosphates, vanates, and arsenates. Actinides easily form strong complexes with oxygen ligands, and thus their species with inorganic ligands such as OH^-, CO_3^{2-}, and HPO_4^{2-} are likely to occur in various environmental conditions. Thus, the distribution of Th and U in the lithosphere is highly controlled by the Eh-pH system.

Actinides, and especially U, exhibit an affinity to be absorbed by organic matter and accumulated by microorganisms. Letunova and Kovalskiy (1978) calculated the factor of bioaccumulation of U by microorganisms at 300 times as compared with its content in soils. Due to easy U sorption by organic matter, its significant accumulation in coal and peat is often reported. Nakashima (1992) described that naturally occurring humic substances such as coals may play important role in the U concentrations at various stages of sedimentary cycles. The average contents of Th and U in coal is estimated as 5.6 and 6.1 mg kg^{-1}, respectively, while in fly ash their contents increased up to above 22 mg kg^{-1} (Llorens et al. 2001).

Among the actinide series, beside Th and U, only Pa occurs as natural unstable isotope. Other transuranic elements are artificial isotopes resulted from Th and U decay reactions.

Americium is a man-made radioactive element that contains four principal isotopes (241, 242m, 242, 243) of very long half-lives from 432 to 7 370 years, and has no stable isotopes. The only commercially important is ^{241}Am and may be found at very low levels, in all environmental media, as a result of fallout from past atmospheric nuclear weapons tests. Since 1973, concentrations of this radionuclide are continuing to decline, especially in air.

II-3.4.2 Production and Use

Information on the production of Th and U is variable, but recent data gave the annual world production of Th to be 700 t in 1984, (Reimann and Caritat 1998) and the production of U to be 35.37 kt in 2003 (WMSY 2004).

The use of actinide isotopes has been quite recent and restricted to scientific experiments and nuclear-power industries. Thorium is used as a coating material for optical lenses and some metallic wires, and as an addition to metallic alloys. Uranium is applied mostly in the nuclear industry and also for some very sophisticated devices. Depleted U (DU) is used in the manufacture of armor-piercing ammunition for the military and its deployment has been undertaken recently in wars (e.g., in Kosovo and Iraq). Compounds of U are used in photography for toning, for special lamps, and in the leather, ceramic and wood industries, for stains and dyes.

Plutonium, ^{239}Pu, has recently received much attention recently because of its use in nuclear power reactions. It has been calculated that since 1980, the world's nuclear power reactors have produced more than 20 t Pu. In addition, Pu isotopes (mainly 239 and 240) are produced for military programs (ATSDR 2002a). This radionuclide is also use in thermo-electric power devices.

Act

II-3.4.3 Soils

Average contents of Th and U reported for different worldwide soils are fairly similar and vary for Th in the range of 3.4–10.5 mg kg^{-1} and for U in the range of 1.2–11 mg kg^{-1} (Table II-3.13). Harmsen and De Haan (*vide* ATSDR 2002a) gave the typical concentration range of naturally-occurring Th in soils as 2–12 mg kg^{-1}, and an average value of 6 mg kg^{-1}. Fairly similar Th contents, in the range of 11–14 mg kg^{-1}, are reported for soils of central California. (Wilson et al. 1994). Data for the reference soils of China gave a bit higher Th contents ranging between 8 and 27 mg kg^{-1} (Govindaraju 1994). In Japanese agricultural soils median concentrations of Th vary from 5.8 to 12 mg kg^{-1}, in Andosols and Acrisols, respectively (Takeda et al. 2004). Swedish arable soils contain Th in the range from 1.6 to 16 mg kg^{-1}, at an average value of 6.1 mg kg^{-1} (Eriksson 2001a).

Soils of the lowland of Poland contain average amounts of Th at the value of 3.23 mg kg^{-1}, and U at the mean value of 1.6 mg kg^{-1}. In the mountain soils (Sudeten), amounts of U increased up to >10 mg kg^{-1} (Strzelecki et al. 1994). Surficial sediments of Finland contain, most commonly, Th in the range of 5–20 mg kg^{-1} and U in the range of 1–5.5 mg kg^{-1} (Koljonen 1992). Yoshida and Muramatsu (1997) reported the lowest contents of both elements for sand dunes and peat soils of Japan (Th 0.3 and U 0.6 mg kg^{-1}) and the highest for red and gray lowland soils (Th 1.4 and U 2.9 mg kg^{-1}). In Japanese agricultural soils median concentrations of U vary from 1.4 to 2.4 mg kg^{-1}, in Gleysols and Acrisols, respectively (Takeda et al. 2004). Swedish arable soils contain U in the range from 1.3 to 16 mg kg^{-1}, at an average value of 4.4 mg kg^{-1} (Eriksson 2001a).

The extensive studies curried out in the USA show relatively small variation in the concentrations of U and Th among soil types, with the exception of the radionuclides ^{238}U and ^{232}Th which are likely to be concentrated in the soils close to aquatic systems (Meriwether et al. 1988). Variable parameters (pH and Eh) of ground water may control the adsorptive reactivity of U species. The hydroxides of Th^{4+} are the dominant species in soil and in aquatic systems, although carbonate complexes are also formed (Zhang et al. 2002).

Table II-3.13. Uranium and thorium in leaves of trees grown on background and contaminated sites of wetland in South Carolina, USA (adapted from Hinton et al. 2005)

Sample	Uranium-238 (mg kg^{-1})		Thorium-232 (mg kg^{-1})	
	BS	CS	BS	CS
Soil	6.7	284	1.3	166
Red Maple, *Acer rubrum* L.	bdl	0.5	0.03	2.6
Water Oak, *Quercus nigra* L.	0.0013	1.6	bdl	0.4
Sweetgum, *Liquidambar styraciflua* L.	0.04	8.2	0.04	5.3

BS – Backgroun site; *CS* – Contaminated site; *bdl* – Below detection limit.

The fate of U in the environment has recently been widely investigated. Although in the soil vadose zone the immobile U^{6+} is predominated species, its reduction to soluble U^{4+} can frequently occur under biological processes and due to changing redox potential and pH (Scheinost et al. 2004). According to Takeda et al. (2005), the water soluble fraction of U in surface soils varies from about 0.05 to below 0.2 $\mu g\, l^{-1}$, being the highest for soil amended with chemical fertilizers and compost.

Several organisms common in soil and subsurface environments have been identified that can enzymatically reduce U^{6+} to U^{4+}, under anoxic conditions (Fendorf et al. 2002). This process, however, is highly controlled by the presence of Fe hydroxides. The reduction of immobile U^{6+} to mobile U^{4+} by microbial activity (mainly by *Thiobacillus ferrooxidans*) has been also observed by Fox et al. (2004). On the other hand, Abdelouas and Grambow (2004) reported that soil microorganisms could enhance the retention of U^{6+} and other radionuclides *via* the precipitation of highly insoluble mineral phases. There is evidence that arbuscal mycorrhizal fungi (AMF) increase concentrations of U in roots of some herbaceous plants and may have a potential role in the phytostabilization of U in soils (Chen et al. 2005). Also Landa (2003) emphasized the role of reducing bacteria in leaching of U from Uranium Mill Tailings. Tsuruta (2003), on the other hand, described possible accumulation of Th by bacteria (*Micrococcus luteus* and *Streptomyces levoris*) that might have a practical use.

Studies on the mobility and phase formation of the DU indicate a low migration of this metal throughout a loamy clay column (Trueman et al. 2004). Grive et al. (2004) emphasized that U cycles are closely associated with Fe cycles in various environments and are also linked to carbonates. Other factors that control the mobility of U are the redox potential, the pH, and CEC. The sorption and co-precipitation of U^{6+} by calcium carbonates and mobilization due to the reduction to U^{4+} is of great environmental importance (Koroleva and Manginii 2004). The increased use of apatite as a soil amendment resulted in the enhanced desorption of Th in soils and sediments. Nevertheless Th predominates in the residual fraction of both apatite and soil amended with apatite (Kaplan and Knox 2004).

There is presently a considerable interest in the behavior of other elements of the actinide group, which are isotopes not occurring naturally but released from the nuclear fuel power industry. This group includes various isotopes of plutonium (Pu), americium (Am), curium (Cm), and neptunium (Np), of which ^{239}Pu and ^{241}Am, long-lived and highly radiotoxic, are of the greatest concern.

Regardless of the forms of Pu and Am entering soils, mainly from the aerosols produced and discharged by the nuclear facilities, their behavior is controlled by various soil factors. Jakubick (1979) described in details the geochemistry of worldwide fallout and discharge of Pu. He has concluded that the behavior of stack discharged PuO_2 from a nuclear facility can give information about the environmental impact of the nuclear industry. Average fallout of $^{239+240}$Pu on soils in the temperate zone of the USA, has been calculated at 7.8×10^4 Bq km^{-2} and of ^{238}Pu at 1.9×10^3 Bq km^{-2} (Hanson *vide* ATSDR 2002a). The concentration of Am in soil in the New York region has been estimated in 1969 at 2 MBq km^{-2}, which may increase up to 33 MBq km^{-2} in the year 2035, resulting from cumulative deposition (ATSDR 2002a). In the period 1969–1983, the mean annual fallout of Am to Antarctica was about 8.3 mBg m^{-2}. The levels of ^{238}Pu in soils of the UK were detected in the range of 0.2–18 nCi kg^{-1}, and of $^{239+240}$Pu in the range of 0.8–83 nCi kg^{-1} (ATSDR 2002a). Amounts of $^{239+240}Pu$ in the surface layer of the contaminated soil near Heidelberg, Germany, were at 0.13–0.45 mCi km^{-2} (Jakubick 1979). Soils of Kazakhstan sampled in 1992 contained $^{239+240}Pu$ in the range of average values from 10 to 93 mCi km^{-2} (Panin 2004).

In soils, Pu occurs mainly as complexes of fulvic and humic acids and in forms absorbed by fine mineral particles, often coated with an organic layer. Oxide compounds, like PuO_2^{2+}, are more stable than nitrate compounds, $Pu(NO_3)_4$, which are likely to occur in soil solution as hydroxides and are easily soluble, and therefore plant-available. Relatively mobile, and thus available PuO_2^{2+} migrates slowly into deeper soil layers, but most of this compound remains within the 10 cm top-soil layer. Plutonium is adsorbed on hematite in larger amounts than on silica and montmorillonite (Runde 2002). Microbial activity may enhance the mobility of Pu, as well as other soil parameters, such as low pH, low clay content, and low organic matter. Zhang et al. (2002) described that surface contamination, rather than uptake by roots, should be considered as primary source of Pu in plants.

Especially ^{241}Am is known to easily form humate complexes. At low humic content and low soil pH (low carbonate content) the adsorption of this radionuclide is reduced and its mobility and phytoavailability increased (Zhang et al. 2002).

To protect soils against contamination with radionuclides, the storage in the form of glass fusion is proposed (Menard et al. 1998). The presence of amorphous silica and phosphates stimulate the processes of the immobilization. Recent studies of Wieland et al. (2004) reveal a great sorption capacity of cementation matrices for radionuclides at the solid/liquid interface. Qinn et al. (2004) reported a great adsorption of YREE (yttrium and rare earth elements) by freshly precipitated hydroxides of trivalent metals (Fe, Al, Ga, and In).

II-3.4.4
Waters

There are very limited data on actinides in waters. Gaillardet et al. (2003) estimated the world average concentration of Th in river waters at 0.041 μg l^{-1} (range 0.0005–1.054 μg l^{-1}) and for U as 0.372 μg l^{-1} (range 0.02–4.94 μg l^{-1}). Levels of U in surficial waters of Germany were reported to range between 0.007 and 52.7 μg l^{-1}, being increased in the vicinities of U mines and in agricultural regions where intensive phosphate fertilization was applied (Birke and Rauche 1997). In the Colorado River, analysis done during the period 1985–1986 showed that levels of total U ranged from 3.4 to 60 pCi l^{-1} (Steward *vide* ATSDR 2002a). The concentration of ^{241}Am in the North Sea and Atlantic Ocean waters is given as 10 Bq m^{-3} in surface seawater (ATSDR 2002a). Szefer (2002a) reported that concentrations of U in Baltic Sea waters range from 0.67 to 1.65 μg l^{-1} and of Th, range between 0.5 and 7.7 ng l^{-1}. Nozaki (2005) cited the mean concentration of U in the North Pacific as 3.2 μg kg^{-1}.

Radioactive actinides, ^{238}U and $^{239+240}Pu$ in waters of the Southern Baltic Sea were at levels 8.37 and 0.0035 mBq l^{-1}, respectively, during the period 1980–1991. After the Chernobyl accident, their increase was not observed (Skwarzec 1997; Skwarzec and Bojanowski 1992).

Unlike PuO_2, $Pu(NO_3)_4$ is an easily soluble compound. However, the presence of complexing agents (e.g., CO_3, PO_4, humic, fulvic acids, etc.) in natural waters are important ligands for Pu, and highly control its solubility (Jakubick 1979). As Yamada et al. (2004) reported, the occurrence of Pu-isotopes in the Pacific Ocean is mainly a consequence of global fallout from atmospheric nuclear weapons testing. Most nuclear power plants release great quantities of radioisotopes, especially of ^{239}Pu into aquatic environment. Bottom sediments can act as both a repository and a source of waterborne actinides.

Drinking waters of the Gdansk agglomeration (Poland 1988) contained (in $mBq\,l^{-1}$): ^{238}U at 2.67 and ^{234}U at 2.86, which gives an average annual intake of both radionuclides at the range of 2.81 Bq per person (Skwarzec et al. 2001b). Drinking waters of the Eastern part of Germany contained U, in the range of average values from 0.29 to 2.4 $\mu g\,l^{-1}$ (Anke et al. 2000). Generally, U found in drinking water is <1 $pCi\,l^{-1}$. The limits for U in domestic water in the USA is <15 $pCi\,l^{-1}$, and the maximum concentrations for ground-water protection is 30 $pCi\,l^{-1}$ (ATSDR 2002a).

Concentrations in ground waters vary in the following ranges: 2–12 and 0.08–2.0 $\mu g\,l^{-1}$ for U and Th, respectively. Their levels are usually lower in zones of the temperate humid climate and higher in areas of hot, dry climate (Kabata-Pendias and Pendias 1999).

Rain water in Norway contained U in the range of 0.01–0.05 $\mu g\,l^{-1}$ at the coastal region and 0.01–0.43 at the inland areas. Fairly similar concentrations of Th are reported for rainwater of other regions (Reimann and Caritat 1998). Rainwater collected in Sweden during 1999 contained U and Th at mean levels of 0.0022 and 0.0023 $\mu g\,l^{-1}$, respectively (Eriksson 2001a).

Aquatic biota. All actinides are easily bioaccumulated in aquatic organisms, primarily at the lower end of the food chain. Contents of $^{239+240}Pu$ in fish and shelfish sampled in the Irish Sea were reported to range from 2×10^{-4} to 2 $pCi\,g^{-1}$ (ATSDR 2002a). Sanchez-Cabeza et al. (2003) have stated, based on enrichment factor values, that discrimination rather than enrichment of Am and Pu in the primary producer trophic chain is observed in the Mediterranean Sea. These authors reported mean concentrations of $^{239+240}Pu$, ^{238}Pu, and ^{241}Am in mesoplankton sampled from <200 m depth as (in $mBq\,kg^{-1}$): 452, 13, and 81, respectively. Mesoplankton collected from a deeper layer contained significantly smaller amounts of these radionuclides.

Bioconcentration factors for radionuclides in fish are fairly high in all tissues, and are the highest for intestines (above 20 000) and for gills (up to 10 000). Nevertheless, the annual doses of U and Pu radionuclides from the consumption of Baltic fish are relatively small (Skwarzec et al. 2001a).

II-3.4.5
Air

Increased concentrations of U and Th in air are associated mostly with coal combustion and nuclear powers and should be considered the most important risk to the environment and health. Thorium abundance in air above the South Pole ranges between 0.02 and 0.08 $pg\,m^{-3}$, and above Greenland it ranges from 20 to 40 $pg\,m^{-3}$. Air above cities of Europe, Japan and the USA is enriched in Th up to 1 000–13 000 $pg\,m^{-3}$. Cawse (1987) estimated atmospheric deposition of Th in urban regions of the UK as <0.5–1 $g\,ha^{-1}\,yr^{-1}$, and Eriksson (2001a) calculated the wet deposition in Sweden at 0.08 $g\,ha^{-1}\,yr^{-1}$.

Act

The maximum concentration of U in air of polluted sites is reported to be <0.5 ng m^{-3} and of Th 1.3 ng m^{-3} (Reimann and Caritat 1998). Therefore the concentration of U in air around urban regions estimated at about 0.5 ng m^{-3} clearly indicates its anthropogenic source.

Concentrations of ^{239}Pu and $^{+240}$Pu and in air near Heidelberg, Germany, during the period 1954–1975 varied from 0.086 to 0.75 fCi m^{-3}, and cumulative total deposition was calculated to range between 0.054 and 2.10 mCi km^{-2} (Jakubick 1979). Contents of both radionuclides in air have stayed presently at an approximate value of <0.04 µBq m^{-3}.

Limits values given by OSHA are 50 µg m^{-3} for soluble U compounds and 250 µg m^{-3} for insoluble U compounds.

II-3.4.6
Plants

Little information is available about the transfer of U, Th, and transuranic radionuclides from soils to plants, although this is very important in environmental researches.

Soluble fractions of these elements in soils are readily available to plants and this is clearly evident from the studies conducted in the U-geochemical provinces where plants accumulate up to 100 times more U than do plants from other areas. Values of the transfer factor for radioactive U and Th from soils around an abandoned U mine to grass pasture indicated a relatively low uptake with an exception of ^{228}Th which was accumulate at a higher concentration in plants (TF 1.6–2.0) than in soils (Vera Tome et al. 2003). Some plants when grown on contaminated sites exhibit a relatively high ability to accumulate both radionuclides (Table II-3.13) and are suggested for the phytoremediation technology (Hinton et al. 2005). Uranium in plants can form complexes with protein and/or be precipitated as mineral compounds (e.g., autunite, $Ca(UO_2PO_4)_2 \cdot 10H_2O$) (Jones et al. 1990).

The gram-positive bacterial strains (*Streptomyces levorie*) can adsorb relatively high amounts of both U and Th from the aqueous solution at pH 3.5: 390 µmol U g^{-1} and 383 µmol Th g^{-1} (Tsuruta 2004). Some plants, e.g., *Brassica* sp. exhibit the capability to absorb large quantity of U, especially with treatment by citric acid (Chang et al. 2003).

Shtangeeva et al. (2005) reported an easy accumulation of Th by wheat seedlings which resulted in the decrease of Ca in leaves as well as in the uptake of other nutrients. A significant decrease of Th in the growth media, both soil and water, up to 5 times and 1.7 times, respectively indicates that Th may be easily phytoextracted (Shtangeeva and Ayrault 2004).

Based on available data on the occurrence of U in plants, following contents can be presented (in µg kg^{-1}): higher vascular plants, 5–60; various vegetables, <5–20; potato tubers, 0.08 µg kg^{-1} (Kabata-Pendias and Pendias 2001). Uranium content in crop plants decreases slightly with the age of plants, e.g., lucerne sampled in May contained U at the level of 9.2, while lucerne sampled in June contained this element at the level of 3.9 µg kg^{-1} (Anke et al. 2000). These authors presented data for U content in several plant foodstuffs in the range from 0.2 to 17 µg kg^{-1} FW in apple and sweet paprika, respectively. Parsley is reported to contain the highest level of U among vegetables

Table II-3.14. Uranium in some foods of the USA (µg kg^{-1} FW) (adopted from ATSDR 2002a)

Plant food	Content	Type of food	Content
Potato, tubers	2.7 – 18	Beef	14
Carrots	7.7	Beef liver	26
Cabbage	4.7	Beef kidney	70
Parsley	60	Poultry	0.1 – 0.4
Red pepper	5	Eggs	0.2 – 9.6
Fresh fruits	0.7 – 1.3	Milk	1 – 2
Fresh vegetables	0.5 – 0.9	Fresh fish	0.4 – 11
Wheat bread	19	Shellfish	9.5 – 31

(Table II-3.14). The average U content in winter wheat grains sampled in Sweden is 0.1 µg kg^{-1} and in barley grains is 0.2 µg kg^{-1}. Fairly similar Th concentrations have been determined in Swedish cereals (Eriksson 2001a). Important sources of certain actinides (Pu, Am, Cm) as well as other radionuclides to plants may be marine algae used as organic fertilizer, particularly for vegetable cultivated in certain areas of the French Atlantic coast (Colle et al. 2002).

Little et al. (1980) studied the behavior of Pu and concluded that its soluble fraction is easily mobile in soils and easily available to grasses. In an area polluted with Pu, much of this radionuclide was attached to the plant surface which indicated its aerial sources (Deinstbach et al. 1983). Accumulation of $^{239+240}Pu$ in mosses of Lithuania is higher in the southwestern part on the country (range 0.4–8.4 Bq kg^{-1}) than in the northern part (range 0.1–1.6 Bq kg^{-1}) (Lukšienė and Druteikienė 2004).

Moss samples from the Scandinavian countries contain U in the range of 2.8–1 300 and Th in the range of 4–5 100 µg kg^{-1} (Berg and Steinnes 1997). Lichens are found to be good monitors of concentrations of U in air and were used to assess the U levels in the Balkan area, after the use of DU during the conflict in 1999. The U contents of lichens varied widely depending on species and sampling sites. The highest levels, from 81 to 306 µg kg^{-1}, seem to be found in *Pseudevernia* sp. collected in Greece. According to the statements of the authors, the average U concentration was in line with the values reported in lichens from other countries and does not indicate any contamination by DU in this area (Loppi et al. 2003).

II-3.4.7 Humans

Actinides are easily accumulated in plants and thus are taken up by humans with vegetation in their diet and by animals with forage. However, presently available data do not indicate any significant biomagnifications in the food chain. Nevertheless, it has been stated that the main source of radionuclides, especially for rural population, is from food chain (Dolin and Bongdarenko 2004). Aerial dust and aerosols also can be a significant source of these elements, especially in areas of the U industry (e.g., mining, production of nuclear fuel).

Act

Easily soluble compounds of U are accumulated in the human body in both, soft organs (mainly in kidneys and liver) and bones. Mean concentrations of U in the organs of persons of all age groups from the USA are as follows (in $\mu g\ kg^{-1}$): lungs, 0.5–1.2; liver, 0.12–0.33; kidneys, 0.4–1; bones, 0.3–1.9; and whole blood, 0.1–0.14. The mean worldwide value for U in blood is 0.58 $\mu g\ kg^{-1}$ (ATSDR 2002a).

Normal levels of dietary $^{234+238}U$ indicate that the intake of these radionuclides by adults might be about 0.3–0.5 pCi d^{-1} for each isotope. Whereas, worldwide intake values for U has been reported to average 1 pCi d^{-1}. Dietary intake of U by the US population varies from 1.3 to 4.4 $\mu g\ d^{-1}$ per person and a mean intake by Japanese people is 0.71 $\mu g\ d^{-1}$ (ARSDR 2002a). Soft tissues of beef organs may contribute relatively high proportion of U in the diet (Table II-3.14). The MPL value of 2 $\mu g\ kg^{-1}\ d^{-1}$ is given for an intermediate-duration of oral exposure to soluble compounds of U, based on renal effects in dogs (ATSDR 2002a).

Plutonium occurs mainly as oxide compound (PuO_2) and thus is less available to humans than U, however, it is considered to be relatively more toxic than U. There is a possibility that Pu can enter the body through ingestion of dust in areas close to nuclear weapon test camps. The gastrointestinal absorption values for actinides, as estimated by ICRP (1993), vary from 0.01 to 0.05% of the dose and are much lower than the absorption due to inhalation.

Plutonium is considered to be the most radiotoxic among the transuranic elements. Several mutagenic, carcinogenic and teratogenic effects, especially pulmonary disorders, after the inhalation Pu compounds, have been frequently reported in recent years (Melo and Burkart 2004). In addition several other health effects on both humans and animals were observed, particularly hepatic degeneration, hematological, neurological, developmental, and musculoskeletal (ATSDR 2002a). Uranium can damage capillary vessels, the nervous system, and chromosomes of white blood cells. Similar toxic effects, as for Pu, were observed at the increased exposure to this radionuclide (Seiler and Sigel 1987).

The concentrations of ^{241}Am are reported to be at the levels of 0.48 and 0.048 mBq kg^{-1} FW in shellfish and in fresh vegetables, respectively. The concentrations of ^{232}Th in vegetables of the USA vary form ≤ 1 to ≤ 9 fCi g^{-1} FW, being the lowest in tomatoes and the highest in celery. In canned fruits, Pu was detected at the concentration up to 0.11 fCi g^{-1} (ATSDR 2002a).

Contents of ^{241}Am in food plants vary greatly and seem to be the lowest in potatoes (0.013 mBq kg^{-1} FW) and the highest in squash and pumpkin (3 mBq kg^{-1} FW) (ATSDR 2002a). The availability of ^{241}Am from the food chain to humans is considered to be low. However, the food chain lichen-reindeer-man might be of concern, since in reindeer livers its content seems to be increased, and in the period 1974–1976, was at 5.4 pCi kg^{-1} FW (Jaakkola *vide* ATSDR 2002a). The overexposure of humans and animals to this radionuclide indicates several health effects such as damage in liver, thyroid, blood, and kidneys. Bones accumulate the majority of the ^{241}Am in all mammalians and this can cause bone fibrosis, cell depletion, and marrow atrophy.

Trace Elements of Group 4 (Previously Group IVb)

The trace elements of Group 4 are titanium (Ti), zirconium (Zr), and hafnium (Hf). Titanium is considered a trace element only because of its low concentrations in plant and animal tissues. The physical and chemical properties of these metals differ (Table II-4.1), but they all are widely distributed in both litho- and biosphere. However, their biochemical functions are hardly known.

II-4.1 Titanium (Ti)

II-4.1.1 Introduction

Titanium shows the strong lithophilic characteristic and is a common constituent of rocks, in the range of 0.03–1.4%. Its average abundance in the Earth's crust is given as 0.4–0.6% (Table II-4.2.). Titanium exhibits variable valences (Table II-4.1), but in minerals occurs mainly in the tetravalent oxidation state as a major component of oxides, titanates, and silicates. The most important and of an economic value are: ilmenite, $FeTiO_3$; rutile, TiO_2; and to a lesser degree, sphene (titanite), $CaTiO \cdot SiO_5$. Anatase and brookite are also composed of TiO_2, but have different crystallographic coordinations. Many of Ti-containing minerals are resistant to weathering and they are often associated with other heavy, resistant minerals that are accumulated in some deposits.

Table II-4.1. Selected properties of trace elements of Group 4

Element	Atomic number	Atomic mass	Atomic radius[a] (pm)	Density (20 °C, g cm^{-3})	Valence[b]	Melting point (°C)
Ti, titanium	22	47.87	200	4.5	+2, +3, **+4**	3287
Zr, zirconium	40	91.22	216	6.5	+2, +3, **+4**	4377
Hf, hafnium	72	178.5	216	13.3	+4	4602

[a] Approximately average values for the main oxidation states.
[b] Values in bold are for the main oxidation state.

Ti

Table II-4.2. Abundance[a] of titanium, zirconium and hafnium in the environment

Environmental compartment	Ti	Zr	Hf
Earth's crust ($mg\ kg^{-1}$)[b]	0.4 – 0.57	100 – 200	3 – 4.9
Igneous rocks($mg\ kg^{-1}$)[b]			
Mafic	0.3 – 2.4	80 – 200	1 – 4.8
Acid	0.3 – 3.4	140 – 240	4.5 – 5
Sedimentary rocks ($mg\ kg^{-1}$)[b]			
Argillaceous	0.6 – 2.7	160 – 200	2.8 – 6
Sandstones	0.15 – 3.6	180 – 250	3 – 6
Calcareous	0.4 – 4	20	0.3 – 0.4
Raw energy material ($mg\ kg^{-1}$)			
Coal, hard	500 – 7400	34 – 50	0.5 – 1
Crude oil	0.1	–	–
Soils ($mg\ kg^{-1}$)[b]	0.33	300	3
Light sandy	0.02 – 1.7	90 – 200	1.8 – 18.7
Medium loamy	0.5 – 1	200 – 550	20.8
Heavy loamy	0.1 – 2.4	330 – 850	–
Calcareous	0.04 – 1	–	–
Organic	0.01 – 0.5	32	0.0X – 0.X[d]
Water ($\mu g\ l^{-1}$)			
Rain[f]	0.12 – 1.2	0.01 – 0.04	0.6 – 1.9[g]
Sea	0.6 – 9	0.1 – 0.3[c]	0.03 – 0.07[c]
River	10	0.2 – 89[c]	0.004 – 0.02[e]
Air ($ng\ m^{-3}$)			
Urban	15 – 25	3.0[c]	18 – 590
South Pole	0.5 – 2.5	–	0.04 – 0.06

[a] Presented are common, possible background, values from various sources as compiled by Kabata-Pendias and Pendias (1999 and 2001), unless otherwise indicated.
[b] Values for Ti are in %.
[c] After Reimann and Caritat (1998).
[d] The orders of magnitude, after Markert and Lieth (1987).
[e] After Gaillardet et al. (2003).
[f] After Eriksson (2001).
[g] Values for Hf in rainwater are in $ng\ l^{-1}$.

II-4.1.2
Production and Use

Annual global production of Ti in 2003 was 4.22 Mt and increased significantly since 1995 when its production was 2.30 Mt (Reimann and Caritat 1998; WMSY 2004). It is obtained mainly from rutile and ilmenite deposits. Additional sources of Ti are byproducts of the production of iron by electronic furnace reduction of ilmenite.

Most of the total quantity of mined Ti is used for the production of white pigments used broadly in various manufactures. It serves as catalyst in the production of polyethylene. It is also used in the aeronautic industry and is added to metal plates, pins, etc.

Additional, but significant use of Ti is in alloys, especially for orthopedic implants and prosthesis. Titanium dioxide is applied in cosmetics, mainly in the therapy of skin disorders. It is an additive to food and toothpaste and is used in variety of tablet-coating formulations.

II-4.1.3 Soils

Titanium contents in surface soils range from 0.02 to 2.4% (Table II-4.2). Calculated mean for worldwide soils is 0.33%. Soils of Kazakhstan contain Ti at the level of 0.2–0.7% of which very low proportion was easily mobile (Grabanov 1970). Swedish arable soils contain Ti in the range from 0.14 to 0.51%, at an average value of 0.37% (Eriksson 2001a).

Since Ti minerals are very resistant to weathering they occur often undecomposed in soils and therefore has been used as an index for soil genesis. Due to its relatively stable content in soil layers, the total Ti (also Zr) contents of soils have been used in several studies of soil genesis and of the continuity of soil profiles.

The solubility of Ti in soils is very limited, therefore in the soil solution it occurs at a relatively low level of about 30 $\mu g\ l^{-1}$. When Ti-bearing silicates are dissolved, the metal is soon transformed into Ti oxide-aquate and transformed to anatase or rutile. Anatase seems to be the most common Ti-mineral in soils. However, Ti can enter into the structure of some layered silicates and is likely to be adsorbed on surface of Fe-Mn concretions.

Soils exposed to effluents or emissions from certain industries (Ti alloys, Ti-paint production) may become contaminated by Ti, however, this element does not create any environmental problems.

II-4.1.4 Waters

Titanium, a slightly mobile element, is present in the hydrosphere in limited amounts. According to data presented by Gaillardet et al. (2003), concentrations of Ti in river waters range between 0.02 and 2.3 $\mu g\ l^{-1}$ and worldwide average is estimated at 0.49 $\mu g\ l^{-1}$. Titanium concentration in stream waters, as given by Reimann and Caritat (1998), varies from 12 to 926 $\mu g\ l^{-1}$, being the highest in India and the lowest in Nova Scotia, Canada. In the US rivers, its concentrations vary from 2 to 107 $\mu g l^{-1}$ (www.inchem.org/documents/ehc).

Titanium in waters of the Baltic Sea drainage basin ranges from 1.2 to 29 $\mu g\ l^{-1}$, and its concentrations in water of the Baltic Sea are between 0.08 and 0.15 $\mu g\ l^{-1}$ (Szefer 2002a). Concentration of Ti up to 9 $\mu g\ l^{-1}$ has been observed in seawater (Nordman and Berlin *vide* Anke and Seifert 2004), whereas Nozaki (2005) reported mean concentration of Ti at 0.007 $\mu g\ kg^{-1}$ for the Northern Pacific.

Rainwater collected in Sweden during 1999 contained Ti at an average value of 0.44 $\mu g\ l^{-1}$ (Eriksson 2001a). Drinking waters in the USA contain Ti from 0.5 to 15 $\mu g\ l^{-1}$ (Anke and Seifert 2004).

Aquatic biota. In seaweeds, Ti is likely to be concentrated at the range of 22–60 mg kg^{-1} (Szefer 2002a). However, higher Ti concentrations, especially in zooplankton also have been reported. Increased level of Ti in some algae may introduce its large quantities into the food chain. Titanium can be concentrated up to 400 mg kg^{-1} in hard tissues of sea animals.

The discharged TiO_2 into water systems can be harmful to some fish.

II-4.1.5 Air

The origin of Ti in the atmosphere is mainly terrestrial. However, anthropogenic sources of Ti are also of an importance as is clearly demonstrated by the increase of its content in the range of 0.5–2.5 ng m^{-3} in air from remote regions to 15–25 ng m^{-3} in urban areas (Table II-4.2). Reimann and Caritat (1998) gave the global median values for Ti in the remote regions as 7 ng m^{-3} and in polluted zones as 85 ng m^{-3}. Concentrations of Ti in air of the USA range from 10 to 100 ng m^{-3}, and in industrial regions can increase up to $\leq$1 000 ng m^{-3} (www.inchem.org/documents/ehc).

II-4.1.6 Plants

Contents of Ti in plants seem to be highly controlled by soil properties and origin. Anke and Seifert (2004) presented the variation in Ti mean contents in various plants as affected by geological origin of soils. The highest contents of Ti, at about 100 mg kg^{-1}, were found in plants from loess and some weathered *in situ* soils, whereas the lowest concentrations were in plants grown on Histosols.

Titanium in food-plants ranges from 0.13 to 6.7 mg kg^{-1}. The lowest values are in prepared cereals and fruits (apples, bananas and oranges) and the highest in lettuce, radish and corn. Contents of Ti in grain crops from Kazakhstan range from of 1.2 to 7 mg kg^{-1} and in wheat flour, from 0.4 to 1.0 mg kg^{-1} FW (Grabanov 1970). In Swedish cereal grains, mean Ti concentrations are 0.17 and 0.25 in wheat and barley, respectively (Eriksson 2001a).

The concentration of Ti in plants decreases with age and may be reduced by one-third in older plants as compared with plants in early growth stage (Anke and Seifert 2004). Some weeds, especially horsetail and nettle are known to accumulate more Ti, up to 80 mg kg^{-1} (Kabata-Pendias and Pendias 2001). The phytoavailability of Ti is low; the TF values are calculated for grass pasture as 0.03–0.04 (Vera Tome et al. 2003).

No clear evidence of a biochemical role of Ti has been given, although there are some suggestions on possible catalytic function in N fixation by symbiotic microorganisms and in photooxidation of N compounds by higher plants (Pais and Jones 1997). These authors described some beneficial effects of Ti-chelates (mainly ascorbates) on photosynthesis and on the molecular N_2 fixation. Also Simon et al. (1990) reported that the addition of Ti stimulates the growth of *Rhizobium* bacteria strains. Only one early report by Wallace et al. (*vide* Kabata-Pendias and Pendias 2001) has described toxicity symptoms at the content of 200 mg Ti kg^{-1} in bush bean, as necrotic and chlorotic spots on leaves.

Concentrations of Ti in moss from Norway average 53 mg kg^{-1} and range from 12 to 310 mg kg^{-1} (Berg and Steinnes 1997). Apparently, lichens and mosses can be bioindicators for Ti aerial pollution.

II-4.1.7 Humans and Animals

Humans. Titanium occurs in all tissues of terrestrial and aquatic organisms. Its contents in soft tissues of mammalian are around <0.3 mg kg^{-1}, whereas in skin and hair it occurs at 0.54 and around 3.0 mg kg^{-1}, respectively (Jørgensen 2000). The highest mean Ti contents in humans are reported for lung and hair, as 3.7 and 10.6 mg kg^{-1}, respectively (Pais and Jones 1997), and it is likely to increase with age, especially in lung and lymph nodes. The calculated Ti content in total soft tissues of humans is 0.26 mg kg^{-1} (Li 2000).

A special source of Ti in human body might be due to its release from some implants made of Ti-alloys. There is evidence that some Ti-implants can cause mild inflammatory reactions. A man exposed to Ti contamination might have 33 and 150 mg of this metal in lung and lymph nodes, respectively (www.inchem.org/documents/ehc).

Food is the principal source of humans exposure to Ti. The intake of Ti varies depending on diet and region, and is calculated, on average as 66–207 μg d^{-1} for women and 90–253 μg d^{-1} for men, being the lowest for vegetarians (Anke and Seifert 2004). Aerial dust can be a significant source of Ti to humans and especially to children. Davis and Waller (1990) calculated that children aged 2–7 years might take up to 250 mg Ti per day (medium 81 mg d^{-1}) derived mainly from the "pica-soil". Daily Ti intake for the world population is reported to vary from 750 to 2 000 μg. Typical American diets were estimated to provide approximately 300 μg Ti d^{-1} (www.inchem.org/documents/ehc).

Although Pais and Jones (1997) described beneficial effects of chelated forms of Ti (mainly ascorbate) on humans and animals, there is no evidence indicating that Ti is an essential element. Still no clear biochemical function has been identified for Ti. However, Anke and Seifert (2004) have stated that the essentiality of Ti to fauna and humans can no longer be excluded, although further investigations are required. They concluded that the daily Ti of 30 μg for humans is likely to meet the Ti requirement. There is not much evidence of Ti hazard to humans. Thus, there is no reason to believe that the exposure to Ti would have any health risk for the general population, with an exception of accidental exposures, especially to $TiCl_4$.

Animals. Relatively high average contents of Ti reported for the house mouse are 1.89 and 3.26 mg kg^{-1} for unpolluted and polluted areas, respectively. Anke and Seifert (2004) calculated that mice and voles ingest 140–170 μg Ti per day, while daily intake by cats via commercial food varies between 17 and 590 μg d^{-1}.

Some evidence has been reported for the carcinogenicity of TiO_2 (mainly due to dust exposure) and for effects on the reproduction of animals (Anke and Seifert 2004).

II-4.2 Zirconium (Zr)

II-4.2.1 Introduction

The crustal abundance of Zr varies from 100–200 mg kg^{-1}, being the lowest in ultramafic rocks and calcareous sediments (Table II-4.2). Zirconium exhibits both oxyphilic and litophilic tendencies and its predominant valency is +4 (Table II-4.1). The main mineral of Zr is zircon, $ZrSiO_4$, but baddeleyite (zirconia), ZrO_2, has also technological importance. It occurs in several complex silicates and is associated with Ti and Hf minerals.

II-4.2.2 Production and Use

Zirconium is obtained as sand deposits which are most often enriched in U and Th what leads to environmental concern. Its production in 2003 was estimated at 1.1 Mt (WMSY 2004). It is also obtained as byproduct of the Ti production.

The principal use of Zr is for atomic energy purposes. Due to its great resistance to heat (melting point 4 377 °C) and to corrosion, it is an important component of alloys, glass and ceramics. It is also applied for water-repellent textiles, dyes and pigments; and utilized in pyrotechnics. Some Zr minerals (zircon and baddeleyite) are used for the gem zircons production and sometimes to imitate diamonds.

The important application of Zr-metal is in medicine as an addition to implant materials for bones and muscles. ZrO_2 is also added to bone cement for better X-ray contrast (Schaller 2004). This metal is a component of some cosmetics and toothpaste.

II-4.2.3 Soils

Parent rocks are the main source of Zr in soils where it is slightly mobile. Some organic acids seem to be the transporting agents for the migration of Zr within soil profiles.

Contents of Zr in soils vary within the range of 30–250 mg kg^{-1}, being the lowest in peat soils and the highest in sandy soils (Table II-4.2). Zicronium distribution in soils seems to be a function of soil texture (Table II-4.2), the lowest contents are in sandy soils (90–200 mg kg^{-1}) and the highest in heavy loams (330–850 mg kg^{-1}). This metal, like Ti, has been used as an index for soil genesis. Especially the Ti to Zr ratio gives reliable results. The greatest abundances of Zr, up to 300 mg kg^{-1}, was found in residual soils derived from Zr-rich rocks, and up to 850 mg kg^{-1} in soils derived from basalts and andesites (Kabata-Pendias and Pendias 2001). Also some chernozems of Russia are reported to contain Zr between 200 and 550 mg kg^{-1} (Protasova and Kopayeva 1985). Median contents of Zr in agricultural soils of Japan are 21 and 100 mg kg^{-1} in Gleysols and Andosols, respectively, and average 85 mg kg^{-1} (Takeda et al. 2004). The mean value of Zr in Swedish topsoil samples is 308 mg kg^{-1} and the maximum content is 757 mg kg^{-1} (Eriksson 2001a).

II-4.2.4 Waters

Due to the limited solubility of most compounds and minerals of Zr, its concentration in waters is relatively low. Water-soluble Zr compounds are converted into insoluble zirconia (ZrO_2) at a broad range of water pH 4–9.5.

Different values for Zr concentration in waters are presented in literature. In stream and river waters its abundance is: *(i)* 0.9–90 μg l^{-1}, median values after Reimann and Caritat (1998), *(ii)* 20–200 μg kg^{-1}, after Schaller (2004), and *(iii)* 0.03–0.98 μg l^{-1}, after Gaillardet et al. (2003). The last authors calculated the global average for Zr in fresh waters as 0.04 μg l^{-1}. The median Zr concentration in sea waters is around 0.03 μg l^{-1} (Reimann and Caritat 1998) and mean Zr content in the North Pacific is 0.015 μg l^{-1} (Nozaki 2005).

Rainwater collected during 1999 in Sweden contained Zr at the average value of 0.024 μg l^{-1} and at the maximum value of 0.04 μg l^{-1}. Its wet deposition has been estimated as 0.25 g ha^{-1} yr^{-1} (Eriksson 2001a).

II-4.2.5 Air

The Zr concentrations in air from polluted regions ranges between 0.7 and 26 ng m^{-3}, and the calculated median value is 3 ng m^{-3} (Reimann and Caritat 1998).

The maximum concentration at the workplace established in Germany (MAK value) for Zr in air is now reduced to 1 mg m^{-3} (Schaller 2004).

II-4.2.6 Plants

The soil Zr is slightly available to plants, and it is taken up presumably as anionic forms, $Zr(OH)_5^-$. Its higher content in roots, especially in nodules and roots of legumes, than in tops, indicates its low mobility within plant organs. The TF value for the accumulation of Zr by grass pasture calculated as 0.09 clearly confirmed its low phytoavailability (Vera Tome et al. 2003).

There are very scanty data on Zr status in plants. The Zr levels in food plants vary from 0.002 to 2.6 mg kg^{-1}, being the lowest in cereals and the highest in beans (Kabata-Pendias and Pendias 2001). Winter wheat and barley grains grown in Sweden contain Zr at the mean value of 0.008 and 0.013 mg kg^{-1}, respectively (Eriksson 2001a). Some herbage, especially legumes, trees and shrubs are likely to concentrate more Zr than other plants. Herb plants sampled in 1999 along highways in Germany contained Zr in the range of 0.07–0.92 mg kg^{-1}, being the highest in dandelion (Djingova et al. 2003).

There is not much agreement among the reported values for Zr (presumably due to low reliability of analytical results), however there is no accumulation of this metal in food plants. This metal can be relatively easily taken up by plants from the deposition of atmospheric particulates, especially in industrial regions. Increased level of Zr in moss (average 0.35 mg kg^{-1}) indicates its source from dust and rainwater (Berg and Steinnes 1997).

Toxic effects of high concentrations of Zr on plants, especially on root growth, are commonly reported. However, a stimulating impact of this metal on the growth of yeast and on protein synthesis of some microfungi has also been described (Pais and Jones 1997).

II-4.2.7 Humans

Zirconium is reported to occur in mammalian tissues at the range from <2 to <5 mg kg^{-1}, and its intake with diet is calculated as 0.1 g d^{-1} by a man of weight 70 kg (Jørgensen 2000). Li (2000) calculated Zr content in total soft tissues of humans as 0.021 mg kg^{-1}, with the highest concentration in lymph, at 0.3 mg l^{-1}. Its increased accumulation in liver may also occur.

Zirconium has been considered as an element of low toxicity. However, the inhalation of Zr compounds by workers exposured to industrial sources of that metal might affect granulomatous changes both in skin and lungs (Schaller 2004).

II-4.3 Hafnium (Hf)

II-4.3.1 Introduction

Crystallographic and chemical properties of Hf are very analogous to those of Zr, especially resemblance of the atomic radius is remarkable (Table II-4.1). However, its terrestrial abundance is much lower than Zr and ranges between 3 and 5 mg kg^{-1}. Some higher Hf contents, up to 6 mg kg^{-1} can be found in sandstones.

Hafnium occurs mainly in minerals of Zr, such as zircon, $(Zr,Hf)SiO_4$ and baddeleyite, $(Zr,Hf,Ti,Fe,Th)O_2$, and may be also associated with other minerals such as biotite and pyroxenes.

II-4.3.2 Production and Use

The global annual production of Hf in 1997 was estimated as 60 t (Kabata-Pendias and Pendias 1999).

Hafnium is used mainly (up to 90% of its production) in the nuclear industry, especially as reactor control rods. It is added to various alloys and applied in photographic flashes. Also biomaterials can contain Hf due to its good biocompatibility and osteoconductivity (Szilàgyi 2004).

II-4.3.3 Soils

Contents of Hf in soils are reported to range between 2 and 20 mg kg^{-1} (Table II-4.2). Reference soils from China contain Hf from 7 to 14 mg kg^{-1} (mean 7.9), and soils of the

USA contain from 2.5 to 17 mg kg^{-1} (mean 7.3) (Govindaraju 1994). The median Hf content of the worldwide soils is 5 mg kg^{-1}, with a range of 1–10 mg kg^{-1}. The mean value of Hf in Swedish topsoil samples is 7.6 mg kg^{-1} and the maximum content is 14 mg kg^{-1} (Eriksson 2001a).

II-4.3.4 Waters

Median concentrations of Hf in stream and river waters range between 8 and 10 ng l^{-1}, after Reimann and Caritat (1998), and from 4 to 110 ng l^{-1}, after Gaillardet et al. (2003). The worldwide average is calculated as 6 ng l^{-1}. Hafnium in ocean water averages 7 ng l^{-1} (Reimann and Caritat 1998) and in the North Pacific its concentration is given as 3.4 ng kg^{-1} (Nozaki 2005).

Somewhat elevated contents of Hf were found in stream sediments in Germany, up to 38 mg kg^{-1} (Reimann and Caritat 1998).

Rainwater sampled during 1999 in Sweden contained Hf at the average value of 1 ng l^{-1} and at maximum of 2 ng l^{-1}, and its wet deposition has been calculated as 11 mg ha^{-1} yr^{-1} (Eriksson 2001a).

II-4.3.5 Air

Hafnium occurrence in air from the remote area (Greenland) ranges between 40–60 pg m^{-3}. Air from industrial regions contains much higher amounts of this metal, up to over 700 pg m^{-3}, that clearly indicates its anthropogenic sources (Kabata-Pendias and Pendias 1999). The hazard level of Hf in air is 0.5 mg m^{-3} (Szilàgyi 2004).

II-4.3.6 Plants

The common range of Hf contents in plants vary from 0.01 to 0.4 mg kg^{-1}. Some parts of plants, e.g., spruce bark, contain a bit higher amounts of Hf, up to 11 mg kg^{-1} (Reimann and Caritat 1998). Lieth and Markert (1985) reported Hf content of blueberry tops to be <1 mg kg^{-1}. Eriksson (2001a) has found the Hf concentrations in cereal grains to be at the mean values of 0.4 and of 0.6 µg kg^{-1} for wheat and barley, respectively. Food plants contain Hf within the range of 0.6–1.1 µg kg^{-1} FW (Oaks et al. *vide* Kabata-Pendias and Pendias 2001). Although Hf is very slightly available to plants, its elevated contents, in plants grown on soil amended with sewage sludges were observed.

II-4.3.7 Humans

Hafnium was detected in mammalian muscles at the content of <40 µg kg^{-1} (Jørgensen 2000), and in human blood at mean concentration of 21 µg l^{-1} (Reimann and Caritat 1998). The distribution of radiohafnium in rat organs was found to be in the following order: skeleton > skin > muscle > liver (Taylor et al. 1983).

The absorption of orally administered ^{181}Hf to hamsters was relatively low, between 0.04 and 0.13% of the dose. The radiohafnium ($^{175+181}$Hf) was found to be bound mainly to blood plasma and in the liver cytosol (Szilàgyi 2004).

Similar concentrations of Hf in plant and animal tissues indicate a lack of the biological discrimination in its easy transport through the trophic chain.

There are no reported toxicity symptoms of Hf, except for an individual allergic sensitivity.

Trace Elements of Group 5 (Previously Group Vb)

Geochemical characteristics of the metals of Group 5, vanadium (V), niobium (Nb) and tantalum (Ta), are variable, although all elements exhibit mainly lithophilic tendencies. Vanadium is widely distributed in geochemical environments where it forms both cationic and anionic compounds. Two elements, niobium and tantalum have very similar atomic radius and resemble each other in geochemical properties (Table II-5.1). Both metals are rather rare in the biosphere (Table II-5.2), are highly associated with Fe, Ti, Mn and Zr, and are likely to be concentrated in polymetallic concretions in the ocean bottom.

II-5.1 Vanadium (V)

II-5.1.1 Introduction

The abundance pattern of V in rocks shows its tendency to be concentrated in mafic igneous rocks and in argillaceous sedimentary rocks (Table II-5.2). Its oxidation state is variable, but under weathering conditions it is mainly V^{5+} (Table II-5.1). In the oxidized state, V^{5+} shows isomorphic relation to other cations such as: As^{5+}, and Mo^{5+}, whereas V^{2+} behaves similarly to Fe^{2+}.

Vanadium does not form many own minerals of which most important are: vanadinite, $Pb_5Cl(VO_4)_3$; roscoelite, $KV_2(OH)_2(AlSi_3O_{10})$; mottramite, $Pb(Cu,Zn)(VO_4)(OH)$; and patronite, of variable composition: VS_2, VS_4, and V_2S_5. Vanadium can form variable

Table II-5.1. Selected properties of trace elements of Group 5

Element	Atomic number	Atomic mass	Atomic radius[a] (pm)	Density (20 °C, g cm^{-3})	Valence[b]	Melting point (°C)
V, vanadium	23	50.9	192	6.1	+2, +3, + 4, **+5**	1 890
Nb, niobium	41	92.9	208	8.6	+2, +3, + 4, **+5**	2 468
Ta, tantalum	73	180.9	209	16.7	+3, +4, **+5**	2 996

[a] Approximately average values for the main oxidation states.
[b] Valence values in bold are for the main oxidation states.

V

Table II-5.2. Abundance[a] of vanadium, niobium and tantalum in the environment

Environmental compartment	V	Nb	Ta
Earth's crust ($mg\ kg^{-1}$)	53 – 60	26	1.5 – 2.2
Igneous rocks ($mg\ kg^{-1}$)			
Mafic	40 – 250	10 – 35	0.5 – 2
Acid	40 – 90	15 – 60	2 – 4
Sedimentary rocks ($mg\ kg^{-1}$)			
Argillaceous	80 – 130	15 – 20	0.8 – 1.7
Sandstones	10 – 60	0.5 – 10	0.5 – 1.5
Calcareous	10 – 45	0.05 – 0.1	0.01 – 0.05
Raw energy material ($mg\ kg^{-1}$)			
Coal, hard	20 – 58	5 – 10	0.1 – 0.3
Crude oil	14 – 90	4	–
Petrol	50	2	–
Soils ($mg\ kg^{-1}$)	60	12	1.1
Light sandy	10 – 260	<4 – 8	1.5
Medium loamy and silty	27 – 110	–	1 – 2
Heavy loamy	20 – 330	–	–
Calcareous	10 – 500	–	–
Organic	10 – 22	–	–
Water ($\mu g\ l^{-1}$)			
Rain[c]	0.28 – 0.94	0.002	0.0002
Sea	2.5[b]	0.01	0.002[b]
River[d]	0.02 – 5.8	0.002 – 0.01	0.0001 – 0.148
Air ($ng\ m^{-3}$)			
Urban	5 – 200	–	0.02 – 0.28
Greenland	0.8 – 1.4	–	<0.01 – 0.03
South Pole	0.0006 – 0.002	–	–

[a] Presented are common, possible background, values from various sources as compiled by Kabata-Pendias and Pendias (1999 and 2001), unless otherwise indicated.
[b] After Reimann and Caritat (1998).
[c] After Eriksson (2001a).
[d] After Gaillardet et al. (2003).

compounds (vanadium oxysalts) of which $(VO_4)^{3-}$ plays significant role due to adsorptive properties. These minerals have complex composition, like e.g., carnolite, $K_2(UO_2)_2(VO_4)_2 \cdot 3H_2O$. Vanadium occurs commonly as admixture in about 70 minerals. Most of the V-containing minerals, however, are of secondary origin. Vanadium is very often associated with primary minerals such as: pyroxenes, hornblende, biotite and magnetite. During weathering, the mobility of V is dependent on the host minerals and on pH-Eh conditions, and finely is incorporated into clay minerals, and Fe-hydroxides. Some V-ions, e.g., $(VO_4)^{3-}$, are easily fixed to the surface of clay particles, and often together with phosphates to secondary apatites.

Vanadium reveals lithophilic affinity, and also is highly concentrated in several bituminous sediments and coals. Its average content in coal is estimated for 58 mg kg^{-1}, and in fly ash for 208 mg kg^{-1} (Llorens et al. 2000). However, some elevated concentrations of V are reported (in mg kg^{-1}) in: *(i)* fly ash up to >1 000, *(ii)* crude oil up to 1 300, *(iii)* crude-oil ash up to 22 000, and *(iv)* asphalt, up to 6 000 (Kabata-Pendias and Pendias 1999). High concentrations of V in organic sediments are effects of several processes like sorption by lipids and cholines, and substitution of Mg^{2+} by vanadyl ions VO^{2+} (Piispanen and Lähdesmäki 1983).

II-5.1.2 Production and Use

Global production of V was around 34 kt in 1990 and 43 kt in 2000 (USGS 2004). Some quantities of V are recovered from petroleum residues, ashes, and catalysts, however it is not possible to calculate this level (ATSDR 2002a).

Various compounds of metallic V are currently used for a wide variety of purposes. A great proportion of this metal (about 85% of mined metal) is utilized as an agent in the steel industry, mainly for heat-resistant and high-strength alloys, especially for high-speed tools in vibration (engines, springs). It is a component of alloys in the manufacture of jet aircraft engines. It is significant corrosion inhibitor.

Small quantities of V play an important role as a catalyst in the production of some chemicals, glass, ceramic and also in electronic industries. Some dyes for textiles contain V compounds. It al also added to some pesticides and is applied as photographic developer.

II-5.1.3 Soils

Levels of V in soils are closely related to the parent rock types. Its world median content has been established at 90 mg kg^{-1} (Reimann and Caritat 1998). Average contents of V in soils vary from 10 mg kg^{-1} in sandy soils to 500 mg kg^{-1} in calcareous soils (Table II-4.2). The reference soils of the USA contain V in the range of 36–150 mg kg^{-1} (Govindaraju 1994). Fairly uniform contents of V, presented as average values from 79 to 91 mg kg^{-1}, are reported for various soils from different regions of Russia (Protasova and Kopayeva 1985). Geometric mean of V in representative soils of Poland is 18.4 mg kg^{-1} (Dudka and Markert 1992). Topsoil samples from Sweden contain V within the range of 28–111 mg kg^{-1}, at average value of 69 mg kg^{-1} (Eriksson 2001a). The median V contents in agricultural soils of Japan vary from 94 mg kg^{-1} in Gleysols to 250 mg kg^{-1} in Andosols (Takeda et al. 2004). Some residual soils derived from rocks enriched in V contain elevated amounts of this metal, from about 70 to 100 mg kg^{-1} (Anke 2004c).

Low V contents are mainly in soils of light texture. Vanadium in soils seems to be associated mainly with Fe hydrous oxides and SOM. In some soils, clay minerals can also control the mobility of V. Apparently, the most common form of V in soils is the vanadyl cation (VO^{2+}), that predominates in acid soils and is likely to be mobilized as complexes with organic acids. Anionic forms of this metal, VO_3^-, VO_4^{2-}, and $H_2VO_4^-$, occur rather in neutral and alkaline soil solutions (Kabata-Pendias and Sadurski 2004). Anionic species of V can be relatively easily sorbed by clay minerals, especially together with Fe cations. This might explain considerable concentrations of V in various concretions and nodules formed in soils: 400 mg kg^{-1} in Fe-rich, and 440 mg kg^{-1} in Mn-rich nodules.

Although there are not many reports on V pollution of soils, it is likely that industrial processing of certain mineral ores as well as burning of coals and oils increase the deposition of V in soil. Increased levels of V are observed in soils of urban and industrial regions, especially around coal-burning plants and oil refinery. Elevated V contents in soils are also noticed in areas with high levels of residual fuel oil consumption. Surface soils polluted by fly ash contain up to 429 mg V kg^{-1} (Morrell et al. 1986).

II-5.1.4 Waters

Average concentration of V in seawaters is established at 2.5 µg l^{-1}, in the range between 1 and 3 µg l^{-1} (Table II-5.2). Waters of the North Pacific contain V at the average value of 2 µg kg^{-1} (Nozaki 2005). Concentrations of V in waters of the Baltic Sea vary from 0.4 to 5 nmol kg^{-1}, whereas in riverine-estuarine sediments of the Baltic catchment V is accumulated and ranged between 53 and 153 mg kg^{-1} (Szefer 2002a). This author reported that also ferromanganese nodules of the Baltic Sea contain increased level of V, up to 400 mg kg^{-1}.

Concentrations of V in river waters range widely from 0.02 to 5.8 µg l^{-1}, and the world average is estimated at 0.71 µg l^{-1} (Gaillardet et al. 2003). Some municipal waters in the USA have been found to contain V levels of between 1 and 6 µg l^{-1}, whereas up to 70 µg l^{-1} was determined in rivers in the U ores regions in the USA (ATSDR 2002a). Waters of rivers polluted by industrial effluents contain V in the range of 4–19 µg l^{-1} (Lis *vide* Kabata-Pendias and Pendias 1999).

The chemistry of V ions in aqueous media is diverse and depends highly on numerous factors involving reactants as well as reaction conditions. In oxidized water conditions, V is likely to occur in forms: VO_2, V_2O_5, VO^{2+}, and $H_2VO_4^-$. However, all V species are easily fixed in bottom sediments. Thus V contents of sediments can be an indicator for the pollution of aquatic systems.

Skjelkvåle et al. (2001) investigated the occurrence of V in the Nordic lakes, and reported its concentrations to vary, at the value of 99.5 weighted percentiles, as follows (in µg l^{-1}): 3.1 in Finland, 2.5 in Sweden, and 2.2 in Norway. Rainwater samples collected in Sweden during 1999 contain V in the range from 0.28 to 0.94 µg l^{-1}, at an average value of 0.5 µg l^{-1} (Eriksson 2001a).

Vanadium concentration in drinking water in Germany varies between 0.18 to 1.1 µg l^{-1}, depending upon the geological origin of bedrocks (Anke 2004c).

Aquatic biota. Both zoo- and phytoplankton accumulate V, and its content in aquatic plants can reach 80 mg kg^{-1} (Szefer 2002a). Vanadium is known to stimulate the photosynthesis in algae, especially blue ones.

II-5.1.5
Air

Vanadium concentrations in atmospheric air vary broadly from 0.0006 ng m^{-3} around South Pole to 200 ng m^{-3} in urban areas (Table II-5.2). Air from Spitsbergen contains V in the range of 0.4–2 ng m^{-3}, and in the remote region of the UK it varies from 5 to 10 ng m^{-3} (Cawse 1987). Concentrations of V in rural air in the USA ranged, during the period 1965–1969, from <1 to 64 ng m^{-3} (ATSDR 2002a). Reimann and Caritat (1998) reported the median values for V in the world air from remote sites at 0.5–08 ng m^{-3}, and from polluted sites at 200 ng m^{-3}.

Vanadium input to the atmosphere from natural sources (soil dusts, marine aerosols) is estimated at 65 kt yr^{-1}, and about the same amounts of V enter the atmosphere from industrial sources (Anke 2004c). Increased V concentrations in air during cold periods, as compared with warm seasons, indicate its sources from burning of fossils fuels. Its emission from the combustion of residual fuel oils in the USA, in 1970 ranged from 14 to 22 kt. Additional V source is from the burning of coal, wood, solid wastes etc. (ATSDR 2002a). The wet deposition of V in Sweden is estimated at 5.7 g ha^{-1} yr^{-1} (Eriksson 2001a).

II-5.1.6
Plants

Contents of V in food plants vary broadly, from about 10 to 700 µg kg^{-1} and do not show any clear regularity in the distribution among plant species and plants organs. The highest reported contents of V are related to the V-polluted areas in Germany (Anke et al. 2003). Several mushroom species contain elevated amounts of V, up to about 2 000 µg kg^{-1} (Table II-5.3).

Mosses collected in Norway during 1990–1995 contained V in the range of <0.01–16 mg kg^{-1}, at the average value of 2.8 mg kg^{-1} (Berg and Steinnes 1997). Vanadium content of moss samples from the vicinity of peat-fired plant was up to 8 mg kg^{-1}, and vegetables grown close to the thermal power station contained up to 4.9 mg V kg^{-1}. Also grasses and clovers, grown in surroundings of the crude oil refinery contain 8 and 13 mg V kg^{-1}, respectively (Kabata-Pendias and Pendias 1999).

Vanadium is relatively easily taken up by plants through root systems from acid soils, and presumably it is absorbed passively as indicated by the linear function of its content in plants *versus* its concentration in soil solution. Apparently cationic species of V are more available to plants than anionic species, however, both species are capable of being chelated and thus could contribute substantially to V uptake by plants from soils. The biotransformation of vanadates (VO_3^-) to vanadyl (VO^{2+}) during uptake by plants has been observed (Morrell et al. 1986).

The essentiality of V for alga species and certain bacteria has been demonstrated, while it is not considered yet to be an essential element for higher plants.

The reduction of elevated V content in plants is of a great significance since V^{5+} is recognized as a potent inhibitor of several enzymes which are largely unaffected by the reduced V^{4+}. There is also some evidence that V is a specific catalyst of N_2 fixation and may substitute for Mo, particularly in *Rhizobium* bacteria. Evidence for this comes

Table II-5.3.
Vanadium in food plants ($\mu g\ kg^{-1}$) (adapted from Kabata-Pendias and Pendias 1999 and Anke et al. 2003)

Plant	Range	Mean
Wheat, grain	7 – 10	9
Winter wheat grain[a]	0.6 – 7.8	2.2
Barley[a]	0.9 – 7.6	3.4
Oats, grain	–	60
Asparagus	52 – 128	–
Lettuce, leaves	280 – 710	300
Onion, bulbs	22 – 58	–
Cucumber	35 – 56	40
Potato, peeled	458 – 606	–
Parsley, roots	–	360
Spinach[c]	533 – 840	–
Wild mushrooms[b]	50 – 2 000	–
Apple, fruits	–	320
Lemon, fruits	–	110
Caraway, seeds	–	430
Black pepper, seeds	–	985

[a] Data for Sweden, after Eriksson (2001a).
[b] Data for the USA (ATSDR 2002a).

from the accumulation of V (up to 12 $mg\ kg^{-1}$) in nodules of several legumes (Dobritskaya *vide* Kabata-Pendias and Pendias 2001).

There is not much evidence for V toxicity to plants. Increased concentration of V (up to 3 $mg\ kg^{-1}$) in the nutrient solution under greenhouse conditions is reported to reduce the length of collard roots by 50% (Gettier et al. 1987). Symptoms of V toxicity (chlorosis and dwarfing) may appear at about its content of 2 $mg\ kg^{-1}$ in some plants.

II-5.1.7
Humans and Animals

Humans. Vanadium is distributed fairly uniformly in mammalian tissues in the range of 20 to <300 $\mu g\ kg^{-1}$ (Jørgensen 2000). In humans, V contents are as follows (in $\mu g\ kg^{-1}$): kidney, 25–28; liver, 61–67; and ribs, 26–40. Li (2000) has calculated V content in the total soft tissues of human as 12 $\mu g\ kg^{-1}$, being the lowest in muscles (10 $\mu g\ kg^{-1}$) and the highest in lung (100 $\mu g\ kg^{-1}$). Anke et al. (2003) have stated that only V content in liver has not been affected by age, while in other organs its content decreases significantly in older people.

Vanadium intake by adults varies from 10 to 60 $\mu g\ d^{-1}$, depending on the diet and regions. The dietary V intake of about 10 $\mu g\ d^{-1}$ seems to meet the requirement (Anke 2004c). The experiments of this author show that goats fed with forage of low V contents have reduced concentrations of this metal, especially in kidneys, uterus, ribs and carpal bones. In serum, V is bound mainly to transferrin and albumines, but its transport within the organism depends highly on the oxidation state (De Cremer et al. 1998). The deficiency of V has not been observed often, however its low supply can affect lipids level in blood and bone mineralization processes (Kabata-Pendias and Pendias 1999).

Biological function of V is related to metabolisms of lipids, glucose and phosphotransferrine (Salifoglou 2003). It was also shown to be a strong inhibitor of Na-K and Ca-Mg balances in the organism. Observed influence of V on metabolic and ionic changes in a heart remains to be understood. Highly oxidized species of V are toxic, especially to nervous system, and breathing and digestion processes. The increased inhalation of V (at 0.1 mg V m^{-3}) resulted in coughing and other signs of respiratory irritation. There were no observed other acute symptoms of its excess (ATDR 2002a).

The common pathway of V exposure for the general population is the ingestion of food, especially that bearing soil residue. Since children have a tendency to ingest soil (pica-soil) and dust during everyday activities, soil containing increased levels of V may result in their increased V exposure.

The threshold level for V toxicity to humans is established at 10–20 mg d^{-1} (Anke 2004c). Airborne V can cause irritation of eyes and respiratory track. The threshold limit for V-pentoxide, V_2O_5, the most toxic compound, in air dust differs in various countries, as follows (in mg m^{-3}): USA, 0.5; Russia, 0.1; and Germany, 0.05–0.1 (Anke 2004c). The guidelines applicable to V in the USA gave most commonly the limit of 0.05 mg m^{-3} (ATSDR 2002a).

Animals. Symptoms of V toxicity in young cattle are observed at intake of 10 mg V kg^{-1} BW, while a rate of 300 mg V kg^{-1} BW may have lethal effects (Hansard et al. *vide* Anke 2004c). There is some evidence that V is involved in antioxidant status of plasma of rats and interacts with Cr in oxidative processes (Ścibior 2003).

II-5.2 Niobium (Nb)

II-5.2.1 Introduction

Niobium, also called columbium (Cb), is relatively common in the terrestrial environment, at an average content of 26 mg kg^{-1} in the Earth's crust. The amounts of Nb increase in acid igneous rocks and in argillaceous sediments (Table II-5.2). The +5 valence state of Nb is the most stable, and its compounds are slightly soluble in both acid and alkaline media, especially in the presence of organic complexing agents.

In the nature, Nb and Ta occur together because of their similarities in some properties and atomic radii (Table II-5.1) and both exhibit an affinity to associate with Fe, Ti, and Zr. It is scattered in several minerals of complex formula, like for example columbite/niobite, $(Fe,Mn)(Nb,Ta)_2O_6$, and pyrochlore, $(Na,Ca)_2(Nb,Ta)_2O_6(OH,F)$. Niobium is also included in the crystal lattices of Ti-minerals and in minerals bearing REEs. Also zircon is known as carrier of Nb in igneous rocks (Fairbridge 1972).

The average Nb content in US coal is estimated to be 2.9 mg kg^{-1}, with the maximum level of 70 mg kg^{-1} (Finkelman 1999).

II-5.2.2 Production and Use

Annual global production of Nb, calculated together as Nb-Ta ores in 1995, was 18.3 kt (Reimann and Caritat 1998). Its annual production reported in the WMSY (2004) is about 27 kt. This metal is used mainly for carbon and stainless still alloys. It is also applied for welding and nuclear-industry tools. Nb-containing alloys are resistant to strength and to cryologic factors, therefore are widely used in pipeline constructions and in aerial transportation industries.

Due to special properties, Nb is added to alloys used for surgical implants as well as in the stomatology.

II-5.2.3 Soils

There are not many reports on the Nb status of soils. Geometric mean value of this element has been calculated for surface soil samples of the USA (in mg kg^{-1}) at 12.5, and for soils of Lithuania and Poland at 3.2 (Dudka and Markert 1992; Kadunas et al. 1999). Topsoil samples from Sweden contain Nb within the range of value from 5 to 17 mg kg^{-1}, at an average of 12 mg kg^{-1} (Eriksson 2001a). The median Nb contents in agricultural soils of Japan vary from 6.3 mg kg^{-1} in Andosols to 14 mg kg^{-1} in Cambisols (Takeda et al. 2004).

Increased levels of Nb, up to 300 mg kg^{-1}, is reported for *in situ* soils derived from various rocks, especially from alkaline basalts and trachyte-basalts (Ure et al. *vide* Kabata-Pendias and Pendias 2001).

II-5.2.4 Waters

Scanty data are presented by Gaillardet et al. (2003) on Nb in river waters, only for the USA, within the range of 1.2–11 ng l^{-1}. The world average for river waters is given by these authors as 1.7 ng l^{-1}. Reimann and Caritat (1998) estimated the mean value of Nb in the world ocean water as 10 ng l^{-1}, and Nozaki (2005) gave the mean concentration at <5 ng kg^{-1} in the North Pacific. In aquatic environments Nb is likely to occur in the ionic form of $Nb(OH)_6^-$.

Elevated concentration of Nb in stream bottom sediments in Europe, at the maximum value of 243 mg kg^{-1}, as reported by Reimann and Caritat (1998), may be an evidence for both anthropogenic sources and easy leaching out from soils. The amount of Nb cycling from the ocean *via* the atmosphere on the continents, as calculated by Korzh (1991) is at the order of 10 t yr^{-1}, and is fairly similar to the most of other easily mobile cationic elements. This might be an evidence of a relatively high mobility of Nb in the environment.

Rainwater samples collected in Sweden during 1999 contain Nb in the range from 0.8 to 7.9 ng l^{-1} (Eriksson 2001a).

The limit for Nb in drinking water was established in the USSR, in 1980, as 10 µg l^{-1} (Goering and Ziegler 2004).

II-5.2.5 Air

Data on Nb concentration in air are lacking, however the maximum allowable level for the hydrate form of this metal has been established at 6 mg m^{-3}, and for the hydrate form, at 10 mg m^{-3} (Goering and Ziegler 2004).

The wet deposition of Nb in Sweden has been calculated to be 23 mg ha^{-1} yr^{-1} (Eriksson 2001a).

II-5.2.6 Plants

The common range of Nb in plants is 0.02–1.1 mg kg^{-1}. Cereal grains of Sweden contain Nb in the range from 0.5 to 1.7 mg kg^{-1} (Eriksson 2001a). Its contents in herbs liquor-ice (*Glycyrhiza uralensis*) from the Ural Mountains range from 0.14 to 1.0 mg kg^{-1} (average 0.4) in root system and from 0.4–2.1 mg kg^{-1} (average 1) in top parts (Grankina et al. 2003). Several native plants have a capability to extract Nb from soils that is enriched in this metal. The evidence for this is the Nb content up to about 10 mg kg^{-1} in plants of *Rubus* species grown in the Nb mineralized area of Komi (Russia) that has served as an exploration indicator for the Nb-ore deposits (Kabata-Pendias and Pendias 2001).

Niobium in moss sampled in Norway in the period 1990–1995 ranged between <0.002 and 0.8 mg kg^{-1} and averaged 0.11 mg kg^{-1} (Berg and Steinnes 1997). Its contents have not been changed much during the recent decades.

II-5.2.7 Humans

Li (2000) calculated the mean content of Nb in total soft tissues of humans at 27 μg kg^{-1}, being the lowest in kidneys (10 μg kg^{-1}) and the highest in lymph (60 μg kg^{-1}).

II-5.3 Tantalum (Ta)

II-5.3.1 Introduction

Mean contents of Ta in the Earth's crust are established at the range of 1.5–2.2 mg kg^{-1}. The highest abundance of Ta is in granites and the lowest in some alkaline igneous rocks (Table II-5.2).

Tantalum exhibits properties of both lithophilic and chalcophlic element and occurs in few minerals, mainly oxides, and in association with Nb, Ti, Y and lanthanides. Most of minerals bearing Ta have complex structure, and several have similar composition to tantalite/columbite, $(Fe,Mn)(Nb,Ta)_2O_6$.

Geochemical properties of Ta are fairly similar to those of Nb, but it is less mobile than Nb because of a lower solubility during weathering processes and a slight stability of organic complexes. It is likely to be concentrated in argillaceous sediments and in Fe-enriched sea bottom sediments.

II-5.3.2 Production and Use

Annual global production of Ta has been calculated to be 1.22 kt in 2000 (USGS 2004).

The major use of Ta is in the production of electronic components that are broadly applied in nuclear, cosmic, and telecommunication industries. Because of high melting point and good ductility, Ta is a component of super alloys used for jet engines, nuclear reactors and missile parts. Thus, Ta is a metal of strategic importance since it is used for rust and acid resistant alloys. Metallic dust of Ta is a fire and explosion hazard.

Tantalum is totally immune to the action of body liquids and therefore is widely used in surgical appliances.

II-5.3.3 Soils

There is a paucity of information on the occurrence of Ta in soils. Its content of reference soils of China is reported to range from 0.8 to 5.3 mg kg^{-1}, and in reference soils of the USA to range between 0.7 and 1.2 mg kg^{-1} (Govindaraju 1994). The 90th percentile for Ta in Finnish soils is about 1.5 mg kg^{-1} (Köljonen 1992). The range of Ta in soils derived from loess deposit in Russia is 1–2 mg kg^{-1} (Vaganov and Nizharadze 1981).

Topsoil samples from Sweden contain Ta in the range of 0.5–1.6 mg kg^{-1}, at an average of 1.1 mg kg^{-1} (Eriksson 2001a). Median levels of Ta in agricultural soils of Japan range from 0.7 to 1.4 mg kg^{-1} being the highest in Gleysols (Takeda et al. 2004). A few reports on Ta in soils from other countries gave the range between 0.2 and 3.9 mg kg^{-1} (Kabata-Pendias and Pendias 2001).

II-5.3.4 Waters

Gaillardet et al. (2003) presented only a few data on Ta concentrations in river waters, in the range between 1 and 148 ng l^{-1}, and the world average at the value of 1.1 ng l^{-1}. Nozaki (2005) cited the mean concentration of Ta in the North Pacific to be <2.5 ng kg^{-1}.

Rainwater samples collected in Sweden during 1999 contained Ta in the range from <0.1 to 0.4 ng l^{-1}, at an average value of 0.2 ng l^{-1} (Eriksson 2001a).

II-5.3.5 Air

The concentration of Ta in air above Greenland is 10–30 pg m^{-3}, and it is substantially increased in urban areas of Japan, up to 100 pg m^{-3} and in atmospheric air of the USA, up to 280 pg m^{-3} (Table II-5.2).

Wet deposition of Ta in Sweden is reported to be 1.5 µg ha^{-1} yr^{-1} (Eriksson 2001a).

II-5.3.6 Plants

In vegetables, Ta ranges from 1 to <6 μg kg^{-1}. Cereal grains (wheat and barley) from Sweden contain Ta in the range of 1.1–5 μg kg^{-1}, with smaller amounts in wheat than in barley (Eriksson 2001a).

II-5.3.7 Humans

Scanty data on the occurrence of Ta in human tissues are presented by Divine and Goering (2004) for lungs, to range from 0.1 to 18 μg kg^{-1}, with the average value of 3.9 μg kg^{-1}.

Human exposure to Ta radionuclides *via* inhalation has been demonstrated a low solubility in human fluids. In workers of the hard-metal industry, high levels of Ta were reported in bronchoalveolar lavage fluids, however no relation between Ta concentration and hard-metal pneumoconiosis was demonstrated (Sabbioni et al. 1994).

There is no defined biological function of Ta in humans and animals. Acute and chronic effects of Ta on humans and animals are related to the impaired impact on respiratory track and cardiovascular functions. When administered orally to rats, it is poorly absorbed and when administered intravenously it is accumulated primarily in kidneys (Wilson *vide* Divine and Goering 2004).

There is also some evidence that Ta compounds can cause tumors and therefore it should be regarded as a toxic element. However, Ta implants are presently considered to be biologically inert and pose little concern for human clinical pathophysiology (Divine and Goering 2004).

The hazardous occupational exposure concentrations of Ta vary depending on its compounds and countries. The limit for metal and oxide dust is 5 mg m^{-3} and for the short-term-exposure is 10 mg m^{-3} (NIOSH, www.cdc.gov/niosh/rtecs/ww53fee8.htlml).

Trace Elements of Group 6 (Previously Group VIb)

The trace metals of the Group 6, chromium (Cr), molybdenum (Mo), and tungsten (W), have strong lithophilic tendencies. They have variable oxidation states, possess bimodal redox behavior in terrestrial and aquatic environments, and are likely to be present in two oxidation states, +3 and +6 (Table II-6.1). Metals of this group have the propensity to form oxyanions and coordinate only weakly with other common environmentally ligands such as Cl^- or OH^-.

The geochemical behaviors of Mo and W are very similar. Furthermore, in biochemical processes of some anaerobic organisms, a substitution of Mo by W in enzymes has been observed (Buc et al. 1999; Makdessi et al. 2001).

II-6.1 Chromium (Cr)

II-6.1.1 Introduction

The distribution pattern of Cr in rocks shows the tendency to its concentration in mafic igneous rocks and in argillaceous sediments (Table II-6.2). In ultramafic rocks its content can be over 1 000 mg kg^{-1} and in felsic ones (e.g., granites) may be less than 20 mg kg^{-1}.

Chromium is precipitated from magma at an early stage, in the chromium spinel mineral group, with the general formula, $(Mg,Fe)O \cdot (Cr,Al,Fe)_2O_4$. All Cr minerals are resistant to chemical weathering and due to their high specific gravity and are usually concentrated in various types of intrusions and deposits. Chromite ores differ in com-

Table II-6.1. Selected properties of trace elements of Group 6

Element	Atomic number	Atomic mass	Atomic radius[a] (pm)	Density (20 °C, g cm^{-3})	Valence[b]	Melting point (°C)
Cr, chromium	24	51.9	185	7.19	+2, **+3**, +4, +5, **+6**	1 857
Mo, molybdenum	42	95.9	201	10.2	+2, +3, +4, +5, **+6**	2 617
W, tungsten	74	183.8	202	19.3	+2, +3, +4, +5, **+6**	3 410

[a] Approximately average values for main oxidation states.
[b] Valence values in bold are for the main oxidation states.

Cr

Table II-6.2. Abundance of chromium, molybdenum, and tungsten in the environment[a]

Environmental compartment	Cr	Mo	W
Earth's crust (mg kg^{-1})	126 - 185	1.1	1
Igneous rocks (mg kg^{-1})			
Mafic	170 - 3 400	0.2 - 1.5	0.3 - 1.1
Acid	10 - 50	1 - 2	1.3 - 2.4
Sedimentary rocks (mg kg^{-1})			
Argillaceous	80 - 120	2 - 2.5	1.8 - 2
Sandstones	20 - 40	0.2 - 0.8	0.3 - 2
Calcareous	5 - 16	0.2 - 0.4	0.3 - 0.6
Raw energy material (mg kg^{-1})			
Coal, hard	10 - 38	3 - 6	0.3 - 3
Fly ash	134.2	14.6	4.5
Crude oil	0.1 - 4	–	–
Petrol	0.001 - 0.3	10	–
Soils (mg kg^{-1})	54	1.8	1.2
Light sandy	2 - 350	0.1 - 3.7	0.7 - 2.7[f]
Medium loamy and silty	10 - 300	0.4 - 6.4	0.6 - 3.4[g]
Heavy loamy	30 - 1 100	0.7 - 7.2	
Calcareous	5 - 150	0.3 - 7.4	–
Organic	1 - 100	0.3 - 3.2	
Water (µg l^{-1})			
Rain	0.2 - 0.5[b]	0.01 - 2.6[b]	–
Sea	0.3[b]	0.01[b]	0.001[b]
River	0.3 - 2.1[c]	0.04 - 1.3[c]	0.1[c]
Air (ng m^{-3})	0.6[e]	<0.2[d]	0.005[d]
Urban/Industrial	1 - 1 100	1 - 10	2.5[b]
Greenland	0.6 - 0.8	0.2[e]	–
South Pole	0.003 - 0.01	–	–

[a] Presented are common, possible background, values from various sources as compiled by Kabata-Pendias and Pendias (1999, 2001), unless otherwise indicated.
[b] After Reimann and Caritat (1998).
[c] After Gaillardet et al. (2003).
[d] Median for World air, remote, after Reimann and Caritat (1998).
[e] Data for Shetland Islands.
[f] Data for various types of soils.
[g] Data for soils of Parana State, Brazil (Licht 2005).

position mainly due to their variable proportion to silicate minerals. Cr-minerals are likely to be associated with pyroxenes, amphibolites and micas. Among these minerals chromite, $FeCr_2O_4$ has an economic importance. Crocoite, $PbCrO_4$, is also relatively common.

The speciation of Cr is Eh-pH dependent under most environmental conditions. The geochemistry of Cr is complex because of its easy conversion from one oxidation state to another. The variable properties of Cr have been described in several fundamental publications (e.g., Bartlett 1999; Barnhart 1997; James et al. 1997; Kotaś and Stasicka 2000).

The geochemical association of Cr with Fe and Mn is responsible for its elevated concentrations in ferromanganese nodules of sea bottom sediments as well as in Fe-concretions of soils.

II-6.1.2 Production and Use

Global production of Cr in 2003 was reported at 15.83 Mt (WMSY 2004). The main use of Cr is in metallurgical, refractory and chemical industries, however, it was first used as a pigment because of its yellow color.

The major proportion of Cr is used for stainless steel and chromate plating. In the chemical industry, Cr (both +3 and +6) is used primarily in pigments, metal finishing, leather tanning and wood preservatives. It is commonly used in the manufacture of green tints for paints, varnishes, glazes and inks. Leather tanning utilizes especially substantial amounts of Cr compounds. This metal is used in various phases of paper manufacture. It is a component of different refractory products, such as refractory bricks.

The main source of Cr pollution is considered to be from dyestuffs and leather tanning when wastes are discharged directly into waste streams, either as liquids or solids. Sewage treatment plants from industrial and residential sources discharge substantial amounts of Cr. Thus, COPR (chromite-ore processing residue) is of the greatest environmental risk in some regions.

II-6.1.3 Soils

The world median content of Cr in soils has been established as 54 mg kg^{-1} (Table II-6.2). Its content in soils is determined mainly by its abundance in the parent material. Since soil Cr is inherited from parent rocks, higher contents are generally found in soils derived from mafic rocks and argillaceous sediments. Soils developed from serpentines are especially elevated in Cr, sometimes to above 100 000 mg kg^{-1}. Chromium occurs in soils mainly (>80% of total content) in the immobile residual fraction (Kabata-Pendias and Pendias 1999).

A positive relationship between Cr and the fine granulometric fraction in soils has resulted in a higher Cr content in silty and loamy soils than in sandy ones. Sandy and light loamy soils contain Cr within the range of 2–350 mg kg^{-1}, whereas in heavy loamy and clay soils Cr ranges from 30 to 1 100 mg kg^{-1} (Kabata-Pendias and Pendias 2001). Swedish arable soils contain Cr in the range from 3 to 50 mg kg^{-1}, at an average value of 22 mg kg^{-1} (Eriksson 2001a). The median Cr contents in agricultural soils of Japan vary from 56 to 70 mg kg^{-1} in Andosols and Acrisols, respectively (Takeda et al. 2004).

The highly variable oxidation states of Cr (Table II-6.1) allow the formation of complex anionic and cationic species, as well as organic complexes. Naturally occurring Cr

Cr

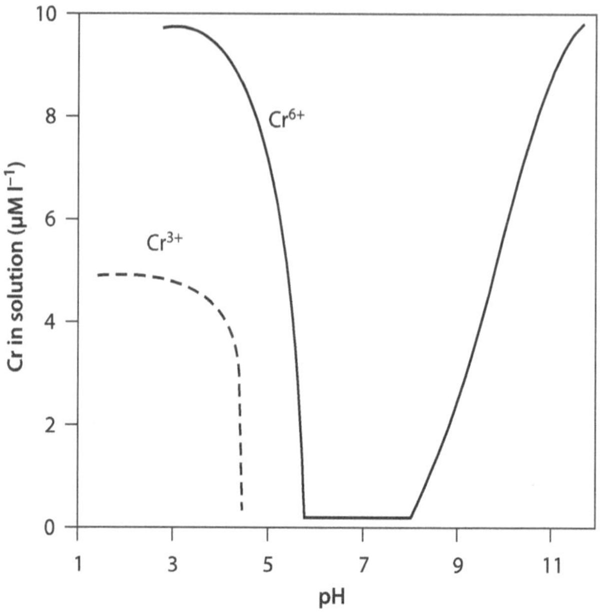

Fig. II-6.1. Solubility of trivalent and hexavalent Cr as a function of pH (based on data after Bartlett and Kimble 1976)

compounds have principal valences of +3 (chromic) and +6 (chromate). Highly oxidized Cr forms are much more mobile than Cr^{3+} species, especially under very acid and alkaline ranges of pH (Fig. II-6.1). Estimated complex ion species mobile (soluble) in soil environments are: $CrOH^{2+}$, CrO_4^{2-}, $HCrO_4^-$, $HCrO_3^{2-}$, $Cr(OH)_4^-$, $Cr(CO_3)_3^{3-}$ (Kabata-Pendias and Sadurski 2004). In the contrary, the species $Cr(H_2O)_6^{3+}$, is considered relatively inert due to the slow exchange of hydration water (Raspor 2004). In addition, Cr reveals a great capability to form various organic complexes. Thus, ionic Cr species are highly susceptible to oxidation-reduction processes that govern the behavior of this metal in soils (Bartlett 1999).

Soil pH has an influence on Cr speciation. In uncontaminated (pristine) soils at neutral pH, Cr is slightly mobile and thus not easily available to plants which may have nutritional implications. However, several soil factors, and especially organic substances have a crucial impact on Cr forms and behavior. Soil pH also has an influence on Cr species and solubility (Fig. II-6.1). As El-Bassam (*vide* Kabata-Pendias and Pendias 2001) observed, under the same redox potential (500 mV), Cr^{3+} predominated at pH 5, $Cr(OH)_3$ was formed between pH 5 and 7, and CrO_4^{2-} occurred at pH >7. The rapid conversion of the Cr oxidation state is closely associated with Fe and Mn oxidation-reduction processes in soils (James et al. 1997; Geelhoed et al. 2003). The reduction of Cr^{6+} can lead to two general types of products: soluble Cr^{3+}-organic compounds or Cr-hydroxide precipitates, $Cr_{1-x}Fe_x(OH)_3$ (Fendorf et al. 2004). However, biological reduction or reduction by organic molecules have the potential to form soluble Cr^{3+}-organic complexes. Biogenic Fe^{2+} is also involved in the reduction of Cr^{6+}. Due to the ease in variability of the oxidation state of Cr, this metal is used for the standard spot plate field test for the oxidation potential in soils (Bartlett 1999).

The sorption of Cr in soil is associated primarily with the clay content, and to a lesser extent with Fe hydroxides and the organic matter content (SOM). The impact of

SOM on the conversion, and particularly on the reduction of Cr (especially Cr^{6+} to Cr^{3+}) is of great environmental significance, and has been studied widely (Bartlett 1999; Barnhart 1997; Wittbrodt and Palmer 1996). James et al. (1997) and James (2001) described the Cr^{3+}/Cr^{6+} balance of exchange as an oscillating "seesaw" with available Mn and organic matter, where the soil pH acts as a controllable master variable that helps to set the oxidizing reactivity of Mn oxide *versus* the reducing properties of organic matter and other compounds (e.g., Fe, S). These authors has also described the variability of Cr ions as affected by the oxidation-reduction potential of soils amended with various wastes, particularly those containing organic matter. Under oxidizing conditions, Cr^{6+} in soil is present mainly as CrO_4^{2-} and $HCrO_4^-$ (James et al. 1997). Under natural soil conditions, the Cr oxidation potential seems to be directly related with Mn-oxide contents (Chung and Sa 2001). Peat exhibits a high sorption capacity for Cr, in the range of values from 24 250 to 52 800 mg kg^{-1}, and bind this metal mainly as an insoluble metal-organic complex (Kyzioł 2004).

Forms and transformation of Cr in soils have great environmental and health implications. Therefore, the speciation of Cr in soils and its solubility in Cr-polluted soils have been widely studied (Barnhart 1997; Rüdel et al. 2001). The conversion of Cr species in soils can be relatively rapid and significant. Dudka et al. (1993) demonstrated, using Tessier's method (Tessier et al. 1979), that organic matter added as sewage sludge to the soil (loamy sand, pH 6.1) caused significant variation in Cr species: exchangeable forms increased from 8 to 52%, and organic matter-bound forms increased from 9 to 31% of the total Cr content. Such concentrations of mobile Cr can be toxic to crop plants. In highly Cr-polluted soils in Mexico, total Cr contents in surface layer vary between 807 and 12 960 mg kg^{-1} and is concentrated mainly in two fractions; reducible hydrous Fe-Mn oxides and oxidizable sulfides and organic phases (Armienta et al. 1996).

In ultramafic soils, where a large part of Cr occurs mainly in insoluble phases, Cr was present in the soil solution, under natural vegetation, in the range of 2.7–10 μg l^{-1}, whereas under crop vegetation, due to effects of phosphorous fertilizers, its concentrations reached up to 700 μg l^{-1} (Becquer et al. 2003).

Readily soluble Cr^{6+} in soils is toxic to both plants and animals. Toxic effects, however, are highly dependent on soil properties, and especially on soil texture and pH. The activities of soil microbial enzymes decrease under increased Cr-levels in soils; dehydrogenase activity and nitrification processes are especially sensitive (Pacha 1989; Rüdel et al. 2001). Hexavalent Cr compounds exhibit a strong mutagenic effect on *Bacillus subtilis*, while Cr^{3+} compounds are only slightly mutagenic to that microbial species (Pacha and Galimska-Stypa (1988). The response of soil bacteria to Cr^{3+} is species dependent, that is, while *Rhizobium* sp. and *Bradyrhizobium* sp. do not tolerate any Cr^{3+}, the growth of other soil bacteria is stimulated by concentrations (0.2–5 mg l^{-1}) of Cr^{3+} (Wyszkowska and Kucharski 2004).

Soil Contamination and Remediation. The Cr content of surface soils is known to have increased due to pollution from various sources, of which the main ones are COPR, pigments and tannery wastes, leather-manufacturing wastes, and municipal wastes. A number of studies have recently been curried out to understand environmental problems and effectives of remedial treatments of Cr-contaminated land (Farmer et al. 1999; Geelhoed et al. 2003; Hiller et al. 2003).

Land contaminated with COPR (millions of tons of COPR in the past have been deposited in lands around Cr industries) has recently become of great environmental concern.

Municipal and industrial wastes from various sources may contain elevated amounts of Cr, up to 10 200 mg kg^{-1}, at an average of 1 993 mg kg^{-1} (Maján et al. 2001). Its contents in sewage sludge (mainly municipal wastes) applied to agricultural soils usually vary between 100 to 200 mg kg^{-1}. The deposition rates of Cr to agricultural soils have been calculated as 0.5–46 g ha^{-1} yr^{-1}, being the lowest in Finland and the highest in Italy (Nicholson et al. 2003). As estimated by these authors, the total annul input of Cr to agricultural land in England and Wales in 2000 was 327 t of which 126 t was from inorganic fertilizers (mainly phosphate), 83 t from atmospheric deposition and 78 t from sewage sludge.

Topsoils around electroplating industries in Portugal contain Cr up to 27 132 mg kg^{-1}, at an average of 1 000 mg kg^{-1} (Morgado et al. 2001). On a worldwide basis, the disposal of commercial products that contain Cr may be the largest contributor, accounting for about 51% of the total Cr released to soil. The disposal of coal fly ash and bottom fly ash contribute more than 30% of Cr to soil (Nriagu and Pacyna 1988).

Farmlands amended with sewage sludges are likely to be enriched in Cr, especially because Cr is usually accumulated at the thin top layer of soils. Chromium, as all microelement-cations, is not distributed uniformly in soils but is likely to be concentrated in nodules and concretions. Nodules separated from the COPR-contaminated soils near Glasgow, Scotland, contain elevated amounts of Cr between 14 700 and 36 400 mg kg^{-1}, whereas, the whole surface soil layer (20–40 cm) averaged 16 200 mg Cr kg^{-1} (Farmer et al. 1999).

The main purpose of remediation treatments of Cr-contaminated soils is the conversion of easily mobile Cr^{6+} to Cr^{3+}. Both Cr species occur in COPR, which has a pH between 11 and 12, and is associated with various mineralogical phases. Most minerals, e.g., carbonates, oxides and silicates exhibit a strong capacity to accommodate significant amounts of Cr. Due to specific geochemical properties of COPR, ferrous sulfate has appeared to be not effective enough in the reduction of Cr^{6+} (Hiller et al. 2003). Geelhoed et al. (2001 and 2003) have emphasized that each Cr-contaminated site has to be studied to evaluate the adequacy treatments designed using the model ORCHESTRA-CHROMIUM, based on chemical equilibrium and the precipitation/dissolution reactions of various mineral phases.

Some recent studies have been focused on the effect of organic compounds (carbon) on the reduction of Cr^{6+} to Cr^{3+}. Tzou et al. (2003) reported that in surface soils this reduction is highly accelerated even when artificial light is available, whereas Tokunaga et al. (2003) indicated an increased reduction of Cr^{6+} due to microbial activities after the addition of organic carbon to heavily contaminated soils. Alcântara and Camargo (2004) reported an increased Cr movement in the A soil horizon, possibly due to the formation of soluble compounds with fulvic acids.

Varieties of treatments have been applied with different effects depending upon specific geochemical conditions of Cr-contaminated sites. Some of the remediation treatments are as follows:

- Immobilization of Cr using materials of high sorption capacity, e.g., smectite clays, coal, bone charcoal (or other sorbents)
- Removal of Cr by electrokinetic techniques
- Reduction of Cr^{6+} by ferrous sulfate and/or sulfate-reducing bacterial biofilms
- Phytoremediation

The suitability of phytoextraction methods is rather limited because the SPTF (Soil-Plant Transfer Factor) for Cr is relatively low and varied for herbaceous plants from 0.1 to 0.6 (Bylińska 1993) and Cr is known to be concentrated mainly in roots. The addition of organic ligands (e.g., picolinic acid) can promote the phytoavailability of Cr, as Simon et al. (2001) have reported for a fodder radish.

The regulatory standards set up for soil Cr in the USA have been based on the inhalation or ingestion exposure and range from 50 to 10 000 mg kg^{-1} (Proctor et al. 1997).

II-6.1.4 Waters

The global median value for Cr in ocean water is given as 0.3 μg l^{-1} (Reimann and Caritat 1998). Nozaki (2005) cited, after Nakayama, the following mean concentrations in the North Pacific Ocean: Cr^{6+}, 0.21 μg kg^{-1}; and Cr^{3+}, 0.002 μg kg^{-1}.

The Cr content in particulate matter of the Baltic Sea varies from 0.3 to 185 mg kg^{-1}, being the lowest in the Southern Baltic and the highest at the Swedish coast (Szefer 2002a). Estimated total input of Cr to the Baltic Sea is relatively high, and Matschullat (1997) calculated this as 540 t yr^{-1}, of which 46% is of anthropogenic origin.

The median Cr content in world river waters is given as 0.7 μg l^{-1}, with the range of 0.04–1.3 μg l^{-1} (Table II-6.2). Average values for Cr in river waters across the eastern UK range from 0.4 to 6.9 μg l^{-1}, reaching the concentration of 46.8 μg l^{-1}, in industrially polluted rivers (Neal et al. 2003).

The chemical mobility of Cr during erosion and transport is relatively low, and calculated as about a 0.02 Cw/Cc ratio (Gaillardet et al. 2003), indicating a low geochemical mobility of the metal. Chromium does not occur as mobile forms in the hydrosphere for a long time period, however, once in marine waters, its residence time is estimated as approximately 10^4 years. Chromium occurs in the oceans as both Cr^{3+} and Cr^{6+} ions, but most commonly in the highest oxidation state, CrO_4^{2-}.

Hexavalent Cr in ground waters is likely to be bound to organic layers (Konek et al. 2004). Under natural conditions in surface waters, aqua/hydroxo complexes of Cr^{3+} are dominant, although various complexes with organic matter are likely to occur. It also may occur in surface waters in forms of ions such as CrO_4^{2-}, $Cr_2O_7^{2-}$, $CrOH^{2+}$, and Cr^{3+} (Witczak and Adamczyk 1995).

The speciation of Cr in wastewaters varies widely since organic and inorganic ligands determine its forms. Hexavalent Cr dominates in effluents from metallurgical, refractory and pigment industries, whereas trivalent Cr is likely to be present in wastewater from tannery, textile and plating industries (Kotaś and Stasicka 2000). Szefer (2002b) reports that the Baltic Sea accumulates 540 t Cr yr^{-1}, 47% of which is from anthropogenic contamination. It has been estimated that electroplating, leather tanning, and textile industries release especially large amounts of Cr to surface waters (ATSDR 2002a).

Farmer et al. (2002) reported that Cr concentrations in groundwater from the vicinity of Cr ore processing residue disposal sites were up to 91 mg l^{-1} and predominantly in the hexavalent form, as CrO_4^{2-}. However, in the presence of organic (humic) matter it is likely to form soluble Cr^{3+}-humic complexes or precipitate as insoluble $Cr(OH)_3$. Thus, the addition of organic matter may be considered for the remediation of Cr-contaminated aquatic sites. Concentrations of Cr in groundwater at a laboratory (Idaho

State), where chromate was used as a corrosion inhibitor ranged from <1 to 280 μg l^{-1} (ATSDR 2002a).

Chromium concentrations in US drinking water are reported to range from 0.4 to 8 μg l^{-1}, with a mean value of 1.8 μg l^{-1} (ATSDR 2002a). Health based limit value is 50 μg Cr l^{-1} (WHO 2004).

Rainwater samples collected in Sweden during 1999 contained Cr in the range of 0.08–0.38 μg l^{-1}, at an average value of 0.23 μg l^{-1} (Eriksson 2001a). Concentration of Cr in rainwater around the Baltic Sea Basin varies from 1.6 to 20 μg l^{-1}, being the highest at the North Sea (Szefer 2002a).

Aquatic biota. Chromium occurs in tissues of all aquatic animals, and its highest content was reported by Lande (*vide* Szefer 2002a) to be 2 mg kg^{-1} FW in haddock (*Gadus aeglefinus*), and 20 mg kg^{-1} in soft tissue of mussels from the Baltic Sea. The content of Cr in seaweeds of the Baltic Sea varies broadly from 0.1 to 15 mg kg^{-1} (Szefer 2002a). Some submerged aquatic plants (e.g., *Vallisneria spiralis*) can accumulate extremely high amount of Cr, above 1 000 mg kg^{-1}, and are proposed for the bioremediation of tannery effluents (Sinha et al. 2002). Trojanowska (2003) reported a close relationship between Cr contents in roots of *Phragmites communis* (macrophyte) and its contents in bottom sediments contaminated by tannery effluent.

Some marine invertebrates (e.g., tunicates, crabs) can accumulate high amounts of Cr, about 1 000 mg kg^{-1}. The toxicity of Cr^{6+} to fish is high and has an impact on some enzymatic activities. Toxicity is, however, very dependent on other factors, mainly water pH (Stoecker 2004).

II-6.1.5 Air

Global median values of Cr in air are reported as 0.5–0.6 ng m^{-3} in remote areas and as 40 ng m^{-3} in polluted regions (Reimann and Caritat 1998). However, Cr in air varies highly from 0.003 ng m^{-3} above the South Pole to over 1 000 ng m^{-3} in industrial areas (Table II-6.2). Continental dust flux is the main natural source of Cr in the atmosphere. Anthropogenic sources, mainly from emissions of the metal industry and from fuel combustion, contribute above 70% of Cr to the atmosphere. The total atmospheric Cr concentration in the USA is <10 ng m^{-3} in rural areas and 10–30 in urban areas. The maximum level of Cr in ambient air sampled during the period 1978–1982 at a site of chromate manufacture in the State of Texas was 5 500 ng m^{-3} (ATSDR 2002a). The global emission of Cr during 1990 has been calculated at 30 kt, most of which was from metal smelters.

In the atmosphere, Cr^{6+} may be reduced to Cr^{3+} by various compounds of V^{2+}, Fe^{2+}, and As^{2+}. Conversely, Cr^{3+} may be oxidized to Cr^{6+} in the presence of at least 1% of Mn oxide (ATSDR 2002a).

Atmospheric deposition of Cr on agricultural land in 1990 was estimated to average (in g ha^{-1} yr^{-1}) as follows: 8.7 in England and Wales, 3.5 in Germany, and 27.8 in New Zealand (Alloway et al. 1999; Gray et al. 2003). However, earlier deposition of Cr in the UK was estimated as 610 g ha^{-1} yr^{-1} (Cawse 1987). Recently, wet deposition in Sweden was calculated to be 2.5 g ha^{-1} yr^{-1} (Eriksson 2001a).

II-6.1.6 Plants

There is no clear evidence to date that Cr is essential in plant metabolisms, and thus, described positive effects of Cr application on plant growth and development may be artifacts.

Chromium is slightly available to plants and not easily translocated within plants, thus it is concentrated mainly in roots. The form most available to plants is Cr^{6+}, which is the very unstable form under normal soil conditions. Also Cr^{3+} and several complex Cr anions (e.g., CrO_4^{2-}) seem to be easy available to plants. Zayed et al. (1998) have shown that patterns of accumulation and translocation of these two Cr species (Cr^{3+} and CrO_4^{2-}) are identical, however, Cr^{3+} is rather benign to most plants and binds strongly to soil solids.

The mechanism of absorption and translocation of Cr in plants is apparently similar to those of Fe, and therefore the capability of roots to convert Cr^{3+} to CrO_4^{2-} is the key process in Cr absorption. This Cr conversion has been observed in spite of the different forms of Cr supplied to plants. There is some evidence that easily available Cr^{6+} is transformed into Cr^{3+} forms in plant cells (Zayed and Terry 2003). Fendorf et al. (2004) concluded, based on results of recent work, that Cr bioaccessibility is a function of soil type and retention time. Plants, in general, accumulate Cr mostly in roots, apparently because of the propensity of Cr^{3+} to bind to cell walls (Zayed et al. 1998). The ratio of the shoot/root Cr concentration varies widely from 0.005 to 0.027 for most vegetable crop plants.

Contents of Cr in plants have recently received much attention due to the knowledge of its importance as an essential micronutrient in human metabolic processes, but also because of its carcinogenic effects. Thus, an adequate rate of nutritional Cr has become an important issue. There is some evidence of Cr phytotoxicity, and Chatterjee and Chatterjee (2000) reported that an excess of Cr in plants caused poor protein formation resulting from the disruption of N metabolism.

In cereal grains and vegetable crops, Cr contents are fairly uniform (Table II-6.3). Contents of Cr in cereal grains seem to be rather stable and vary either from 0.01 to 0.35 mg kg^{-1} (Kabata-Pendias and Pendias 2001) or from 0.07 to 0.41 mg kg^{-1} (Bratakos et al. 2002). Winter wheat and barley grains grown in Sweden contain Cr at the mean values of 0.01 and 0.02 mg kg^{-1}, respectively (Eriksson 2001a). Cereal grains sampled in France in 1998 contained Cr in the range from 0.04 to 0.25 mg kg^{-1} which was influenced mainly by soil types (Baize et al. 2003). The highest content of Cr is in carrot (0.13 mg kg^{-1}), onion (0.16 mg kg^{-1}), and cabbage (0.13 mg kg^{-1}). Leafy vegetables and fruits contain Cr in the range of 0.04–0.08 mg kg^{-1}. Relatively higher amounts of Cr are found in nuts (0.16 mg kg^{-1}) and almonds (0.11 mg kg^{-1}).

Chromium contents in moss samples from Norway, collected during the period 1990–1995, averaged 2.6 mg kg^{-1}, in the range of 0.21–290 mg kg^{-1}, across all sampling sites (Berg and Steinnes 1997). Lichens from Wisconsin State, USA sampled during the period 1987–2001 contained Cr at the average value of 1.2 mg kg^{-1}, in the range from 1.1 to 1.4 mg kg^{-1} (Bennett and Wetmore 2003). These data indicated a rather stable emission of Cr to the atmosphere in these regions.

The range of average contents of Cr in the mushroom, bay bolete (*Xerocomus badius*) sampled in the period of 1993–1998, from the northern part of Poland, were 0.22–0.75 and 0.20–0.71 mg kg^{-1}, in caps and stalks, respectively (Malinowska et al. 2004).

Cr

Table II-6.3. Chromium in food plants (μg kg^{-1}) (compiled from Kabata-Pendias and Pendias 1999, 2001, unless otherwise indicated)

Plant	Range	Mean
Wheat, grains[a]	4 – 0.20	10
Barley, grains[a]	10 – 20	20
Oats, grains	40 – 600	90
Rye, grain	40 – 90	60
Carrots[b]	50 – 210	130
Onion, bulbs	20 – 240	160[b]
Maize (corn), cobs[b]	220 – 290	–
Cabbaged	50 – 210	130
Potatoes, unpeeled	20 – 50	40[b]
Legumes	500 – 160	90
Lettuces	30 – 60	50
Spinach	69 – 110	80
Cucumber	60 – 90	70[??]
Tomato, fruits	70 – 130	90[b]
Apple, fruits	10 – 100	80
Lemon, fruits	30 – 50	40
Hazelnuts[c]	90 – 240	160
Almonds[c]	80 – 150	110
Grains and cereals[d]	40 – 220	–

[a] After Eriksson (2001a).
[b] After Czekala (1997).
[c] After Bratakos et al. (2002).
[d] After ATSDR (2002a).

II-6.1.7 Humans

Contents of Cr in mammalian tissues vary between 30 and 290 μg kg^{-1}, being the highest in skin and the lowest in heart and liver (Jørgensen 2000). Its concentrations in soft tissues of humans range from 5 to 500 μg kg^{-1}, in muscles and lungs, respectively (Li 2000). Normal endogenous Cr levels for the USA population have been reported as 0.01–0.17 μg l^{-1} in serum, 0.24–1.8 μg l^{-1} in urine, and 234 μg kg^{-1} in hair (ATSDR 2002a). In workers exposed to Cr^6, the mean level of total Cr body's fluids significantly increases.

Amounts of Cr in many human organs decrease with aging, especially in the liver, whereas its concentration in the lungs is likely to increase with age. The Cr concentration in human milk is especially variable, with a range of 0.4–80 $\mu g\,l^{-1}$. This can be related to both different dietary habits as well as to genetic factors (Stoecker 2004).

Chromium is an essential micronutrient for normal energy metabolism of humans and animals. It is reported to control the metabolism of glucose and lipids. It is a part of the glucose tolerance factor (GTF) and appears to affect some of the enzymes that regulate cholesterol synthesis. The beneficial impact of Cr on cholesterol fractions and the ratio of LDL and HDL (low and high density lipoproteins) have been observed. Although severe Cr deficiency is rare, marginal Cr deficiency is relatively common and results in:

- Increased cholesterol levels
- High blood sugar levels
- Coronary dysfunction
- Some changes in aortas
- Some abnormalities of nerve stimulation of the extremities

However, Cr^{6+} also exerts toxic effects and any exposure (oral, dermal and inhalation) to increased concentrations of Cr can be harmful (ATSDR 2002a). Mechanisms of Cr^{6+} induced carcinogenesis have been widely investigated (Anderson 1997; Klein 1996; Stoecker 2004). Excessive doses of Cr may result in liver and kidney failure, anemia, muscle breakdown and abnormalities in blood clotting. When an excess of Cr compounds are inhaled, lung, nasal and possible stomach cancer may develop (ATSDR 2002a). Also dermal allergy and asthma may occur under the exposure to Cr compounds.

Toxic effects in humans, and occupational exposure limits (OELs) for Cr^{6+} have been reported by Cross et al. (1997). When Cr compounds in air are poorly soluble the exposure limit of 50 $\mu g\,m^{-3}$ may be adequate, however for a safe protection TWA (Time-Weighted Average) limits have been set at 10–25 $\mu g\,m^{-3}$. These authors presented the quantitative risk assessment for lung cancer related to occupational exposure over a working lifetime and have stated that mortality from lung cancer is apparently influenced by cumulative exposure.

The primary sources of dietary Cr are vegetables, whole cereal grains and nuts, (Table II-6.3) followed by egg yolks, brewer's yeast, liver, and seafoods (Stoecker 2004). Recently, a dietary supplement of Cr picolinate (picolinic acid is the body's prime natural chelator) has been proposed. Ascorbic acid (vitamin C) and glutathione are found to reduce Cr^{6+} to Cr^{3+} and thus may reduce toxic effects of the excess of Cr^{6+}.

Several oral studies reported no signs of Cr^{6+} toxicity following administration of 70–100 $mg\,kg^{-1}$ in the diet or drinking water (Cross et al. 1997). In the USA, daily dietary intake of Cr by adults varies in the range of 25–224 μg, at an average value of 75 $\mu g\,d^{-1}$ (ATSDR 2002a). Anderson (1989), on the other hand, estimated the average Cr intake by population in the USA at about 50 $\mu g\,d^{-1}$ and concluded that it might not cover the requirements of some organisms. This author reported that the RDI is estimated for adults as 50–200 μg in different countries.

The regulatory limit for Cr in drinking water is 0.05 $\mu g\,l^{-1}$, for dietary intake is 3 $\mu g\,kg^{-1}\,d^{-1}$, and in air is from 0.5 to 500 $\mu g\,m^{-3}$, depending on the Cr species, compounds and time of exposure (ATSDR 2002a).

II-6.2 Molybdenum (Mo)

II-6.2.1 Introduction

The average content of Mo is given for the upper Earth's crust as 1.1 mg kg^{-1} (range 1–2 mg kg^{-1}), and its concentration is elevated in granitic rocks and argillaceous sediments, up to 2.5 mg kg^{-1} (Table II-6.2).

Molybdenum exhibits unusual geochemical behavior since in most environmental compartments it is likely to form oxyanions, however, under reducing conditions it has a high affinity for sulfur. The most common Mo mineral is molybdenite, MoS_2, which occurs in several geological formations. Less frequent are: molibdite, MoO_3; wulfenite, $PbMoO_4$; and powellite, $CaMoO_4$. Several other Mo minerals, mainly oxidates, are known to be associated with various sediments. Especially Mn-Fe rich nodules have the ability to concentrate Mo up to 2 000 mg kg^{-1} (Fairbridge 1972). Ferromanganese oxides in ocean sediments are known to accumulate high amounts of Mo (Anbar 2004; Arnold et al. 2004). These concretions seem to occur under anoxic S-rich conditions (Lei and Boström 1995; Erickson and Helz 2000; Arnold et al. 2004). Szefer (2002a) gives the range of 20–507 mg kg^{-1} for Mo in ferromanganese nodules of the Baltic Sea. The Mo content in soil Mn-nodules is reported to reach 410 mg kg^{-1} (Kabata-Pendias and Sadurski 2004).

II-6.2.2 Production and Use

Molybdenum is mined as a primary ore deposit, mainly as molybdenite, and as a by-product, generally from copper mines. Annual production of Mo in 2003 was 127.4 kt (WMSY 2004). Its main use is in metallurgy for the hardening of alloys. This metal is also widely used in electronics and chemistry (e.g., catalysts).

II-6.2.3 Soils

For soils, the Global mean of 1.8 mg kg^{-1} for Mo has been calculated. However, its occurrence in soils ranges from 0.1 to above 7 mg kg^{-1}, and its concentrations resemble its abundance in parent rocks (Table II-6.2). The Mo means for different soil units vary from 1.3 mg kg^{-1} for sandy podzols to 2.8 mg kg^{-1} for loamy cambisols. The 90th percentile value for Mo in Finnish soils is 1.5 mg kg^{-1} (Köljonen 1992), whereas the median content of Mo in Lithuanian soils is reported to be 0.62 mg kg^{-1} and does not show any distinct association with soil texture (Kadunas et al. 1999).

Soils in arid and semiarid regions, especially ferrasols, usually contain relatively high amounts of Mo. These soils, and soil contaminated with Mo, need amelioration to depress the Mo availability to plants. Increased contents of Mo, up to 35 mg kg^{-1} are reported for the vicinity of Mo processing plants in the USA, and up to 38 mg kg^{-1} in Chile (*vide* Kabata-Pendias and Pendias 2001). Also high concentrations of Mo were determined in soils of British Columbia, up to 24 mg kg^{-1}, on which vegetables were commercially grown (Pieri et al. 1996). Swedish arable soils contain Mo in the range from 0.1 to 4 mg kg^{-1}, at an average value of 0.6 mg kg^{-1} (Eriksson 2001a).

Molybdenum behaves both like a chalcophile and a lithophile element and its chemistry in soils is related mainly to anionic species. The MoO_4^{2-} anion dominates in the neutral and moderate alkaline pH range, whereas $HMoO_4^-$ occurs at lower pH values. Both anions occur in soil solutions (Kabata-Pendias and Sadurski 2004). The behavior of Mo in soils differs greatly from that of other trace metals, and it is slightly mobile in acid soils (pH <4–5) and readily mobilized in alkaline soils (pH >6.5). However, relatively easily soluble Mo anions are readily coprecipitated by organic matter, $CaCO_3$, and, in addition to Fe and Mn, by several other cations (e.g., Cu, Zn, Pb). It is especially readily adsorbed by freshly precipitated $Fe(OH)_3$. Molybdenum is likely to form various complex compounds and/or minerals in soil, as for example $PbMoO_4$ (wulfenite), $Fe_2(MoO_4)_3 \cdot 8H_2O$ (ferrimolybdite), and other semicrystalline Fe-Mo forms. Under reducing conditions Mo can easily form soluble thiomolybdates (e.g., MoS_4^{2-}, $MoO_2S_2^{2-}$). All these reactions are highly dependent on Eh-pH conditions of soil, and thus may be quite variable.

II-6.2.4 Waters

Reimann and Caritat (1998) calculated the median concentration of Mo in global ocean waters at 10 μg l^{-1} and a similar value is given by Nozaki (2005) for the North Pacific Ocean. Waters of the Baltic Sea contain Mo in the range from 1.6 to 6.3 μg l^{-1}, as recalculated from values given in nmol kg^{-1} (Szefer 2002a).

Concentrations of Mo in river and lake waters are highly variably between <0.2 and 60 μg l^{-1}, from clean to polluted waters, respectively (Kabata-Pendias and Pendias 1999). The world average of Mo in rivers is estimated at 0.42 μg l^{-1}, whereas the mean value for the Amazon River is given as 0.18 μg l^{-1} (Gaillardet et al. 2003).

Concentrations and behavior of Mo in waters, especially in seawaters, have recently received much attention due to its unique properties. Under oxygenated conditions its concentrations, nominally do not change, whereas in anoxic waters it is readily removed and precipitated in bottom sediments, so its enrichment is considered diagnostic of reducing conditions (Anbar 2004).

The speciation of Mo in waters is affected by the pH-Eh regime. The ion MoO_4^{2-} predominates in water with pH >5, H_2MoO_4, whereas $HMoO_4^-$ occurs at pH <5, and such species as $Mo_7O_{24}^{6-}$ or $Mo_8O_{26}^{4-}$ predominates at pH <6, at a high Mo concentration (Anbar 2004).

The background level of Mo in ground waters is calculated at 5 μg l^{-1}, and the recommended threshold Mo concentration in irrigation waters is 10 μg l^{-1} (Vermes 1989).

Concentrations of Mo in atmospheric precipitations over the Baltic Sea are in the range of 0.08–0.27 μg l^{-1} (Szefer 2002a). Rainwater samples collected in Sweden during 1999 contain Mo in the range between 0.03 and 0.06 μg l^{-1} (Eriksson 2001a). The wet deposition of Mo in this country was determined as 0.4 g ha^{-1} yr^{-1}.

Health based limit value for Mo in drinking water is 70 μg l^{-1} (WHO 2004).

Mo

II-6.2.5 Air

The background concentration of Mo in air above the Spitsbergen and Shetland Islands ranges between <0.05 and <0.6 ng m^{-3}, whereas above urban regions of Europe the concentration of this metal can reach 10 ng m^{-3} (Table II-6.2).

II-6.2.6 Plants

Molybdenum is an essential nutrient to all plants, but the requirement for this element by most plants is relatively low.

The average Mo content of cereal grains is 0.5 mg kg^{-1} and does not show a great variation neither between cereals nor between field conditions (Table II-6.4). For example, the lowest average Mo content of barley, 0.16 mg kg^{-1}, was estimated in Norway and the highest, 0.72 mg kg^{-1} – was in Russia (Kabata-Pendias and Pendias 2001). Winter wheat and barley grains grown in Sweden contain Mo at the mean values of 1 and 0.5 mg kg^{-1}, respectively (Eriksson 2001a). Staple foods from Germany, analyzed during the period 1990–1994, contain following amounts of Mo (in mg kg^{-1}): rye grains, 0.54; wheat grains, 0.40; and unpeeled potato tubers, 0.21 (Brüggemann et al. 1996).

Molybdenum content of forages is of a special concern in animal nutrition. Therefore, most available data are on the Mo status in grasses and legumes. In general, contents of this metal greater that 10 mg kg^{-1} can cause molybdenosis, a serious problem to most ungulate livestock. However, in areas where toxicities of Mo were observed in livestock, its contents were reported as low as 1.5 mg kg^{-1} in grasses and 5.2 mg kg^{-1} in

Table II-6.4. Molybdenum in food plants[a] (mg kg^{-1})

Plant	Range	Mean
Wheat, grain[b]	0.2 – 2.4	1.0
Barley, grain[b]	0.2 – 1.2	0.5
Rye, grain	0.12 – 1.3	0.45
Oats, grain	0.28 – 1.9	0.55
Pea, seeds	1.2 – 1.8	–
Kidney bean, seeds	0.9 – 1.6	–
Carrot, roots	–	0.04
Onion, bulbs	0.16 – 0.24	–
Potato, tubers	0.1 – 0.25	–
Tomato, fruits	–	0.82
Apple, fruits	–	0.07

[a] Presented are common, possible background, values from various sources as compiled by Kabata-Pendias and Pendias (1999, 2001), unless otherwise stated.

[b] After Eriksson (2001a).

Table II-6.5. Mean contents of molybdenum in grasses and legumes from areas where Mo toxicity was observed and not observed in grazing animals ($mg\ kg^{-1}$) (values from various sources as compiled by Kabata-Pendias and Pendias 1999, 2001)

Country	Grasses		Legumes	
	Toxicity	No toxicity	Toxicity	No toxicity
Canada	4.0	1.3	5.2	–
Finland	–	0.45	–	0.7
Great Britain	1.5	0.56	–	–
Japan	–	0.72	–	0.9
Poland	–	0.33	26.6	0.5
USA	3.7	–	–	1.8
Russia	10 – 50	–	–	2.3

legumes (Table II-6.5). Increased Mo contents of grasses (up to 36 $mg\ kg^{-1}$) have been observed in surroundings of various industries and especially near Mo ore-processing plants (Kabata-Pendias and Pendias 1999).

The translocation of Mo in plants is moderate, and there is the assumption that Mo-S amino acid complexes are involved in this process. Chemical similarities between MoO_4^{2-} and SO_4^{2-} have led to the suggestion that the biological assimilation of Mo may be inhibited by sulfur compounds (Marinio et al. *vide* Anbar 2004).

The role of Mo as a component of enzymes involved in the N metabolism, mainly nitrogenase and nitrate reductase, has been the subject of several studies. Also its inhibitory and regulatory functions in the activity of DNA and RNA have been emphasized (Ivchenko 1988). Two Mo-containing enzymes in N metabolism are involved in either N_2 fixation or in NO_3 reduction. Therefore, the requirement of plants for Mo is related to N supply; plants supplied with NH_4-N have less need for Mo than those utilizing NO_3-N. Molybdenum is an essential micronutrient to most microorganisms and is obviously functional to Rhizobium bacteria and other N-fixing microorganisms. Some fungi and free-living bacteria can concentrate high amounts of Mo, up to 100 $mg\ kg^{-1}$.

Mobile species of Mo in soils (MoO_4^{2-}, and $HMoO_4^-$) are easily available to plants, however, Mo content of plants is not a simple function of its abundance in soil. The uptake of Mo by plants is, in most cases, controlled by the soil pH. Molybdenum availability from reducing environments, where thiomolybdates are formed (MoS_4^{2-}, $MoO_2S_2^{2-}$) is apparently due to the oxidizing ability of the root surface (Kabata-Pendias and Pendias 2001).

Usually a positive correlation is observed between Mo contents in plants and the soil pH. Thus, plants (mainly native) grown on neutral or alkaline soils can accumulate extremely high amounts of Mo (up to 52 $mg\ kg^{-1}$, average 11 $mg\ kg^{-1}$) whereas on Mo-rich acid soils, Mo concentrations in the same varieties of plants average 0.9 $mg\ kg^{-1}$ (Doyle et al. *vide* Kabata-Pendias and Pendias 2001). Contents of Mo do not exceed 1 $mg\ kg^{-1}$ in most plants, with the exception of legumes, and grassland vegetation on soils with increased levels of Mo (Tables II-6.4 and II-6.5).

In general, Mo contents in plants, in the range from 0.1 to 0.5 $mg\ kg^{-1}$, are sufficient for plant metabolism, whereas concentrations of Mo in the range of 10–50 $mg\ kg^{-1}$ are toxic to most plant species. Only some leguminous and cruciferous crops require more Mo. Deficiencies of Mo are likely to occur on acid soils (pH <5.5) and soils with a low

Mo content and high Fe oxide levels. The most sensitive to Mo deficiency are some vegetables, e.g., cauliflowers, that exhibit deficiency symptoms (mainly chlorosis) at a Mo content of <0.4 mg kg^{-1} in leaves. Also on organic-rich soils (peat soils) plants can suffer from Mo deficiency due to its slow release from organic bound forms. Liming of acid soils is a common practice to increase the Mo availability to plants. However, an inadequate rate of liming may decrease its mobility. The application of Mo fertilizers has to be based on a proper assessment of Mo doses and is the subject of publications (Mortvedt et al. 1991; Pierzynski and Jacobs 1986; Wang et al. 1994).

Where Mo-rich sludges (with Mo up to 50 mg kg^{-1}) are used for soil amendments, crops should be monitored to ensure that contents of this metal are within safe limits, especially for grazing animals. Agricultural utilization of alkaline-stabilized biosolids (ASB) poses a special potential risk to ruminant animals due to the easy transfer of Mo from ASB to forage (Stehouwer and Macneal 2004).

Predicting the phytoavailability of Mo to various crops is not easy and a case study should always be undertaken for the assessment of its status in food and forage plants.

II-6.2.7
Humans and Animals

Humans. Molybdenum occurs in all human and animal tissues. Its range in human organs is given by Li (2000) to be from 0.001 to 0.4 mg kg^{-1}, being the lowest for blood and the highest for kidneys and liver. Its average content in soft tissues of "reference man" is calculated by this author as <0.075 mg kg^{-1} and in skeleton as <0.48 mg kg^{-1}. Average values of Mo concentrations in human whole blood range from 3.4 to 14.9 μg l^{-1} and seem to be a little bit lower in women than in men (Monèiloviã and Ivičić 2003). Jørgensen (2000) gave the range of Mo in mammalian tissues from <0.2 to 2.8 mg kg^{-1}; the lowest in muscles and the highest in liver.

Molybdenum is an essential micronutrient being a component of several enzymes involved in oxydation-reduction processes and regulates the metabolism of purines and fats. The oxomolybdenum enzymes belong to the xanthine oxidase, sulfide oxidase and reductase families. It is relatively easily absorbed at all levels of intake and its whole blood levels vary directly with the dietary Mo intake. There is only one human case of the Mo deficiency described, that responded to Mo therapy. Symptoms included tachycardia, headache, nausea and vomiting (Combs 2005).

The main sources of Mo in human diet are legumes, pulses, dark leafy vegetables and organ meats (Combs 2005). The minimum daily requirement for Mo by adults is estimated at about 0.02 mg (Turnlung et al. 1993). The upper tolerable intake of Mo by children is established as 0.3–1.7 mg d^{-1}, depending on the age and by adults as 2 mg d^{-1} (WHO/IPCS 2002). Daily intake of Mo by adults in Germany (in the period 1990–1994) was about 100 μg, and the RDI for adults is established at 150–500 μg for adults, 50–100 μg for children, and 40–80 μg for infants (Brüggemann et al. 1996). Daily intake of Mo by adults in the UK was estimated at 113 μg, and in the USA at 120–140, whereas in Poland the Mo intake varied highly from 37 to 137 μg d^{-1}, in different regions of the country (Kabata-Pendias and Pendias 1999).

The excess Mo in the human diet influences its accumulation in serum, urea, and hair. Apparently, it is also related to cancer of the stomach (Xianmao et al. 1990).

Animals. The deficiency of Mo in animals is seldom observed when low content Mo feed is supplied. Chickens seem to be most sensitive and show low growth and feather quality. There is some evidence that Mo is involved in the metabolism of Ca and F in bones; its excess can lead to osteoporosis and bone deformation (Kabata-Pendias and Pendias 1999).

An excess of Mo, especially in cattle, is widespread in the world. It can affect growth and health of animals, and in that particular of ruminants. Frank (2003) and Frank et al. (2002) described molybdenosis symptoms in the Swedish moose. They presented a negative relationship between concentrations of Mo and Cu in the liver from moose; at the lowest Cu content of about 20 mg kg^{-1} FW was the highest Mo content of about 55 mg kg^{-1} FW.

The function of Mo in animals is closely related to interaction reactions with Cu and S. The dietary ratio of Cu/Mo for ruminants should range from >4 to 5, and this ratio affects the metabolism of sulfur. Any variation in the ratio between Cu-Mo-S in ruminant feed can affect a secondary deficiency of Cu, and some times impaired metabolism of Fe. The Mo-induced Cu deficiency has been relatively often observed (Jones 2005).

II-6.3 Tungsten (W)

II-6.3.1 Introduction

The distribution of W in the Earth's crust shows that its concentration seems to increase with increasing acidity of magmatic rocks and with increasing clay content of sedimentary rocks (Table II-6.2). The common concentration of W in granite rocks and argillaceous sediments varies from 1 to 2 mg kg^{-1}, and in mafic rocks and in limestones, from 0.5 to 1 mg kg^{-1}. Its content in coals is relatively high and averages 0.5–1 mg kg^{-1}, and in some oil shales can reach up to 2.5 mg kg^{-1} (Kabata-Pendias and Pendias 1999; Li 2000). In fly ash samples from coal-fired power plant in Indiana State, USA, contents of W vary in the range of 1.04–1.17 mg kg^{-1} (Brake et al. 2004).

Geochemical properties of W are similar to those of Mo and in several minerals the substitution of W for Mo is observed. Predominate minerals contain the anionic form of the metal, $(WO_4)^{2-}$. The most common minerals are wolframite, $(Fe,Mn)(WO_4)$ and sanmartine, $(Zn, Fe)(WO_4)$. However, several tungstanites, anhydrous and hydrous, in coordination with different metals are known. Under most geochemical conditions, minerals and compounds of W are slightly soluble, but some complex compounds, especially those associated with Mo are likely to be mobilized under specific conditions.

II-6.3.2 Production and Use

Global production of W was estimated as 50 kt in 2003 (WMSY 2004). It is used mainly in metallurgy as an addition to alloys (steel) for abrasives and cutting tools. Tungsten is broadly utilized for rocket engines, X-ray tubes and various electric tools and in the lighting industry (e.g., light bulbs). It is also used for catalysts in automobile exhaust systems, and as pigments in dyes and inks. Tungsten carbide (WC) has a hardness of that of diamond.

II-6.3.3 Soils

The distribution of W in soils has not been widely studied. Compiled data for the occurrence of W in non-polluted soils indicate the following ranges for W (in mg kg^{-1}) in soils of different countries:

- Brazil: 0.62 – 3.4 mean 1.38 (Licht 2005)
- China: 0.95 – 85.5 (Govidaraju 1994)
- Finland: 2.23 90th percentile range (Köljonen 1992)
- Japan: 0.9 – 1.8 (Takeda et al. 2004)
- Scotland: 0.68 – 2.7 (Ure et al. *vide* Kabata-Pendias and Pendias 2001)
- Sweden: 0.4 – 2.4 mean 1.3 (Eriksson 2001a)
- USA: 0.5 – 5.0 (Govindaraju 1994)

Soils surrounding an W ore-processing plant contain total W in the range of 100–200 mg kg^{-1}, of which 30% is water soluble, 15% bound to Fe oxides, and 5% fixed by organic matter (Panova 1988).

Fly ash can be a serious source of W in soils; added at the concentration of 20%, by weight, has increased W level from 0.01 mg kg^{-1} to 0.28–0.52 mg kg^{-1} in different soil treatments (Brake et al. 2004). Also sludges used for amending soil can add W, as sludges often contain elevated amounts of this metal, up to 1.2–2.5 mg kg^{-1} (AW) (Furr et al. *vide* Kabata-Pendias and Pendias 2001).

II-6.3.4 Waters

Although most of W minerals are considered to be slightly soluble, its global mobility, as calculated by Gaillardet et al. (2003), is relatively high (Cw/Cc ratio is about 0.9).

Data for W concentrations in water are very scant. The concentration of W in seawaters is estimated as 1 µg kg^{-1} (GERM database: http://www.earthref.org). However, Reimann and Caritat (1998) have cited 0.1 µg l^{-1} as a median value of W in ocean waters and Nozaki (2005) gave the value of 0.01 µg kg^{-1} in the North Pacific Ocean. In sea water W, similarly as Mo, is likely to occur in the oxide species of WO_4^{2-}.

In stream waters of various continents, W concentrations vary from 0.002 to 0.89 µg l^{-1}, with the median values in the range of 0.01–0.05 µg l^{-1} (Reimann and Caritat 1998). Rainwater samples collected in Sweden during 1999 contain W in the range from 4.2 to 23 ng l^{-1} (Eriksson 2001a).

Tungsten is readily accumulated and concentrated in polymetallic nodules of the ocean bottom. In ferromanganese nodules of the Baltic Sea, W contents vary from 0.4 to 285 mg kg^{-1} (Szefer 2002a).

II-6.3.5
Air

Natural concentrations of W in air above the Spitsbergen range from 0.004 to 0.08 ng m^{-3}, and above urban areas of the UK, between <0.5 and <1 ng m^{-3} (Kabata-Pendias and Pendias 1999). Reimann and Caritat (1998) gave the range of 0.005–0.02 ng m^{-3} and median value of 2.5 ng m^{-3} for W concentrations in air from remote and polluted regions, respectively.

II-6.3.6
Plants

There is some evidence that W might have a biological function in plants. However, it displays competitive inhibition of Mo in the enzyme nitrate reductase, reducing that enzyme's catalytic activity (Pais and Jones 1997). Makdessi et al. (2001) reported that the function of W as a component of two dehydrogenases of some bacteria (e.g., *Eubactrium acidaminophilum*) has now been fully recognized. The behaviors of Mo and W are very similar in biochemical processes and some substitution by W for Mo has been reported (Buc et al. 1999; Makdessi et al. 2001; Tajima 2005). There have been some observations on the antagonistic interactions between W and Mo causing a slower N-fixation process that is governed by pH of the media (Zajic *vide* Kabata-Pendias and Pendias 2001).

The common range of W in terrestrial plants is established at the range of <1–150 μg kg^{-1}, whereas Markert (1992) has proposed 200 μg kg^{-1} for W in the standard plant. Contents of W in vegetables vary within the range from <1 to 350 μg kg^{-1}. Wheat and barley grains from Sweden contain W at averages values of 6 and 5 μg kg^{-1}, respectively (Eriksson 2001a).

In mosses sampled in Norway in the period 1990–1995, amounts of W varied from <2 to 1 500 μg kg^{-1} (Berg and Steinnes 1997). Concentrations of W in mosses from Germany range from 10 to 2 500 μg kg^{-1} (Reimann and Caritat 1998).

There are confusing reports on the phytoavailability of W. Results of studies with radionuclide ^{185}W have indicated its easy accumulation by plants. Elevated levels of W, up to 100 mg kg^{-1} AW, in some trees and shrubs grown in the Rocky Mountains in Colorado, USA were reported by Shacklette et al. (*vide* Kabata-Pendias and Pendias 2001). Wilson and Cline (*vide* Kabata-Pendias and Pendias 2001) reported that barley removed a large proportion of radioactive W from polluted soils, and suggested that it is taken up as the anionic form, WO_4^{2-}.

Panova (1988) found W contents in semiarid herbs grown in the vicinity of the W industries to decrease from 600 mg kg^{-1} AW to 50 mg kg^{-1} AW with increasing distance from the pollution source. This observation, however, might be associated also with the atmospheric source of W. Brake et al. (2004), on the other hand, suggested that the absence of W in plants grown on soils enriched in this metal indicates its low bioabsorption. However, Furr et al. (*vide* Kabata-Pendias and Pendias 2001) found W to range from 0.7 to 3.5 mg kg^{-1} in vegetables growing on soil amended with fly ash. Apparently, these observations could be related to species of W that are present in polluted soils.

II-6.3.7 Humans and Animals

Humans. The only available data gave concentrations of W in human fluids as follows (in ng l^{-1}), after Li (2000): blood, 1; and urine, 5; and after Reimann and Caritat (1998): blood, 390; and in urine, 320. The reference value for W in urea has been estimated by Kraus (2004) as 860 ng l^{-1}.

The biological functions of this metal are not known well, though naturally occurring enzymes using W seem most prevalent among anaerobic bacteria (Vorholt and Thauner 2002; Laukel et al. 2003). Easily soluble W (e.g., from sodium tungstate, Na_2WO_4) is also easily absorbed and is harmful to the nervous system. Dust exposure of man to W is associated with tool-manufacturing industries. The MAK value in Germany has been set at 1 mg m^{-3} for soluble W compounds.

Animals. Information on adverse and stimulating effects of W in animals are confusing and need more studies (Kraus 2004). The toxicity of W depends on its compounds and is the highest for sodium tungstate and the lowest for tungsten metal. The intake of W by individual animals is estimated as 13 μg d^{-1}. Fibrotic lung changes are observed in animals under exposure to tungsten-carbide dust (Kraus 2004).

Trace Elements of Group 7 (Previously Group VIIb)

The elements of Group 7 are manganese (Mn), technetium (Tc) and rhenium (Re). Only Mn is essential to living organisms. Technetium is a by-product of nuclear fuel processing and occurs as unstable radionuclides with variable but mainly short half-lives (from seconds to years). Rhenium is highly dispersed in the environment. All elements of this group have variable valences (Table II-7.1) and exhibit both lithophile and chalcophile characters.

II-7.1 Manganese (Mn)

II-7.1.1 Introduction

Manganese is one of the most abundant trace elements in the lithosphere. Its common occurrence in rocks ranges from 350 to 2 000 mg kg^{-1} and higher concentrations are associated with mafic rocks (Table II-7.2).

Manganese is a member of the iron family and is closely associated with Fe in geochemical processes. Thus, the Mn cycle follows the Fe cycle in various geochemical environments. During weathering, Mn is oxidized in minerals, under atmospheric conditions and may be mobilized. The Mn oxides are reprecipitated and readily concentrated in forms of secondary Mn minerals, often in forms of concretions and nodules. Under weathering in tropical and subtropical conditions, Mn is concentrated in residual deposits, while under humid colder climate, Mn is leached by acid solutions from sediments and soils.

Table II-7.1. Selected properties of trace elements of the Group 7

Element	Atomic number	Atomic mass	Atomic radius (pm)	Density (20 °C, g cm^{-3})	Valence	Melting point (°C)
Mn, manganese	25	54.9	179	7.44	+2[a]	1 244
Tc, technetium	43	98.9[b]	–	11.5	+4, +6, +7	2 172
Re, rhenium	75	186.2	197	21.0	+7[c]	3 180

[a] Variable valences from +2 to +7.
[b] For the most stable isotope ^{98}Tc.
[c] Variable oxidation states from –1 to +7.

Mn

Table II-7.2.
Abundance[a] of manganese and rhenium in the environment

Environmental compartment	Mn	Re
Earth's crust ($mg\ kg^{-1}$)	716 – 1 400	0.4[b]
Igneous rocks ($mg\ kg^{-1}$)		
Mafic	850 – 2 000	0.4[b]
Acid	350 – 1 200	0.6[b]
Sedimentary rocks ($mg\ kg^{-1}$)		
Argillaceous	400 – 850	0.5[b]
Sandstones	100 – 500	0.1[b]
Calcareous	200 – 1 000	0.1[b]
Raw energy material ($mg\ kg^{-1}$)		
Coal, hard	50 – 100	<0.1
Fly ash	325 – 360	–
Crude oil	0.1 – 1	–
Petrol	0.005 – <0.1	–
Soils ($mg\ kg^{-1}$)		
Light sandy	7 – 2 000	–
Medium loamy and silty	50 – 9 200	–
Heavy loamy	100 – 3 900	–
Calcareous	50 – 7 750	–
Organic	10 – 2 200	–
Water ($\mu g\ l^{-1}$)		
Rain	–	–
Sea	0.2 – 04	4[c]
River	0.02 – 130[d]	<10[c]
Air ($ng\ kg^{-3}$)		
Urban/Industrial	50 – 900	–
Greenland	2.8 – 4.5	–
South Pole	0.004 – 0.02	–

[a] Presented are common, possible background, values from various sources as compiled by Kabata-Pendias and Pendias (1999 and 2001), unless otherwise indicated.
[b] Unit: $\mu g\ kg^{-1}$.
[c] Unit: $ng\ l^{-1}$.
[d] After Gaillardet et al. (2003).

Complex behavior of Mn in sedimentation processes resulted in the accumulation of this metal in different layers of sediments. Polymetallic concretions in deep-sea sediments are effects of specific Mn properties. Manganese is the dominating metal of these concretions and its concentration can reach from 15 to over 40% (Kotlinski 1999). The genesis of deep-sea concretions is not fully understood, but recently the influence of microbes and organic matter has been recognized. Occurrence of Mn is highly associated with activities of microbes in both sediments and soils.

The redox state of Mn is very variable (Table II-7.1) and is mediated by both geochemical and biological processes. The most common cation is Mn^{2+} that readily occupies the sites of other divalent cations (e.g., Fe^{2+}, Mg^{2+}). Several complex anions and cations of Mn as well as hydroxides of variable composition are also common.

The complex mineralogical and chemical behavior of Mn and its participation in oxidation-reduction processes resulted in the formation of the large number of oxides and hydroxides of variable stability and properties. The physical features of Mn compounds and minerals, especially the small size of crystals and large surface areas of amorphous forms have important geochemical and pedochemical implications.

Both Mn and Fe oxides are considered to be the most abundant compounds of the Earth's surface that can serve as anaerobic terminal electron acceptors in microbial metabolism. Thus, it has recently been recognized that microorganisms play a key role in Mn cycling (Megonigal et al. 2003). Tebo et al. (2003) have described the Mn-oxide biomineralization and the role of those Mn oxides in the transformation/degradation of organic and inorganic compounds in various environmental compartments.

Manganese micro-nodules found on ancient brick walls have resulted, as stated by López-Arce et al. (2003), from the impact of water enriched in Mn compounds or due to oxidation processes linked to algae biofilms.

II-7.1.2
Production and Use

Global annual production of Mn in 2003 was 24.35 Mt (WMSY 2004). Manganese ores are widely distributed in various geological and climatic environments and concentrated mainly in residual deposits, as various oxides and hydroxides; to a lesser extend as carbonates and silicates.

Manganese is used mainly in the metallurgical industry for steel and various alloys to provide hardness and toughness, and as an anti-oxidant. Alloys with Mn are used in the electrical industry (e.g., dry-cell batteries). It is also widely applied in the production of pigments, ceramics, and glass. Manganese sulfate ($MnSO_4$) is used as fertilizer and as a fodder supplement. It is relatively broadly used as fungicides, mainly for seed treatment (e.g., *Maneb*, manganese ethylene-bis-dithiocarbamate).

Since 1970, Mn has been used in gasoline, first in Canada and the USA, as an anti-knock agent (as replacement for Pb in unleaded fuel), in the form of an organic compound – methylocyclopentadienyl manganese tricarbonyl (MMT), which has been already approved for use in several countries (Howe et al. 2004).

Potassium permanganate ($KMnO_4$) has long been used as an oxidizing catalyst in chemical processes, and in disinfection in medicine, as well as in the purification of various waters. Also several other Mn compounds (e.g., gluconate, sulfate) are used in pharmaceutical production.

II-7.1.3
Soils

Contents of Mn in soils are highly diverse and range approximately from 10 to 9 000 mg kg^{-1} (Table II-7.2). The world grand mean Mn content is calculated (in mg kg^{-1}) as 437, whereas for soils of the USA it is estimated at 495, and for Finnish soils at 600 (Kabata-Pendias and Pendias 2001). The variation of Mn contents of surface soils seems rarely to be correlated with soil classification, but is positively associated with clay contents. However, higher Mn levels are often found in soils derived from mafic rocks, in soils rich in Fe and/or organic matter, and in soils from arid or semiarid climatic regions.

Manganese is not distributed uniformly in soil substrata and, in addition to various nodules, is also concentrated in certain spots and veins. It is likely to occur in soils as oxides and hydroxides in the forms of coatings on soil particles and as nodules of different diameters. The nodules often exhibit a concentric layering that suggests a seasonal growth. The Mn concretions and various colloidal Mn hydroxides exhibit a great affinity for the adsorption of Fe and several cationic and anionic forms of trace metals (Table II-7.3). On the other hand, due to both reducing and oxidizing properties, Mn oxides can increase the mobilization of some metals under specific soil conditions. Therefore, Mn has a strong influence on the mobility of trace inorganic pollutants through precipitation-dissolution processes. Synthetic HMnO reveals a great capability to bind several bivalent cations such as Ba, Ca, Cd, Co, Cu, Mg, Mn, Ni, Pb, Sr, and Zn, that is under the influence of redox conditions and cation exchange reactions (Tonkin et al. 2004).

Table II-7.3. Trace elements in manganese oxides (mg kg^{-1}) (data sources: Kabata-Pendias and Pendias 2001, Bartlett 1999)

Element	Mn nodules	Mn minerals from soils[a]	Mn minerals from sediments[b]
Mn (%)	0.4 – 50	78 – 60	28 – 60
Ba	14 – 2300	32000 – 54000	110000 – 128000
Cd	8	–	–
Ce	720	–	–
Co	82 – 3000	4500 – 12000	140 – 12000
Cr	14 – 120	–	–
Cu	2600	–	130 – 12600
I	120 – 900	–	–
Li	–	300 – 700	2 – 53400
Mo	410	–	–
Ni	40 – 4900	100 – 3400	120 – 10900
Pb	34 – 870	–	–
V	88 – 440	–	–
Zn	30 – 710	–	50 – 3800

Identified minerals:
[a] Lithiophorite, birnessite, and hollandite.
[b] Psylomelane, cryptomelane, and pyrolusite.

Although Mn occurs in soils predominantly as amorphous oxides and hydroxides, some crystalline minerals of Mn are also present. These are mainly: pyrolusite, β-MnO_2; manganite, γ-MnOOH; hausmanite, Mn_3O_4; lithiophorite, (Al, Li) $MnO_2(OH)_2$ (likely to occur in acid and neutral soils); and birnessite, $Na_xCa_yMn_7O_{14}(2.8H_2O)$ of unconfirmed composition, apparently associated with alkaline soils.

The ionic species of Mn commonly occurring in soil solutions are as follows (Kabata-Pendias and Sadurski 2004):

- Cationic forms: Mn^{2+}, $MnOH^+$, $MnCl^+$, $MnHCO_3^+$, $Mn_2(OH)_2^{2+}$, Mn_2OH^{3+}
- Anionic forms: MnO_4^-, $HMnO_2^-$, $Mn(OH)_3^-$, $Mn(OH)_4^{2-}$

The redox behavior of Mn in soils has been broadly studied and described by Bartlett (1999). As Sparks (*vide* Bartlett 1999) has stated: "Soil Mn, or its lack, appears to be the key to the entire spectrum of soil redox behavior and status". Recent results presented by Nelson and Lion (2003) verify that biogenic, amorphous Mn oxides exhibit greater specific surface areas than typical abiotic Mn oxides. The biogenic Mn oxides exhibit very high adsorption capacity for Pb; especially at a pH about 8 (Fig. II-7.1).

Biotic and abiotic processes in the oxidation and reduction of Mn have been broadly described by Megonigal et al. (2003). Based on their data, the following sequence for Mn-redox cycling can be presented:

- The reduction: $Mn^{3+} \rightarrow Mn^{2+}$, abiotic and biotic, by Fe^{2+}, Cr^{3+}, S, phenols, organic compounds (humic substances), and reducing bacteria
- The oxidation: $Mn^{2+} \rightarrow Mn^{3+}$, Mn^{4+}, can occur under both aerobic and anaerobic conditions and is biologically mediated or autocatalytic
- The Mn^{3+} is an extremely reactive redox species and quickly disappears, either by accepting or donating an electron
- The Mn^{2+} is either adsorbed by MnO_2, or oxidized to Mn^{3+}, or to Mn^{4+}
- The organic and phosphate ligands are also involved in the Mn-redox cycling

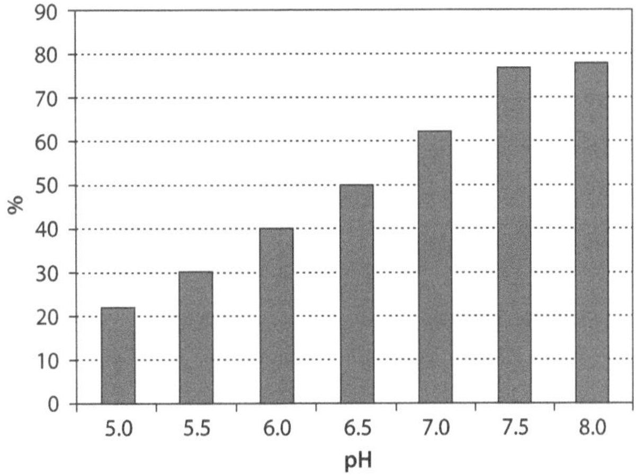

Fig. II-7.1. The Pb adsorption (in % of initial Pb concentrations in solution) on natural biofilm (Mn^{2+} biologically oxidized by *Leptothrix discophora* SS-1) as a function of pH (based on data from Wilson et al. 2001)

Megonigal et al. (2003) emphasized that biologically mediated processes are the most significant in the Mn-redox cycling in soils. This statement has been supported also by several other findings. Tani et al. (2003) reported that Mn oxides enzymatically produced by a Mn-oxidizing fungus reveal specific adsorption properties to Co, Ni, Zn, and As. Glasauer and Beveridge (2003) described the dissimilatory reduction of Mn by bacteria attached to the surface of Mn-oxide minerals. Some organic acids secreted by microorganisms can release Mn from synthetic MnO_2 (Jones 1998).

Mn-oxides have relatively high total surface (32–300 $m^2 g^{-1}$) and CEC values (150–320 cmol(+) kg^{-1}). Metal species bound to Mn oxides in metal-polluted soils made up <1–30% of total Cu, 1–26% of Pb, 3–27% of Zn, <1% of As, and <1% of Fe (Karczewska 2004; Kabala and Wilk 2004). The sorption of Co by Mn oxides, resulting in changes of Co valences and in its unavailability to plants, has been of great importance in Australian sheep farming regions (Fleming *vide* Kabata-Pendias and Pendias 1999).

The distribution of Mn among soil fractions was investigated using ^{54}Mn tracer (Goldberg and Smith 1984). The results have shown that the distribution of Mn in soil various fractions varies depending on soil types (pedological origin), however, the fraction of resistant Mn minerals (oxides) contains usually the highest proportion of the metal, about 50% of the total contents.

All Mn compounds are very important constituents of soil because this metal is essential in plants nutrition and controls the microbial activity of soils. The mobility of Mn in soils is highly dependent on the pH and redox potential. The predicted Mn solubility, based on mathematical models usually does not correspond to the actual Mn mobility in soils because multiple factors influence the Mn behavior, such as metastable and/or mixed composition of Mn oxides and hydroxides, organic complexes of Mn and variable Eh-pH condition. Even small changes in Eh-pH conditions might affect Mn contents of a soil solution.

Concentrations of Mn in soil solutions vary highly from 25 to 2 000 $\mu g\, l^{-1}$, depending on soil kinds and techniques used for obtaining solutions (Kabata-Pendias and Pendias 2001). The mobility of Mn is especially affected by soil pH. Concentrations of Mn in solutions of metal-polluted soils vary highly depending on soil pH. These changes are by a factor from 7.6 to 270 of the ratio of acid soil (pH 2.5) to alkaline soil (pH 9–10), depending on soil properties (Karczewska 2002).

The solubility of soil Mn is very significant since the plant supply of Mn depends mainly on the soluble Mn pool in soils. Several soil factors influence the Mn availability to plants. Among abiotic and biotic soil parameters, the complexing by root exudates and cross-interactions with Fe oxides play a crucial role. The solubility of soil Mn is of significance since plants can uptake mainly soluble fractions of Mn. However, the ability of Mn to form anionic complexes and organic ligands may contribute to increased Mn solubility, even at the alkaline pH range. In general, Mn is highly mobile at the acid range of pH. In specific soil redox conditions, as e.g., poorly aerated soil at pH about >6.0 or well-drained soils, at pH <5.5, the Mn mobility increased, and Mn became easily available to plants.

Manganese budgets (input/output ratios) in soils of various ecosystems indicated a predominance of leaching processes over atmospheric input. In pine, spruce and birch forests its leaching from soil profiles accounts from 360 to 11 000 g $ha^{-1} yr^{-1}$, whereas in other ecosystems some accumulation (91–191 g $ha^{-1} yr^{-1}$) of Mn was observed (Kabata-Pendias and Pendias 2001).

The major anthropogenic sources of Mn are: municipal wastewaters, sewage sludge, and metal smelting processes. The combustion of fuel additives (MMT) is of a lesser importance. However, in some regions (e.g., Mississippi River Delta) alluvial sediments can concentrate Mn up to >1 000 mg kg^{-1} from MMT fuel use (Mielke et al. 2002).

II-7.1.4 Waters

Manganese is relatively easily mobile in the terrestrial environment, and its Cw/Cc ratio is about 0.5 (Gaillardet et al. 2003). Nevertheless its concentration in suspended matter and bottom sediments is much higher that in waters.

Ambient Mn concentrations in sea water have been reported to range from 0.4 to 10 µg l^{-1}, with an average of about 2 µg l^{-1} (ATSDR 2000). However, Reimann and Caritat (1998) calculated the median Mn concentration in the world ocean waters at 0.2 µg l^{-1}. The Mn range of sea waters given by Kitano (1992) is from 0.02 to 0.1 µg l^{-1}, whereas mean Mn concentration after GERM (www.earthref.org) is 0.01 µg l^{-1}, and after Nozaki (2005) in the North Pacific waters is 0.02 µg kg^{-1}. In waters of the Baltic Sea, Mn contents range from <1 to 3 µg l^{-1} (Szefer 2002a).

The Mn concentrations in sea basins vary with the water depth; its highest content (450 µg l^{-1}) was noticed at the depth of 30–40 m of the Black Sea (Konovalov and Korenova *vide* Kabata-Pendias and Pendias 2001). The Mn fraction in suspended particulate matter in waters of the Gotland Deep is reported to increase also with the depth of water column and to reach over 150 µg l^{-1} at the depth of over 200 m (Brügmann et al. 1998). Suspended matter of the Baltic Sea contain extremely variable amounts of Mn, from 20 to 73 000 mg kg^{-1} (Szefer 2002a).

Manganese input to the World Sea by rivers is calculated by Kitano (1992) as 400–1 000 kt yr^{-1}, whereas the global riverine flux is estimated by Gaillarted et al. (2003) at 1 270 kt yr^{-1}.

Skjelkvåle et al. (2001) investigated the occurrence of Mn in Nordic lakes, and reported its concentrations to vary, at the value of 99.5 weighted percentiles, as follows (in µg l^{-1}): 316 in Finland, 474 in Sweden, and 101 in Norway. The Mn range reported for US river waters is between 11 and 51 µg l^{-1} (ATSDR 2000). Rainwater samples collected in Sweden during 1999 contain Mn in the range of 0.41–3.1 µg l^{-1}, at an average value of 1.7 µg l^{-1} (Eriksson 2001a).

In various surface water conditions, several species of Mn are likely to occur, such as: Mn^{2+}, $MnOH^{+}$, $MnCl^{+}$, $MnHCO_3^{+}$, $Mn(OH)_3^{-}$, and $MnSO_4^{0}$ (Witczak and Adamczyk 1995).

The precipitation of Mn in bottom sediments of seas and ocean's basins is controlled mostly by seasonal variations of the redox conditions within the water column and the sediments. There are different forms, mainly of secondary Mn and Fe formations. Bacteriogenic Mn oxides occur commonly as nanoparticles and grain coatings in sediments of marine and fresh water environments. Hydrated biogenic Mn oxides produced by spores of marine and fresh water bacteria have flexible structure and are strong scavengers of several metals, particularly Co, Cu, Zn, and U (Bargar et al. 2004). Hlawatsch et al. (2002) stated that there is not observed a long-term accumulation of trace metals in Fe-Mn nodules, with the exception of Zn.

Concentration of Mn in polluted river waters can reach the value >1 000 µg l^{-1} (Kabata-Pendias and Pendias 1999). The highest content of Mn is, however, in bottom sediments of polluted rivers and/or water basins. Salomons and Förstner (1984) reported the content of Mn in sediments of Rhine River at the value of 960 mg kg^{-1} to be natural, whereas the highest Mn concentration in sediments of Odra River was 770 mg kg^{-1} (Kucharzewska et al. 1991). The distribution of Mn in the surface sediments of the Baltic Sea varies from 120 to 2 290 mg kg^{-1}, and the highest concentrations, up to 7 260 mg kg^{-1}, are in sediments of gulfs and lagoons (Szefer 2002a). The author reported Mn concentrations in bulk precipitation over the Baltic Sea to range from 4 to 76 µg l^{-1}. In groundwater, its concentrations can differ highly, from <1 µg l^{-1} in European Russia up to 550 µg l^{-1} in Japan (Uchvatov and Baszkin 1997).

Manganese concentration in drinking waters generally varied from 5 to 25 µg l^{-1}. The recommended Mn concentration in drinking water is estimated by the EU at 50 µg l^{-1}, health based limit value is established at 400 µg l^{-1} (WHO 2004), and the ceiling value is proposed as 500 µg l^{-1} (Schäfer 2004).

Aquatic biota. Some reports have shown that Mn in waters can ameliorate the toxicity of other metals, and especially of Cd, to microalgae. However, Mn may also induce Fe deficiency in some algae. Water hardness significantly affects Mn toxicity on aquatic invertebrates as well as on fish (Howe et al. 2004).

Manganese is likely to concentrate, up to about 23%, in shells of mollusks, in the form of layers coating the shell's surface. Extremely high Mn contents were found in shells of mollusks from the Baltic Sea, in the range of 7.5–233 mg kg^{-1}, and in some species (cockle), to over 34 000 mg kg^{-1} (Szefer 2002a).

II-7.1.5
Air

The origin of Mn in the atmosphere is from both terrestrial and industrial sources. Its concentrations in air vary from <0.02 ng m^{-3} above the South Pole up to 900 ng m^{-3} above industrial regions of the USA (Table II-7.2). Its contents in remote areas of the UK, in 1970, varied from 7–28 ng m^{-3} (Cawse 1982), and in Montreal, during the period 1993 to 2000, was relatively stable at about 10 ng m^{-3}, but then a decreasing trend was observed (Bankovitch et al. 2003). Median world concentration of Mn in air has been estimated at 2 ng m^{-3} (Reimann and Caritat 1998).

The global atmospheric emission of Mn has been calculated by Livett (1992) as 38.27 kt yr^{-1}, with the highest contribution from ferric metal smelting, and oil combustion. Since the addition of MMT to unleaded fuel, the increase of atmospheric Mn concentration has been expected. However, results of the investigations of Bankovitch et al. (2003) have not indicated a significant impact of the combustions of MMT used in gasoline to increasing Mn levels in the urban atmosphere, nevertheless its concentration in areas with very high traffic density could be higher than the US EPA reference concentration of 50 ng m^{-3}. Loranger et al. (*vide* Bankovitch et al. 2003) reported a correlation between the atmospheric Mn concentration and traffic density. Relatively stable Mn contents in lichen (*Hypogymnia physodes*) collected from Apostle Islands WI, USA, during the period 1987–2001, indicated rather the decrease of this metal (by 14.5%), and did not show any aerial pollution by Mn (Bennett and Wetmore 2003).

Manganese occurs in aerial particles mainly as different oxides that easily react with SO_2 and NO_2 and became soluble in rainwater. Concentrations in rainwater samples of the remote area of the Kola Peninsula ranged from 0.45 to 6.1 $\mu g\,l^{-1}$ (Reimann and Caritat 1998), and in rainwater of Sweden ranged between 0.5 and 3.1 $\mu g\,l^{-1}$ (Eriksson 2001a). These authors reported the bulk deposition of Mn in Germany to range from 30 to 720 $g\,ha^{-1}\,yr^{-1}$, and its wet deposition in Sweden to average 17 $g\,ha^{-1}\,yr^{-1}$.

Relatively high Mn average content of moss from Norway, 310 $mg\,kg^{-1}$, and the broad range, from 28–2 100 $mg\,kg^{-1}$ have indicated an impact of the aerial pollution (Berg and Steinnes 1997).

II-7.1.6 Plants

Contents of Mn in plants are results of both the plants characteristics and the pool of available Mn in soil, which is highly controlled by soil properties. Although the Mn uptake is metabolically controlled, its passive absorption, especially at high and/or toxic ranges, is also observed.

Concentrations of Mn fluctuate greatly within plant types, plant parts and during the vegetative period. Usually concentrations increase in older parts of plants. Among food plants, higher Mn concentrations are in cereal grains; average 27–50 $mg\,kg^{-1}$ (Table II-7.4). Relatively low contents in fruits, average 1.3–1.5 $mg\,kg^{-1}$, are affected by the slight Mn transport through the phloem vessels due to complexation with large organic molecules (Kabata-Pendias and Pendias 2001). In vegetables grown in Poland, Mn contents vary from 0.6 $mg\,kg^{-1}$ FW in tomatoes to 20 $mg\,kg^{-1}$ FW in green peas (Kunachowicz et al. 1998). Extremely high Mn contents in various types of black tea leaves, with average values from 560 to 1 366 $mg\,kg^{-1}$, apparently are a result of its easy bioavailability from very acid soils of tea plantations (Smiechowska et al. 2004).

Grasses and clovers from various countries contain relatively high and fairly similar amounts of Mn; mean range for grasses: 71–127 $mg\,kg^{-1}$, and for clover: 25–89 $mg\,kg^{-1}$ (Table II-7.5). The elevated average contents of Mn in grasses were given only for three countries: Australia, Hungary and Japan.

The most important Mn function in plants is related to the oxidation-reduction processes. The role of Mn has been already widely reviewed and summarized in several books and articles (Crowley et al. 2000; Nelson and Lion 2003; Rengel 2004; Schäfer 2004). The Mn^{n+} cations in enzymes can catalyze several redox reactions. The Mn^{2+} cation is a specific component of two enzymes, arginase and phosphotransferase. However, this metal can also substitute for Mg in other enzymes. The mechanism by which Mn activates several oxidases is not yet known precisely, but it appears to be related to the valence change between Mn^{3+} and Mn^{4+}. The role of Mn in the NO_2^- reduction step is also not clear. Apparently it is related to the indirect relationship between the Mn activity and N assimilation by plants. Manganese participates in the O_2-evolving system of photosynthesis and also plays a basic role in the photosynthetic electron transport system. Bartlett (1999) described Mn as "a key to life" due to its function in the photosynthesis – the vital link in life's most basic processes.

Mn

Table II-7.4. Manganese in food plants grown in various countries[a] (mg kg^{-1})

Plant	Range	Mean
Wheat, grains	16 – 103	40
Wheat, grains[b]	24 – 29	27
Wheat, grains[c]	8 – 53	25
Barley, grains[c]	12 – 34	18
Rye, grains	10 – 87	30
Oats, grains	17 – 121	50
Pea, yellow, seeds	–	21
Bean, pods[b]	4 – 25	21
Carrot, roots	9 – 28	15
Carrot, roots[b] (FW)	1 – 1.9	1.3
Onion, bulbs	16 – 24	19
Beet (sugar and red), roots	36 – 113	80
Potato, tubers	4 – 15	9
Potato, tubers[b] (FW)	0.4 – 0.9	1.2
Tomato, fruits	–	12
Apple, fruits	1.3 – 1.5	–
Strawberry, fruits[b] (FW)	1 – 1.9	1.3
Nuts and almonds[d]	10 – 42	22

[a] Presented are common, possible background, values from various sources as compiled by Kabata-Pendias and Pendias (1999, 2001), unless otherwise indicated.
[b] After Szteke et al. (2004); range and median values of 130 samples from Poland.
[c] After Ericksson (2001a); cereal grains from Sweden.
[d] After Jedrzejczak (2004); from markets in Poland.

Adequate levels of available Mn in growth media are necessary in plants nutrition. Skinner et al. (2005) reported that its transport across the soil-root interface is in the reduced, Mn^{2+}, state. Complex interactions between roots and microorganisms may affect the phytoavailability of Mn (Marschner and Rengel 2005). Deficiency of Mn is relatively common in certain crops grown on neutral and calcareous soils. In such soils, the interactions of plant roots with microorganisms (in the rhizosphere) can facilitate the oxidation of soluble Mn^{2+} into unavailable Mn^{3+} or Mn^{4+} (Rengel 2004).

Most crop plants do not suffer from Mn deficiency at concentrations above the range of 15–25 mg kg^{-1}. Rengel (2004) reported that Mn concentrations in the phloem sap range from 2 to 910 $\mu M\,l^{-1}$, whereas the required concentration, to keep the minimal levels of Mn in grains (10–20 mg kg^{-1}), ranges between 25 and 100 $\mu M\,l^{-1}$.

Chloroplasts are the most sensitive of all cell components to Mn deficiency and react by showing structural impairment. The deficiency symptoms occur first in younger

Table II-7.5. Manganese in grasses and clover at the immature grown stage from different countries (mg kg^{-1}) (values from various sources as compiled by Kabata-Pendias and Pendias 1999, 2001)

Country	Grasses		Clover	
	Range	Mean	Range	Mean
Australia	67 – 187	120	33 – 43	38
Czech Republik	24 – 130	71	17 – 42	25
Finland	41 – 144	77	34 – 140	87
Germany	35 – 128	76	24 – 420	62
Hungary	67 – 309	161	55 – 126	82
Ireland	77 – 116	86	18 – 39[a]	26[a]
Japan	20 – 330	127	15 – 436	89
Poland	20 – 665	98	16 – 260	58
Russia	26 – 43	44	19 – 165	70

[a] Alfalfa.

leaves as interveinal chlorosis. At further stages, necrotic, reddish and brownish spots on leaves and browning of roots appear. The growth of Mn-deficient plants is retarded, the turgor is reduced and plant resistance to diseases and frost is lower. The most sensitive to the Mn deficiency are oats (gray speck symptoms), peas (marsh spot symptoms) sugar beet, and some fruit trees. The optimum rates and the methods of Mn applications, as soil fertilizers and foliar application, have been extensively reviewed in several textbooks. However, a high rate of Mn application or use of inappropriate forms, as well as variable soil conditions, can easily result in toxic effects on plants.

An excess of phytoavailable Mn is associated with several soil properties, particularly with a high acidity (pH <5.5), anaerobic condition, and poor aeration. The Mn toxicity in plants can occur also on highly alkaline soils, at about pH 8, where complex Mn anions can be formed. There is a great difference between plant species and genotypes in the sensitivity to Mn. Among crop plants, cereals, legumes and potatoes are the most sensitive to the Mn excess in growth media. The physiology of Mn toxicity is associated mainly with impaired impact on the activities of some enzymes and hormones, as well as on amino acid synthesis. Plants resistant to Mn excess have an ability to accumulate Mn in roots and/or to precipitate MnO_2 within the epidermis. Generally, most plants are affected by Mn contents around 500 mg kg^{-1}. However, the accumulations above 1 000 mg kg^{-1} in some resistant plant species or genotypes, and up to 10 000 mg kg^{-1} in hyperaccumulators have been reported (Greger 1999).

Foy (*vide* Kabata-Pendias and Pendias 2001) described the Mn tolerance as related to several plant characteristics and metabolic processes:

- Oxidizing power of plant roots, possible oxidation of Mn to MnO_2, which is stored as inactive fraction
- Complexation of Mn by low-molecular-weight compounds produced by plant roots or soil microorganisms
- Mn absorption rate and Mn entrapment in nonmetabolic forms
- Interactions with other elements, especially with Fe, Co, Ca, Al, and Si

The Mn biochemistry is controlled by interactions with several elements, in particular with Fe and Co. In general, Fe and Mn are interrelated in their metabolic functions, and their appropriate levels and ratio (the Fe/Mn ratio should range from 1.5 to 2.5) is a prerequisite of healthy plants. Several interactions between Mn and other trace metals have been observed, but are not fully confirmed, and in most cases might be cross-linked with other elements.

Such a relationship can also vary, as in the case of P and Ca impacts, affecting either a Mn deficiency or elevated Mn uptake, depending on plant characteristics and/or soil properties (Kabata-Pendias and Pendias 1999).

II-7.1.7
Humans and Animals

Humans. Manganese occurs in all tissues of terrestrial and aquatic organisms. Common ranges of Mn in soft mammal organs vary, depending on the source of data, from 0.2 to 3.8 mg kg^{-1} (Jørgensen 2000) or from 0.04 to 1.3 mg kg^{-1} (Li 2000), being always the lowest in muscles and the highest in kidneys. Also data on mean Mn concentrations in human fluids vary greatly; for blood from 0.008 to 0.05 mg l^{-1}, and for milk from 0.0032 to 0.12 mg l^{-1}. Higher contents of Mn are found in bones of terrestrial mammals (up to 10 mg kg^{-1}) and in hard tissues of aquatic animals.

The physiological function of Mn is closely associated with some enzyme activities, (e. g., superoxidase dismutase-SOD, arginase) and with metallothionein (Brandi et al. 2004). It is also known that Mn^{2+} is involved in gene expression processes and stabilizes the DNA structure (Munno De et al. 1996). The metabolism of Mn in the body is closely related to the metabolism of Fe that is manifested by a higher Mn absorption by women with low ferritin (Fe-protein) contents. This indicates an impact of Fe status of the organism on the Mn metabolism (Momčilović 2004). In general, Mn is an activator of different enzymes that control the metabolisms of carbohydrates, proteins and lipids (including cholesterol), and nitrogen metabolism. A great proportion (over 50%) of the body Mn is located in the hepatocyte nuclei of liver what can indicate some functions of this metal in genetic regulation (Leach and Harris *vide* Schäfer 2004b).

Symptoms of the Mn deficiency in humans are very rare. The Mn contents in human tissues, especially in bones, decrease with age, what can be associated with bone fractures (osteoroposis), dermatitis and hypochotesterolemia. Skeletal deformities and testicular dysfunctions can result from a Mn deficiency (Plumlee and Ziegler 2003). A low or absence of Mn-SOD activity in human cancer cells indicates a possible tumor-suppressive action of this enzyme (Schäfer 2004b). At the normal, balanced diet, the Mn supply is adequate to meet its requirement.

Average contents of Mn in different foods and beverages ingested by adults in Germany varied for women from 6.8 to11 mg kg^{-1}, and for men from 5.9 to10 mg kg^{-1}, depending on location and year (Schäfer et al. 2002). The contribution of various food groups to oral Mn exposure highly differs and is the highest for bread and cereals (49% of total Mn intake), vegetables (18%), and beverages (16%) (Nkwenkeu et al. 2002). The infant food in Poland contains Mn at average 0.97 mg kg^{-1} FW, and one meal covers about 65% of the daily requirement (Marzec et al. 2004). The intake of Mn by adults varies from 0.1 to 10 mg d^{-1} in different countries (Table II-7.6). The estimated safe

Table II-7.6. Average daily intake of Mn (mg d^{-1}) by population in different countries

Country	Adults	Women	Men	Source
Australia	2.9	–	–	Nkwenkeu et al. (2002)
Germany	–	0.9 – 1.3	2.7 – 3.4	Schäfer et al. (2004)
Germany	0.1 – 10	–	–	Brüggeman et al. (1996)
Belgium	2.6	–	–	Nkwenkeu et al. (2002)
Poland	–	6.7	4.1	Skibniewska et al. (2004)
Serbia	–	4.3 – 5.7	5.4 – 6.1	Djujić et al. (2003)
UK	4.6	–	–	Nkwenkeu et al. (2002)
USA	2.4	–	–	Nkwenkeu et al. (2002)

and adequate daily intake of Mn by adults was previously given to be between 2 and 6 mg d^{-1} whereas the recent EPA recommendation is up to 10 mg d^{-1} for a 70 kg BW (Djujić et al. 2003; Nkwenkeu et al. 2002). The upper tolerable intake of Mn by children is established as 2–9 mg d^{-1}, depending on the age (WHO/IPCS 2002).

Some kinds of vegetarian diet, as well as Fe deficiency, anemia, and some hepatic dysfunction can promote chronic manganese toxicity. This can affect the Mn accumulation in brain that can be associated with neurological symptoms. Manganese, either ingested or inhaled is likely to concentrate in brain and causes a manganism – acute toxic symptoms of psychiatric disorders. Normandin et al. (*vide* Bankovitch et al. 2003) have shown that the solubility of Mn in the respiratory tract influences the brain Mn concentrations.

Oral Mn poisoning has not been observed often and is mainly related to drinking water. The ingestion of excess Mn can result in liver cirrhosis (Plumlee and Ziegler 2003). The harmful excessive Mn exposure in the working place was recently recognized. The recent use of MMT in gasoline has prompted debates on the Mn potential health risk, especially since several studies have proven that the inhalation of Mn-rich dust by humans can increase susceptibility of the respiratory tract to infection and can induce Mn-pneumonitis and some damage to the central nervous system (Plumlee and Ziegler 2003; Schäfer 2004b). Neurotoxicity of inhaled Mn depends on its particle size and species. In Germany, the MAK value of Mn has been set at 0.5 mg m^{-3}. The highest Mn concentration in drinking water, established by the WHO, is 0.5 mg l^{-1}. A much lower concentration of Mn, 0.05 mg l^{-1}, has been recently proposed in the USA and Europe (Schäfer 2004b).

Animals. In animals, Mn deficiency has been manifested relatively often. Initially, symptoms of Mn deficiency in farm animals were recognized in the area with soils derived from calcareous rocks. These symptoms vary depending on animals, and are always related to the impaired functions of some enzymes (as in the human body). The Mn requirement for animals varies from 20 to 60 mg kg^{-1} BW, being the highest for lactating ruminants and the lowest for pigs (Schäfer 2004b). Requirements for some domestic animals vary from <10 to 60 mg kg^{-1} feed (DW) and most commonly is 40 mg kg^{-1} BW (Jones 2005). The Mn deficiency in animals is affected, in most cases, by interactions between elements (e.g., Mg, Ca, P, Fe, Co) that influence a low Mn availability from the digestion track. A low or absence of Mn-SOD activity in various rodent cancer cells indicates, as in humans, a tumor-suppressive action of this enzyme (Schäfer 2004b).

High levels of dietary Mn intake are tolerated by most animals. However, in experimental animals, the increased Mn exposure resulted in damage to the central nervous system and in alterations in carbohydrate homeostasis (Schäfer 2004b). Increase Mn intake by animals (e.g., fodder Mn around or above 100 mg kg^{-1}) can affect deficiency of some bio-elements, (i.e., Fe, Cu, Co, P).

II-7.2 Technetium (Tc)

II-7.2.1 Introduction

Technetium is a by-product of nuclear fuel processing, and the only artificial metal that is used in metal processing (Heiserman 1997). Metallic Tc, mainly used in alloys with Mo or Nb exhibits electrical superconductivity. Very resistant to oxidation, Tc is added to some Fe-alloys, to inhibit the corrosion.

Minute quantities of the isotope ^{99}Tc were found in uranium ores (Fairbridge 1972). This isotope is widely used in medicine for the diagnostic of thyroid disorders.

II-7.2.2 Occurrence in the Environment

Geochemical properties of Tc are similar to those of Mn and Re. In compounds, Tc occurs mainly as heptavalent (Table II-7.1) and in water forms the complex anion TcO_4^-.

The source of ^{99}Tc is mainly global fallout. According to Tagami and Uchida (1998) contents of this radionuclide in paddy soils of Japan vary from 6.1 to 110 mBq kg^{-1}, being accumulated in surface soil layers. The soil to rice-plant transfer of Tc is much lower in flooded soil than in non-flooded soil. Apparently it is controlled by the ratio of two Tc species, soluble TcO_4^-, and insoluble TcO_2 (Yanagisava and Muramatsu 1995). Adsorption of pertechnetate (TcO_4^-) on mineral surface is weak and tends to be strongly affected by pH and other anions, while reduced Tc is readily sorbed by soils constituents and also forms complexes with organic matter (Zhang et al. 2002).

The transfer factor of Tc from soil to edible parts of vegetables ranges and is the lowest for tomato (BCF – 0.3) and the highest for spinach (BCF – 17), on a dry weight basis (Yanagisava and Muramatsu 1993). Recent studies have indicated that the Tc bioavailability in natural soils environments is lower than expected (Tagami and Uchida 2005). The phytoavailability of reduced Tc is much less than that of pertechnetate (Zhang et al. 2002). In plants, Tc is relatively immobile, therefore the highest Tc concentrations were found in older plant tissues.

Denys et al. (2005) reported that the correlation between TcO_4^- and NO_3^- might be useful to assess the fate of ^{99}Tc in soils. Organic and nitrogen fertilizers inhibit its phytoavailability since they stimulate changes of the mobile TcO_4^- to slightly soluble TcO_2 and TcS_2 (Echevarria et al. 1997).

The disposal of Tc-enriched wastes is an environmental hazard since it is easily included in the biocycle. It is subjected to microbial oxidation and reduction processes and is easily taken up by plants, possible as the anion TcO_4^- (Echevarria et al. 1997).

Tagami and Uchida (2005) reported that the dominant species of Tc in natural aqueous environments is TcO_4^-. However, Hu et al. (2005) concluded that in reduced groundwaters this radionuclide is likely to occur in the slightly mobile species, $TcO_2 \cdot nH_2O$.

Ingested Tc localizes in the thyroid gland and gastrointestinal tract. Its biological half-life in the body is approximately 60 hours (Zhang et al. 2002).

II-7.3
Rhenium (Re)

II-7.3.1
Introduction

Rhenium is highly dispersed in the Earth's crust, in the range from 0.4 to 0.6 μg kg^{-1} (Table II-7.2). It occurs primarily in minerals of other elements and is commonly associated with molybdenite, zircon, and gadoline. The mineral dzhezkazganite, of possible formula $CuReS_4$, containing 40–50% of Re, occurs rather seldom.

The annual global production of Re in 2003 was 39.3 t (WMSY 2004). The use of Re is mainly in productions of lamps, photographic flashes, and thermostats. It is also used as catalyst in some chemical reactions. This metal is added to alloys used for medical tools and in jewelry production. The supplementation of Re compounds has been recently used in some cancer patient treatments (Collery et al. 2004).

The most common oxidation state is +7, but other oxidation states are also known (Table II-7.1). Two ionic forms Re^{4+} and Re^{6+} reveal similarity to Mo cations and are likely to substitute for this metal in geochemical processes. Thus, molibdenite can contain Re up to around 2%.

II-7.3.2
Occurrence in the Environment

During weathering, Re is readily soluble, especially in oxidizing and acidic conditions, mainly as the anionic form, ReO_4^-. It is precipitated or sorbed by various sediments, such as bituminous copper shales, black piritic shales, as well as some coals, up to 10 mg kg^{-1}. Swedish arable soils contain Re at average value of <0.04 mg kg^{-1} (Eriksson 2001a).

The world mean Re in river waters has been calculated by Gaillardet et al. (2003) at 0.4 ng l^{-1}. Much higher Re concentrations found in stream water of India, up to 10 ng l^{-1}, are apparently associated with geochemical conditions. The world mean in seawater is estimated by Reimann and Caritat (1998) at 4 ng l^{-1}. Anbar et al. (1992) gave the range of its average values as 7.4–9.1 ng kg^{-1} in the Pacific Ocean, and the range of 4–6.9 ng kg^{-1} in the Atlantic Ocean. The concentration of Re in rainwater collected in Sweden during 1999 averaged 0.17 ng l^{-1} (Eriksson 2001a).

Soluble forms of Re are apparently relatively easily available to plants. Barley and wheat grains from Sweden contained Re at the average value of <0.1 μg kg^{-1} (Eriksson 2001a). The highest Re contents are observed in vegetation of areas surrounding uranium ore deposits.

Rhenium supplied to human organisms is concentrated mainly in the thyroids. Its cationic form ($ReCl_3$) is more toxic than the anionic form (K_2ReO_4). Symptoms of the Re toxicity is associated with impairments of nervous and circulation systems (Collery et al. 2004).

Trace Elements of Group 8 (Previously Part of Group VIII)

According to the previous nomenclature system of the Periodic Tables, the elements of the three groups of the Table: 8, 9, and 10 were included in one VIII group. The elements of the first subgroup (triad) were the so-called: iron family metals (Fe, Co and Ni). Two next triads contained noble metals of the platinum-group (Ru, Rh, Pd, Os, Ir, Pt). The old system referred to chemical and geochemical properties of the elements and therefore it was easier to compare and describe these elements. However, the new nomenclature system, as obligatory, will be followed in the text.

Group 8 of the New Periodic Table of Elements (after the latest IUPAC recommendation) is composed of three elements: iron (Fe), ruthenium (Ru), and osmium (Os). These elements exhibit very variable states of oxidation (Table II-8.1).

II-8.1 Iron (Fe)

II-8.1.1 Introduction

Iron is the most important metal and one of the major constituents of the lithosphere. Its average content of the Earth's crust is about 5% (Table II-8.2). The global abundance of Fe is calculated to be around 4.5% and it is not considered a trace element in rocks and soils. However, Fe plays a special role in the behavior of several trace elements and is in the intermediate position between macro- and micronutrients in plants, animals and humans.

Table II-8.1. Selected properties of trace elements of Group 8

Element	Atomic number	Atomic mass	Atomic radius[a] (pm)	Density (20 °C, g cm^{-3})	Valence[b]	Melting point (°C)
Fe, iron	26	55.8	172	7.87	+2, **+3**, +4, +6	1535
Ru, ruthenium	44	101.1	189	12.41	**+3**, +4[c]	2450
Os, osmium	76	190.2	192	22.61	**+3**, +4, +6, +8	3050

[a] Approximately average values for the main oxidation states.
[b] Valences value in bold are for the main oxidation states.
[c] Oxidation stages range between 0 and +8.

The geochemistry of Fe is very complex in the terrestrial environment and is largely determined by the easy change of its state of oxidation in response to physicochemical conditions. Iron is very chemically reactive and is similar in chemical behavior to other metals, especially to Co and Ni. Its behavior is also closely linked to the cycling of O, S, and C. In variable environmental conditions, Fe reveals various characteristics and can be siderophilic, chalcophilic or lithophilic. In most minerals formed near the Earth's crust surface, Fe occurs as the ferric (Fe^{3+}) ion, whereas in deeper rocks the ferrous (Fe^{2+}) ion predominates.

There are recognized two cycles of Fe: *exogenic* that takes place at the surface Earth's crust, and involves the action of water and air, and *endogenic* – taking place beneath the surface of the Earth, and includes geological processes, such as e.g., melting, metamorphism.

Chief Fe ore minerals are mainly ferric oxides (e.g., hematite) or hydrated ferric oxide (goethite), and various dimorphous minerals including those that are commonly occurring in soils. Important Fe ore deposits can also be composed of other Fe minerals, like siderite, $FeCO_3$; pyrite, FeS_2; and ilmenite, $FeO \cdot TiO_2$. "Bog ores" are small deposits of $FeCO_3$ resulted from the precipitation from soluble $Fe(HCO_3)_2$ in lakes, swamps and shallow shelf regions.

The fate of Fe in processes of weathering is dependent largely on the Eh-pH system and on the stage of oxidation of the Fe compounds involved. The general rules governing the behavior of Fe are that oxidation and alkaline conditions promote the precipitation of Fe, whereas reducing and acid conditions promote the mobilization of Fe compounds. The complex geochemical behavior of Fe associated with its participation in oxidation-reduction processes, resulted, similarly as in the case of Mn, in the formation of great number of oxides and hydroxides. Several minerals and amorphous compounds of Fe have a large surface area that is involved in the reactions of various cations and anions in terrestrial environments. Iron complexes with organic ligands have a significant impact on the metal's fate in the environment. A variety of humic and nonhumic organic compounds are involved in the Fe mobility in various geochemical and pedochemical environments.

Megonigal et al. (2003) have broadly reviewed the cycling of Fe in relation to abiotic and biotic processes. The authors have emphasized (after Lovley) that dissimilatory Fe^{3+} reduction is the dominant form of anaerobic carbon metabolism in many ecosystems, and may be one of the earliest forms of microbial metabolism to evolve on the Earth or elsewhere. Microbial oxidation of Fe that can occur under both aerobic and anaerobic conditions stimulates the precipitation of poorly crystalline Fe oxides, which have significant impact on the behavior of several other cations. Microorganisms play a key role in Fe cycling, which is closely associated with organic carbon oxidation *via* Fe^{3+} reduction and with sulfide and sulfate compound transformations. About eighteen Fe minerals, oxides, hydroxides, carbonates, phosphates, sulfates and sulfites are listed as possibly biologically-induced (Frankel and Bazylinski 2003).

II-8.1.2
Production and Use

Since about 3 000 years BC iron has been used by man and has played a dominant role in human civilization. It is the most commonly used metal in all civilizations. Iron mining and processing is mentioned in the Old Testament.

Global mining of Fe has not changed significantly during recent decades and has been given by various sources as 550–870 Mt yr^{-1}. The calculation for 2000 gave the annual Fe production as 571 Mt (USGS 2004). The Fe-processing industry is spread out in all countries and approximately 30% was produced in the former Soviet Union.

Over 90% of "pig iron" obtained from smelting furnace is processed to steel and various alloys. As a very chemically reactive metal, Fe tarnishes rapidly in air or water. Additions of several other metals to alloys increased Fe resistance to corrosion. However, effects of corrosion processes, on a global scale, are a serious source of this metal in different environmental compartments.

The main use of Fe is for various tools, and in the transport and construction industry.

II-8.1.3 Soils

The common range of Fe contents in soils is between 0.1 and 10% and its distribution in soil profiles is variable and controlled by several soil parameters. Soil texture seems to be highly correlated with Fe concentrations that resulted in increased Fe contents with increasing amounts of fine granulometric fractions (<0.02 mm). Mean Fe contents in the surface layer of sandy soils of Poland have been determined to be 0.57% and in heavy loamy soils to be 1.2% (Kabata-Pendias and Pendias 1999). Contents of Fe in soils are both inherited from parent materials and/or resulted from soil processes that are controlled by climatic factors.

In soils, Fe occurs mainly in forms of oxides and hydroxides, as amorphous compounds, small particles, fillings in cracks and veins and coatings on other minerals or particles. The accumulation of Fe in concretions and nodules is most commonly observed. Also Fe frequently forms chelates with humic and nonhumic organic substances, especially in topsoil layers.

The transformation of Fe compounds and minerals, both biotic and abiotic, play an important role during weathering and soil formation. The vertical Fe translocation influenced mainly by organic ligands is a process called podzolization.

The Fe minerals that occur in soil and are also formed pedogenically and biologically, and are used as a key for soil and soil horizon characterization. Soil Fe minerals, associated with pedogenic processes are:

- Hematite, α-Fe_2O_3, occurs in soils of arid, semiarid, and tropical regions and is most often inherited from parent rocks
- Maghemite, γ-Fe_2O_3, is formed in highly weathered soils of tropical zones and often occurs as concentrations associated with hematite, magnetite, and goethite
- Magnetite, Fe_3O_4, is mostly inherited from parent materials, and is often accompanied by maghemite
- Goethite, α-FeOOH, is the most common mineral in soils over broad climatic regions (temperate and tropical). The crystallinity and composition of this mineral vary and reflect the environment in which it has formed
- Lepidocrocite, γ-FeOOH, is common in poorly drained soils (e.g., paddy soils) and in soils of humid temperate regions. A low pH value, low temperature, and absence of Fe^{3+} are favorable conditions for its formation

Fe

- Ferrihydrite, $Fe_2O_3 \cdot nH_2O$, is not stable but relatively common form in mineral soils. It is easily transformed to hematite in warm regions and to goethite in humid temperate zones
- Ilmenite, $FeTiO_3$, does not occur commonly in soils. It is resistant to weathering, and usually is inherited from parent materials
- Pyrite, FeS_2; ferrous sulfide, FeS; and jarosite, $KFe_3(SO_4)_2(OH)_6$, are widely distributed in submerged soils containing S-compounds
- Vivianite, $Fe_3(PO_4)_2 \cdot 2H_2O$; and hydrous ferric phosphate, $FePO_4 \cdot nH_2O$ are formed mainly under the influence of bacterial processes

Mineral and organic compounds of Fe are easily transformed in soils, and both organic matter and microorganisms appear to have a significant influence on the formation of Fe oxides, as well as on their forms (crystalline, semicrystalline, amorphous). Especially some bacteria (e.g., *Thiobacillus, Metallogenium* sps.) are involved in the accumulation and cycling of Fe (Megonigal et al. 2003). Tan (1998) has emphasized that various binding mechanims between living organisms, like bacteria and fungus, and soil clay and humus matter play a crucial role in the Fe distribution and forms. The precipitation of Fe due to the accumulation of $Fe(OH)_3$ at a bacteria cell surface is relatively common processes in soils (Paul and Clark 2000).

Goethite is apparently the most frequently occurring Fe-mineral in soils, and its sorption properties have been studied broadly (Gerth et al. 1993). Sorption processes are complex and involve: *(i)* absorption of metals on external surface, *(ii)* solid state diffusion of metals, and *(iii)* metal binding and fixation at position inside the mineral particles. A strong ability of goethite to bind selected metals might suggest that this

Fig. II-8.1. Influence of soil pH on the activity of Fe^{3+} (log concentration in solution) maintained by Fe oxides present in soils: **a** goethite; **b** $Fe(OH)_3$, amorphous; **c** $Fe(OH)_3$, soil constituent; **d** Fe concentration in soil solution (various sources, compiled by Kabata-Pendias and Pendias 2001)

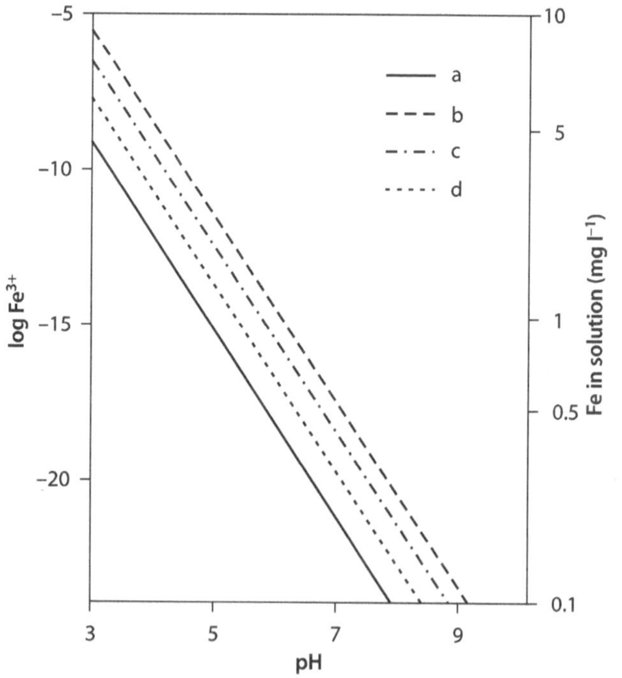

mineral as well as other Fe hydroxides could be used to ameliorate metal-polluted soils. Different authors have presented several orders of preferential sorption of metals by goethite. Most commonly, however, metal ion affinity for the Fe-oxide surface has been presented as follows: Cu > Zn, Co, Pb > Mn > Cd (Kabata-Pendias and Pendias 2001). The negative correlation between Se grains and Fe in soils is suggesting a possible immobilization of Se by secondary Fe minerals (Johnson et al. 2000). It has been confirmed by Cheney et al. (2004) who studied the adsorption of selenite and iodate onto goethite.

Results presented by Karczewska (2004) indicated that amounts of metals bound to both amorphous Fe oxides and crystalline iron oxides in metal-polluted soils are fairly similar and governed mainly by soil properties. Relative amounts of total metal contents bound to Fe oxides were as follows:

- By amorphous Fe oxides: 3–24% of Cu, 1–54% of Pb, 3–25% of Zn, and 43–90% of As;
- By crystalline Fe oxides: 5–29% of Cu, 3–53% of Pb, 15–34% of Zn, and 5–46% of As

Amounts of crystalline Fe oxides in soils irrigated with municipal wastewater were relatively high, within the range from 22 to 66% of the total Fe contents, whereas Fe species bound to Mn oxides were rather negligible, from 1 to 8% of the total Fe (Kabała and Wilk 2004).

The distribution of Fe minerals and compounds in soil profiles is highly variable and reflects several soil processes. The color of soils is largely associated with amounts and forms of Fe compounds. Therefore the content and profile distribution of Fe compounds has been used for the description of soil processes and for the soil classification (Zonn 1982).

Many reactions are involved in the solubility of Fe in soil, but hydrolysis and formation of complexed species appear to be most important. The solubility of Fe^{3+} and Fe^{2+} amorphous hydrous oxides especially control the mobility of soil Fe. However, the formation of other Fe compounds, such as phosphates, sulfides, and carbonates greatly modify the solubility of Fe.

Contents of easily soluble and exchangeable fractions of Fe are very low in comparison with the total Fe content. Easily soluble species of Fe, determined as an operational group by sequential extraction, account from 0.01 to 0.1% of the total Fe, and exchangeable species account from 0.05 to 0.21% (Kabała and Wilk 2004). The concentration of Fe in soil solution, at the common soil pH, ranges from 30 to 550 $\mu g\ l^{-1}$, but in very acid soils it can exceed 2 000 $\mu g\ l^{-1}$. Thus, in calcareous and loamy soils (pH range 7–7.8), Fe contents vary from 100 to 200 $\mu g\ l^{-1}$, whereas in light acid sandy soils (pH range2.5–4.5), Fe contents range from 1 000 to 2 223 $\mu g\ l^{-1}$ (Kabata-Pendias and Wiącek 1985).

The ionic species of Fe commonly occurring in soil solutions are as follows (Kabata-Pendias and Sadurski 2004):

- Cationic forms: Fe^{2+}, $FeCl^{+}$, $Fe(OH)_2^{+}$, $FeH_2PO_4^{+}$
- Anionic forms: $Fe(OH)_3^{-}$, $Fe(SO_4)_2^{-}$, $Fe(OH)_4^{2-}$

The redox behavior of Fe in soils has been broadly studied and described in several textbooks, and recently by Alloway (1995), Bartlett (1999), and Sparks (1995).

Fe

Table II-8.2. Abundance[a] of iron the environment

Environmental compartment	Fe	Environmental compartment	Fe
Earth's crust (%)	5.0		
Igneous rocks (%)		Soils (%)[b]	3.5
Mafic	3.7 – 8.7	Light sandy	0.1 – 1.0
Acid	1.4 – 2.7	Medium loamy	0.8 – 2.8
		Heavy loamy	1.5 – 3.0
		Calcareous	0.5
		Organic[c]	0.03 – 4.9
		Lateritic[c]	0.4 – 33.3
Sedimentary rocks (%)		Water ($\mu g\ l^{-1}$)	
Argillaceous	3.3 – 4.7	Rain[c]	11 – 40
Sandstones	10 – 3.0	Sea[c] (World median)	2
Calcareous	0.4 – 1.0	River[d] (World median)	66
Raw energy material (%)		Air ($ng\ m^{-3}$)	
Coal, hard	0.8 – 19	Urban/Industrial	130 – 14000
Fly ash[e]	12 – 15	Greenland	166 – 171
Crude oil[f]	11[g]	South Pole	0.5 – 1.2
Petrol[f]	2.5[g]		

[a] Presented are common, possible background, values from various sources as compiled by Kabata-Pendias and Pendias (1999 and 2001), unless otherwise stated.
[b] Median for World soils.
[c] After Reimann and Caritat (1998).
[d] After Gaillardet et al. (2003).
[e] After Llorens et al. (2000).
[f] After Li (2000).
[g] Values in $mg\ kg^{-1}$.

In waterlogged soils, the reduction of Fe^{3+} to Fe^{2+} affected by both abiotic and biotic factors results in an increase of its mobility. Also some anthropogenic factors, such as acid rain, acid fertilizers, high input of organic matter, may increase the Fe mobility in most soils. In consequence, geochemical and pedochemical Fe cycling associated with its mobility and phytoavailability may be significantly changed (Zonn 1982).

Although soil Fe is slightly mobile under most soil conditions, it also exhibits a tendency to form mobile organic complexes and chelates. These compounds are largely responsible for the Fe migration between soil layers and for leaching of Fe from soil profiles; they are also important in the supply of Fe to plant roots. The budget of Fe in soils calculated as input/output differences indicates losses of Fe mainly from soils under the spruce forest, and this phenomenon is particularly associated with acid and light soils (Kabata-Pendias and Pendias 2001).

Areas of Fe deficiency in soils for certain crop plants are relatively common, however, in most cases this deficiency is related to a low concentration of soluble Fe species.

Table II-8.3. Trace elements in iron oxides of soils (mg kg^{-1}) (data source: Kabata-Pendias and Pendias 2001)

Element	Fe nodules	Goethites	Fe-Mn concretions
Fe (%)	40.1	52 – 62	2 – 15
Mn (%)	3.4 – 5	0.3 – 0.5	16 – 40
Ti (%)	0.13	–	31 – 39[a]
Ba	890 – 1790	–	–
Co	300	80[b]	70 – 390
Cu	100	800[b]	80 – 450
Ni	100	170[b]	80 – 680
Pb	160 – 2720[a]	–	120 – 11910
V	400	17000	–
Zn	400	17300 – 23500	350 – 1800

[a] Data for Ti-Fe-rich nodules.
[b] Data for magnetite separated from soils.

Most of Fe deficient soils occur under arid climates and are associated with calcareous, alkaline, or other specific soils. In the humid climatic zone, with the predomination of acid soils, Fe deficiency is most unlikely to occur. However, some activities of humans, like improper liming and disturbing the soil chemical balance by remediation action, may affect a low Fe availability to plants. Bocanegra et al. (2004) reported that iron-humate chelates might be a source of easily available Fe to plants.

Soil testing and the correction of the Fe deficiency have been the subject of many studies, however, there is still a lack of suitable methods to assess the Fe availability to plants (Mortvedt et al. 1991). The chelating agents (DTPA and EDTA) appear to be most often recommended for measuring the availability of Fe to plants, however, Kreij et al. (1996) reported that the correlation between Fe in all substrate extracts (also H_2O and $CaCl_2$) and Fe in the test plants was poor. Recently Houba et al. (1999) showed that 0.01 M $CaCl_2$ is relatively simple and suitable for the extraction of the available pool of Fe.

Iron is an excellent metallic material for environmental remediation because it is a strong and nontoxic reducer, effective in coprecipitating and sorbing inorganic pollutants (Kaplan et al. 2002). Several Fe compounds for the remediation of soil contaminated with trace metals have been proposed. Also steel shots containing mainly Fe oxides are effective in the immobilization of soil trace metal excess. The method presented by Mench et al. (1998) is based on the ability of Fe and Mn oxides to absorb and occlude several trace elements from the ambient solution. Zero-valent Fe decreases the solubility mainly of As and Cr, is less effective in the immobilization of Zn and seems to increase the labile pool of Cu (Kumpiene et al. 2005). Steel shots are also little successful in the stabilization of Cd. However, several authors suggested that the zero-valent Fe applied as remediation treatment, gives beneficial effects in trace metal stabilization, in a long-term perspective.

The increased ratio of magnetic Fe to regular Fe in soils, as compared with mother rocks, indicates the impact of industrial pollution. Dust from metal processing exhibits greater magnetic susceptibility than Fe minerals. Measurement of the anomaly in magnetic reactivity of soils is proposed as a tool for the monitoring of soil pollution (Magiera et al. 2004).

II-8.1.4 Waters

According to Gaillardet et al. (2003), the global mobility index (Cw/Cc ratio) of Fe is low (<0.01). These authors calculated the world average of Fe in the dissolved load (<0.2 μm) of rivers as 66 $\mu g\, l^{-1}$, and its concentrations in the range of 11–739 $\mu g\, l^{-1}$.

Iron concentrations in seawaters range broadly and depend on the source, these values are (in $\mu g\, l^{-1}$):

- 2 – 20 world ocean (Fairbridge 1972)
- 25 – 743 world ocean (Reimann and Caritat 1998)
- 0.3 – 35 Baltic Sea (Szefer 2002)
- 0.03 $\mu g\, kg^{-1}$, North Pacific (Nozaki 2005a)

Iron plays an important role in marine environments as a micronutrient for organisms and its low bioavailability may limit growth of phytoplankton that are critical for the oceans primary production (Achterberg et al. 2004).

Concentrations of Fe in surface waters can be quite variable under the impact of different factors. McKnight and Duren (2004) reported that the Fe^{2+} oxidizing reactions in stream water including photochemical and/or microbial reactions might constrain the mid-day Fe^{2+} maxima concentrations. The importance of these reactions depends upon the sources and concentrations of DOM, and dissolved Fe.

Most of Fe transported by rivers reaches the ocean, however, it is also partly precipitated in bottom sediments. Riverine flux of Fe to the ocean has been calculated by Kitano (1992) as 7 $Mt\, yr^{-1}$, whereas this value estimated by Gaillardet et al. (2003) is 2.47 $Mt\, yr^{-1}$.

The behavior and chemistry of Fe in water systems are very complex and controlled by several parameters of which redox potential is the most important. The predominated forms of Fe are colloids, but several hydrous ions of Fe^3 and Fe^2 can also dominate. Most of Fe compounds are easily soluble in water at pH <7, however, under oxidizing conditions they are precipitated in various oxide forms. Due to reactive surface properties, hydrous Fe oxides are the most important factor controlling the behavior of dissolved metals in water. The colloidal Fe oxides play a dominate role in the sorption and coagulation of other colloidal substances and ions. Dissolved Fe compounds readily precipitate in most aquatic environments and form various multimetallic concretions in bottom sediments. Bacteriogenic Fe oxides occurring in ground waters, composed mainly of amorphous hydrous ferric oxides, Mn oxides, and bacterial organic matter, play an especially important role in the transport and fate of dissolved metals in subterranean water systems (Ferris et al. 2000).

Iron concentrations in landfill leachate and ground water can create various technologic and environmental problems. Redox conditions of in landfill leachates are very complex, but reducing conditions may predominate. Therefore concentrations of Fe in groundwater around waste disposal sites may vary in the broad range from 3 to 5 500 mg l^{-1} (Bjerg et al. 2003), whereas Fe contents of groundwater in unpolluted zones of temperate humid climate average 0.75 mg l^{-1} (Kabata-Pendias and Pendias 1999). Recently, several technologies based on the precipitation of metallic pollutants on ferric iron band have been proposed for the remediation of contaminated waters.

Rainwaters contain Fe in the range of 11–40 µg l^{-1}, the lowest value for coastal region of Norway and the highest for polluted region of the Kola Peninsula (Reimann and Caritat 1998).

Aquatic plants. Zhu et al. (2005) described that Fe plaque is commonly formed on the root surface of aquatic plants, mainly due to the oxidation of Fe^{2+} to Fe^{3+}, and the precipitation of Fe oxide on the root surface. This plaque is composed predominantly of ferrihydrite (63%) with lesser amounts of goethite (32%) and a minor level of siderite (3%).

II-8.1.5 Air

The origin of Fe in the atmosphere is from both terrestrial and industrial sources. Its concentrations in air vary from 0.5 to 1.2 ng m^{-3} above the South Pole, from 13 to 40 ng m^{-3} at Spitsbergen, and from 166 to 17 ng m^{-3} above Greenland (Table II-8.2). Contents of Fe in the atmosphere of different cities, from various continents, range broadly from 50 to 14 000 ng m^{-3}, and are clearly associated with industrial activities (Kabata-Pendias and Pendias 1999). In world remote regions, median Fe concentrations vary from 60 to 70 ng m^{-3}, whereas in polluted areas range between 130 and 14 000 ng m^{-3}, with a median of 2 500 ng m^{-3} (Reimann and Caritat 1998).

Aerial dust of urban region is composed, of 33–38% (weight), of Fe which the annual deposition was estimated at 16 800–43 200 g ha^{-1} yr^{-1} (Manecki et al. 1988). The deposition of Fe in inhabitant areas of Europe ranges from 300 to 5 700 g ha^{-1} yr^{-1} (Kabata-Pendias and Pendias 1999).

II-8.1.6 Plants

The mechanisms of Fe uptake by plants and transport within plants, as well as its metabolic functions have received much study because Fe is the key metal in energy transformation needed for syntheses and other life processes of the cells. The present knowledge on this topic has been extensively reviewed by several scientists and compiled by Kabata-Pendias and Pendias (2001).

The main features of the Fe uptake by plants and Fe transport between plant organs may be summarized as follows:

- Various species may be absorbed, mainly as Fe^{2+}, but also Fe^{3+}, and Fe chelate
- Plant roots may reduce Fe^{3+} to Fe^{2+}, which is fundamental in the Fe absorption by most plants
- At Fe deficiency conditions, roots (especially of cereals and other *Gramineae*) release mugineic acids that are effective in mobilizing Fe

- In xylem exudates, Fe occurs mainly in unchelated forms
- The Fe transport is mediated largely by citrate chelates and by soluble ferritins (transferrins)

At conditions of Fe deficiency, roots of some plants can develop various mechanisms responsible for enhanced Fe solubility through the reduction of Fe^{3+} and/or chelation in phytosiderophore forms that are efficient in mobilizing Fe (Marschner et al. 1989). Phytosiderophores released by the roots of *Poaceae* into the rhizosphere especially influence the availability and uptake of Fe and, probably, other metals (Reichman and Parker 2005). Boganegra et al. (2004) reported that Fe-humate complexes are a good source of this metal for plants.

The essential role of Fe in plant biochemistry can be summarized as follows:

- Several Fe-proteins, mainly transferrins, ferritins, and siderophores, are involved in transport, storage and binder systems
- Fe occurs in hem and non-hem proteins and is concentrated mainly in chloroplasts
- Chlorophyll formation is influenced by Fe
- Organic Fe complexes are involved in the mechanisms of photosynthesis electron transfer
- Non-hem Fe proteins are involved in the reduction of nitrites and sulfates
- Fe is directly implicated in the metabolism of nucleic acids
- Both cations, Fe^{3+} and Fe^{2} may also play a catalytic role in various reactions

Interactions between Fe and other trace and major elements are very complex and may occur within plants as well as at the root surface. In most cases these are antagonistic interrelationships, and can be presented as follows:

- Excess amounts of Mn, Ni, and Co cause a reduction in the absorption and translocation of Fe that resulted in a decrease of chlorophyll
- High levels of Fe compounds in growth media result in decreasing uptake of several trace metals, mainly of Mn, Ni, and Co
- Interactions between Fe and Mn in both growth medium and plant tissues have the most important impact on plant metabolism
- The interaction of Fe-Zn seems to be associated with the precipitation of franklinite, $ZnFe_2O_4$, that depresses the availability of both metals
- Antagonistic interactions between Fe and Si, and Fe and Se have been observed
- Fe-P interactions, in both plants and soils, are related mainly to the precipitation of $FePO_4 \cdot 2H_2O$. Furthermore, P anions compete with plants for Fe and interfere with Fe uptake and its transport. The appropriate P/Fe ratio is fundamental to plant health
- In K-deficient plants, Fe may become toxic due to disrupted oxidation of Fe^{2+} to Fe^{3+}
- Fe-Ca interactions, in both plant and soil, are very complex, but the significant impact of Ca is to suppress Fe availability, and thus may lead to Fe chlorosis in plants on calcareous soils
- Fe-S interactions cause a low Fe availability at low soil S, whereas a high S content may also result in a low Fe solubility, depending on soil environments

The physiology of Fe toxicity and of plant resistance to the excess of this metal can be summarized as follows:

- Plants high in nutrients, especially in Ca and SiO_2, can tolerate increased levels of Fe
- Rice roots (also roots of other plants) are able to oxidize Fe and deposit it on the root surface
- Mycorrhizas reveal a great capability to bind Fe either at the root surface or in root cells
- Root damage by various factors destroys the oxidizing power of roots and aggravates Fe availability and thus Fe toxicity
- Plant resistance to excess Fe is associated with different reactions like oxidation, immobilization and exclusion of mobile Fe forms by roots
- Plants adapted to waterlogged conditions are commonly more tolerant to high Fe levels than plants grown in well-aerated soils

The above summaries are oversimplified and very generalized, but give some information showing that in addition to the active Fe roles in redox reactions of plant cells, Fe also performs other fundamental functions in plants. Plant response to Fe toxicity or deficiency is highly variable among genotypes and plant species. Therefore, genetic transformations in plants seem to be most promising in resolving problems of Fe in plant nutrition.

The appropriate content of Fe in plants is essential to both, the plant metabolism and the nutrient supply to humans and animals. The plant ability to absorb Fe is variable and is affected by changing conditions of soil and climate and by stages of plant growth and specific genotype properties. Some plants, like e.g., *Alyssum bertoloni*, exhibit a special capability to absorb Fe and can accumulated this metal to about 4 000 mg kg^{-1} in roots and 1 300 mg kg^{-1} in leaves (Brooks 1998).

Various cereal grains do not differ much in their Fe concentrations (Table II-8.4). Mean grain Fe amounts vary from 31 to 98 mg kg^{-1}, being higher in oats and barley than in wheat and rye. The grand mean of 48 mg kg^{-1} for Fe in grains has been estimated (Kabata-Pendias and Pendias 2001). The range of mean Fe contents in belowground parts (roots, bulbs, tubers) of selected food plants vary from 34 to 46 mg kg^{-1}, and from 3.3 to 3.7 mg kg^{-1} FW. Green, aboveground parts of vegetables (especially spinach) contain higher amounts of Fe than other parts of plants. Phytoferritin, a metalloprotein, is a soluble and available (to animals and humans) form of Fe in plants, concentrated mainly in the shoots, root apex, and seeds.

Iron contents of forage plants (grass and clover) vary greatly and indicate the impact of soil and climatic conditions (Table II-8.5). Mean Fe concentrations in grasses range from 43 to 376 mg kg^{-1}, being the highest for Australia, Hungary and New Zealand. Mean Fe amounts in clover vary from 117 to 400 mg kg^{-1} and are higher in the same countries as Fe in grasses. The nutritional requirement of grazing animals is usually met at the Fe contents ranging from around 50 to 100 mg kg^{-1}.

Contents of Fe in moss growing in Norway vary highly, from 120 to 21 000 mg kg^{-1} and average 660 mg kg^{-1}, suggesting the impact of atmospheric deposition (Berg and Steinnes 1997). Results of the research of Nieminen et al. (2004) indicated that Fe from aerial deposition dominates (Fe in pine needles: 883 mg kg^{-1}) over root uptake pathway of this metal (Fe in pine needles: 69 mg kg^{-1}).

Fe

Table II-8.4.
Iron in food plants grown in various countries[a] (mg kg^{-1})

Plant	Range	Mean
Wheat, grains	17 – 50	31
Wheat, grains[c]	31.5 – 8.7	34
Rye, grains	34 – 43	38
Barley	33 – 218	98
Oats, grains	54 – 140	93
Bean, pods[b] (FW)	17 – 89	46
Carrot, roots	16 – 67	3.3
Carrot, roots[c] (FW)	3 – 3.4	34
Onion, bulbs	33 – 50	77
Beet (sugar and red), roots	71 – 82	40
Potato, tubers	21 – 58	3.7
Potato, tubers[c] (FW)	3.4 – 4.0	3.1
Strawberry, fruits[c] (FW)	2.7 – 3.7	15
Nuts and almonds[b]	9 – 67	–
Nuts and almonds[d]	11.1 – 47	–
Sunflower, fruits[d] (FW)	1 – 11	–

[a] Presented are common, possible background, values from various sources as compiled by Kabata-Pendias and Pendias (1999, 2001), unless otherwise stated.
[b] Jedrzejczak (2004), from markets in Poland.
[c] Szteke et al. (2004), range and mean of median values of 130 samples from Poland.
[d] Ensminger et al. (1995), data for USA.

II-8.1.7 Humans and Animals

Humans. The content of total Fe of soft tissues of a "reference man" is estimated at 55 mg kg^{-1}. Skin tissues contain only 29 mg kg^{-1}. In human fluids the Fe concentrations vary from 0.14 to 0.45 mg l^{-1} in urine and milk, respectively (Li 2000).

Iron is an essential nutrient for all organisms and each species regulate its own Fe metabolism. Free ionic Fe is very reactive, being involved in the generation of free-radicals that can be very toxic. It is, therefore, commonly stored in inert forms, e.g., bound to serum transferrin. Ferritin was first discovered in the mucous membrane of the stomach and small intestine, as well as in the liver and spleen. It is a metalloprotein involved in both the transport and storage of Fe. Iron binds to transferrin that plays a fundamental role in Fe metabolism of cells (Schümann and Elsenhans 2004). By chelating some divalent cations (e.g., Cd, Co, Cu, and Zn,), ferritin is able to protect the cells from metal toxicity (Sczekan and Joshi *vide* Prasad 2004a).

Table II-8.5. Iron in grasses and clover at the immature grown stage from different countries (mg kg^{-1}) (values from various sources as compiled by Kabata-Pendias and Pendias 1999, 2001)

Country	Grasses		Clover	
	Range	Mean	Range	Mean
Australia	228 – 264	245	252 – 357	285
Great Britain	73 – 154	103	122 – 132[a]	–
Finland	39 – 49	43	–	–
Germany	79 – 430	157	116 – 253	175
Hungary	133 – 923	376	118 – 535	346[b]
Ireland	34 – 78	55	6 – 85	74[a]
Japan	55 – 157	106	76 – 136	152
Poland	60 – 140	92	76 – 136	117
New Zealand	69 – 1510	320	105 – 1700	400
USA	18 – 320	73	–	–

[a] Alfalfa.
[b] Various legumes.

A peptide called hepcidin, recently isolated from human urine and blood, appears to be capable of regulating Fe absorption. It exhibits antimicrobial activity (Mazur et al. 2003; Kośla et al. 2004). As a hormone, hepcidin seems to be a key regulator of Fe metabolism. As such it is involved in anemia (Niçolas et al. 2002). An insufficient production of hepcidin is accompanied by Fe overloading, whereas overproduction of this hormone causes Fe deficiency and anemia. Hepcidin is, therefore, considered a primary factor in regulating Fe homeostasis (Gehrke et al. 2003).

Both non-heme and heme Fe forms are adsorbed in variable proportion depending on the body Fe status and dietary bioavailability, and vary from 0.1 to >35% to 20–50%, respectively (Antal 2006). The author suggests that the Fe supply of the population should be provided primarily by nutrition. Any forms of Fe supplementation are justified only for patients with a confirmed Fe-deficiency.

Iron deficiency fosters anemia and reduces the availability of other nutrients. Iron-deficiency anemia, particularly in young children, produce pallor, fatique and susceptibility to infections. In most instances, however, it results from the disturbed absorption and not from the low intake of Fe. The availability of Fe depends on its species in the diet. Commonly, from 5 to 20% of its total content is easily available. The most highly available Fe is protein-bound, such as occurs in women's milk, meat, and especially in liver. It is much less readily available in plant food, particularly in plants rich in cellulose.

Excessive ingestion of Fe is mainly accidental and can be associated with Fe-containing medications. It can result in haematochromatosis, siderosis, and heart and liver damage (Plumlee and Ziegler 2003). Some persons absorb unduly high amounts of Fe, due to hederitary haemochromatosis. They have too much Fe (not bound to transferring) in their plasma, a condition that produces a significant health risk. Increased levels of serum ferritin and transferrin are indicators of excessive Fe intake and Fe overload. According to Schümann and Elsenhans (2004), an overdose of oral Fe-containing substances results in corrosion of the gastric and intestinal mucosae, particularly in children. The clinical effects from Fe overload are increased risk of cardiomyopathy (and possible coronary heart disease), hepatic cancer, and tumor of other organs (WHO/IPCS 2002).

Home-brewed beer, fermented in Fe drums (in some region of Africa) is a source of highly available Fe (50–100 mg d^{-1}) and may induce hepatic cirrhosis and diabetes. Excessively high Fe content in human plasma is associated with myocardial infarction, corrosion of gastric and intestinal mucosa, hepatic cirrhosis, and risk of cancer. However, the epidemiological evidence of neoplasmic tumours and the association with increased Fe accumulation in bodies has not been, so far reported, and therefore, it is still controversial. (Schümann and Elsenhans 2004).

Interactions between several elements may determine the availability of Fe to humans. Antagonistic relationships have been observed between Fe, on one hand, and Cd, Zn, Mn, Pb, and P, on the other. The interaction between Fe and Cu is complex and is associated with oxidation-reduction processes.

The Fe content in food has been investigated in many countries. The recommended daily dose for adults varies from 10 to 18 mg and might be higher for pregnant women and some other persons. The world mean daily uptake of Fe has been estimated as 13 mg (Hamilton *vide* Kabata-Pendias and Pendias 1999). The mean Fe content of daily home-made meals in Poland is calculated to be 14.9 mg. The food rations served to prisoners contain only 2.5 mg (Marzec et al. 2004b). An increased dose of Fe (e.g., 60–100 mg d^{-1}) may cause nausea, vomiting, heartburn, and epigastric discomfort. Doses of 180–300 mg kg^{-1} BW can be lethal (Schümann and Elsenhans 2004).

Animals. Iron occurs in all tissues of organisms and seems to be more concentrated in marine animals, as a whole, than in terrestrial mammals. According to Jørgensen (2000), common Fe content in soft mammalian organs varies from 140 to 520 mg kg^{-1}, being the lowest in muscles and the highest in liver.

There are very few instances of Fe deficiency or toxicity in animals. Anemia associated with the lack of Fe was reported for baby pigs (Anke 2004b). This author reported that normative Fe requirements of farm animals range from 40 to 75 mg kg^{-1} feed (DW) for hens and piglets, respectively.

II-8.2
Ruthenium (Ru)

II-8.2.1
Introduction

Ruthenium is a member of the group of platinum metals and is considered to be the least abundant of the metals of this group. Its mean concentration in the Earth's crust is estimated at 1 μg kg^{-1}, whereas its contents in rocks vary from 0.01 to 60 μg kg^{-1}. The occurrence of Ru is often associated with Os and Ir, as well as with some base metals such as Fe or Cu. Some sulfide (e.g., laurite, RuS_2) and arsenosulfide minerals of Ru have been recently found at very low concentrations in other sulfide and arsenosulfide minerals (Fairbridge 1972). Naturally occurring isotopes are mainly ^{102}Ru and ^{104}Ru.

The deposits of Ru are closely associated with Pt and Pd. Pure Ru is a very hard metal and not subject to attack by acids, but reacts with alkalis, especially under oxidizing conditions. Like other Pt-group metals, Ru does not easily combine with other elements. This metal is extracted mainly at the refinery of Ni and Cu. It is used as a hardener for Pt and Pd alloys and to increase the resistance to corrosion of some other metal alloys. It is also used as a substitute for Pt in jewelry. Ruthenium is used in some electronic devices and for special ceramic paints, optical instruments, and glass of a required quality. This metal exhibits catalytic properties in the synthesis of some hydrocarbons.

II-8.2.2
Soils and Plants

Recent data on the distribution of Ru in Sweden are presented by Eriksson (2001a) as follows: *(i)* <40 μg kg^{-1} in topsoils, *(ii)* <1 ng l^{-1} in rain, *(iii)* <1 μg kg^{-1} in cereal grains, and *(iv)* <11 mg ha^{-1} yr^{-1}.

Concentrations of Ru are given as 0.7 ng l^{-1} in seawaters, 20 ng l^{-1} in Norwegian lakes, 100 ng l^{-1} in stream waters of India (Reimann and Caritat 1998), and 0.005 ng kg^{-1} in the North Pacific (Nozaki 2005a).

The average Ru content in various plants is given as 5 μg kg^{-1} (Kabata-Pendias and Pendias 2001). Plants exposed to automobile exhausts contain Ru in the range from 0.3 to 0.9 μg kg^{-1}, with the highest value for moss and the lowest for mushrooms (Djingova et al. 2003). The concentrations of both radioisotopes, ^{103}Ru and ^{106}Ru, in the lichen (*Stereocaulon vesuvianum*), sampled in Italy in autumn 1986, indicated an increase with elevation of Mt. Vesuvius, from 90 to 110 Bq kg^{-1} and from 320 to 520 Bq kg^{-1}, respectively (Adamo et al. 2004).

The radionuclides ^{103}Ru and ^{106}Ru are released during nuclear reactions. An elevated content of ^{106}Ru was found in bottom sediments of Danube River, below the nuclear reactor (Kabata-Pendias and Pendias 2001). Both radionuclides, and particularly ^{106}Ru, were deposited on soil by the fallout after the Chernobyl accident. According to Schimmack et al. (1989) this radionuclide is rather mobile and easily migrates down soil profiles.

Plants relatively easily take up the ^{106}Ru radionuclide, but a large proportion is concentrated in root. According to Grogan et al. (1987), this radionuclide is not likely to be easily transferred from soil to the food chain, however, under specific conditions it might be significantly increased in plants.

II-8.2.3
Humans

Humans absorb Ru relatively easily from the diet. However, it does not create any health risk since this metal is concentrated mainly in bones. Veronese et al. (2001) investigated the absorption rate of the stable isotope ^{101}Ru into the circulatory system of humans and concluded that it seems to be lower than predicted from the data recommended by ICRP (International Commission on Radiological Protection). Studies on the kinetics of Ru in humans have demonstrated that complexed Ru, particularly as a citrate compound, is cleared from blood very rapidly while inorganic Ru remains longer in the circulatory system (Veronese et al. 2004).

Some Ru complexes are presented as potential anti-cancer drugs, similarly as Pt-based anti-tumor medications (Pieper and Keppler 1998). The volatile compound, RuO_4, may irritate some mucous membranes and, at an elevated concentration, may disturb the metabolism of Ca.

II-8.3 Osmium (Os)

Osmium is a member of the platinum metal group, but in contrast to other metals of this group, it does not concentrated preferably in the nickel-iron phase of chondrites. This metal is notable for its great density and high melting point (Table II-8.1). Its abundance in the Earth's crust has been calculated at 1 $\mu g\,kg^{-1}$. Like other Pt-group metals, Os occurs mainly in native alloys of variable composition. There are no Os minerals, but this metal is likely to be associated (at about 3%) with laurite, $(Ru,Os)S_2$. Some other mineral species of Pt metals exhibit an affinity for increased amounts of Os, up to about 30%, e.g., iridosmine, and osmiridium – native alloys of variable composition.

Unlike other Pt metals, Os easily reacts with oxygen and may occur at various oxidation states, from –1 to +8 (Table II-8.1). This metal is resistant to acids but is susceptible to reactions with alkalis.

There is small production of Os, estimated as 300 t yr^{-1} in 1995 (Reimann and Caritat 1998). The metal is used mainly for the production of hard-wearing alloys and for electronic devices. During metal processing, various Os compounds are formed, such as, OsF_6, $OsCl_4$, and volatile OsO_4. Unlike metallic Os, these compounds are strongly toxic to humans.

Osmium concentrations in river waters range from 3.3 to 42.1 $pg\,l^{-1}$, and average 9 $pg\,l^{-1}$ (Gaillardet et al. 2003). The mean concentration of Os in ocean waters is given as 2 $pg\,l^{-1}$ (Reimann and Caritat 1998; Nozaki 2005a).

There are seven stable isotopes of Os, of which three Os (186, 187, and 192) have been determined in river waters. The total concentrations of these isotopes were estimated (in $fm\,mol\,kg^{-1}$) in waters of Columbia River as 15, of Mississippi River as 34–45, and of Vistula River as 44 (Sharma and Wasserburg 1997; Kabata-Pendias and Pendias 1999).

Soluble Os^{4+} species are easily mobile and thus may be available to plants. Osmium metal has no biological role and is unreactive. However, all its compounds, and particularly OsO_4, is highly toxic to humans and animals, and can cause lung congestion, skin damage, and severe eye damage. Longer exposure to increased levels of some Os compounds is hazardous to marrow, liver, and kidneys.

Trace Elements of Group 9 (Previously Part of Group VIII)

The elements of Group 9 are cobalt (Co), rhodium (Rh), and iridium (Ir). Cobalt belongs to the iron family of metals that include also Fe and Ni, whereas Rh and Ir are noble metals of the platinum group (PGMs).

Based on the IUPAC recommendation, these elements will be described under one group. All three metals exhibit variable properties, siderophilic and/or lithophilic characters, and variable oxidation states (Table II-9.1).

II-9.1 Cobalt (Co)

II-9.1.1 Introduction

Cobalt, in the Earth's crust, is highly concentrated in mafic rocks, up to 200 mg kg^{-1}, when compared to its content in acid igneous rocks, 1–15 mg kg^{-1}, and to its abundance in the upper continental crust, 12 mg kg^{-1} (Table II-9.2). It is also likely to concentrate in black shales.

As Co reveals siderophilic character it is likely to form minerals with S, As, and Se. Its common rock-forming minerals are: cobaltite, CoAsS; smaltite, $CoAs_{2-3}$; linneite, Co_3S_4; and arsenosulfide, CoAsS. In the hypergenic zones and in soils, the formation of erythrite, $Co_2(AsO_4)_2 \cdot 8H_2O$ is likely to occur.

The geochemical cycle of Co closely resembles Fe and Mn cycling, and is likely to be associated with several minerals of these metals. During weathering processes, Co^{3+}

Table II-9.1. Selected properties of trace elements of Group 9

Element	Atomic number	Atomic mass	Atomic radius[a] (pm)	Density (20 °C, g cm^{-3})	Valence[b]	Melting point (°C)
Co, cobalt	27	58.93	167	8.9	**+2**, +3, +4	1495
Rh, rhodium	45	102.9	183	12.4	+2, **+3**, +4, +6	1964
Ir, iridium	77	192.2	187	22.4	+2, +3, **+4**, +6	2443

[a] Approximately average values for the main oxidation states.
[b] Valence values in bold are for main oxidation states.

Table II-9.2. Abundance[a] of cobalt in the environment

Environmental compartment	Co	Environmental compartment	Co
Earth's crust (mg kg^{-1})	10 – 12		
Igneous rocks (mg kg^{-1})	–	Soils (mg kg^{-1})[b]	8
Mafic	35 –200	Light sandy	5.5
Acid	1 – 15	Medium loamy/ silty	7.0
		Heavy loamy	12
		Calcareous	7.0
		Organic	4.5
Sedimentary rocks (mg kg^{-1})	–	Water (µg l^{-1})	–
Argillaceous	14 – 20	Rain	0.025[c]
Sandstones	0.3 – 10	Sea	0.002[c]
Calcareous	0.1 – 3	River	0.15[d]
Raw energy material (mg kg^{-1})	–	Air (ng m^{-3})	0.05[e]
Coal, hard	5 – 40	Urban/Industrial	0.13 – 6.8
Fly ash	30	Greenland	0.07 – 0.15
Crude oil	0.05	South Pole	0.0001 – 0.12
Petrol	0.2		

[a] Presented are common, possible background, values from various sources as compiled by Kabata-Pendias and Pendias (1999 and 2001), unless otherwise indicated.
[b] Arithmetic mean for World soils.
[c] After Reimann and Caritat (1998).
[d] After Gaillardet et al. (2003).
[e] Median value for World air in remote regions, after Reimann and Caritat (1998).

is unstable in the aqueous phase and precipitates as sulfides and/or adsorbed by hydrous oxides. In soils, however, Co behavior seems to be strongly influenced mainly by the Mn oxide phase formation. The absorption capacity of Co by Mn oxides, of both biogenic and abiogenic origin, is very high (Tani et al. 2003).

Relatively easy interactions of Co with all metals that are associated geochemically or biochemically with Fe, have a significant impact on its behavior in soils and its phytoavailability. Especially exhibit are unique antagonistic reactions of Fe and Mn arise from their ability to occupy the same sites in crystalline structures and from the similarity of their metallo-organic compounds.

II-9.1.2
Production and Use

Global annual mining of Co in 2000 was 35.6 kt (USGS 2004) and in 2003 was 43.03 kt (WMSY 2004). Other sources give the current annual production on the order of 15 kt (Reimann and Caritat 1998; Hoppstock and Sures 2004). Most of Co is used for special alloys. The application of Co is very broad and includes several industries such as stainless steel production, aerospace equipments, electroplating, and magnetic devices. This metal is also used in chemical catalyses and synthesis of fuels, in paints, and as a plastic hardener.

A significant branch of the pharmacy uses Co for medical and veterinarian drugs. The radionuclide ^{60}Co has been used for some medical treatments as a source of gamma radiation. This radionuclide has also been used in special equipments in steel production.

It is estimated that about 2% of Co used in the metallurgical industry is discharged to sewage sludge.

II-9.1.3 Soils

The content of Co in soils is inherited mainly from parent materials. The range of world mean values of Co in surface soils is between 4.5 and 12 mg kg^{-1}, being the highest for heavy loamy soils and the lowest for light sandy soils and organic soils. A fairly similar range is reported for Japanese soils, from 7.1 to 23 mg kg^{-1}, with the highest content in Andosols and the lowest in Acrisols (Takeda et al. 2004). The range of Co in reference samples of the USA is from 5.5 to 29.9 mg kg^{-1} and in Chinese soil, in the range of 5.5–97 mg kg^{-1} (Govindaraju 1994).

The content and distribution of Co in soil profiles are dependent on soil-forming processes and therefore differ for soils of various climatic zones. Usually, higher Co contents of surface soils are found in arid and semiarid regions, e.g., Egyptian soils contain Co from 16.5 to 26.8 mg kg^{-1} (Nasseem and Abdalla 2003), while low soil Co has been reported for the glaciated regions of northern parts of different continents, as well as for the Atlantic Coastal Plain of the USA. Light soils developed from glacial deposits under temperate humid climate usually contain small quantities of Co. Thus, median value for Co in Lithuanian soils is reported to be 3.4 mg kg^{-1}, and in soils of the northern region of Russia is 5.5 mg kg^{-1} (Kabata-Pendias and Pendias 1999). Swedish arable soils contain Co in the range of 0.4–14 mg kg^{-1}, at the average value of 7.1 mg kg^{-1} (Eriksson 2001a). Surface horizons of Russian chernozems contain Co in the range of mean values from 9.2 to 10.8 mg kg^{-1} (Protasova and Kopayeva 1985). Naturally high Co contents are observed in soils over serpentine rocks, 10–520 mg kg^{-1}, and in soils around ore deposits, 13–85 mg kg^{-1} (Kabata-Pendias and Pendias 2001).

Several soil factors govern the Co distribution in soils. However, hydrous oxides of Fe and Mn are of the greatest importance. Oxides of Fe have strong affinity for the selective adsorption of Co. This is well reflected in the Co distribution in soil profiles that shows a close relation between the levels of Fe and Co in a particular soil horizon. Amounts of Co are always higher in the B-soil horizons where Fe is concentrated. Iron nodules may accumulate up to 300 mg Co kg^{-1} and Fe-Mn concretions contain Co in the range of 70–390 mg kg^{-1} (Table II-8.3).

The speciation of Co in soils depends on several factors, among which the Eh potential plays a significant role. Both the reduction of soil Eh and a decrease in soil pH may result in the solubilization of precipitated or adsorbed Co. The oxidation of Co^{2+} to Co^{3+}, especially by Mn oxides, is a common process in soils. Interchange reactions are involved in the sorption of Co by Mn oxides, leading to the formation of hydroxyl species, $Co(OH)_2$ that precipitated at the oxide surface.

Different redox mechanisms have been proposed for the sorption of Co by Mn oxides. In general, these include: *(i)* oxidation of Co^{2+} to Co^{3+} at the oxide interface, *(ii)* reduction of Mn^{4+} to Mn^{3+} in the oxide crystal lattice, and *(iii)* replacement of Mn^{4+} or Mn^{3} by Co^{3+} (Fleming *vide* Kabata-Pendias and Pendias 2001). In soils, the sorption of Co by Mn increases greatly with pH and is controlled by redox soil potential (Fig. II-9.1). Soil Mn nodules and concretions may adsorb up to 12 000 mg Co kg^{-1} (Table II-7.3).

Soil texture and SOM play also an important role in the behavior of Co in soils. Contents of Co in mineral soils indicate a positive relationship with the fine granulometric fraction (Table II-9.2). Montmorillonite and amorphous Fe-Al hydroxides especially exhibit a great sorption capacity for Co (Kabata-Pendias 1973). The impact of SOM on the Co behavior is variable and depends on the kind of organic matter and pH. Although soils rich in SOM usually have low Co contents and low Co phytoavailability, Co-organic chelates may be readily available. Some organic ligands stimulate the sorption of Co on Fe-coated silica compounds, but increase Co mobility when Fe hydroxides are absent (Brooks and Herman 1998). In most soils, Co is rather slowly mobilized and thus its concentrations in soil solution range from 0.3 to 87 μg l^{-1} (Kabata-Pendias and Pendias 2001). Common ionic species occurring in soil solutions are: Co^{2+}, Co^{3+}, $CoOH^{+}$, and $Co(OH)_3^-$. Also complex compounds, such as $Co(OH)_2^0$, of a relatively high stability constant, may occur in soil solution (Kabata-Pendias and Sadurski 2004).

An adequate supply of Co to grazing animals (especially ruminants) has been for years of a great challenge. Acetic acid-soluble and EDTA extractable Co usually corresponds to the readily available pool of the metal. Pasture soils, of the texture of loamy sands and clays, from the North Western Coast of Egypt contain the AAAc-EDTA extractable Co from 3.60 to 4.69 mg kg^{-1} (Nasseem and Abdalla 2003). The authors sug-

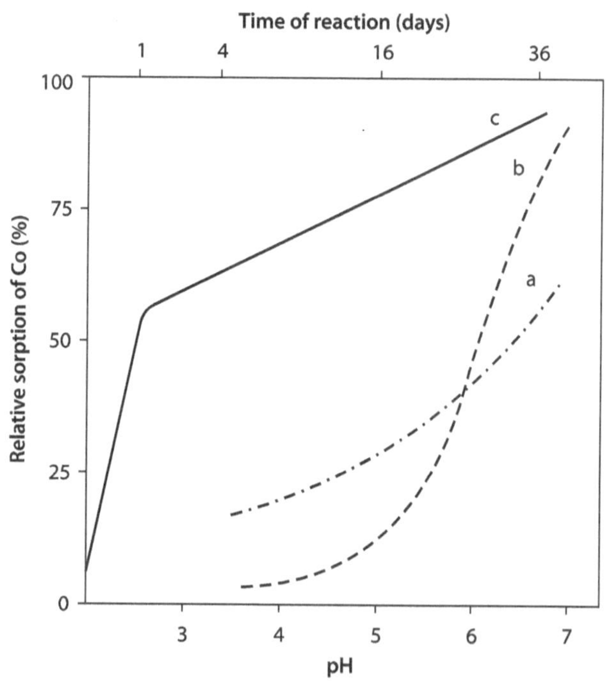

Fig. II-9.1.
Effects of pH on the sorption of Co by Mn nodules (**a**), goethite (**b**), and effects of time on the sorption of Co by birnessite (**c**) (adapted from McKenzie 1975)

gested that an extractable Co content less than 2.45 mg kg^{-1} in pasture soils may cause a deficiency of Co in animals. The application of Co sulfate compounds or EDTA chelates to soil are the most common practices in controlling the Co deficiency in ruminants.

There is some evidence of Co enrichment in contaminated soils. In surface soils around metal processing industries, Co can reach up to 154 mg kg^{-1}, which has been reported by Smith and Carson (*vide* Kabata-Pendias and Pendias 2001). Some increase in Co contents has also been observed in sludge-applied farm soils as well as in roadside soils.

II-9.1.4 Waters

Cobalt is slightly mobile in the terrestrial environment, and its Cw/Cc ratio is about 0.08 (Gailardet et al. 2003). However, it does not remain long in aquatic environments since soluble species are easily adsorbed by suspended organic matter and by Fe-Mn hydroxides.

Concentrations of Co in river waters range from 0.02 to 0.43 µg l^{-1}, and the world average amount is calculated at 0.15 µg l^{-1} (Gaillardet et al. 2003). Rivers of Western Siberia contain Co in the range of 0.2–2 µg l^{-1} (mean 1 µg l^{-1}), and its input to Baikal Lake has been estimated as 3 t yr^{-1} (Niechayeva 2002; Vietrov 2002). Relatively low Co concentrations are reported by Pekka et al. (2004) for dissolved and suspended phases in the Kola River (Russia), in the range from 0.03 to 0.83 µg l^{-1} and from 0.003 to 0.190 µg l^{-1}, respectively. In heavily polluted rivers of Poland, Co concentrations average 2.2 µg l^{-1} (Lis *vide* Kabata-Pendias and Pendias 1999).

In ocean waters Co ranges from 0.001 to 0.02 µg l^{-1} (Reimann and Caritat 1998) and in the North Pacific it averages 0.001 µg kg^{-1} (Nozaki 2005). In waters of the Baltic Sea, Co contents vary from 0.001 to 0.07 µg l^{-1} (Szefer 2002a). The world riverine flux of Co has been estimated by Gaillardet et al. (2003) as 5.5 kt yr^{-1}, whereas this value for the Baltic Sea has been given by Matschullat (1997) as 220 t yr^{-1}, of which 44% is from anthropogenic sources.

In waters, Co occurs at two oxidation states (+2 and +3); Co^{3+} is very reactive and is rapidly adsorbed and deposited in bottom sediments. These processes have a significant impact on the accumulation of Co in concretions and nodules of ocean-sediments, where maximum concentration of this metal can reach 0.8% (Kotlinski 1999).

Several factors, like the presence of organic ligands and various anions, as well as redox potential and pH control the speciation of Co in waters and sediments. The percent of Co species in a Welsh lake was found to be: Co^{2+}, 76%; $CoCO_3$, 9.8%; $CoHCO_3^+$, 9.6%; humate complexes, 4%; and $CoSO_4$, 0.4%. However, in another models for Co species in fresh water, the speciation of Co was completely different, with the predomination of either $CoCl^{2+}$ and Co^{2+} or $CoCO_3$ and Co^{2+} (ATSDR 2002a).

Concentrations of Co in surface water and groundwaters seem to be fairly similar and range from 0.03 to 10 µg l^{-1} (ATSDR 2002a). Extremely high Co concentrations (5–60 µg l^{-1}) in ground waters of the European part of Russia can be related to the impact of both geochemical and anthropogenic factors (Uchvatov and Bashkin 1997). High levels of Co, in the range of 4.5–6.5 mg l^{-1}, were reported at the sites near a copper mine and smelter in the USA (ATSDR 2002a). Skjelkvåle et al. (2001) investigated the occurrence of Co in the Nordic lakes, and reported its concentrations to vary, at the value of 99.5 weighted percentiles, as follows (in μg l^{-1}): 2.4 in Finland, 1.4 in Sweden, and 0.96 in Norway.

In atmospheric precipitation (bulk rain) over the Baltic Sea, Co concentrations are between 0.8 and 12.5 μg l^{-1} whereas over the Kola Peninsula rain water contains Co within the range from 0.02 to 11.8 μg l^{-1} in remote and polluted regions, respectively (Reimann and Caritat 1998; Szefer 2002a). Rainwater collected during 1999 in Sweden contained Co at the average value of 0.01 μg l^{-1} and at the maximum value of 0.03 μg l^{-1} (Eriksson 2001a).

Water pollution with Co is usually reflected in its accumulation in bottom sediments. The average Co content of river bottom sediments has been estimated at 13 mg kg^{-1}, but in industrially polluted rivers its concentrations can vary from around 20 to 50 mg kg^{-1} (Kabata-Pendias and Pendias 1999). About 60% of the ^{60}Co was bound to the hydrous Fe-Mn oxide phase of the streambed sediments of White Oak Creek Watershed, USA (Zhang et al. 2002). Some nuclear reactors release ^{60}Co picolinate complex that is relatively stable in waters. It is possible, therefore, that soluble species of radiocobalt will increase in the future (ATSDR 2002a).

For drinking water, no limits have been set in most countries, but in Russia the MAC was established as 100 μg l^{-1} (Reimann and Caritat 1998). The median Co concentration in US drinking water is <2 μg l^{-1}, however increased levels up to 107 μg l^{-1} has been also reported (Smith and Carson 1981).

Aquatic biota. Fish from the Baltic Sea accumulate rather small amounts of Co, in the range from 1 to 10 μg kg^{-1} FW, whereas in soft tissue of mussels its contents vary from 100 to 900 μg kg^{-1} FW, and in plankton, from <100 to 2 100 μg kg^{-1} (Szefer 2002a,b).

A number of aquatic microorganisms and blue-green algae synthesized vitamin B_{12} that is released into the water when organisms die. Thus, relatively increased level of cobalamins may be expected in some coastal waters (Smith and Carson 1981).

II-9.1.5
Air

The median Co concentration in air of the world remote regions is 0.05 ng m^{-3}, and of world polluted regions is 3 ng m^{-3} (Reimann and Caritat 1998). However, Co concentrations in air highly vary from 0.0001 to around 6 ng m^{-3}, above the South Pole and urban regions, respectively (Table II-9.2). Mean Co levels in air at unpolluted sites are generally <1 to 2 ng m^{-3}, but in open-ocean environments, Co levels of 0.0005 ng m^{-3} were recorded (ATSDR 2002a). At the site of a Ni refinery in Wales, Co concentrations averaged 48 ng m^{-3} (Smith and Carson 1981).

Natural aerial deposition of Co into the Baltic Sea has been estimated as 3 t yr^{-1}, whereas the deposition from anthropogenic sources is 20 t yr^{-1} (Matschullat 1997). Atmospheric input of Co on inhabitant regions of the UK was calculated to be between 1.5 and 6 g ha^{-1} yr^{-1} (Cawse 1987), and in Germany (Schwarzwald) was reported to be 5.6 g ha^{-1} yr^{-1} (Zöttl et al. *vide* Kabata-Pendias and Pendias 2001). According to Reimann and Caritat (1998), throughfall deposition of Co in Germany varies from 13 to 27 g ha^{-1} yr^{-1}. Wet deposition of Co in Sweden has been calculated at 0.19 g ha^{-1} yr^{-1} (Eriksson 2001a).

II-9.1.6
Plants

Contents of Co in plants are highly controlled by both soil factors and the ability of plants to absorb this metal. The relatively large variations in Co levels, as reported by different authors for cereals and food plants, apparently reflect the impact of various factors, among which plant genotypes and environmental conditions play significant roles. Sillanpää and Jansson (1992) evaluated results of analysis of wheat (before shooting stage, $N = 1\,723$) and of corn (6-leave stage, $N = 1\,892$) from 30 countries, and stated that soil texture is the most significant parameter controlling the Co level in plants. Mean values of Co in cereal grains, in the range of 5–270 µg kg^{-1}, show that its lowest concentrations are in cereals from Norway and the highest in cereals from Egypt (Table II-9.3). Wheat and barley grains from Sweden contain Co at mean values of 4.5 and 7.6 µg kg^{-1}, respectively (Eriksson 2001a).

Contents of Co in plant foodstuffs vary from 8 to 170 µg kg^{-1}, being the lowest in apples (Table II-9.3.). In general, legumes accumulate more Co than grasses. Mean Co contents of clovers from various countries range from 100 to 570 µg kg^{-1}, whereas these values for grasses are between 60 and 270 µg kg^{-1} (Table II-9.4).

Cobalt in nutrient or soil solution is very easily available to plants. The relationship between Co in red clover and its concentration in the nutrient solution is very close (Fig. II-9.2). Also contents of Co species soluble in AAAc-EDTA extractant are highly correlated with Co levels in crop plants (Sillanpää and Jansson 1992).

Table II-9.3.
Cobalt in food plants grown in various countries[a] (µg kg^{-1})

Plant	Range	Mean
Wheat, grains (Norway)	1.5 – 14	4.7
Wheat, grains (Sweden)[b]	1.1 – 18	4.5
Wheat, grains (Australia)	13 – 231	82
Wheat, grains (Egypt)	160 – 380	270
Barley, grains	4.4 – 44	18
Oats, grains	10 – 300	96
Cabbage, leaves	100 – 160	–
Carrot, roots	37 – 120	–
Onion, bulbs	28 – 80	–
Tomato, fruits	62 – 20	–
Cucumber, fruits	87 – 170	–
Potato, tubers	37 – 160	–
Apple, fruits	8 – 16	–
Orange, fruits	19 – 45	–

[a] Presented are common, possible background, values from various sources, as compiled by Kabata-Pendias and Pendias (1999, 2001), unless otherwise indicated.
[b] After Eriksson (2001a).

Co

Table II-9.4.
Cobalt in grasses and clover at the immature grown stage from different countries (μg kg^{-1}) (values from various sources as compiled by Kabata-Pendias and Pendias 1999, 2001)

Country	Grasses		Clover	
	Range	Mean	Range	Mean
Australia	–	–	70 – 530	190
Great Britain	<30 – 100	270	60 – 1700	570
Finland	40 – 80	60	–	–
Germany	30 – 220	90	60 – 210	120
Sweden	10 – 400	60	80 – 300	150
Japan	10 – 510	120	20 – 750	200
Poland	10 – 240	80	50 – 260	100
New Zealand	30 – 150	80	–	–
USA	<40 – 390	80	150 – 270	190

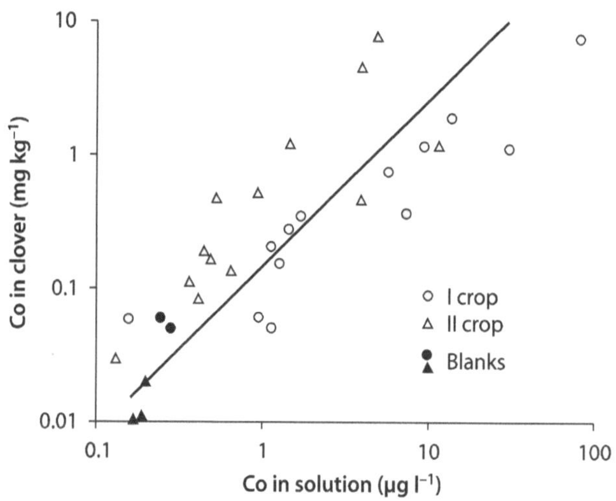

Fig. II-9.2.
Cobalt content of red clover as a function of its concentration in the solution containing clay minerals impregnated with this metal (Kabata-Pendias 1973)

On Co-rich soils, such as serpentine soils or soils over Co ore bodies, some plants can accumulate much Co; for example *Astragalus* sp. may contain up to 100 mg kg^{-1}. The range of Co in native vegetation from serpentine soils is reported to vary from 17 to 540 mg kg^{-1} AW and in vegetation grown around Co-Cu ores, was found up to about 4 000 mg Co kg^{-1} (Brooks 1998). Several plant species, mainly from *Cruciferae*, *Leguminosae*, *Nyssaceae* families reveal the ability to accumulate extremely high amounts of Co; e.g., *Hausmaniastrum* sp. up to 4 000 mg kg^{-1}, *Nyssa sylvatica* above 200 mg kg^{-1}, without any symptoms of the toxicity. However, when Co is taken up in excess by roots of plants that have not developed a mechanism of the tolerance to Co, it usually follows the transpiration stream, resulting in an enrichment of Co at leaf margins and tips. Therefore, white, dead leaf margins and tips are symptoms of the Co toxicity, as well as interveinal chlorosis of new leaves, which is closely linked with the Fe chlorosis.

The essentiality of Co to microorganisms, as well as to higher plants, has not been yet fully understood, although there is some evidence of a favorable effect of Co on plant growth and especially on N-fixing processes. Studies on functions of Co in N_2 fixation indicated that cobamide coenzymes are involved in the H atom migration during the formation of the NH_3 compound by rhizobia (Nicholas *vide* Kabata-Pendias and Pendias 2001). The deficiency of Co in legumes inhibits the formation of leghemoglobin and hence, inhibits N_2 fixation. However, the Co requirement for this process is low. In natural conditions, Co deficiency is not known to retard the growth of either legumes or nonlegumes.

The Co deficiency in herbage has been considered mainly from the viewpoint of ruminant requirement since Co is a component of a precursor of the vitamin B_{12} for these animals. It has been observed that grasses grown on soils with the Co content less that 5 mg kg^{-1} may be Co-deficient for the normal growth of animals. The critical Co levels in ruminant diet is around 80 to 100 μg kg^{-1} and thus common Co contents in grasses and clovers met this requirement (Table II-9.4).

Moss samples collected in Norway in the period 1990–1995 contain Co in the range of 0.045 to 0.7 mg kg^{-1}, at the average value of 0.37 (Berg and Steinnes 1997).

II-9.1.7
Humans and Animals

Humans. Cobalt occurs in all mammalian tissues and its contents vary from 5.5 to 230 μg kg^{-1}, with the highest value in the liver and the lowest in the brain (Jørgensen 2000). Average contents of Co in human soft tissues has been estimated at <20 μg kg^{-1} (Li 2000). According to data presented by the ATSDR (2002a), mean concentrations of Co in some tissues of unexposed population from various countries are (in μg kg^{-1}): liver, 17–120; kidneys, 12; pectoral muscles, 16; nails, 40–170; and hair, 20–180. Mean concentrations of this metal in human fluids are given (in μg l^{-1}) as follows: blood, 0.39; serum, 0.21; milk, 0.27; and urine, 0.57 (Reimann and Caritat 1998). Somewhat different Co concentrations (in μg l^{-1}) are given by Schrauzer (2004) for unexposed humans: blood, 0.09; serum, 0.11; and urine, 0.25; whereas hard-metal workers may contain much higher levels of Co, up to 245 μg l^{-1} in blood and up to 303 μg l^{-1} in urine. The biological half-life of ^{60}Co in the body is 9.5 days (Zhang et al. 2002).

Cobalt is essential for humans and for most animals as a component of the vitamin B_{12}. In the form of inorganic metal, Co is required for the synthesis of bacterial vitamin B_{12} by ruminants. Although inorganic Co (non-B_{12}) is present in organs and fluids of a body, its other physiological functions are unknown. Cobalt is likely to be bound by some proteins and to replace other divalent cations (e.g., Zn, Mn) in various enzymes, without any effects. Some organic Co compounds are apparently involved in processes of stabilizing the DNA structure (Munno et al. 1996).

The deficiency of Co may affect anemia and anorexia. Anemic patients show increased requirements for Co. Presumable Co binds to Fe-transport proteins and is involved in the formation of hemoglobin. The excessive ingestion of Co may cause polycythemia (increased red blood cells), cardiomyopathy, hypothyroidism, pancreas failure, bone marrow hyperplasia, and some types of cancer (Plumlee and Ziegel 2003).

Human dietary intakes of Co vary from 5 to 40 μg d^{-1} and is mainly from the ingestion of foods, particularly from livers and meet products (Table II-9.5). Mean intake of

Co

Table II-9.5.
Cobalt levels in food from various countries ($\mu g\ kg^{-1}$ FW) (compiled from ATSDR 2002a)

Food item	Range	Median/maximum
Bakery good/cereals	–	10.9/75.7
Fats and oils	–	5.9/9.1
Beef	–	1
Beef liver	43 – 74	–
Beef kidneys	8 – 10	–
Pork	1 – 12	–
Pork liver	10 – 23	–
Pork kidneys	4 – 11	–
Meat and poultry	–	<5.5/38.2
Fish, Sweden	–	–/20
Fish, Canada	<1 – 8	18.6/–
Milk and milk products	14.3 – 29	<1.4/18.9
Sugar and candy	–	<0.4/3.5
Vegetables	–	2.4/18.1

Co by the selected population groups of different age in Canada is (in $\mu g\ d^{-1}$): children, 7–10; adults, 9–15; and elders, 8–10 (Dabeka and McKenzie *vide* ATSDR 2002a).

Most of Co is ingested in inorganic forms, while vitamin B_{12} accounts for a very small fraction of the total Co intake. Only a small fraction of Co is inhaled from air. When some implants in humans contain Co, it may be a significant source of this metal; in hip arthroplasty patients increased levels of Co in urine, up to 3.8 $\mu g\ l^{-1}$, are reported (ATSDR 2002a).

Animals. Animals require very small amounts of Co. However, the Co deficiency syndromes in ruminants have been spread through several continents, and the most severe were observed, at the beginning of the last century, in Australia. According to Schrauzer (2004a), it is likely that Co deficiency will become a problem in the future, since the natural Co content of soils is low, and the depletion of this metal occurs through agricultural practices and natural soil leaching processes. The deficiency of Co can be controlled by the application of Co salts to the soil, and the effect of such treatment may last for several years. There are differences of opinion, however, as to whether it is better to apply Co to the soil or give directly to livestock.

There are no reports on the Co toxicity to animals, attributed to the consumption of natural feedstuffs. However, in certain geochemical areas, and under the influence of man-induced pollution, an excess of Co in plants may be a health risk.

Stemme et al. (2002) have investigated the effect of oral Co supplies on the vitamin B_{12} status of calves. They concluded that relatively low Co content in fodder (130 $\mu g\ kg^{-1}$) seems to be sufficient for pregnant cows, and that only about 3% of orally intake Co is involved in metabolic processes.

II-9.2 Rhodium (Rh)

II-9.2.1 Introduction

Data on Rh are too scarce to estimate its environmental abundance. Some authors consider Rh as the least frequent element of the PGMs. The average content of Rh in the bulk continental crust is given as 0.06 μg kg^{-1}, and its range in different rocks from 0.01 to 20 μg kg^{-1}. In coal, Rh was found at the concentration of <0.1 mg kg^{-1}.

Oxidation stages of Rh may vary, however, most commonly it occurs at the +3 valence. In minerals, Rh is associated with other PGMs. They are mainly arsenosulfides, e.g., hollingworthite, (Rh,Pd,Pt)AsS, and irarsite (Ir,Ru,Rh,Pt)AsS. The Pt-mineral, sperrylite, can contain a large proportion of Rh and some of Fe and Cr minerals may host Rh.

The melting point of Rh is much higher than that of Pd and Pt (Tables II-9.1 and II-10.1) and this metal is also less subjected to chemical reactions than PGMs. Rhodium is used mainly for the production of jewelry and of special tools, but also as a catalyst in organic synthesis.

Its global production has been estimated at 300 t yr^{-1} in 1995 (Reimann and Caritat 1998).

II-9.2.2 Occurrence in the Environment

Data on Rh in soils are too scarce to evaluate its mean abundance in soils. Eriksson (2001a) reported for Swedish arable soils the content of Rh at the mean value of <0.04 mg kg^{-1}.

The median Rh concentration in ocean waters is estimated as 0.08 ng l^{-1}. This metal has not been commonly determined in waters and the only data are on Rh concentration at the value of <100 ng l^{-1}, in stream waters of India (Reimann and Caritat 1998). Nozaki (2005) cited the mean concentration of Rh in the North Pacific to be 0.08 ng kg^{-1}. It seems to be relatively easily bioaccumulated by crustaceans (isopods), up to 12 μg kg^{-1}, from polluted freshwaters (Moldovan et al. 2001).

Rainwater collected during 1999 in Sweden contained Rh at the average value of <1.0 ng l^{-1}, and wet deposition is calculated to be <11 mg ha^{-1} yr^{-1} (Eriksson 2001a). This author estimated the mean concentration of Rh in cereal grains at <1 μg kg^{-1}.

There has been an increased release of Rh noticed, together with other PGMs, from catalytic converters of cars. The Rh contents in dust samples collected in 1999 along the highways in Germany ranged from 30 to 42 μg kg^{-1}, whereas plant samples from those sites contained Rh (in μg kg^{-1}), as follows: dandelion, 2.2; plantain, 1.3; and moss, 5.4 (Djingova et al. 2003).

Compounds of Rh are not encountered by most people, but should be considered to be toxic and carcinogenic. Some Rh compounds are suspected human carcinogens mostly at high levels, however the MTD depends on its compounds and sources. Several water-soluble Rh^{3+} compounds are mutagenic in bacteria and in mammalian cells and induce chromosomal aberrations. The exposure of humans to the increased inhalation of Rh can be hazardous. Recommended limits for Rh in air is established at 0.1 mg m^{-3} for soluble, and 1 mg m^{-3} for insoluble Rh compounds (Hoppstock and Sures 2004).

II-9.3 Iridium (Ir)

II-9.3.1 Introduction

Mean content of Ir in the bulk continental crust is estimated at 0.05 $\mu g\ kg^{-1}$, and in the upper continental crust as 0.02 $\mu g\ kg^{-1}$. It is concentrated in the Fe-Ni core and in some meteorites. Like other PGMs, Ir occurs as natural alloys, mainly with Rh and Os, in arsenosulfide minerals.

Increased content of Ir in sedimentary rocks, at the border layer between Cretaceous and Tertiary deposits, is apparently related to an effect of the meteorite collision about 5 million years ago (Dai et al. 1998). It is concentrated in some coals, up to <200 $\mu g\ kg^{-1}$.

The melting point of pure Ir is very high (Table II-9.1) and its resistance to corrosion is the greatest among the PGMs. Due to a great mechanical strength at high temperatures, Ir is used for special alloys and also as a catalyst in different chemical processes. Previously Ir was commonly used for tips of ink pen.

Its production has been estimated as 300 $t\ yr^{-1}$ in 1995 (Reimann and Caritat 1998).

II-9.3.2 Occurrence in the Environment

According to Reimann and Caritat (1998) the median Ir concentration in ocean waters is 0.13 $pg\ l^{-1}$. Anbar et al. (1996) reported its concentrations in the Baltic Sea at the range from 0.32 to 1.17 $pg\ kg^{-1}$ (calculated from values: $10–39 \times 10^8$ atoms kg^{-1}). In stream and lake waters, Ir ranges from 6 to 100 $ng\ l^{-1}$ (Reimann and Caritat 1998). Rivers supply more dissolved Ir to the oceans than do extraterrestrial sources (Anbar et al. 1996). Rainwater, collected during 1999 in Sweden, contained Ir at the average value of <1 $ng\ l^{-1}$ (Eriksson 2001a).

Swedish arable soils contain Ir at the average value of <40 $\mu g\ kg^{-1}$ (Eriksson 2001a). Increased levels of Ir, up to around 5 $\mu g\ kg^{-1}$, have been reported for humus layer (topsoil) from the Kola Peninsula. Increased Ir levels in soils adjacent to a motorway are attributed to motor vehicles with catalysts (Zereini et al. 1997). In dust collected along streets in Germany, Ir contents vary from 1.2 to 3.5 $\mu g\ kg^{-1}$, and in plants grown along those roads; it ranges from <0.02 to 0.4 $\mu g\ kg^{-1}$, being the highest in dandelion (Djingova et al. 2003).

Iridium is apparently easily taken up by plants and concentrated in leave margins. Wheat and barley grains from Sweden contain Ir at the average value of <0.1 $\mu g\ kg^{-1}$ (Eriksson 2001a).

When ingested, Ir is absorbed by animals, and at higher doses is accumulated in soft tissues but does not have any biochemical implications. Its concentrations in human fluids are 7.4 $ng\ l^{-1}$ in blood, and 18 $ng\ l^{-1}$ in urea (Hoppstock and Sures 2004).

Trace Elements of Group 10 (Previously Part of Group VIII)

Group 10 contains three elements: nickel (Ni), palladium (Pd), and platinum (Pt). Nickel belongs to the iron family that also includes Fe and Co, whereas Pd and Pt are noble metals of the PGMs.

Based on the IUPAC recommendation, these elements will be described under one group. All three metals reveal variable properties (Table II-10.1), siderophilic and/or lithophilic characters, and a tendency to occur under various oxidation states. All metals readily form sulfide and sulfarsenide compounds.

II-10.1 Nickel (Ni)

II-10.1.1 Introduction

In the Earth's crust, the mean Ni abundance has been estimated at around 20 mg kg^{-1}, whereas in the ultramafic rocks Ni ranges from 1 400 to 2 000 mg kg^{-1}. Its concentrations decrease with increasing acidity of rocks, down to the range of 5–20 mg kg^{-1} in granites. Sedimentary rocks contain Ni in the range from 5 to 90 mg kg^{-1}, with the highest range being for argillaceous sediments.

Nickel reveals both chalcophilic and siderophilic affinity and readily combines with metallic Fe. Therefore, Ni-Fe compounds are common in the Earth's core and meteorites. The Ni-Fe alloy of the Earth's core (called barysphere or NIFE) is composed of Fe/Ni within the ratio of 11:1 (Fairbrigde 1972). Also the great affinity of Ni for S accounts for its frequent association with segregates of S bodies. The principal Ni metallic ore

Table II-10.1. Selected properties of trace elements of the Group 10

Element	Atomic number	Atomic mass	Atomic radius[a] (pm)	Density (20 °C, g cm^{-3})	Valence[b]	Melting point (°C)
Ni, nickel	28	58.69	162	8.90	**+2**, +1, +3, +4	1 454
Pd, palladium	46	106.42	178	12.02	**+2**, +1, +3, +4	1 555
Pt, platinum	78	195.07	183	21.45	**+4**, +2, +3, +5	1 768

[a] Approximately average values for the main oxidation states.
[b] Valence values in bold are for the main oxidation states.

composed mainly of pentlandite, $(Ni,Fe)_9S_8$ and pyrrhotite, $Fe_{1-x}Ni$). In rocks, Ni occurs primarily as sulfides, (millerite, NiS), arsenides (niccolite, NiAs), and antimonides (ullmanite, NiSbS). Nickel often forms sulfides and sulfarsenides together with Fe and Co, and is associated with several Fe minerals.

After weathering, most Ni is coprecipitated with Fe and Mn oxides, and becomes included in goethite, limonite, serpentinite, as well as in other Fe minerals. It is also associated with carbonates, phosphates, and silicates. Mn-nodules of the Pacific Ocean floor may contain up to 1% Ni.

Organic matter exhibits a strong ability to absorb Ni, thus it is likely to be highly concentrated in coal and oil. This concentration is apparently an effect of the precipitation of Ni as sulfides in sediments rich in organisms and under reducing conditions. Therefore, a significant proportion of Ni loads to the environment is from biolite combustion. Global emission of Ni from coal combustion is calculated to be around 20 kt yr^{-1}, and from oil products combustion it is estimated to range from 10 to >40 kt yr^{-1} (Kabata-Pendias and Pendias 1999).

II-10.1.2
Production and Use

Global Ni production in 1995 was estimated to be 1.14 Mt in 2000 and 1.2 Mt in 2003 (WMSY 2004; USGS 2004).

Nickel has been broadly used in several industries. A great proportion of Ni goes into the production of variety of metal alloys for the aircraft and the plating industries. Most of Ni is used in various alloys (e.g., hastelloy, nickelbase, nickel-silver) that provide oxidation and corrosion resistance for use with acids and salts. Ni-alloys are also high-temperature and creep and stress resistant. Stainless steel usually contains Ni at 8–10%, however amounts up to 30% are also reported (ATSDR 2002a).

It is widely used for magnetic components (e.g., magnetic tapes, permanent magnets) and electrical equipments. Ni-alloys are used for different tools and vessels used in medicine and in food technology, as well as for kitchen equipment (e.g., cutlery). Its compounds are utilized as dyes in ceramic and glass manufactures, and in batteries containing Ni-Cd compounds. Nickel has been a common catalyst for hydrogenation of fats and oils, and for the oxidation of various organic compounds.

II-10.1.3
Soils

The status of Ni in soils is highly dependent on its contents in parent rocks. However, the concentration of Ni in surface soils reflects the additional impact of both soil-forming processes and pollution.

Soils throughout the world contain Ni in the very broad range from 0.2 to 450 mg kg^{-1}. The common background range of mean Ni contents varies between 19 and 22 mg kg^{-1} (Table II-10.2), but various values, ranging from 20 to 40, have been cited (Adriano 2001). The geometric mean of Ni contents in surface (0–20 cm depth) agricultural soils of Poland (calculated for 48 590 soil samples) is 6.2 mg kg^{-1} (Terelak et al. 2000). Sandy soils of Poland contain, on average, 8 mg Ni kg^{-1} and loamy soils 18 mg Ni kg^{-1} (Kabata-

Table II-10.2. Abundance[a] of nickel in the environment

Environmental compartment	Ni	Environmental compartment	Ni
Earth's crust ($mg\ kg^{-1}$)	20		
Igneous rocks ($mg\ kg^{-1}$)		Soils ($mg\ kg^{-1}$)[c]	19 - 22
Ultramafic	1400 - 2000	Light sandy	7 - 33
Mafic	130 - 160	Medium loamy	11 - 25
Acid	5 - 20	Heavy loamy	23 - 50
		Calcareous	18 - 92
		Organic	4 - 12
Sedimentary rocks ($mg\ kg^{-1}$)		Water ($\mu g\ l^{-1}$)	
Argillaceous	40 - 90	Rain	0.3 - 0.7[g]
Sandstones	5 - 20	Sea	0.7[d]
Calcareous	5 - 20	River	0.8[h]
Raw energy material ($mg\ kg^{-1}$)		Air ($ng\ m^{-3}$)	
Coal, hard	14 - 340[b]	Urban/Industrial	4 - 120
Fly ash	88[e]	World remote areas	0.9[g]
Crude oil	9.4[f]	Shetland Island	4
Petrol	10[f]		

[a] Presented are common, possible background, values from various sources as compiled by Kabata-Pendias and Pendias (1999 and 2001), unless otherwise indicated.
[b] After Finkelman (1999).
[c] Range of means for World soils.
[d] After Szefer (2002a).
[e] After Llorens et al. (2000).
[f] After Li (2000).
[g] After Reimann and Caritat (1998).
[h] World average (Gaillardet et al. 2003).

Pendias and Pendias 1999). The geometric means Ni concentrations in soils of two-background-sites in the State of Idaho, USA, are 11.8 and 23.4 $mg\ kg^{-1}$ (ATSDR 2002a). The median Ni contents in Japanese soils range from 25 to 31 $mg\ kg^{-1}$, with the lowest value for Acrisols and the highest for Andosols (Takeda et al. 2004). These authors have indicated slightly elevated Ni contents in agricultural soils as compared with non-agricultural ones. Swedish arable soils contain Ni in the range of 1–30 $mg\ kg^{-1}$, with an average value of 13 $mg\ kg^{-1}$ (Eriksson 2001a).

Senwo and Tazisong (2004) reported a close relationship between Ni in surface soils of Alabama and CEC and clay content. The lowest Ni level (7.6 $mg\ kg^{-1}$) was in soil with 0.9% clay content, and the highest (53.8 $mg\ kg^{-1}$) in soil with 4.6% clay content. Data presented by Bettinelli et al. (2000), 88 mg Ni kg^{-1} in calcareous loam and 27 mg Ni kg^{-1} in light sandy soil, also show the impact of soil texture on the soil Ni status. Niechayeva (2002) reported Ni contents in soils of the taiga zone of Western Siberia to be in the range of 20–100 $mg\ kg^{-1}$ (mean 40). In surface horizons of Russian chernozems, mean Ni contents range from 28 to 34 $mg\ kg^{-1}$ (Protasova and Kopayeva

1985). Naturally elevated Ni contents are observed in soils over basic and volcanic rocks and especially in soils derived from serpentine rocks, where Ni ranges from 770 to 7375 mg kg^{-1} (Kabata-Pendias and Pendias 2001).

Soil Ni is strongly associated with Fe and Mn oxides. Iron oxides and minerals contain Ni within the range from 100 to 170 mg kg^{-1}, whereas Fe-Mn concretions accumulate this metal up to 680 mg kg^{-1} (Table II-8.3). Manganese oxides in soils accumulate Ni in the range from 40 to 4900 mg kg^{-1}, and Mn minerals present in sediments contain this metal from 120 to 10900 mg kg^{-1} (Table II-7.3). Also clay minerals, in particular montmorillonite, exhibit a great capability to bind this metal (Dähn et al. 2003).

Nickel in soils is slightly mobile and occurs mainly in the residual fraction, in over 50% of its total contents in loamy sandy soil, and in about 70% of its occurrence in estuary muck (Kabata-Pendias and Pendias 2001; Zhang et al. 2004). However, in surface soil horizons Ni occurs mainly in bound organic forms, a part of which are relatively easily soluble chelates. Bonding of Ni to organic ligands may be very strong, and can highly affect its mobility. Conversely, the remobilization of Ni from solid phases appears to be possible in the presence of FA and HA. Thus, Ni may be quite mobile in soils with high complexation ability (e.g., organic-rich and polluted soils). Generally, the mobility of Ni is inversely related to the soil pH. This is clearly demonstrated by Siebielec and Chaney (2006) who reported a drastic decrease of the Ni extractability at soil pH above 6.5 (Fig. II-10.1).

The forms of Ni in soils are diverse and range from highly mobile to ones that have no reactivity. Several soil properties, particularly clay and SOM contents, and pH, control Ni behavior, and its phytoavailability. Massoura et al. (2005a) reported that Ni associated with phyllosilicates is more easily taken up by plants than that associated with Fe oxides. Isotopic exchange kinetics, based on the soil-solution distribution coefficient of the radionuclide, ^{63}Ni, is proposed as an easy method to measure the available fraction of Ni in soils (Echevarria et al. 1998; Massoura et al. 2005b).

Concentrations of Ni in soil solution vary highly, from 3 to 150 μg l^{-1}, depending on soil type and techniques used for obtaining the soil solution (Kabata-Pendias and Pendias 2003). Species of Ni in soil solution have been specified as follows: cationic: Ni^{2+}, $NiOH^+$, $NiHCO_3^+$, and anionic: $HNiO_2^-$, $Ni(OH)_3^-$. Also complex compounds like $Ni(OH)_2^0$ and $NiSO_4^0$, of a relatively high stability constant, are known to occur in soil solution (Kabata-Pendias and Sadurski 2004; Uren 1992).

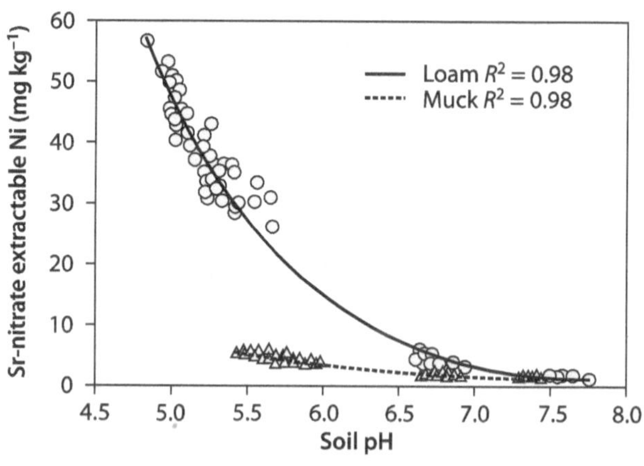

Fig. II-10.1. Effect of harvest soil pH on Ni extractable with 0.01 M $Sr(NO_3)_2$ from two different soils (loam and muck) contaminated by emissions of a Ni refinery at Port Colborne, Ontario (after Siebielec and Chaney 2006, with the authors' permission)

Soil contamination. Nickel has recently become a serious pollutant that is released from metal processing operation and from the increased combustion of coal and oil. Also some sewage sludges and phosphate fertilizers may be importance sources of Ni in agricultural soils. The Ni balance in soils of crop farms varies significantly (in $g\,ha^{-1}\,yr^{-1}$) from –3.3 in Denmark to 33 in France (Eckel et al. 2005). The highest contamination by Ni, up to 26 000 mg Ni kg^{-1}, has been reported for topsoils near the Ni-Cu smelter at Sudbury, Canada (Freedman and Hutchinson 1980). The impact of the Ni industry of Russia, has been reported to affect elevated levels of Ni in soils of Norway (Almås et al. 1995). The concentrations of HNO_3-extractable Ni varied in surface agricultural soils of this region from 6.25 to 136.88 mg kg^{-1} (mean 30.43), whereas the background content was estimated as 25 mg kg^{-1}.

Some municipal sludges are enriched in Ni, and for American cities its content is reported to range between 29 and 800 mg kg^{-1}, with a median value of 195 mg kg^{-1} (ATSDR 2002a). The presence of sludge in soils increases the mobility of Ni, which may be because of complexation with dissolved organic compounds. However, land application of Ni-contaminated sludge did not result in an increased level of Ni in groundwater (ATSDR 2002a).

II-10.1.4 Waters

Nickel is slightly mobile in the terrestrial environment, and its Cw/Cc ratio is about 0.08 (Gaillardet et al. 2003). However, it does not remain long in the aquatic environment as soluble species since it is easily adsorbed by suspended matter and by Fe-Mn hydroxides and deposited in bottom sediments. These processes have a significant impact on the Ni accumulation in concretions and nodules of ocean-sediments, where maximum concentration of this metal, in certain types of nodules, can reach 1.35% (Kotlinski 1999). According to Vignati (2004), the Ni distribution in freshwaters between the particulate, colloidal and truly dissolved phases, depends on the variation in master variables by Mn controlling the oxygen saturation.

Concentrations of Ni in river waters range from 0.15 to 10.39 $\mu g\,l^{-1}$, and the world average is 0.8 $\mu g\,l^{-1}$ (Gaillardet et al. 2003). Rivers of Western Siberia contain Ni within the range 1–15 $\mu g\,l^{-1}$ (mean 5 $\mu g\,l^{-1}$), and its input to the Baikal Lake has been estimated as 30 t yr^{-1} (Niechayeva 2002; Vietrov 2002). Relatively low Ni concentrations are reported by Pekka et al. (2004) for dissolved and suspended phases in Kola River (Russia), within the ranges from 0.68 to 4.62 $\mu g\,l^{-1}$, and from 0.06 to 1.58 $\mu g\,l^{-1}$, respectively.

The world riverine flux of Ni has been estimated by Gaillardet et al. (2003) as 305 kt yr^{-1}, whereas this value for the Baltic Sea has been given by Matschullat (1997) as 400 t yr^{-1}, of which 57% is of anthropogenic sources.

In seawaters Ni concentrations ranges (in $\mu g\,l^{-1}$) vary, depending on data sources:

- 0.48 – 1.7 (Reimann and Caritat 1998)
- 0.1 – 3.0 (ATSDR 2002a)
- 0.48 mean in the North Pacific (Nozaki 2005)
- 0.09 – 1.08 mean 0.7 in the Baltic Sea (Szefer 2002a)
- 0.12 – 0.33 mean 0.2 in the Sargasso Sea (ATSDR 2002a)

Ni

In atmospheric precipitation (bulk rain) over the Baltic Sea, Ni concentrations are between 0.9 and 19.7 $\mu g\,l^{-1}$, whereas over the Kola Peninsula rain water contained Ni within the range from 0.09 to 57 $\mu g\,l^{-1}$ in remote and polluted regions, respectively (Reimann and Caritat 1998; Szefer 2002a). Rainwater collected during 1999 in Sweden contained Ni at the average value of 0.34 $\mu g\,l^{-1}$ and at the maximum value of 0.56 $\mu g\,l^{-1}$ (Eriksson 2001a). Rainwater collected in Canada has been reported to contain Ni at the concentration of ≤1.5 $\mu g\,l^{-1}$ (ATSDR 2002a). Nickel in aerobic waters is likely to occur as the hexahydrate species, $Ni(H_2O)_6^{2+}$, but also easily makes complexes with common anions, such as OH^-, SO_4^-, HCO_3^-, and Cl^-.

Background concentrations of Ni in ground water, in temperate humid zones, average 2.5 $\mu g\,l^{-1}$, while the global Ni average has been estimated as 15 $\mu g\,l^{-1}$ (Reimann and Caritat 1998). Data presented by the ATSDR (2002a) gave median Ni concentrations in both surface and groundwater in the range of about 0.5–6 $\mu g\,l^{-1}$. Waters of the industrial region of Uzbekistan contain following amounts of Ni (in $\mu g\,l^{-1}$): surface waters, 30; well water at 8 m depth, 120; well water at 70 m depth, 80 (Galiulina et al. *vide* Kabata-Pendias and Pendias 1999). In ground waters of the European part of Russia, Ni varies from 2 to 45 $\mu g\,l^{-1}$, however in some regions (e.g., tributary of Oka River) the concentration of this metal has reached 300 $\mu g\,l^{-1}$ (Uchvatov and Bashkin 1997). Skjelkvåle et al. (2001) investigated the occurrence of Ni in the Nordic lakes, and reported its concentrations to vary, at the value of 99.5 weighted percentiles, as follows (in $\mu g\,l^{-1}$): 7.1 in Finland, 5.7 in Sweden, and 3.2 in Norway. The authors cited that content of Ni in the lakes of the Russian Kola Peninsula is 50.4 $\mu g\,l^{-1}$, at 99.5 weighted percentiles.

Water pollution with Ni is usually reflected in its accumulation in bottom sediments. Background Ni content in bottom sediments of Rhein River has been established as 46 $mg\,kg^{-1}$, and of Vistula River as 40 $mg\,kg^{-1}$ (Kabata-Pendias and Pendias 1999). The fine granulometric fraction of bottom sediments of polluted rivers may contain highly elevated amounts of Ni, e.g., up to around 300 mg Ni kg^{-1} in rivers of industrialized Silesia region of Poland (Bojakowska et al. 1992).

For drinking water, various limits (MAC) have been set in different countries (in $\mu g\,l^{-1}$): 50 in Norway, 100 in Russia, and 20 in the WHO recommendation (Reimann and Caritat 1998). The NIOSH regulations vary from 500 to 1 700 $\mu g\,l^{-1}$, for children and adults, respectively (ATSDR 2002a). The value of MCL for Ni established by EPA as 100 $\mu g\,l^{-1}$ has been discussed recently.

Drinking water of the USA analyzed during the period of 1969–1970 contained Ni at levels (in $\mu g\,l^{-1}$) as follows: 22% of samples, <1; 43% of samples, 1–5; 26% of samples, 6–10; 9% of samples, 11–20; and 1% of samples, >20. The highest level of Ni, 68 $\mu g\,l^{-1}$, was reported for municipal drinking water in Sudbury, Ontario, Canada (ATSDR 2002a).

Elevated Ni levels may exist in tap waters as a result of the corrosion of Ni-containing alloys used in some water distribution systems.

Aquatic biota. Fish from the Baltic Sea contain Ni in muscle tissue within the broad range from <0.04 to 0.8 $mg\,kg^{-1}$ FW, whereas in soft tissue of mussels its contents vary from <1 to 13.8 $mg\,kg^{-1}$ FW, and in plankton from 0.16 to 16.6 $mg\,kg^{-1}$ FW (Szefer 2002a). The accumulation of Ni in muscles of flounder from a clean area close to Finland was found to be 0.2 $mg\,kg^{-1}$, whereas from polluted area was 0.6 $mg\,kg^{-1}$ (Mukherjee 1999).

II-10.1.5 Air

The median Ni concentration in air of remote world regions is 90 ng m^{-3}, and of polluted world regions 60 ng m^{-3} (Reimann and Caritat 1998). The Ni concentrations in air vary highly, from 0.9 to 120 ng m^{-3} in control and urban regions, respectively (Table II-10.2). Ninety-nine percent of the urban and non-urban samples collected in the USA, during 1977–1979, did not exceed 68 and 52 ng m^{-3}, respectively. Later determinations suggest that atmospheric Ni concentrations in the USA have been declining (ATSDR 2002a).

It is estimated that 8.5 kt of Ni are emitted into the atmosphere from natural sources each year, whereas the anthropogenic emissions are calculated to be 55.6 kt (ATSDR 2002a). The main natural Ni sources are windblown dust (56%) and volcanoes (29%), while anthropogenic sources are principally burning residual and fuel oil (62%), and Ni metal refining (17%).

Global atmospheric Ni deposition has been estimated by Livett (1992) as 55.7 kt yr^{-1}, of which the major proportion is related to oil (27 kt yr^{-1}) and coal (13.8 kt yr^{-1}) combustion. The bulk aerial Ni deposition (in g ha^{-1} yr^{-1}) has been reported to range in Germany from 20 to 510, in Great Britain from 59 to 110, and in the industrial area of Poland up to 150 (Kabata-Pendias and Pendias 1999; Reimann and Caritat 1998). Nicholson et al. (2003) reported that the European average Ni deposition amounts 10 g ha^{-1} yr^{-1}, and varies from 1.5 (in Finland) to 36 (in Italy). The wet deposition of Ni in Sweden has been calculated as 3.6 g ha^{-1} yr^{-1} (Eriksson 2001a).

The natural aerial deposition of Ni on the Baltic Sea has been estimated as 5 t yr^{-1}, whereas deposition from anthropogenic sources as 100 t yr^{-1} (Matschullat 1997).

Rocher et al. (2004) have estimated Ni loads in roof runoff in central Paris as 340 mg ha^{-1} d^{-1} that is relatively highly increased as compared with its atmospheric deposition. This elevated amount of Ni is due to both the material used for roof covering and the vicinity of the chimney stacks of heating boilers.

The Ni contents of forest moss (*Pleurozium schreberi*) sampled in 1975 and 2000 clearly have decreased 2.8 folds and clearly indicate its smaller global emission (Rühling and Tyler 2004).

II-10.1.6 Plants

The content of Ni in plants seems to be highly controlled by soil properties and origin, however, plant abilities to absorb Ni (e.g., accumulators and hyperaccumulators) also has a strong influence.

Average contents of Ni in cereal grains from different countries vary from 0.34 to 14.6 mg kg^{-1}, and show the highest contents in oat grains, especially oats grown in Finland (Table II-10.3). Barley and wheat grains from Sweden contain Ni at mean values of 0.15 and 0.16 mg kg^{-1}, respectively (Eriksson 2001a). The Ni content of plant foodstuffs varies from 0.06 to 2 mg kg^{-1} with the lowest value in apples and the highest in cucumber fruits (Table II-10.3). The mean Ni contents of clover from various countries range from 1.2 to 2.7 mg kg^{-1}, whereas these values for grass are between 0.13 and

Ni

Table II-10.3.
Nickel in food plants grown in various countries[a] ($mg\ kg^{-1}$)

Plant	Range	Mean
Wheat, grains	0.17 – 0.67	0.34
Barley, grains	0.10 – 0.67	0.41
Oats, grains	0.20 – 8.0	1.28
Peas and beans, seeds[b]	0.2 – 0.25	–
Cabbage, leaves	0.6 – 3.3	1.03
Carrot, roots	0.26 – 0.98	–
Onion, bulbs	0.59 – 0.4	–
Tomato, fruits	0.43 – 0.48	–
Cucumber, fruits	1.3 – 2.0	–
Potato, tubers	0.29 – 1.0	–
Apple, fruits	–	0.06
Orange, fruits	–	0.39
Hazelnuts and walnuts[b]	1.6 – 1.7	–
Cocoa, powder[c]	5 – 10	–

[a] Presented are common, possible background, values from various sources as compiled by Kabata-Pendias and Pendias (1999, 2001), unless otherwise indicated.
[b] After Przyslawski and Duda (2000).
[c] After Sunderman (2004).

1.7 $mg\ kg^{-1}$ (Table II-10.4). A much higher Ni content has been reported for meadow grass (13–75 $mg\ kg^{-1}$) and forest grass (10–100 $mg\ kg^{-1}$) from taiga zone of Western Siberia (Niechayeva 2002). In spite of the impact of various factors, among which plant genotypes and environmental conditions play significant roles, the Ni contents in food and forage plants are fairly similar.

Nickel is very easily extracted from soils by plants and the Ni contents of plants are simple functions of the Ni forms in soils (Fig. II-10.2). Plants more readily absorb in the ionic Ni^{2+} form than when it is chelated. Both plant and pedological factors affect these processes and the most pronounced factor is soil pH. Berrow and Burrige (*vide* Kabata-Pendias and Pendias 2001) found that increasing soil pH from 4.5 to 6.5 decreased the Ni content of oats by a factor of about 8.

Stauton et al. (1999) noted that the secretion of organic anions, and modification of soil pH by roots and fungi in the rhizosphere may decrease Ni adsorption by soils components, and thus increase its plant availability. In plants, Ni is very mobile and is readily accumulated in both leaves and seeds.

There are a great number of plants, 317 taxa of 37 families, which have the ability to accumulate over 1 000 mg Ni kg^{-1} (Greger 1999). The most common Ni accumulators are *Alyssum* sps., and Chaney was one of the first to suggest the possible phytoextraction of Ni from polluted soils, using these species that can accumulate over 400 kg Ni from one hectare (Chaney et al. 1999a). The mechanism of the Ni hyperaccumulation is not

Table II-10.4.
Nickel in grasses and clover at the immature grown stage from different countries[a] (mg kg^{-1})

Country	Grasses		Clover	
	Range	Mean	Range	Mean
Belgium	0.9 - 1.3	1.1	–	–
Great Britain	0.8 - 2.4	–	1.3 – 6.2	–
Finland	0.15 – 1.1	0.4	–	–
Germany	1.3 - 2.5	1.7	1.0 – 1.3	1.2
Hungary	–	–	1.4 – 2.4	1.9
Ireland	0.8 - 2.2	1.4	–	2.7
Lithuania	2.4 - 4.8	–	–	2.6
Poland	0.01 – 24.1[b]	0.76[b]	0.2 – 8.2	2.5
USA	<0.07 – 0.7	0.13	<0.5 – 5[c]	1.5[c]

[a] Values from various sources as compiled by Kabata-Pendias and Pendias (1999, 2001), unless otherwise indicated.
[b] After Kabata-Pendias and Terelak (2004).
[c] Calculated from AW basis for alfalfa.

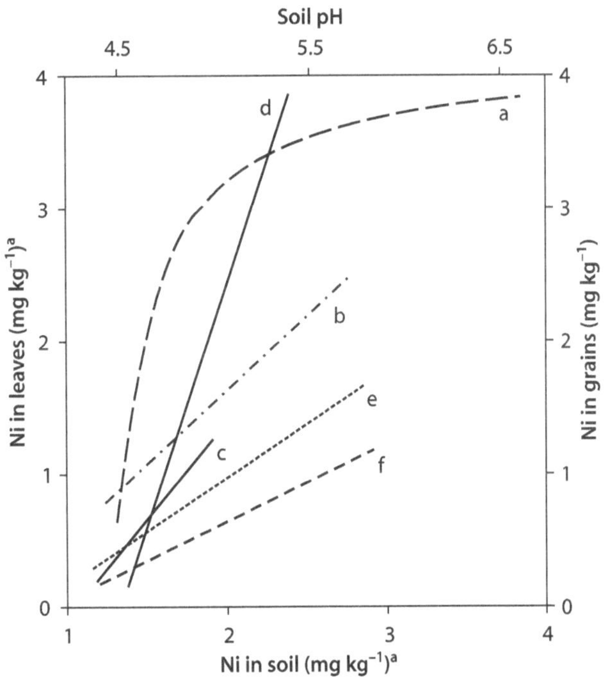

Fig. II-10.2.
Nickel concentrations in plants as a function of Ni contents of soils: **a** leaves of *Alyssum* sp., hyperaccumulator; **b** *Alyssum montanum*, nonaccumulator; **c** wheat leaves; **d** barley grains; **e** carrot roots; and **f** radish roots. [a]Ni in soils, leaves and roots are given in powers of ten (modified from Kabata-Pendias and Pendias 2001)

yet well-understood, but evidently it is associated with organo-metallic complexes. An association has been found between Ni concentration in the sap of several genotypes of hyperaccumulators and concentrations of some organic acids (polycarboxylate), especially of citric acid (Brooks 1998). Apparently these complexes react as transport-ligands for Ni within plants which have accumulated the metal. Pohlmeier (1999) broadly reviewed the mechanisms of the complexation of Ni^{2+} in plants and listed several organic ligands involved in these processes. Krämer et al. (1996) reported that exposing a species of *Alyssum* to Ni elicits a large increase in the levels of free histidine in a Ni-accumulator phenotype. The authors concluded that an enhanced production of histidine is responsible for the Ni hyperaccumulation in these plants. Nicks and Chambers (1998) have described a great variation in Ni content (3 280–7 820 mg kg^{-1}) of individual plants of one species of the Brassicaceae family that is endemic to serpentine soils in California.

Several investigators have suggested that Ni might be essential for plants. The essentiality of Ni for some biosynthesis of a number of bacteria has been proven. However, there is no evidence yet of a metabolic role of Ni in higher plants, although reported beneficial effects of Ni on plant growth have stimulated the speculation that this metal may have some function in plants. Its role in the nodulation of legumes and effects on the nitrification and mineralization of N has been described. Thus Ni is considered as essential for legumes in which is involved in the urease metabolisms (Eskew et al. 1983).

Although Ni is already considered as an essential element for higher plants, its toxicity to plants has been most frequent studied. Restricted growths of plants and injuries caused by an excess of Ni have been observed for quite a long time. The mechanism of Ni toxicity and biological effects are highly related to its species. The cationic form, Ni^{2+}, is more readily absorbed and more toxic than its complexed forms. Molas (1997a,b) observed that Ni-glucose chelate was more toxic to cabbage than Ni-EDTA complex, and described various morphological and anatomical changes caused by an *in vitro* excess of Ni.

The most common symptom of Ni phytoxicity is chlorosis followed by yellowing and necrosis of leaves. This is induced mainly by a Ni-Fe interaction; low foliar Fe levels are closely associated with toxic Ni concentration in the growth media. Plants under the Ni stress exhibit retarded absorption of nutrients, impaired root development, and metabolic dysfunctions.

The phytotoxic Ni concentrations vary widely among plant species and cultivars, and have been reported to range from 40 to 246 mg kg^{-1} (Kabata-Pendias and Pendias 2001). However, some plants are very sensitive to increased levels of Ni in the growth media, and small Ni concentration in nutrient solution, 1 mg Ni l^{-1}, has been reported as toxic to hybrid poplars (Punshon and Adriano 1999). Chang et al. (1992) attempted to establish the phytotoxicity threshold for Ni and determined that the PT_{50} leaf tissue corresponds to 3 mg Ni kg^{-1} and is much lower than Ni concentrations reported to be associated with its phytotoxicity levels. These authors suggested that the typical plant tissue Ni concentration is approximately 10 mg kg^{-1}. Generally, the range of excessive or toxic amounts of Ni in most plant species varies from 10 to 1 000 mg kg^{-1}. However, several species are known for their great tolerance and hyperaccumulation of Ni. These are mainly species of the Boraginacea, Cruciferae, Myrtaceae, Leguminosae, and Caryophyllacea families.

Under natural conditions, Ni toxicities are associated with serpentine or other Ni-rich soils. Some native plants from such locations accumulate leaf concentrations of over 6 000 mg Ni kg^{-1} (Kabata-Pendias and Pendias 2001). Baker et al. (*vide* Greger 1999) described that hyperaccumulators may concentrate Ni in leaves over 1 000 mg kg^{-1}. The pronounced ability of some plants to extract Ni from soils over Ni ore bodies may make them useful biogeochemical indicators.

The easy phytoavailability of Ni is of real environmental concern. Increased Ni levels in food plants are associated with health hazards. The ready transfer of Ni from soils to vegetables has been illustrated by Frank et al. (1982) in the notable accumulation of this metal, up to 4 000 mg kg^{-1}, in cabbage heads and celery roots and tops planted on soil around a nickel smelter (Niagara Peninsula) with Ni contents between 500 and 1 500 mg kg^{-1}.

II-10.1.7
Humans and Animals

Humans. The average Ni content in human soft tissues is estimated as 88 μg kg^{-1} (Li 2000). Its concentrations in soft human organs vary greatly, with the mean values highest for the lung (173 μg kg^{-1}) and lowest for the pancreas (34 μg kg^{-1}) (Rezuke et al. *vide* Sunderman 2004). In human fluids, mean Ni concentrations are (in μg l^{-1}): blood, 2.3; serum, 1.2; and urine, 0.9 (Reimann and Caritat 1998).

The Ni deficiency seems unlikely to occur in humans (Sunderman 2004). However, Anke et al. (1995b) have established the Ni requirements by adults as 25–35 μg d^{-1}. Lower Ni intake can induce some dysfunction of fat metabolism, but human diets generally contain sufficient amounts of Ni.

Average dietary Ni uptake by the German population has been estimated as 144 μg d^{-1} for women and 166 μg d^{-1} for man (Pilz et al. 2004). Fairly similar values, average of 168 and 145 μg d^{-1}, are reported for the USA and the UK, respectively (Sunderman 2004). Pais and Jones (1997) have calculated the average intake of Ni by adults to range from 300 to 500 μg d^{-1}. A market basket survey in England, in 1984, estimated a dietary intake of 154–166 μg d^{-1} for a 70-kg person (ATSDR 2002a). According to this source of data, the oral RfD for Ni is proposed as 20 μg kg^{-1} d^{-1}. The adsorption of Ni is relatively low, when administrated with drinking waters was about 27% of the dose whereas Ni mixed with food was absorbed only around 0.7% of the dose (Sunderman 2004).

Hundreds of foods from various European countries have been analyzed for Ni, and its common contents were established as <0.5 mg kg^{-1} (Sunderman 2004). This value corresponds well to Ni contents of food plants presented in the Table II-10.3.

The toxicity of Ni compounds to animals and humans has been broadly review by Sunderman (2004). Investigations on the toxicity of Ni have indicated various effects of its excess, among which the most important are developmental, genotoxic, neurological, reproductive, and carcinogenic. Generally, the soluble compounds are more toxic than the less-soluble compounds. However, slightly soluble Ni compounds are likely to be carcinogenic at the site of deposition (ATSDR 2002a). Nickel tetracarbonyl, $Ni(CO)_4$, has been found to be the most toxic, as reported for industrial workers (Sunderman 2004).

Nickel should be used with great precautions in industry since is exceedingly toxic when inhaled. Lung and nasal cancer were the forms of cancer commonly reported in the nickel-refinery workers. Up to 55% of the inhaled less-soluble Ni is retained in the lungs. However, when it has been absorbed into the blood stream is primarily excreted in the urine. Significant correlations have been found between exposure to Ni and its levels in urine and serum. Correlations between exposure concentration and the level of Ni in hair were not reported (ATSDR 2002a).

The atmospheric limits (expressed as an 8-h TWA) for soluble Ni compounds established in different countries vary from 0.05 to 1 mg m^{-3}. A much lower value, 0.005 mg m^{-3} is reported for Russia (Sunderman 2004). Regulations for Ni concentrations in air vary, depending primarily on its compound, from 0.02 to 1 mg m^{-3}. The average Ni intake by inhalation in US cities is calculated to range from 0.1 to 1 μg d^{-1}. Smokers inhale from 2 to 12 μg of this metal, for each pack of cigarettes (ATSDR 2002a).

Animals. Nickel occurs in mammalian tissue within the range from 8 to <300 μg kg^{-1}, being fairly uniformly distributed, with an exception of the skin where its lowest content is reported (Jørgensen 2000).

Based on reports of experimental induction of dietary Ni deficiency in animals, this metal appears to be essential. However, at the established dietary requirements for ruminants around 0.1 mg Ni kg^{-1}, deficiency is not likely to occur under normal conditions (Pais and Jones 1997). Symptoms of induced Ni deficiency in experimental animals are associated with the growth retardation of hemoglobin, hematocrit and erythrocytes.

Increased administration of water-soluble Ni compounds ($NiCl_2$) to animals resulted in acute toxicities, such as: *(i)* immune inhibition of spleen cells, *(ii)* hyperglycemia, *(iii)* nephrotoxicity, *(iv)* hypothermia, *(v)* hepatotoxicity, *(vi)* coronary vasoconstriction, *(vii)* bronchoalveolar hyperplasia, and *(viii)* increased heme oxydase activity in kidneys and liver. Several other chronic effects have been noted, such as: pulmonary damage, renal dysfunction, arteriosclerotic lesions, and allergy. Carcinogenic effects, and particularly reproductive and teratogenic disorders, have been described (ATSDR 2002a).

II-10.2
Palladium (Pd)

II-10.2.1
Introduction

The abundance of Pd in the Earth's crust is given as 10 μg kg^{-1}, and in the upper continental crust, 1 μg kg^{-1}, but it may be concentrated in ultramafic rocks, some coals, black shales and phosphorites. Higher concentrations, from 0.5 to 0.6 μg kg^{-1}, are found in basalts and shale schists. The arithmetic mean of Pd in US coal is <1 μg kg^{-1} (Finkelman 1999). Manganese ores and Mn concretions are known for their capacity to accumulate Pd, which ranges from 3.7 to 11.4 μg kg^{-1}. Palladium exhibits a relatively strong chalcophilic character. In general, Pd is considered as a greater potent threat to the environment than the other PGMs.

Palladium is more reactive chemically than other PGMs, and therefore it is known in more compound mineral forms that others. Among better-known minerals are: potarite, PdHg; stibiopalladinite, Pd_3Sb; and arsenopalladinite, Pd_3As. Also relatively common are sulfide minerals, like braggite (Pt,Pd,Ni)S and cooperite (Pt,Pd)S. This metal is relatively common in several sulfide minerals, at the approximate concentration range from 0.7 to 10 mg kg^{-1}.

II-10.2.2 Production and Use

The global production of Pd, in the period 1999–2002, has varied from 140 to 230 t yr^{-1} (Hoppstock and Sures 2004), whereas its use in 2000 was estimated as 238 t (Ravindra et al. 2004). This metal is widely used for electronics and jewelry production and in dentistry. Alloys with Pd are also used for bearings and springs in watches, as well as for chemical and medical instruments. Its compounds are used in photography, electroplatings, and as a catalyst, mainly in organic synthesis.

Above 50% the Pd supply was used in the year 2000 for automotive catalytic converters. The use of Pd in car converters has increased 200-fold between 1989–1997 (Stilianakis 2005). Palladium, as other PGMs (Pt and Rh) used to reduce hydrocarbon, carbon monoxide and nitrous oxide in automotive emissions pose a potential risk to humans and the environment (Zereini and Alt 2006).

II-10.2.3 Soils

There are only few reports on Pd contents in soils. Contents of Pd in surface soils have been found by different authors to range between 0.5 and 130 μg kg^{-1} (Kabata-Pendias and Pendias 2001). An association of concentrations of Pd with the clay fraction and SOM has been observed. Swedish arable soils contain Pd in the range of <40–150 μg kg^{-1}, at an average value of 40 μg kg^{-1} (Eriksson 2001a). In the Report of the UNEP/WHO (2002), its mean content is given as 40 μg kg^{-1}. Soils near German highway contain Pd in the range between 1 and 47 μg kg^{-1}, at an average value of 6.6 μg kg^{-1} (Zereini et al. 1997; Ravindra et al. 2004). Pd-hydroxides and complex compounds are relatively mobile in soil.

Concentrations of Pd in roadside dust have been found to range from 1 to 26 μg kg^{-1} near Frankfurt and from 20 to 115 μg kg^{-1} near San Diego, California (UNEP/WHO 2002). The highest contents of Pd, however, are reported for road tunnel dust, from 20 to 297 μg kg^{-1} (Helmers et al. 1998). Contents of Pd in soils from the Napoli metropolitan area vary from 8 to 110 μg kg^{-1}, with a mean value of 12.7 μg kg^{-1} and indicate a positive relationship with the traffic density (Cicchella et al. 2003). Since 1990, Pd has been added instead of Pt to some vehicle catalytic converters, which has resulted in increased amounts of Pd released to roadside areas (Helmers and Krümmerer 1999).

II-10.2.4 Waters

Palladium ranges in river waters from 1 to 30 ng l^{-1}, and its world-average concentration has been calculated by Gaillardet et al. (2003) as 2.8 ng l^{-1}. In waters of Rhein River it is estimated by Ravindra et al. (2004) to be 0.4 ng l^{-1}. Contents of Pd in waters of the Pacific Ocean vary from 0.04 ng l^{-1} to 0.06 ng kg^{-1} (Ravindra et al. 2004; Nozaki 2005). Rainwaters collected during 1999 in Sweden contained Pd at the average value of <1 ng l^{-1} and its wet deposition is calculated to be <11 mg ha^{-1} yr^{-1} (Eriksson 2001a).

According to the UNEP/WHO (2002) data, common ranges of Pd in waters are as follows (in ng l^{-1}): 0.4–22 in fresh water; 19–70 in salt water; and <24 in drinking water.

Concentrations of Pd in sewage sludges of Southern Australia vary from 18 to 153 $\mu g\,kg^{-1}$ in industrialized areas and from 9.4 to 20.4 $\mu g\,kg^{-1}$ in rural areas. Lottermoser (1995) has stated that these amounts of Pd in waste dumped on land or at sea, might pose a threat to the environment and human health. Recently, increased application of Pd has been reflected in its concentration in sewage sludge ashes from the municipal purification plant of Stuttgart. Its concentration below 100 $\mu g\,kg^{-1}$ in the late 1980 has rapidly increased and reached 600 $\mu g\,kg^{-1}$ in 1992 (Helmers et al. *vide* Pyrzyńska 2000).

Aquatic biota. Palladium seems to be relatively easily bioaccumulated by crustaceans (isopods) up to 155 $\mu g\,kg^{-1}$ from polluted freshwater (Moldovan et al. 2001). Also hyacinth accumulates Pd from water, up to 785 $\mu g\,kg^{-1}$ (Abbasi 1987).

II-10.2.5
Air

Palladium contents in air samples seem to range commonly from <0.07 to 280 $pg\,m^{-3}$. However, mean Pd concentrations of 12 700 $pg\,m^{-3}$ for Chicago, USA and 56 600 $pg\,m^{-3}$ for Chernivtsi, Ukraine, have been reported (Ravindra et al. 2004). According to data of UNEP/WHO (2002), average Pt concentration in air in USA is <1 $pg\,m^{-3}$, and varies from <0.06 to 110 $pg\,m^{-3}$, in unpolluted and urban areas, respectively.

The environmental exposure limit of Pd in air has been established at 5 $mg\,m^{-3}$ (Hoppstock and Sures 2004).

II-10.2.6
Plants

Palladium seems to be easily available to plants when it occurs in soluble forms. The accumulation of Pd by plants takes place in both experimental and natural conditions. Its contents in various shrubs and trees are in the range of 30–400 $\mu g\,kg^{-1}$ AW. In twigs from birch grown close to the mineral deposits of PGMs in Canada, Pd is accumulated up to 4 014 $\mu g\,kg^{-1}$ (UNEP/WHO 2002). Barley and wheat grains from Sweden contain Pd at a mean value of <1 $\mu g\,kg^{-1}$ (Eriksson 2001a). Its contents in staple foods vary from <0.1 to 3 $\mu g\,kg^{-1}$ FW, being the highest in nuts and bread (Table II-10.5).

Table II-10.5. Mean palladium contents of selected food ($\mu g\,kg^{-1}$ FW) (data from UNEP/WHO 2002)

Food group	Pd	Food group	Pd
Green vegetables	0.6	Meat products	0.6
Potatoes	0.5	Poultry	<0.3
Other vegetables	0.5	Fish	2
Canned vegetables	0.4	Oil and fats	0.4
Fresh fruits	0.4	Eggs	0.4
Cereals, miscellaneous	0.9	Milk	<0.3
Bread	2	Dairy products	0.4
Nuts	3	Sugar	0.5

Plants exposed to automobile exhausts contain Pd in the range from 0.43 to 2.4 μg kg^{-1}, with the highest value for moss and the lowest for plantain (Djingova et al. 2003). Farago and Parsons (1994) reported that Pd is relatively toxic to water hyacinth which has a strong ability for its accumulation.

Palladium is likely to be bound to high molecular-weight proteins and is able to replace Mn in some metalloenzymes due to similar ionic radii. Phytotoxic effects in terrestrial plants at Pd concentrations from about 2 to 60 mg l^{-1} of the nutrient solution have been reported (UNEP/WHO 2002).

II-10.2.7 Humans

The occurrence of Pd in animal and human tissues is almost unknown. Jørgensen (2000) has cited the Pd content in muscles of mammals as 2 μg kg^{-1}. Ravindra et al. (2004) compiled data for Pd in urine in German population; the general population contained Pd in the range from 0.01 to <0.036 μg l^{-1} and occupationally exposed population – in the range from 0.08 to 3.4 μg l^{-1}. The population is preliminarily exposed to Pd through dental alloys, jewelry, and automobile catalytic converters. People working in the catalyst production are exposed to increased levels of Pd; their urinary Pd concentrations are reported to be from 0.2 to 1 μg l^{-1} (Hoppstock and Sures 2004). According to data of the UNEP/WHO (2002) its concentrations in urine of adults range between 0.0006 and <0.3 μg l^{-1}. It is likely to accumulate in kidneys.

Exposure of humans to increased levels of Pd may be associated with iatrogenic effects of dentist therapy. Sensitization effects of increased levels of Pd can be observed in some individuals as dermatitis, stomatitis, and oral lichen planus. Excess of Pd causes histopathological effects in liver, kidneys and spleen. It may inhibit most major cellular functions and is likely to bind to amino-acids, DNA, and other macromolecules (e.g., vitamin B_6). Palladium is considered to be a relatively strong allergen, and there has been a suggestion for no dental use of Pd-Cu-alloys (Zinke *vide* Hoppstock and Sures 2004). Human average dietary intake of Pd appears to be up to 2 μg d^{-1} and varies from <1.5 to 15 μg d^{-1} (UNEP/WHO 2002).

Palladium can form complexes with nucleic acid components that behave similarly as Pt complexes but are much more active. These complexes could find their application in the treatment of different types of tumor (Rau and Eldik 1996). Several Pd compounds have been found to have antiviral, antibacterial and fungicidal properties.

II-10.3 Platinum (Pt)

II-10.3.1 Introduction

The bulk continental abundance of Pt is given in the range form 0.4 to 5 μg kg^{-1}. The range of its content in rocks has been estimated to vary from 0.05 in limestones to 3 μg kg^{-1} in ultramafic rocks (Table II-10.2). As a very non-reactive metal with other elements, Pt occurs mainly in alloys with other PGMs. Its more common minerals are:

sperrylite, $PtAs_2$; cooperite, (Pt,Pd)S; and braggite, (Pt,Pd,Ni)S. Several sulfide minerals are possible hosts for Pt which has an proximate range of concentrations from 0.2 to 2 mg kg^{-1}.

According to Finkelman (1999), the arithmetic mean of Pt in US coal is <1 µg kg^{-1}, whereas Li (2000) reported 230 µg Pt kg^{-1} in the reference coal sample (SRM1632).

Although Pt does not represent an acute danger, due to its low concentrations in the environment at the present, long-term monitoring studies are prerequisite for the assessment of the impact of its increased levels to the environment and human health.

II-10.3.2 Production and Use

The annual global production of Pt in the period 1997–1983 ranged between 200 and 203 t (WHO/IPCS 1991) and in 2000 it was 365 t (USGS 2004). The highest proportion of Pt is presently used for automotive catalytic converters and for jewelry (Table II-10.6). It is also broadly used in various industries, mainly as catalyst in chemical processes. In different alloys, Pt is used in dentistry and in medicine for neurological prostheses and some long-term electrode implantations. Pt-alloys are increasingly used in glass production and in dental medicine.

II-10.3.3 Soils

The common range of Pt in surface soils vary from <20 to 75 µg kg^{-1}, although concentration of up to 150 µg kg^{-1} has also been reported (Kabata-Pendias and Pendias 2001). Swedish arable soils contain Pt at the mean value of 40 µg kg^{-1} (Eriksson 2001a). Laterite soils of Australia contain Pt in the range from <5 to 70 µg kg^{-1} and its contents in soils from the Napoli metropolitan area vary from 1.6 to 52 µg kg^{-1}, with the mean value 4.2 µg kg^{-1} (Reimann and Caritat 1998; Cicchella et al. 2003). The average content of Pt in soils near Sudbury, an industrial region of Canada, is <0.8 µg kg^{-1} (WHO/IPCS 1991).

Table II-10.6. Platinum use to various types of industry, after the introduction of automotive catalytic converters, in the USA and in Western-World countries (kg yr^{-1}) (adopted from WHO/IPCS 1991)

Industry	USA, 1987	Western-World, 1986
Automobile	18 817	41 107
Chemical	1 920	5 528
Petroleum	739	3 260
Dental and medical	479	-
Electrical	1 821	3 969
Glass	285	3 685
Jewellery	177	36 996
Miscellaneous	1 430	3 260
Total	25 668	97 805

The distribution of Pt in surface soils shows positive correlation with major traffic flow. Its highest concentrations in the soil occur directly at the edge of the motorway and decrease with increasing distance from the highway. Along highways in Germany, Pt content ranges from 0.35 to 330 $\mu g\,kg^{-1}$, whereas in soils of botanic garden in the UK it varies from 0.8 to 1.6 $\mu g\,kg^{-1}$ (Ravindra et al. 2004). Farago et al. (1998) presented mean concentrations of Pt in soils and road dust in the UK from 2.29 to 20.8 $\mu g\,kg^{-1}$, being the lowest for minor roads and the highest for major road intersections.

The estimation of Pt emission from traffic varies and recent calculations showed that it might be between 0.5 and 0.8 $\mu g\,km^{-1}$ (Pyrzyńska 2000). Based on Pt concentrations found in soils near roadsides and on volumes of traffic, Zereini et al. (2001) calculated this value at an average of 0.27 $\mu g\,km^{-1}$ in the vicinity of Frankfurt, Germany. Although Pt emitted from the catalytic converters are mainly in the form of nanocrystals of metallic Pt, some oxidized species have been identified (Nachtigall et al. *vide* Pyrzyńska 2000). Such Pt species could be easily mobilized by complexation, particularly when complexed with chlorides and organic matter (e.g., humic and fulvic acids) and/or undergo biotransformation (e.g., biomethylation). Increased Cl levels in soil solution especially promote its mobility (WHO/IPCS 1991).

Platinum content in sewage sludge ashes from the municipal purification plant of Stuttgard, Germany, has been monitored during the period 1972–1994. Since the early 1980s it has rapidly increased, from $<100\,\mu g\,kg^{-1}$ to around 300 $\mu g\,kg^{-1}$ (Helmers et al. *vide* Pyrzyńska 2000).

II-10.3.4 Waters

Platinum concentrations in seawaters vary, depending on the location and water depth, from 37 to 332 $pg\,l^{-1}$ (Ravindra et al. 2004). The median value for the North Pacific is reported to be 50 $pg\,kg^{-1}$ (Nozaki 2005). A much higher Pt concentration, up to 20 $ng\,l^{-1}$, has been found in lake water of Norway (Reinmann and Caritat 1998). Rainwater collected during 1999 in Sweden contained Pt at the average value of 22 $ng\,l^{-1}$ with the maximum value of 39 $ng\,l^{-1}$, and its wet deposition is calculated to be 250 $mg\,ha^{-1}\,yr^{-1}$ (Eriksson 2001a).

Concentrations of Pt in sewage sludges of Southern Australia vary from <0.5 to 157 $\mu g\,kg^{-1}$ in industrialized areas and from <0.5 to 6.1 $\mu g\,kg^{-1}$ in rural areas. Lottermoser (1995) has stated that increased amounts of Pt in waste dumped on the land or at sea, might pose a threat to the environment and human health.

The recent discharge of Pt from some hospitals should be of environmental concern. The total Pt emission for all German hospitals was calculated as 28.6 $kg\,yr^{-1}$ (Krümmerer and Helmers 1997). It has been shown that 70% of this metal, administrated during anti-neoplastic drug therapy, was excreted, ended in the hospital effluents, and was transferred into the aquatic environment (Ravindra et al. 2004).

Aquatic biota. Platinum seems to be relatively easily bioaccumulated by crustaceans (isopods), up to 38 $\mu g\,kg^{-1}$ from polluted freshwater (Moldovan et al. 2001). In marine algae its concentrations vary from 0.08 to 0.32 $\mu g\,kg^{-1}$ (WHO/IPCS 1991).

II-10.3.5 Air

Recently, Pt concentrations in airborne samples, especially from urban and highway areas, have been investigated. Ravindra et al. (2004) reported the following data for various locations (in pg m^{-3}): California, <0.05; Rome, 7.8–38.9; Madrid, <0.1–57.1; and Copenhagen, (heavy traffic) 25–2 740. According to data of the WHO/IPCS (1991), Pt contents in air near Frankfurt vary from <1 to 13 pg m^{-3}, whereas in rural areas it is between <0.6 and 1.8 pg m^{-3}. Road dust from Munich (1997/1998) contained an average of 730 µg Pt kg^{-1}.

Maximum Pt concentration in air recommended in Germany for occupational exposure is 2 µg m^{-3} (Hoppstock and Sures 2004).

II-10.3.6 Plants

Although Pt is strongly held in soils, apparently plants easily take up mobile complexes of Pt (mainly associated with SOM), as of other PGMs. Plants grown in a medium enriched in Pt compounds, absorb this metal which accumulates mainly in roots. Several Pt-binding proteins were isolated from those plants (Hoppstock and Sures 2004). Pyrzyńska (2000) cited, after Helmers, that values of the TF for Pt, from soil to plants, are relatively high and similar to those for Cu. Wheat and barley grains from Sweden contain Pt at mean values of 0.2 and 0.9 µg kg^{-1}, respectively (Eriksson 2001a).

Grasses from remote locations contain Pt in the range of 0.2–0.5 µg kg^{-1} (Hoppstock and Sures 2004). Plants (collected in Germany) exposed to automobile exhausts contain Pt in the range from 3.6 to 30 µg kg^{-1}, with the highest content in moss and the lowest in plantain (Djingova et al. 2003). Roadside grasses (collected also in Germany) contained Pt between 5 and 8 µg kg^{-1} (Ravindra et al. 2004). The highest Pt content in grass, in the range between 17 and 95.6 µg kg^{-1}, was found in Germany along the motorway near Siegen (Hees et al. *vide* Pyrzyńska 2000).

Some plants collected from a roadside in France are reported to contain 100–830 µg Pt kg^{-1} (Cabrera et al. 1996). According to these authors, mean Pt concentration of wine from different regions of France averages around 10 µg l^{-1} and does not represent any imminent toxicological risk. According to Leiterer et al. (2004), the main part of Pt emitted from catalytic converters is deposited as suspended particles on the plant surface.

II-10.3.7 Humans

There is very scant information on Pt occurrence in human tissues. Ravindra et al. (2004) compiled some data as follows: unexposed persons of Austria contain Pt in liver in the range of 0.05–0.24 µg kg^{-1}, and in blood, 0.1–2.8 µg l^{-1}. These authors have cited that the physiological levels of Pt in blood of unexposed persons of Germany range from 0.032 to 0.078 µg l^{-1}. According to Farago et al. (1998), mean concentrations of Pt in blood of different populations vary as follows (in µg l^{-1}): precious metal workers,

0.246; motorway maintenance workers, 0.145; and Imperial College staff, 0.129. These authors have reported the linear relationship between Pt in blood and Pt in urine. Intake of Pt by population in the UK is given as 1 μg d^{-1} $person^{-1}$ (WHO/IPCS 1991).

Metallic Pt is considered to be biologically inert. Most probably, metallic and/or oxides forms emitted from car exhausts are not toxic. However, emitted Pt compounds, when ingested or inhaled, became available and form compounds that have mutagenic and carcinogenic effects. Platinum may readily form complexes with proteins and cause a reduction in enzymatic activity. The toxicity of Pt depends considerably on its species and is reported to decrease in the following order: $PtCl_4 > Pt(SO_4)_2 \cdot 4H_2O > PtCl_2 > PtO_2$. Methionine seems to be a primary binding compound for Pt (WHO/IPCS 1991).

The Pt-DNA complexes reveal high cytotoxicity and antitumor activity (Farrell 1996). Platinum anticancer drugs based on cis-Pt compound (cis-[$Pt(NH_3)_2Cl_2$]) have been widely used. A new generation of Pt-drugs, carboplatin and lobaplatin show increased activity and reduced toxicity (Gauchelaar et al. *vide* Pyrzyńska 2000). Kümmerer and Helmers (1997) have estimated that during the therapy one patient exudes in urine from 2 to 330 g Pt yr^{-1}, depending on the kind of therapy and hospital.

The excess of cis-Pt compounds administered to mammals causes nausea and vomiting, anorexia, nephrotoxicity, and some neuropathy symptoms. Applications of antitumor Pt compounds have been broadly investigated, and the results should be still reexamined (Farrell 1996; Arpalahti 1996; Hoppstock and Sures 2004).

Some Pt compounds that may occur during the refining processes of Pt, may cause allergy sensitization. Halogenated Pt salts are especially known to cause allergic asthma and dermatitis in occupational individuals (Hoppstock and Sures 2004). However, opinions on physiological effects of Pt-halogenated complexes are variable (Farago et al. 1998).

It is now under the discussion, whether this metal is toxic for living organisms and humans (Bencs et al. 2005; Bowles 2005). Metallic Pt is inert as far as biological reactions are concerned. However, its chloro-compounds and complexes with organic acids become mobile in the environment and may be of potential health risk for individuals living in an urban area and/or along major highways.

Platinum released during the past decades by vehicle exhaust catalysts, has been significantly accumulated in abiotic and biotic matrices and organic bound species that are relatively easily bioavailable. There is no evidence, as yet, of adverse health effects of the great accumulation of Pt near roads (during 5 years increased up to 90-fold above natural background levels). However, the situation warrants continued monitoring and health risk assessment (Hutchinson and Pearson 2005).

Trace Elements of Group 11 (Previously Group Ib)

The Group 11 is composed of copper (Cu), silver (Ag), and gold (Au). Silver and Au belong to the so-called noble metals. All these metals occur in variable oxidation stages, mainly +1, and +2, (Table II-11.1). Copper and Ag reveal chalcophilic tendencies whereas Au is siderophilic.

The terrestrial abundance of Cu is very well documented, however, data for Ag and Au occurrence in some environmental compartments are rather scarce and presented values vary, especially for the bulk continental crust (Table II-11.2). The co-occurrence of As is used (e.g., in Albania) as a geo-indicator element in Au and Cu surveys due to their frequent geochemical association (Lazo et al. 2007).

For nearly nine millennia, Cu and Au were the only metals known to man (Aitchison 1960; Tylecote 1976) and they thus are very closely associated with the progress of human civilization. Copper was known already at the end of Stone Age and has been used by human beings since 4000 BC. Romans exploited Cu from the island Cyprus, and the metals was named *cuprum*, after the island name.

II-11.1 Copper (Cu)

II-11.1.1 Introduction

Copper occurs in the Earth's crust at concentrations between 25–75 mg kg^{-1}. Its abundance pattern in rocks shows the tendency for the concentration in mafic igneous rocks

Table II-11.1. Selected properties of trace metals of the Group 11

Element	Atomic number	Atomic weight	Atomic radius[a] (pm)	Density ($g\ cm^{-3}$)	Valence state[b]	Melting point (°C)
Cu, copper	29	63.54	157	8.96	+1, **+2**	1083
Ag, silver	47	107.87	175	10.5	**+1**, +2, +3	962
Au, gold	79	196.97	179	19.3	0, **+1**, +3[c]	1064

[a] Approximate average values for the main oxidation states.
[b] Valence values in bold are for main oxidation states.
[c] Oxidation states +2 and +5 are also reported.

Cu

Table II-11.2. Abundance[a] of elements of Group 11 in the environment

Environmental compartment	Cu	Ag	Au
Earth's crust ($mg\ kg^{-1}$)	25 – 27	0.07 – 0.08	3 – 4[c]
Igneous rocks ($mg\ kg^{-1}$)	50	0.07	1.1[c]
Mafic	10 – 120	0.07 – 0.1	0.5 – 5[c]
Acid	5 – 30	0.05 – 0.07	1.2 – 1.8[c]
Sedimentary rocks ($mg\ kg^{-1}$)			
Argillaceous	40 – 60	0.07 – 0.1	2.5 – 4[c]
Sandstones	5 – 30	0.05 – 0.25	3 – 7[c]
Calcareous	2 – 10	0.1 – 0.15	2 – 6[c]
Raw energy material ($mg\ kg^{-1}$)			
Coal	12 – 280	0.01 – 19	10 – <50[c]
Fly ash	72	–	–
Crude oil	0.2 – 1	0.06 – 0.3	–
Petrol	0.05 – <1	–	–
Soils ($mg\ kg^{-1}$)	20	0.06 – 0.4	1 – 2[c]
Light sandy	1 – 70 (13)[b]	–	1 – 5[c,d]
Medium loamy	4 – 100 (23)	–	–
Heavy loamy	7 – 140 (24)	–	–
Calcareous	7 – 70 (23)	–	–
Organic	1 – 115 (16)	2 – 5	1 – 8[c]
Waters ($\mu g\ l^{-1}$)			
Rain	0.02 – 0.3	–	–
Ocean, sea	<1 – 8[g]	10 – 300	0.46[f]
River	0.27 – 3.53	<5 – 210[g]	0.2 – 0.7[e]
Air ($ng\ m^{-3}$)	150 – 1600[i]	–	–
Urban (USA)	1.0 – 0.20	>10	440[h]
Rural (USA)	0.01 – 0.06	0.04	0.05 – 0.01[j]
Background (Finland)	0.03 – 0.06	0.012[k]	–
South Pole	–	–	–

[a] Presented are common, possible background, values from various sources as compiled by Kabata-Pendias and Pendias (2001), unless otherwise indicated.
[b] Arithmetic means are given in parenthesis.
[c] Values are $\mu g\ kg^{-1}$.
[d] Data for agricultural soil of Egypt (Rashed and Awadallah 1998).
[e] Data in $ng\ l^{-1}$ for Uzbekistan (Kist 1994).
[f] Data in $ng\ l^{-1}$ for Northern Atlantic Ocean, surface layer (Gordeyev et al. 1997).
[g] After Eisler (1998).
[h] Data for aerial dust near Frankfurt (Messerschmidt et al. 2000).
[i] Data for UK (Cawse 1987).
[j] After McDoland and Duncan (1979).
[k] After Jalkanen and Häsänen (1994).

and in argillaceous sediments, however, it is rather excluded from the carbonated rocks (Table II-11.2).

Copper reveals a strong affinity for sulfur; hence its principal minerals are chalcopyrite, $CuFeS_2$; bornite, Cu_5FeS_4; chalcocite, Cu_2S; and covellite, CuS. The second group contains oxidized copper minerals, which are mainly formed during the weathering of copper sulfides exposed to erosion. Here, Cu is incorporated in oxide and carbonate minerals of which cuprite, Cu_2O; tenarite, CuO; malachite, $Cu_2CO_3(OH)_2$, and azurite, $Cu_2\ (CO_3)_2\ (OH)_2$ are the most common. Copper is often associated with sphalerite, ZnS; pyrite, FeS; and galena, PbS. Its deposits are commonly found in acid igneous rocks and various sedimentary deposits.

II-11.1.2 Production and Use

World primary production of Cu has increased from 9.2 Mt in 1990 to 13.7 Mt in 2003 (USGS 2004). In the 20th century its utilization increased rapidly and now the mining and refining of Cu takes place on all six continents.

Due to its versatile properties, Cu has a wide range of applications. It is used for the production of conductor materials, wire, rod and bar; it is also used for current generation, transmission and conduction, and in the manufacture of motors and electrical equipment. Other applications of Cu include: fabrication of household articles, coins, art objects and ammunition. In the automobile industry, Cu has been partly replaced by Al. Copper wires and cables used for telecommunications are being nowadays displaced by microwave technology and fiber optics.

Copper is also widely used in agriculture (fertilizers, pesticides etc.). Due to its bacteriostatic properties, it is also used as a feed additive in livestock and poultry nutrition.

II-11.1.3 Soils

The general values for the average total Cu contents in soils of different types all over the world are reported to range between 20 and 30 mg kg^{-1} (Alloway 1995). However, values below 10 mg kg^{-1} are very often cited and thus, the global range for mean Cu concentration is cited to be from 8 mg kg^{-1} in acid sandy soils to 80 mg kg^{-1} in heavy loamy soils (Kabata-Pendias and Pendias 2001). Other sources (Aaseth and Norseth 1986; ATSDR 1990) gave the global range for Cu in soils as 2–250 mg kg^{-1}. Generally, Cu is accumulated in the upper few centimeters of soils, but it has a tendency to be adsorbed by organic compounds, carbonates, clay minerals, and oxyhydroxides of Mn and Fe also in deeper soil layers (Janssen et al. 1997).

Contents of Cu are closely associated with soil texture and usually are the lowest in light sandy soils and the highest in loamy soils (Table II-11.2). In addition to soil texture, several other soil parameters in particular soil pH and SOM control its distribution and behavior. Average contents and ranges of Cu in arable top-soils of different EU countries are fairly similar (in mg kg^{-1}):

Cu

- Finland: 13 – 29 means for mineral and organogenic soils, respectively (Mäntylahti and Laakso 2002)
- Germany: 1 – 130, mean 14.4 (Alloway 1995)
- Italy, mean: 51 (Arduini et al. 1995)
- Poland: 1 – 100, mean 14.6 (Kabata-Pendias and Pendias 1999)
- Sweden: 2 – 45, mean 17 (Eriksson 2001a)
- UK: 1.2 – 1 508, mean 23 (Alloway 1995)

Copper occurs in most soils as the $Cu(H_2O)_6^{2+}$ ion adsorbed on clay minerals or coprecipitated on other mineral and organic soil components. Depending upon soil pH, Cu may occur in various forms: Cu^{2+}, Cu^+, $Cu(Cl_2)^-$, $CuSO_4^0$, $Cu(OH)_2^0$, $CuCO_3^0$, $CuCl^0$, and (McBride 1981; Alloway 1995). The behavior of Cu in soils depends highly on its compounds and species (Fig. II-11.1). Concentrations of Cu in soil solution range from 0.5 to 135 $\mu g\ l^{-1}$, depending on techniques used for the solution extraction and on soil types. In soil solution it may occur as cations: Cu^{2+}, $CuOH^+$, $Cu_2(OH)_2^{2+}$, and as anions: $Cu(OH)_3^-$, $Cu(OH)_4^{2-}$, and $Cu(CO_3)_2^{2-}$ (Kabata-Pendias and Sadurski 2004). Ponizovsky et al. (2006) calculated that most of the added Cu is sorbed in soil, and only about 1% of the total amount remained in soil solution. These authors adapted a model for a prediction of dissolved Cu concentration in soils at low moisture content.

The behavior, phytobioavailability and toxicity of Cu are influenced by its species, and are not a function of its total concentration (Allen 1993). Several soil variables control the Cu solubility and thus bioavailability; these include: pH, oxidation and reduction potential, organic matter, soil texture, mineral composition, temperature, and water regime. The mobility of Cu is especially reduced in the presence of large mineral colloids with Fe-Al-oxyhydroxide coatings by oxyhydroxide particles of Al, Mn, Fe, and by organic matter (Kabata-Pendias and Sadurski 2004). Dissolved organic matter (DOM) has a great affinity to fix Cu and thus to inhibit its sorption in soils. These phenomena are attributed to the formation of soluble Cu-organic complexes (Zhou and Wong 2003). In mineral soils, natural attenuation of Cu occurs as an effect of Cu substituting for Ca in calcites present in calcareous soils and precipitation of $Cu(OH)_2$ and/or $Cu_2(OH)_2CO_3$ in other soils (Ma et al. 2006).

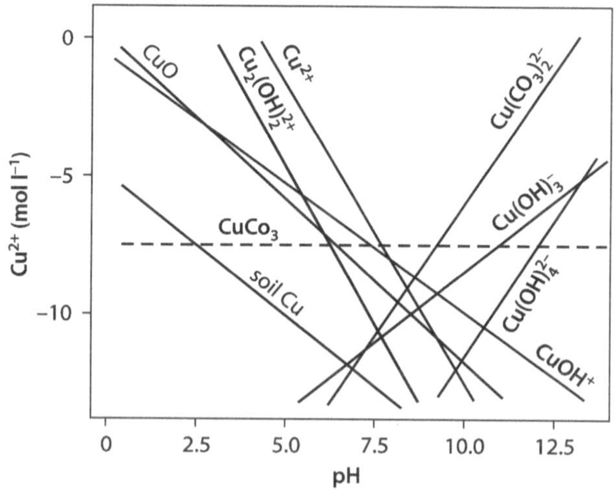

Fig. II-11.1. Schematic diagram of solubility of Cu compounds and species as a function of soil pH. (adopted from Lindsay and McBride *vide* Kabata-Pendias and Pendias 2001)

The Cu deficiency in crop plants due to low contents of Cu in soils has been reported for many countries since the 19th century (Gartrell 1981; Alloway 1995). However, Cu deficiency in world agriculture is also closely associated with soil conditions and management practices. Copper deficiency most commonly occurs in soils which are: *(i)* coarse structured and high pH, *(ii)* formed from on calcareous parent material, and *(iii)* rich in organic matter. Thus, the deficiency can be a consequence of the low content of total soil Cu, or the condition of the soil reducing the mobility and thus availability of Cu. Among different exctractants used for the diagnosis of Cu deficiency, the solution of AAAcEDTA has been recently used most commonly, however EDTA and DTPA also have been used for long time. The interpretation of these values has been broadly discussed by Alloway (1995). However, it may be generally stated that total Cu contents below 10 mg kg^{-1} in different soils may indicate deficiency.

Soil contamination by Cu compounds has been the subject of detailed studies for several decades and a large database has been already collected and presented in a number of monographs and papers. Several significant sources such as fertilizers, sewage sludge, manures, agrochemicals, industrial by-product wastes and the quality of irrigation waters have contributed to increased Cu levels to agricultural soils. Highly increased levels of Cu are observed in soils surrounding Cu mines and smelters (Table II-11.3). As Cu is only slightly mobile under most soil conditions, its elevated contents may persist for a long time (Hutchinson 1979).

Input of Cu to soils of livestock farms in the EU member countries vary greatly (in g $ha^{-1} yr^{-1}$): from 29 in Czech Republic to 2 771 in Italy. Minus load charges were calculated only for Switzerland (–17) and for Norway (–138). Copper balance in crop farms is much lower and in several countries show some loss of Cu from top-soils, up to 54 g $ha^{-1} yr^{-1}$ in Norway. However, the balance of Cu for vineyard soils and sludged soils of Italy have been estimated at 13 923 and 3 905 (g $ha^{-1} yr^{-1}$) respectively (Eckel et al. 2005). The maximum allowable loading of Cu with wastes to arable soils of the EU countries is estimated at 12 kg $ha^{-1} yr^{-1}$. Ranges of MAC values commonly cited in literature range from 20 to 100 mg kg^{-1} (Kabata-Pendias and Sadurski 2004). Precau-

Table II-11.3. Copper in contaminated soils (mg kg^{-1})

Site, pollution source	Country	Mean or range	Reference
Vineyard soils	France	100 – 1 500	Besnard et al. (1999)
Coffe plantation	Kenya	883	Lepp and Dickson (1994)
Near copper facility	USA	7 000	USEPA (1980)
Near copper facility	Poland	70 – 1 600	Kabata-Pendias and Pendias (1999)
Contaminated sites	UK	200 – 500	Alloway (1995)
Heavily contaminated	UK	500 – 2 500	Alloway (1995)
Mine dump	Galicia (NW Spain)	273 – 5 241	Álvarez et al. (2003)
Salez soil (no waste)	Switzerland	750	Keller et al. (2001)
Soil + municipal waste + sewage sludge		250	Keller et al. (2001)

tionary values for Cu in Germany are established for soils of various texture as follow (in mg Cu kg^{-1}): clay, 100; loam, 60; and sand, 30 (Eckel et al. 2005).

Among various methods used to remediate Cu-contaminated soils, a relatively common technique is immobilization by organic matter. For example, Kikkkilä (2002) studied the remediation techniques for polluted soils, 0.5 km away from a Cu-Ni smelter in Finland, by mulching with organic matter (mixture of compost and woodchips). The organic mixture converted the Cu into less exchangeable and thus less bioavailable forms. This technique has resulted in a low Cu in soil solution and in the subsequent reduction of the toxic effects on microbiota.

II-11.1.4 Waters

Nozaki (2005) gave the mean Cu concentration in the North Pacific Ocean as 0.15 μg kg^{-1} (Table II-8.2). The median for world ocean waters is estimated at 0.25 μg kg^{-1} (Reimann and Caritat 1998).

The variation in Cu concentrations in seawater is greatly dependent on the water pH. For example, $Cu(OH)_2$ concentration increases by 90% with a pH change from 7 to 8.6 (USEPA 1980). The prevailing Cu species in seawater over the entire ambient range are $Cu(OH)_2$, $CuCO_3$ and Cu^{2+} (USEPA 1980). In surface waters, Cu is likely to occur as Cu^{2+}, $Cu(OH)^+$, $CuCO_3$, and $CuSO_4$ (Witczak and Adamczyk 1995).

Waters in sea basins receive Cu from various sources. Matschullat (1997) calculated the Cu input to the Baltic Sea as follows (t yr^{-1}):

- Transport with river water: 1 300
- Atmospheric deposition: 1 200
- Total input: 2 500 of which 82% has been established to be from anthropogenic sources

World average of Cu in river waters is estimated at 1.48 μg l^{-1} and its riverine flux at 55 kt yr^{-1} (Gaillardet et al. 2003). However, its concentration in surface and ground waters differ significantly (Table II-11.4).

The speciation and chemical forms of Cu in waters are important in controlling biogeochemical processes and its bioavailability (Hodson et al. 1979). In freshwater, the solubility of Cu salts is decreased under reducing conditions, which is governed by factors such as water pH, temperature, hardness, density of suspended materials and many others parameters (Luoma 1983; Eisler 1998). In sea water, the major chemical species of Cu are Cu(OH)Cl and $Cu(OH)_2$. Brügmann, (*vide* Szefer 2002b) reported that Cu in seawater occurs in both organic and inorganic fractions in equal amounts, while in pore water samples, Cu is present mainly in inorganic forms.

The concentration of Cu in water is in a dynamic equilibrium with its content in surface bottom sediments. Generally, Cu in bottom sediments is associated with mineral matter or is tightly bound to organic material (Kennish 1998). Helios-Rybicka (1993) investigated phase-specific bonding of metals in sediments of the Vistula River and found that in most cases the moderately reducible fraction of Cu predominated (up to 50%) over other fractions and that a relatively very small proportion of Cu occurred

Table II-11.4. Copper in waters in selected areas of North America

Source of water	Location	Concentration	Remarks	References
Private wells	Nova Scotia	0.04 – 0.2	Running tap water	Maessen et al. (1985)
	Four communities	0.13 – 2.45[a]	Standing water	
River	Canada (National Survey)	≤0.005 – 0.53	Raw water	Meranger et al. (1979)
		≤0.005 – 0.1	Treated water	
		≤0.005 – 0.22	Distributed water	
Groundwater	New Jersey	0.005 (median)		Page (1981)
Surface water	USA	0.004	53 862 samples	Eckel and Jacob (1988)
Lakes	Canada	0.001 – 0.008	Acid sensitive lakes	Reed and Henningson (1984)
	Great Lakes	0.63 – 0.83	Lake Superior	Nriagu et al. (1996)
		0.7 – 1.06	Lake Erie	
		0.54 – 1.1	Lake Ontario	

[a] Levels of Cu in 53% of samples are greater than 1 mg l^{-1}.

in the organic fraction. The distribution of Cu fractions was not related to Cu contents in sediments that varied from 48 to 2 800 mg kg^{-1}. The available data suggests that Cu is generally in the bound forms in natural waters and sediments.

The threshold values for Cu in drinking water vary in different countries from 0.04 µg l^{-1} to 3.0 µg l^{-1}. However, in most European countries it is established as 0.05 µg l^{-1} (WHO 1998). The limit value of 1.3 µg l^{-1} proposed by the US EPA (2002) may be exceeded in some waters. Higher values of Cu in drinking water may be due to variation of pH and hardness; Cu may also be taken up from the plumbing pipes through which it is distributed.

Bottom sediments. In many areas, sediments of lakes and/or rivers may be sinks for Cu, with little release to the water body. Usually increased concentrations of Cu are observed near smelter and mining areas, and near sewage outlets. Increased Cu contents in surface layers of bottom sediments are reported as follows (in mg kg^{-1}):

- India, rivers near the megacity Chenai (formerly Madras): 760 – 939 (Govindarajan and Rao 1992)
- Poland, highly industrialized region: 17 – 278 (Aleksander-Kwaterczak and Helios-Rybicka 2004)
- Russia, lake near Cu-Ni smelter: 905 (Lukin et al. 2003)
- Sweden, lake near smelter area, up to 2 000 (Bradley and Morris 1986)
- UK, southwest industrialized region: 590 (Bryan and Langston 1992)

Copper contents in sediments of lakes and rivers of uncontaminated regions are reported to range (in mg kg^{-1}) between:

- 36 and 74 in Lake Asososca, Nicaragua (Cruz et al. 1994)
- 18 and 39 in Lake Rożnowskie, Poland (Skwarzec 2005)
- 7 and 70 in streams of Wales (Thornton et al. 1979)
- 57 and 107, during wet and dry season, respectively, karstic river Nahr Ibrahim, Lebanon (Korfali and Davies 2005)
- 31 mean, Lower Ganges, India (Datta and Subramanian 1998)

Aquatic Biota. Copper is essential for the growth and development of aquatic species and its accumulation rate and retention are based on both biological and abiotic variables.

Copper concentrations in the tissues of aquatic biota (Table II-11.5) depend on a number of factors such as: temperature, pH, salinity and depth, chemical species of Cu, organic and inorganic compounds, and chelators (Eisler 1998). Biological variables affecting Cu build up in marine organisms include age, the size and development stage of a particular organism, and physiological adaptation for Cu substrates (Eisler 1979).

Table II-11.5. Copper content in selected aquatic biota (mg kg^{-1} FW)

Species and organ	Country, site, period	Mean, range	Reference
Invertebrates, contaminated *vs.* uncontaminated	Canada, Lake Ontario, 1988	89	Miller et al. (1992)
Alga ,whole Oyster,soft part Fish, muscle	India, Chenai (Madras)	12.3 4.2 <0.09	Miller et al. (1992)
Mollusks, soft parts Bivalves Gastropods African sharp-tooth catfish, liver	Israel, Acre Valley, 1988–1991	9.4 - 13.3 31 - 48 <92.0	Fishelson et al. (1994)
Perch, liver Perch, muscle	Norway, small lakes, 1991	1.6 - 1.5 0.16	Baekken (1994)
Crab Fish, four species, liver	Wal, coastal area, 1989	58.0 1.6 - 4.4	Morris et al. (1989)
Crayfish whole	United States	56.2 - 77.7	Morrissey and Edds (1994)
Mussel, soft parts	Italy, Goro Bay, 1991–1992	6.9	Fagioli et al. (1994)
Cod, muscles Cod, liver Herring, muscles Herring, liver	Southern Baltic, 1964–1980	0.01 - 1.3 0.3 - 1.5 8.83 5.35	Szefer (2002a)

In Cu contaminated river waters, detritovores contained Cu as high as 102 mg kg^{-1}, whereas predators and omnivores contained between 54 and 43 mg kg^{-1}, respectively (Cain et al. 1992).

In general, trace elements in aquatic fauna are higher in the coastal waters than in the middle of the Baltic Sea (Miettinen et al. 1985; Voigt 1997, 2000; HELCOM 2001; Szefer 2002a).

The European Council guideline for Cu in waters inhabited by salmon and trout is 0.04 mg l^{-1} (Ruf 1981). Apparently the Cu species, Cu^{+2}, $CuOH^+$ and $Cu_2(OH_2)^{2+}$ are more toxic to aquatic biota than the species, $CuHCO_3^+$ and $CuCO_3$ (Meador 1991).

II-11.1.5
Air

Copper content in the air varies greatly, from 0.03–0.06 ng m^{-3} around the South Pole and up to 4900 ng m^{-3} in industrialized regions of Germany (Kabata-Pendias and Pendias 2001). Livett (1992) proposed a concentration of 4 ng m^{-3} as the Cu background value in the atmosphere and Reimann and Caritat (1998) calculated the median Cu in world-remote regions at 2.6 ng m^{-3}.

Global emission of Cu in 1983 was calculated to range from 20 to 51 kt but by 1995 had decreased to 35 kt (Nriagu and Pacyna 1988; Pacyna and Pacyna 2001). Due to the strict pollution control regulation, and use of the best available technology (BAT), emissions of Cu have recently been significantly reduced. This is confirmed by the

Table II-11.6. Copper deposition rate on agricultural soils in some European countries (adopted from Nicholson et al. 2003)

Country	Deposition (g ha^{-1} yr^{-1})
Austria	100
Belgium	7
Denmark	8
Finland	5
Germany	53
Hungary	62
Ireland	13
Italy	60
Norway	12
Poland	40
Sweden	15
Switzerland	18
The Netherlands	27
UK[a]	25
European average	34

[a] After Cawse (1987).

observations of Berg and Steinnes (1997) who reported Cu content in moss sampled in Norway during the period 1990–1995 to average only 7 mg kg^{-1} whereas mosses sampled earlier contained this metal at levels above 20 mg kg^{-1}; even in Greenland, it was measured at 11.2 mg kg^{-1} (Rühling et al. 1987).

The Cu deposition rate in European countries has varied from 5 to 100 g ha^{-1} yr^{-1} (mean 34), being the lowest in Finland and the highest in Austria (Table II-11.6).

Sources of anthropogenic Cu in the atmosphere include smelting of ferrous and non-ferrous ores, heat and electricity production by combustion of fossil fuels, foundries, kiln operation for the production of cement, application of fertilizers and municipal sludge to the land and industrial manufacturing (Ruskin et al. 1994; Cunningham et al. 1997; Pacyna et al. 2001). Natural sources of Cu are: weathering of rocks, volcanoes, thermal springs and blown dust from terrestrial components.

Copper is likely to form oxides that are often associated with dust particles and are relatively easily dissolved in rainwater. Oxides of Cu react easily with SO_x and therefore in air around Cu smelters (e.g., in Arizona, USA) it is strongly correlated with sulfur (Schroeder et al. 1987).

II-11.1.6
Plants

Copper contents in plants vary greatly and are controlled by several factors of which the pool of mobile Cu in soils and plant properties play significant roles. Crop plants from various countries contain fairly similar amounts of Cu, e.g., mean contents in wheat grains vary from 3.8 to 6.7 (mg kg^{-1}) and leafy vegetables 3–8 mg kg^{-1} contain Cu at the range of 3–8 mg kg^{-1}. Average Cu contents in various food plants show a relatively small variation between countries (Table II-11.7). Higher contents of Cu are in grasses, in the range of 2–10 mg kg^{-1} and in clovers, from 7 to 15 mg kg^{-1} (Table II-11.8).

A number of projects have been carried out recently on Cu levels in edible plants, since about 30% of daily Cu intake by adults is from food plants. The reference, after Ensminger et al. (1995) for food composition in the USA, gives the following values for Cu in some categories of plant food (in mg kg^{-1} FW):

- vegetables, range 0.1 – 3.2 lowest in celery roots, highest in garlic cloves
- fruits, range 0.3 – 4 lowest in grapes, highest in avocados
- cereals, range 0.3 – 13 lowest in oats, highest in rye
- nuts, range 0.2 – 23.8 lowest in coconut, highest in Brazil nuts

Copper, the essential metal in plants, is a constituent of several "key" enzymes, and plays important functions in physiological processes, such as:

- photosynthesis and respiration
- carbohydrate and nitrate metabolisms
- water permeability
- reproduction
- disease resistance

Table II-11.7. Copper in food plants grown in various countries[a] (mg kg^{-1})

Plant	Range	Mean
Wheat, grains[b]	1.3 – 10	4.7
Barley, grains	4 – 15	5.5
Rye, grain	–	4.5
Oats, grains	2 – 6	3.6
Cabbage	3 – 4	–
Lettuce	6 – 8	–
Carrot, roots	4 – 8.4	5.7
Onion, bulbs	4 – 6	4.9
Tomato, fruits	6 – 9	–
Potato, tubers	3 – 6.6	4.4
Tea, leaves[c]	–	20

[a] Presented are common, possible background, values from various sources, as compiled by Kabata-Pendias and Pendias (1999, 2001), unless otherwise indicated.
[b] After Kumpulainen (1996) and Eriksson (2001a), for wheat from seven countries.
[c] Data for Kenya, after Wong et al. (1998).

Table II-11.8. Copper in grasses and clover at the immature grown stage from different countries (mg kg^{-1}) (values from various sources as compiled by Kabata-Pendias and Pendias 1999, 2001)

Country	Grasses		Clover	
	Range	Mean	Range	Mean
Canada	–	5	–	–
Finland	3.8 – 4.8	4.3	–	–
Germany	7.4 – 15	10	7.6 – 15	10.5
Hungary	3.6 – 8.4	5	4.2 – 16.2	10.5
Japan	1.3 – 33	6.9	2 – 12.5	6.5
Poland	2.2 – 21	6	4.2 – 21	11.3
New Zealand	7.3 – 13.4	10.5	8.1 – 17.5	11.7
USA	1.5 – 18.5	9.6	10 – 29	16.2
Russia	–	–	–	–
Meadow bog	1.1 – 3.8	1.8	8.8 – 20	12.5
Forest steppe	1.1 – 3.9	2.6	5.1 – 24	14.8

The deficiency of Cu strongly affects physiological processes and therefore plant production. Cu-deficiency in crops is widespread and often it is in a hidden form. As Alloway (1995) has stated this form of Cu-deficiency in crop plants is the most ubiquitous and economically important, and is a more serious problem in Europe than in many other parts of the world. This is associated mainly with soil types, climatic conditions and crops grown. Common crop plants that are highly sensitive to Cu deficiency are: wheat, oats, sunflower, alfalfa, carrots, lettuce, spinach, onion, and citrus trees.

Copper is slightly mobile in plants as it is strongly bound by nitrogen and proteins. With Cu deficiency, a plant may grow normal vegetation for a short time, but there will be a failure in the redistribution of Cu from old leaves and roots causing chlorosis and tip necrosis in new leaves, known as Cu deficiency symptoms.

Plants growing on Cu-polluted sites tend to accumulate increased amounts of this metal, especially near industrial areas, and in soils treated with Cu-bearing herbicides (Read and Martin 1993; Fishelson et al. 1994; Reimann et al. 1999). Thus, coffee trees from African plantations using Cu herbicides contain the following amounts of Cu (in mg kg^{-1}): bark, >1 000; leaves 400; and roots 150 (Lepp 1994). Forest trees, e.g., pine trees, grown near a Cu-smelter in Finland contain also very high amounts of Cu (in mg kg^{-1}): bark, 1 500; phloem, 66; wood, 11; and needles, 500 (Heliövaara and Väisänen 1990, 1991).

In the root or rhizosphere, Cu is in complex form, but its uptake and translocation is a function of the activity of Cu^{+2}. When the total soil concentration of Cu is between 25–40 mg kg^{-1} at below 5.5 pH, then a soil solution level of Cu^{2+} can be toxic to many higher plants (Baker et al. 1994). Hence, soil pH plays an important role in the availability and toxicity to many terrestrial plants.

Some plants of the mint Family (e.g., *Aeolanthus biformifolius* and *Becium homblei*) are good indicators of Cu deposits and, when grown on mineralized land contain Cu at level of 13 700 and 1 000 mg kg^{-1}, respectively (Hutchinson 1979; Jenkins 1980). Cu-tolerant plant communities and the role of Cu-hyperaccumulators in geochemical prospecting have been described widely by several authors (Kovalevsky 1991; Brooks 1998; Vernadsky 1998).

Despite general Cu tolerance in plants, this element is considered to be quite toxic. The effects of Cu excess such as decrease of photosynthesis and yield are well known. Sensitive plants are already affected at Cu contents of 15 to 20 mg kg^{-1} in the growth media (Kloke et al. 1984). The most common characteristic symptoms of the Cu^{2+} and Cu^{+} toxicity are: *(i)* root malformation, *(ii)* damage to membrane permeability, *(iii)* inhibition of photosynthetic electron transport, *(iv)* damage to DNA, and *(v)* immobilization of Cu in membranes and protein complexes.

II-11.1.7
Humans and Animals

Humans. According to Jørgensen (2000) the average contents of Cu in mammalian tissues range from 1.7 to 196 mg kg^{-1}, with the lowest value in skin and the highest in liver. The range of Cu concentrations in human tissues is given by Li (2000) to be between 0.7 and 7.8 mg kg^{-1}, with the lowest value for muscles and the highest for liver; mean Cu level in tissues of the "reference man" is estimated at 1 mg kg^{-1}. Copper concentration of blood varies in the range of 0.8–1.6 mg l^{-1} and usually is higher in women than in men (Yunice and Hsu *vide* Kabata-Pendias and Pendias 1999).

Copper is an essential element that is a component of several proteins and metalloenzymes. It is prerequisite in oxidation-reduction reactions and hemoglobin synthesis. The liver is a real storage site for Cu that can be mobilized in case of a negative Cu balance (Momčilović 2004). Metallothionein is the main intracellular Cu-protein involved in Cu transport within the body. Copper is associated mainly with three distinctive phases, ceruloplasmin, albumins, and amino acids (Kleczkowski et al. 2004).

Copper deficiency in humans may affect several diseases such as: slow growth, anemia, and fertility, hair and weight loss, disorders of central nervous system, cardiovascular problems, osteoporosis and several other metabolic dysfunctions (ATSDR 1990; Percival 1995). Menke's syndrome, a result of the Cu deficiency, causes brain diseases, and low concentration of Cu in liver and other tissues (Aaseth and Norseth 1986). Chronic electroconvulsive shock induces an increase, by 36%, in the Cu levels in the hippocampus and by 16%, in the cerebellum (Schlegel and Nowak (2000). Schlegel-Zawadzka et al. (1999) reported elevated Cu levels in the serum of depressed patients (1.2 mg l^{-1}) as compared with its level in the control persons (0.95 mg l^{-1}).

An excess of Cu has an adverse affect on several physiological processes such as:

- impairment of membrane properties,
- impairment of DNA and enzyme functions, especially those containing sulfhydryl groups,
- cellular injury due to the generation of hydroxyl radicals OH^- and OH^+ from hydrogen peroxide H_2O_2 which can lead to oxidative stress

An excess of Cu may result in Wilson's disease, mostly ending in death (USEPA 1980; Aaseth and Norseth 1986). The increased level of Cu in tissues of humans with Wilson's disease as compared with healthy adults is, after the data of ATSDR (1990), as follows (in mg kg^{-1} for healthy *versus* ill individuals):

- Bone: 2.9 *vs.* 31.0
- Brain: 5.4 *vs.* 54.9
- Cornea: 3.8 *vs.* 35.1
- Kindeys: 2.8 *vs.* 36.2
- Liver: 7.8 *vs.* 99.2

The main source of Cu for humans and animals are all kinds of food and water. However, the inhalation of atmospheric fine particles (<10 μm) which are carriers of Cu, can also be a significant source of Cu, in Cu-rich regions. Copper in human diets does not generally exceed the human requirement. Several European studies indicated that intake of Cu is rather low among the populations of many European countries. Daily intake of Cu by adults varies for European countries from 0.9 to 1.8 mg, being the lowest for Poland and the highest for Austria (Jędrzejczak and Ręczajska 1996).

In the USA, adults' Cu intake from food is 0.93–1.3 mg d^{-1} (i.e., 0.013–0.019 mg kg^{-1} d^{-1} considering 70 kg body weight) (ATSDR 2002a), whereas a dietary allowance (RDA) of 0.9 mg d^{-1} (0.013 mg kg^{-1} d^{-1}) has recently been proposed. Daily intake of Cu by two groups in Ankara (Turkey) was 1.78 mg d^{-1} and 1.87 mg d^{-1}. This value is lower than the estimated safe and adequate daily dietary intake (ESADDI) value of 2.6–5.0 mg d^{-1} (Aras and Kumpulainen 1996). The general estimation by Jørgensen (2000) gives 2–5 mg Cu per day for a man (70 kg weight). However, dietary habits including different agricultural and food processing practices can greatly influence these values. Brüggemann et al. (1996) estimated that most adults in Europe intake on average 33% of the daily Cu dose from cereals and potatoes.

Copper is absorbed mainly through the gastrointestinal tract (WHO 1998). The requirement of adults for Cu is established as 1.5–4 mg d^{-1} and is practically covered by the normal diet of the population in most countries. The FAO/WHO (1998) have adopted the provisional dietary requirement of Cu to be 0.05 mg kg^{-1} BW, and toxicological intake as 0.5 mg kg^{-1} BW. There are a few adopted levels by the FAO/WHO (2003) for Cu in food that is 5 mg kg^{-1} FW for fruit juices, and 0.1–0.4 mg kg^{-1} FW for fats and oils. The maximum limits for Cu established by the Ordinance of the Ministry of Health and Human Welfare of Poland are 6.0 mg kg^{-1} in wheat grain, rye grain and flour, and 5.0 mg kg^{-1} in powdered milk (Wojciechowska-Mazurek et al. 1996).

Animals. Copper levels in different tissues from selected wild and domestic animals indicate fairly similar distributions, being always the highest in the liver and/or in kidneys and the lowest in muscles (Table II-11.9). Liver of both, ducks and various mammals, contain the highest amounts of Cu, up to 200 mg kg^{-1} FW whereas other tissues contain Cu at the level <10 mg kg^{-1} FW (Falandysz et al. 1994a).

Important agronomic problems in certain regions are related to secondary Cu deficiency in cattle and other ruminants due to the presence of Mo and S in diets which are responsible for the decreased availability of Cu caused by the formation of the unavailable Cu-thiomolybdate complex (Aaseth and Norseth 1986). Cattle may die if diets contain low Cu and high Mo concentrations, 2–20 mg kg^{-1}, (Eisler 1989). It is believed that ruminants are more susceptible to Cu deficiency than nonruminant animals because of a difference in a sensitivity rate of Cu absorption from fodder. On the other hand, non-ruminant animals are extremely resistant to Cu toxicity (Momčilović 2004).

The deficiency of Cu in mammals is relatively often observed and has been reported for several animals, i.e., domestic swine, camels, sheep, and mink. Swine need a higher Cu-containing diet than mink. Swine have a low hemoglobin and growth rate, when the animals' diets contain 15–36 Cu mg kg^{-1} BW, for only seven days (ATSDR 1990). Chickens may suffer from anemia, poor growth, cardiovascular disease and high mortality due to Cu deficiency, at <2.7 mg kg^{-1} in fodder (Carlton and Henderson 1964). However, Cu is ubiquitous in common food and fodder hence its deficiency is not normally the result of continually consuming a Cu-poor diet (Percival 1995).

Many animals reveal a relatively high tolerance to increased levels of Cu in fodder. Apparently some enzymatic systems are involved in the reduction of metal toxicity, but these processes are still unclear (Augustyniak and Migula 2000). The most tolerant are chicken and swine. In Cu contaminated regions domestic animals may have elevated contents of this metal, as for example near a copper foundry in Poland, cattle livers contained more Cu (35–140 mg kg^{-1} FW) than livers of cattle grazed on agricultural lands (7–32 mg kg^{-1} FW) (Falandysz et al. 1994c). Gummow et al. (1991) suggested that high concentrations of Cu, up to 141 and 444 mg kg^{-1} FW in kidneys and liver, respectively, were responsible for the death of some wild ruminants.

The concentrations of Cu in birds of colonies in the Barents Sea vary between species (Savinov et al. 2003). The hepatic Cu concentration in muscle tissues of common eider was twice as high (227 mg kg^{-1}) as that of the king eider (84.9 mg kg^{-1}). The

Table II-11.9. Copper in tissues of livestock and poultry[a] ($mg\ kg^{-1}$)

Animal	Liver	Kidneys	Muscle
Cow	8.15	8.15	1.41
Lamb	89.8	5.39	1.47
Chicken	4.6	3.07	0.67
Turkey	7.14	3.68	0.83
White-tailed deer[b]	122	29	–
Rabbit[b]	19.2	21.5	11.9

[a] After Coleman et al. (1992), unless otherwise indicated.
[b] After Storm et al. (1994).

authors confirmed that different levels of Cu in the liver of sea birds are results of differential intake of Cu containing food. Copper concentration in liver may also be increased due to the starvation of birds (Parker and Holm 1990).

II-11.2 Silver (Ag)

II-11.2.1 Introduction

The geochemical characteristics of Ag are similar to those of Cu, but its concentration in the Earth's crust is about 1 000 times lower than Cu and ranges around 0.08 $mg\ kg^{-1}$ (Table II-11.2). In some sandstones and calcareous rocks its abundance can be enriched up to about 0.2 $mg\ kg^{-1}$. After weathering it is relatively easily precipitated in alkaline reduction-potential media and media rich in S compounds, due to its chalcophilic properties.

The most common Ag minerals are: argentite, Ag_2S; cerargyrite, AgCl; arsenide, Ag_3As; pyrargyrite, Ag_3SbS_3; proustite, Ag_3AsS_3; and polybasite $(Ag,Cu)_{16}Sb_2S_{11}$. Silver is easily combined with minerals of other trace elements, such as Se, Te, As, Sb and Bi (Etris 1997). Minute concentrations of Ag impurities occur often in sulfide minerals, e.g.: sphalerite, ZnS; pyrite, FeS_2, chalcopyrite $CuFeS_2$, and galena PbS.

In 1978, the disposal of Ag to ecosystems (air, water and land) of the USA had been estimated at the range of 2.39–2.54 kt (Smith and Carson 1977a). In northern Europe, the annual loss of Ag to the environment has been calculated as follows: to land (FI), 23.3 t, and to water (FI, DK, N, SE), 3.2 t (Mukherjee 1994,1997).

II-11.2.2 Production and Use

The global annual production of Ag in 2000 was estimated at 320 kt (USGS 2004). There are about 20 top Ag-producing countries in the world. The highest amount of Ag (2 852 t) was produced in 2002 in Mexico and the lowest (115 t) in South Africa (The Silver Institute 2003). The specific metallurgical processes applied to Ag-bearing mineral concentrates depend on whether the primary metal is Cu, Zn, Pb or Au. However, most often it is recovered as byproducts at Cu and Zn ores exploration.

Silver was discovered about 4000 BC, when it was used as jewelry and a medium of exchange. Presently, the major part of Ag production is used in the manufacturing and photographic industries. It can be recycled from spent photographic processing solutions and photographic and X-ray films. In the former case, electrolytic method for the recovery of Ag is generally adopted (Kodak Publication 1983). Its demand is also in different other sectors, including: batteries, coins, jewelry, silverware, brazing and soldering, catalysts, electronics and electroplating,

Accordingly to data of The Silver Institute (1994), Ag consumption in 2002 was in various manufacturing as follows:

- industrial application: 39.67% (10.63 kt)
- photographic sector: 23.75% (6.37 kt)
- jewelry and silverware markets: 30% (8.04 kt)
- other uses (medical applications, mirrors, coatings): 6.58% (1.76 kt)

II-11.2.3
Soils

The common range of mean contents of Ag in soils is between 0.06 and 0.4 mg kg^{-1} and is likely to be increased in some organic soils (Table II-11.2). Smith and Carson (1977a) have estimated soil Ag concentration to be between 0.03 and 0.09 mg kg^{-1}, whereas the background concentration of Ag was reported by (US EPA 1981a) to range from 0.1 to 0.2 mg kg^{-1}.

In soils, Ag is present primarily as sulfides associated with Fe, Pb. Also MnO_2 has a strong affinity for Ag, and appears to be the most significant sorbant of this metal. Silver may occur also as complex ions associated with chlorides and sulfates which are mostly insoluble or sparingly soluble (Etris 1997; Cappel 1997). Humic substances, known to absorb and complex Ag, control its mobility in soils (Sikora and Stevenson 1988).

Common species of Ag in soils are simple cations, Ag^+, Ag^{2+}, and AgO^+; and complex anions that predominate in soil solution: $AgCl_2^-$, $AgCl_3^{2-}$, $Ag(SO_4)_2^{3-}$ (Kabata-Pendias and Sadurski 2004).

Soils from mineralized areas are usually enriched in Ag and may contain up to 2.5 and 3.2 mg kg^{-1} in the USA and Canada, respectively (Shacklette and Boerngen 1984). Near Ag-Pb mining areas in the UK, soils contain this metal at the range of <0.1–18 mg kg^{-1} (Jones et al. 1985). Lockhart (1983) described that Ag may be extracted from soil containing about 1 000 mg Ag kg^{-1} (e.g., in Mexico and western USA).

Rasmussen et al. (2001) reported for garden soil and street dust in Canada contents of Ag at 0.4 and 0.2 mg kg^{-1}, respectively. According to these authors, Ag concentration in household vacuum dust is 4.5 mg kg^{-1}. Elevated Ag contents, in the range between 15 and 120 mg kg^{-1}, in soils near industrial areas and landfills also have been reported.

Ewell et al. (1993) studied survival and growth of earthworms (*Lumbricus terrestris*) in soil with the addition of Ag_2S. Results indicate the growth reduction (62 mg Ag kg^{-1}, NOEC) due to contact with Ag_2S, but there was no bioaccumulation of Ag. In general, Ag and its compounds are toxic to meso and microbiota.

II-11.2.4
Waters

Data for mean Ag concentrations in ocean waters vary; after Andren and Bober (2002), the range is <0.024 to 0.56 ng l^{-1}, and after Reimann and Caritat (1998) the range is 2–40 ng l^{-1}. In the North Pacific Ocean waters it averages 2 ng kg^{-1} (Nozaki 2005). In surface waters AgS concentrations are higher and vary (in ng l^{-1}) from <0.01 to 140 in rivers and from 0.2 to 0.5 in Great Lakes water (Andren and Bober 2002). Drinking water in Finland contains Ag at the level of 30 ng l^{-1} (Mukherjee 1997).

According to Andren et al. (1995), data for Ag in waters prior to "clean technology" (e.g., before 1980) are no longer valid. Shafer (1995) confirmed the above statement, indicating that levels of Ag in Lake Michigan dropped during the last three decades. Mean Ag concentrations in surface waters have changed in recent years, as follows (in ng l^{-1}):

- 1972: 300 (Copeland and Ayers 1972)
- 1984: 57 (Rossman 1984)
- 1995: 0.2 – 0.5 (Shafer 1995)

Silver in waters occurs in several ionic species, mainly associated with Cl and S. Common species that control Ag solubility are: Ag^+, $AgCl_2^-$, $AgCl_3^-$, $Ag(MS)_2^-$, and $Ag_2S_3M_2^{2-}$. Also some inorganic and organic complexes are known to control Ag concentrations and behavior in waters. The model for Ag transfer in fresh water is depicted in Fig. II-11.2.

Total Ag discharged from photographic processing facilities, prior to sewage treatment plant, was mainly in the form of Ag thiosulfate that is not toxic in concentrations of mg l^{-1}. After the treatment, the effluent discharged to the open waters contained Ag at the levels below μg l^{-1} (Wood et al. 1999). Due to marketing of digital cameras, Ag release to the aquatic environments will be significantly lower. Silver is transported in the

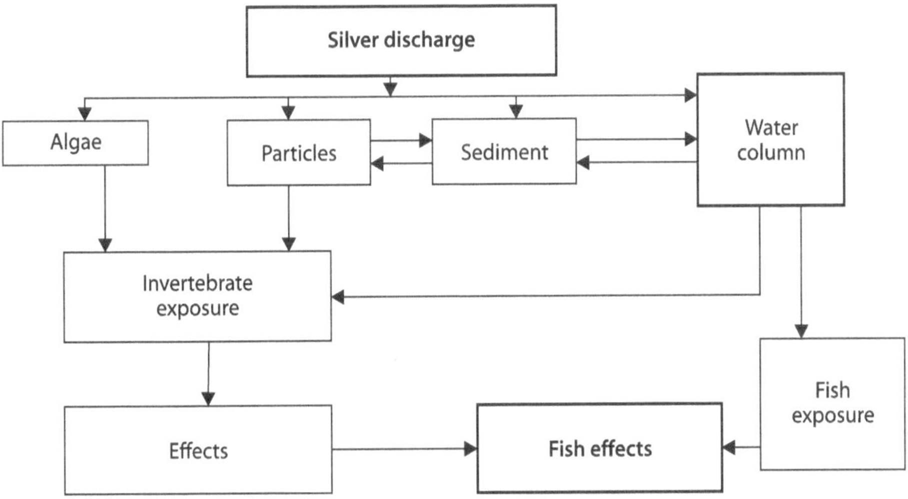

Fig. II-11.2. Conceptual model for silver transfer in the aquatic systems (reproduced with permission from the SETAC)

Table II-11.10. Silver in bottom sediments in different countries (mg kg^{-1}) (after Andren and Bober 2002, unless otherwise indicated)

Country, location	Environment	Range
New Zealand, Maratoto	Stream	0.75 - 7.0
USA, South Plate River, WY, NB, CO	River	0.2 - 5.1
USA, Long Island Sound, CT – NY	Estuary	0.006 – 25
USA, Puget Sound, WA	Estuary	0.015 - 0.71
USA, San Francisco Bay, CA	Estuary	0.09 - 1.10 0.04 - 0.83
USA, New Haven Harbor, CT	Estuary	0.06 - 33.9
USA, Quinnipiac River Marsh, CT	Salt marsh	0.1 - 31.6
USA, MA and Cape Cod Bays	Coastal	0.03 - 6.7
USA, Pacific Coast[a]	Coastal	1.5 - 3.5
Canada, St. Lawrence River	River-Estuary	0.04 - 1.1
Australia, Brisbane River	Mangrove Swamp	<0.6 - 2.8
UK[b]	Estuary	0.07 - 4.13
Finland[c]	Stream	<0.65

[a] After Eisler (1996).
[b] After Bryan and Langston (1992).
[c] After Mukherjee and Lahermo (1995).

colloidal phase or in fixed forms with organic and mineral substances and is readily scavenged by suspended sediments (Sañudo-Wilhelmy and Flegal 1992; Smith and Flegal 1993).

Silver concentrations in surface waters vary greatly, depending on distances from the industry or discharged sources. Its level in waters near a point source discharge is between 0.8 and 100 ng l^{-1}, whereas far from a point source discharge it is 0.58 ng l^{-1} (Andren and Bober 2002).

Bottom sediments are principal sinks for Ag that may cause large environmental risk (Luoma et al. 1995; Purcell and Peters 1998). Concentrations of Ag in sediments vary significantly upon site and pollution source (Table II-11.10). Depending on several properties of the sediment, e.g., salinity, pH, hydroxides of Fe and Mn, and organic matter, Ag may become slightly mobile and remains for over 100 years (Wingert-Runge and Andren 1994). The speciation of Ag and geochemistry of bed-rocks also significantly control its behavior (Call et al. 1999).

Aquatic biota. At the beginning of the 1990s, research on the toxic effects of Ag in the aquatic environment had been initiated by the Silver Council and the Silver Council of the USA (Gorsuch and Klaine 1998). The bioaccumulation of Ag in marine food chains and its transfer to human foods are subjects of a great interest (Fisher and Rainfelder 1995; Fisher and Wang 1998).

Silver is a highly reactive element in the aquatic environment, however, only the free ionic form, Ag^+, is highly toxic at low concentrations. This cation can exist in waters for quite a short time due to reactions with inorganic and organic ligands (Lytle 1984; Purcell and Peters 1998). In seawater, Ag toxicity to fish is caused by the Ag^+ cation that attacks fish gills or other parts of fish, causing osmoregulatory failure; hence fishes appear to die from elevated plasma Na^+ and Cl^- concentrations and dehydration (Hogstrand and Wood 1998).

The toxicity of Ag species in seawater is still poorly understood. It is 10–100 times less toxic in marine water than in fresh water (Cappel R. pers. comm.; Wood and Hogstrand 1998; Wood et al. 1999). According to Andren and Bober (2002), the toxic concentration of Ag to fresh water organisms is between 0.85 and 1 542 $\mu g\, l^{-1}$ whereas to marine organisms it varies between 13.3 and 2 700 $\mu g\, l^{-1}$.

Silver accumulations by algae play a crucial role in its cycling in the aquatic environments and especially in its transfer to food chain. The bioaccumulation of Ag in marine food chains and its transfer to human foods are subjects of a great interest (Fisher and Reinfelder 1995; Fisher and Wang 1998).

Silver is relatively easily accumulated by aquatic biota that play a crucial role in its cycling in aquatic environments and especially in its transfer to the food chain (Fisher and Reinfield 1995). Silver concentrations in the zooplankton range from 10 to 20 $mg\, kg^{-1}$ AW and in the phytoplankton it averages 4 $mg\, kg^{-1}$ AW (Kabata-Pendias and Pendias 1999). The concentrations Ag in algae and periphyton grown in river water with $AgNO_3$ addition were 3 and 0.1–3.6 $mg\, kg^{-1}$, respectively (Forsythe et al. 1996; Di Toro et al. 1996). Mussels accumulate more Ag at low salinities than at higher salinities, due to decreased Ag influx and reduced filtration activity of mussels (Langston and Burt 1994). Its content in shells of mollusks varies depending on the area. In mollusks collected from the Baltic Sea, Ag varies in the range <0.01–4.0 $mg\, kg^{-1}$ WW, and from the coast of the United States, it ranges from 0.002 to 0.09 $mg\, kg^{-1}$ WW (Szefer 2002a). Silver content in organs of cutthroat trout (*Oncorhynchus mykiss*) in vicinities of urban areas was as follows (in $\mu M\, kg^{-1}$): bone, 26–44; liver, 14–23; and skin, 2–3 (Cain and Luoma 1990; Berthet et al. 1992). Silver is the most easily available to trout (Ratie 1999).

The toxicity of Ag depends upon chemical forms and availability of free Ag^+ cation (Wood et al. 1994). Silver has been stamped as a toxic element by the US EPA (1999) based on water hardness. The equation proposed by for the prediction of the Ag toxicity in aquatic systems in relation to water hardness has been widely discussed, because several other factors also have a significant impact on the Ag toxicity (Bills 1996; Purcell and Peters 1998; Karen et al. 1998).

II-11.2.5 Air

The world median Ag concentration has been estimated at 0.044 $ng\, m^{-3}$ (Reimann and Caritat 1998). In the USA, the background value of Ag in the atmosphere has been calculated to be <1 $ng\, m^{-3}$, whereas in urban areas it can be >10 $ng\, m^{-3}$. In industrialized cities of Europe (e.g., Heidelberg and Glasgow), mean Ag concentrations in the air were reported to range from 4.2 to 2.7 $ng\, m^{-3}$, respectively (McDonald and Duncan 1979). The background value of Ag concentrations measured in summer, autumn and winter at Ähtäri, Finland, were 0.014, 0.010, and 0.011 $ng\, m^{-3}$, respectively (Jalkanen and Häsänen 1994).

In the atmosphere, Ag occurs manly in forms of halides, sulfides, bromides or oxides (US EPA 1973). Cawse (1987) calculated the atmospheric deposition of Ag to be <0.7–1 g ha^{-1} yr^{-1} in rural regions of the UK. The wet Ag deposition in Sweden, during 1999, was 0.4 g ha^{-1} yr^{-1} (Eriksson 2001a).

Coal combustion may be a significant source of Ag. There is an estimation that about 50% of Ag occurring in coal (average value of Ag in coal: 0.5 mg kg^{-1}; range: 0.02–2.0 mg kg^{-1}) is released to the atmosphere from thermal power plants (Clarke and Sloss 1992; Mukherjee 1994).

II-11.2.6
Plants

Contents of Ag in plants vary greatly; the lowest is in cereals grains and the highest in grasses and fungi (Table II-11.11). Concentrations of Ag in plants may be highly elevated when grown in Ag-contaminated soils (Klein 1978; Hirsch 1998).

Roots of grasses and agricultural crops accumulate more Ag than other parts of the plant. Silver is apparently deposited in root tissues in the forms of Ag_2S and/or metallic, and therefore is excluded from metabolic processes (Kabata-Pendias and Pendias 2001). Grasses grown on Ag-contaminated soils accumulate large amounts of Ag in roots, up to 64 mg kg^{-1}, whereas above ground parts of grass contained only 1.2 mg kg^{-1} (Ratie 1999). Ward et al. (1979) showed that 90% of Ag in

Table II-11.11. Silver in plants and fungi from uncontaminated and contaminated sites (mg kg^{-1})

Plant	Range, mean	Reference
Unconntaminated sites		
Cereal grains	<0.0005 – 0.0027	Eriksson (2001a)
Cereal products	0.008 – 0.14	Cunningham and Stroube (1987)
Vegetables	0.007 – 0.039	Cunningham and Stroube (1987)
Fruit	<0.05	Cunningham and Stroube (1987)
Grasses	0.06 – 5.04	Cunningham and Stroube (1987)
Ag-contamianted sites		
Trees		Matuk *vide* Mukherjee (1994)
– Leaves	1	
– Twigs, 25 m above	13	
– Bark, 15 m above	5.3	
– Wood; 22 m above	2.4	
Fungi (various)[a]	119	Eisler (1996)
Mushrooms[b]	120 – 150	Falandysz et al. (1994a)

[a] Vicinity of AgI-seeding generator site.
[b] Soil with addition of Ag_2S.

rye grass and white clover was immobilized in the root system for 10 days, under laboratory study.

The resistance of plants to increased Ag contents varies greatly. Some crop plants were not affected when they were grown in the sludge amended soils containing Ag_2S at the concentration of 106 mg Ag kg^{-1} soil, whereas the yield of soybean plants grown in the same condition was decreased (Hirsch 1998). In most cases, no significant toxicity and growth reduction have been reported for higher plants. However, in sensitive plants, growth and germination may be reduced at the concentration of 7.5 mg $AgNO_3$ l^{-1} in nutrient solution (the most toxic Ag form to plants) (Ratie 1999).

II-11.2.7 Humans and Animals

Humans. Jørgensen (2000) reported that normal contents in human tissues vary (in mg kg^{-1}) between 0.06 and 0.08. The toxic level is 60, and the lethal level is 1 300. Similar Ag concentration, 0.04 mg kg^{-1}, in "reference man", is given by Li (2000). Human fluids contain Ag at following mean concentrations (in μg l^{-1}): blood, 0.37; serum, 0.18; and urine, 0.46 (Reimann and Caritat 1998).

The total (with food and water) daily intake of Ag by adults ranges from 35 to above 100 μg. Also inhalation of air, especially in urban areas, is a source of this metal (Table II-11.12). Chronic exposure of humans to Ag and its compounds has caused argyria (a slate-gray pigmentation of the skin and hair caused by Ag). It is rare and it occurs when Ag compounds enter into the broken skin or mucous membranes (Smith and Carson 1977a). However, it has been reported for workers engaged in Ag-industries: mining, smelting operation, mirror plating, soldering, drilling or silver nitrate manufacturing. The MAK-value at the workplace has been set at 0.1 mg Ag m^{-3} for metallic forms and 0.01 mg Ag m^{-3} for Ag salts (Doherty 2004).

However, some authors concluded that Ag exposure is not expected to cause any significant risk to human health (Eisler1996; Campbell et al. 2002). Argyria, a cosmetic pigmentation of the skin does not occur as a result of environmental exposure. Hence human beings seem to be excluded from the risk of increased Ag levels in the ecosystem.

Animals. There is no report on the toxicity of Ag to grazing animals. However, rats and mouse exposed to Ag deposits had higher Ag content in tissues and developed some neurological disorders and argyria, e.g., a bluish gray coloration of the skin (Walker 1971).

Table II-11.12. Silver exposure of adults from ingestion and inhalation (as compiled from various source by Andren and Bober 2002)

Root of exposure	Estimated intake (μg day^{-1})
Ingestion-diet	35 – 88
Drinking water	0.2 – 20[a]
Inhalation (urban area)	0.05 – 0.19[b]

[a] Ingestion of water = 2 l day^{-1}.
[b] Inhalation rate = 1.8 m^3 day^{-1}.

II-11.3 Gold (Au)

II-11.3.1 Introduction

The abundance of Au in the continental crust averages 3 μg kg^{-1} and its concentration in sedimentary rocks is usually lower than in magmatic rocks (Table II-11.2). It was one of the first metals to be recognized by humans and due to its chemical and physical properties, it became very important throughout human civilization.

Gold has a tendency to occur in the free state and tends to remain in this form due to its great chemical resistance. It occurs most commonly in the form of the native metal. Relatively common Au minerals are: calverite, $AuTe_2$; montbrayite, Au_2Te_3; petzite, Au,$AgTe_2$; and sylvanite, $AuAgTe_4$. It also may be a minor constituent of some types of ores.

There are several oxidation states of Au (Table II-11.1), however the occurrence of Au^{+}, Au^{2+}, Au^{3+}, and Au^{5+} in the environment or in biological systems has not been very often reported (Shaw 1991).

II-11.3.2 Production and Use

Accurate information on Au production is difficult to obtain, but it has been estimated that the global Au production has increased from 2 180 t in 1990 to 2 590 t in 2000 (USGS 2004). In Brazil and many other countries, Au production is often under-reported (Porvari 1995).

The oldest method of metallic Au exploration is washing out from sediments such as sands, sandstones, gravels, and conglomerates. Gold in residual sediments tends to collect in various bottom sediments as well as in some alluvial soils. The flotation method of Au extraction is based on the amalgamation with Hg and therefore creates a great environmental risk (Jernelöv and Ramel 1994). The use of Hg in the small-scale Au mining (ASM) industry is quite common in 62 countries, including, South America, USA, Canada, Tanzania, Indonesia, India, Vietnam, Zambia, Zimbabwe and many others (Veiga and Meech 1995; Veiga and Baker 2004).

Recently, Au became an important by-product from non-ferrous Cu and Pb ores. In a modern Au-recovery plant, it is extracted by the cyanide process wherein finely ground ore is agitated in a diluted solution of sodium cyanide and calcium hydroxide (Adams et al. 1999). These technologies create health risks because of contact with toxic chemicals. The inhalation of Hg at ASM method of Au mining affects Au-dealers and also surrounding inhabitants (Veiga and Baker 2004).

The bio-methods of dissolving Au from solution cause less risk and environmental problems for microorganisms and other aquatic biota (Ting et al. 1995; Ting and Mittal 1999; Niu and Volesky 1999, 2000). In 1990, the bacteria *Thiobacillus ferrooxidans* was used for the recovery of 30 t of Au from a complex ore in Chile, under laboratory conditions (Maturana et al. 1993).

Gold is widely used in jewelry, coinage, electroplating, electronics, dental alloys, and anti-arthritic drugs. It is also an essential metal in information processing, telecommunications, and the military and space industries. The biggest advantages of this metal are its resistance to oxidation and corrosion. Pure Au is soft and therefore is alloyed with Ag, Pd, Cu, Ni and Zn for industrial applications (Ishida 2004).

II-11.3.3 Soils

Common average content of Au in soils is between 1 and 3 μg kg^{-1} and usually is elevated in organic soils (Table II-11.2). The Au status in reference soils is in the range from 0.6 to 20 μg kg^{-1} (Govindaraju 1994). The background contents of Au in soils are proposed as 0.8–8 μg kg^{-1} (Roslanov *vide* Kabata-Pendias and Pendias 2001). Agricultural soils of Sweden contain Au at level <5 μg kg^{-1} (Eriksson 2001a).

Its distribution in soils depends on the origin of the soil material and the weathering stage and in most cases its content is inherited from parent rocks. In most soils, Au forms several complex ions such as $AuCl_2^{2-}$, $AuBr_4^-$, AuI_2^-, $Au(CN)_2^-$, $Au(CNS)_4^-$, $Au(Se_2O_3)_2^{3-}$ which are relatively mobile. Gold complexes such as $Au(OH)_3 \cdot H_2O^0$, $AuClOH^-$, and $Au(OH)_2FA$ are known to be formed in tropical lateritic soils. These complexes are very mobile and increase Au phytoavailability (Davies 1997).

Some alluvial soils, soils with postmining Au deposits, and soils amended with Au-enriched sludge may contain much higher Au levels that are reported to range from 43 to 473 mg kg^{-1} (Williams et al. 1995; McBride et al. 1997; Rashed and Awadallah 1998).

In rain forest lateritic soils, Au complexes may be easily mobilized and thus readily available. In the Au mining area of Sri Lanka, high concentration of Au in peat and algal mats was increasing with depth and correlated with Fe, Mn, Co, Zr, Na, Mg and K (Dissanayake and Kritsotakis 1984).

The global balance of the accumulation of Au in sewage streams is estimated at 360 t yr^{-1} (Eisler 2004). Sewage sludge may be a major source of Au when applied to soils. Elevated contents of Au have been reported as follows (in mg kg^{-1}):

- 0.18 – 2.35 soils at Au mining area in Australia (Lottermoser 1995)
- 0.5 – 4.5 sewage sludge, Germany (Eisler 2004)
- 0.5 – 3.0 sewage sludge, USA (Lottermoser 1995)

II-11.3.4 Waters

World median Au concentration in the oceans is estimated to be 4 ng l^{-1} (Reimann and Caritat 1998). Cidu et al. (1994) reported contents of Au in seawaters and river waters to vary from ≤0.03 to 119 ng l^{-1}, depending on sampling sites. Nozaki (2005) cited Au at the mean concentration of 0.02 ng kg^{-1} in the North Pacific Ocean. Gordeyev et al. (1997) reported that in 1989, surface water of NE Atlantic Ocean Au at the level of 2–7 ng kg^{-1}.

The speciation of Au in seawater may be $AuCl_2^-$, and in smaller amounts $AuClBr^-$. Several bromo-, iodo-, hydroxyl complexes of Au^+ also are likely to occur. In the bottom water samples of the mid-Atlantic ridge, Au was found in different oxidation states: Au^0, Au^+, and Au^{+3} (Karamushka and Gadd 1999). It is known that there are Au-rich sea-floor deposits in the SW Pacific Ocean and the sea-floor hydrothermal system of this region contains 28.7 mg Au kg^{-1} at an average of 3.1 mg Au kg^{-1} (Eisler 2004).

In marine sediments, Au is associated with other trace elements (e.g., As, Sb, and Pb). However, the correlations between these elements and Au are variable (Herzig et al. 1993). High Au concentrations in sediments around Soda Island in the Sea of Japan resulted from auriferous mineralization and mining activities (Terashima et al. 1991). According to these authors, Au was supplied to the bottom sediments by river and seawaters, as dissolved forms and as minerals.

Aquatic biota. Some aquatic biota (e.g., brown algae) may accumulate considerable amounts of Au. Occasionally very high concentrations of Au in marine species have been reported near the mining and smelting areas. Gold concentration in the soft parts of Common mussel (*Mytilus edulis*) was cited to be 2–38 µg kg^{-1}, whereas in fish, mackerel, (*Pneumatophorous japonicus*) it was 0.12 µg kg^{-1} (Eisler 1981).

II-11.3.5 Air

Concentrations of Au in air from remote regions vary from 0.08 to 1.5 pg m^{-3}, whereas at polluted sites its content can reach 300 pg m^{-3} (Reimann and Caritat 1998). In the 1970s, Au contents in Great Britain were reported to range from 5 to 10 pg m^{-3} and its deposition was estimated between <0.07 and <0.2 g ha^{-1} yr^{-1} (Cawse 1987). Its wet deposition in Sweden has been calculated at 0.07 g ha^{-1} yr^{-1} (Eriksson 2001a).

Atmospheric particulates contain relatively increased amounts of Au when compared with mean crust abundance. Volcanoes can emit large amounts of Au in gaseous and aerosols forms and it has been estimated that Mt. Etna, Italy emitted 880 kg yr^{-1} and Mt. St. Helen, USA 1 100 kg yr^{-1} of Au (Reimann and Caritat 1998).

Elevated contents of Au, in the range from 224 to 7 450 mg kg^{-1}, in lichens collected from the vicinity of an Au mine in New Zealand indicated aerial pollution (Williams et al. 1995). Gold content in lichens collected from nearby Au mine accumulate from 0.22 to 1.45 mg Au kg^{-1} whereas the average background value in this area is 0.12 mg kg^{-1} (Limbong et al. 2003).

II-11.3.6 Plants

Plants grown in uncontaminated sites usually contain Au in the range 0.5–10 µg kg^{-1}, and in most cases it is concentrated more in roots than in above-ground parts. Greger (1999) calculated the content of Au in the reference plant to be 1 µg kg^{-1}. Accumulator plants (e.g., *Artemista persia*, *Prangos popularia*, *Stripa* sp.) may contain Au at levels of 0.1–100 mg kg^{-1} (Table II-11.13). Cyanogenic plants and some deciduous trees can accumulate above 10 mg Au kg^{-1}.

In most cases, microorganisms in the plant rhizosphere are responsible for the mobility and the phytoavailability of this metal (Eisler 2004). However, this process is

Table II-11.13. Gold in plants growing in various sites (μg kg^{-1})

Plant, site, country	Mean or range	Reference
Various plants		Davies (1997)
– Uncontamined sites	<5.0	
– Gold mining area	3 – 19	
Faba bean, contaminated sites		Rashed and Awadallah (1998)
– Leaves	170	
– Stems	50	
– Pods	40	
– Pericarp	36	
– Testa, seeds, cotyledon	<7	
Various plants, mineralized areas	0.7 – 6.5	Girling et al. (1979)
Horsetail species, Alaska, USA	0.1 – 0.5 (FW)	Cannon et al. (1968)
Poplar, roots, Germany	2 – 28	Messerschmidt et al. (2004)
Accumulator plants	>100 – 100000	Sadler (1976)

still not fully understood. The bioaccumulation of Au is also observed in stems and needles of Corsian pine tree (*Pinus laricio*) and in other plants growing in soils with Au levels of 1–25 mg kg^{-1} (Pyatt 1999). Higher concentrations of Au in seeds than in leaves and stems have been reported for planted fenugreek and lupin (both of Leguminosae family).

The concentration Au in plants and shrubs grown in various place has been used extensively for the biogeochemical survey for Au mineralized areas, especially in Russia (Vernadsky 1998). Some plants (e.g., *Sedum aceotatum*, *Phaselia servicea*, *Artemisia* sp.) have been used successfully for biogeochemical prospecting (Girling et al. 1979). Promising results were reported also for the use of Au hyperaccumulation for phytomining (Anderson et al. 1999).

Some microfungi (e.g., ectomycorrhizal and terrestrial saprobes) are reported to accumulate Au between 20 and 30 μg kg^{-1} (Borovička et al. 2005). According to these authors, Au levels in *Boletus edulis* (ectomycorrhizal sp.) may be up to 235 μg kg^{-1}. Increased Au concentrations in microfungi are controlled by several factors of which humidity and pH are of real importance.

II-11.3.7 Humans

Jørgensen (2000) gave the Au content in mammalian tissues in the range <0.1 to about 500 μg kg^{-1}, being the lowest in liver and the highest in kidneys and brain. In the human liver, Au concentration averages 0.72 μg kg^{-1} that is more than the concentration in pig liver, 0.43 μg kg^{-1} (Chen et al. *vide* Ishida 2004).

Gold concentrations in human fluids, as compiled by Ishida (2004), have been reported to range (in μg l^{-1}): from 0.125 to 0.413 in blood and from 0.02 to 0.13 in urine.

Mean Au concentration in human urine is given at 0.07 $\mu g\,l^{-1}$ (Reimann and Caritat 1998). It may vary, however, upon people's occupations. Students contain Au at mean level (in $\mu g\,l^{-1}$) 0.03, construction workers 0.03, and dental technicians 0.19 (Begerow et al. 1999).

Content of Au in human hard tissues may be of diagnostic potential. It ranges (in $\mu g\,kg^{-1}$) in fingernails of adults of Niger from 8–39 (mean 20) and in hair from 6 to 880 (mean 47) (Oluwole et al. 1994). Hairs of goldsmiths contain 1 440 μg Au kg^{-1} (Caroli et al. 1998). People dealing in articles of Au have higher Au contents in urine than students or road construction workers (Begerow et al. 1999). Gold concentration in hair of goldsmiths in Italy has been reported to be 1 400 $\mu g\,kg^{-1}$, and healthy adults of Nigeria to range between 6 and 800 $\mu g\,kg^{-1}$ (Oluwole et al. 1994). It has also been reported that the population of Uzbekistan has high Au concentration in blood compared to the rest of the world. This has resulted in more frequent hypertension and anemia among this population than among others people (Eisler 2004; Zhuk et al. 1994).

An estimate of dietary Au intake by adults is $\leq 7\ \mu g\,d^{-1}$. However the absorption of Au compounds from the alimentary tract and skin is very poor (Ishida 2004). Gold poisoning is not common but it may occur, as a result of therapeutic overdose or as a side effect of chrysotherapy, e.g., the use of chemicals containing Au for treatment, especially rheumathoidal diseases (Ishida 2004). However, due to unique Au properties, its safe application has been documented in medicine. There are observations that in cancerous tissues and parkinsonian brain, the Au content may be elevated (and Zecca et al. *vide* Ishida 2004).

Ishida (2004) reported that some Au complexes, especially of Au^{3+}, may be toxic to mammals due to the rapid reduction to Au^{+} that is involved in the metabolism of sulfide proteins.

Trace Elements of Group 12 (Previously Group IIb)

The Group 12 consists of zinc (Zn), cadmium (Cd), and mercury (Hg). These metals have quite a low abundance in the Earth's crust. These metals form compounds in which their oxidation states are usually not higher than +2 and easily form metal-metal ($^+M{-}M^+$) bonds (Table II-12.1). The strength of the bond increases down the group, in the following order: Hg < Cd < Zn. The Zn_2^{2+} and Cd_2^{2+} ions are highly unstable, however, the +1 state of Hg is quite stable compared with the other two elements. The toxicity of Cd and Hg is well known, whereas Zn has enormous biological importance.

II-12.1 Zinc (Zn)

II-12.1.1 Introduction

Zinc has been known as the metal since the Middle Ages, but industrial extraction and Zn refining only began in Europe in the late 18th Century.

Zinc is quite uniformly distributed in magmatic rocks, whereas in sedimentary rocks is likely to be concentrated in argillaceous sediments (Table II-12.2). It is very mobile during weathering processes and its easily soluble compounds are readily precipitated by reaction with carbonates, or it is sorbed by minerals and organic compounds, especially at the presence of sulfur anions. Some of the most common compounds are zinc chloride, $ZnCl_2$; zinc oxide, ZnO; zinc sulfate, $ZnSO_4$; and zinc

Table II-12.1. Selected properties of trace elements of the Group 12

Element	Atomic number	Atomic mass	Atomic[a] radius (pm)	Density (20 °C, g cm^{-3})	Valence[b]	Melting point (°C)
Zn, zinc	30	65.38	153	7.13	**+2**	419.55
Cd, cadmium	48	112.41	171	8.65	**+2**	321.05
Hg, mercury	80	200.59	176	13.55	+1, **+2**, +3	−38.85

[a] Approximately average values for the main oxidation states.
[b] Valence values in bold are for the main oxidation state.

Zn

Table II-12.2. Abundance of trace metals of Group 12 in the environment[a]

Environmental compartment	Zn	Cd	Hg
Earth's crust (mg kg^{-1})	52 – 80	0.1 – 0.2	0.02 – 0.09
Ingeous rocks (mg kg^{-1})			
Mafic	40 – 120	0.03 – 0.22	0.004 – 0.01
Acid	40 – 100	0.05 – 0.2	0.03 – 0.08
Sedimentary rocks (mg kg^{-1})			
Argillaceous	80 – 120	0.2 – 0.3	0.2 – 0.4
Sandstones	15 – 30	<0.04 – 0.05	0.01 – 0.05
Calcareous	10 – 25	0.04 – 0.1	0.02 – 0.05
Raw energy material (mg kg^{-1})			
Coal, hard[b]	52 – 19000	0.47 – 170	0.17 – 10
Fly ash[c]	194	1.3	0.01
Crude oil	0.5 – 85	0.003 – <1	0.002
Petrol	<0.2 – 3	>0.001	<0.02
Soils (mg kg^{-1})	63	0.5	0.07
Light sandy	31 – 61	0.01 – 0.2	0.08 – 0.7
Medium loamy	47 – 61	0.08 – 0.3	0.01 – 1.2
Heavy loamy	35 – 75	0.2 – 0.5	0.02 – 1.5
Calcareous	50 – 100	0.4 – 0.8	0.01 – 0.5
Organic	57 – 100	0.2 – 2.5	0.04 – 1.12
Waters (μg l^{-1})			
Snow	2.6 – 3	0.2 – 0.7	–
Rain	1.2 – 6.6	<0.01 – 0.05	<0.02[e]
Ocean, sea	0.04 – 4.9	0.07 – 0.11	0.03 – 0.05
River	3.3 – 10.3	0.03 – 0.05	0.005 – 0.01
Air[d] (ng m^{-3})	0.03 – 15	0.2 – <0.4	0.01 – 0.06
Urban/industrial	550 – 16000	0.5 – 620	0.17 – 11.2
Rural	10 – 200	0.1 – 4	0.01
South Pole	0.002 – 0.05	0.015	–
Greenland	18 – 41	0.003 – 0.6	0.04 – 0.08

[a] Presented are common, possible background, values from various sources as compiled by Kabata-Pendias and Pendias (1999, 2001), unless otherwise indicated.
[b] Finkelman (1999).
[c] Llorens et al. (2000).
[d] Reimann and Caritat (1998).
[e] Eriksson (2001a).

sulfide, ZnS. This means that the metal has the potential for forming a variety of compounds with organic and inorganic groups. Most Zn minerals contain above 50% of pure metal (Table II-12.3).

Table II-12.3.
Zinc contents in common Zn minerals (adopted from Landner and Lindeström 1998)

Mineral	Chemical composition	Zn (%)
Sphalerite	ZnS	67.0
Smithsonite	$ZnCO_3$	52.0
Hemimorphite	$Zn_4Si_2O_7(OH)_2 \cdot H_2O$	54.2
Zinc bloom	$Zn_5(OH)_6(CO_3)_2$	56.0
Zincite	ZnO	80.3
Willemite	Zn_2SiO_4	58.5

II-12.1.2 Production and Use

Global production of Zn in 2000 was 8.73 Mt and in 2003 was 9.17 Mt (WMSY 2004; USGS 2004). Compared with the production in 1998, it has increased by 15%. The principal ores of Zn are composed of sphalerite and wutzite, ZnS, of the same chemical formula but with various crystallographic structures, cubic and hexagonal, respectively. Relatively common is also smithonite, $ZnCO_3$. Zinc ores contain different elements such as Pb, Cu, Ag and Cd of which Cd is most closely associated with Zn.

Zinc is used in many industries, mainly as corrosion protection on steel components and other metals. It is an important component of various alloys and is widely used as catalyst in different chemical production (e.g., rubber, pigments, plastic, lubricants, and pesticides). Due to its versatile properties, its use has been documented in different sectors such as batteries, automotive equipment, pipes and household devices. Different compounds of Zn have dental and medical applications.

II-12.1.3 Soils

Zinc is unevenly distributed in soils and its concentration ranges between 10 to 300 mg kg^{-1}, with a mean of about 50 mg kg^{-1} (Malle 1992). Mean Zn for worldwide soils is calculated as 64 mg kg^{-1} (Table II-12.2). The highest background Zn contents are observed in calcareous soils and organic soils. Mean contents of Zn in soils reported for several European countries vary between 7 and 89 mg kg^{-1}, being the lowest for Denmark and the highest for Italy (Angelone and Bini 1992; Kabata-Pendias and Pendias 2001). Agricultural practices are known to increase Zn contents of surface soils. Japanese agricultural soils contain Zn in the range of mean values from 59 to 99 mg kg^{-1}, with the lowest value for sandy Acrisols and the highest for Andosols (Takeda et al. 2004). Zinc contents in agricultural soils of Sweden ranges from 6 to 152 mg kg^{-1}, with mean value of 65 mg kg^{-1} (Eriksson 2001a). In the USA, the geometric mean of Zn in cultivated soils was reported to be 43 mg kg^{-1}, in the range of $<$3–264 mg kg^{-1} (Holmgren et al. 1993).

Although Zn is very mobile in most soils, clay fractions and SOM are capable of holding Zn quite strongly, especially at neutral and alkaline pH regimes (Kabata-Pendias and Pendias 2001; Peganova and Edler 2004; Shtangeeva et al. 2005b). It has been calculated that the clay fraction controls up to about 60% of Zn distributed in soils (Kabata-Pendias and Krakowiak 1995). Apparently, there are two different mechanisms of Zn sorption: one in acid media related to cation exchange sites, and the other in alkaline media that is considered to be the chemisorption and is highly influenced by organic ligands. Also hydroxides of Al, Fe, and Mn appear to be of importance in binding Zn in some soils.

The behavior and phytoavailability of Zn are controlled by several soil parameters and they vary greatly in different soils (Mukherjee and Hartikainen 2002). Silicates, carbonates, phosphates, oxides and organic matter may contribute to the metal retention in soils (McBride et al. 1997). In the soil solution, Zn occurs in forms of free and complexed ions; as cations: Zn^{2+}, $ZnCl^+$, and $ZnOH^+$, $ZnHCO_3^+$, and as anions: ZnO_2^{2-}, $Zn(OH)_3^-$, and $ZnCl_3^-$ (Kabata-Pendias and Sadurski 2004). According to Tyler and Olsson (2002) the concentration of Zn in cultivated soil solution was 78 (12–223) μg l^{-1}, whereas in natural soil it was 35 (13–72) μg g^{-1}. However, depending on techniques used for obtaining soil solution its concentrations vary between 21 and 570 μg l^{-1} (Kabata-Pendias and Pendias 2001).

Main Zn sources are related to nonferric metal industry and to agricultural practices. Some fertilizers and in particular super phosphate can significantly contribute to Zn levels in soils (Table II-12.4).

It is a very difficult task to estimate Zn pollution in soils. Anthropogenic Zn in forest soils is mostly of industrial origin, whereas in agricultural soils it may originate from numerous other sources. The fate of the Zn from various sources (e.g., atmospheric deposition, fertilizers, pesticides, sewage sludge, leaching from galvanized materials, manure, waste, slag and ashes) differs depending upon its chemical species and their affinity to soil and soil parameters (Kabata-Pendias et al.1992; Ross 1994).

Near to nonferric smelters and contaminated sites, levels of Zn along with other metals may be quite high, from 443 to 1 112 mg kg^{-1} (ATSDR 1994). High levels of Zn in soils around mining areas have been reported by Angelone and Bini (1992) for different countries, as follows (in mg kg^{-1}): the Netherlands, 1 020; Austria, 8 900; and Greece, 10 547.

In a recent report (EC 2003), the average concentration of Zn in sludge for the EU countries was reported at 811 mg kg^{-1}. The lowest concentration of Zn in sludge occurred in the Netherlands (131 mg kg^{-1}, in 2000) whereas Portugal had relatively high

Table II-12.4. Trace elements of Group 12 in fertilizers and slurry[a] (mg kg^{-1})

Element	NPK fertilizer	Ammonitrate fertilizer	Super phosphate triple	Pig slurry	Cow slurry	Poultry slurry
Zn	61 – 200	<2	141 – 625	919	580	495
Cd	3 – 24	<1	43 – 53	0.5	0.4	–
Hg	–	–	–	0.05	0.17	–

[a] Data for six members of the EU: Belgium, Germany, France, the Netherlands, Sweden and the United Kingdom (Etat de l'environment wallon 1995).

Table II-12.5.
Maximum allowable loading of metals of Group 12 (kg $ha^{-1}yr^{-1}$) (after Kabata-Pendias and Sadurski 2004)

Element	USA	EU-15	Poland
Zn	140	30	10
Cd	1.9	0.15	0.2
Hg	0.85	–	–

Zn concentration (1 670 mg kg^{-1}, in 2000). In some countries (e.g., Sweden), public opinion and farmer's attitudes are quite important regarding the use of sludge in agricultural soils. Regulations and loading rates for trace metals exist for the developed world, although variation is quite noticeable (Table II-12.5). Recently, the report of AROMIS has summarized that the input of Cd, Cu, Pb and Zn in 2000, in some countries (UK, CH, NL and DE) was lowest in the UK (454 g ha^{-1} yr^{-1}) and the highest (1 468 g ha^{-1} yr^{-1}) in the Netherlands (Eckel et al. 2005). Apparently, the highest Zn value for the Netherlands is due to considerable high Zn input from corrosion, dredging and hunting. Zinc loads, with various materials used for soil amendments vary in the following order (in g Zn ha^{-1}): compost, 1 190; pig manure (solid), 1 080; triplesuperphosphate, 54 (Döhler et al. 2002).

The Zn balance in surface soils has been summarized for different ecosystems and it has been observed that this metal exceeds its output due to both leaching and the production of biomass. However, in most ecosystems the anthropogenic source of Zn may highly influence its elevated contents, especially in surface soils. The Zn balance in crop-farm soils vary (in g ha^{-1} yr^{-1}) from –115 in Germany to 838 in France (Eckel et al. 2005).

II-12.1.4
Waters

World average of Zn in river waters has been calculated at 0.6 μg l^{-1} and its riverine flux at 23 kt yr^{-1} (Gaillardet et al. 2003). The median Zn has been estimated for world ocean water from 0.5 to 4.9 μg l^{-1} (Reimann and Caritat 1998). However, Zn concentrations in open oceans, lakes and rivers vary considerably (Table II-12.6). In Europe, water quality has been studied for 550 rivers and of these 176 stations were analyzed for Zn. At 44 stations, samples contained 5 μg l^{-1} whereas in other stations, the Zn concentration was 10 μg l^{-1} (EEA 1995).

In lakes, rivers and estuaries, Zn is bound by hydroxides, clay minerals, and other sediment materials. Zinc is present in aquatic systems in the divalent state and the chemical speciation of Zn^{2+}, by analytical-chemical methods remains indispensable (Table II-12.7). The speciation of Zn varies with pH; between pH 4 and 7, Zn exists in freshwater as aqua ion, at pH 7, the dominant forms of Zn are the free ion (98%) and $ZnSO_4$ (2%), and at pH 9, the main species of Zn are the monohydrooxide ion (78%), $ZnCO_3$ (16%) and the free ion 6% (Witczak and Adamczyk 1995).

Zinc may enter into waters from numerous sources including mine drainage, industrial and municipal wastes, urban runoff, but the largest input occurs from the erosion of soil particles containing Zn (US EPA 1980). Due to the strict regulation, in recent years (1987–1997) Zn discharge to the river Rhine from industrial and

Table II-12.6. Zinc concentrations in different water systems (μg l^{-1})

Sites	Mean and range	References
Open ocean	0.002 – 0.1	US EPA (1987)
Baltic Sea	1.4 – 5.4	Szefer (2002)
North Pacific	0.35	Nozaki (2005a)
Coastal/inland sea	0.5 – 1.0	van Assche et al. (1996)
Low land rivers, Europe	5 – 40	van Assche et al. (1996); EEA (1995)
US Rocky Mountain stream	10	van Assche et al. (1996)
Great Lakes, USA	0.09 – 0.3	van Assche et al. (1996)
Drinking water	<0.0002	US EPA (1980)
Water after passing through corrosive pipe	0.001 – 0.005	Hoell et al. (1986)

Table II-12.7. Chemical forms of dissolved zinc in surface waters (after Florence 1980)

Form	Compound	Diameter (nm)
Particultae matter	Not retained by 0.45 μm filter	>450
Hydrated metal ion	$Zn(H_2O)_6^{2+}$	0.8
Labile inorganic complexes	$Zn(H_2O)_5Cl^+$, $Zn(H_2O)_5OH^+$	1
Labile organic complexes	Zn-citrate, Zn-glycinate	1 – 2
Stable inorganic complexes	ZnS, $ZnCO_3$, Zn_2SiO_4	1 – 2
Stable organic complexes	Zn-humate, Zn-cysteinate	2 – 4
Adsorbed on inorganic colloids	Zn^{2+}-Fe_2O_3, Zn^{2+}-SiO_4	10 – 500
Adsorbed on organic colloids	Zn^{2+}-humic acid, Zn^{2+}-organic detritus	10 – 500

wastewater treatment plants has decreased by 20.4% and 35.0%, respectively. The largest (28%) Zn discharge to the aquatic environment in the European Communities occurs from the manufacturing of basic industrial chemicals. Matschullat (1997) estimated the total load of Zn to Baltic Sea at 11 000 t yr^{-1} of which 81% is of anthropogenic source. During the flood of the Odra River in 1997, the load of Zn into the Baltic Sea was calculated at 168 t yr^{-1}, of which about 70% was of geogenic source (Helios-Rybicka et al. 1998).

Zinc is easily accumulated in bottom sediments. Its background contents in sediments of rivers are estimated at 110 and 115 mg kg^{-1} for Vistula and Rhein River, respectively. However, in polluted rivers (or parts of rivers flow) its contents may reach up to 2 000 and 14 000 mg kg^{-1}, as reported for these two rivers (Helios-Rybicka et al. 1998; Leńczowska-Baranek *vide* Kabata-Pendias and Pendias 1999).

Aquatic biota. Contents of Zn in aquatic organisms vary greatly and are influenced mainly by sites of sampling. In the Baltic Sea, fish contain Zn in the range of 1.4–48 mg kg^{-1} FW, soft tissues of mussels contain from 50 to 600 mg kg^{-1}, and plankton, from 11 to 673 mg kg^{-1} (Szefer 2002a). Also the rhizosphere of aquatic plants can adsorb relatively high amounts of Zn in root tissues (Séguin et al. 2004).

Zinc in waters is not very toxic to the biota, however, concentrations above 240 µg l^{-1} may have adverse effects on some sensitive organisms, as for example salmons (Weis and Weis 1991). Depending on the hardness of fresh water, the limit for safe Zn concentration is established at 180–570 µg l^{-1}, for soft and very hard water, respectively (Peganova and Eder 2004).

II-12.1.5
Air

Generally, levels of Zn in the air are low and fairly constant. Median Zn concentrations in world air are estimated at 7 and 900 ng m^{-3} for remote and polluted sites, respectively (Reimann and Caritat 1998). Zinc concentration in rural air varies between 10–200 ng m^{-3} whereas in urban air, it can reach 16 000 ng m^{-3} (Table II-12.2). Aerosol samples over the Southern Bight of the Northern Sea collected between September 1988 and October 1989 averaged 67 ng m^{-3}, in the range of 3–220 ng m^{-3} (Barrie and Hoff 1985; Injuk et al. 1992).

Zinc enters into the atmosphere in the vapor as well as particulate forms from industrial processes, incineration of waste, cement plant, and from fuel fired power plants. Combustion temperature, process parameters, composition of raw materials and emission control technologies influence the load of Zn in the atmosphere.

In the Netherlands, 3.8 kt yr^{-1} of Zn was deposited to the soil from diffuse sources in the 1980s (RIZA 1989). There are data that in the Netherlands, 575 t yr^{-1} entered directly the soil and groundwater.

Considering the dry and wet deposition, the later process dominates supplying by 60–90% of Zn deposition over the soil (Pattenden et al. 1982). There is no recent information on the total global deposition of Zn in soil. Previously, Nriagu and Pacyna (1988) estimated global deposition of Zn in soils, for the 1980s, to vary between 683 and 1 954 kt. The largest quantity stemmed from coal fly ash and bottom ash generated during the coal combustion process for heat and electricity.

II-12.1.6
Plants

Although the Zn contents of plants vary considerably, reflecting impact of different factors of the various ecosystems and of genotypes, in certain food and fodder plants its content is fairly similar (Siegert et al. 1986). In cereal grains from worldwide countries, the mean Zn concentrations vary between 18 and 33 mg kg^{-1}, being the lowest in rice and the highest in oats (Table II-12.8). Elevated Zn contents are observed only in cassava grain from Nigeria and in rice from industrial regions of India. In vegetables, Zn also does not show any accumulation in a specific plant, but is relatively very low

Zn

Table II-12.8.
Zinc concentration in cereals from various countries[a] (mg kg^{-1})

Cereal	Range[b]	Mean[c]
Wheat	23 – 37	24
Barley	20 – 30	26
Oats	29 – 37	33
Rye	29 – 31	28
Rice	1 – 41	18
Rice[d]	26 – 33	–
Cassava (Nigeria)	–	129

[a] Sources: Kabata-Pendias and Pendias (2001), Puwastien et al. (2000), Chon and Lee (2004), Chukwuma (1995), and Eriksson (2001).
[b] Range of mean values as given for different countries.
[c] Mean of range values.
[d] Rice grown in India at 21 km of and near mining area, respectively.

Table II-12.9.
Zinc in food plants[a] (mg kg^{-1})

Plant	Range	Mean
Sweet corn, grains	25 – 36	–
Been, pods	32 – 38	–
Cabbage, leaves	24 – 31	27
Lettuce, leaves	44 – 73	–
Carrot, roots	21 – 27	24
Beet, roots	28 – 46	–
Onion, bulbs	22 – 32	25
Potato, tubers	10 – 26	17
Tomato, fruits	17 – 26	–
Apple, fruits[b]	–	1.2
Banana, fruits[b]	–	2.8

[a] Presented are common, possible background, values from various sources as compiled by Kabata-Pendias and Pendias (1999, 2001), unless otherwise indicated.
[b] Peganova and Eder (2004).

in fruits (Table II-12.9). Zinc in grains is bound strongly, which is attributed to high phytate content, as well as to high level of fiber, hemicelluloses, and amino acid-carbohydrate complexes and this causes decrease in its availability to animals (Case and Hambidge 1980). In grasses its mean contents vary from 25 to 47 mg kg^{-1} and is fairly similar to those in clover (Table II-12.10). Studies by Eisler (1997) indicated that Zn concentration in plant parts follows the pattern: roots > foliage > branch > trunk. Small lateral roots retain more Zn than other vegetation components.

Table II-12.10. Zinc in grass and clover from different countries (mg kg^{-1}) (values from various sources as complied by Kabata-Pendias and Pendias 1999, 2001)

Country	Grasses		Clovers	
	Range	Mean	Range	Mean
Bulgaria	24 – 50	34	–	–
Czech Republic	15 – 35	25	–	–
Germany	15 – 80	47	20 – 50	24
Great Britain	22 – 54	33	–	–
Hungary	21 – 26	27	30 – 126	39
Japan	18 – 36	28	23 – 44	34
New Zealand	16 – 45	28	20 – 49	27
Poland	12 – 72	30	16 – 86	37

Zinc plays essential metabolic roles in plants, being an active component of a variety of enzymes, such as dehydrogenase, proteinases, peptidases, and phosphohyrolases. Thus, the basic Zn functions are related to the metabolism of carbohydrates, proteins, and phosphate and also to auxins, RNA, and ribomosome formations. It influences the permeability of membranes and stabilizes cellular components. There are evidences that Zn stimulates the resistance of plants to dry and hot weather and also to bacterial and fungal diseases. Von Wiren et al. (1996) studied Zn^{2+}-phytometallophore complexes and observed that these complexes were readily absorbed by maize roots where the maize root-cell plasma membrane binding sites were not highly specific for the Fe^{3+}-phytometallophores.

Zinc deficiency in plants is generally observed when the plant contains less than 20 mg kg^{-1} and toxic effects are expected when the concentration exceeds 300–400 mg kg^{-1} (Boawn and Rasmussen 1971; Vitosh et al. 1994). These values may, however, vary significantly because: "the Zn deficiency reflects both the requirements of each genotype and effects of the interactions of Zn with other elements within the plant tissues" (Kabata-Pendias and Pendias 2001). Due to Zn deficiencies, large crop loses have been cited for citrus grown in California, peach trees in Texas and for various crops in Australia. Zinc is not translocated readily in plants; deficiency occurs generally first in younger leaves. Due to the Zn deficiency, some plants remain short and underdeveloped which is the result of an inadequate supply of the growth hormone, indole acetic acid.

Zinc toxicity and Zn tolerance in plants have been of concern because the prolongated use of Zn fertilizers, as well as its input from sewage and from other pollution sources, that enhances Zn content of surface soils. Zn-phytotoxicity is reported relatively often, especially for acid and heavily sludged soils. The toxicity limit for Zn depends on the plants species and genotypes, as well as on the growth stage. Sensitive terrestrial plants die when soil Zn concentration exceeds 100 mg kg^{-1} and photosynthesis is stopped when the content is more than 178 mg Zn kg^{-1} (Eisler 1993). Some species are known to hyperaccumulate Zn, as for example *Thlaspi* species that can contain above 10 000 mg Zn kg^{-1} and is applied for the phytoremediation of contaminated soil (Greger 1999). These plants can extract of

Zn from a soil in the range of 14–444 kg ha^{-1}, depending on species and soil parameters (McGrath 1998).

The mobility of Zn within plant highly varies depending on species and Zn nutrition status. In most cases, however, Zn is likely to concentrate in mature leaves and in roots. The rhizosphere plays a significant role in the concentration of Zn in roots. Séguin et al. (2004) found water extracted Zn from inner, and outer rhizosphere and bulk saturated soils to be (in mg kg^{-1}): 3.77, 1.83 and 1.0, respectively. This indicates increased content of Zn in rhizosphere as compared with its content in surrounding soil. However, other authors reported that the concentration of water-soluble Zn in the rhizosphere is lower relative to that of bulk soils (Sarong et al. 1989). Nevertheless, more opinions suggest that the rhizosphere contains elevated amounts of metals that are complexed with organic matter (Gobran and Clegg 1996; Courchesne et al. 2001).

Plants cultivated close to metal smelters may contain elevated amounts of Zn. Its contents in seeds of oil crops grown in the vicinity of Non-Ferrous Metal Works near Plovdiv, Bulgaria varied from 13.9 to 66.9 mg kg^{-1} whereas in crops at 15 km distance from the Works it was at the range of 8.7–33.6 mg kg^{-1} (Angelova et al. 2004).

II-12.1.7
Humans and Animals

Humans. Contents of Zn in mammalian tissues vary between 13 and 210 mg kg^{-1}, being the highest in kidneys and liver and the lowest in skin (Jørgensen 2000). Its concentrations in soft tissues of humans range from 10 to 57 mg kg^{-1}, in lung and liver, respectively. The average content in tissues of the "reference man" is 33 mg kg^{-1} (Li 2000). Its mean concentrations in human fluids have been reported (in mg l^{-1}) as: 6.3 in blood, 0.9 in serum, and 0.5 in urine (Reimann and Caritat 1998). The highest Zn accumulation (about 130 mg kg^{-1} FW) was found in prostate gland and retina (Peganova and Eder 2004).

The serum Zn ranged between 0.39 and 1.5 mg l^{-1} with a mean value of 0.77 mg l^{-1}. The difference in the Zn blood value between men and women was quite narrow (Kabata-Pendias and Pendias 1999). Schuhmacher et al. (1994), however, observed higher Zn values for females, although many other authors found no correlation of serum Zn between the sexes (Songchitsomboom and Komindr 1996).

Zinc plays several fundamental functions in metabolic processes, mainly of proteins, carbohydrates and DNA (Prasad 1980; Apgar and Everell 1988). It is essential for growth, development, and reproduction. Its intake from food varies highly and depends on several factors, but often on intereactions between other metals. Important antagonistic relationship is between Zn-Cd and Zn-Cu. Also increased levels of Ca and Mg in food inhibit its availability (Kabata-Pendias and Pendias 1999).

Dietary reference values for Zn intake (in mg d^{-1}) as established by the WHO in 1996 (*vide* Peganova and Eder 2004) are as follows: 5.5–7.5 for children (1–10 years), 12–13 for children (11–18 years), 9.5 for adult man, and 6.5 for adult female.

Meat products are a better source of Zn than plants. Meat contains Zn at the level of 19 to 90 mg kg^{-1}, in pork and pork-liver, respectively. The highest Zn concentration (1 430 mg kg^{-1}) is in oysters (Peganova and Eder 2004).

Zinc is relatively slightly toxic and therefore its deficiency occurs most commonly. Most serious symptoms of its severe deficiency are according to Maher (*vide* Peganova and Eder 2004):

- Frequent infections
- Diarrhea
- Compromised immune function
- Alopecia
- Delayed sexual and bone maturation
- Mental disturbance

In addition, several other dysfunctions like impaired wound healing and skin lesions, enlarged spleen and liver, impaired taste and smell, and night blindness, may occur. Schlegel-Zawadzka et al. (2000) observed a negative correlation between depression symptoms and Zn level in the serum. Zinc deficiency also cause growth retardation and various blood disorders. At risk of the Zn deficiency are mainly infants and children, but also older adults, non-healthy adults, and some vegetarians.

The ingestion or inhalation of larger doses of Zn, especially in forms of inorganic compounds, can be harmful to individuals. It can damage alimentary tracts, and affect diarrhea and fever. A short-term illness called *metal fume fever* can result if workers breath very high levels of zinc dust or fumes at the working place. It may cause chills, fever, excessive sweating and weakness.

Animals. Zinc is very essential for to animals and because it is not stored in the body; it must be supplied continuously with the diet. The required amounts of Zn vary highly depending on various parameters (species, age, health condition) and in the most cases its supply is sufficient with normal feeding. Thus, in farm animals, few cases of Zn deficiency have been reported. Symptoms of Zn deficiency, such as decreased growth, testicular atrophy, alopecia, and dermal lesions have been observed mainly in young animals.

Farm animals are rather tolerant to high Zn levels in the diet. However, animals grazing on forage in the vicinity of Zn smelters may suffer from Zn toxicity. Horses are the most sensitive to the Zn excess which symptoms are: lameness, osteoporosis, and lymphoid hyperplasia of spleen and lymph node (Peganova and Eder 2004). However, some other pathological changes of different organs may also be observed. Near a zinc and lead smelter, dairy cattle did not show abnormal level of Zn in blood and hair although their daily Zn intake was 5.6 mg kg^{-1} BW against 1.2 mg kg^{-1} BW for cattle in a control area (Milhaud and Mehannaoui 1998).

Accumulation of Zn in seabirds has been studied and it is considered to be a good bioindicators of Zn pollution (Kim et al. 1998). Its levels in marine birds normally range between 6.4 and 150 mg kg^{-1} FW in eggs and liver, respectively (Eisler 1997). Both environmental quality and health condition of birds may highly influence Zn levels in body (Debacker et al. 2000).

Cd

II-12.2 Cadmium (Cd)

II-12.2.1 Introduction

The average content of Cd is given for the Earth's crust as 0.1–0.2 mg kg^{-1} and its abundance is fairly similar in both igneous rocks and sedimentary rocks (Table II-12.2). The metal occurs rarely in nature in the pure form. Its common minerals are: greenockite, CdS; octavite, CdSe; and monteponite, CdO. Its host minerals are Zn and Pb ores, sphalerite, biotite, and amphiboles. Sphalerite is its principal carrier in geochemical processes (Fairbridge 1972; Garrett 1996).

Cadmium and Zn have similar ionic structures and electronegativities and belong to the chalcophile group (Table II-12.1). However, Cd has stronger affinity for S than Zn and thus its mobility in an acidic environment is higher than that of Zn (Alloway 2005). Although Cd has the same valence and similar ionic radius as Ca, it does not substitute for Ca in minerals.

During weathering processes, Cd forms simple compounds, such as CdO, $Cd(OH)_2$, $CdCl_2$ and CdF_2 that are easily mobile and follow Zn, especially in sedimentation processes.

Cadmium is considered as being one of the most ecotoxic metals that exhibits adverse effects on all biological processes of humans, animals, and plants. This metal reveals its great adverse potential to affects the environment and the quality of food.

II-12.2.2 Production and Use

The global production of Cd decreased from 20 kt in 1990 to 19.2 kt in 2000, and to 16.8 kt in 2003 (USGS 2004; WMSY 2004).

Cadmium is produced mainly as a by-product in mining and refining of Zn. Therefore its supply is more dependent on the Zn production than on the Cd demand. Generally, 3 kilograms of Cd is produced from one ton of the Zn production (OECD 1994). Cadmium is also recovered from recycled materials such as Zn-Cd batteries. Approximately 10–15% of the Western World production stems from secondary materials.

The main use of Cd is in the sector of Ni-Cd and Ag-Cd battery production (Table II-12.11). Relatively high amounts of Cd are used as pigments (yellow) and for

Table II-12.11. World use pattern of cadmium in 2000

Use sector	Tons (% of total)
Batteries	12390 (70%)
Pigments	2301 (13%)
Coatings	1416 (8%)
Stabilizers	1239 (7%)
Alloys and other	354 (2%)
Total	17700

coatings and stabilizers. Due to its unique physical and chemical properties, this metal is also added to alloys as stabilizers for various plastics, mainly for a purpose to provide protection from sunlight, chemical attack or heat degradation.

II-12.2.3
Soils

The global average soil Cd concentration is estimated to range between 0.06 and 1.1 mg kg^{-1}, and to average 0.5 mg kg^{-1} (Table II-12.2). In uncontaminated soils, its contents are highly governed by soils texture and averages from 0.22 to 0.51 mg kg^{-1} in light sandy and heavy-loamy soils, respectively (Kabata-Pendias and Pendias 1999).

In most soils, 99% of Cd is associated with soil colloids, thus very proportion occurs in the soil solution (Christensen and Huang 1999). Cadmium may form several types of complex ions and organic chelates. Its concentrations in the soil solutions may vary, depending on techniques used for obtaining solution, from 0.2 to 300 μg l^{-1}. It may occur as cationic species: $CdCl^+$, $CdOH^+$, $CdHCO_3^+$, $CdHS^+$ and as anionic species: $CdCl_3^-$, $Cd(OH)_3^-$, $Cd(OH)_4^{2-}$, $Cd(HS)_4^{2-}$ (Kabata-Pendias and Sadurski 2004). Sposito and Page (1984) have predicted, based on a model, following species of Cd in oxic soil: Cd^{2+}, $CdSO_4^0$, and $CdCl^+$, in acidic solution, and Cd, $CdCl^+$, $CdSO_4^0$, and $CdHCO_3^+$ in alkaline solution. Taylor and Percival (2001) studied soil solution Cd that is readily available to plants. Between 55 to 90% of Cd in soil solution is present as free metal ion Cd^{2+}, although speciation of a metal in the solution depends on concentrations of ligands and metal complexes. According to Welch and Norvell (1999), in addition to inorganic species, Cd in the soil solution also occurs in complexes with various organic acids such as aminoacids, humic and fulvic acids.

Basta et al. (1993) demonstrated the importance of organic matter to the Cd and Pb adsorption, however, a simple correlation analysis indicated that only pH is significantly correlated with metal absorption. The pH is a controlling variable of metal complexation by organic matter and it is not possible to decouple pH from the organic matter.

The oxidation-reduction reactions and formation of complexes are important for the Cd mobility in soils (Herrero and Martin 1993). Lamy et al. (1993) explained that the pH (4.4–7.0) dependent Cd sorption capacity involves competition between the metallic cation and protons for reactive sorption on soil constituents. Increasing pH results in increased sorption of Cd on humic material. The precipitation of Cd may occur at high Cd^{2+} activities under alkaline conditions (pH >7.0) Under anaerobic conditions, the Cd in soil solution is governed by sulfide precipitation (Table II-12.12).

Table II-12.12. Solubility products of Cd compounds and calculated minimum anion activity controlling the Cd concentration (from Christensen and Haung 1999)

Cd compounds	Solubility product, pK_{SO}	Anion activity at Cd concentration of 10^{-7} M
CdS	27.9	$S_2 = 10^{-20.9}$
$CdCO_3$	12.1	$CO_3^{2-} = 10^{-5.1}$
$Cd(OH)_2$	13.7	$OH^- = 10^{-3.35}$
$Cd_3(PO)_4$	32.6	$PO_4^{3-} = 10^{-5.8}$

The mobility and plant availability of Cd in wetland soil differs from that of upland soils. The metal is strongly retained in wetland soils, but it is available to plants under upland conditions (Grambrell 1994). Cadmium is very mobile in oxidizing conditions: pH <3, in low contents of Fe-rich particles and at pH >5; and slightly mobile in the presence of oxidizing particulates and pH >5. However, in reducing conditions and in the presence of hydrogen sulfide and pH >5, high mobility of Cd is also reported (Kabata-Pendias and Sadurski 2004).

Soil contamination and remediation. Anthropogenic source of Cd in soils, and in particular in arable soils are of a great concern. The deposition rate of this metal in Europe has been cited as 3.0 g $ha^{-1} yr^{-1}$ (Pacyna 1987) whereas the smallest deposition rate 0.15 g $ha^{-1} yr^{-1}$ has been reported for Finland (MAF 1997). In Sweden, the wet deposition has been calculated at 0.39 g $ha^{-1} yr^{-1}$ (Eriksson 2001). In USA, about 300 t of Cd (22% of the total release) was deposited on the soil in 1996 (TRI96 1998). It is believed that the deposition rate of Cd in the industrialized countries will be further reduced due to strict environmental regulations and decreased Cd production (Beavington et al. 2004).

Considering all sources of Cd in soils, the major sources of pollution are atmospheric deposition and P-fertilizers (Table II-12.13). Most of Cd pollution, up to 90%, from various sources remains in the top, 15 cm depth layer of soil.

In Sweden, Cd concentration in topsoil has increased by 33% during 1900–1999 years. After this period, its concentration decreased by 0.21% due to: *(i)* reduced trace metal emissions to the atmosphere, *(ii)* less use of P-fertilizers, and *(iii)* leaching (Eriksson

Table II-12.13. Anthropogenic input of cadmium from different sources to soil (from Alloway and Steinnes 1999)

Cd source	Cd in soil (mg kg^{-1})	Cd input (g $ha^{-1} yr^{-1}$)
Atmospheric deposits		
Wet/dry deposition	–	<1.1 – 9.0
Wet/dry deposition, smelter	–	25 – 1 000
Street dust	1 – 12.0	–
Rubber tire wear	20 – 90	–
Fly ash during incineration	2.6 – 68	–
Direct application		
P-fertilizers	0.2 – 345	0.3 – 8.9
By-products gypsums	<6.0	–
Sewage sludge	<1 – 3 410	Up to 150
Composts	0.26 – 11.7	Up to 150
Fly ash[a]	1.5 – 13	–

[a] From other source.

Table II-12.14. Cadmium status and sources in arable soils of Sweden and Finland (based on data summarized by Jansson 2002)

Bilance	Sweden, 1990	Sweden, 2000	Finland, mean of few years
Input (g ha^{-1} yr^{-1})			
P-fertilizer	0.72	0.12	0.10
Manure	–	–	0.22
Deposition	0.78	0.70	0.15
Lime	0.02	0.02	0.03
Sludge	–	–	0.02
Total input	1.52	0.84	0.52
Removal (g ha^{-1} yr^{-1})			
Crops	0.21	0.23	0.14
Leaching	0.06	0.40	0.06
Runoff	–	–	0.11
Total removal	0.27	0.63	0.31
Accumulation	1.25	0.21	0.21
Change in topsoil (%)	0.21	0.03	–

2001b). The study indicated that Cd accumulation in soils of Sweden and Finland were about equal (Table II-12.14). The EU regulations and national policies encourage use of compost materials on agricultural lands when Cd content in composts varies between 0.2 to 1.3 mg kg^{-1} (Eckel et al. 2005).

In New Zealand, the Cd input to soils from P-fertilizer production has been cited at 8.9 g ha^{-1} yr^{-1} whereas average value of EC countries is at 2.5 g ha^{-1} yr^{-1} (Bramley 1990). In New Zealand, the input of fertilizers to agricultural soils has caused double increase of Cd concentration compared to their non-agricultural soils (Roberts et al. 1994). In Switzerland, Cd concentrations in soils decreased from 25.2 mg kg^{-1} in 1975 to 1.7 mg kg^{-1} in 1999 (Herter and Kuelling 2001), whereas in the EU countries, it has decreased from 9 mg kg^{-1} in 1982/1983 to 1.6 mg kg^{-1} in 1996/1997 (Smith 1996; Genderbien et al. 1999).

The Cd content in phosphate rocks depends on the geological nature. Its concentrations in P-fertilizers used in the EU are rather high. Also elevated Zn content is observed in these fertilizers (Table II-12.15).

Phosphate fertilizer application causes increased in the Cd concentration in soil solution. However, the increase of Cd can also be due its desorption from the soil matrix that occurs at low soil pH (Alloway 1995; Taylor and Percival 2001).

Industrial Cd emission is well documented as its significant source in agricultural lands of several countries. The highest Cd concentrations (in mg kg^{-1}) are reported for soils in the vicinity of metal-processing industries, e.g., in Belgium, 1 781; in Poland, 270; and in the USA, 1 500 (Kabata-Pendias and Pendias 2001). The Cd balance in crop-farm soils vary (in g ha^{-1} yr^{-1}) from 0.3 in Denmark to 4.4 in France (Eckel et al. 2005).

Cd

Table II-12.15. Cadmium and zinc concentrations in mineral fertilizers (mg kg^{-1}) (after Eckel et al. 2005)

Fertilizer	Cd, median and range	Zn, median and range
N fertilisers	0.9 (0.01–6.9)	5.0 (0.6–39)
P fertilisers	13 (0.11–47)	236 (24–557)
K fertilisers	0.5 (0.1–2.6)	6.6 (0.7–11)
NPK fertilisers	2.5 (0.1–12)	166 (15–240)
Lime fertilisers	0.2 (0.03–1.0)	22 (2.5–166)

Table II-12.16. Factors affecting Cd uptake from soil by plants (after McLaughlin et al. 1995)

Factors	Cd uptake by plants
pH ↓	Incseased
Soil salinity ↑	Increased
Concentration of Cd ↑	Increased
Metal sorption by soil ↑	Decreased
Cation exchange capacity ↑	Decreased
Clay, Fe and Mn oxides ↑	Decreased
Macronutrients ↑	Incseased/decreased
Aeration	Decreased
Micronutrients (e.g. Zn) ↓	Increased

Increase: ↑; decrease: ↓.

Several methods, including incineration of soils, adsorption and desorption processes, solubility, chemical treatment and biodegradation have been applied for soil remediation (Rulkens et al. 1993; Rulkens and Honders 1996; Ottosen et al. 1997; Cunningham and Berit 2000).

Liming is an old practice that reduces uptake of Cd, as well as other metals by of crop plants. The effectiveness of liming on its adsorption by plants depends greatly on the relative change in pH and Ca^{2+} concentration in soil solution (Adriano et al. 2004). Even coal fly ash (if it has alkaline properties) may decrease the Cd availability to plants.

In recent years, the phytoremediation technique has been applied on Cd and other metals and organics (Huang and Cunningham 1996; Li and Chaney 1998; Ma et al. 2001; Tu and Ma 2003). Although many soil parameters control Cd uptake by plants (Table II-12.16), some plant species reveal unusual ability to its uptake from soils (Table II-12.17). When plants are watered with EDTA, the uptake of trace metals from contaminated soils may be increased (Jørgensen 1993). Many different plants can be used to clean contaminated soils, but growth rate is very important when considering phytoextraction. It has been observed that the transfer of Cd is much lower to barley grains than to straw (Fig. II-12.1).

Recently, another technique known as "inoculating soils with viable microorganisms" has been introduced, that depends on the bioaccumulation of metals from soils

Table II-12.17. Crops accumulating cadmium from contaminated soils (after Felix et al. 1999)

Plant	Harvest yield ($t\ ha^{-1}$)	Cd in plants ($mg\ kg^{-1}$)
Alyssum murale	0.8	34
Thaspi caerulescens	16	12
Nicotiana tabacum	9	10
Zea mays (B37 inbred line)	10	8
Brassica juncea	21	3
Salix viminalis	10	22

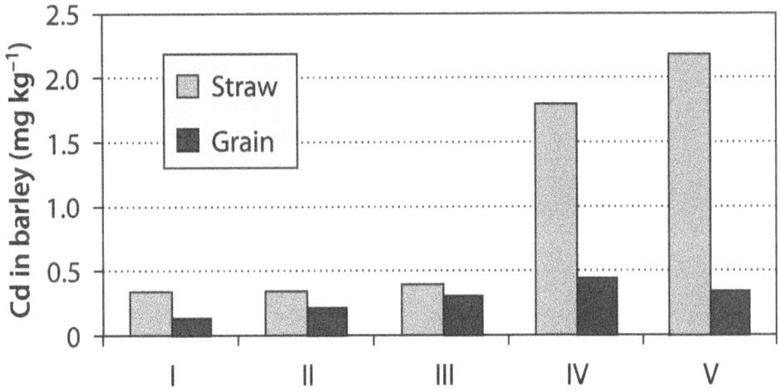

Fig. II-12.1. Impact of Cd added to the soil with Zn and Pb smelter flue-dust on its concentration in spring barley. Cd contents ($mg\ kg^{-1}$) in soil: *I*, 0.7; *II*, 3.7; *III*, 7.7; *IV*, 50.7; *V*, 100.7 (Kabata-Pendias and Piotrowska 1999)

Table II-12.18. Cadmium biosorption by microorganisms (after Lebeau et al. 2002)

Species	Solution pH	Cd in solution ($mg\ l^{-1}$)	Cd sorption (%)	Cd sorption ($mg\ g^{-1}$)
Actinomycete R27		110	100	1 120
Fomitopsis pinicola CCBAS 535	6.2	1 120	51	130
Pseudomonas aeruginosa	6	170	99	58
Pseudomonas fluorescens 4F39	5 – 7	95	–	28
Bacterium ZAN-044	6	11	–	1.1

(Table II-12.18). The soil microorganisms will accumulate high Cd levels (Ledin et al. 1999; Burd et al. 2000).

II-12.2.4 Waters

The median Cd concentration in world ocean waters has been estimated to range from 0.07 to 0.11 $\mu g\ l^{-1}$ (Reimann and Caritat 1998). However, Cd concentrations in open oceans, lakes and rivers vary considerably (Table II-12.19). The residence time of Cd in ocean water has been calculated at 0.7×10^4 to 25×10^4 yr (Boyle et al. 1976; Nriagu 1980).

Cd

Table II-12.19. Cadmium in waters ($\mu g\ l^{-1}$)

Sites	Range, mean	Reference
Ocean	<1.0	Salomons and Kerdijk (1986)
Open sea	0.02 – 0.1	Elinder (1992)
Surface water of the open ocean	<0.005	Eisler (1985)
Deep ocean water	0.12	Elinder (1992)
Baltic Sea (whole)	0.06 – 1.99	Szefer (2002)
North Pacific Ocean	0.07	Nozaki (2005)
Coastal sea water	0.05	Sprague (1986)
Rivers and estuaries	10 – 1 000	Salomons and Kerdijk (1986)
River water, Poland	<0.15	Roszak (1991)
Mississippi River[a]	0.2 – 6.0	IARC (1993)
Rhine River, Germany	0.39	Taylor et al. (1990)
Drinking water	1.0	Eisler (1985)
Groundwater, Russia	0.05 – 8.0	Uczvatov and Bashkin (1997)

[a] Data for samples collected during May–June 1988.

World average of Cd in river waters has been calculated at 0.08 $\mu g\ l^{-1}$ and its riverine flux at 3 kt yr^{-1} (Gaillardet et al. 2003). Nriagu (1980) has estimated the atmospheric Cd input to the oceans at 2.4 kt yr^{-1} while the annual input via stream waters at 7.5 kt yr^{-1}.

Cadmium may enter into water systems from various sources of which smelting of nonferrous metal ores are considered to be the largest one. In the EU-15 states, 11.33 t of Cd and its compounds entered into the aquatic environment in 2001 of which 8.1 tons stemmed from the ferrous and non-ferrous metal industries (www.eper.cec.eu.int/eper/).

Cadmium is easily mobile in seawaters, but much less soluble in other waters (Sundby et al. 2004). In fresh water, Cd is present as Cd^{2+}, $Cd(OH)^{+}$, and $CdCO_3^0$, and also as inorganic compounds and organic complexes (NTP 1991).

Different studies suggest that the Cd concentration in river systems depends on the discharge that is shown for the river Rhine to be affected by the following parameter:

- Dilution as the volume of discharge increases
- Increased erosion, maximum flows will contaminated fluvial particulates with soil particles
- Grain size difference (at a low discharge suspended particulates are finer)
- Difference of residence time – the residence time of particulates will be low when river discharges high volume of water compared to low

In the USA, the mean Cd level in 2 595 samples of drinking water was reported at 3 $\mu g\ l^{-1}$, but the authorities believe that the general value of Cd is less than 1.0 $\mu g\ l^{-1}$. The Cd concentration may increase up to 10 $\mu g\ l^{-1}$ as a result of industrial discharge or

Table II-12.20. Cadmium in different aquatic species (mg kg^{-1}) (after Jeng et al. 2000, unless otherwise indicated)

Organism, site	Mean, range
Sword prawn (*Parapenaeopsis cornuta*), Taiwan	0.090
Swimming crab (*Portunus hastatoides*), Taiwan	0.120
Rock-shells (*Thais clavigera*), Taiwan	1.15
Green mussel (*Mytilus smarrangdinus*), Taiwan	1.21
Oyster (*Crassostrea gigas*), Taiwan	1.76
Flounder (*Psettodes erumei*), Taiwan	0.030
Flounder (*Platichthys*), Tvärminne, Baltic Sea[a]	2.48 (liver)
Green mussel (*Mytilus smarrangdinus*), Baltic Sea[b]	0.98
Shells of blue mussels (*Mytilus edulis*), Baltic Sea[c]	0.002 – 1.62
Macoma, Baltic Sea[b]	0.64

[a] Voigt (2004).
[b] Voigt (2003).
[c] Szefer (2002a).

leaching from metal or plastic pipes (IARC 1993). The WHO (1984) guideline for Cd in all kinds of drinking water is 5.0 µg l^{-1}, whereas Cd in drinking water and the permissible level in bottle water in the USA is 10 µg l^{-1} (IARC 1993). The toxicity of Cd in waters decreases with increasing water hardness.

Aquatic biota. Marine biota bioaccumulates Cd from sediment and water and they may act in the biomonitoring of aquatic pollutants (Table II-12.20). Several aquatic organisms may have specific physico-chemical properties that they can detoxify metals or can reduce the effects of the exposure to high concentrations of metals (Phillips 1995). Bivalves (such as oyster) and mussels (green mussels) are used as bioindicators of pollutants in coastal environments. Oysters collect high concentrations of trace metals. Near the Machu Islands, southern China, Hung et al. (1997) observed 6.82 mg kg^{-1} Cd (geometric mean value) in oysters from water basin where heavy metal concentration in the area was quite high in mud, sand and water.

In 17 species of macrophytes from Lake Maggiore, Italy, Cd contents ranged from 0.20 to 9.60 mg kg^{-1} (average 0.33) for *Nympheea*, the species with lowest collection capacity, and average 3.93 mg kg^{-1} for *Ceratophyllum*, the species with highest collection capacity (Gommes and Untau 1981). In the same lake, Gommes and Muntau (1981) observed higher concentration of Cd (15.70 mg kg^{-1}) in the soft tissue of *Unio mancus* (bivalve) whereas in *Viviparus ater* (Gastropod) only 0.75 mg kg^{-1}.

Effects of Cd on green algae (*Chlorella* spp.) indicate that growth stopped at 2.0 mg kg^{-1} whereas Hutchinson (1979) observed more sensitive algae species (*Chlorella vulgaris*) whose cell division started to decline at 0.05 mg kg^{-1}.

Several studies have shown that the Cd concentration in plankton varied from lake to lake. In Lake Washington, USA, it was noted to be 15 mg kg^{-1} whereas in Lake Alpaca, Switzerland, the mean Cd concentration was 5.6 mg kg^{-1}. Some plankton species have evolved mechanisms to cope with relatively high concentrations of Cd in growth media.

Aquatic plants take Cd from the aquatic environment and the body burden of Cd is many times higher than the level of Cd in water. The Cd threat to aquatic plants may be due to the Cd concentration in the surrounding medium but also due to its uptake and accumulation (Prasad 1995).

II-12.2.5
Air

Cadmium concentration in air of Greenland varies between 0.003 and 0.6 ng m^{-3} (Table II-12.2). Its concentration in rural areas ranges from 0.1 to 4.0 ng m^{-3} whereas in urban/industrial areas it is in the range of 2–150 ng m^{-3}, after OECD (1994), and between 0.5 and 620 ng m^{-3}, after other sources (Table II-12.2). According to data presented by Reimann and Caritat (1998), the worldwide median Cd concentration in remote regions ranges between 0.2 and 0.4 ng m^{-3} whereas in polluted sites is at 20 ng m^{-3}.

Cadmium and its compounds have negligible vapor pressures due to which they may exist in the atmosphere as suspended particulate matter as well as a vapor (Wilber et al. 1992). For this reason, small particulates of Cd may cross the national boarder and travel for a long distance. Particles greater than 10 µm may fall onto soils and vegetation near the released sources by wash out and self cleaning processes. It has been cited that metallic Cd and its oxides are the prevailing forms discharged from industrial sources. These two forms together with $CdCl_2$ released from waste incineration are toxic to the environment (AMAP 1998).

Most significant sources of Cd in the atmosphere are from human activity. However natural sources, such as volcanic activity, forest fire, wind-blown soil and rock dusts may also contribute to its concentration in air (Alloway and Steinnes 1999). The industrial Cd sources are associated with many industrial and manufacturing activities such as: ferrous and non-ferrous metallurgical facilities, waste incineration, cement kiln, fossil fuel combustion, wear of vehicle tires, pigment production, and metal scrap melting and refining. It has been suggested that the coal and oil fired power plants and metal industries are responsible for major Cd emissions to the atmosphere. In the EU-15 states, total Cd emission from anthropogenic sources was 24 t in 2001 of which 13.84 t stemmed from metal production and refining. In 1996, 25.0 t of Cd was released to the atmosphere of the United States from manufacturing and processing industries (TRI96 1998).

The deposition rate of Cd in Europe between 1970s to 1990s has been estimated by many authors and the summarized value varies between <1.1 to 12.5 g ha^{-1} yr^{-1} (Alloway and Steinnes 1999). Its deposition in the EU-15 countries has been reduced in recent years. Unfortunately, in many areas of Europe (for example south of Sweden), elevated concentration of trace metals due to transboundary emission and deposition is observed. It has been calculated that for long-term protection of the environment, the Cd atmospheric deposition should be reduced by 50% (Johansson 1996).

II-12.2.6
Plants

Plants have no metabolic requirement for Cd, however its relatively easy availability to plants reveals serious health risk. Its concentration in food, especially in food and feed plants is of great concern. Contents of Cd in food plants varies in the range of 5–400 μg kg^{-1} and is a little bit higher in leafy vegetables and in roots than in other parts of plants (Tables II-12.21, II-12.22). Its common range in grasses is 50–320 μg kg^{-1}

Table II-12.21.
Cadmium in food plants[a] (μg kg^{-1})

Plant	Range and mean
Cereal, grains[b]	5.6 – 32
Wheat, grains	20 – 70
Barley, grains	13 – 22
Legumes	1 – 30
Broccoli	10
Cabbage	5 – 10
Lettuce	29 – 400
Carrot	30 – 240
Onion	80
Potato	16 – 300
Tomato	30
Spinach	43 – 150

[a] Presented are common, possible background, values from various sources as compiled by Kabata-Pendias and Pendias (1999, 2001), and Herber (2004), unless otherwise indicated.
[b] After Eriksson (2001a), mean values for barley and wheat, respectively.

Table II-12.22.
Cadmium content in selected foods in USA (μg kg^{-1}) (after Gartrell et al. 1986)

Type of food	Average
Potatoes	42.1
Leafy vegetables	32.8
Grains, cereal products	23.7
Root vegetables	15.9
Garden fruits	17.1
Oils and fats	10.8
Sugar and adjuncts	10.9
Meat, fish, and poultry	5.7
Legume vegetable	4.4
Dairy products	3.5
Fruits	2.1
Beverages	1.3

and similarly in clover, from 20 to 350 μg kg^{-1} (Kabata-Pendias and Pendias 2001). Thus, it seems that its availability to plants depend rather on soil parameters than on plants species. However, there are several indications that biotic factors such as plant species, root activity, and crop cultivation may have an impact on Cd uptake by plants (Guo et al. 1996; Welch and Norvell 1999; Lebeau et al. 2002).

Plant species, root activity and rooting pattern may affect the accumulation of Cd that can result in its elevated accumulation in the leaves. The Cd contents of cereal grains, as illustrated for spring barley, are not increased significantly under its elevated amounts in soil (Fig. II-12.1). Usually, Cd concentration is the highest in roots and decreases towards the top of plants (Welch and Norvell 1999). In the USA, 29 rice samples were analyzed and geometric mean concentration of Cd in sample was at 7.43 ng Cd g^{-1} (IARC 1993). In cereal grains, Cd distribution follows in the following order: wheat > oat > barley (Eriksson et al. 1996). In Sweden, Cd concentration for wheat grains is below 0.2 mg kg^{-1} and 0.1 mg kg^{-1} for other cereals, such as oats and rye. These data are below the EU's limit value. In Swedish potatoes and carrots, Cd concentration was reported at 0.049 mg kg^{-1} and 0.30 mg kg^{-1}, respectively (Olsson et al. 2005).

The absorption of Cd via roots and transport within the plants is highly affected by soluble species of Cd^{2+} in soils. Organic Cd complexes such as phytometallophore may play a very significant role. However, it is believed that there is no direct evidence in support of Cd binding with phytometallophores during its transport in root cells (Singh and McLaughlin 1999). Nowadays it has been suggested that Cd-binding complexes similar to metallothionein exists in several higher plants such as tomato, bean, tobacco, soybean, and rice (Bartolf et al. 1980; Kaneta et al. 1983).

There is a general agreement, over many years, that an increase in the solubility and hence availability of metals go together with a decrease in soil pH that affects Cd concentration in the soil solution. Hence, the composition of soil solution is a good indicator for the Cd concentration in crops (Wolt 1994). Although several factors may control the uptake of Cd by plants such as soil properties, weather, and plant species/cultivars, the soil Cd and soil pH are probably the most important (Eriksson et al. 1996).

Several interactions of Cd with other elements have a great impact on its uptake by plants and its fate. Commonly described interactions are related to some metals of which Zn is the most significant (Kabata-Pendias and Pendias 2001). This interaction appears contradictory since both depressing and enhancing effects of Zn on Cd have been reported. Recent studies indicated that there is no effect on inhibition of Cd accumulation by plants due to increase Zn supply to the growth media (Eriksson 1990; Grant and Bailey 1997; Turner 1997). When there is Zn deficiency in a plant, increasing Zn supply decreases Cd accumulation in various plant organs. However, addition of Zn further will promote Zn accumulation in shoots and leaves of plants. The build up of Cd in seeds and grains is possible often, but not always reduced by increasing the Zn supply to plants (Oliver et al. 1994; Grant and Bailey 1997). The behavior of Cd in plants is closely related to Zn as both metals have affinity for sulfur, particularly sulphydryl groups (Garrett 1996).

The symptoms of Cd toxicity in plants generally are stunting and chlorosis. The chlorosis may be due to the Fe deficiency or it may be due to direct or indirect interaction of Cd with foliar Fen (Das et al. 1997). Some researchers have reported on the association of Cd with other elements in plants (e.g., Zn, Cu, Se, P, Cl) and on the role of pH, carbonates, organic carbon, (Page et al. 1987; Jackson and Alloway 1992; McLaughlin et al. 1994).

Cadmium is an effective inhibitor of photosynthesis and inhibited the net photosynthesis of several plants by increasing stomatal and mesophyll resistance to carbon dioxide uptake (Ferretti et al. 1993; Greger et al. 1994). In presence of Cd, the inhibition of stomatal openings in plants depends on Cd concentration and exposure time. Damage due to presence of Cd has been cited (Prasad 1995). In pigeon pea (*Cajanus cajan*), Cd reduced photosynthesis indirectly, stomatal conductance and the electron transport system.

Generally, Cd decreases water stress tolerance of plants, causing the increase of bulk elastic modules and decreased cell wall electricity (Becerril et al. 1989; Prasad 1995). Cadmium induced cell wall degradation in xylem, which is due to reduced water transport (Barceló and Poschenrieder 1990).

Several native plants, in particular trees, shrubs, fern herbs and mosses may accumulate variable and considerable amounts of Cd that ranges between 7 and 2 700 μg kg^{-1} with an average value of 529 (Aloway and Steinnes 1999). In some tree species (*Salix aurita* and *Populus tremula*), average concentration of Cd was less in leaves (1 500 μg kg^{-1}) than in twigs (2 250 μg kg^{-1}). In dwarf shrubs, Cd concentrations varied between 600–1 000 μg kg^{-1} that was 25 times higher than other shrubs species. The highest Cd accumulation, above 100 mg kg^{-1}, has been reported for *Thlaspi caerulescens* (Greger 1999). Felix et al. (1999) reported extremely high Cd-content (560 mg kg^{-1}) in this plant, grown on contaminated soil, which suggest its suitability for phytoremediation.

Nuorteva (1990) observed high concentrations of Cd in the phloem of conifers, bark beetles, honeydew of the aphids, formica ants, forest humus, earthworms, and mushrooms. The increased Cd level in roots is not only the cause of forest decline in Europe but also of deleterious effects of Cd and other trace elements on forest systems.

The accumulation or exclusion of Cd by plants differs at the family levels. *Brassicaceae* and *Fabaceae* species are the highest and the lowest tolerant to Cd, respectively (Rauser 1990). It is often observed that the resistance of plants functions normally even at the presence of high concentration of Cd, as for example in the mining or contaminated sites by Cd waste which is generally accomplished strategies of avoidance or tolerance of metal toxicity (Kanerva et al. 1988). This type of plant can be used as indicators for exploration of metal ores.

II-12.2.7
Humans and Animals

Humans. Contents of Cd in mammalian tissues vary between <0.06 and 1 300 mg kg^{-1}, being the highest in kidneys and the lowest in muscles (Jørgensen 2000). Its concentrations in soft tissues of humans range from 0.03 to 14 mg kg^{-1}, in muscles and kidneys, respectively. The average content in tissues of the "reference man" is 0.17 mg kg^{-1} (Li 2000). Its mean concentrations in human fluids have been reported (in μg l^{-1}) as 0.6 in blood, 0.2 in serum, and 0.9 in urine (Reimann and Caritat 1998).

Cadmium is one of the most toxic metal to humans, mainly due to the ability to combine with sulfhydryl groups and thus disturb function of several SH-group enzymes that leads to changes in proteins. Recognized in 1960s in Japan, so-called *Itai-*

Itai disease has increased the attention paid to environmental Cd pollution and its impact on health. *Itai-Itai* disease had occurred due to run off Cd containing mining tailings in the paddy fields. Thus, Cd uptaken by rice was subsequently ingested by humans.

The Cd excess caused mainly the damage to skeletal structure, especially in elderly women. This phenomenon has brought important information on the Cd transfer in food-chain.

Main toxic effects of the excess of Cd on humans are related to:

- renal damage
- hypertention
- emphysema
- carcinogenic changes, mainly of kidney and prostate
- skeletal deformation due to impaired Ca metabolism
- low reproduction functions

Cadmium intake by humans is mainly by inhalation and ingestion of food and drink. For inhalation exposure, particle size and solubility in biological fluids (in contrast to solubility in water) appears to be the important detriments of toxicokinetics (Hirano et al. 1989; Oldiges and Glaser 1986). Other mechanism is "hand-to-mouth" which generally occurs at the working place. The dietary exposure of population varies highly between countries and is affected by local conditions and occupations. The highest source of dietary Cd is from plant food, up to 75% (Tables II-12.21, II-12.22), among which cereals and particularly rice, and also potatoes, may contribute above 50% of total intake (Louekari et al. 2000).

The current PTWI is 7 µg kg^{-1} BW, i.e., 400 to 500 µg per adult per week (WHO 2001). However, most data on Cd intake is given in µg d^{-1} $person^{-1}$. These values, for different countries, are as follows: Japan, polluted sites, 59–113 µg; Belgium, 16–30; Sweden, 8–15, and Finland, 9–16 (compiled from various sources by Kabata-Pendias and Pendias 1999).

The primary mechanisms of Cd exposure are: inhalation and hand-to-mouth factor where hygienic practices are not followed (Yost 1986; NTP 1991). The high temperature processes generally generate Cd-oxide fumes that are absorbed well through the lungs (IARC 1993). Also the smoking is a significant root of the Cd exposure to general population. A person may absorb 1 µg Cd d^{-1} from smoking 20 cigarettes that is about the as same daily absorption of Cd in Sweden (Friberg et al. 1986). In most countries, Cd may be inhaled by the general population at 50 µg Cd d^{-1}, at a variation from 25 to 75 µg Cd d^{-1} (Friberg et al. 1976). The 10–40% of inhaled Cd may be retained in the body depends on the particle size of Cd whereas ingested Cd is retained by 4 to 10% (Yost 1986).

It has been estimated that the intake of Cd (in µg d^{-1}) by humans from different sources follows the order: diet, 20–50 > smokers, 2–4 µg > air intake, ca 2 µg (Yost 1986). The primary human health concern is now related to the long-term chronic exposure.

The concentration of Cd in the atmosphere does not appear to pose a great threat to human unless one lives close to a zinc-lead smelter. But still, physicians are concerned about Cd exposure in Pb-Zn smelter, Ni-Cd battery factories or other manufacturings that use Cd compounds, especially when the operation involves heatings. These persons can build up Cd level in the liver over time. Cadmium is not very well absorbed (it may be about 20%), but still it is high comparing with other metals in the environment. Children living in the vicinity of the Pb-smelter contained elevated BCd levels up to 8.9 ng l^{-1}. As Carvahlo et al.(1989) reported, houses of these children were contaminated with Pb-smelter dust.

The FAO/WHO expert committee jointly proposed tolerable intake of Cd at about 70 µg Cd d^{-1} (WHO 1992). Other sources recommend safe daily intake of Cd by adults at 40–80 µg (Louekari et al. 2000). For smokers and anemic persons the critical level of dietary Cd intake should be lower.

The acute toxicity of Cd is rather rare. With the Zn deficiency, more Cd is stored in human body. The highest chronic exposure is generally found in worker populations in facilities where Cd is used as a feed material or is present in by-product stream (Yost 1986). With long body retention time, the concentration of Cd in target organs will increase in time. The target organ is generally kidney or more specifically the renal cortex. The renal cortex dysfunction generally occurs when Cd concentration is about 200 mg kg^{-1} (Kjellstrom and Nordberg 1978). Recent suggestions, however, consider that <50 mg Cd kg^{-1} FW in renal cortex should be accepted (Louekari et al. 2000). Elderly women living in the industrialized urban areas often suffer from lowered kidney function (Yost 1986).

Animals. Cadmium is concentrated mainly in animal kidneys; however, other soft tissues may also contain relatively high amounts of this metal (Table II-12.23). Eisler (1985) reported that: *(i)* Cd level is higher in older organisms than in younger ones, *(ii)* marine biota contained more Cd than their freshwater or terrestrial counterparts, and *(iii)* it is accumulated mostly in liver and kidneys.

Table II-12.23. Cadmium in liver and muscles of animals (mg kg^{-1}) (adoped from WHO (2001)

Animal	Organ	Year 2000	Year 2001	Predicted, year 2002
Bovines	Liver	3368	1230	1164
	Muscles	3046	1230	1164
Pigs	Liver	1416	790	781
	Muscles	1061	790	781
Sheep and goats	Liver	942	490	495
	Muscles	700	490	495
Horses	Liver	462	300	300
	Muscles	473	300	300
Poultry	Liver	371	507	514
	Muscles	–	507	514
Rabbit	Liver	49	30	30
Wild game	Liver	74	100	100

Cadmium is easily accumulated, especially in animal kidneys and thus it may serve as a sensitive bioindicator for the Cd contamination. Especially horse kidneys are good indicators (Szkoda and Żmudzki 2000). As these authors reported, the median content of Cd in young horses is 0.02 mg kg^{-1} FW, whereas in old ones (above 2 years old) it increased to 30.72 mg kg^{-1} FW. The level of Cd in soft tissues of cattle in Poland was at about similar levels during the period 1995–1997 (in mg kg^{-1} FW): kidneys, 1.0; liver, <0.2; and muscles, 0.005. However, in kidneys of hunting animals, Cd contents increased during that period from 1.5 to 2.5 mg kg^{-1} FW, but remained constant in other soft organs (Żmudzki and Szkoda 2000). These authors concluded that due to highly elevated Cd in kidneys of most animals, these organs should be excluded from the food products.

According to data presented by the WHO (2001), Cd levels in both, kidneys and liver of different animals is about equal. Wild game (not specified species) seems to contain Cd in liver at lower levels than farm animals (Table II-12.23).

The toxicity of Cd is to animals is similar to those described for humans. Its acute toxicity is not common and is observed mainly in fish and other aquatic organisms (Louekari et al. 2000). The main concern related to increased contents of Cd in animal tissues is related to its transfer in the food-chain.

II-12.3
Mercury (Hg)

II-12.3.1
Introduction

Mercury has been known since prebiblical times and this element occupies a unique place in the human civilization. It has been known by alchemists in China, perhaps since 2 000 years BC. For over a century it has been known as an environmental pollutant.

Mercury abundance in the Earth's crust is very low, in the range of 0.02–0.06 mg kg^{-1}, but is it likely to be concentrated in argillaceous sediments and in coal (Table II-12.2). The most important mineral, cinnabar, HgS, has been mined continuously since 415 BC (Clarkson and Marsh 1982). There are several other Hg minerals, mainly simple compounds with some anions and cations, but of much less importance (Fairbridge 1972). Thus, it can occur in rocks in forms of oxides, sulfides, halides, organic complexes and native metal (Mitra 1986). There are several Hg host minerals such as amphiboles, sphene, sphalerite, and other sulfides. It may evaporate from some compounds release to various ecosystems.

Mercury exists in several states, like Hg-metallic liquid, Hg-vapor metallic, inorganic Hg^{+} (mercurous salts), inorganic Hg^{2+} (mercuric salts) and gaseous Me-Hg (Drasch et al. 2004). It is chalcophilic and shows the affinity to several metals, such as Ag, Zn, Cd. The mercuric salts (such as $HgCl_2$) are soluble in water whereas organomercurics are neither soluble nor do they react with weak acids and bases due to the low affinity of Hg for oxygen bound to carbon (US EPA 1997a).

The characteristics of Hg in the environment are in a number of chemical and physical forms, including Hg^0, $HgCl_2$, HgO, HgS, $CH_3HgCl$and $(CH_3)_2Hg$ (Schroeder and Munthe 1998), and each species behaves differently in various ecosystems. In the 1950s, it was established that emissions of Hg to the environment could have serious effects on human health. Concerns arise when Hg-contaminated fish and wildlife from ecosystems are passed into the human food chain (D'Itri 1972; Friberg and Vostal 1974). The tragic case of Hg poisoning known as a Minamata Disease was a fundamental information for the scientific community and decision makers on its use and discharge to the environment (Tsubaki and Takahashi 1986). In the United States, the 1990 US Clean Air Act Amendments have identified Hg as one of the trace elements listed in the legislation as "hazardous air pollutant" because of its toxicity, availability, and potential bio-accumulation within the environmental compartments (Petersen et al. 1998).

Increased Hg load into the environment is additionally due to its use in Au-mines for amalgamation process became of a real risk of its transfer to the food chain. This is a serious threat in the Amazonian region and in the Siberian subregion.

II-12.3.2 Production and Use

The global production of Hg has decreased from 5.36 kt in 1990 to 2.2 kt in 1999, whereas virgin Hg production in EU-25 was about 0.55–0.68 kt in 1999. An approximate value of the Hg annual production is given at 1.8 kt in the year 2000 (Drasch et al. 2004). However, Jernelöv and Ramel (1998) have emphasized a disagreement between the established global Hg production and the use of Hg that was about twice higher, during the period 1991–1995.

Due to toxic properties of Hg and strict regulations, primary Hg production and its uses have been restricted in many countries. In Finland, virgin Hg is recovered from roaster gases by precipitation-filtration methods at the Kokkola Zinc Plant. In the West, secondary Hg production (e.g., recycling) can be accomplished by chemical or thermal treatment. The most common method is thermal treatment where Hg-containing scrap is sized and heated in furnaces or retorts at about 538 °C to vaporize the Hg. The vapor is then condensed and collected under water (Reisdorf and D'Orlando 1984; US EPA 1984). In the EU-25 states, Spain and Finland still produce virgin Hg.

In some countries, the Hg production has been lowered or completely stopped (Ferrara 1999). The old Hg mine in Almadén, Spain, operating since 4th Century BC, the largest Hg producer in the world, has been stopped in recent years.

Main uses of Hg are in the caustic soda and chlorine production and in the Au-ore processing (amalgamation). There is a tendency to replace Hg by other metals or compounds, as for example, in Northern Europe; an organic compound has replaced Hg-amalgam for teeth feelings. Also Hg free thermometers or Hg free batteries are quite common in the western countries. However, Hg is still used in many products including: fluorescent lamps, batteries, instruments, manometers, electric conductors, metal recovering processes, dental fillings, bactericides, fungicides, insecticides, pharmaceuticals and also in paints (Mitra 1986; US EPA 1997b). The use of Hg in both interior and exterior paints is prohibited in the USA and in many other countries in Europe (particularly in the north Europe).

II-12.3.3 Soils

Background levels of Hg in soils are not easy to estimate due to the widespread Hg pollution that has a strong affinity for most soils, which act as a vast reservoir (Meili 1991; Swedish EPA 1991). When added to soil, both as elemental Hg or inorganic and organic complexes, it is likely to be strongly bond.

Natural Hg contents in soils are fairly low and seldom exceed 1 mg kg^{-1} (Table II-12.2). Its highest levels are always in organic soils (histosols) than in sandy or loamy soils. In the reference soils of China, Hg contents vary between 0.015 and 0.294 mg kg^{-1} and average 0.14 mg kg^{-1} (Govindaraju 1994). However, most top soils contain increased amounts of this metal, especially near mining and smelting areas (Table II-12.24).

The behavior of Hg in soils depends highly on its species, which is a function of several soil parameters and on interactions with other ions (Yin et al. 1997). In soils, the Hg^{2+} species undergo biological and chemical reaction forming different inorganic and organic compounds that are controlled by such factors as temperature, pH and SOM content (Schuster 1991). In the surface soil, about 1–3% of total Hg is in the Me-Hg form and the rest may be considered as Hg^{2+} complexes (Revis et al. 1990).

Table II-12.24. Mercury in soils of various countries (mg kg^{-1})

Sites and country	Total-Hg	Reference
Discovery Park, Seattle, WA, USA	0.029 – 0.133	Lindqvist et al. (1991)
Control soil, NY, USA	0.117	Cappon (1981)
Garden soil, New York State	0.406	Cappon (1987)
Typical US soils	0.008 – 0.117	Cappon (1987)
Podzols and sandy soils, Holland	0.01 – 0.70 (0.04)[a]	Ferara et al. (1991)
Agricultural surface soils, Poland	0.003 – 0.28 (0.06)	Kabata-Pendias and Pendias (2001)
Monte Amiata, mining area, Italy	90 – 212 (182)	Ferara et al. (1991)
Siberia, tectonic fault zone; sandy soils, Russia	3.59; 1.24	Laperdina et al. (1999)
Paddy soils, Japan	0.15 – 0.76	Kitagishi and Yamane (1981)
Forest soils (raw humus layers), Sweden	0.24	Andersson (1987)
PVC-SODA factory, Vlora, Albania	0.4, 2.5, 20[b]	Reif et al. (2004)
Chlor-alkali plant, Kazakhstan	17 100	Randall et al. (2004)
Volcanic areas		
– Mt. Etna, Italy	0.04 – 7.45	Compiled by Kabata-Pendias and Pendias (2001)
– Mt. St. Helen, USA	0.001 – 0.23	

[a] In parenthesis: mean values.
[b] At A, B, C soil horizons, respectively.

Some specific properties of Hg species in soils are as follows:

- easy volatilization: Hg^0 and $(CH_3)Hg$
- easy solubilization: $HgCl_2$, $Hg(OH)Cl$, and $Hg(OH)_2$
- low mobility: CH_3Hg^+ and CH_3HgS^-

Its concentration in the soil solution is very low, at about 2.5 µg l^{-1}, and preferable as cationic species: Hg_2^{2+}, $HgCl^+$, and $HgCH_3^-$. However, it can form also anions; $HgCl_3^-$ and HgS_2^{2-} (Kabata-Pendias and Sadurski 2004).

Mercury enters soils from several sources: atmospheric fall out and rainfall, sewage sludge application, Hg-based pesticides, disposal of industrial, domestic solid waste products, and municipal incinerator ash.

Mercury contents in topsoils may greatly differ depending on the soil origin and pollution sources. Its highest accumulation is reported for soils from the vicinity of chlor-alkali plant (Table II-12.24). Often these are cumulative effects of the anthropogenic deposition of Hg over the last 100 years (Fitzgerald 1994; Expert Panel on Mercury Atmospheric Processes 1994).

The phytoavailability and toxicity of Hg in soil-plant systems depends on its forms in soils. Simple Hg salts or metallic Hg may create hazardous effects on plants and soils biota due to the toxic nature of Hg vapor. Leinonen (1989) and Nuorteva (1990a,b) presented results of studies on effects of tilling of soil on the Hg concentration in leaves of *Vaccinium myrtillius*. The Hg concentration in leaves of plant grown in untreated and clear-cut forest plots was at 0.02 mg kg^{-1}, whereas plants from tilled plots contain a bit more Hg, up to 0.03 mg kg^{-1}. These findings indicate that the mobility of Hg in soils is very labile and controlled by several soil and climatic factors.

Due to specific properties and behavior of Hg in soils, remediation of Hg contaminated soils is complicated and costly. However, increasing Hg pollution in some regions, affected mainly by Hg uses in the Au-amalgamation processes in Au-mines, has stimulated intensive studies on this topic (Ebinghaus et al. 1999). Lacerda and Salomons (1991) calculated that during the Au-Hg amalgam processes, mean total Hg losses of 1.32 kg are for 1 kg of Au produced.

The vaporization of Hg from soils is generally accelerated by microbial and non-microbial processes and the volatilization rate is probably maximum when microbial processes are most intense (Andersson 1979). Several studies have demonstrated that the addition of water significantly increases Hg release from soils and the magnitude of its release correlates with the Hg concentration in the soil (Lindberg et al. 1999; Frescholtz and Gustin 2004). The release mechanism of Hg from soil is assumed to be due to the displacement of Hg by the more polar water molecules (Engle et al. 2001).

The soil pools of Hg and Me-Hg are the potential source of Hg accumulated in the freshwater food chain. Bishop and Lee (2004) observed high concentrations of Me-Hg (2–20 mg kg^{-1}) in stream-bank sphagnum mosses and (Me-Hg: 5–10 mg kg^{-1}) in swampy ground at the depth 0–20 cm. Their study confirmed that the Me-Hg is not only from atmospheric sources but also from the methylation of inorganic Hg in soil/soil water system, especially in wetlands. Bishop et al. (2004) demonstrated that boreal wetlands may bear important source of Me-Hg to the watershed.

Hg

II-12.3.4 Waters

Concentrations of Hg in seawaters vary highly depending on sampling sites; water from open sea contains Hg in the common range of 0.5–3.0 ng l^{-1}, whereas coastal and bay waters contain from 2 to 65 ng Hg l^{-1}. Its median concentration in worldwide ocean water is calculated at 30 ng l^{-1} (Reimann and Caritat 1998). Terrestrial waters, and in particular, mines and drainage waters might be enriched in Hg (Table II-12.25).

Mercury occurs in waters in several forms: Hg^0, Hg^+, Hg^{2+}, $HgCl_2^0$, HgS_2^-, and various methylated species (Me-Hg). It is partitioned between suspended particulate matters, organic matters and DOC. About 25–60% of Hg^{2+} and Me-Hg are particle-bound in the water and the rest is in dissolved and DOC-bound phases (Nriagu 1979; Bloom et al. 1991; Witczak and Adamczyk 1995). Phenyl-Hg, or alkoxyalkyl-Hg can be converted into Me-Hg compounds by natural processes (Jernelöv 1969). Mercury methylation and its bioaccumulation in the aquatic ecosystems depend on Hg loadings, microbial activity, nutrient content, pH and redox potential, sediment load, temperature and other factors (NAS 1978; Compeau and Bartha 1984; Callister and Winfrey

Table II-12.25. Mercury in waters (ng l^{-1}) (as summarized by US EPA 1997a)

Sites and country	Total-Hg	Me-Hg
Lakes: 8 sites, Sweden	1.35 – 15	0.04 – 0.8
Riwers: 8 sites, Sweden	2.9 – 12	0.08 – 0.73
Runoff: 8 sites, Sweden	2 – 12	0.04 – 0.64
Lake Cresent, WA, USA	0.163	<0.004
Lake Michigan, USA	8.0; 6.3a	n.d.
Worldwide, rivers and streams	1 – 7	n.d.
Open ocean	0.5 – 3.0	n.d.
– Coastal sea water	2 – 15	
– Coast at Italy	1.7 – 12.2; 0.3 – 80[b]	
Baltic Sea[c]	0.5 – 65	n.d.
North Pacific Ocean[d]	0.14	n.d.
Domestic well, NJ, USA	>2 000	n.d.
– Well water, WA, USA	0.3	
– Tap water	0.3 – 25	
Rainwater, Sweden[e]	<2	n.d.

n.d. – No data.
[a] At 0.3 and 10 m depth, respectively.
[b] Dissolved and particulates, respectively.
[c] After Szefer (2002a).
[d] After Nozaki (2005).
[e] After Eriksson (2001a).

1986; Wastras and Bloom 1998). There are two processes, biotic and abiotic, which are guiding the Me-Hg formation mechanisms (Celo et al. 2004). Mercury in the water may revolatilize, deposit on sediments and be uptaken by aquatic biota.

Aquatic biota. The Me-Hg is highly bioavailable and easily accumulates in fish (Bloom et al. 1991). Planktivorus and piscivorus fish uptake Me-Hg through their food chain. Relatively high Me-Hg concentration in fish tissues is reported in acidic lakes (Winfrey and Rudd 1990). The Hg levels in different fish species highly depend on their nutritional habits (Fig. II-12.2). Fish take Hg straight from the water as well as from their

Fig. II-12.2. Average Hg contents in fish species according to their alimentary habits (reproduced with permission from Brasil et al. 2004)

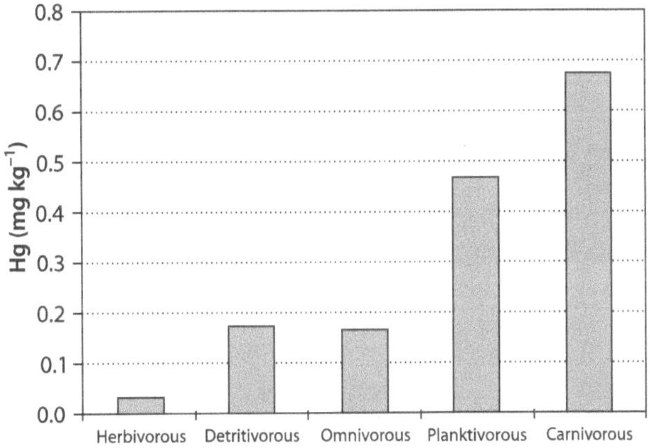

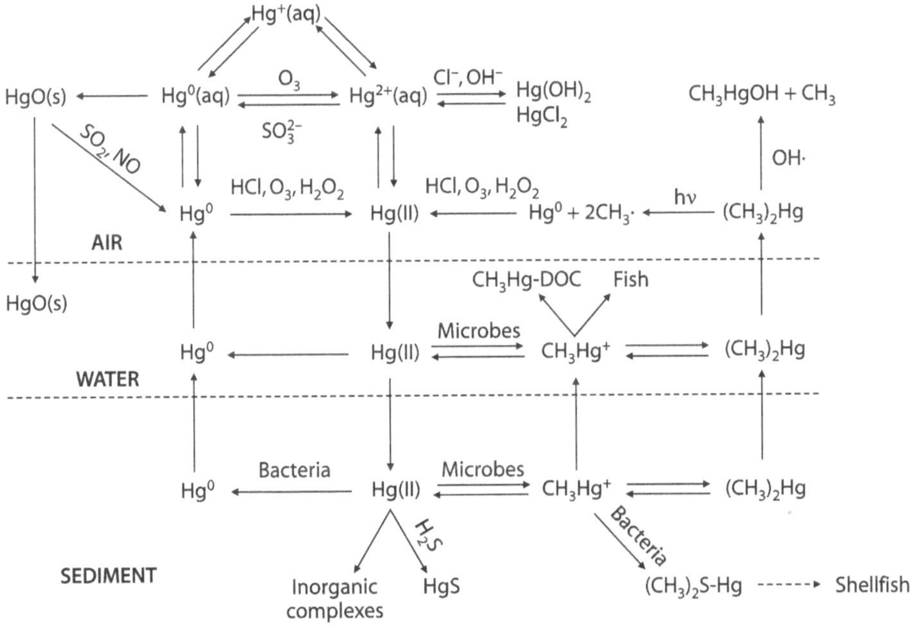

Fig. II-12.3. Transformation of mercury in air, water and sediment (based on data from ATSDR 1999a). *aq* – associated with aqueous phase; *s* – solid fraction

Hg

fodder. Mercury discharged to the Amazon River is easily bioaccumulated in aquatic biota, whereas its increased content in bottom sediments of the Baikal Lake (at a level of percentage) seems to not affect its concentration in fish tissues (Jernelöv and Ramel 1998).

The Hg concentrations in pike have risen from 0.05–0.3 mg kg^{-1} to 0.5–1.0 mg kg^{-1} in rivers of southern and central Sweden. Due to a high acidity, many lakes have been black listed in Sweden and Finland because of elevated concentrations of Hg in fish that is observed at low water pH. This is not a general rule, however, as in some lakes positive correlation with water pH has been observed and a negative in others (US EPA 1993).

The geometric mean concentration of all freshwater fish (in USA, $N = 112$ samples) was 0.11 mg kg^{-1} FW, in the range of 0.01–0.77 mg kg^{-1} FW (Lowe et al. 1985). Bahnick et al. (1994) measured Hg levels from 314 sampling sites of the United States, in five bottom feeders (e.g., carp) and five game fish (e.g., bass). Mercury was detected at 92% of the sampling sites. Average concentration of Hg in fish tissues was 0.26 mg kg^{-1} FW. This difference may be also due to sampling position, the size and age of fish.

A great part (>95%) of total Hg in both fresh and saltwater fish is in the form of Me-Hg (Bloom 1992). Voigt (2004) reported that Hg contents in various organs of 30 fish species sampled around the Archipelago Sea, Gulf of Riga and at Firth of Kiel were, in most cases, the highest in muscles and the lowest in gonads (Table II-12.26). However, Hg concentrations in fish species varied according to sampling sites, uptake behavior of food, age, and deep or shallow water fish.

In crucian carp (*Carassius carassius*) from ponds located near Chernobyl, Hg concentration was, after the nuclear accident, in the range of 0.03–1.28 mg kg^{-1} (Jagoe et al. 1997). Fresh water fish sampled in the 1970s in Poland contained Hg at the range of 0.014–0.464 mg kg^{-1} FW and fish from the Baltic Sea had this metal at the levels of 0.007–0.215 mg kg^{-1} FW (Nabrzyski 1996). The highest mean Hg levels in commercial fish in USA are reported to range between 0.73 and 1.45 mg kg^{-1} FW in makrel king and tilefish, respectively, and its lowest range varies from 0.01 to 0.11 FW in salmon and cod, respectively (US Dept. of Health and Human Services 2004).

Mercury contamination of fish from gold mining areas in south America, Asia, and Africa have been cited by some authors (Castilhos et al. 2004a–c; Veiga and Baker 2004). In most studies, fish type, weight and length, age, and sex of the fish and sampling sites have been considered. The Hg level in fish is quite high near mining areas in the Brazilian Amazon – São Chico. The Hg values lay between 2.21 mg kg^{-1} for Carás (non-carnivorous) and 6.11 mg kg^{-1} for Trairas (carnivorous), whereas the limit content of Hg in fish is 0.5 mg kg^{-1} (Castilhos et al. 2004a–c).

Table II-12.26. Mercury content in fish organs collected from the Baltic Sea near Tvärmine, Finland (mg kg^{-1}) (after Voigt 2004)

Fish species	Muscle	Liver	Gonad/F	Gonad/M
Esox lucius	0.33	0.13	0.03	0.02
Pearch	0.22	0.10	0.05	0.06
Platichthys	0.13	0.09	0.03	0.04
Osmerus	0.12	0.05	0.02	0.02
Clupea	0.10	0.12	0.03	0.03

F – female; *M* – male.

II-12.3.5 Air

The worldwide mean Hg concentration in the atmosphere ranges between 0.01 and 0.06 ng m^{-3}, whereas in urban/industrial areas it can increase up to the range of 0.17 to 11.2 ng m^{-3} (Table II-12.2). According to Reimann and Caritat (1998), Hg ranges from 0.01 to 0.06 ng m^{-3} in air from remote world regions and from <0.09 to 38 ng m^{-3} in air from polluted areas.

The Hg sources and its flux from the atmosphere to land or watershed generally stem from the natural global cycling and from anthropogenic sources. On a global scale, the atmospheric Hg cycle is dominated by elemental Hg vapor (>95%), whereas its deposition to terrestrial and aquatic environments depends strongly on the Hg speciation released from different industrial sources to the atmosphere (Ebinghaus et al. 1999). Estimation made by Nriagu (1979) indicates that ocean sediments may contain the total Hg at about 10^{17} g, soils and freshwater sediments 10^{13} g, and the atmosphere 10^{8} g. In recent years, Fitzgerald (1994), Mason et al. (1994) and Pacyna and Pacyna (2001a) estimated global atmospheric emission for the 1990s and values lay between 5 kt yr^{-1} to 2.14 kt yr^{-1}, which is over 60% less than the estimated value for the 1970s. In all these estimations, there is a broad range of uncertainties and the Expert Panel on Mercury Atmospheric Processes (1994) believed that one third of the estimated amount has been added from pre-industrial time and the rest is from present anthropogenic sources to the atmosphere. There is estimation that the global anthropogenic Hg fraction constitutes 40 to 50% of its total current (Horvart et al. 1993). However, only half of the anthropogenic Hg released enters into the global cycle, the rest is being removed through local or regional cycles. The Me-Hg compounds are very important in the overall Hg loadings in terrestrial and aquatic ecosystems (Hultberg et al. 1995).

The atmospheric Hg emission was estimated in 1995 for Europe and North America at 249.7 t and 213 t, respectively (Pacyna and Pacyna 2002). During the same period, Mniszek (1996) calculated the Hg atmospheric emission in Poland at 40 t yr^{-1} of which 44% and 18.3% were produced from burning hard coal and lignite, respectively.

A wide range of mining and mineral processing activities are classified as artisanal mining where miners apply low-tech manual processes. In artisanal gold mining operations in South America, 200 t of Hg are discharged to the environment, during recovering 115–190 t of Au (Veiga 1997). A part of the released Hg enters into the atmosphere and the rest into the aquatic environment and the food chain of the local population (Veiga 1997; Veiga and Baker 2004). The speciation of Hg from artisanal mining differ due to high temperature processes. Mercury vapor emitted from gold mining is transported only for the distance of 2–3 km (Borochoff 2001).

The chemical forms of Hg in the atmosphere are still under discussion, although it has been generally accepted that the major form of Hg in the atmosphere is elemental (Hg^0) followed by minor concentrations of Me-Hg, which concentration in the atmosphere increases with increase in latitude (Bloom and Fitzgerald 1988). Various forms of Hg behave differently. Metallic Hg^{2+} is generally deposited by both dry and wet fallout within the distance of 100 km from sources, whereas Hg^0 is insoluble in water and thus is transported mainly as particles.

Table II-12.27. Mercury in air and in precipitation in Europe (based on the data from Working Group of Mercury 2001)

Hg species	Hg in air	Hg in precipitation (ng l^{-1})
Total mercury	1.2 – 3.7 (ng m^{-3})	5 – 80
Elemental mercury Hg (O)	1.0 – 3.6 (ng m^{-3})	<0.005
Reactive gaseous mercury (RGM)	1 – 50 (pg m^{-3})	5 – 50
Total particulate mercury (TPM)	1 – 50 (pg m^{-3})	5 – 50
Methyl mercury, MeHg	1 – 20 (pg m^{-3})	0.005 – 0.5

In the atmosphere, Hg is present mainly in three different forms: Hg^0, Reactive Gaseous Mercury (RGM) and Total Particulate Mercury (TPM). Only elemental Hg vapor (Hg^0) has been identified whereas the other two mentioned species are still to be identified, in respect of their physical and chemical structure (Edner et al. 1989).

Based on sediment Hg measurements in different regions of lakes in Sweden and State of Wisconsin, USA, it has been estimated that the atmospheric Hg burden has increased by a factor of 2 to 5 since the beginning of the industrial period. Mercury in air and in the precipitation in Europe varies highly (Table II-12.27). However, it can be stressed that the Hg dispersion from a point source is governed by a series of atmospheric processes and chemical interactions in which the changing of Hg speciation is of a central importance and is on the current research agendas (Petersen et al. 1996).

II-12.3.6 Plants

The concentration of Hg in plants, especially food and fodder plants, has recently received much attention, because of its pathway into the food chain. Especially Hg in potatoes and cereal grains, as its main dietary sources, has been intensively studied. The background value of Hg in vegetation has been reported to be 100 μg kg^{-1} (Lindqvist 1991). However, its contents vary highly, depending on plants species and growth conditions and is lower in cereals than in leafy vegetables (Table II-12.28). The baseline Hg in indicator plants from forests in Finland is established at the range between 10 and 330 μg kg^{-1} (Nuorteva 1990a). Fodder plants grown in Poland contain Hg in the range of mean values at 12–142 μg kg^{-1}, being the highest in beets (Dąbrowski 1992).

Plants differ in their ability to take up Hg and can develop a tolerance to its high concentrations when grown in contaminated sites. Plants seem to absorb Hg easily from solution, however there is not much evidence that increasing soils Hg generally causes an increase in the Hg content of plants, partly due to its great accumulation in roots (Kabata-Pendias and Pendias 2001). There is evidence, however, that increased soil Hg will cause increase Hg in roots of plants and its translocation will occur via roots to other parts of plants (Siegel and Siegel 1988; Ferrara et al. 1991). Increased Hg content in vegetation from contaminated sites indicates that it is taken up from soil, however, its atmospheric source is known to contribute also significantly to Hg levels in plants. In recent years, the attention has been paid to Hg uptake from the

Table II-12.28. Mercury in food plants (μg kg^{-1})

Source	Range and mean	Reference
Wheat	<0.1 – 34	Wiersma et al. (1986)
Barley	1.1 – 34	Wiersma et al. (1986)
Oats	<0.1 – 22	Wiersma et al. (1986)
Maize	1.7 – 73	Szymczak and Grajeta (1992)
Lettuce, greenhouse	38.5	Wiersma et al. (1986)
Tomato, greenhouse	22.0	Wiersma et al. (1986)
Cucumber, greenhouse	7.7	Wiersma et al. (1986)
Carrot	16.9	Wiersma et al. (1986)
Potato	13.5	Wiersma et al. (1986)
Apples	6.3	Wiersma et al. (1986)
Spinach	68.5	Wiersma et al. (1986)
Lemon	4.3[a]	Dassani et al. (1975)
Tea, leaves	34 – 46	Inarida et al. (1984)
Leafy vegetables, N.Y.	64 – 139	Cappon (1987)
Herbs, garden products	130[b]	Temmerman et al. (1986)

[a] At FW basis.
[b] Conversion to dry wt. assuming 90% water by wt.

atmosphere. In some cases, the major part of Hg in above-ground parts of trees seems to be associated mainly with its atmospheric deposition (Ericksen et al. 2003). Certain plant species, such as lichens, carrots, lettuce, and mushrooms in particular, are likely to take up more Hg than other plants grown at the same sites (Falandysz et al. 2001, 2003).

In the atmosphere, Hg is present mostly as elemental Hg^0 that is readily converted into Hg^{2+} in the plant interior. Inorganic Hg can be translocated from leaves to fruits, from potato leaves to tubers and from rice leaves to grains (Lagerwerff 1972). Studies (by spiking soil with ^{203}Hg) indicated that 90–95% of Hg is associated with leafy parts of plants and with roots only 30–60% (Mosbaek et al. 1988). The methylation process occurs in plants, whereas the Me-Hg uptake from soil is not confirmed. There are some plants (*Elodea densa*) where movement of Me-Hg and inorganic Hg is in opposite directions. In this particular species, Me-Hg moves towards the young shoot apex whereas inorganic Hg is moving towards older (lower) parts of shoots (Cuzba and Mortimer 1980).

Plants grown in contaminated regions may accumulate much higher than common amounts of Hg. Especially increased Hg contents are reported for mushrooms that are grown in the vicinity of Hg mining (mean 38 mg kg^{-1}) and for those from the region of chloralkali works (72–200 mg kg^{-1}) (Kosta et al. 1974; Lodenius and Herranen 1981).

Lichens sampled from a chlorakali plant region, in Finland, contained the mean Hg at 36 mg kg^{-1}. Vegetation grown at the distance of 1 km away from the lead smelter in Czechoslovakia contained Hg between 0.3 and 12.0 mg kg^{-1} (Kalac et al. 1991). Mercury contents in moss samples from remote regions of Norway ranges from 0.002 to 0.48 mg kg^{-1} and averages 0.08 mg kg^{-1} (Berg and Steinnes 1997a,b).

Increased contents of Hg may be toxic to both higher plants and microorganisms. The symptoms of Hg toxicity are commonly: *(i)* stunting of growth of seedlings and roots, *(ii)* inhibition of photosynthesis, *(iii)* inhibition of K uptake, and *(iv)* reduction of yield. The most sensitive plants to the Hg excess are sugar beets, corn, and roses. Some microorganisms may suffer from the soil Hg concentration >10 mg kg^{-1}.

II-12.3.7
Humans and Animals

Humans. Contents of Hg in mammalian tissues vary between 0.02 and 0.25 mg kg^{-1}, being the highest in kidneys and the lowest in muscles (Jørgensen 2000). Its mean concentrations in human fluids have been reported (in µg l^{-1}) at 5.3 in blood, 2.1 in serum, and 3.5 in urine (Reimann and Caritat 1998).

The biochemistry of Hg has received significant attention because of the toxicity of Me-Hg, its accumulation in biota and biomagnification in the human food chain (Horvat et al. 2003). The Me-Hg can be taken up from the digestive tract and transported by the blood to other parts of the human body, causing major damage (Magos 1997).

Both forms of Hg, elemental and organic have acute or chronic effects on humans. Either excessive amounts of Hg are ingested or inhaled; most of metabolic functions are inhibited that finally leads to death (Drasch et al. 2004). In a brief summary, Hg toxicity to humans is related to:

- Mutagenic, carcinogenic, and teratogenic effects
- Immunology impaired
- Coronary disease
- Central nervous system damage
- Deformation of DNA
- Low reproduction
- Disturbance of the development of a child

The Me-Hg compounds irreversibly damage neurons of the central nervous system. At high sublethal Hg doses in man, it may also cause: nervous disorders, brain damage, cancer, cerebral palsy, gross motor and mental impairment, speech disturbances, blindness, deafness, microcephaly, intestinal disturbances, tremors, and tissue pathology (Drasch et al. 2004).

The Hg concentration in fish tissues is of great health risk. Dietary intake of Hg has been studied recently for 14 countries, including the USA. Galal-Gorchev (1993) calculated that 20–35% of the Hg is contributed by fish diet for populations of these countries. Intake of Hg depends, to a great extend, on dietary habits. In the Nordic countries, especially in Sweden and Finland, high intake of Hg has been observed.

Several human populations, around the world, have been identified as being in high risk groups due to the consumption of fish with elevated Me-Hg contents. However, higher consumption of cereals and other foods contribute more or less about the same amount of Hg intake as that from fish. As a result of dietary habits, fish cannot be blamed as the highest source of Hg for human diets. Nevertheless, fish food is usually considered to contribute a great proportion of Hg in a diet. However, this can highly differ. For example, in France, the Netherlands, the UK, and the USA, fish consumption contributes 35% of Hg in a diet, whereas in Finland the intake percent through fish is the highest, at 85% (Galal-Gorchev 1993).

The intake of Hg via the food by the population of the USA was: 126 ng kg^{-1} d^{-1} (range 5.7–3 131) and for females and 123 ng kg^{-1} d^{-1} (range: 3.1–2 367) for males (Gunderson 1988). The youth (>14 years old) of Poland takes up Hg at the range of mean values at 7–114 μg per week in rural and urban regions, respectively (Ludwicki and Wiadrowska 1992).

It is advisable for pregnant women to limit their consumption of shark and swordfish as these contain higher concentrations of Me-Hg than other fish (FDA 1998). The concentration of Me-Hg in fish is of a special concern. It is regulated by restrictions of the consumption, especially for pregnant woman and children, when the concentration in fish is higher than 1 mg Hg kg^{-1}. The tolerable intake of Me-Hg has been established by the FDA in 1969 at 1.6 μg kg^{-1} BW. At Amazon, Brazil, 80% of the people consume fish daily and on average two meals a day. The mean concentration of Hg in hair of the studied population was 17.8 mg kg^{-1}, at the range of 6.3–36.9 mg kg^{-1}. According to the WHO, the limit value of Hg in hair is 6 mg kg^{-1}. These results suggest that the population of Amazonian riverine communities is at risk from the Hg contamination due to high Hg fish ingestion (Cavalcante de Oliveria et al. 2004).

Since the 1950s, when Hg compounds were commonly used as fungicides, several Hg intoxications have been observed in humans. It was known as *Minamata* disease and has occurred in Japan, after the consumption of fish and shellfish with increased Hg contents, in the range of 2.6–6.6 mg kg^{-1} (Adriano 2001). In the 1970s, in Iraqui, people consumed cereals that were dressed with Hg-fungicides and were also heavily intoxicated with this metal (Mitra 1986; Takeuchi et al. 1989; Tsubaki and Takahashi 1986).

Animals. Mercury is present in tissues of all organisms and usually is more concentrated in sea animals than in terrestrial ones. Ringed seal (*Phoca hispida*) contained 27–187 mg Hg kg^{-1} FW, and this meat is a traditional and common food of the coastal Inuit people (Eaton et al. 1980). Elevated levels of Hg in hair (109 mg kg^{-1}) and blood (0.037 mg l^{-1}) of Inuits were reported, but they had no symptoms of Hg poisoning (Eisler 2000).

The biomagnification and transfer of Hg through mammalian food chains is well documented, however it can vary considerable (Galster 1976; NAS 1978; Eaton et al. 1980; Eisler 1981; Wren 1986). There are different Hg concentrations in herbivorous mammals and carnivorous mammals. The former species (e.g., mule deer, moose) contain <1.0 mg Hg kg^{-1} FW in liver and kidney, whereas the later species (e.g., marten, polecat, and red fox) frequently contain >30 mg kg^{-1} (NAS 1978).

Trace Elements of Group 13 (Previously Group IIIa)

The Group 13 consists of five elements: boron (B), aluminum (Al), gallium (Ga), indium (In), and thallium (TI) (Table II-13.1). These elements are characterized by having three electrons in their outer energy levels, but reflect a wide range in the occurrence and behavior. The geochemistry of these widely distributed elements is quite complicated. Among them, only B is metalloid and plays a significant role in plants. Aluminum, being one of the basic constituents of the lithosphere, reveals amphoteric properties. Three other trace metals are widely distributed in the lithosphere and, due to their special properties, share an important position in the field of electronic industry.

II-13.1 Boron (B)

II-13.1.1 Introduction

The chemistry of B is extremely complex due to which a special attention has been paid to it by the scientific community. Boron and its compounds have a special application in various military sectors and in organic syntheses. The importance of B compounds in bonding theory and organic chemistry has been presented by two Noble prize-winners (Power and Woods 1997).

Table II-13.1. Selected properties of trace elements of the Group 13

Element	Atomic number	Atomic mass	Atomic radius[a] (pm)	Density (20 °C, g cm^{-3})	Valence[b]	Melting point (°C)
B, boron	5	10.81	117	2.34	+3	2079
Al, aluminum	13	26.98	143	2.69	+3	660
Ga, gallium	31	69.72	181	5.90	+1, +2, **+3**	29
In, indium	49	114.81	200	7.31	+1, +2, **+3**	156
Tl, thallium	81	204.38	208	11.84	**+1**, +2, +3	303

[a] Approximately average values for the main oxidation states.
[b] Valence values in bold are for the main oxidation states.

B

Table II-13.2. Abundance of elements of Group 13 in the environment[a]

Environmental compartment		B	Al	Ga	In	Tl
Earth's crust (mg kg^{-1})		10	8.2[b]	15 – 19	0.25	0.85 – 1
Igneous rocks (mg kg^{-1})	Mafic	5 – 20	7.3 – 8.8	15 – 24	0.02 – 0.22	0.1 – 1.4
	Acid	10 – 30	6 – 8.3	16 – 20	0.04 – 0.2	0.6 – 1.8
Sedimentary rocks (mg kg^{-1})	Argillaceous	120 – 130	7 – 10[b]	15 – 25	0.05 – 0.07	0.5 – 2
	Sandstones	30 – 35	2.5 – 4.3	5 – 12	0.02	0.4 – 1
	Calcareous	20 – 30	0.4 – 1.3	1 – 3	0.03	0.01 – 1.5
Raw energy material (mg kg^{-1})	Coal, hard[c]	49 (1 700)[d]	1.5 (10.6)[b,d]	5.7 (45)[d]	0.1 – 0.2[a]	1.2 (52)[d]
	Fly ash[c]	50 – 509	13.23	40.8	–	1.4
	Crude oil	<0.2	–	0.02	–	–
	Petrol	0.02 – 0.1	–	–	–	–
Soils (mg kg^{-1})		–	1 – 4[b]	28	–	–
	Light sandy	10 – 22	–	<3 – 30	<0.2 – 0.5	0.014
	Medium loamy and silty	15 – 35	–	5 – 50	0.03 – 4.1[e]	0.18 – 2.8[e]
	Heavy loamy	15 – 40	–	5 – 70	–	–
	Calcareous	10 – 100	–	5 – 30	–	–
	Organic	20 – 54	–	7 – 50	<0.1 – 2.6	0.17 – 0.22
Water (µg l^{-1})	Sea[f]	4500	0.03 – 2	0.001	0.0001	0.019
	River[g]	10 – 100	32	0.03	–	0.008 – 0.04
Air (ng m^{-3})		–	–	–	–	0.06 – 0.22[h]
	Urban/industrial	–	150 – 15 000	0.23 – 1	20 – 1 200	–
	Remote regions	–	240 – 380	<0.001	<0.08	–
	South Pole	–	0.32 – 0.8	<0.14	0.05	–

[a] Presented are common, possible background, values from various sources as compiled by Kabata-Pendias and Pendias (1999 and 2001), unless otherwise indicated. [b] Data for Al are in %. [c] After Finkelmen (1999) and Llorens et al. (2000). [d] Arithmetic mean and maximum concentrations. [e] Data for reference soils, after Govindaraju (1994). [f] After Reimann and Caritat (1998). [g] After Gaillardet et al. (2003). [h] Data from ATSDR (1992).

Boron is widely but not uniformly distributed in the environment. Its content in igneous rocks increased with the acidity of the rocks, while in sedimentary rocks it is closely associated with the clay fraction (Table II-13.2). Its common minerals are: borax, $Na_2B_4O_7 \cdot 10H_2O$; colemanite, $Ca_2B_6O_{11} \cdot 5H_2O$; ulexite, $NaCaB_5O_9 \cdot 8H_2O$; kernite, $Na_2B_4O_6 \cdot 3H_2O$; and tourmaline minerals of a very complex composition. Boron might be associated with feldspars and micas.

Boron has a high affinity for oxygen, due to which it occurs in the nature mainly bound with oxygen (B-O compounds) (Wyness et al. 2003). Its chemistry is rather complex and resembles that of Si (Cotton and Wilkinson 1989). It is relatively widely distributed in the lithosphere and hydrosphere (Table II-13.2).

II-13.1.2
Production and Use

About 0.8 Mt of commercial borate products (B_2O_3) were manufactured in 1994 from B minerals. In 2003, the global B production was 4.4 Mt (WMSY 2004).

Boron is produced by the chemical reduction of B-compounds with reactive metals, either by thermal decomposition or by nonaqueous electronic reduction (US Bureau of Mines 1989). Borate deposits are quite rare and generally found in arid regions, with a history of volcanism or hydrothermal activities. The United States is the world's major producer of borates (WHO 1998). Boron is a highly refractory metalloid and the production of pure B is rather difficult due to its high melting point and corrosive properties in the liquid form.

Borates have varied uses (Table II-13.3), and the principal sector (56%) is fiberglass and heat resistant borosilicate glass (e.g., Pyrex) production. It is used in flame retardant tools and textiles, agricultural fertilizers and pesticides, cosmetics, antiseptics, and leather tanning. Its great proportion is added to laundry products (US Bureau of Mines 1989; WHO 1998).

Table II-13.3. Use pattern of boron compounds (kt)

Sector	USA in 1992[a]	Europe in 1993[b]
Fiberglass	129	44.6[c]
Textile-grade fiberglass	78.5	27.1
Detergents and soaps	38.6	142.5
Borosilicate glass	34.4	12.2
Fire retardants	13.4	–
Agriculture	11.1	–
Enamels, ceramic glass	9.3	3.5
Metallurgy	3.7	–
Nuclear application	0.9	–

[a] After Lyday (1993).
[b] After ECETOC (1997).
[c] Does not include minerals.

B

II-13.1.3
Soils

The B content in soils is associated with the clay fraction and with SOM. The B status of arable soils has been extensively investigated and shows a very broad range from 1 to over 400 mg kg^{-1}, at a common range between 10 and 100 mg kg^{-1} (Table II-13.2). Its highest concentrations were reported for lateritic and calcareous soils. Boron is inherited mainly from parent rocks, but industrial pollution, sludge and coal fly ash application to soils, as well as B-fertilizers are significant sources of this element (Butterwick et al. 1989; WHO 1998; Matsi and Keramidas 2001).

Boron is relatively strongly sorbed by both inorganic and organic soil components and increases with soil pH, reaching a maximum level at the alkaline pH range of 9 (Goldberg et al. 1993a, 1996). Than, its sorption decreases at the pH range of 10–11.5 (Goldberg and Glaubig 1986). The pH is a guiding factor for the mobility and phytoavailability of B in soil. Several other soil parameters, such as SOM, clay minerals, Fe- and Al-hydroxides, and carbonates may also highly control its behavior. The mechanism of B adsorption on Al- and Fe-oxides is based on ligand exchange with reactive surface hydroxyl groups. The B adsorption creases for Al_2O_3 at pH range of 6–8, and for FeO at a pH between 7 and 9 (Goldberg and Glaubig 1986; Su and Suarez 1995; Toner and Sparks 1995). Goldberg et al. (1993b) reported that the B adsorption decreases in soils dominated by crystalline minerals, when temperature varies between 10 and 40 °C.

In soils, B is considered to be the most mobile element among the micronutrients, although its water-soluble fractions is relatively low and varies from about 3 to 5% of the total content (Table II-13.4). Although B is rather deficient micronutrient in most soils, some soil of arid and semi-arid regions, soils overfertilized with B, and soil amended with sewage sludge and fly ash may contain hazardous amount of this element (Zhang et al. 2004a). Surface water contaminated with B may create agricultural problems in some areas, especially when it is used for the irrigation (Nicholson and Wood 1997).

Table II-13.4. Total and water-soluble boron content of soils in China (mg kg^{-1}) (from Liu Zheng et al. 1980)

Region, soil	Parent material	Total B	Water-soluble B
South of Yangtze, acrisols	Granite	4 – 16	0.06 – 0.29
	Rhyolite	13 – 19	0.21 – 0.24
	Phyllite	15 – 40	0.02 – 0.19
	Red sandstone	25 – 28	0.0 – 0.11
	Gneiss	13 – 22	0.0 – 0.13
North of Yangtze, different soils	Loess	55 – 87	0.14 – 0.70
Central China, different soils	Quaternary red clay	48 – 96	0.08 – 0.37
	Limestone	49 – 145	0.14 – 0.34

II-13.1.4 Waters

The median concentration of B in worldwide ocean water is calculated at 4 500 $\mu g\,l^{-1}$ (Reimann and Caritat 1998). Mean content of B in river water is 10.2 $\mu g\,l^{-1}$, in a broad range of 1.5–150 $\mu g\,l^{-1}$ (Gaillardet et al. 2003). These authors estimated the annual riverine flux of B to be 380 kt. In the USA, B in surface waters ranges from 1 to 5 000 $\mu g\,l^{-1}$ (US EPA 1986; Butterwick et al. 1989).

Recent investigations on B in the EU countries have indicated its concentrations between <10 and >1 000 $\mu g\,l^{-1}$. Wyness et al. (2003) reported B concentrations in European rivers to range (mean values) from 17 to 632 $\mu g\,l^{-1}$, being the lowest in England and Finland and the highest in Germany and Portugal. Rivers of Asia contain B in the range of <10–7 000 $\mu g\,l^{-1}$. Mean B values for river water in Japan is 10 $\mu g\,l^{-1}$ and in South Africa is 300 $\mu g\,l^{-1}$ (Neal et al. 1998). This author explains such variations due to both natural and anthropogenic factors. High content of water-soluble B in fly ash may be easily leached into the aquatic environment depending on the ash to water ratio (Eisler 1990). At the presence of soluble Al compounds, B forms an insoluble aluminum-borate complex, which reduces its solubility by about 90%.

Boron concentration in rainwater sampled in Sweden in 1999 varied between 1.3 and 2.9 $\mu g\,l^{-1}$, and its wet deposition is calculated at 18 $g\,ha^{-1}\,yr^{-1}$ (Eriksson 2001). Median B concentrations in rainwater from the Kola Peninsula range from 0.5 to 0.7 $\mu g\,l^{-1}$ in remote and polluted regions, respectively (Reimann and Caritat 1998).

In surface waters, B can occur mainly as boric acid H_3BO_3 and anion $B(OH)_4^-$. Besides, it forms polyborate and fluoroborate complexes. Boron may also occur in complexes with transition metals and organic compounds (Bassett 1980).

Boron is released into surface waters through weathering processes, volcanic emission, and, in variable amounts, from anthropogenic sources that include sewage outfalls, especially with detergent products, leaching salt deposits, and B-fertilizers. The yearly input of B in surface waters of the USA is calculated at 32 Mt, and this creates environmental problems, especially to agricultural soils (Nicholson and Wood 1997).

In sewage, B occurs primarily as an undissociated boric acid. The concentration of B in sewage of various countries is relatively high (in $mg\,l^{-1}$):

- USA: 0.4–1.5; max. 4, Butterwick et al. (1989)
- Europe: mean 2; max. 5, Butterwick et al. (1989)
- Spain: mean 1.45, Navarro et al. (1992)
- Sweden: 0.34–0.436, Ahl and Jönsson (1972)
- Egypt: 0.32–0.38, El-Hassanin et al. (1993)

Boron in bottom sediments may persist for a long period; desorption and leaching of B from sediment may occur when the system reaches equilibrium at the sediment-water interface. The primary desorption processes are based mainly on the diffusion.

Maximum adsorption generally occurs at pH 7.5–9.0 (Keren and Mezuman 1981; Keren et al. 1981).

The B concentration in groundwater may be of a real agricultural concern in some regions. Its worldwide contents in groundwater vary between <0.3 and >100 mg l^{-1} whereas in Europe it is in the range of 0.3–1.5 mg l^{-1}, where the highest concentration was noted for Italy and Spain, and the lowest for countries in the northern Europe (ECETOC 1997).

The concentration of B in tap water has been reported to range from 0.007 to 0.2 mg l^{-1} in the USA and England. A survey of 969 public water supply systems of the USA showed B concentration at <1 mg l^{-1} (ATSDR 1992). Boron is not removed from groundwater and surface water used for drinking purposes and it is not of a health concern.

Aquatic biota. Boron is not toxic to aquatic biota, but some of its compounds, like boric acid and borax may be harmful to sensitive organisms (e.g., trout). Fish and amphibians can tolerate up to 10 mg B l^{-1}, for some periods without any adverse effects, but it has been suggested that concentration >0.1 mg l^{-1} may affect reproduction of rainbow trout (*Oncorhynchus mykiss*) and >0.2 mg l^{-1} may affect the survival of other fish species (Eisler 1990).

Aquatic plants of three rivers in the USA, polluted from industrial and agricultural sources contain B in the ranges of 26.3–170 mg kg^{-1}, and 11.3 to 57 mg kg^{-1}, depending on plant species (Adams et al. 1973). In marine algae and phytoplankton, B values ranged from 4.2 to 14.9 mg kg^{-1} (Yamamoto et al. 1973). Studies of B in plants, insects and fish have shown that the B does not biomagnify in aquatic species and in aquatic food-chain (Table II-13.5).

Table II-13.5.
Boron concentration in aquatic organisms (mg kg^{-1})

Organism (year of sampling)	Range, mean
Lake trout, muscle[a]	0.2 – 0.6 FW
Aquatic macrophytes[b]	2 – 19
Waterweed, whole[a]	18 – 44
Yellow pond lily, whole[a]	23 – 31
Aquatic plants[c] (1983)	270 – 510 (382)
Aquatic insects[c] (1983)	36 – 54 (45)
Widgeongrass[d], whole (1984)	120 – 780 (371)
Widgeongrass[d], seeds (1984)	450 – 3 500 (1 860)
Seaweeds[e], whole, 41 species	16 – 319 (106)

[a] Jenkins (1980).
[b] Ahl and Johnsson (1972).
[c] Ohlendorf et al. (1986).
[d] Schuler (1987).
[e] Eisler (1981).

II-13.1.5 Air

Natural emissions of B occur from the sea and volcanic activities. Its industrial sources include: mining operations, glass and ceramics industries, chemicals production, and coal fired power plants. It may occur in forms of aerosols or in particulate (<1–45 μm in size), as borides, boron oxides, borates, boranes, organoboron compounds (e.g., borazines). During coal combustion, about 70% of the coal B enters into the atmosphere (Kuhn et al. 1980; Pagenkopf and Connolly 1982). Borates are soluble in water and can be removed from the atmosphere by wet precipitation or by dry deposition (US EPA 1987).

The global anthropogenic emission of B varied between 180 and 650 kt yr^{-1} whereas B from marine sources has been estimated at the range of 0.8–4 Mt yr^{-1} (Anderson et al. 1994). This author has calculated the total global release of B to the atmosphere at the range of 2.0–7.2 Mt yr^{-1}. Considering the available emission data, the mean global B concentration range in the air is: >0.5–80 ng m^{-3}, but over the continents it is 20 ng m^{-3} (WHO 1998). The half-life of airborne B particles has been estimated to be in the order of days depending on the particle size and weather conditions.

Increased B amounts may be released from ceramics industries, as it has been reported for Spain (Riog-Navarro et al. 1997). According to these authors, 30% of samples from the working atmosphere contained B above the recommended TLV, established at 1 mg B m^{-3} (ACGIH 1994).

II-13.1.6 Plants

Boron is an essential element needed for the growth and development of vascular and some aquatic plants while bacteria, fungi, and green algae do not require this element (Loomis and Durst 1992). Its deficiency in crop plants is widespread (Gupta 1993a). However, in arid and semi-arid irrigated regions, particularly in humid and temperate climatic zones, B toxicity does also occur. The water-soluble B pool is readily available for plant uptake (Tsadilas et al. 1994; Xu et al. 2001).

The B contents of plants grown under natural conditions widely vary for plant species and kind of soil. Since dicot plants have a higher B requirement they contain a higher B amounts (Table II-13.6). The lowest B contents have been always found in seeds and grains. The highest B level, >100 mg kg^{-1}, has been often cited for sugar beet leaves. The critical B levels for most plants ranges at 5–30 mg kg^{-1} (Kabata-Pendias and Pendias 2001).

Despite of the essentiality of B for higher plants, its biochemical role is still not well understood. Boron has a complexing ability to trigger the movement of sugars and other materials, and it is involved in cell-wall bindings, conversion of glucose-l-phosphate sugars to starch, and in the metabolism of nucleic acids (Goldbach and Amberger 1986). Recent finding have emphasized the particular importance of B interactions with biological membranes, and with variety of biomolecules. The deficiency of B in crop plants affects seed yield reduction, low male fertility, an impaired reproduction,

B

Table II-13.6. Boron in food and forage plants ($mg\ kg^{-1}$) (after Kabata-Pendias and Pendias 1999, 2001, unless otherwise indicated)

Plant	Mean
Wheat, grains[a]	0.69
Barley, grains[a]	0.83
Rye, grains	4.3
Oats, grains	3.1
Bean, pods	13
Cabbage, leaves	14
Carrot, roots	9.9
Onion, bulbs	10
Potato, tubers	6
Tomato, fruits	6
Apple, fruits	8.3
Orange, fruits	8.4
Grasses	5 – 26
Clovers	14 – 78

[a] After Eriksson (2001a).

and disruption to the ascorbate metabolism (Dell and Huang 1997; Dixon 1998). Its deficiency can be detected visually in dicots, maize, and wheat, at tissue concentrations of less than 20–30, 10–20 and 10 $mg\ kg^{-1}$, respectively (Gupta 1993b). Dixon (1998) described limiting impact of B on Clubroot pathogens of brassica plants.

Boron deficiency often occurs in plants growing in sandy soils because the availability of B from coarse textured soil is less than that from fine textured soil (Fleming 1980). One of the causes of B deficiency in plants is dry soil. Plant roots cannot extract moisture from lower depths during dry conditions and the diffusivity of B decreases with decreasing water content (Scott et al. 1975).

Soluble forms of B are easily available to plants that can take up undissociated boric acid as well as other B species present in the ambient to root solution. Brown and Shelp (1997) have classified plants into two categories: *(i)* species with restricted B mobility, and *(ii)* species with significant B mobility. In the former case, water is the translocation agent for the movement of B. The B uptake is therefore proportional to its concentration and the water flow. In the later case, B shows rapid and significant phloem mobility in species for which sorbitol is the primary photosynthetic product (Brown and Hu 1996). According to Kohl and Oertli (1961): "the pattern of B distribution in plants and symptoms of B toxicity are correlated with leaf ventilation and are consistent with the hypothesis that B moves with the transpiration stream".

Oertli and Richardson (1970) concluded that B in leaves can be mobile and accumulated at the sites of the termination of leaf veins. Boron immobility, due to a kind of fixation, has also been observed in the leaf of walnut (Oertli 1993, 1994). Eaton (1944)

Table II-13.7. Boron removal by crops (from Shorrocks 1997)

Crop	Yield (t ha^{-1})	B removed by crop (g ha^{-1})
Sugar beet	50	300
Oilseed rape	4	80
Cotton	3	150
Sunflower	3.5	100
Apple	40	250
Lucerne	7	350
Wheat	3	25
Grape	6	120
Maize	6	30

has explained that B may react with organic compounds which prevent its translocation to other tissues. However, there are several theories on the mobility of B in plants (Campbell et al. 1975; Hanson and Breen 1985; Shelp and Shattuck 1987).

Most crops remove less than 100 g B ha^{-1}, but there are some exceptions such as lucerne and sugar beet, which often remove about 300 g B ha^{-1} (Table II-13.7). Boron removal by cereals and grains is quite low but still additional B fertilizer is needed to repair B deficiency in some soils, although significant amounts of B may be supplied through rainfall (Liu et al. 1980).

The deficiency of B has been reported in 132 crops and 80 countries, including several European countries (Shorrocks 1997). A global assessment of B in soils and plants has indicated that the B deficiency in crops is possible in any country, but its deficiency is noticeable in the most Asian and African countries (Sillanpää 1982).

Boron is an essential element for higher plants, but it is toxic to most plant species when accumulated at higher concentration. The increased B content of irrigation water may be especially toxic to crop plants grown in arid regions. The ability of a plant to adopt to high concentrations of B may depend on the germplasma and may be controlled by both physiological mechanisms and genetic diversity of species.

Bañuelos et al. (1999) and Ayars et al. (1990, 1994) observed plant resistance to B-enriched waters in the arid regions of California and Chile. Toxic effects may occur in some B-sensitive crops such as beans, carrots, and lettuce when the concentration of B in water exceeds 4 mg l^{-1} (Maas 1986; Shennan et al. 1995). Boron tolerance has been reported for non-horticultural plants such as salt grass (*Distichlis spicata*), Bermuda (*Cynodon dactylon*) and mesquite (*Prosopis chilensis*) growing in the arid regions of Chile.

II-13.1.7 Humans

Contents of B in mammalian tissues vary between <0.2 and <0.5 mg kg^{-1}, being the highest in kidneys and liver and the lowest in skin (Jørgensen 2000). Its concentra-

B

Table II-13.8.
The dietary boron intake by adults (mg d^{-1}) (after Coughlin and Nielsen 1999)

Country	Male	Female
United States	1.11	0.89
Germany	1.72	1.62
Great Britain	1.30	1.14
Mexico	2.12	1.75
Kenya	1.95	1.80
Egypt	1.31	1.24

Table II-13.9.
Boron in some foods (mg kg^{-1} FW) (after Mosonen 1994)

Food	Mean
Human milk	0.80
Cows milk	0.20
Egg yoke	0.008
Egg white	0.14
Barley	2.3
White floor	0.45

tions in soft tissues of humans range from 0.06 to 0.6 mg kg^{-1}, in brain and kidneys, respectively. The average content in tissues of the "reference man" is 0.3 mg kg^{-1} (Li 2000). This author has cited the B mean concentrations (in mg l^{-1}) at 0.13 in human blood, and at 0.71 in urine.

In 1996, the WHO Expert Committee on Trace Elements in Human Nutrition concluded that B is "probably essential". Penland (1994) suggested that B may play a role in cell membrane function, mineral and hormone metabolisms, and enzyme reactions. Nielsen (1991, 1994, 2004) reported that B is required or beneficial in humans and animals for many life processes, such as embryo development, bone growth and calcification, immune function, psychomotor skills, and cognitive function.

The common total intake of B by adults from food is calculated at about 1.2 mg d^{-1} (WHO 1998). It varies a little bit among the countries and sex (Table II-13.8). Vegetables and fruits are better source of the dietary B than milk and eggs (Table II-13.9). Tolerable intake (TI) by adults has been estimated for B at 0.4 mg kg^{-1} BW, thus an acceptable intake of B has been proposed for adults as 18 mg d^{-1} (Murray 1995). The WHO (1998) guideline value in drinking water is 0.3 mg B l^{-1}, whereas Murray (1995) calculated that since drinking water consumption is about 2 l d^{-1}, a person could drink water containing up to 8.25 mg B l^{-1}.

Boron is not very toxic, especially when ingested. Symptoms of acute toxicity, are: nausea, vomiting, diarrhea and dermatitis. In animals, toxicity occurs after the dietary B exceeds 100 mg kg^{-1}.

II-13.2 Aluminum (Al)

II-13.2.1 Introduction

Aluminum is the third most abundant element in the Earth's crust, occurring at about 8% (Table II-13.2). It reveals lithophile properties, one oxidation state (+3) and is slightly mobile in hypergenic zones. However, in an acidic environment, its solubility increases and Al becomes toxic to terrestrial and aquatic species. Naturally occurring Al compounds exist mainly in undissolved forms.

Aluminum is either main or secondary component of many minerals, especially of all silicates. Minerals composed only of Al are: boehmite, γ-AlOOH; diaspore, α-AlOOH; corundum, Al_2O_3; and hydrargilite/gibbsite, γ-$Al(OH)_3$ that is the main component of bauxite rocks. Possible host minerals of Al are: feldspars, micas, and all layer silicates.

Aluminum reacts with mineral acids and strong alkalis due to its amphoteric character. It exhibits a strong relation to electronegative groups (e.g., OH^-) and to some ions (e.g., F^-). Its partitioning between solid and liquid phases has been cited as results of reactions between water molecules and electron-rich anions, e.g., chloride, fluoride, sulfate, nitrate, and negatively charged functional groups of humic materials and clays (WHO 1997).

In the 1970s and 1980s, forest began to die due to root damage in Europe, the USA, and Canada which was found to be due to acid rain liberating soil metals, especially Al (Rosseland et al. 1986; Wells et al. 1986). Nuorteva (1990a) assessed forest damage to be due to a high Al-induced Ca deficiency syndrome. The toxic forms of Al have been found to be mainly inorganic monomeric Al (Berggren and Fiskesjo 1987), but Kennedy (1986) mentioned that Al^{+3} and Al-hydroxide, $Al(OH)_2$ are the most usual toxic forms of Al and it has been observed that Al toxicity is rather a complex phenomenon which is influenced by the hardness, pH, and DOC of the water (Freda 1991).

II-13.2.2 Production and Use

World production of Al was 139 Mt in 2000 and 148 Mt in 2003 (WSMY 2004). In 1998, global recycling of Al scrap was about 11.6 Mt, which fulfilled close to 40% of the global demand. In Europe, average recycling of Al cans and other scrap is about 40%. In some countries, such as Sweden and Switzerland, Al scrap recycling is over 80%.

Aluminum ores, most commonly bauxites, contain up to 55% of $Al_2O_3 \cdot H_2O$ and occurs mainly in tropical and sub-tropical areas. There are also some bauxite deposits in Europe. The commercial deposits are mainly gibbsite, and boehmite. Primary aluminum production facilities are located all over the world, often in areas where there are abundant supplies of inexpensive energy, such as hydro-electric power (Dinman 1983; IARC 1984).

Scrap Al has significant value and can be recycled repeatedly. The recycling of Al requires only 5% of the energy as compared to primary production and as such generates only 5% of the "green house" gas emissions (http://www.world-aluminium.org/production/recycling/index.html).

Due to versatile properties of Al, it is used in different industrial sectors including: metallurgical, packaging, transportation, construction, and electrical and chemical productions. Its compounds are used in paper manufacturing, water purification, sugar refining, wood preservation, leather tanning, waterproofing textiles and many others. Aluminum is a component of cathode ray tube heating elements, is used as a corrosion inhibitor, cement accelerator, and has several others applications (HSDB 1995).

II-13.2.3 Soils

Aluminum is an abundant element in soils and its contents vary commonly between 1 and 4% (Table II-13.2). The total Al content of soils is inherited from parent rocks and its species and distribution in soil profiles are highly governed by several soil properties of which the soil pH is especially significant. Only easily mobile and exchangeable fractions of Al play an important role in soil fertility, however, the dynamic equilibrium among various Al species and insensitivity in the transformation are difficult and almost impossible to identify correctly, even using sophisticated modern technologies (Boudot et al. 1996; Smith et al. 1999; Umemura et al. 2001, 2003).

Aluminum species in soil solution highly depend on the pH. The complexation also increases Al solubility and mobility. In the absence of ligands (mainly DOC), pH value controls Al speciation; following species are likely to predominate at given pH values: <5.0, Al^{3+}; 5.0, Al^{3+}, $AlOH^{2+}$, $Al(OH)_2^+$; 5.5, $Al(HO)_2^+$; and >6.2, $Al(HO)_4^-$ (Kinraide 1991). Some colloidal forms of Al compounds and organic complexes also occur in aqueous phase.

The solutions of neutral soils contain Al at about 400 μg l^{-1}, while its concentration in the soil solution at pH 4.4 was reported to be 5 700 μg l^{-1} (Gough et al. 1979). The concentration of organic Al in soil solution linearly correlates with concentration of carbon and increases with depth, from 3.3–9.8 μmol l^{-1} at 5 cm to 115 μmol l^{-1} below 55 cm (Nilsson and Bergkvist 1983).

The behavior of Al is often modified by SOM, and the organically bound Al forms in solid and liquid soil phases are resulting from different equilibrium processes (Berggren and Mulder 1995; Drabek et al. 2003). Results of studies carried out by Walker et al. (1990) confirmed conclusions that the equilibrium solubility of Al is dependent on pH and the degree of SOM saturation with this metal. Temperature dependence and rate studies indicated that the Al solubility is governed by an ion exchange reaction between H^+, Al and SOM (WHO 1997). The influence of soil acidity on Al binding in sandy soils with low humus content (<2%) has indicated that binding capacity of Al decreases slowly from very strong at pH 5.5–7.0 to very weak at pH 2.5, but increases with increased SOM content.

During weathering of parent materials, at acid environment, highly reactive Al compounds and poorly-ordered alumino-silicates (including organic-alumunic compounds) are formed (Álvarez et al. 2002). These authors studied the role of tree vegetation (oak, pine and eucalypt) and types of parent materials (granodiorite, shale and limestone) in the chemical properties of forest soils, with a reference to Al speciation in solid and liquid phases. The results have indicated that plant species have effective impact on the concentration and mobilization of Al in soils, and thus may affect the formation of Al-organic ligands and Al phytotoxicity. The acidification of soils of coniferous forests in the Scandinavian countries highly controls the Al speciation in soil solution. Its total concentration in the solution increases with the increase of soil acidity (McBride 1994; Lundström and Gieslet 1995; Hees et al. 2001).

The Al mobilization in acid soils together with the release of basic cations has been often reported for Swedish forest (Bergkvist 1987; Driscoll et al. 1989). Leaching of Al has been observed in all soil kinds, but the maximum leaching was noticed from podzols (Bergkvist 1987). Also, soil acidification due to the atmospheric deposition of S (mainly SO_2) increases the Al solubility. The oxidation of sulfides present in soils (e.g., pyrites) produces H_2SO_4, which interact with other minerals to yield Al ions. The mobile Al in acid soils can be taken up by plants and it creates a problem of chemical stress in plants. Foy (*vide* Kabata-Pendias and Pendias 2001) reported that Al toxicity in subsoils is particularly harmful because it causes shallow rooting, drought susceptibility, and poor use of subsoil nutrients.

II-13.2.4 Waters

Aluminum concentration in ocean and seawaters depends on the salinity and is relatively low, varying commonly between 0.03 and 2 $\mu g\,l^{-1}$ (Table II-13.2). River waters contain much higher Al amounts that range broadly from about 2 to over 1 000 $\mu g\,l^{-1}$, depending on surroundings rocks and soils. Its average content is calculated at 32 $\mu g\,l^{-1}$, and the global riverine flux at 1 200 kt yr^{-1} (Gaillardet et al. 2003). Shiller (1988) reported Al concentrations in rivers of the USA to vary between 95 to 2 370 pmol kg^{-1}.

In the surface layer of seawater it may be at a higher level due to the atmospheric deposition and higher scavenging rate. In the North Atlantic Ocean, Al level is on the lower side due to a low atmospheric input (WHO 1997). In North Sea, the measured value of Al for 1988 was cited at the range of 10.2–49.2 nmol l^{-1} (Hydes and Kremling 1993).

The speciation of Al in the aquatic systems depends on several parameters, such as pH, DOC, and contents of fluoride, sulfate, phosphate, silicate, and suspended particular matter. In acidic environments, the concentration level of Al may rise to 500–1 000 $\mu g\,l^{-1}$ (WHO 1997). According to thermodynamic modeling, there is considerable uncertainty regarding presence of $AlOH^{2+}$, $Al(OH)_2^+$ and $Al(OH)_3^0$ which are important species at the pH between 4.5 and 6, the most dominant form of Al species being Al^{3+} at pH <5 (Dobbs et al. 1990).

The presence of organic ligands, such as salicylate and oxalate, has important impact on Al complexes, especially at the pH range 4–5 (Plankey and Patterson 1987). The nadir of aqueous Al solubility is at pH 6.2 (Yokel 2004). The Al species in waters may be quite variable (Table II-13.10).

Moiseenko et al. (1995) studied Al and other trace elements released from metal industries in the northeastern part of Fennoscandia, located above the Arctic (Polar) Circle and have indicated that sulfate concentrations in the surrounding lakes are causing acidification which triggered the toxicity of Al. Around Al smelter, the Al concentrations in waters and sediments increase, as well as its mobilization. At the distance of 20 km away from the smelter, Al concentrations were between 15 and 63 mg Al l^{-1}, and did not decrease at 80 km from the smelter, where it was still at the value of 70–90 mg Al l^{-1}.

Table II-13.10. Aluminum speciation in waters from different sources (data from Cañizares et al. 1994, Mitrovic et al. 1998, and Quintela et al. 1993)

Species	Water type
$Al(OH)^{2+}$, $Al(SO_4)^+$, $Al(OH)_4^-$	Natural
Fast reactive	Tap
Free, labile, hydroxopolymers, Al-humate	Aqueous soil extracts, humic waters
Nonlabile monomeric, reactive	River, tap
Non-labile monomeric, labile monomeric, anionic	Bottled, tap, spring, well, river

Rainwater from the Kola remote region contain Al at 2.9 μg l^{-1} and from polluted region 105 μg l^{-1} (Reimann and Caritat 1998).

The Al concentration in drinking water of the USA is mainly below 100 μg l^{-1} (Schenck et al. 1990). Acceptable level of Al in drinking water is established in Europe at 50 μg l^{-1}, and by the WHO (1997) at 200 μg l^{-1}. The US EPA accepted a broad level for Al in drinking water at the range of 50–200 μg l^{-1} (Yokel 2004). At neutral pH, the Al level in groundwater wells can be less than 100 μg l^{-1}, but in acid waters (pH 4.4), it is often above 5 000 μg l^{-1}.

Aquatic biota. There is major concern regarding the Al toxicity in freshwater systems. The toxicity of Al to fish in acidic waters, especially of lakes and rivers, has been well documented (Godal et al. 1995; Camilleri et al. 2003) and it is believed that aqueous Al is the main toxicant killing fish in these waters (Howells et al. 1994; Gensemer and Playle 1999).

Respiratory problems in fish generally occur due to the Al toxicity at pH >5.5, when the disturbance of ion regulatory is resulting in a decrease in plasma Na^+ and Cl^- (Neville 1985; Gensemer and Playle 1999). High concentrations of water Ca may decrease Al toxicity in fish (Wood and McDonald 1987; Playle et al. 1989). Poléo and Hytterød (2003) have stated, based on results of the laboratory experiment, that there is no mortality at Al concentration of 350 μg Al l^{-1} to Atlantic salmon for three weeks when water pH is 9.5. It also has been documented that rainbow trout exposed to Al at pH >10 survived for a few weeks (Yesaki and Iwama 1992; McGeer and Eddy 1998).

There is evidence that in freshwater, 40–50% of Al is present as particulate/colloidal forms while in seawater Al occurs mainly as reactive Al species. This suggests that increased salinity transformed Al from nonreactive to reactive form that has toxic effects to fish. There is also a contradiction regarding the presence of base cations (Ca^{2+}, Mg^{2+}, Na^+, K^+) and Al toxicity on fish mortality (Alstad et al. 2005). Generally, increased water ionic strength (due to presence of Cl^+ and Na^+) mitigates the effect on fish exposed to toxic Al, by reducing the Al concentration on gill surfaces (Dietrich et al. 1989; Lydersen et al. 2002).

II-13.2.5 Air

World median Al concentration in air from remote regions is calculated by Reimann and Caritat 1998 at the range of 46–70 ng m^{-3}, and in air from polluted areas at

1 300 ng m^{-3}. Other estimations gave somewhat higher amounts of Al in air from urban/industrial regions (Table II-13.2). According to data of the WHO (1997), the atmospheric concentration of Al is 0.5 ng m^{-3} over Antarctica and up to >1 000 ng m^{-3} in industrial areas.

Aluminum dust in the atmosphere is found as aluminosilicates associated with crustal particulate matter. This dust is from both natural and industrial sources. Most of particle-borne Al dust stems from volcanic eruption, wind blown soil dusts, and Sahara sands.

The information from the European Aluminum Association (Nordheim@eea.be) suggests that Al released as Al_2O_3 from the European Al-industry is 0.21 kg t^{-1} of Al mined, whereas total dust emission containing Al_2O_3, AlF_3 and Na_3AlF_6 is about 2.33 kg t^{-1} of Al production.

Aluminum is generally associated with large particles (>2 μm diameter) that are deposited in the vicinity of a source. Rain (1981) calculated Al input to the Arctic Ocean due to the rainfall at 30 kt yr^{-1} that is much less than recently estimated global riverine inputs.

II-13.2.6 Plants

Aluminum contents in food plants vary from 7 to 104 mg kg^{-1}, being the lowest in apples and the highest in spinach. Its much higher concentration is in fodder legumes and grasses, up to above 3 000 mg kg^{-1} (Table II-13.11).

The physiological function of Al in plants is not clear, although there is some evidence that low levels of Al may have a beneficial effect on plant growth (Kabata-Pendias

Table II-13.11. Aluminum in food and forage plants (mg kg^{-1}) (after Kabata-Pendias and Pendias 1999, 2001, unless otherwise indicated)

Plant	Mean range
Wheat, grains[a]	31
Barley, grains[a]	38
Rye, grains[a]	70
Oats, grains[a]	47
Cabbage, leaves	8.8
Spinach, leaves	104
Onion, bulbs	63
Potato, tubers	76
Tomato, fruits	20
Apple, fruits	7.2
Orange, fruits	15
Grasses	7 - 3410
Clovers	85 - 3470

[a] Data for Poland, after Indeka and Karaczun (1993).

and Pendias 2001). The most important problem is associated with the Al toxicity, as one of the major factors which limits the growth and yield of plants cropped on acid mineral soils (Matsumoto et al. 2001).

During acidification of soil, there is an increased concentration of Al^{3+} ion in the soil solution, which is the most phytotoxic form of Al (Jones and Kochian 1997). It has been demonstrated that under low pH, polynuclear Al, such as Al_{13} may exist in solution, which is more toxic than monometric Al^{3+} (Parker et al. 1989). However, its existence in soil solution is still under question. A number of reports suggest that free Al is toxic to plant roots and very little proportion of Al is transported or adsorbed to the aerial plant parts (Mendoza and Borie 1998; Rufyikiri et al. 2000). Apparently, a large part of the Al is associated with cell walls and plasma membrane because the cell wall pectin and plasma membrane are negatively charged (Delhaize and Ryan 1995).

The main toxic effect of Al is the inhibition of root growth that has been shown on Al-sensitive maize and many other plants, such as tobacco and tea plants (Lewis 1990; Sivaguru and Horst 1998; Blancaflor et al. 1998). Additional effects of Al toxicity include: poor root hair development and brittle roots due to decreased translocation of water, nutrients and phosphate from roots to shoots (Delhaize and Ryan 1995). One of proposed mechanisms of the Al toxicity is rapid displacement of Ca by Al from cell walls, especially the Ca bound to pectin. This phenomenon causes cell wall rigidity due to which cells cannot expand (Watmough and Dickinson 1995). Another important phenomenon is the accumulation of callose (β-1,3-glucane) which has been considered as one of the Al toxicity sensitizing effects (Horst et al. 1997).

Plant species and even cultivars of the same species differ considerable in their tolerance to the excess of Al and their ability to its uptake and translocation. It is not very clear yet to what extent root-food crops and leafy vegetables can uptake this metal. There is calculation that uptake factors (ratio of plant Al/soil Al) for leafy vegetables, fruits and tubers are 0.004 and 0.00065, respectively (Kabata-Pendias and Pendias 1999). The tolerance of plants to the excess of Al seems to be based on avoiding metal accumulation. This can be accomplished by giving off organic acids such as oxalate and malate from roots which chelate the toxic Al rendering it non-toxic form to plants (Miyasaka et al. 1991; Ma 2000). In addition, plants may develop internal resistance mechanisms against Al toxicity, these include *(i)* release of organic ligands such as malate acid, *(ii)* increase of rhizopheric pH by roots, and *(iii)* increased binding capacity to cell walls (Matsumoto et al. 2001).

Much work has been focused on the Al toxicity in terms of nutrients such as Ca, Mg, K, P and N. In practice, the uptake of cations by plants is reduced with increased Al levels (Kabata-Pendias and Pendias 2001). Effects of Al on the Ca^{2+} transport are more prominent in the Al-sensitive wheat than in Al-tolerant one. The Al-induced inhibition of Ca^{2+} uptake and its translocation from root apex to the basal part of the plant can be reduced when the Al is removed from the soil solution (Matsumoto et al. 1996).

The intensity of the toxicity of different metal ions seems to be dissimilar due to ion radius, valence and organic complexes which may effect plants, including alteration of plasma membrane properties (Ernst 1998a). Aluminum as well as other trace metals can affect the plant water relation at multiple levels based on stomatal effects and photosynthesis. Nowadays, it is known that roots can regulate shoot water relation not only by hydraulic signals but also by chemical messages (Poschenrieder and Barceló 2004).

II-13.2.7 Humans and Animals

Humans. Contents of Al in mammalian tissues vary between 0.9 and 4.4 mg kg^{-1}, being the highest in skin and the lowest in heart and brain (Jørgensen 2000). Very high Al contents, up to 30 mg kg^{-1}, are reported for hair. Its concentrations in soft tissues of humans range from 0.4 to 18 mg kg^{-1}, in kidneys and lung, respectively. The average content of Al in tissues of the "reference man" is 2.6 mg kg^{-1} (Li 2000). Its mean concentrations in human fluids have been reported (in µg l^{-1}) at 0.6 in serum, and 11 in urine (Reimann and Caritat 1998). Yokel (2004) cited that healthy humans contain Al at the level of 3 and 9 µg l^{-1} in serum and urine, respectively. The Al toxicity is seen when its level in serum is >200 µg l^{-1}.

There is increasing awareness and concern of the Al toxicity to world population and ecosystems due to acid rain/precipitation (Matsumoto 2000). Although there is much debate on Al effects on humans, so far no positive conclusions have been made. Aluminum is a well-known neurotoxic metal (Sjögren and Elinder 1997). Islam et al. (1999) described neurodegenerative disorder known as Motor Neurone Disease (MND) among young people (age 17–30) due to the excess of Al and other trace elements. It has been cited that there is a risk of Alzheimer's disease (AD) when one is exposed to Al-dusts at work (WHO 1997; Polizzi et al. 2002). According to Bjernes (1993), elevated content of Al in damaged brain cells of individuals suffered from this disease is a secondary effect.

The Al cation reacts with HPO_4^{2-} and form slightly soluble $Al(OH)_2H_2PO_4$, decreasing availability of P compounds. Such effects are observed at patients after long-time application of drugs on the base of $Al(OH)_3$ (Matczak-Jon 1995). The Al level in blood plasma of AD patients is higher (mean 9.9 µg l^{-1}) than in healthy persons (mean 2.3 µg l^{-1}) (Bartoń et al. 2003).

Considering daily intake of Al, at the limit value of 1 mg Al kg^{-1} BW (WHO 1989), no risk is expected from eating food cooked in Al-pots and/or Al-foil (Ranau et al. 2001). A minimal risk level for oral exposure of 2 mg Al kg^{-1} BW per day has been set up in the USA (Yokel 2004). Aluminum is relatively easily excreted from human body, however at kidneys' disorders, Al may be accumulated in brain cells and in bones.

Table II-13.12. Aluminum in liver of animals from Norway (mg kg^{-1} FW) (after Godal et al. 1995)

Sampling area	Reindeer		Moose		Sheep	
	Range	Mean	Range	Mean	Range	Mean
Jarfjord	0.38 – 2.0	0.75	0.02 – 0.10	0.04	0.03 – 0.10	0.07
Lower Pasvik	0.11 – 0.71	0.35	0.01 – 2.2	0.08	0.05 – 0.14	0.09
Upper Pasvik	–	–	0.02 – 0.25	0.06	–	–
Neider	–	–	0.01 – 0.21	0.06	–	–
Finnmark	0.28 – 1.35	0.59	0.02 – 0.50	0.05	0.05 – 0.30	0.10
All results	0.11 – 2.0	0.56	0.01 – 2.2	0.06	0.03 – 0.30	0.09

Table II-13.13. Aluminum in tissues of animals near the aluminum smelter in Slovakia (mg kg^{-1}) (after Mankovská and Steinnes 1995)

Species and limit	Organ	Mean
Roe deer	Liver	0.74
	Kidney	0.65
	Meat	1.29
Red deer	Liver	1.18
	Kidney	0.82
	Meat	0.92
Wild boar	Liver	0.86
	Kidney	2.27
	Meat	3.44
Hygienic limit[a]	Viscera	10
	Meat	10

[a] As given by the authors, source not identified.

Animals. There is little information on the requirement for Al in the biological systems (Eichenberger 1986). Godal et al. (1995) studied 213 liver samples from reindeer, moose and sheep and reported the ranges of all mean values to be from 0.04 to 0.75 mg kg^{-1}, being the lowest for moose livers and the highest for reindeer livers (Table II-13.12). Similar data have been reported by Jorhem et al. (1989) for the Al distribution in sheep and moose. These results, as well as data presented by Godal et al. (1995) have indicated that there is no relationship between the acidification of soils and Al in livers of ruminants. Somewhat higher Al contents, especially in liver and kidneys, have been reported for animals from a vicinity of the Al-smelter in Slovakia (Table II-13.13).

II-13.3 Gallium (Ga)

II-13.3.1 Introduction

Gallium is one of the few metals with a melting point close to room temperature. In its compounds, the valency of Ga is generally 3 but the oxidation states of +1 and +2 may also occur (Table II-13.1). It reveals chalcophile properties and resembles the behavior of Zn. The geochemistry and aquatic chemistry of Ga are similar to those of Al, but Ga is less reactive than Al (Orians and Bruland 1988). Because Ga^{+3} is slightly larger than Al^{+3} its content of aluminous minerals might increase during geochemical differentiation. The distribution of Ga in various igneous rocks and in rock forming minerals depends primarily on the magmatic melt. Its concentration is lower in ultrabasic rocks than in alkali rocks.

The Ga contents in the Earth's crust range between 15 and 19 mg kg^{-1}. Its increased contents in argillaceous sediments show an association with clay minerals. The only own Ga minerals are: gallite, $CuGaS_2$ and söhngeite, $Ga(OH)_3$. It may be associated with feldspars, amphiboles and micas. Due to the ability to combine with O and S, it has formed different complexes with some anions, e.g., O, F, and S. The electronegativities and ionic potential suggest that Ga, similarly like Al and Fe, possess amphoteric characteristics, as a result of which they are placed in the group of complex elements.

II-13.3.2
Production and Use

World production figures for Ga are not available, nevertheless there is an estimation of its production in 1999 at 70 t, and the primary Ga production capacity in 2003 at 203 t (USGS 2004).

Gallium occurs in minute concentrations in some ores and is generally produced as a by-product of treating bauxite during the production of Al. Its level varies in bauxite mined from different regions, the highest concentration (0.08%) being reported for the bauxite from tropical regions. It is also obtained at sphalerite (ZnS) and other sulfide metals exploration. The world supply of recoverable Ga in bauxite is estimated at 1.6 Mt, based on bauxite reserves and the Ga concentration.

The of Ga became important in the 1970s, due to its semi-conducting properties. About 44% of Ga is used in electronic devices such as light emitting diodes (LEDs), photo detectors, and solar cells. About 55% of Ga demand is for integrated circuits. A small amount of Ga is also used as an alloying element and for laboratory purposes. Different Ga compounds are applied in the cancer therapy.

II-13.3.3
Soils

The common range of Ga in soils is between 3 and 70 mg kg^{-1}, and its mean content is calculated at 28 mg kg^{-1} (Table II-13.2). The Ga distribution in soils reflects the positive correlation with clay fraction, and Fe and Mn hydroxides. The abundance of Ga in soils of various countries are reported, as follows (in mg kg^{-1}):

▪ USA, San Joachim Valley, range:	16–35	(Wilson et al. 1994)
▪ China, range:	10–40	(Govindaraju 1994)
▪ Japan, range:	18–23	(Takeda et al. 2004)
▪ Sweden, range:	3.4 - 16	(Eriksson 2001)
▪ Sweden, forest soil (O horizon), average:	0.41	(Tyler 2005a)

Concentrations of Ga in rock stream-bed and in top soils seem to be not closely associated (Table II-13.14). Due to the association of Ga with Al, its higher contents in soils than in the continental crust is observed some times (Henderson 1982).

Gallium is relatively slightly mobile in soils, partly due to a low solubility of $Ga(OH)_3$, its common compound. Therefore it is considered to be a useful marker for the characterization of soil genesis and parent materials, for example based on Ga *vs.* Zr and Zr *vs.* Ga ratios (Muggler 1998).

Galium is emitted from Al works and during coal combustion. Its elevated contents in effluents, sludge and composts may be a source of soil contamination (Table II-13.15).

Ga

Table II-13.14. Gallium in Californian stream-bed, top-soils and rocks ($mg\,kg^{-1}$) (after Shiller and Frilot 1996)

Major lithology	Rock	Stream-bed	Top-soil
Sand/mudstone	11	13	15
Graywacke	10	12	13
Quartzite	20	18	16
Argillaceous sandstone	16	15	15
Granodiorite	16	17	14
Basalt/diabase	15	22	33
Basalt	15	12	15
Basalt/rhyolite	18	17	18
Diabase	9.5	14	22
Serpentinite	1.7	6.6	4.6
Schist	11	14	15
Sandstone	16	20	14
Graywacke	12	14	13
Meta-graywacke	3	8.5	15

Table II-13.15. Gallium and indium in selected Danish waste products (data for 2001, from the Danish-EPA 2003)

Waste	Ga	In
Household compost ($\mu g\,kg^{-1}$)	1900	20
Compost from garden waste ($\mu g\,kg^{-1}$)	1600	14
Landfill leachate ($\mu g\,l^{-1}$)	0.22 – 0.25	<0.04
MSW[a] incinerator, semi dry gas cleaning ($\mu g\,m^{-3}$) Landfill leachate ($\mu g\,l^{-1}$)	<0.11 0.19	<0.045 <0.04
MSW[a] incinerator, wet gas cleaning ($\mu g\,m^{-3}$) Landfill leachate ($\mu g\,l^{-1}$)	<0.2 0.28	<0.02 <0.04
WWTP[b], effluent ($\mu g\,l^{-1}$)	<0.02 – 0.12	<0.009
WWTP[b], sludge ($\mu g\,kg^{-1}$)	2200 – 6400	34 – 94
Motorway ($\mu g\,kg^{-1}$)	4300 – 5200	61 – 68

[a] *MSW* – Municipal solid waste.
[b] *WWTP* – Waste water treatment plant.

II-13.3.4 Waters

Bowen (1979) presented data for average Ga concentrations in seawater at $0.03\ \mu g\,l^{-1}$ and in fresh water at $0.09\ \mu g\,l^{-1}$. According to (Nozaki 2005) Ga in the Pacific Ocean, averages at $0.0012\ \mu g\,l^{-1}$. Gallium in worldwide rivers ranges between 0.001 and

0.022 μg l^{-1} (mean 0.03), and its annual riverine flux is calculated to be 1.1 kt yr^{-1} (Gaillardet et al. 2003). In the Northeast Pacific Ocean, Ga concentration at a depth of 250 m was measured to be 0.5 pmol kg^{-1}, whereas at the surface of the Sargasso Sea, it was about 45 pmol kg^{-1} (Shiller 1988). This author reported Ga levels in river waters in the USA to vary between 68 and 250 pmol kg^{-1}.

In fresh and sea waters, Ga occurs as $Ga(OH)_3^0$ and $Ga(OH)_4^-$ (Danish EPA 2003). Its solubility depends upon the pH value, $Ga(OH)_4^-$ being soluble at low pH value. Shiller (2003) estimated residence time of Ga in deep water to vary between 100 to 700 years depending on the location, which is 3–5 times greater than the residence time of Al. Due to variable sources of Ga in aquatic systems, its levels and the mode of residence time in the Norwegian Sea and Atlantic Ocean are different.

Rainwaters sampled in Norway contained Ga at the range of 0.03–0.04 μg l^{-1} (Reimann and Caritat 1998), whereas these sampled in Sweden, in 1999, contained Ga between 0.002 and 0.011 μg l^{-1} (Eriksson 2001a).

Aquatic biota. In the USA, toxicity of $GaCl_3$ on rainbow trout was tested for 28 days, and at the concentration of 3.5 mg l^{-1} was harmful (US EPA 2000).

II-13.3.5 Air

Data on Ga occurrence in the atmosphere are very scanty. Its concentrations above remote regions vary from <0.001 to <0.14 ng m^{-3} (Table II-13.2). It is emitted from Al works and coal combustion plants, thus its contents over industrial areas may reach up to 1 ng m^{-3}.

Due to high levels of Ga in various waste products, as reported for Denmark (Table II-13.15), it is quite possible that the emission of Ga will increase at local, regional, and global levels. According to data given by Bertine and Goldberg (1971) the Ga emission from coal burning is about 1 t yr^{-1}.

Relatively high content of Ga, ranging between 0.14–16 mg kg^{-1}, in moss sampled in Norway during the period 1990–1995, may reflect the atmospheric deposition (Berg and Steinnes 1997).

II-13.3.6 Plants

Although Ga is commonly found in plants, and there is no evidence neither for its necessity nor for the toxicity. The Ga contents in wheat and barley grains from Sweden average about 0.001 mg kg^{-1} (Eriksson 2001). Its higher amounts, from 0.02 to 5.5 mg kg^{-1}, were found in native herbages Russia (Gribovakaya et al. *vide* Kabata-Pendias and Pendias 2001). Elevated levels of Ga in mosses, lichens and bryophytes indicate apparently its aerial sources. A higher ratio of Ga/Al in terrestrial plants than in the soils in which the plants grown can reflect the selective uptake of Ga by plants.

Tyler (2004) selected a site untouched by mechanical disturbances for the last 50 years and also unpolluted from industrial sources in south Sweden, where Ga concentrations in ectomycorrhizal beech roots and in organic soil horizon were at 0.06 and 0.41 mg kg^{-1}, respectively.

II-13.3.7
Humans

Content of Ga in most mammalian tissues is <40 μg kg^{-1} (Jørgensen 2000). Its concentrations in soft tissues of humans range between 0.3 and 5 μg kg^{-1} in muscles and lung, respectively (Li 2000). There is the estimation that a person with a mass of 70 kg contains 0.7 mg Ga in the body, what is equal to the content at 10 μg kg^{-1} (Danish EPA 2003). Its mean concentrations in human fluids have been reported (in μg l^{-1}) at 0.3 in blood, and 0.5 in urine (Reimann and Caritat 1998).

The physiological function of Ga has not been understood yet. There are no data, which suggest that Ga is harmful to humans (Danish EPA 2003). However, the information on its nephrotoxicity and on irritation of skin and mucous membranes are reported. Gallium has been shown to have significant effects at the molecular level; its increased doses to rats inhibit some enzyme and DNA synthesis (Ziegler et al. 2004).

II-13.4
Indium (In)

II-13.4.1
Introduction

The abundance of In in the Earth's crust is reported to range between 0.11 and 0.25 mg kg^{-1} and is higher in igneous rocks than in sedimentary ones (Table II-13.2). Available data on In occurrence in terrestrial material might be not very precise because of difficulties during analyses of geological samples (Yi et al. 1995).

Indium is easily oxidize to +3 and usually precipitates under condition that form Fe and Mn hydroxides. Thus, it is associated with Fe-bearing minerals, mainly due to sorption processes. Indium exhibits a chalcophilic behavior and therefore forms sulfide minerals, such as: indite, $FeInS_4$; and roquésite, $CuInS_2$. However, selenide and telluride minerals are also common. Its sources of economic importance are mainly: sphalerite, ZnS; galena, PbS; and chalcopyrite, $CuFeS_2$. The In concentrations in these minerals may be very high, up to (in mg kg^{-1}): 10 000 in sphalerite, 1 500 in chalcopyrite, and 100 in galena. However, generally its contents in these minerals vary between 10 and 20 mg kg^{-1} (Noël 1999).

II-13.4.2
Production and Use

Galena and sphalerite are the most important minerals for the recovery of In. However it is often recovered from residues, slag, dusts and ashes, during production of Zn, Pb and Sn. The global production of In was about 335 t in 2002 (http://minerals.usgs.gov/minerals/pubs/of 01-06/).

Indium is generally used in alloys, solders and electronics. At the end of 1980s, its use as a thin-film display (TPD) in the electronic industry was developed and aroused much interest. An alloy consisting of 80% Ag, 5% Cd and 15% In is used in thermal-neutron absorbers for nuclear reactors. Indium is also used in Na-lamps and its compounds (hydroxide, acetate and sulfate) have replaced Hg in alkaline batteries (Danish EPA 2003).

II-13.4.3
Soils

Worldwide mean contents of In in uncontaminated soils are reported to range from 0.01 to 0.5 mg kg^{-1}. Its contents in soils from Sweden average <0.04 mg kg^{-1}, with the highest concentration up to 0.06 (Eriksson 2001a). Japanese agricultural soils contain in top layer In at the range from 0.08 to 0.1 in Acrisols and Cambisols, respectively (Takeda et al. 2004). Fergusson (1990) reported mean concentrations of In in soils from Brazil and Canada, at 0.15 and 1.71 mg kg^{-1}, respectively. Surface horizon of forest soils in Sweden contains 0.007 mg In kg^{-1} (Tyler 2005). Increased In contents, at the range of 0.11–1.92 mg kg^{-1}, are reported for soils from the vicinity of Zn-Pb smelter (Asami et al. 1990). Its elevated contents in sludge and composts may be a source of soil contamination (Table II-13.15).

Indium distribution among soils indicates the association with clay fraction. It is also likely to be sorbed by Fe and Mn hydroxides. Soluble species of In in soils are mainly $InCl^{2+}$ and $In(OH)^{2+}$ that are mobile at acid and neutral pH range. Above pH 9.5, the anion $In(OH)_4^-$ is likely to occur.

II-13.4.4
Waters

Mean In concentration in ocean waters is reported to be low, at 0.1 ng kg^{-1} and it is considered as the least soluble elements (Nozaki 2005). Alibo et al. (1999) reported In levels in the Atlantic Sea to vary between 6.25–10.71 pmol kg^{-1} at the depth of 20–1 000 m, which is higher by a factor of 2 to 10 than in the North Atlantic Sea.

In freshwater, In may be present as $InOH^{2+}$ whereas in natural waters and sea waters, the dominant species are $In(OH)_3$ and $InCl_2^+$ (Fergusson 1990; Alibo et al. 1999). At decreased chloride concentration (<10^{-2} mol l^{-1}), In may be present as In^{3+} and/or as hydroxide species, which have relatively short oceanic residence time (less than a few hundred years) and it is removed from seawater by scavenging on particulate matter and deposited in bottom sediments (Alibo and Nozaki 1999).

II-13.4.5
Air

The concentration of In in the atmosphere from remote areas varied between 0.054–78 pg m^{-3} (Fergusson 1990). Air from urban/industrial areas contains much higher amounts of this metal, up to 1 200 ng m^{-3} (Table II-13.2). This indicates its anthropogenic sources, mainly from waste and coal combustion, where In may accumulated up to 0.09 and 0.2 mg kg^{-1} (Tables II-13.2 and II-13.15). Considering these values, the global emission of In from coal and waste combustion seems to be insignificant. There may be atmospheric emissions of In from other sources, but no reports on this topic are available today (Clarke and Sloss 1992).

As Rühling and Tyler (2004) reported, the increase of In in litter (2.3 µg kg^{-1}) as compared with its concentration in leaves (0.2–0.5 µg kg^{-1}) is due to organic matter decay. The content of In in moss (*Pleurozium schreberi*) decreased significantly from

Table II-13.16. Indium concentration, stemming from rock and soil dusts, in Taylor Dome ice samples (after Matsumoto and Hinkley 2001)

Sample depth (m)	Approximate age (years BC)	Dust in ice (ng g^{-1})	Calculated In in ice from dust present (pg g^{-1})
87	1 300	14	0.00069
131	2 200	6	0.00029
250	5 800	16	0.00078
280	7 000	14	0.00069
335	10 000	29	0.0014
340	10 400	32	0.0016
368	13 400	31	0.015
386	27 200	300	0.015
488	72 900	20 000	0.98

0.011 in 1975 to 0.0013 mg kg^{-1} in 2000. This seems to indicate that the emission of In has been decreased during that period. Decreased In contents in younger layers of ice cores and snow is an important information of its atmospheric deposition affected by decreased volcanic emission through the prehistoric time (Table II-13.16).

II-13.4.6 Plants

Although In is known to be readily available to plants, its accumulation in plants has not been observed. Its common contents in plant tissues vary between 1 and 2 mg kg^{-1}. Few data on In in some plants are reported by Fergusson (1990), as follows (in µg kg^{-1} FW): beets 80–300, leaves of fruit trees 0.64–1.8, tomato leaves 0.64–1.8, and vegetables 30–710. Cereal grains sampled in Sweden contain In at <5 µg kg^{-1} (Eriksson 2001a). Beets grown in soil amended with sewage sludge contain much higher amounts of this metal, in the range of 80–300 µg kg^{-1}. Unwashed grass samples from industrial regions contain In at the range of 8–2 100 µg kg^{-1} FW (Furr et al. *vide* Kabata-Pendias and Pendias 1999).

The concentration of In at 1–2 mg l^{-1} in culture solution induces the toxicity, observed mainly in roots. Microorganisms reveal a higher resistance to In than do higher plants, however at its contents of soil in the range of 5–9 mg kg^{-1}, nitrification processes are inhibited (Kabata-Pendias and Pendias 1999).

II-13.4.7 Humans

The only data for In in human tissues are given for muscles at the concentration of 16 µg kg^{-1} (Jørgensen 2000) It is reported to average in human urine at 0.15 µg l^{-1} (Reimann and Caritat 1998).

The In toxicity depends on the solubility of its compounds. It can hinder protein synthesis causing neurological effects such as limb paralysis, and renal and hepatic damage. Its chloride compounds are nephrotoxic, whereas colloidal forms are hepatotoxic (Madden et al. 2004). Appetite problems in humans have been observed, and death of a person from In poisoning has been reported (Danish EPA 2003).

Several studies have shown that the excess of In may cause teratogenic, embroytoxic, and reproductive dysfunction effects in animals. The chronic inhalation exposure affects inflammatory changes in lung (Madden et al. 2004).

II-13.5 Thallium (Tl)

II-13.5.1 Introduction

The range of Tl content in the Earth's crust is calculated at 0.85–1 mg kg^{-1}. Its content seems to increase with increasing acidity of igneous rocks and with increasing clay contents of sedimentary rocks (Table II-13.2).

Thallium has lithiophilic properties and occurs mainly in three oxidation states, but preferable in +1 oxidation state (Table II-13.1). It is associated with K-minerals (e.g., feldspars, micas) and with S-minerals (e.g., pyrite, sphalerite).

Thallium exhibits higher electronegativity than any of the alkali elements. It occurs in several minerals, but only in some is present at higher amounts: lorandite, $TlAsS_2$; and avicennite, Tl_2O_3. Its other minerals, are rather rare and without any potential mining, such as: crookesite, $(Cu,Tl,Ag)_2Se$; and hutchinsonite, $Pb,TlAs_5S_9$.

II-13.5.2 Production and Use

The global production of Tl has been rather stable during the recent decades. At the end of the 1980s, production of Tl was 17 kt yr^{-1} and has not changed significantly till the year 2000 (WMSY 2004). The recent worldwide industrial production of Tl is estimated to be in the range of 10–15 kt yr^{-1} (Bertram and Bertram 2004).

Thallium is generally recovered during the production of Zn since its production from primary products is not profitable. Many Zn ores contain >2 mg kg^{-1} of Tl which goes into the residues and dusts in Zn industries. Minor amounts of Tl are also obtained from Pb, Cu and Fe smelters, and from sulfuric acid plants (Nriagu 1998). It can also be obtained during the recovery of Au and U (Twidwell and Beam 2002).

Commercial use of Tl as an alloying element is quite limited due to its toxic properties. In recent years, its largest user (60–70%) has been the electronic industry. It is added to the production of special refractive glass. Thallium is also used as radioisotopes in medicine for scintigraphy and the diagnosis of melanoma. Its previous use in pesticides (rodenticides and insecticides) has been recently strictly avoided due to the toxicity (ATSDR 1992).

Tl

II-13.5.3 Soils

Thallium content of soils is controlled by parent material as well as by contamination. Its background abundance varies from 0.01 to 2.8 mg kg^{-1}, being increased in loamy and organic soils (Table II-13.2). Its occurrence in uncontaminated soils from different countries has been reported as follows (in mg kg^{-1}):

- Canada, garden soils, range 0.17 – 0.22 (Fergusson 1990)
- China, range 0.21 – 2.4 (Govindaraju 1994)
- France, mean–max. 1.54 – 55 (Tremel et al. 1997a)
- Japan, range 0.39 – 0.59 (Takeda et al. 2004)
- Poland, range 0.01 – 0.41 (Łukaszewicz et al. 1996)
- Sweden, range <0.04 – 0.52 (Eriksson 2001)
- Sweden, forest soil (O horizon), average 59 (Tyler 2005)
- USA, range 0.18 – 0.64 (Govindaraju 1994)

Anthropogenic sources of Tl in soils are related to coal combustion, heavy metal smelting, refining processes, and cement industry that is considered to be the largest emitter of this metal and is clearly supported by data presented in Table II-13.17.

Soils derived from Tl-sulfide mineralized parent material and soils from industrial regions contain its increased levels. In China (Lanmuchang area), Tl levels in various samples from the vicinity of mining region range as follows (in mg kg^{-1}): soils, 40–124; wash materials, 20–28; alluvial deposits, 14–62; undisturbed natural soils, 1.5–6.9; and background soils, <0.2 to 0.5 (Xiao et al. 2004a,b). Slag dump of mining Tl-rich pyrite (China, Guangdong province) contains this metal from 30 to 70 mg kg^{-1} (mean 47). Because H_2SO_4 is used during the production, there is a great risk of Tl leaching that affect soil and groundwater contamination (Yang et al. 2004; Xie et al. 2000; Wu et al. 2000).

Yang et al. (2005) and Xiao et al. (2004) concluded that Tl may pose a problem in the environment if Tl-rich minerals and Tl compounds containing waste are not disposed of correctly. In southern Poland, waste tips from the Zn and Pb mine contain Tl at mean level of 39 mg kg^{-1} (max. 76) (Wierzbicka et al. 2004).

Thallium can be immobilized and oxidized to Tl^{+3} by sulfides in the terrestrial environment (Sager 1998). Its mobility in soils was reported due to acid rain. The Tl concentration in German soils, at the range of 0.2–4.0 mg kg^{-1}, has decreased by

Table II-13.17. Thallium in soils surrounding cement plant and in its products (mg kg^{-1}) (after Scholl 1980, data for cement plant in Baden, Germany)

Sample (number)	Range	Mean
Non-contaminated soil (478)	0.1 – 2.2	0.4
Soils near cement plant (575)	0.1 – 15	1.4
Soils near old mines (221)	0.1 – 73	14.7
Cement raw materials (11)	0.13 – 2.6	0.45
Flue dust from cement plants (21)	42 – 2370	428

10–15% of the HNO_3-extractable fraction at pH <7.0, within 14 years (in the 1970s) (Crössmann 1994).

Microorganisms are relatively sensitive to Tl, and therefore the inhibition of nitrate formation in Tl-contaminated soils may have an agronomic impact.

II-13.5.4
Waters

In the North Pacific Ocean, Tl mean concentration is calculated to be 13 ng kg^{-1} (Nozaki 2005). According to Cheam et al. (1995) its average amounts in seawaters range between 10 and 15 ng l^{-1}. These authors have reported concentrations in unpolluted and polluted freshwater samples to ranges between 5–10 and 20–25 ng l^{-1}, respectively. Lin and Nriagu (1999) reported Tl concentrations in river waters of different countries to vary from 5 to 1 350 ng l^{-1}, being the lowest in Poland and the highest in Japan. Excluding its very elevated contents, apparently in polluted rivers, its concentration varies between 5.1 and 56 ng l^{-1}.

Rainwater samples collected in Sweden during 1999 contained Tl at the range of 3.5–16 ng l^{-1} (Eriksson 2001a). The concentration of Tl in tap water of most cities of Poland is reported to range between 5 and 24 ng l^{-1} (Łukaszewicz et al. 1996). Safe limit for Tl in drinking water is established at 2 000 ng l^{-1} (US EPA 2003). According to other source, exposure limit values for ground and drinking water are set in the USA as 500 ng l^{-1} (Bertram and Bertram 2004).

Based on the thermodynamic calculation, it may be predicted that Tl in seawaters occurs mostly as $TlCl_6^{3-}$. The Tl species in two river waters (Raisin and Huron, USA) are: Tl^{3+} at the concentration of 11–13 ng l^{-1}, and Tl^{+} at 6 ng l^{-1} in both rivers. The main Tl^{3+} hydroxyl species is $Tl(OH)^{2+}$ and is likely to form at the pH range of 4–6, whereas $Tl(OH)_2^{+}$ predominates at pH around 7, and $Tl(OH)_3$ at pH 7.5–8.8 (Lin and Nriagu 1998). As these authors have stated, the hydrolysis of Tl is generally governed by the pH of the aqueous solution. Also the redox potential governs the chemical speciation of Tl. Considering the standard reduction potential, Tl^{3+} has a limited stability field. According to some authors, Tl^{+} is the dominant species under most environmental conditions (Flegal and Patterson 1985; Vink 1993).

Compounds of Tl are readily soluble in water and the metal has a tendency to be sorbed to bottom sediments. Sediments of a polluted river in Germany, contain about 20 mg Tl kg^{-1}, which is mobilized at a pH <2, whereas complexing agents had no influence on the mobility of Tl in the river sediments (Gunther et al. 1987).

Thallium in tailing ponds or in waters of mining areas may be quite high, and has been reported to reach 1 620 µg of dissolved Tl l^{-1}. Xiao et al. (2004b) cited Tl levels in deep groundwater of the industrial region of China to reach 1 100 µg l^{-1}, whereas the background value is <0.005 µg l^{-1} (Table II-13.18).

Aquatic biota. The bioconcentration of Tl in aquatic plant species have been reported by some authors (Barrows et al. 1978; Ewers 1988), however its toxicity, mobility, bioavailability, and distribution are still under a discussion (Flegal and Patterson 1985; Lin and Nriagu 1998). According to Flegal and Patterson (1985), following contents of Tl are in aquatic species from the Central Pacific Ocean (in mg kg^{-1}): phytoplankton, 0.02–0.8; zooplankton, 0.03–0.5; and ichthyoplankton, 0.1.

Tl

Table II-13.18. Thallium in water samples collected from different sources in the Lanmuchang area, China (μg l^{-1}) (after Xiao et al. 2004a)

Sample	Range	Mean
Deep groundwater	13 – 1100	62
Shallow groundwater	0.005 – 0.75	0.04
Well water	0.01 – 0.38	0.2
Background groundwater	–	<0.005
Stream water (base flow regime)	0.09 – 31	1.8
Stream water (flood flow regime)	0.07 – 4.5	1.5

II-13.5.5 Air

The concentration of Tl in the air in Europe has been reported to be 0.06 ng m^{-3} and in North America 0.22 ng m^{-3} (ATSDR 1992).

Thallium enters into the atmosphere mostly from industrial sources, such as cement plants, coal combustion, ferrous and non-ferrous metallurgical plants, and burning of solid waste. As a result of its physicochemical properties and volatile nature, the metal cannot be retained either by an electrostatic precipitator (ESP) or by other gas cleaning equipment. Cement dust and fly ash are considered to be the main sources of Tl pollution (Sager 1998).

The deposition of Tl on soils and waters occurs generally by dry and wet fall down. Its wet deposition in Sweden has been estimated at 0.083 g ha^{-1} yr^{-1} (Eriksson 2001a).

II-13.5.6 Plants

The Tl content of plants seems to be a function of the concentration of its mobile fraction in soils. It is relatively easily taken up by plants, especially by tree roots from acidified soil (Tyler 2005). Its concentration in roots up to about 2 mg kg^{-1}, can inhibit the germination, plant growth, and chlorophyll content (Kabata-Pendias and Pendias 2001).

The reference content of Tl in plants has been calculated by Markert (1992) at 0.05 mg kg^{-1}. Smith and Carson (1977) gave Tl levels in various plants as follows (in mg kg^{-1}): vegetables, 0.02–0.13; clover, 0.008–0.01; and meadow hay, 0.02–0.025. Wheat and barley grains sampled in Sweden contain average Tl amounts at 0.0002 and 0.0011 mg kg^{-1}, respectively (Eriksson 2001a). Its increased levels (up to 5.5 mg kg^{-1}) have been sometimes in mushrooms (Kabata-Pendias and Pendias 1999). The maximum Tl content in some food plants (vegetables and fruits) has been set in Germany as 0.1 mg kg^{-1} FW, however it is not a legal value (Bertram and Bertram 2004).

Some plants, especially *Cruciferae* and *Gramineae* families, exhibit a special ability to take up this metal. Anderson et al. (1999) and Leblanc et al. (1999) discovered hyperaccumulator plants *Brassicaceous iberis intermedia* Guersent and *Biscutella laevigata* L., grown in mining areas in southern France that contain Tl at levels of 1.4 and 0.4%, respectively. These authors have cited the field trials in New Zealand that show that the *Iberis* species may produce about 700 kg ha^{-1} of bio-ore containing 8 kg of Tl. Other plants, that can be used for Tl phytomining and the cleaning of contaminated soils, are *Biscutella laevigata* and *Iberis intermedia* accumulating this metal up to 13 768 and 4 055 mg kg^{-1}, when grown in soils containing Tl at levels of 28 and 16–18 mg kg^{-1}, respectively (Anderson et al. 1999).

Xiao et al. (2004a) studied Tl in crops in southwest Guizhou, China, and reported the concentration of Tl decreasing in the following order: green cabbage > carrot > chili > chinese cabbage > rice > corn. Thallium in green cabbage was at 500 mg kg^{-1}, which is more than its concentration in soils.

In the EU region (Germany, Poland and the Czech Republic), the moss, *Pleurozium schreberi* contained 0.04–0.13 mg kg^{-1} and *Polyrichum formosum* 0.01–0.05 mg kg^{-1} in less polluted areas; in the study areas Tl in topsoil was about 0.5 mg kg^{-1} (Heim et al. 2002). In south Sweden, Rühling and Tyler (2004) measured Tl in the carpet of forest moss (*Pleurozium schreberi*) sampled in 1975 and 2000. They have indicated that cross boundary deposition of Tl in the remote areas has decreased from 0.152 to 0.066 mg kg^{-1} in 1975 and 2000, respectively.

II-13.5.7 Humans and Animals

Humans. Contents of Tl in most mammalian tissues is reported to be <200 µg kg^{-1}. Its increased level, at 500 µg kg^{-1}, is in skin (Jørgensen 2000). Its concentrations in soft tissues of humans range from <1 to 9 µg kg^{-1} (Li 2000) and mean concentrations in human fluids have been reported (in µg l^{-1}) as: 0.4 in blood, 0.2 in serum, and 0.4 in urine (Reimann and Caritat 1998). Bertram and Bertram (2004) cited, based on several sources, the median contents of Tl in kidneys and hair of healthy unexposed individuals at 0.5 and <1 µg kg^{-1}, respectively. These authors reported Tl to range in whole blood and urine of adults (in µg l^{-1}) from 0.05 to 0.39 and from 0.02 to 0.82, respectively.

The toxicity of Tl has not been greatly studied, but its harmful impact has been observed to both humans and animals. It has been confirmed that Tl is toxic to humans and animals, and plays no role in their metabolisms. Considering its toxicity, Tl ranks alongside Pb, Hg, and Cd (Tremel et al. 1997; Wierzbicka et al. 2004). Mechanisms of Tl toxicity are related to: K-substitution, enzyme inhibition, lipid peroxidation, peripheral neuropathy, and antagonism to Ca in heart function (Bertram and Bertram 2004).

Toxic effects of increased amounts of Tl are: sleeping disorder, headache, fatigue, polyneuritic symptoms, and baldness (Brockhaus et al. 1981). The mechanism of Tl poisoning is not very clear, but similarly to other trace metals, Tl binds sulfhydryl groups of proteins and membranes, thereby inhibiting a range of enzyme reactions thus causing poisoning (Ramsden 2002).

Its toxicity to humans generally occurs from skin contact, as it is easily absorbed through the skin (Ewers 1988). However, ingested and inhaled Tl is also harmful to organisms. General exposure occurs through the food chain, especially from fruits and vegetables what has been observed at the population living in the vicinity of a cement plant that discharges high-Tl containing dusts (Dolgner et al. 1983). Other significant Tl sources are from coal combustion, roasting of sulfide concentrates and manufacturing of Tl-containing products (Kazantzis 2000). In the Ti-Hg-As mineralization areas of Guizhow Province, China, a number of causalities were reported from thallotoxicosis in the 1960s; still today high concentrations of Tl (153–2 668 μg l^{-1}) in urine among the population of that area are detected (Zhang et al. 1998; Xiao et al. 2004b).

In the Lanmuchang Tl mineralized areas of China, the estimated daily intake of Tl by village people, via local planted crops, was reported to be 1.9 mg $person^{-1}$ (Xiao et al. 2004a). This high ingestion rate of Tl is 1 000 times higher than the world average intake by adults, 2 μg d^{-1} (Sabbioni et al. 1984) and also far above the element's "oral reference dose" of 56 μg d^{-1}.

Animals. In the 1960s, Tl toxicity has been reported among dog, cats, red foxes and wild martens due to the use of rodenticides and insecticides containing Tl. Due to its toxicity, Tl in rodenticide has been banned, but in developing countries it is still used which is of a great concern. Sheep and horses died without any signs of disorder were reported to lose their hair due to high Tl concentrations in the environment (Schoer 1984).

Trace Elements of Group 14 (Previously Group IVa)

Trace elements of the Group 14 are: silicon (Si), germanium (Ge), tin (Sn) and lead (Pb). Silicon, in combination with oxygen, is the basic non-metallic component of all rocks and is considered as a trace element only in respect to its biochemical role. It is the second most abundant element in the lithosphere. It performs an important role in connective tissues, especially in bone and cartilage and it is believed to have played, together with carbon, the crucial function in processes of the first synthesis of proteins (Sedlak 1967). The next three elements, Ge, Sn, and Pb are metals, which show chalcophilic properties and occur in the environment in two oxidation states, +2 and +4 (Table II-14.1).

II-14.1 Silicon (Si)

II-14.1.1 Introduction

Silicon is the second most abundant element, after oxygen, in the Earth's crust and occurs mainly as SiO_2, in considerable amounts in nature; it has three oxidation states of which +4 is the most common (Table II-14.1). It is most abundant and also most stable element in the Earth's crust. However, under specific conditions its compounds can be dissolved and transported, mainly in colloidal phase.

Table II-14.1. Selected properties of trace elements of the Group 14

Element	Atomic number	Atomic mass	Atomic radius[a] (pm)	Density (20 °C, g cm^{-3})	Valence[b]	Melting point (°C)
Si, silicon	14	28.08	117	2.33	+2, −4, **+4**	1410
Ge, germanium	32	72.61	152	5.23	+2, **+4**	937
Sn, tin	50	118.71	172	7.31	**+2, +4**	232
Pb, lead	82	207.20	181	11.35	**+2**, +4	327

[a] Approximate average values for the main oxidation states.
[b] Valence values in bold are from the main oxidation states.

Table II-14.2. Abundance of silicon, germanium, tin, and lead in the environment[a]

Environmental compartment	Si	Ge	Sn	Pb
Earth's crust (mg kg^{-1})	26 – 29[b]	1.4 – 1.6	2.3 – 2.5	15
Igneous rocks (mg kg^{-1})				
Mafic	20 – 30[b]	0.7 – 1.5	0.3 – 1.5	0.1 – 8.0
Acid	31 – 34	1.0 – 1.5	1.3 – 3.6	10 – 25
Sedimentary rocks (mg kg^{-1})				
Argillaceous	25 – 28[b]	1 – 2.5	6 – 10	14 – 40
Sandstones	30 – 40	0.8 – 1.4	0.5	5 – 10
Calcareous	5 – 30	0.1 – 0.3	0.3 – 0.5	3 – 10
Raw energy material (mg kg^{-1})				
Coal, hard[c]	1.5 – 20[b]	1 –780	1 – 140	10 – 1900
Fly ash	–	6.6 –250	8.7 – 30	52
Crude oil	–	–	–	1 – 4
Petrol	–	–	–	2 – 650
Soils (mg kg^{-1})	54[bx]	2.0	3.5	15 – 28
Light sandy	–	0.6 – 2	<0.1 – 2	5 – 40
Medium loamy and silty	–	1 – 2.5	0.3 – 2	10 – 50
Heavy loamy	–	1 – 2	0.3 – 4	10 – 90
Calcareous	–	0.5 – 1.3	<0.1 – 1.8	17 – 63
Organic	–	0.8 – 1.6	0.2 – 5	2 – 80
Water (µg l^{-1})				
Sea	2.8[d]	0.05 – 0.5	0.004 – 0.05[f]	<0.03 – 0.27
River[e]	1.3[f]	0.003 – 0.08	0.005 – 1.2[f]	0.04 – 3.8
Air (ng m^{-3})	20 – 100[f]	–	–	–
Urban/industrial	>1000	–	2 – 800[f]	120 – 13000
Remote regions	–	–	10 – 70	10 – 22
South Pole	–	–	–	0.2 – 1.2

[a] Presented are common, possible background, values from various sources as compiled by Kabata-Pendias and Pendias (1999 and 2001), unless otherwise indicated. [b] Data for Si are in %. [bx] Estimation for SiO_2. [c] After Finkelmen (1999) and Llorens et al. (2000). [d] Data for Si in water are in mgl^{-1}. [e] After Gaillardet et al. (2003). [f] After Reimann and Caritat (1998).

Quartz, SiO_2, is very resistant mineral in all terrestrial compartments and occurs primarily as silica in the forms of quartz, flint and sand. Most silicate minerals are built of a fundamental unit, SiO_4, called tetrahedron. It is the basic component of kaolinite, $Al_2O_3 \cdot 2SiO_2 \cdot 2H_2O$; several other clay minerals, and of all silicate minerals. Non silicate mineral containing Si is carborundum, SiC.

II-14.1.2 Production and Use

Annual Si production in 1995 was calculated at 3.1 Mt (Reimann and Caritat 1998). Silicon is commercially produced by heating SiO_2 and carbon in an electric furnace, using carbon electrodes. There are other methods by which amorphous Si can be prepared as a brown powder that can be easily melted or vaporized. Hyperpure Si can be produced by the thermal decomposition of ultra-pure trichlorosilane in a H-atmosphere, and by the vacuum zone process. (http://pearll.lanl.gov/periodic/elements/14.html).

Due to its versatile properties, Si is used in many different sectors of industry. It is used in the electronics to produce transistors, computer chips, and solar cells, because it has good semiconductor properties. It is commonly used for high temperatures devices, because of its high melting point (Table II-14.1). Silicon is a useful refractory material for high temperature processes and as silicate, is used in the production of bricks, pottery, enamels, etc. Silica (sand) is the main ingredient of glass and silicon monoxide and is widely used as a protective coating. It is also used as an alloying element in the metallurgical industry, especially in the steel and aluminum industries.

II-14.1.3 Soils

Silicon is the second most abundant element in soils. It is one of the basic components of the geosphere and at the same time it occurs in trace amounts in the biosphere (Zielecka 1996). The common form of Si in soil is quartz, SiO_2. It is also known to occur in noncrystalline form, opal, which is believed to have had a biological origin.

The soil solution contains Si as silicic acid, H_4SiO_4 in the range of 1 to 200 mg l^{-1} (Kabata-Pendias and Pendias 2001). Weathering of aluminosilicate minerals is the main source of dissolved Si in soil solution (D'Haese et al. 2004). The pH level has especially important effect on the Si mobility, although several other soil and climatic factors influence its behavior. The reaction of silicate anions is enhanced at high pH (>9.5) and also the dissociation of H_2SiO_4 is enhanced.

The main source of Si, besides geogenic, is wind blown dusts composed mainly of $SiO_2 \cdot nH_2O$ (Epstein 1994). There is evidence that Si reacts with other ions such as P, Al and Ca, and Fe that modify its behavior. In such cases, silicate and phosphate precipitate in acid solution and may fix other cations such as Fe and Al oxides which have the capability to absorb dissolved Si as H_4SiO_4. Often the content of Si in soils is associated with types of parent material. Takeda et al. (2004) described that in agricultural soils of Japan, levels of Si increase from mafic (Andosols) to felsic (Acrisols) rocks (Table II-14.3).

Table II-14.3. Mean contents of SiO_2, Sn, and Pb in agricultural soils of Japan ($g\,kg^{-1}$) (after Takeda et al. 2004)

Soil group (sample numbers)	SiO_2	Sn	Pb
Andosols (220)	430	2.0	23
Cambisols (70)	560	3.0	28
Gleysols (79)	630	2.6	23
Acrisols (53)	660	3.1	25
All soils (514)	530	2.4	24

II-14.1.4 Waters

The average worldwide dissolved Si in the ocean, river and freshwaters has been reported by different authors to be (in $mg\,l^{-1}$): 2 (Treguer et al. 1995), 2.8 (Nozaki 2005), 4 (Conley 1997), and 6 (Willén 1991).

Silica in waters occurs mainly in form of silica acid, H_2SiO_4, which resulted from inorganic and biochemical processes. Both processes may lead to the formation of amorphous Si that is accumulated in bottom sediments. Biomass cycling, mineral breakdown and leaching of minerals are main sources of Si in the aquatic environment (Zevenhoven-Onderwater 2001; Conley 2002). Treguer et al. (1995) studied the biogeochemical cycling of Si, during the 1990s, and calculated that the global net river input of Si is 145.45 $Mtyr^{-1}$ and input from other sources 30.9 $Mt\,yr^{-1}$. This means that 80% of Si input to ocean stems from rivers and is in colloidal forms and/or biochelates. Thus, the gross oceanic BSi production can be estimated at 6740 $Mt\,yr^{-1}$ and in permanent burial of BSi at 176.35 $Mt\,yr^{-1}$.

Neal et al. (2005) studied Si concentration in the UK waters, for a wide range of catchment systems (pristine, rural, agricultural and urban impact systems). Mean Si concentrations in waters range from 0.06 to 2.58 $mg\,l^{-1}$, being the lowest in rainwater and the highest in groundwater (Table II-14.4).

Hartikainen et al. (1996) studied P and Si in lake sediments in southern Finland. Both elements have similar sorption mechanism, however, the P is retained at a larger pH range than Si. The sequential fractionation analysis indicated that P is bound by a ligand exchange mechanism onto oxide surfaces (35–40%) and this process also dissolves half of the BSi. Due to chemical competition, silicate may depress the retention of phosphate. These authors have hypothesized that silicate may contribute to phosphate release, especially in shallow eutrophic lakes where high algal production maintains a high pH in the water.

II-14.1.5 Air

In the atmosphere, it is found as Si compounds, mainly oxides or silicates, stemming from natural and industrial sources. Wind blown dust, especially dust from the Sahara desert, may carry huge amounts of Si into the atmosphere. Abundance of the Si in moss samples may indicate its aerial transport. Its lower content in moss sampled during 2000 than in moss of 1975 shows a lower Si air pollution (Table II-14.5).

Table II-14.4. Silicon in various waters of the United Kingdom (mg l^{-1}) (after Neal et al. 2005)

Source and region	Maximum	Average
Rainwater	4.0	0.06
Cloud water	1.5	0.07
Throughfall	0.55	0.20
Stemflow	1.25	0.22
Streams	5.60	1.45
Groundwater	7.20	2.58
Streams[a]	4.50	1.35
Streams[a]	3.43	1.70
Lakes[a]	2.15	0.60

[a] Data for Welch, after Neal et al. (1998).

Table II-14.5. Mean concentrations of Si, Ge, Sn, and Pb in moss (*Pleurozium schreberi*) collected from south Sweden in 1975 and 2000 (mg kg^{-1}) (after Rühling and Tyler 2004)

Element	1975	2000
Si	470	248
Ge	0.091	0.015
Sn	0.49	0.09
Pb	47.5	4.2

II-14.1.6 Plants

Silicon is a common mineral constituent of plants, however, it is not defined to be essential. Its amounts vary greatly, mainly upon plant species. For example, mean Si contents in grasses range between 0.3 and 1.2%, whereas its concentrations in leguminous vary from 0.05–0.2%. Some species may accumulate much higher amounts of Si, up to >10%, as for example: sedges, nettles, horsetails, and diatoms. Silica in the form of H_4SiO_4 (known as dissolved silicate), is readily absorbed by plants and is deposited as a solid amorphous SiO_2. Körös (*vide* Kabata-Pendias and Pendias 1999) reported that Si may be taken up in a form of monomeric silicic acid and/or as organic complexes. It is deposited as amorphous SiO_2, as well as a special silicon complex that has been isolated from *Thuja plicata*.

Rice plants are also known as accumulators of Si and can contain it up to 10% in hulls and up to 4% in grains, where is concentrated mainly in chloroplasts (Alyoshin et al. 1988). There is evidence for two mechanisms of Si uptake; passive in grasses and several other plants and active in rice. Silicon is absorbed from the soil solution as monosilicate acid or amorphous silica, and its absorption depends on the concentration of Si, the water regime, and the pH (Richmond and Susman 2003). Ma et al. (2001) investigated more than 150 plant species and divided mechanisms of the Si uptake into three different modes: active, passive and rejective.

Silicon has not yet been known as an essential element for higher plants because its role in plant biology is poorly understood (Epstein 1999). However, recent studies confirmed that Si is an important element that increases the tolerance of plants against environmental stress (Savant et al. 1999). It is also believed that Si is necessary for plant growth, development, and structure strength.

In plants, this element can improve resistance or tolerance to Al, Mn, salt toxicity (Yeo et al. 1999; Rogalla and Römheld 2002; Liang and Ding 2002; Liang et al. 2001; 2003). Chen et al. (2000) indicated that furnace slag is more effective in suppressing Cd uptake by rice and wheat due to increased pH and availability of the Si from the furnace slag. A general rule is that elevated pH reduced Cd uptake *via* Cd immobilization in the presence of a Si source.

In most rice-growing regions, there is a problem of severe heath blight due to the deficiency of phytoavailable Si in soils (Ou 1985; Rush and Lee 1992; Roy 1996). The most important diseases affecting rice are blast, brown spot, stem rot, scald and grain discoloration, which is due to sheath blight disease. These effects can be minimized by using Si-fertilization (Seebold et al. 2000, 2001). Rodrigues et al. (2003) have confirmed positive effects of available Si in great plantations of rice in Brazil. It has also been cited that Si could increase the drought tolerance of plants (Hattori et al. 2001; Lux et al. 2002). Gong et al. (2005) have indicated effects of Si on improved drought tolerance of wheat plants, increasing antioxidant defense abilities and on maintaining of many physiological processes such as photosynthesis under drought conditions.

II-14.1.7
Humans and Animals

Humans. Silicon is quite abundant element in mammalian tissues in the range from 70 to 450 mg kg^{-1}, being the lowest in liver and brain and the highest in skin (Jørgensen 2000). Its highest contents, up to 500 mg kg^{-1} are also in bones, tendons, skin and lungs. The concentration of Si in tissues of a "reference man" is calculated at 260 mg kg^{-1} (Li 2000). Much lower amounts are in soft tissues (2–10 mg kg^{-1}) and in blood (<1 mg kg^{-1}). Its lowest concentrations, between <7.12 and 2 421 µmol l^{-1}, are reported for serum (Bercowy et al. 1994). The serum Si level is rather constant over the age of between 18–59; it may decrease when age is over 60 years (Jugdaohsingh et al. 2002). Some authors believe that hormonal activity plays an important role in the serum Si level of woman (Bissé et al. 2005). The Si concentration of human urine vary significantly upon kidney function, from 140 to 4 600 µg l^{-1}, in healthy individuals and in person with kidney failure and/or hemodialyzed, respectively (D'Haese et al. 2004).

Silicon is not easily available to the humans due to its low water solubility, however, it is an essential nutrient (Zielecka 1996). It is required for bone, cartilage and connective tissue formation and is also necessary for proper development of skin and hair. Thus, Si has been reported to be a major ion in osteogenic cells (Carlisle 1986; Bissé et al. 2005).

Excess of Si in diets may have an adverse effect on the metabolism of fats and phosphates. Save doses of Si for adults in food and water are proposed at 5–10 mg d^{-1} and 3–5 mg d^{-1}, respectively (Suslikov *vide* Kabata-Pendias and Pendias 1999). Most forms of Si, amorphous and crystalline, are harmless to organisms, however, some silicates of needle-like structure, as e.g., asbestos, may have very adverse impact on cells due to physical irritation. The excessive inhalation of Si-containing dust may affect chronic bronchitis and silicosis as well as tuberculosis and lung cancer. There is also reported its indirect beneficial effects, since Si has a preventive role in senile dementia, reducing the absorption of Al (D'Haese et al. 2004).

Animals. Most fodder plants contain Si in variable amounts (at common range of 0.05 to >1%) and it absorbed into tissues and fluids of (Carlisle 1982). The highest concentration of Si is generally observed in the epidermis, appendages and connective tissues, and which is generally derived from silicic acid. There is also observed that a low Si level in forage can induce a development of osteoporosis at some animals (Birchall and Espie 1986).

Silicon plays a significant role in the growth of hair and feather. The use of dietary Si supplement has been reported for cows, but very little effect regarding concentration of Si in cow's milk has been reported (Carlisle 1982).

II-14.2 Germanium (Ge)

II-14.2.1 Introduction

Germanium concentration in the Earth's crust averages 1.5 mg kg^{-1}. The distribution of Ge in major rock types resembles that of Si, but at much lower contents, being the lowest in calcareous sediments and mafic magmatic rocks (Table II-14.2).

Germanium and Si have similar crystal structures and their atoms are similarly arranged at their surface due to which their chemical characteristics are similar and both occur preferable in the +4 oxidation stage (Table II-14.1). It can readily substitute Si in silicate minerals (Martin et al. 1996). Germanium exhibits siderophilic properties, but can also reveals lithophilic and chalcophilic properties (Vlasov 1966). Its common minerals are: germanite, $Cu_3(Ge,Fe)S_4$; argyrodite, Ag_8GeS_6. It is likely to be associated with Pb-Zn-Cu sulfide ores.

During weathering, Ge is easily mobilized, mainly as $Ge(OH)_2$ and transported in aquatic systems, but then it is readily fixed by clay minerals, Fe hydroxides, and organic matter. Apparently, this might explain its elevated contents in several coals (Table II-14.2). In most coals Ge ranges from about 1.6 to 13.4 mg kg^{-1} in samples from the EU-15 and South Africa, respectively (Meij and Winkel 2004). In coal fly ash its concentration averages about 95 mg kg^{-1}. Font et al. (2004) reported that Ge recovery from coal fly ash is associated with the occurrence of sphalerite enriched in this metal.

II-14.2.2 Production and Use

Total world production of Ge (metallic Ge, $GeCl_4$, GeO_2) has been rather constant, at between 65 and 70 t yr^{-1} (USGS 2004). Generally, it is recovered from smelter dusts during the Zn production, from Pb-Zn-Cu sulfide ores as well as from the by-products of combustion of certain coals. Its presence in coals insures a large reserve of the element.

Germanium is economically and strategically an important metal. Due to special properties, its major part is used in fiber-optic systems (50%), polymerization catalysts (20%), infrared optics (15%), electronic/solar electrical applications (10%) and other uses in metallurgy, and chemotherapy (5%) (USGS 2004). There are thousands of electronic applications for Ge doped with minute amounts of As, Sb, Ga, In and P. This element is a very competitive sister for In and Ga, within the global electronic market (Moskalyk 2004).

II-14.2.3
Soils

There is not much information available regarding Ge in soils. It is reported to range in various soils of the USA from <0.1 to 2.1 mg kg^{-1}, and from <1 to 95 mg kg^{-1} in topsoils of Sweden (Skacklette and Boerngen 1984; Eriksson 2001a). Tyler and Olsson (2001, 2002) found concentrations of Ge in soil and soil solution at 0.44 mg kg^{-1} (range 0.05–0.64) and 0.06 μg l^{-1} (range 0.01–0.12), respectively. Apparently, the level of Ge in soil solution might be less due analytical errors during the HNO_3 digestion. Germanium contents in beech forest growing on a podzolized soil vary from 1–6 to 27 μg kg^{-1} in fresh leaf litter and old leaf litter, respectively (Tyler 2005).

Kurtz et al. (2001) described the Ge behavior in soil weathering processes, using the Ge/Si system as a tracer of silicate weathering processes. They indicated that the Ge concentration in soils increases with Si, and decreases with Fe. However, the Ge/Si fractionation is not related to Ge-Fe or Ge-organic complexes. Sequential extraction experiments indicated that most of the soil Ge is associated with a secondary silicate fractions.

II-14.2.4
Waters

El Wardani (1958) measured Ge in ocean waters to be 0.05–0.07 μg l^{-1}, whereas Braman and Tompkins (1978) observed Ge concentration in fresh and estuarine waters at levels of 0.016 and 0.029 μg l^{-1}, respectively. In the Pacific Ocean, Ge is reported to average 0.006 μg kg^{-1} (Nozaki 2005). The median value of Ge in the worldwide ocean water is reported at 0.05 μg l^{-1} (Reimann and Caritat 1998). In river waters, Ge ranges from 0.0014 to 0.08 μg l^{-1}, at the world mean concentration of 0.007 μg l^{-1} (Gaillardet et al. 2003). These authors calculated the riverine flux of Ge at 0.25 kt yr^{-1}.

Apparently, in addition to cationic species (metallic), Ge is likely to occur in aquatic environment as complex anions, such as $HGeO_2^-$, $HGeO_3^-$, and GeO_3^{2-} (Lewis et al. 1988). Different organo-Ge species were also identified in ocean waters. Unlike metallic Ge, its methylated forms (Me-Ge, MM-Ge, and DM-Ge) are non-reactive in estuaries and oceans. The MM-Ge and DM-Ge species are uniformly distributed in oceans depending on the salinity of water. It is believed that organo-Ge in ocean waters comprises >70% of the total Ge concentration. The formation of Me-Ge species and their occurrence in seawaters may also be due to the biomethylation processes that take place at the presence of microbes. It has been cited that in pristine rivers, MM-Ge and DM-Ge species are present in very minute levels, which are barely detectable.

II-14.2.5 Air

Germanium is generally released in to the atmosphere from coal fired power plants, metallurgical industry and waste incineration.

Rühling and Tyler (2004) reported for moss samples from Sweden, collected in 1975 and 2000, Ge average levels at 0.091 and 0.015 mg kg^{-1}, respectively (Table II-14.5). Moss sampled in Norway, during the period 1990–1995 contained Ge at the range of 0.24–11 mg kg^{-1} (mean 1.1) (Berg and Steinnes 1997). These differences are apparently related to variable long-distance aerial transport and to decreased emission of the metal.

In the UK, it has been estimated that Ge is discharged to the environment from stack gases, flue dust and ashes from coal burning, which amounted to about 2000 t yr^{-1}, with the ash containing 20–280 mg Ge kg^{-1} (http://www.speclab.com/elments/germanium.html).

II-14.2.6 Plants

Germanium is known to occur in all plants, but it is not known to have any physiological functions. In different plant species it levels range from 50 to 754 mg kg^{-1} (Table II-14.6). Its elevated contents, up to 100 mg kg^{-1} in leaves of hazel grown in Wroclaw, indicate polluted sites due to a high industrial emission (Sarosiek et al. 1998). Connor and Schacklette (1975) reported the Ge level in plant ash to average 20 mg kg^{-1}.

Table II-14.6. Germanium in selected plants (mg kg^{-1}) (source: www.oxygentimerelease.com/A/Therapies/Germanium/b8.htm)

Plant	Mean content
Aloe	77
Comfrey	152
Chlorella	76
Garlic	754
Bandai moss	255
Ginseng[a]	250
Pearl barley	50

[a] Plants from Shimane Prefecture, Japan.

Ge

II-14.2.7 Humans and Animals

Germanium occurs in all mammalian tissues, in common range of 0.07–1.5 mg kg^{-1} and is likely to be in higher amounts is aquatic organisms than in terrestrial ones. In human urine Ge is reported to occur at 1 mg l^{-1} (Li 2000).

There is no evidence that Ge is essential to humans or animals. In general, the toxicity of Ge is quite low except for the tetrahydride Ge that causes toxic effects in humans. Organic Ge-compounds claim to be destroyers of cancer cell and reveal antineoplastic properties in man (Gerber and Léonard 1997). However, there is information on the death of a Japanese woman from taking a compound composed mainly of inorganic Ge, who had suddenly taken ill and died. Inorganic forms of Ge, especially when inhaled, may be toxic in high doses, but organic Ge is quite safe (http://www.positivehealth.com/permit/Articles/Nutrition/Germanium/chapter11.htm.

Some foodstuffs such as seafood, oats, tomato juice, and garlic may content substantial amounts of Ge (Table II-14.7). Human intake of Ge through foodstuffs varies between 0.4 to 1.5 mg d^{-1} (Vouk 1986), whereas according to Glei (2004), its daily intake varies between 0.4 and 3.5 mg. Schauss (1991) estimated its average dietary intake by adults at about 1.5 mg d^{-1}.

Cigarettes, containing mean amount of Ge at 0.61 mg kg^{-1} FW, may be a substantial source of inhaled metal (Schroeder et al. 1968). However, there is evidence that the toxic effect of dusts containing Ge is not due to Ge, but to other dust constituents, such as SiO_2. The most harmful are Ge hydride and chloride compounds. During the course of the Ge production, the exposure to air contaminated with Ge fumes and dusts may occur. Workers exposed to Ge compounds show an increased urinary excretion without any other symptoms. In users of Ge preparations (e.g., as antioxidant), increased contents of this element are observed in spleen, renal cortex, brain, and skeletal muscles (Glei 2004).

Certain Ge compounds have low mammalian toxicity but there is a stimulated interest and marked activity in their use in pharmaceutical products, against some bacteria (http://www.ilo.org/encyclopaedia/. Only a few experimental studies on animals indicated toxic effects of Ge (Schauss 1991).

Table II-14.7. Germanium in foodstuffs (mg kg^{-1} FW) (after Schroeder et al. 1968)

Food species	Mean	Food species	Mean
Seafood (21)[a]	0.87	Meats (6)	0.35
Dairy products (6)	0.73	Grains (14)	0.36
Wheat (8)	0.54	Vegetables (19)	0.61
Nuts (7)	0.41	Beverages (10)	0.44
Fruit (5)	0.41	Oils (2)	0.17
Condiments and spices (18)	0.46	Canadian rat food	0.32

[a] In paretheses numbers of analyzed samples.

II-14.3
Tin (Sn)

II-14.3.1
Introduction

The distribution of Sn in the Earth's crust averages at above 2 mg kg^{-1} and shows an increase in acid igneous rocks and in argillaceous sediments (Table II-14.2). Tin is a component of only few minerals, of which the most important is cassiterite, SnO_2. It may combine with other metals in such minerals as, stannite, Cu_2SnFeS_4; teallite, Cu_3SnS_4; and montesite, $PbSn_4S_3$. The ability of Sn to form complexes with organic substances may affect its enrichment in biolites.

Tin was mined in Spain, from the beginning of the Bronze Age and at that time the main suppliers of Sn were the Phoenicians who carried on trade from their mines and smelters situated in the Iberian Peninsula. Hence it is one of the earliest metals known to and used by humans, mainly as a component of bronze alloys.

Tin exhibits two oxidation states: Sn^{2+} (stannous) and Sn^{4+} (stannic). The former is basic and behaves like a metal, whereas the latter form is amphoteric and may behave in an acidic manner in an alkaline solution. Most organotin compounds have tin in the 4+ oxidation state. Common organotin compounds, OTCs, are composed of MBT, DBT, and TBT (Table II-14.9).

II-14.3.2
Production and Use

The primary world Sn production was 231 kt in 2002 and 258 kt in 2003 (WMSY 2004). Its secondary production of Sn in the United States touched 9.0 kt at that time (USGS 2004). The annual global production of organotin is about 50 kt (Mercier et al. 1994).

Tin is generally extracted from the ore of cassiterite. Other Sn-containing minerals are of much smaller commercially importance. The production of pure Sn requires the following steps: roasting, acid leaching, smelting and electrorefining.

Most of Sn is used as a component of protective coatings and as an alloying element. Due to a large temperature difference between melting (232 °C) and boiling (2 270 °C) points, Sn can be used as an alloy without loss through the vaporization. The metal has a large number of organometallic derivatives, which are used commercially more than any other element. There is observed a significant increase of its use in the South-East Asian region and in the developing countries (Fent 1996b).

Its uses are documented in the coating of steel (e.g., sheet metals) and for solder, glass-making, PVC stabilizers (mainly organotin), pesticides and wood preservatives. Organotin is a common component of ship paints that has significant environmental effects (Hoch 2001).

II-14.3.3
Soils

The average values of Sn in 15 different types of soils in the USA range between 0.6 and 1.7 mg kg^{-1} (Shacklette and Boerngen 1984). Eckel and Langley (1988) estimated the mean background Sn concentration in the US at 0.89 mg kg^{-1}. Mean Sn contents in agricultural soils of Japan range from 2 to 3.1 mg kg^{-1} in Andosols and Acrisols, respectively (Takeda et al. 2004). According to Eriksson (2001a), Swedish topsoils contain Sn in the range of 0.4–8.6 mg kg^{-1} (mean 1.8).

In uncontaminated soils, Sn is largely derived from its content in the bedrocks, and occurs at the background level. In polluted sites, however, its concentration may be highly elevated, as for example up to 1 000 mg kg^{-1}, near smelter areas (Schafer and Fembert 1984).

The presence of organic Sn compounds (OTCs) in the terrestrial systems stems mainly from landfills, as it has been stated by Huang and Matzner (2004). It has been also cited that the deposition of OTCs and Sn_{tot} on forest soils occurs via throughfall or literfall and there is evidence that the phytouptake of OTC occurs from soils. The OTCs are regarded as contaminants in the environment. They are associated mainly with the clay fraction, which is an important control factor regarding the distribution and fate of OTCs in soils. Both clay minerals and metal oxides, bearing a net negative charge, are very effective sorbents for several OTCs under various conditions (Hoch 2001). Soil parameters, and in particular CEC, pH, salinity, and kind of minerals, are important in controlling the sorption of OTCs (Batley 1996; Weidenhaupt et al. 1997).

II-14.3.4
Waters

The mean Sn concentration in worldwide ocean water is calculated at 0.004 μg l^{-1} (Reimann and Caritat 1998). Lower concentration in the North Pacific Ocean, at 0.0005 μg kg^{-1}, has been reported by Nozaki (2005). Increased Sn levels, up to 0.43 μg l^{-1} has been detected in waters of Swedish west coast during 1986–1988 (Szefer 2002a). However, Sn concentration in unpolluted waters has been often below the detection limit (WHO 1980). Earlier citations gave higher values for Sn in waters; in Canadian rivers its levels were given to range from 1.3 to 2.1 μg l^{-1}, whereas in the seawater it was between 0.2 to 3.1 μg l^{-1} (WHO 1980).

Tin may be present in aqueous solution in either divalent or the tetravalent oxidation states, depending on the oxygen pressure and pH of the system. In natural waters, OTCs are present mostly as neutral species or as cations (Weidenhaupt et al. 1997). At different pH, forms of OTCs in the aquatic environment differs considerably (Table II-14.8). Methylated OTC from the river estuaries and the transformation of ionic OTC into volatile species are significant sources of Sn in the marine environment (Feldmann and Hunter 1995; Tessier et al. 2002). These organic species are potentially volatile and the mobilization of Sn species in the water column may be possible (Amouroux et al. 2000).

Table II-14.8. Forms of OTCs in waters at different pH (after Blunden and Evans 1990)

pH	Species
<4	DMT, Me_2Sn^{2+}
6–8	DMT, $Me_2Sn(OH)_2$
<5	TMT, Me_3Sn^+
>5	TMT, Me_3SnOH

Tin may enter into the aquatic systems from various sources. However, very significant pollution with different Sn compounds, including OTCs, has occurred in harbors, from shipping and shipbuilding (Table II-14.9). Here, the water receives antifouling paints, which are important factors influencing the concentration of Sn compounds (Koivisto 2003).

In the late 1980s and at the beginning of 1990s, the use of TBT-based antifouling paints was banned in many western countries as a result of which the concentration of *n*-TBT in waters and sediments is on the lower side. In the open Arachon Bay, France, the concentration of TBT in water was measured to be only 1 ng Sn l^{-1} (Ruiz et al. 1996). The quality standard for waters at 2 ng TBT l^{-1} has been recommended in the UK (Alzieu et al. 1991).

Table II-14.9. Butylintin concentrations in water (ng Sn l^{-1})

Site, country	MBT, monobutyltin	DBT, dibutyltin	TBT, tributyltin
Sado Estuary, Portugal[a]	18 – 60	52 – 160	39 – 870
San Pedro River, Spain[b]	6.9 – 14.4	5.5 – 14.3	9.3 – 16.3
Cadiz Bay, Spain[b]	8.9 – 41	8.3 – 68	8.3 – 488
Guadalete River, Spain[b]	9.8 – 25	5.7 – 39.5	9.9 – 116
Antwerp Harbour, the Netherlands[c]	51 – 76	217 – 283	765 – 1000
Ganga Plain, India[d]	2 – 70	2 – 101	3 – 20
Baltic Sea, German coast[e]	–	–	10 – 300

[a] Quevauviller and Donard (1990).
[b] Gomez-Arza et al. (1995).
[c] Dirkx et al. (1993).
[d] Ansari et al. (1998).
[e] Szefer (2002b).

Table II-14.10. Concentration of butyltin in sediments (µg Sn kg^{-1}) (collected from different sources)

Location	Depth (cm)	MBT	DBT	TBT
San Pedro River, Spain	0 – 10	1.9 – 6.1	1.5 – 8.7	0.7 – 11.1
Guadalete River, Spain	0 – 10	16.1 – 129	20.5 – 510	26.5 – 601
Cadiz Bay, Spain	0 – 10	1.2 – 31	1.1 – 52	1.6 – 225
Arcachom Bay, France	0 – 51	6 – 156	5 – 141	16 – 161
San Diego Bay, CA	0 – 6	2 – 185	2 – 265	2 – 242
Pearl Habor, Hawaii	0 – 6	5 – 533	4 – 367	4 – 2830
Leman Lake, Switzerland	n.d.	186	295	627
Toronto, Canada	n.d.	580	248	539
Vancouver, Canada	n.d.	3360	8510	10780

In the aquatic environment, OTCs have a tendency to be adsorbed onto suspended particulate matter. Through this mechanism, there is scavenging of OTC by sediments in which Sn-chemicals and degradation products can occur (Table II-14.10). Sometimes, biomethylation of OTCs occur due to very fast photochemical and biological processes in the sediments, the product being, e.g., mixed butylmethyltin species, $BuMe_3Sn$; up to 540 µg Sn kg^{-1} (Jantzen 1992). In addition to OTCs, the mobile inorganic Sn^{4+} is also found in sediments that is due to discharge of chemicals from industry (Kuballa et al. 1996).

Aquatic biota. Elevated concentrations of Sn and especially of organotins affect the growth and reproduction of both zoo- and phytoplanktons. Experimental data showed that TBT at concentrations <1 ng l^{-1} modifies behavior of marine gastropods. Organotins are easily accumulated in aquatic birds. In some cormorans in Lake Biwa, Japan, the TBT levels were at 42–160 µg kg^{-1}, which is a result of the ingestion of fish containing 10–55 µg Sn kg^{-1} (Thompson et al. 1990).

The OTCs are considered to be one of the most hazardous known pollutants in the aquatic systems (Fent 1996b). Due to the extensive use of OTCs, significant concentration of these chemicals has been observed in different compartments of the aquatic environment. The highest contents of Bu_3Sn are found in fish and mammals (Fig. II-14.1).

Table II-14.11.
Tin species in green algae (*Enteromorpha* sp.) (mg kg^{-1}) (after Donard et al. 1987)

Sn species	Sn content
Total-Sn	4.4
Monomethyltins	0.5 (FW)
Dimethyltins	0.5 (FW)
Trimethyltins	<0.001 (FW)
Monbutyltins	0.4
Tributyltins	0.6

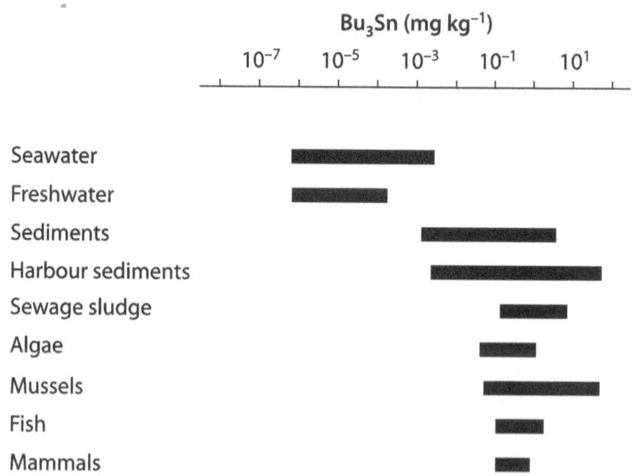

Fig. II-14.1.
Ranges of tributyltin (Bu_3Sn) levels in various samples of aquatic systems (adapted from Hoch 2001)

Marine plants are also important in the cycling of Sn. The total Sn concentrations in marine algae and macrophytes vary between 0.5 and 101 mg kg^{-1} and clearly demonstrate that most species of aquatic flora accumulate this metal from seawater (Eisler 2000). Living algae are useful in immobilizing Sn from seawater and regulating the formation and degradation of toxic methyl-Sn compounds. Dead and decaying algae accumulate inorganic and organo-Sn compounds, and ultimately removes Sn from the estuary to the atmosphere by formation of tetramethyltins (Donard et al. 1987).

Concentrations of toxic OTCs have often been reported for some water system (e.g., Mediterranean Sea) that resulted in a significant increase of Sn and OTCs in tissues of aquatic animals (Table II-14.12). A very high concentration of Σbutyltins (sum of TBT, DBT, and MBT), up to 10 000 μg kg^{-1}, was found in aquatic species from the Seto Inland Sea, Japan (Tanabe et al. 1998).

In the deep-seas the butyltin compounds affect aquatic species in a similar manner to the affect on living organisms in the coastal areas. Takahashi et al. (1997) observed butyltin compounds in the Suruga Bay, Japan at 21 to 980 ng kg^{-1} and suggested that there is a less decomposition rate of TBT due to lower temperature, less sunlight penetration and the lack of phytoplankton in mid sea. The concentration of TBT in Australian oysters after the banning of OTCs, was found to be close to the detection limit 0.2 μg Sn kg^{-1}. But still there are problems with this compound in the aquatic environment of many countries, even though its use has been restricted.

Table II-14.12. Concentration of OTCs in aquatic organisms (μg Sn kg^{-1})

Area	Animal (tissue)	MBT	DBT	TBT
Italian coastal waters[a]	Bottlenose dolphins (blubber)	55	16	41
	Bottlenose dolphins (liver)	150	800	250
	Bluefin tunas (liver)	38	125	46
	Bluefin tunas (muscle)	15	8.6	39
Californian coastal waters[b]	Sea otters (liver)	<7 – 360	21 – 5 820	19 – 3 020
	Sea otters (kidney)	<7 – 61	3.7 – 200	4 – 210
	Sea otters (brain)	<2.4 – 24	1.8 – 105	2.7 – 140
Lake, The Netherlands[c]	Zebra mussels (body)	21 – 120	20 – 160	180 – 2 500
	Eel (body)	13 – 63	9 – 40	50 – 390
	Roach	7 – 34	20 – 210	160 – 2 500
Indonesia[d]	Fish	0.93 – 14	<1.5 – 18	1.4 – 52
Malaysia[e]	Fish	2.5 – 7.4	<1.3 – 13	2.4 – 190
Bangladesh[f]	Fish	<5.6 – 170	0.36 – 15	0.47 – 3.0
India[f]	Fish	<5.6 – 78	0.36 – 0.65	<0.13 – 0.9
Australia[f]	Fish	<4.0 – 42	<0.36 – 0.98	0.13 – 13

[a] Kannan et al. (1996).
[b] Kannan et al. (1998).
[c] Stäb et al. (1996).
[d] Sudaryanto et al. (2005).
[e] Sudaryanto et al. (2004).
[f] Kannan et al. (1995).

Sn

II-14.3.5
Air

Air from remote regions contains Sn at the level of 10–70 ng m^{-3}, whereas in polluted areas its concentration may be up to 800 ng m^{-3} (Table II-14.2). There are some reports on Sn concentrations in the atmosphere of the United States that have indicated values from below the detection limit to 800 ng m^{-3}; the average concentration was at 100 ng m^{-3}. Tin in a highway tunnel exhaust in Hamburg, was measured in 1988–1989, at the value of 10 900 ng m^{-3} (Dannecker et al. 1990).

Industrial emissions of Sn and its compounds, and especially the emission from some coal burning plants, may be a significant source of Sn pollution. Fly ash released from the coal combustion contains a substantial level of Sn, at 7–19 mg kg^{-1} (WHO 1980). In the 1980s, worldwide industrial Sn emission into the atmosphere was estimated at 1.47 to 10.8 kt (Nriagu and Pacyna 1988). Organotin may be release into the atmosphere from the incineration of organic waste, uses of fertilizers and pesticides, from antifouling paints and from glass manufacturing facilities. However, due to its low vapor pressure and rapid photodegradation, its release into the atmosphere is less important (Blunden et al. 1984; Fent 1996a).

Rühling and Tyler (2004) observed that the concentration of Sn in the moss carpet (*Pleurozium schreberi*) decreased from 0.49 mg kg^{-1} in 1975 to 0.09 mg kg^{-1} in 2000 (Table II-14.5). Mosses sampled in Norway in the period 1990–1995 contained Sn in the range of <0.008–25 mg kg^{-1} (mean 0.25) (Berg and Steinnes 1997).

Huang and Matzner (2004) reported that the precipitation of Sn_{tot} is about 10 times higher than of OTCs. The deposition of OTCs, expressed as Sn, was 170 mg ha^{-1} yr^{-1} that indicates significant emissions of OTCs into the atmosphere. The same authors estimated the emission of OTCs in the atmosphere to range between 2 000 to 6 000 kg yr^{-1}. These authors concluded that this is a crude estimation, but it still suggests that OTCs emission and deposition in terrestrial and aquatic environments are substantial and have been underestimated so far. A high deposition of Sn, at 60 g ha^{-1} yr^{-1}, has been calculated for agricultural soils of Sweden (Eriksson 2001a).

II-14.3.6
Plants

There is no evidence that Sn is either essential or beneficial to plants. It is considered as toxic to both higher plants and fungi. However, when it exists in easily available forms in soils, such as cations Sn^{2+} and Sn^{+4}, it is absorbed by roots, where is likely to be accumulated.

Common range of Sn in food plants is reported to be between <0.04 and <0.1 mg kg^{-1}, and in grass from 0.2 to 1.9 mg kg^{-1} (Kabata-Pendias and Pendias 2001). Its contents in wheat grains are given for various countries to be (in mg kg^{-1}): 0.01–0.12 in Sweden, 5.6–7.9 (FW) in USA, and 0.5 (FW) in Japan (Eriksson 2001a; Jenkins 1980). These great differences should be apparently related to different analytical methods. Wood contains 1.8 and 1.5 mg Sn kg^{-1}, respectively of 40 and 80 years old trees from Vermont, USA (Jenkins 1980).

According to Tyler (2005), the R/S ratio (root/soil) for Sn is about 0.11 (concentration of Sn in root: 74 and in soil: 651 mg kg^{-1}). In Swedish beech forest growing in podzolized soil, ranges of Sn in various species were (in mg kg^{-1}) from 0.2–0.3 in *Lactarius blennius* to 1.0–1.8 in *Collybia peronata* (Tyler 2004).

Plants growing in mineralized or contaminated soils accumulate Sn to high levels; Jenkins (1980) reported Sn concentrations between 338 and 2 165 mg kg^{-1} and Peterson et al. (1976) up to 2 000 mg kg^{-1} in vegetation near Sn-smelters.

II-14.3.7
Humans

Mammalian tissues contain Sn at the range of <0.2–0.85 mg kg^{-1}, being the lowest in brain and the highest in liver and kidneys (Jørgensen 2000). According to Li (2000) total soft tissues of human contain Sn at 0.1 mg kg^{-1}, and its content of all tissues of a "reference man" averages at 0.24 mg kg^{-1}. Information on OTC contents in human tissues is quite rare. Takahashi et al. (1999) reported concentrations of OTCs between <2 and 78 μg kg^{-1}, in tissues of people of Japan.

Humans are exposed to Sn from ingestion, inhalation, and dermal adsorption. However, the main source of Sn uptake is from food, with an exception of industrial areas where its concentrations in water and air are elevated (Biégo et al. 1999).

Tin in the inorganic form is considered as being non-toxic, whereas some organotins are mainly toxic, especially triethyltin (Et_3SnOAc). The toxicity of OTCs decreases in the following order: $R_3Sn^+ > R_2Sn^+ > RSn^{3+}$. This toxicity may be inhibited by the phosphorylation processes (Huang and Matzner 2004).

Leaching of OTCs from tin-coated cans and other food containers, as well as from tap water pipes may contaminate foodstuff and beverages. The OTCs are often used in household commodities (e.g., a siliconized baking pan contains 260 μg MBT kg^{-1}) (Takahashi et al. 1999). Also higher levels of Sn may be in some processed food due to the addition of Sn-based preservatives and stabilizers such as stannous chloride (Anger 2004).

Canned foods, especially fruits and vegetable products are considered to be the main source of Sn in the diet. Some canned fruits and juice may contain Sn at the range from 141 to 2 000 mg kg^{-1} (FAO/WHO 2002). Organotins present in wine, at levels: MBT (1.7–20 μg l^{-1}), DBT (0.3–160 μg l^{-1}) and TBT (0.8–1.6 μg l^{-1}), are most probably resulted from leaching of containers (Forsythe et al. 1994).

The RDI value for adults is 120 mg $person^{-1}$ (FAO/WHO 2002). The provisional maximum tolerable daily intake is 2 mg kg^{-1} BW (Anger 2004). Some individuals, however, can tolerate up to 700 mg kg^{-1}, and this is highly variable depending on the specific Sn chemical form. Mean Sn uptake by adults in different countries varies between 1.7–3.1 mg d^{-1}.

Increased Sn concentration in food may cause acute gastric irrigation, impaired reproductivity, and bone strength failure. It is also considered to be genotoxic. Some organotins are highly dermal irritants.

II-14.4 Lead (Pb)

II-14.4.1 Introduction

Of the known environmental pollutants, Pb has few competitors as a persistent pollutant, which causes harmful effects to humans, animals and ecosystems. However, it is also a useful metal. It occurs in the environment mainly as Pb^{2+} (Table II-14.1) and its compounds are mostly insoluble in waters. There are still large quantities of Pb, mainly as $PbSO_4$ and $PbCO_3$ released into the environmental compartments. This metal is remobilized in ecosystems causing toxic effects to terrestrial and aquatic species.

In the 1970s and 1980s, most Pb pollution was emitted from the use of leaded gasoline, and the emission of Pb was higher from anti-knock additives in gasoline than it was from pollution point sources. In the 1990s, strict regional regulation has virtually eliminated the use of Pb in gasoline in the developed countries. However, it is still use in several African countries, such as in Nigeria, where Pb in gasoline is about 0.6 g l^{-1} that is the highest concentration noted in the world.

II-14.4.2 Production and Use

The global production of refined Pb, as given by the WMSY (2004) in 2000, was 3.1 Mt and in 2003, was 2.9 Mt. However, other sources, e.g., Smith (2004) reported much higher production, up to 6.82 Mt in 2003, which include both primary and secondary (from scrap, etc.) Pb production.

The most important Pb ore is the heavy black mineral galena, PbS. Other common minerals are: anglesite, $PbSO_4$; cerussite, $PbCO_3$; minium, Pb_3O_4; pyromorphite, $Pb_5(PO_4)_3Cl$; and mimetesite, $Pb_5(AsO_4)_3Cl$. Lead ores are widely distributed throughout the world, but only fifty countries have Pb-ores that can be mined commercially. Normally, Pb is obtained from galena deposits. The Pb-concentrate is roasted at a temperature of 600 °C in a limited supply of air and the product, PbO, is mixed with coke and limestone (as a flux) and reduced to liquid Pb in the blast furnace at 1 250 °C. The product contains different metals as impurities: Cu, Ag, Au, Zn, Sn, As, and Sb, which must be removed to obtain 99.99% of pure Pb.

In the USA, 92% of all Pb is produced from secondary sources, i.e., Pb scrap from spent lead-acid batteries (Platts Metals Week 2003b). In the near future, the EC will adopt a new battery directive, which requires collection of all used batteries in the EU and the recovery of metal used in car and industrial-type lead-acid batteries. This proposal will ban landfilling and incineration of all types of batteries, and create the EU-wide framework for national battery collection and recycling systems (Platts Metals Week 2003a).

The largest use of Pb worldwide is for lead-acid batteries known as the SLI battery (starting, lighting and ignition). In addition, it is used in solders, alloys, chemicals, cables, lead weights, lead wool, and for many other purposes, due to its special physical and chemical properties. Nowadays, the use of Pb in petrol as an anti-knock additive in developed countries has been phased out in an effort to reduce the atmospheric Pb pollution. Hence the overall market for tetraethyl and tetramethyl lead (TEL and TML) has declined considerably, but still there is some market for these chemicals in the developing countries.

There are strict regulations in many countries on the release of Pb to the environment, but still it continues to be one of the most serious global human hazards (United Nations 1996; Brännvall et al. 1999).

II-14.4.3 Soils

The natural Pb content in soils stems from parent rocks. Its abundance in sediments is a function of clay fraction content and thus argillaceous sediments contain more Pb than sands, sandstones and limestones (Table II-14.2). Soil Pb status has been intensively studied (Table II-14.13). However most of presented data in this table are not comparable because of very different methods used for the preparation of soil samples and for the metal extraction. Only the values obtained from the HF/$HClO_4$ fusion and aqua regia extraction can be considered as compatible for, so called, the total content of Pb. The overall mean value of Pb for different soils has been calculated to be, in the average, at 25 mg kg^{-1} (Kabata-Pendias and Pendias 2001). Angelone and Bini (1992) cited unpolluted soil Pb concentration as being less than 100 mg kg^{-1}. In Greenland

Table II-14.13. Different forms of lead in worldwide soils (mg kg^{-1})

Location	Sample no.	Range	Mean	Method of sample preparation or extraction	Reference
Ireland	1305	0.5 – 138	–	0.05 M EDTA	Dickson and Stevens (1983)
Norway	512	–	40	HNO_3	Steinnes et al. (1997)
Sweden, remote areas	–	–	27	HNO_3	Tyler (2005)
UK, Somerst	174	8 – 10000	183	HNO_3	Davies (1995)
Wales	1521	4.5 – 2900	39.8	HNO_3/$HClO_4$	Archer and Hodgson (1987)
	5692	3 – 16338	–	Aqua regia	McGrath and Loveland (1992)
USA,					
high risk	75	50 – 10900	1275	HNO_3	Holmgren et al. (1993)
low risk	25	50 – 700	205	HNO_3	
various soils	3045	<1 – 135	12.3	HNO_3	
Poland, agricult. soils	40400	0.1 – 5000	13.7	Aqua regia	Terelak and Piotrowska (1998)
Australia					
Wollongong	80	1 – 187	9.9	0.5 N Acetic acid	Lottermoser (1997)
Port Macquarie	30	6 – 108	20	HF/HNO_3/HCl/$HClO_4$	
Mort Bay	27	14 – 780	256.2		
Adelaide	69	5 – 1450	97	0.1 M EDTA	Tiller (1992)

soils, Pb contents were found to vary between <12 and 13.8 mg kg^{-1}. Thus, at the end of 1990s, the Pb contents in arctic soils were at the values below the stringent guideline value of 25 mg kg^{-1}.

Contents of Pb in soils show a positive correlation with contents of fine granulometric fractions. Its lowest amounts, up to 40 mg kg^{-1}, are in light sandy soils, and the highest, up to 90 mg kg^{-1}, in heavy loam soils. Some increased Pb contents have also been reported for calcareous soils and organic ones (Table II-14.2).

Lead is not distributed uniformly is soil horizons and reveal a great association with hydroxides, especially of Fe and Mn. Its concentrations in Fe-Mn nodules and concretions may be very high, up to 20 000 mg kg^{-1} (Kabata-Pendias and Sadurski 2004). It also can be concentrated in carbonate or phosphate particles. Lead is generally accumulated near the soil surface, mainly due to its sorption by SOM. The mobilization of Pb is usually slow, but some soil parameters, such as increased acidity, formation of Pb-organic complexes may increase its solubility. The Pb concentrations in soils solution are relatively low, and depending on the methods used for obtaining soils solution, vary from <1 to 60 µg l^{-1}. It occurs in soil solution as cationic species: Pb^{2+}, $PbCl^+$, $PbOH^+$, and anionic species: $PbCl_3^-$, $Pb(CO_3)_2^{2-}$ (Kabata-Pendias and Sadurski 2004). Lead desorbed into the soil solution may easily move from the upper horizons to lower causing pollution of groundwater (Alumaa et al. 2002).

Some authors have pointed out that the Pb fixation by SOM is more important than fixation by hydrous oxides (Li and Shuman 1996). However, it is not yet known exactly how much of the soil-Pb is available to plant. Davies (1995) reported that only about 0.005–0.13% of Pb in the soil solution is available to plants. However, the Pb uptake by plants depends on several soils properties, such as SOM kind and content, granulometric composition, CEC, pH, as well as root surface area, and root exudates (Sillanpää and Jansson 1992).

Contamination and remediation. Among various natural and anthropogenic sources of Pb contamination, the impact of industrial emissions and previously used leaded petrol are considered to be of the greatest environmental risk. It has been estimated that Pb loading rate in soil exceeds its natural removal by 20-fold or more (Nriagu 1988).

Soil pollution due to Pb from mining and industrial activities is not a new problem. Historical evidence suggests that the problem of Pb in the environment began when our ancestors learned the use of fire. Marked soil-Pb contamination occurs in the vicinity of mining and industrial activities, in urban areas, and along high-traffic roads. The old industrial Pb contamination has been a subject of several studies, as for example those conducted by Davies (1995).

At high temperature processes in smelters, Pb is emitted as PbS, $PbCO_3$, $PbCl_2$, ($PbO \cdot PbSO_4$), Pb^0, and PbO, while Pb from automobile exhausts is mainly in forms of halide salts, e.g., PbBr, PbBrCl, Pb(OH)Br, and $(PbO)_2PbBr_2$. It has been estimated that the atmospheric half-life for organic Pb vapor is somewhat less than 12 hours that is a great concern for the populations living in megacities (Neilsen 1982).

In the 1980s and 1990s, many studies have reported on Pb contamination in soils due to the huge atmospheric emission of Pb (global estimated emissions 332 kt yr^{-1}; in Europe 85 kt yr^{-1}) (Pacyna 1995). As a result of strict regulations and campaigns against Pb hazards in society, Pb emissions in most European countries and in the North America have decreased considerably. Nevertheless, in several countries, especially those where the Pb emission is not restricted controlled (e.g., Newly Developed Countries), this metal is still of a great concern. Such situations in vicinities of nonferric metal processing industries have been reported, e.g., by Kachur et al. (2003) for the far East Russia, by Shomar et al. (2004) for Lebanon, and by da Silva et al. (2004) for Portugal. In mining areas, Pb may be dispersed due to the erosion and chemical weathering of tailings. The severity of these processes depends on chemical characteristics, and the minerals present in the tailings (da Silva 2004).

Increased Pb level in top soils may also be due to Pb-shots that has been reported by Darling and Thomas (2003) for some hunting areas (in over 211 locations) in Ontario. A similar problem has been reported by other authors (Gzyl 1995; Wolz et al. 2003; Agbenin 2002). In Tasmania and South Australia, Pb contamination of surface soils has been described in orchard soils as a result of the use of arsenate insecticides (Merry et al. 1983). Other important sources of Pb in home-garden soils are Pb-containing paints. The geometric mean of Pb in such soils of the UK was calculated at 298 mg kg^{-1} (Thornton 1990).

The Pb contamination of roadside soils has been for quite a time of a real environmental concern. Though Pb additives in petrol are banned in most countries, increased Pb levels in soil along high traffic road (up to 7000 mg kg^{-1}) have been still observed (Kabata-Pendias and Pendias 2001). Especially organolead compounds are still in these soils. The available data suggest that tetraethyl and tetramethyl Pb are converted into sparingly water-soluble Pb compounds in soil, however, recent laboratory tests indicated that tetraethyl and tetramethyl Pb do not leach significantly through soil profile (Ou et al. 1995). Exhaust Pb particles are unstable and readily converted into oxides, carbonates, and sulfates.

Some chelating agents and organic matter ligands can modify the behavior of Pb in soils (Cunningham and Berti 2000). Due to a great sorption capacity of organic matter, peat bogs reveal a great capacity to sorb Pb. The level of Pb is used for the assessment and dating the pre-anthropogenic and anthropogenic pollution using isotope ^{210}Pb (Shotyk 1996). The author has estimated that the present Pb concentrations in recent peats are about 100 times higher than the pre-industrial time pollution.

The remediation of Pb-contaminated soils has become recently broadly investigated. Although Pb in most soils is slightly available to plants, there are several plants, such as *Thlaspi* sp., corn, sunflower, that accumulate larger amounts of this metal. Greger (1999) cited that *Minuartia verna* may concentrate in leaves >1000 mg Pb kg^{-1}. Increasing Pb mobility by using various chelators, such as EDTA and DTPA, can stimulate its phytoavailability, and thus the phytoremediation (Jarvis and Leung 2002; Lim et al. 2004). The effectiveness of chelating agents increases Pb accumulation in several plants such as Indian mustard or some shrubs (Huang et al. 1997). The laboratory test conducted by Chandra Sekhar (2005), in the presence of other metals Ni, Zn, Cr and Cd, indicated that Pb uptake by *Hemidesmus indicus* decreased by 50% at the presence of Ni, but there were only marginal effects from the presence of Zn, Cr, and Cd (Ebbs et al. 1997; Garcia et al. 2002).

Lepp and Dickinson (2003) studied trace metals in an urban woodland ecosystem (Pb concentration in soils varied between 348 to 76,410 mg kg^{-1}) in St. Helens, Merseyside, UK. They believe, that the closed and stable vegetation cover, could prevent re-entrainment of Pb to plant species. According to their statement, at the absence of human intervention, the flexibility of the natural vegetation may be responsible for the bioremediation of polluted soils in St. Helens.

Adriano et al. (2004) have carried out the broad survey on the natural remediation of soil contaminated with trace metals. The addition of Fe-rich lime stabilized biosolids and of $CaCO_3$ raise the soil pH to 7, that decreased the uptake of Pb by tall fescue (*Festuca arundinacea*). In addition to the impact of high pH, the competition of Pb with other cations also affect its phytoavailability.

II-14.4.4
Waters

Worldwide reports confirm that there are great variations in Pb concentrations in seawaters. The median Pb content in worldwide ocean waters has been calculated at 0.03 $\mu g\, l^{-1}$ (Raimann and Caritat 1998). The Pb mean concentration in the North Pacific Ocean is given at 0.003 $\mu g\, kg^{-1}$ (Nozaki 2005). The US EPA (1982) report has estimated this value for seawater at 0.005 $\mu g\, l^{-1}$. The Baltic Sea waters contain Pb at the range from 0.05 to 3.6 $\mu g\, l^{-1}$, being the highest at coastal waters and in gulf areas (Table II-14.14). In remote ocean waters, there is no immediate aquatic input of Pb, but only a modest atmospheric deposition of the metal and therefore its concentration is low, at <0.1 $\mu g\, l^{-1}$ (Prospero et al. 1996).

Concentrations of Pb in river waters vary highly from 0.007 to 3.8 $\mu g\, l^{-1}$, being the lowest for Mississippi tributaries and the highest for Harz Mt. regions, Germany (Gaillardet et al. 2003). These authors calculated the worldwide average in river waters at 0.08 $\mu g\, l^{-1}$, and the Pb riverine flux at 3 kt yr^{-1}. Kitano (1992) has estimated the global discharge of Pb to marine basins at 40 and 20 kt yr^{-1} with river waters and atmospheric precipitations, respectively.

Relatively, high Pb concentrations, between 5 and 30 $\mu g\, l^{-1}$, have been measured, in 50 000 samples of surface waters and groundwater in the USA (US EPA 1998). In the European part of Russia, groundwater contains Pb at the range of 0.1–8 $\mu g\, l^{-1}$ (Uchvatov and Bashkin 1997).

Table II-14.14. Lead in waters of the Baltic Sea ($\mu g\, l^{-1}$)

Region	Year	Sample depth (m)	Range or mean	Reference
Baltic Sea (whole)	1972–1973		0.4 – 2.1 0.64	Bürgmann (1979)
East Gotland	1996	10	0.083[a]	Pohl and Hennings (1999)
Gulf of Finland	1991	53	0.1 – 3.6	Kravtsov and Emelyanov (1995)
Baltic Sea	1988	9	0.075	Brügmann et al. (1991)
German coast	1977	5	0.053	Mart and Nürnberg (1986)
Northern Baltic	1985/1986	10	0.17 – 0.19[b]	Szefer (2002a)

[a] In mMol kg^{-1}.
[b] In fraction <0.45 μm.

The chemistry of Pb in aqueous systems is highly complex because this element exists in multiple forms. The Pb concentrations in surface waters depend especially on the pH and dissolved salt contents of the water. Additional factors, such as pollution sources, sediment Pb content, temperature, and organic matter kinds and amounts have also significant impact on the Pb status in waters. In seawaters, Pb occurs mainly in forms of $PbCO_3$ and $PbCl_2$. In surface and ground waters, its species are: Pb^{2+}, $PbOH^+$, $PbHCO_s^+$, and $PbSO_4^0$ (Witczak and Adamczyk 1995). In most aquatic environments, it forms one of the less mobile species, which eventually will precipitate on the bottom sediments. Surface waters are generally neutral or alkaline which results in Pb being associated with less mobile forms (Laxen and Harrison 1983).

In the Northern Hemisphere, a great attention has been paid to the potential effects on the Baltic Sea and the North Sea in a respect to atmospheric deposition of pollutants from industrial sources as well as coastal pumping of treated and untreated effluents. The Baltic Sea is one of the largest brackish water regions of the world. There is a Baltic Monitoring Programme (BMP), which operates on an international basis due to the discharge of pollutants through rivers, and high transportation of saline water from the North Sea into the Baltic Sea. Mean concentration of Pb in nine Polish rivers discharging into the Baltic Sea has been estimated at 2.72 $\mu g\,l^{-1}$ within the range of 0.1–5.5 $\mu g\,l^{-1}$ (HELCOM 1998). In the course of downstream dispersion, a part of Pb amounts in river waters is removed by the precipitation to bottom sediments of riverbeds. The studies conducted within the projects of HELCOM (1987, 1991, 1993, and 1998) indicated that the total Pb input via rivers and atmosphere was 3 200 t, 2 894 t and 975 t for the years 1987, 1990 and 1995, respectively. Matschullat (1997) calculated the total load of Pb to the Baltic Sea at 2 800 t yr^{-1}, in about equal proportions from river flux and atmospheric fallout. Above 90% of Pb discharged to this basin is of anthropogenic origin. Data presented by some authors may suggest that Pb concentrations in the Baltic Sea have decreased a little bit during last two decades (Table II-14.14).

The ratio of Pb isotopes in the northeast Pacific Ocean waters was used for the identification of pollution of sources of Pb; the surface layer of water contains Pb of industrial dust from Japan whereas deeper layers were polluted by North American emissions (Flegal et al. 1986).

During the course of high flow, Pb entered into sediments may become resuspended and enters as solid particles into the water column. Lead concentration in river water depends on local sources and the geochemistry of the area. In a Pb mineralization area, the concentration of Pb may be ten fold that of an unmineralized area. Near to point sources, elevated Pb concentration in river water is to be expected. Water of the polluted Ogunpa River (Nigeria), is reported to contain Pb at 9.8 $\mu g\,l^{-1}$, in the range of 1.3–46 $\mu g\,l^{-1}$ (Monbesshora et al. 1983).

Due to the decreasing concentration gradient, part of the Pb is removed to bottom sediment, which is the long-term sink for the metal and in which the stored Pb is in an unavailable form (OECD 1993). Thus, the best indicator of Pb pollution of river waters is its accumulation in bottom sediments where a great proportion of this metal is deposited. Its background contents in bottom sediments vary from 30 to 45 mg kg^{-1}. However, in polluted rivers, especially by industrial and mines discharges, Pb contents range from 700 to 2 600 mg kg^{-1} in rivers of Poland and of the UK, respectively (Kabata-Pendias and Pendias 1999).

The Pb concentration in drinking water is of special concern. The recommended maximum level of Pb in drinking water is 15 μg Pb l^{-1}, if it is over this level, then further treatment of tap water is necessary, in accordance with the regulations in the USA (US EPA 1999). This document confirmed that over 219 million people in the USA drink tap water containing <5 μg Pb l^{-1}. The Pb concentrations in tap and well waters of the Democratic Republic of Congo (previously Zaire) and Nigeria were reported to be at 2.0 and 0.9 μg l^{-1}, respectively (Kapu et al. 1989). In most countries, drinking water contains Pb at mean levels in the range of 2–3 μg l^{-1} (various sources *vide* Kabata-Pendias and Pendias 1999).

Aquatic biota. Although high concentrations of Pb, up to 3 000 mg kg^{-1} in bryophytes, have been reported, lethal poisoning of fish is quite limited (Chimie et Ecologie 1998; Pain and Donald 2002). It is accepted that exposure to Pb may cause mucous formation, causing coating the gills, which damages respiratory system. In the USA it has been observed that the Pb concentration in fish has been declining steadily, after the reduction of Pb in gasoline and due to the strict regulation on the pumping of untreated effluents into the aquatic environment (Schmitt and Brumbaugh 1990).

Lead contents in muscles of herring (*Clupea harengus*) from the Baltic Sea vary broadly from 0.01 to 1.9 mg kg^{-1} FW, and in shells of mollusk (Blue mussel, *Mytilus edulis*) range between <0.5 and 5.3 mg kg^{-1} (Szefer 2002a).

II-14.4.5
Air

Concentrations of Pb in air vary highly from about 1 ng m^{-3} around the South Pole to above 10 000 ng m^{-3} in urban/industrial areas (Table II-14.2). The common range of natural Pb concentrations is calculated at the range of 0.5–10 ng m^{-3} and most often aerial pollution – at the range of 70–8 000 ng m^{-3} (Livett 1992). Reimann and Caritat (1998) calculated the worldwide median for remote regions at 10 ng m^{-3}. The Polish industry has contributed about 2.5% of the total European Pb emissions and its contains in air are commonly reported to be <200 ng m^{-3} (Gzyl 1997).

The indicator for changes in the atmospheric Pb deposition is its concentration in the glacier in Greenland that has increased during about 100 last years from 0.05 to 0.5 μg l^{-1} (Buat-Ménard 1984). Nriagu and Pacyna (1988) estimated the Pb deposition in agricultural land areas to be about 50 g ha^{-1} yr^{-1}. However, Pb inputs were not uniform in the studied area. Its wet deposition on agricultural soils of Sweden has been calculated to average at 17 g ha^{-1} yr^{-1} (Eriksson 2001a).

Lead enters into the atmosphere from natural and different industrial sources, including coal and oil combustion in electric power stations, heating and industrial plants, gasoline combustion, roasting and smelting processes, and incineration of waste and cement production. On a global basis, anthropogenic inputs of Pb predominate over natural sources and are 96% of the total emission that has averaged at about 334 kt yr^{-1} (Nriagu 1989; Nriagu and Pacyna 1988).

The chemical forms of Pb emitted into the atmosphere depend on sources. From the coal combustion, Pb enters into the atmosphere as $PbCl_2$, PbO, PbS, and Pb, whereas at oil combustion, it is mainly in the form of PbO. Other Pb compounds, including alkylated Pb species, are also emitted. The size of Pb particles depends on its sources. For example, particle size of Pb from automobile exhaust varies between 0.1 and 1.0 µm. In the atmosphere, Pb particles/aerosols are transformed by chemical and physical processes and ultimately removed from the atmosphere by dry and wet depositions in terrestrial or aquatic ecosystems (Mishra et al. 2004).

In the past, up to 95% of Pb in the Western world stemmed from automobile sources (Hanger 1999, 2000). As a strategy to reduce automobile emissions of Pb, leaded fuel was banned in the USA in the1970s, whereas in Europe it was completely banned from January 1, 2000 (Spokes et al. 2001). Thus, the estimated global anthropogenic Pb emission value of 344 kt yr^{-1} (Nriagu and Pacyna 1988) is no longer valid, due to phasing out of leaded gasoline from many countries. In Europe, the estimated Pb emission was at 17.3 kt in 1998, excluding vehicular emission (EMEP 2000), whereas according to Pacyna and Pacyna (2001), the anthropogenic Pb emission in Europe has decreased from 62.53 kt in 1955 to 28.39 kt in 1995. However, there are still Pb vehicular emissions in many regions of the world, especially in Africa, where the contamination of crude oil is at 10–15 mg Pb l^{-1} (Pacyna 1998). According to Adriano (2001), the Pb emissions in some regions, mainly in the newly developed countries, are still increasing.

Lichens and mosses are very sensitive indicators of Pb aerial pollution. It has been observed a continuous decrease of its level during recent decades. In mosses sampled in Sweden during the period 1969–1985, the content of this metal decreased from about 50 to 20 mg kg^{-1} (Rühling et al. 1987). The Pb levels in mosses sampled during the period 1990–1995 range between 0.88 and 59 mg kg^{-1} (mean 8.6) (Berg and Steinnes 1997). In lichen (*Parmelia sulacata* Taylor) collected from olive tree bark at the vicinity of incinerators near Lisbon, Portugal, Pb was recorded at 25–50 mg kg^{-1} (Freitas et al. 2000).

II-14.4.6 Plants

Lead in plants is inherited from both, soil and atmospheric pollution. The variation of Pb contents of plants is highly influenced by environmental factors. When plants are grown in uncontaminated sites, its level is relatively stable. Its lower contents (<1 mg kg^{-1}) are in cereal grains and fruits and higher (>1 mg kg^{-1}) in roots and tubers. Also leafy vegetables and fodder plants contain elevated amounts (>2 mg kg^{-1}) (Table II-14.15). Very low Pb contents have been reported by Eriksson (2001a) for wheat grains, at the range of 0.005 to 0.011 mg kg^{-1} (mean 0.007), and for barley grains, at the range of 0.007–0.028 mg kg^{-1} (mean 0.013).

Lead occurs naturally in all plants, however is has not been shown yet to play any essential roles in their metabolism. There has been some discussion on this topic, and final conclusion was that if Pb would be necessary to plants it could be at the low level of 2–6 µg kg^{-1} (Kabata-Pendias and Pendias 2001). The absorption of Pb by root is passive and thus the rate of its uptake from soils is rather low. Its transfer in plants is very

Pb

Table II-14.15.
Lead in food and forage plants ($mg\ kg^{-1}$) (data compiled from various sources by Kabata-Pendias and Pendias 1999, 2001)

Plant	Range	Mean
Wheat, grains	0.2 – 0.6	0.5
Barley, grains	0.1 – 0.4	0.2
Sweet corn, grains	<0.3 – 3	1.4
Bean, pods	1.5 – 2	–
Carrot, roots	0.5 – 3	1.7
Lettuce, leaves	0.7 – 3.6	2.4
Cabbage, leaves	1.7 – 2.3	–
Onion, bulbs	1.1 – 2	1.5
Potato, tubers	0.5 – 3	–
Tomato, fruits	1 – 3	1.7
Apples, fruits	0.05– 0.2	–
Grasses	0.4 – 4.6	2.1
Clovers	1.3 – 8	3.8

limited and its great proportion is accumulated in roots. However, increased level of Pb in some plants has been used as a very good indicator for the geochemical prospecting.

The bioavailability of soil Pb to plants is limited, but may be enhanced by reduced soil pH, organic matter and inorganic colloids, iron oxides and phosphorus content. The binding and exchange capacity of soils are crucial in determining the Pb availability to plants (Zimdahl and Koeppe 1979).

Airborne Pb is taken up by plants through foliage. Thus, its concentrations always are higher in older parts of plants than in shoots or flowers (Bunzl and Kracke 1984). There is evidence, based in the experiments with isotope ^{210}Pb, that the atmospheric deposition of Pb may be its significant source in above-earth plant parts, in Pb polluted regions (Chamberlain 1983; Dalenberg and van Driel 1990). Levels of Pb in leaves, needles and sprouts of a plant correlate often with its atmospheric concentration, as for example, Pb content in poplar leaves in Spain correlates positively with local traffic density (Toro et al. 1995).

Elevated Pb contents were recorded in various plants in vicinities of following sites: metal smelters, roadsides soils contaminated with Pb, natural ore deposits, and Pb recycling factories. Gzyl (1997) observed high concentrations of Pb in carrot (3.24 $mg\ kg^{-1}$) and red beet (3.11 $mg\ kg^{-1}$) grown in heavily industrialized region of Poland, Silesia. Plants grown in contaminated soils of India contain extremely high Pb levels, especially in roots (Table II-14.16). In the mining areas of Portugal, plants grown in soils with very high Pb concentrations (mean about 2 694 $mg\ kg^{-1}$, at the range of 234–12 218 $mg\ kg^{-1}$), contain this metal in following ranges (in $mg\ kg^{-1}$): 23.5–84.8 $mg\ kg^{-1}$ in *Juncus conglomerates*, 8.5–22.4 in *Juncus effusus*, and 8.0–51.7 in *Scirpus holoschoenus* (Freitas et al. 2004).

Table II-14.16. Lead in different food crops grown in polluted and unpolluted soils in India (mg kg^{-1}) (data for 10–20 samples; after Srikanth and Reddy 1991)

Plant, soil	Range	Mean
Polluted soil	80 – 280.4	183.5
Unpolluted soil	0.5 – 7.8	3.4
Plants in polluted soil		
Spinach leaf	5.5 – 22.6	14.9
Spinach root	6.0 – 23.6	16.1
Amaranthus leaf	7.50 – 14.5	12.2
Amaranthus root	3.0 – 68.2	38.6
Cabbage leaf (inner leaves)	5.50 – 10.14	7.2
Cabbage root	9.0 – 15.6	12.4
Plants in unpolluted soil		
Spinach leaf	0.03 – 0.2	0.05
Spinach root	0.08 – 2.5	1.4
Amaranthus leaf	0.09 – 1.6	0.1
Amaranthus root	0.94 – 2.3	1.4
Cabbage leaf (inner leaves)	0.19 – 1.0	0.6
Cabbage root	0.98 – 3.5	2.7

The Pb deposition, investigated in the forest ecosystem of northern Germany, between 1983 and 1985, was the greatest in higher parts of trees; over 50% of atmospheric Pb was filtered by the top layer of spruce trees (Schultz 1987). Hagner (1999) reported that in the industrial areas in Germany, a significant decrease of Pb concentration in spruce (*Picea abies*) and poplar (*Populus nigra*) during the period between 1985 and 1994, is in a line with the phase out of leaded gasoline in central Europe.

Liu et al. (2003) studied the uptake of Pb by paddy rice and its effect on different rice cultivars. Results of the pot experiment show a great variation in its distribution among different parts of plant, as follows (mean value, mg kg^{-1}): root (4 300) > stem (358.6) > leaf (71.8) > panicle (16.4). According to these authors, the soil Pb level at 1 000 mg kg^{-1} affected the growth of rice plant significantly and reduced the crop yield by 11.5%. Similar results are published by Gu et al. (1989). However, there is observed a great different in the tolerance to high Pb concentration, for example, the hybrid rice cultivar *Shan you* 63 was not affected by 2 500 mg Pb kg^{-1} in soil (Xie and Huang 1994).

The toxicity of Pb depends not only on the total concentration but also on forms of Pb. The variable plant resistance is of a great importance. Plant toxicity levels of Pb in soils are not easy to evaluate, but it is still generally agreed that a soil Pb concentration ranging from 100 to 500 mg kg^{-1} is considered as excessive (Kabata-Pendias and Pendias

2001). The toxicity of organic Pb species, especially alkylated Pb, has been of a special interest. Röderer (1984) reported that the toxicity of organolead compounds exceeds the toxicity of inorganic species, however this highly differs upon several other factors. Common symptoms of the excess of Pb in plants are dark green leaves, wilting older leaves, and brown short roots.

II-14.4.7
Humans and Animals

Humans. Lead is an abundant element in mammalian tissues and ranges from <0.2 to 4.8 mg kg^{-1}, being the lowest in muscles and highest in liver and kidneys (Jørgensen 2000). Its highest contents, up to 35 mg kg^{-1} are reported for hair, and serves as a useful bioindicator for the Pb exposure of human, especially children (Chlopicka et al. 1998). The concentration of Pb in tissues of a "reference man" is calculated at 1.7 mg kg^{-1}, and its distribution is between 0.18 and 11.0 mg kg^{-1} in total soft tissues and skeleton, respectively (Li 2000). Human fluids contain Pb at the mean levels (in µg l^{-1}): 158 in blood, 17 in urine, and 0.3 in serum (Reimann and Caritat 1998). However much lower blood Pb has been reported recently, e.g., for children in Sweden about 20 µg l^{-1}, and for females and males in Germany, 26 and 36 µg l^{-1}, respectively (Becker et al. 2002).

The exposure pathway of Pb for the adult population is from food, water, and air. The Pb inhalation is the major pathway for workers of various industries, particularly of persons working at Pb-Zn smelters and mines.

The ingestion of soil and house dust is a major pathway for the exposure of young children to Pb, due to hand-to-mouth activity (pica-soil). Amounts of Pb intake depend on several factors, such as children behavior and age, and the bioavailability of Pb compounds. A child of age 1–6 years may intake 0.2 g of Pb containing dust per day, and occasionally 5.0 g d^{-1} (ATSDR 1999b). Lead in drinking water may be its significant source that is especially important for both children and pregnant women. The Pb concentration in drinking water is generally safe when at <5 µg l^{-1}. However, sometimes in many old houses, water pipe lines are still made of Pb that may increase its content in tap water above the maximum WHO recommendation value of 10 µg l^{-1}, and even be above the EU statutory limit of 50 µg l^{-1}. Occasionally the Pb content in tap water can be as much as 100 µg l^{-1} (LDAI 2002).

Various foodstuffs have been, and still may be, a serious source of Pb in diets. However, the Pb contamination of most foodstuffs has recently declined substantially. Previously, leaded solders in food cans and wine bottles contributed to the Pb in foods. Also some ceramic articles may contain leaded glazers. Luckily, now it is reported that leaching of Pb from ceramic tableware is acceptably low (Tukker et al. 2001).

The toxicity of Pb and its adverse effects have been known for many centuries. Even the Romans were familiar with some of the toxic effects of Pb. The excess of Pb to the Roman population was from broadly used leaded glazers, and in addition, from the sweet precipitation of Pb acetate, $Pb(CH_3COOH)_2 \cdot 3H_2O$, that was consumed (Kabata-Pendias and Pendias 1999).

The worldwide average dietary Pb intake by adults has been estimated at 440 μg d^{-1} (Hamilton 1987). Recent data gave lower values for the daily Pb intake by humans: mean 128 μg in Germany, and 107–121 μg in Uzbekistan (Brüggemann et al. 1996; Kavas-Ogly et al. 1996). It has been reported that adults may absorb about 5–15% of the ingested Pb, of which less than 5% is retained. Children absorb and retain much more Pb due to physiological and metabolical differences (American Council on Science and Health 1997). Young children can absorb up to 50% of the Pb from the diet whereas the adult population takes only about 10% (WHO 1995). The absorption of Pb by humans is a function of both, forms of Pb and kind of a diet. When the diet is low in Fe, Ca, and proteins then absorption of Pb is higher.

The absorbed Pb is distributed in blood, soft tissues, and in particular in bones, including teeth. The human body cannot differentiate between Pb and Ca; hence the majority of Pb is stored in the bones and teeth, where up to 90% of absorbed Pb may be accumulated. The excess of Pb may cause several health effects:

- Damage to nervous systems
- Inhibition of heme formation
- Kidney damage
- Anemia (inhibitiona hem synthesis)
- Impaired mental development of young children
- Carcinogenicity and genotoxicity
- Impaired reproductivity

It has been reported that neurobehavioral effects impaired academic performance even after the blood Pb levels returned to normal. Although there is no threshold limit for this effect, the available evidence suggests that Pb toxicity may occur at blood Pb levels of 100–150 μg l^{-1} or even less (ATSDR 1988; Gerhardsson 2004). The author reported impaired effects of Pb on several enzymes at the B-Pb level of about 100 μg l^{-1}.

During recent years, non-occupational Pb poisoning in the EU is quite rare. Between the years 1981–1996, there was only one death, of a two years old child, reported due to Pb poisoning (Elliott et al. 1999). In the 1970s, over 100 children were admitted to hospitals in England and Wales each year with suspected of Pb poisoning (HMSO 1974).

Due to phasing out of leaded gasoline and several other technological restrictions in the Western Europe and the USA, the average B-Pb levels of the major part of the population of Western Europe and the USA are below 100 μg l^{-1}. A biological limit B-Pb value of 300 μg l^{-1} has been suggested for both males and females (Gerhardsson 2004).

Some other Pb sources, such as soft water and Pb paints has still affected increased Pb exposure of the population in some areas of the Western Europe. In the USA, for example, increased B-Pb level is linked with the low-income group. Flecks of paint from older houses in the USA and France have caused B-Pb problems among dwellers (LDAI 2002). Outside Europe, there are many countries where B-Pb levels of the population are quite high. In South Africa, with decrease Pb concentration in gasoline from 0.8 to 0.4 g l^{-1}, B-Pb levels of urban and rural athletes declined (in μg l^{-1}) from 55.5 to

13.0, and from 17.7 to 8.5, respectively (Grobler et al. 1996). Due to poor standards of health and a low safety at working sites, higher B-Pb levels may be expected rather in developing countries than in developed ones.

Animals. Animal studies have indicated that Pb exposure may cause a low resistance of some animals to various infection diseases. In vertebrates, kidney failure, and anemia are common (Gainer et al. 1974). Several factors, e.g., low water hardness and pH, and elemental interactions may greatly affect the Pb toxicity.

Grazing animals, such as sheep, cattle and horses, ingest large amounts of soil through their food habits. In some areas Pb content of soils is quite high that affects its increased intake by grazing animals than by non-herbivorous animals. Poisoning by Pb among domestic animals has been mentioned by some authors (McEvoy and McCoy 1993; Gerhardsson 2004). The toxicity of Pb occurs in adult cattle at its intake at 96–120 g for about 10 days (Humphreys 1991 *vide* Adriano 2001). Near a Pb-smelter, high Pb blood levels in grazing cattle has been reported, that decreased with distance from the smelter (Newman and Dollhopf 1992).

The highest Pb level in cows milk in Sweden was 0.22 mg kg^{-1} when cows were exposed to increased Pb for 18 weeks. After six weeks under the normal exposure, a lower Pb value, <0.03 mg kg^{-1}, was again observed in cows milk (ATSDR 1999).

The toxic impact of Pb is clearly recognized in the case of bird mortality due to the exposure to Pb shots from fishing or shooting (Académie des Sciences 1998). This problem has been reported especially for the USA and the Nordic countries. In Japan, Kim et al. (1999) observed large sea eagles, such as Steller's sea eagles (*Haliaeetus pelagicus*), suffering from the ingestion of Pb shots. Due to the ingestion of Pb shot, symptoms of anorexia and ataxia, weakness, and high mortality are generally observed among the bird population. In the USA, about 2% of waterfowl die each year, due to the ingestion of Pb shot (LDAI 2002).

Trace Elements of Group 15 (Previously Group Va)

The Group 15 is composed of semi-metallic elements including arsenic (As), antimony (Sb) and bismuth (Bi). The elements have similar structure and the same symmetry and some similar properties (Table II-15.1). Due to decreased electronegativity, proper metallic character has not been bestrode on As and Sb for which they are often referred to as metalloid which means that these elements have both properties of metals and non-metals. In many ways, As is similar to phosphorous (P), especially in aerated systems.

II-15.1 Arsenic (As)

II-15.1.1 Introduction

Arsenic is widely distributed in the environment. It occurs in the Earth's crust at levels between 0.5 and 2.5 mg kg^{-1} and is likely to be concentrated, up to 13 mg kg^{-1}, in argillaceous sediments (Table II-15.2). Arsenic rich pyrite is common in the sedimentary formations rich in organic matter, especially shale, coal, and peat deposits (Nordstrom 2000; Bhattacharya et al. 2002a). Its concentration in coals causes environmental problems and it exhibits extreme variability among coals of different origin. It has been reported that it varies between 0.5 to 80 mg kg^{-1} (average 10 mg kg^{-1}) (Clarke and Sloss 1992), although higher values have been cited occasionally. The As^{5+} form is adsorbed in a greater extent by sediments than other As ions (Welch et al. 1998).

Table II-15.1. Selected properties of trace elements of Group 15

Element	Atomic number	Atomic mass	Atomic radius[a] (pm)	Density (20 °C, g cm^{-3})	Valence[b]	Melting point (°C)
As, arsenic	33	74.92	139	5.78	−3, **+3**, +5	817
Sb, antimony	51	121.76	153	6.69	−3, **+3**, +5	630.9
Bi, bismuth	83	208.98	170	9.75	**+3**, +5	271.3

[a] Aproximately average values for the main oxidation state.
[b] Valence values in bold are for the main oxidation state.

Table II-15.2. Abundance of arsenic, antimony and bismuth in the environment[a]

Environmental compartment	As	Sb	Bi
Earth's crust (mg kg^{-1})	0.5 – 2.5	0.1 – 0.9	0.06 – 0.13
Igneous rocks (mg kg^{-1})			
Mafic	0.5 – 2.5	0.1 – 1	0.001 – 0.15
Acid	1 – 2.5	0.2 – 2	0.01 – 0.12
Sedimentary rocks (mg kg^{-1})			
Argillaceous	5 – 13	0.8 – 4	0.05 – 0.5
Sandstones	0.5 – 1.2	0.05	0.1 – 0.2
Calcareous	1 – 2.5	0.15 – 0.3	0.1 – 0.2
Raw energy material (mg kg^{-1})			
Coal, hard[b]	7 – 24	0.6 – 1.2	0.7 – 5
Fly ash[c]	60	3.8	–
Crude oil	0.005 – 0.14	0.003 – 0.5	–
Petrol	0.02 – 2	>0.005	–
Soils (mg kg^{-1})			0.04 – 1.5
Light sandy	<0.1 – 30	0.05 – 1.33	–
Medium loamy	1.3 – 27	0.05 – 2	–
Heavy loamy	2 – 23	0.3 – 0.6	–
Calcareous	–	–	–
Organic	<0.1 – 67	0.08 – 0.6	–
Waters (µg l^{-1})			
Snow[e]	0.1	0.03 – 0.05	<0.02
Rain[e]	0.3 – 0.4	0.02 – 0.05	<0.01
Ocean, sea[e]	1.2 – 3.7	0.24	0.02
River[d]	0.11 – 9.4	0.005 – 0.1	0.006 – 0.04[e]
Air (ng m^{-3})			0.2 – 3
Urban[e]	2 – 53	1 – 55	–
South Pole	0.007	0.003	–
Greenland/Shetland Islands	0.6	0.002 – 0.22[f]	–

[a] Presented are common, possible background, values from various sources as compiled by Kabata-Pendias and Pendias (1999, 2001), unless otherwise indicated.
[b] Data after Finkelman, geometric and arithmetic mean, respectively (1999).
[c] After Llorens et al. (2000).
[d] Data after Gaillardet et al. (2003).
[e] After Reiman and Caritat (1998), for remote and polluted regions.
[f] Data for Spitzbergen.

Arsenic is associated with several ores of metals, most often as arsenite minerals where occurs as anions, As^{2-} and As^{3-}. It is likely to combine with several metals (e.g., domeykite, Cu_3As; nikieline, NiAs; sperrylite $PtAs_2$). Its common minerals are: arsenopyrite, FeAsS; orpiment As_2S_3; realgar, AsS; arenolite, As_2O_3, and some other polymorphous forms of minerals. Arsenopyrite can be easily oxidized by both O_2 and Fe^{3+} (Sracek et al. 2001). Due to similar ionic radii, the geochemical behavior of As^{5+} is very similar to that of P, especially in phosphates (Goldschmidt 1958).

II-15.1.2 Production and Use

Global production of As was estimated at 27.9 kt in 2000 (WMSY 2004). However, Zevenhoven et al. (2007) cited the world As production in 1995 at 38.8 kt which was mostly derived from Cu-ore processing.

Arsenic is generally recovered from sludge and flue dust during smelting of Cu, Zn, Pb, Au, and Ag ores that are usually enriched in As. Its presence in non-ferrous ores is generally regarded to be an environmental problem, and not a benefit (Loebenstein 1994). It has been estimated 6.5 kg of As per ton of Cu production and calculated the annual accumulative post-industrial world production of As to be at 3 Mt in year 2000.

Until the 1970s, about 80% of As was used in the manufacturing of pesticides. Due to its toxicity, the use of As in pesticides has decreased to about 50%, but organic As compounds still dominate the pesticides production by 90% (Matschullat 2000). The rest of the amount is used in wood preservation, photoelectric devices, glassware, and Pb-acid batteries. It is also used to improve corrosion resistance and tensile strength in Cu alloys (Zevenhoven et al. 2007).

II-15.1.3 Soils

The recent estimation of As in global soils was at 5.0 mg kg^{-1} (Koljonen 1992). Topsoil samples from Sweden contain As in the range of 0.4–10.5 mg kg^{-1}, at an average value of 3.9 mg kg^{-1} (Eriksson 2001a). The geometric mean for As in surficial materials of the USA is reported to be 5.8 mg kg^{-1} (Shacklette at al. *vide* Kabata-Pendias and Pendias 2001). Chernozems of western Siberia contain As concentrations from 18 to 32 mg kg^{-1}, of which up to 90% is strongly fixed by the fine granulometric fraction (Ilyin and Konarbayeva 1995). The range of As in soils of Poland is 0.9–3.4 mg kg^{-1}, with the highest values in soils derived from shales. However, the concentration of As over 650 mg kg^{-1} also has been reported for polluted soils of some industrial regions of Poland (Kabata-Pendias and Pendias 1999). In other regions in Europe (e.g., Kärnten, Austria) As concentrations in soils also are highly elevated and range from 100 to 115 mg kg^{-1} (Göd 1994; Göd and Heiss 1996). Due to the release of As from industrial sources and in mining areas, its concentration in soils can reach 20 000 mg kg^{-1} (Smith et al. 1998).

Agricultural practices may be a significant source of As, as its contents may be elevated in pesticides, fertilizer, sludge and manure. Thus, increased contents of As in agricultural soils, have recently become a real problem. This is especially emphasized in the case of paddy soils that are highly modified by anthropogenic activities. Paddy soils are often contaminated by fertilizers and organic chemicals, especially in those areas where rapid economic development is needed (Cao and Hu 2000; Wang et al. 2003). Especially the irrigation with As loaded groundwater increases its level in soils (Bhattacharya et al. 2002a). The As contents in soils of Bangladesh range between 2 and >80 mg kg^{-1} (Huq and Naidu 2005). These authors estimated the maximum concentration of As in the top layer (0–15 cm) of the Gangetic alluvium soil at 20 kg As ha^{-1}.

All soil properties, and especially Eh-pH regime influence the sorption capacity to As (Bissen and Frimmel 2003). The As speciation in the soil has been widely reviewed, indicating that in acid soils predominate $H_3As_3O_3^0$ and at higher pH values (>9.0), $H_2As_3O_3^-$ is likely to form. In the oxidation environment, As^{5+} is present mainly as $H_2AsO_4^-$ (pH <7.0) or as $HAsO_4^{2-}$ (pH >7.0) (Montgomery 1985; Cotton and Wilkinson 1989). The As^{3+} is more toxic and mobile in soils than As^{5+}. Ferric hydroxide generally plays an important role in controlling the concentration of As in soils and soil solutions. Anionic forms of this element, AsO_2^-, $HAsO_4^{2-}$, and $H_2AsO_3^{2-}$ predominate in soil solutions, although the cations As^{3+} and As^{5+} can also occur (Kabata-Pendias and Sadurski 2004). Cationic species of As, especially of As^{5+} and As^{3+} can be relatively easily sorbed onto clay minerals and onto oxy/hydroxides of Fe and Mn, as well as on organic compounds that can highly control its behavior in soils (Bhattacharya et al. 2002b). The maximum adsorption of As^{5+} occurs at pH 7.0, whereas of As^{3+} at pH 4.0. Complexes of As with P, Fe, Al, Ca, and Mn have a significant impact on its behavior in soils (Lund and Fobian 1991). In acid soils, dominant forms are Al-As and Fe-As compounds whereas in alkaline conditions predominate Ca-As species.

Some bacteria in the soils are known to accelerate the oxidation of arsenites, As^{3+} to arsenates, As^{5+}, and also to stimulate the methylation of As. Hence, microbiota may play a crucial role in the As behavior in soils (Dudas 1987).

Forms of As from all sources, e.g., irrigation waters, industrial wastes and pesticides (especially Pb-arsenates and Cu-acetate-arsenates) may need decades to be converted to non-phytotoxic forms in soils. The EU Community recommends that As in agricultural soil fertilized with sewage sludge should not exceed 20 mg kg^{-1}, whereas in the UK permissible level is 50 mg kg^{-1} (Stoeppler 2004).

II-15.1.4 Waters

Lide (1996) estimated the global median concentration of As in seawaters at 3.7 μg l^{-1} and Nozaki (2005) gave the mean As concentration in the North Pacific Ocean at 1.2 μg kg^{-1} (Table II-15.2).

The world average of As in river waters is estimated at 0.62 μg l^{-1} and its riverine flux at 23 kt yr^{-1} (Gaillardet et al. 2003). Skjelkvåle et al. (2001) investigated the occurrence of As in the Nordic lakes, and reported its mean concentrations to vary, at the value of 99.5 weighted percentiles, as follows (in μg l^{-1}): Norway, 2.02; Finland, 2.4; and Sweden, 4.6. Water of polluted Lake Erie (USA) contains As up to 300 μg l^{-1}, and in Tocone River in Chile, As was reported to reach 800 mg l^{-1} (*vide* Kabata-Pendias and Pendias 1999).

In marine water, As is present as trivalent, pentavalent and methylated forms (Matschullat 2000). In deep sea water, it may be present as As^{5+} and $HAsO_4^{2-}$ species and its average concentration is at 1.7 μg l^{-1} at a salinity of 3.6% (Bruland 1983). In all aquatic environments, As occurs generally as pentavalent arsenate ($H_nAsO_4^{n-3}$) under oxidizing conditions (Robertson 1989), whereas trivalent arsenite ($H_nAsO_3^{n-3}$) predominates under reducing conditions (Pokrovski et al. 1996). It has been reported that As^{3+}

in aquatic environments is more mobile and toxic for living organisms than As^{5+} (Guha Mazumdar et al. 1992; Bhattacharya et al. 1996, 1997; Ahmed et al. 1999). In surface water, As reacts with phytoplankton and phosphate, converting As^{5+} to arsenite As^{3+}, methylarsenate and dimethylarsenate. The methylated As is about 10% of the total As in the euphotic zone in most oceans.

Arsenic levels in ground and spring waters of geothermal fields are highly variable and influenced by geochemical properties of geologic formations of various regions (Nordstrom 2000). In several parts of the world, e.g., USA, Argentina, Chile, New Zealand, Taiwan, India, and Bangladesh, elevated concentrations of As in groundwater have been reported due to weathering and leaching of As from As-rich geological formations, mine tailings and wastes, and thermal springs. Its concentrations may highly vary and the highest values are reported by Bhattacharya et al. (2002b), Burgess and Ahmed (2006), and Mandal et al. (1996), as follows (in $\mu g\, l^{-1}$):

- Thailand, shallow wells 120 – 6 700
- Argentina, shallow wells 100 – 4 800
- China, shallow and deep wells <100 – 1 860
- Taiwan, deep wells <1 800
- West Bangal, shallow and deep wells 10 – 3 700
- Bangladesh, shallow wells <1 – 3 000
- Mexico, shallow and deep wells 300 – 1 100
- Chile, shallow and deep wells 100 – 1 000

In some countries, As enriched ground waters have affected very large areas and may cause severe problems due to its toxic effects on human health.

The Eh and pH are the most important parameters that control the speciation of As in ground waters (Smedley and Kinniburgh 2002). According to these authors, the dominant As species, at the mean concentration of As at 0.07 $\mu g\, l^{-1}$, was (in 100%) $H_3AsO_3^0$, under the mean Eh of 93.76 mV (range 158.0–7.0) and mean pH of 4.7 (range: 5.9–3.3). Low pH and reduced Eh increase the mobility of As whereas under strongly reducing conditions, the formation of sulfide minerals control As concentrations (Bissen and Frimmel 2003).

In aqueous environments, anaerobic bacteria such as prokaryotes and eukaryotes reductively biomethylate inorganic As to dimethylarsenic acid [DMAA, $(CH_3)_2AsO(OH)$] and monomethylarsenic acid [MMMA, $CH_3AsO(OH)_2$]. The toxicity of methylated forms is low and the volatile forms are not stable under oxidizing conditions. The biomethylation causes degradation of organic matter and conversion of As^{5+} to As^{3+} and its mobilization (Routh et al. 2000). The presence of Fe and Mn hydroxyoxides, as well as organic matter, have great impact on the behavior of As in all aquatic environments (BGS 2001; Smedley and Kinniburgh 2002; Bhattacharya and Mukherjee 2002).

Exposure of people to increased levels of As in waters of Bangladesh represents an example of hazard related to geogenic As as well as to anthropogenic impact. Naidu and Nadebaum (2003) have presented a concise description of this problem and several methods for treatment technologies for the removal of As from water: *(i)* coagulation, *(ii)* use of activated alumina- or Fe-oxides, *(iii)* ion exchange, *(iv)* reverse osmosis, *(v)* biotic-abiotic oxidation, and *(vi)* passive reactive barriers and oxidic soils.

Aquatic biota. Planktivorous species may contain more As than ominivorous or piscivorous species. In fish species, As is present as the organoarsenic form that is essentially non-toxic. Result of studies in the Netherlands suggested that sea fish contain inorganic As (0.1–41%) (Vaessen and van Ooik 1989), but the Canadian studies indicated that sea fish contain 74–91% of As in organic forms (Lawrence et al. 1986). McSheehy et al. (2001) identified in the oyster reference material several inorganic and organic As species. The most common As species found in fish and algae are DMA (dimethylarsinic acids) and MMA (monomethylarsenous acid) and also As-sugars (Stoeppler 2004).

Some marine plants are likely to concentrate a great proportion of As from water. The predominant As species in those sea plants, so called "arsenosugars", are easily converted in the human body to dimehtylarsinic acid, which has carcinogenic potential (McSheehy and Szpunar 2000).

II-15.1.5
Air

Concentrations of As in air vary highly from about 0.007 ng m^{-3} at the South Pole to above 50 ng m^{-3} in urban regions (Table II-15.2). According to Reimann and Caritat (1998) median worldwide As concentrations in remote areas range from 0.5 to 1 ng m^{-3} and in polluted regions it is 15 ng m^{-3}.

Natural sources of As in the atmosphere are soil-derived dusts, volcanic eruptions, sea salt aerosols and forest fires. Volcanic eruptions can account for 20–40% of total the natural As emission, whereas sea salt aerosols may contribute for <10% As emission (Nriagu 1998). The wet deposition of As in Sweden has been estimated at 1.8 g ha^{-1} yr^{-1} (Eriksson 2001a).

Both, natural and anthropogenic emissions may significantly affect As concentrations in air. Emissions of As to the atmosphere from industrial sources in EU-15 countries have decreased from 650 t in 1995 to 176.6 t in 2000 (Pacyna 1998; Pacyna and Pacyna 2001b). The major source of As to the atmosphere is coal combustion. Zevenhoven et al. (2006) summarized the As emission data for 2000, for the ten newly joined EU countries and it was about 83.1 t. It means that total emissions of As from EU-25 states for the 2000 was at 260 t. On the other hand, Pacyna and Pacyna (2001) estimated world As emissions to the atmosphere at 5 011 t in 2000 of which 68.9% stemmed from non-ferrous metal production followed by 16.14% from fossil fuel fired power plants (Pacyna and Pacyna 2001).

Arsenic and some of its compounds are easily volatile and generally escape to the gas and aerosol phases during coal combustion. It has been calculated that about 50%

Table II-15.3. Concentrations of As related to ash contents in coal (based on data from Coleman and Bragg 1990)

Kind of coal, ash content (%)	Mean ash content (%)	Mean As content (mg kg^{-1})
Low ash (0–8)	5.24	≈4
Medium ash (8–15)	10.87	≈7
High ash (>15)	23.2	≈10

of As in the forms of AsO and As_2O_3, enters the atmosphere as gaseous phase at low temperature combustion (1 000–1 200 °C), whereas only the As_2O_3 species are released at high temperature (1 200–1 600 °C) (Yan et al. 2001).

The world average As contents in bituminous and lignite coals (Clarke's value) are 9.0 and 7.4 mg kg^{-1}, respectively (Yudovich and Ketris 2005). However, the distribution of As in coal is strongly affected by the ash and S contents (Table II-15.3).

II-15.1.6 Plants

Arsenic is a common constituent of most plants, but little is known about its biochemical role. Apparently plants take up As passively with the water flow. Some plant species (e.g., Douglas fir) reveal a great capability to accumulate As and often are useful guides to geochemical prospecting of subsurface mineralization. Contents of As in food plants vary highly, most commonly in the range from 10–60 µg kg^{-1} and seem to be increased in green vegetables (cabbage) and in plants grown in contaminated soils (Tables II-15.4, II-15.5). The use of As-containing pesticides has been forbidden in several, mainly European countries, since the 1970s.

Excessive uptake of As, in both forms, as arsenites and arsenates, is toxic to plants, reacting with many enzymes and also decoupling P flow in the plant system causing disruption of energy flows in cells (Dixon 1997; Mukherjee and Bhattacharya 2001). The inhibition of root growth and its ultimate death also have been cited, under the As stress (Meharg and Macnair 1991; Paliouris and Hutchinson 1991).

Some plants can be resistant to the excess As which has been observed in some plant species, such as *Holcus lanatus*, *Calluna vulgaris* and *Silene vulgaris* (Meharg 1994). It has been also observed that the reduction of arsenate to arsenite in plants may trigger methylated species causing oxidative stress, as reported for some fungus (*Scopulariopsis brevicaulis*) and for phosphate-starved tomato plants (Zaman and Pardini 1996). Ma et al. (2001) and Tu et al. (2002) have found Chinese brake fern (*Pteris vittate*) extremely efficient in As uptake from soil. These authors claim that brake fern has great potential to clean up As contaminated soils. They have demonstrated that this plant accumulated 6 805 mg As kg^{-1} from contaminated soil (400 mg As kg^{-1}) during six weeks.

Leaves of fruit trees injured by the As excess contain this element in the range from 2.1 to 8.2 mg kg^{-1}. In general the tolerance for As in plants is established as 2 mg kg^{-1}. In rice plants, however, the critical value is as high as 100 mg kg^{-1} in tops and 1 000 mg kg^{-1} in roots (*vide* Kabata-Pendias and Pendias 2001).

Very crucial is the distribution of As in plants from various countries where the problem of As in ground waters has occurred. Food plants gown in Bangladesh regions with contaminated sites contain As in the rage from 1 to 153 mg kg^{-1}, whereas the same plants from uncontaminated sites contain As from 0.1 to 1.56 mg kg^{-1} (Table II-15.5).

The crucial problem in Bangladesh is rice cultivated in irrigated fields. Rice needs 1 000 mm of irrigation water per growth season. The load of As with this water varies between 1.36 and 5.5 kg ha^{-1} yr^{-1}. Winter wheat needs only 150 mm water, that resulted in the addition of As between 0.12 and 0.82 kg ha^{-1} yr^{-1}. Rice and wheat take up As from soils and waters in the following order: (average values in mg kg^{-1}): root, 23 > stem, 4 > leaf, 1.5 > grain, 0.2 > husk, 0.1. Hence, highest amounts of As are

As

Table II-15.4.
Arsenic contents in food plants

Country, plants	Range	Mean
England[a] (µg kg^{-1})		
Carrots	9.8 – 10.4	10.1
Radish	14.9 – 16.5	15.7
Potatoes	9.1 – 9.3	9.2
Parsnips	6.7 – 8.1	7.4
Spring onion	13.6 – 20.3	17
Broccoli	14 – 15.4	14.7
Cabbage	53 – 68.5	60.8
Leek	10.9 – 11.1	11
Bangladesh[b] (mg kg^{-1})		
Rice from polluted areas	–	46.0
Rice from low levels of As in groundwaters	–	10.0
Chile[c] (µg kg^{-1}, FW)		
Broad bean	–	112/26[e]
Potato	–	864/241[e]
Potato skin	–	101/55[e]
Maize	–	1 848/–[e]
Onion	–	–/7.2[e]
China[d] (mg kg^{-1})		
Rice	0.49 – 0.93	–
Rice husk	–	4.11
Sorghum flesh	–	2.22
Sorghum husk	–	11.55
Maize	–	0.12
Beans	–	1.33
Capsicum	–	1.30
Soybean	–	0.79
Peanut	–	0.27

[a] Data for vegetables of various origin, from Leicester city marked, after Al Rmalli et al. (2005).
[b] After Meharg and Mazibur (2003).
[c] After Queirolo et al. (2000).
[d] For Hunan mining area, after Liu et al. (2005).
[e] For Socaire/Talabre areas, respectively.

accumulated in roots and stems (Huq and Naidu 2005). According to Khan et al. (2005) rice yield decreases with increased As concentration in irrigation water. These authors presented another order of As concentrations in different parts of rice (in mg kg^{-1}): rice straw, 2.74–24.9 > rice husk, 1.0–2.72 > rice grain, 0.26–1.23. The problem of As in groundwater used for irrigation is also serious in several other regions, especially of Asia, where the As level in ground water is increased.

Table II-15.5. Arsenic in different plants growing in contaminated and uncontaminated soils in Bangladesh (mg kg^{-1}) (adopted from Huq and Naidu 2005)

Plants	Uncontaminated site	Contaminated site
Green papaya	0.46	2.22
Arum (*Colocassia antiquorum*)	0.30	153.2
Bean	0.092	1.16
Indian spinach	0.15	1.0
Long bean	0.30	2.83
Potato	0.62	2.43
Bitter gourd (*Momordicum charatia*)	1.56	2.12
Aubergine (*Solanum melongena*)	0.23	2.30
Chili	0.41	1.52

[a] Plant grows in water culture.

II-15.1.7 Humans and Animals

Humans. Arsenic occurs in mammal tissue in the range from 10 to 500 μg kg^{-1}, being the lowest in heart and the highest in liver (Jørgensen 2000). Li (2000) reported much lower As concentrations in soft tissues of human and estimated its average at 3.6 μg kg^{-1}. In human fluids, mean As concentrations are (in μg l^{-1}): blood, 7.9 and urine, 16.7 (Reimann and Caritat 1998). However, in regions contaminated with As much higher concentrations of this element in human tissues are observed (Table II-15.6).

Arsenic is known to be highly toxic to humans and animals. Both cations, As^{3+} and As^{5+} may cause similar toxicological effects, but the former is considered to be more mobile and toxic for living organisms (NRC 1999). However, both specie are metabolized in humans, forming methylated As species which are excreted readily into the urine. Hence, urine As is a good indicator to assess the As exposure (Akter and Naidu 2006).

Since the 13th Century, As has been used as a medicine while at the same time it is known to be poisonous. Several compounds of As may interfere with activities of some enzymes, especially those with sulfhydryl groups. It is also known to be involved in the Krebs cycle (*vide* Kabata-Pendias and Pendias 1999). The toxicology of As is quite complex and is highly dependent upon its species. Species of As^{3+} and inorganic compounds are more toxic than As^{5+} and organic compounds.

Recently, the most topical problems are associated with increased levels of As in ground waters and drinking waters that has been observed and investigated in several countries such as Argentina, Bolivia, Chile, Taiwan, India, Bangladesh and others (Das et al. 1996; Mukherjee and Bhattacharya 2001; Bejarano-Sifuentes and Nordberg 2003). In Bangladesh over 50 million people are at risk due to As in waters. It is reported by the WHO that 1.0 mg of inorganic As per day may give rise to skin lesions within a few years. There is an estimation that 13 persons out of 100 may have a risk of developing

As

Table II-15.6.
Arsenic in human tissues in the As-affected areas of Bangladesh and As thresholds values

Human sample	As concentration[a]	Limit value[b]
Urine	0.05 – 9.42 mg l^{-1}	0.005 – 0.04 mg day^{-1}[c]
Hair	1.1 – 19.84 mg kg^{-1}	0.08 – 0.25 mg kg^{-1}
Fingernails	1.3 – 33.98 mg kg^{-1}	0.43 – 1.08 mg kg^{-1}
Skin	0.6 – 53.59 mg kg^{-1}	–

[a] After Chowdhury et al. (1999), and Karim (2000).
[b] After Samanta et al. (1999), and Karim (2000).
[c] Acceptable daily excretion of As from a body.

cancer in their lifetime at 50 µg As ingested by adults per day from drinking water (Mukherjee et al. 2006). In the affected areas of Bangladesh and West Bengal, As concentrations in ground waters vary between >50 and 1 000 mg l^{-1}, whereas the maximum permissible level of As in drinking water in Bangladesh is 50 µg l^{-1}.

Arsenic toxicity from food and drinking water can cause several health problems (Lai et al. 1994; Rahaman et al. 1998, 1999; Smith et al. 2000): mutagenic, carcinogenic and teratogenic effects that affect various disfunctions:

- Neurological effects
- Cardiovascular disease
- Respiratory disorders
- Hypertension
- Diabetes mellitus
- Keratosis

Arsenic toxicity is generally noticeable within a period of 2–6 years depending on the concentration of As in drinking water and daily intake with food (Chowdhury et al. 2000). The daily acceptable maximum intake of As by a healthy 70 kg BW person, from all sources: air, water and food is as follows: air, 0.8–30 ng m^{-3}; water, 10–50 µg l^{-1}; and food, 0.8–120 µg kg^{-1} (FAO/WHO 1998). Due to drinking well water of high As concentrations (>3 400 µg l^{-1}) elevated contents of As in humans have been reported in many areas of Bangladesh, especially in village areas (Table II-15.6).

Fresh water fish may be serious dietary source of As in some regions. The EPA has set As tissue residues of 1.3 mg kg^{-1} FW in those fish as the criterion for human health protection (Eisler 1994). However, most of As in seafood is in organic compounds that are less toxic than inorganic species.

Some metallurgic industries (e.g., brass production) are also known as a source of increased concentration of As in workers; their hair, fingernail and toenails were reported to contain As (in mg kg^{-1}) 10.8–76.1, 15–188, and 37.1–60.9, respectively (Vahter 1988). Soil-Pica habits can result in the intake of As between 0.05–0.57 mg kg^{-1} d^{-1} by children (ATSDR 2002a).

Chronic oral exposure of As by adults at the LOAEL is probably about 10–20 µg kg^{-1} d^{-1} and that for the NOAEL is between 0.4 and 0.9 µg kg^{-1} d^{-1} (ATSDR 2002a). The cancer risk of workers has been estimated for the inhalation by a person at 50 mg As kg^{-1} d^{-1}, assuming that 30% of the inhaled As is absorbed (US EPA *vide* Stoeppler 2004).

Animals. Arsenic contents of muscles and liver were determined in several species of seabirds from the Barents Sea Region (Norderhaug et al. 1977; Savinov et al. 2003). These authors observed a significant impact of geographic locations. Contents of As vary (in mg kg^{-1} FW) in liver from 0.79 to 18.7 and from 0.17 to 4.21 in muscles. Jiménez et al. (2005) studied presence of As in bird species in Baja California, México. They observed a higher As level (46 mg kg^{-1} FW) in the liver of ground doves than in house sparrow (25 mg kg^{-1} FW). Organic As in bird species may not present any problem as this compound is more easily excreted and less toxic than inorganic species (CEPA 1993; Neff 1997). However, As at 50 mg kg^{-1} in the diet was reported to cause decreased growth in chicks, after three weeks of exposure (Howell and Hill 1978).

Episodes of wildlife poisoning occur relatively often. Death of white-tailed deer from eating sodium arsenite due to licking trees and death of snowshoe hares affected by As compounds as a result of silviculture practices have been reported (NAS 1977; Hood 1985). However, nowadays As poisoning is relatively rare in wildlife.

Domestic livestock are often fed arsenic acid derivatives in food to promote growth and retard disease. Some commercial pet food contains about 2.3 mg As kg^{-1} (NRCC 1978). The lethal dose of As in domestic animals ranges from 1 to 25 mg kg^{-1}. Symptoms of As poisoning are variable; abdominal pain, vomiting, diarrhea, staggering gait, hypothermia, and at acute poisoning – necrosis and perforation of the stomach and intestine (Stoeppler 2004).

II-15.2 Antimony (Sb)

II-15.2.1 Introduction

Antimony exhibits chalcophilic properties for which it combines readily with sulfides and occurs mainly at 3+ and 5+ oxidation stages (Table II-15.1). Its contents in igneous rocks range between 0.1 and 0.9 mg kg^{-1} and in sediments are likely to be increased, up to 2 mg kg^{-1} in argillaceous rocks (Table II-15.2).

Common Sb minerals are stibnite, Sb_2S_3 and valentinite, Sb_2O_3. Antimony may substitute for As in several minerals, although there is a difference in ionic radii (Sb^{3+}, 76 pm; As^{3+}, 58 pm). Common minerals also are: pyrargyrite, Ag_3SbS_3; and bourmite, $PbCuSbS_3$. It also easily forms soluble thiocomplexes, like SbS_2^{2-} and SbS_4^{3-}, depending on pH. Antimony is associated with over 100 Sb-bearing minerals. Therefore, increased levels of Sb are observed in many regions of the world, especially around smelters, chemical facilities, and mining, and mineralized areas, particularly with ores of Cu, Ag, and Pb. It is often used, together with As and Bi, as path-finders for the geochemical exploration (Hale 1982).

In recent years, Sb and its compounds are considered to be an important pollutant by the US EPA and the EU. Especially, Sb released from coal combustion is an important pollutant due to its high solubility and reactivity; it is carried on small sized aerial particles and deposited, which has been documented by Shotyk et al. (2004) in profiles of ombrotrophic peat in Switzerland.

Chemical and toxicological properties of Sb are similar to those of As. There is not much information on the toxicity of Sb and on its effects on the environment and human health. In ancient times, Sb compounds were used for medical purposes.

II-15.2.2
Production and Use

The global production of Sb was about 150 kt in 2001 and 155 kt in 2003 (WMSY 2004). The main source of Sb is stibnite, Sb_2O_3. Antimony is also recovered from primary concentrates, recycled scrap, and antimonial Pb-battery scrap. Primary production of Sb has been ceased in many countries due to its low price (http://minerals.usgs.gov/minerals/pubs/commodity/antimony/).

Antimony increases the strength and hardness of Pb alloys and is used as an alloying element for Pb-acid batteries. It also is used in solders and bearings, as alloying element. Hard Pb (3–9% Sb) is used in the grids and Pb-oxide paste for automobile batteries. In recent years, its uses in high-tech applications are increasing. Sb-containing lead lining is used to protect pipes, valves, and pumps in the chemical industry. This metal is used for the manufacturing of flame-retardants in plastics, vinyl and synthetic fibers.

II-15.2.3
Soils

The background level of Sb in surface soils ranges from 0.05 to 4.0 mg kg^{-1} (Table II-15.2). In agricultural soils of Sweden its contents vary from 0.07 to 0.41 mg kg^{-1}, and average 0.25 mg kg^{-1} (Eriksson 2001a). In Japanese soils the median Sb contents vary from 0.7 to 1.0 mg kg^{-1}, for Cambisols and Acrisols, respectively (Takeda et al. 2004). The 90 percentile for the total Sb in Finnish soil is 0.6 mg kg^{-1}, and ranges between <0.2 and 0.9 mg kg^{-1} (Köljonen 1992). Govindaraju (1994) reported that Sb in reference soils of China range from 0.5 to 60 mg kg^{-1} (mean 3.6) and in soils from the USA, range from 0.11 to 14 mg kg^{-1} (mean 2.4). In the State of Idaho, soils Sb ranges from 2.3 to 9.5 (Flynn et al. 2003), whereas acid sandy soils of Great Britain contain only 0.7 mg Sb kg^{-1} (Ainsworth et al. 1990).

Antimony is slightly mobile in soils and sediments since it is attached strongly to particles that contain Fe, Mn, and Al. It is also easily adsorbed when occurs as oxyanions, especially at low pH values (Bhattacharya et al. 2002; Filella et al. 2002; Smedley and Kinniburgh 2002).

In contaminated soils, near industrial sites, Sb is quite high in surface layers and decreases with depth which indicates that it is rather non-reactive and mostly immobile and will not contaminate the groundwater (Prüess 1994). Nakamaru et al. (2005) showed very low mobility and reactivity of Sb in Japanese soils. However, under specific conditions, mobility and leaching of Sb from the mineral stibnite, near a Sb-smelter in New Zealand has been observed (Wilson et al. 2004). Also Ashley et al. (2003) confirmed the mobility of Sb in Fe-free or low-Fe systems. In soils surrounding Sb smelters and Sb mining, it is mainly bound to Fe-Mn-oxides and organic matter, its mobility is not based on the pH value of soils and contents of mobile species range between from 0.007 to 0.03 mg kg^{-1} (Li and Thornton 1993; Prüess 1994).

The contamination with Sb is often reported, mainly for soils at hazardous waste sites and at Sb-processing sites. Data presented by Flynn et al. (2003) gave the following high concentrations of Sb in topsoils (in mg kg^{-1}):

- 65 – 109 Cu-smelter, USA, East of Tacoma
- 49 – 204 Cu-smelter, USA, Maury Island
- 280 Au-refining, USA, Yellow Knief
- 360 Sb-smelter, UK
- 1 565 Sb-mining areas, China (He and Yang 1999)
- 109 – 2 550 various Sb-smelter (ATSDR 2002a)

II-15.2.4 Waters

Average Sb concentration the North Pacific Ocean water is estimated at 0.2 μg kg^{-1} (Nozaki 2005). Range of Sb in seawaters is between 0.15 and 0.44 μg l^{-1} (Reimann and Caritat 1998). The common range of Sb in river waters varies from 0.005 to 0.27 μg l^{-1}, and the mean concentration is 0.07 μg l^{-1}, with the world average riverine flux of 2.6 kt yr^{-1} (Gaillardet et al. 2003). Accordingly to several other sources, Sb concentrations in surface waters vary from 0.16 to 0.98 μg l^{-1}, being the highest in USA rivers and the lowest in the Rhine River estuary.

Rainwater collected during 1999 in Sweden contained Sb at the average value of 0.07 μg l^{-1} and at the maximum value of 0.14 μg l^{-1}. Its wet deposition was estimated at 0.8 g ha^{-1} yr^{-1} (Eriksson 2001a).

There is disagreement about its reactivity in the open ocean waters and many authors believe that it behaves conservatively (Middelburg et al. 1988; van der Weijden et al. 1990; Takayanagi et al. 1996). In most aquatic environments, Sb occurs in two oxidation states, and its behavior may be altered by the redox potential. Thermodynamically unstable Sb^{3+} has been detected in marine waters, groundwater and rainwater (Andreae 1983). However, in the contrast with thermodynamic equilibrium predictions, it has been suggested that Sb^{5+} concentrations should exceed Sb^{3+}, by a few orders of magnitude.

Waters of polluted aquatic environments may contain Sb at concentrations of over 100 μg l^{-1} (Filella et al. 2002), whereas, its concentration in hot springs and geothermal waters may reach 500 mg l^{-1} (Sabadell and Axtmann 1975). He and Yang (1999) reported the following concentrations of Sb in various waters in Chinese mining areas (in mg l^{-1}): river waters, 0.07–0.04; waste waters, 1.3–21.8; and polluted well-waters, 24–42.

The WHO (2004) has set the provisional guideline for Sb in drinking water at 5 mg kg^{-1}.

II-15.2.5 Air

Natural concentrations of Sb in the atmosphere vary between 0.002 and 0.25 ng m^{-3} (Table II-15.2), whereas in air of urban and industrial regions it can be elevated to about 1 000 ng m^{-3} (ATSDR 2002a). In Buenos Aires City, the aerial content of Sb was

Sb

cited to be 3.3 ng m^{-3} (Smichowski et al. 2005), and in the Helsinki area, urban and rural concentrations of Sb were estimated at 0.77 and 0.36, respectively (Pakkanen et al. 2002).

Antimony and its compounds are easily volatile and released from both natural and industrial sources. The main natural sources are volcanic eruptions, sea salt spray, forest fires, and wind blown dusts. Industrial sources include fossil fuel combustion, incineration of wastes, cement kilns, and metallurgical industries. In the 1980s, Nriagu and Pacyna (1988) estimated global anthropogenic and natural Sb emissions as 3.5 kt and 2.6 kt, respectively. But these values should be used cautiously because the industrial production of Sb increased from 80 kt in the 1980s to 155 kt in 2003. Also the pattern of Sb use has changed significantly.

Antimony enters the atmosphere mainly from high temperature processes, as oxides, sulfides, and elemental Sb. It is often strongly correlated with Pb, As, Cd and Zn in aerosols and is readily transported over long distances – from central Europe to Norway (Steinnes et al. 1997). In addition, sources of Sb are from the abrasion of Sb containing brake-linings of vehicles. In the region of Bavaria, Germany, 6–8 kt of Sb containing dust is released annually, but its concentration in dust is not known (Wegenke et al. 2005).

II-15.2.6 Plants

Antimony is not essential to plants, however it is known to be easily taken up by plants if present in soluble forms in growth media. Its concentration varies from <20 to 1 130 µg kg^{-1} in food plants that grown in soils of historical mining areas (Table II-15.7). In these soils, Sb occurs mainly in the sulfide form, SbS, that is slightly phytoavailable, at the range of 0.06–0.59% of the total Sb content (Hammel et al. 2000).

Table II-15.7. Antimony in food plants from historical mining areas (µg kg^{-1}) (after Hammel et al. 2000)

Plant	Range, mean
Wheat	<20
Barley	<20 – 20
Rye	<20
Oat	<20 – 60
Carrot	<20 – 30
Red beet, roots	<20 – 90
Potato, tubers	<20 – 20
Tomato	<20
Spinach	150 – 1 130
Parsley	50 – 420
Maize, shoot	60 – 350

Commonly reported Sb contents in agricultural plants range from <2 to 29 μg kg^{-1}, whereas its concentrations in trees and shrubs growing in mineralized areas can range between 7 and 50 mg kg^{-1} (Kabata-Pendias and Pendias 2001). Winter wheat and barley grains grown in Sweden contain Sb at the mean values of 0.5 and 1 μg kg^{-1}, respectively (Eriksson 2001a). Usually its contents in plants decrease in the following order: roots > leaves and shoots > grains (Adriano 2001).

There is a lack of reports the Sb toxicity to plants. Only results of the experiment curried out by He and Yang (1999) show that Sb^{3+}, at the highest dose (1 000 mg kg^{-1}), significantly reduced the yield of rice, in the pot experiment, as compared with the impact of Sb^{5+} addition.

The distribution of Sb in Norwegian mosses analyzed in 1995 ranges from 8.3 to 4 300 μg kg^{-1} (average 150) and does not reflect any significant effect of aerial pollution (Berg and Steinnes 1997).

II-15.2.7
Humans and Animals

Humans. Antimony occurs in human tissue in the range from 5 to 10 μg kg^{-1}, being the lowest in heart and the highest in muscles, its average for total soft tissue is estimated to be 9.4 μg kg^{-1} and its content in "reference man" is calculated at 30 μg kg^{-1} (Li 2000).

Antimony is a cumulative poison (Shotyk et al. 2004). Average contents of Sb in food products are between 0.2 and 1.1 μg kg^{-1}. Normal daily intake of Sb by adults from both water and food is calculated for the USA to be 5 μg (ATSDR 2002a). The US EPA and WHO (*vide* Rish 2004) suggested TDI of Sb to be 0.4 μg d^{-1} and 0.86 μg d^{-1}, respectively.

Most common health effects of Sb intoxication, as cited by Abbaspour and Baramakeh (2005), are:

- Carcinogenic and mutagenic problems
- Diarrhea
- Nephropathy
- Encephalopathy
- Muscles/joint pain
- Gastrointestinal problems
- Anemia
- Heart problems

The general population is rarely exposed to Sb, but in smelter and mining regions some individuals and especially workers can be affected *via* skin and inhalation of Sb. In Sb-mining smelting areas of China, some diseases, such as dermatitis and pneumoconiosis occur in over 20% of workers (He and Yang 1999).

Animals. Acute effects of the overdose of Sb in animal nutrition are probably similar to those observed in humans. However, there are not set any exposure limits for animals. An adverse effect was not observed when mice were fed with forage containing 6 700 mg kg^{-1} Sb_2O_3 (Hammel et al. 2000).

II-15.3 Bismuth (Bi)

II-15.3.1 Introduction

Bismuth is crystallo-chemically very closely related to As, Sb, and Pb, and its most common oxidation number is +3 (Table II-15.1). At the +5 oxidation state, Bi is a strong oxidizing agent, such as $NaBiO_6$ or BiF_5 (Krüger et al. 1985).

In the Earth's crust its concentrations range between 0.06 and 0.13 mg kg^{-1}. In sedimentary rocks it is distributed rather uniformly at the range of 0.05–0.5 mg kg^{-1} and it is likely to be concentrated in hard coal, up to 5 mg kg^{-1} (Table II-15.2).

Bismuth reveals both chalcophilic and lithophilic properties. During weathering it is readily oxidized, and when it becomes carbonated (e.g., as $Bi_2O_2CO_3$), it is relatively stable (Neagoe 2001). Its common minerals are: bismuthinite, Bi_2S_3; and bismite Bi_2O_3. Possible host minerals are galena, sphalerite and chalcopyrite.

It has been suggested that Bi can be a good tracer of volcanic emissions. In Greenland, Bi concentrations on ice and snow indicated that its source (total deposition 0.14 t) was the volcanic eruption of Laki (1783–1784) (Ferrari et al. 2000).

II-15.3.2 Production and Use

The annual world production of Bi in 2003 was estimated, depending on the data source, at 3.75 kt (USGS 2004) and at 4.47 kt (WMSY 2004). It is associated with several ores (e.g., Cu, Hg, Sn, Ag, Au, W) and is often obtained as a by-product of various ore processing, mainly of Pb ores (in the USA), and of W ores (in China).

Bismuth is widely used in low-melting alloys, fuses, sprinklers, glass and ceramics production, and also as a catalyst (e.g., production of rubber and acrylic fibers). It is used in other metallurgical industries, as well as in different electronic devices.

Due to restrictions on the use of Pb, the outlook for Bi uses for plumbing fittings for drinking water systems is increasing. As a non-toxic replacement for Pb, Bi is used in several types of food processing equipments.

Due to its versatile properties, other uses are in pharmaceutics and cosmetics production, as well as in dentistry and medicine.

II-15.3.3 Soils

There are not much data on the Bi content of soils (Table II-15.8).

Some ferralictic soils may concentrate up to 10 mg Bi kg^{-1} (Fergusson 1990). Tyler (2005) reported that Bi levels in ectomycorrhizal beech roots and raw humus soils are 12 and 92 mg kg^{-1}, respectively. The calculated root/soil (R/S) ratio of Bi is about 0.13, and seems to be related to the ionic potential and the ionic charge of the cation.

Table II-15.8. Bismuth in various soils ($mg\,kg^{-1}$)

Country	Soil, site	Range	Mean	Source
Uncontaminated soils				
World-wide	Various	0.1 – 13	0.2	Bowen (1979)
Canada	Garden	1.33 – 1.52	<1	Fergusson (1990)
China	Reference	0.04 – 1.2	–	Gowindaraju (1994)
Sweden	Arable	0.03 – 0.47	0.16	Eriksson (2001a)
Scotland	Arable	0.13 – 0.42	0.25	Ure et al.[a]
USA	Reference	0.03 – 0.69	–	Gowindaraju (1994)
Contaminated soils				
Japan	Cu smelter	0.45 – 122	4.2	Asami et al. (1992)
Korea	Au-min. area	1 – 1 200	138	Park et al.[b]
UK	Mining area	–	16.4	Li and Thornton (1993b)

[a] Ure et al. cited after Kabata-Pendias and Pendias (2001).
[b] Park et al., cited after Veiga and Baker (2004).

Fertilizers, sewage sludge and especially some phosphate waste solutions are significant sources of Bi in soils. Eriksson (2001a) reported the following Bi concentrations (in $mg\,kg^{-1}$ of phosphate content) in material used for soil amendment: sewage sludge, 0.73; manure, 0.72; P-fertilizers, 0.18; and NPK-fertilizers, 0.02.

The presence of radionuclide, Bi-214 in P-fertilizers depends on the quality of phosphate rocks. In a cement plant in Lebanon, sea sediment (50 m from the plant) and fertilizers contained this radionuclide at 579 and 401 $Bq\,kg^{-1}$, respectively (Brigden et al. 2002).

II-15.3.4 Waters

The Bi concentration in the North Pacific Ocean is estimated at 0.03 $ng\,kg^{-1}$ (Nozaki 2005). According to other sources, Bi concentrations in surface ocean water range from 20 to 40 $ng\,l^{-1}$ (Goldberg 1965) and in world ocean water its mean is estimated at 20 $ng\,l^{-1}$ (Reimann and Caritat 1998). In surface terrestrial waters, Bi contents vary between 10 and 50 $ng\,l^{-1}$ (Kabata-Pendias and Pendias 1999).

The species of Bi in seawater are mainly: BiO^+, $BiCl^-$ and BiOCl, whereas in fresh water, principal species are: BiO^+, $Bi(OH)_2^+$ and $B_6(OH)_{12}^{6+}$ (Stumm et al. 1975).

Rainwater collected during 1999 in Sweden contained Bi at the average value of 30 $ng\,l^{-1}$ and the maximum value of 52 $ng\,l^{-1}$. Its wet deposition has been estimated at 33 $mg\,ha^{-1}\,yr^{-1}$ (Eriksson 2001a).

Bismuth is likely to be associated with Fe hydrous oxides and is commonly concentrated in polymetallic concretions of the sea bottom.

II-15.3.5 Air

There is a scarcity of data on Bi in the atmosphere. Its global concentration apparently varies from 0.2 to 3 ng m^{-3} (Kabata-Pendias and Pendias 1999). Most Bi emission originates from mining and smelting of metal ores, chemical plants, coal combustion, and waste incineration. Natural Bi emissions to the atmosphere are from geothermal power plants and volcano eruptions.

The annual wet deposition of Bi in Sweden has been estimated at 0.033 g ha^{-1} yr^{-1} (Eriksson 2001a). Atmospheric deposition of Bi in Solling Mountains (Germany) is calculated to be at the mean value of 0.4 g ha^{-1} yr^{-1} (*vide* Kabata-Pendias and Pendias 1999).

II-15.4.6 Plants

Bismuth is not essential to plants and there is no evidence of its toxicity. Although Bowen (1979) showed weight loss of some plants in the solution culture with increased Bi concentration, the mechanism of its toxicity has not been presented (Efroymson et al. 1997).

The standard Bi content in plants is estimated at 10 µg kg^{-1} (Markert 1992). The range of Bi in various plants is from <20 to 800 µg kg^{-1} (Kabata-Pendias and Pendias 1999). However in some plants, especially in trees, its contents can be elevated up to about 15 mg kg^{-1} AW (Fergusson 1990). Winter wheat and barley grains grown in Sweden contain Bi at the mean values of 3 and 5 µg kg^{-1}, respectively (Eriksson 2001a). Berg and Steinnes (1997) found Bi in Norwegian mosses collected in 1995 to range from 1 to 800 µg kg^{-1} (average 33).

Plants grown in some contaminated sites contained Bi in concentrations of less than 20 µg kg^{-1} (Li and Thornton 1993).

II-15.4.7 Humans and Animals

Humans. Bismuth occurs in mammalian tissue in the range from <0.03 to <0.1 mg kg^{-1}, being the lowest in skin and the highest in brain (Jørgensen 2000). Li (2000) reported other ranges of Bi in human tissues, from 0.007 to 0.4 mg kg^{-1}, the lowest in muscles and the highest in kidneys. According to Fergusson (1990), the body burden of Bi is 13 mg per 70 kg and its distribution among human tissues is following: blood, <0.3 mg l^{-1}; kidney, <0.1–1.0 mg kg^{-1} FW; liver, <0.1–0.5 mg kg^{-1} FW; muscle, 0.01 mg kg^{-1} FW; and bone, <0.2 mg kg^{-1}. The mean concentration of Bi in human blood and urine is reported to be 0.49 and 1.2 µg l^{-1}, respectively (Reimann and Caritat 1998).

The toxicity of Bi depends on its digested or inhaled species. For example, citrate compounds increase the uptake of Bi (Slikkerveer et al. 1992; Niebore et al. 1995). It reveals an affinity to combine with sulfhydryl groups and thus can be harmful to some metabolic processes and to the nervous system (Thomas 1991). Neurotoxic effects and kidney disorders of humans may occur from intake of either insoluble and or soluble compounds of Bi (Slikkerveer and de Wolff 1989). In the 1970s, an epidemic of Bi poisoning occurred in France and symptomatic patients had a wide range of Bi blood at concentrations from 0.01 to 4.6 mg l^{-1} (Martin-Bouyer et al. 1978). The total daily intake of Bi by adults, combined with inhalation, is estimated at 5–20 μg (Neagoe 2004).

Since ancient times, Bi has been used in medicine, recently often in combination with an antibiotic. However, presently it is applied mainly as a homeopathic remedy. The common compounds used in medicine are Bi-citrate and Bi-subsalicytate. It is known to cause constipation. According to Neagoe (2004), an important source of exposure to Bi of some individuals is the therapeutic use of its compounds.

Animals. Bismuth levels in land animal tissues are reported to range between <4 and 20 μg kg^{-1} (Neagoe 2004).

Acute local effects of inhalation of Bi compounds by animals were observed mainly as kidney damage (Neagoe 2004). The accumulation pattern of Bi in mice is following: lungs > spleen > liver > brain > muscles (Slikkerveer and de Wolff 1989).

The addition of Bi to Pb pellets as a non-toxic alternative for shotguns became a significant source of this metal in some animal tissues (Ringelman et al. 1993). Mean concentrations of Bi in liver and muscle of ducks and geese hunted with Pb shot shells range from 0.05 to 0.15 mg kg^{-1} and from 0.06 to 0.29 mg kg^{-1} (Jayasinghe et al. 2004; Pamphelet et al. 2000). Concentrations of Bi in muscles on hunted birds are a bit higher than in liver. It seems to be evident that increased level of Bi in muscles of hunted birds originated from Pb pellets that may contain Bi at the level of >5.5 mg kg^{-1}.

Trace Elements of Group 16 (Previously Group VIa)

The Group 16 of the Periodic Table is composed of only two stable trace elements: selenium (Se) and tellurium (Te) which resemble S in a number of geochemical properties, however their behavior in the environment is more complex. Polonium (Po), a natural isotope of the U-Ra transformation chain also belongs to this group.

These elements are highly mobile in the Earth's crust and they are considered to exhibit both chalcophilic and siderophilic properties. All of these elements exist in variable valence states from –2 to +6 (Table II-16.1).

II-16.1 Selenium (Se)

II-16.1.1 Introduction

The concentration of Se in the Earth's crust is between 0.05 and 0.5 mg kg^{-1}. It is slightly more concentrated in mafic rocks, but rarely exceeds 0.1 mg kg^{-1}. In sedimentary rocks, Se is associated with the clay fraction and thus its contents in argillaceous sediments (0.3–0.6 mg kg^{-1}) are significantly higher than in sandstones and limestones (Table II-16.2). Enriched Se concentration in Cretaceous rocks (up to above 100 mg kg^{-1}) was derived from volcanic gases and dust brought down by rain into the Cretaceous sea. However, some sedimentary rocks formed in non-volcanic periods are also enriched in Se, probably in the normal course of weathering.

Table II-16.1. Selected properties of trace elements of the Group 16

Element	Atomic number	Atomic mass	Atomic radius[a] (pm)	Density (20 °C, g cm^{-3})	Valence[b]	Melting point (°C)
Se, selenium	34	78.96	122	4.28; 4.79[c]	**–2**, **+2**, +4, +6	217; 50[c]
Te, tellurium	52	127.60	142	6.24	–2, **+4**, +6	449.5
Po, polonium	84	209[d]	153	9.32 (alpha)	–2, **+4**, +6	254

[a] Approximately average values for the main oxidation states.
[b] Valence values in bold are for the main oxidation states.
[c] Grey and red species, respectively.
[d] Average atomic mass.

Se

Table II-16.2. Abundance of selenium in the environment[a]

Environmental compartment	Se	Environmental compartment	Se
Earth's crust (mg kg^{-1})	0.05 - 0.50		
Igneous rocks (mg kg^{-1})		Soils (mg kg^{-1})	0.33
Ultramafic and mafic	0.01 - 0.12	Light sandy	0.25
Acid	0.01 - 0.05	Medium loamy	0.35
		Heavy loamy	0.40
		Calcareous	0.20 - 0.22
		Organic	0.30 - 1.00
Sedimentary rocks (mg kg^{-1})		Water (µg l^{-1})[d]	0.07
Argillaceous	0.3 - 0.6	Rain[e]	0.03 - 1.75
Sandstones	0.01 - 0.08	Sea[f]	0.2
Calcareous	0.03 - 0.10	River	0.06 - 22
Raw energy material (mg kg^{-1})		Air (ng m^{-3})	0.2
Coal, hard	0.21 - 10.7[b]	Urban/Industrial	0.03 - 30
Fly ash	1.5[b]	Shetland Island	0.5
Crude oil	0.01 - 1.4	South Pole	0.06
Petrol	0.001		

[a] Presented are common, possible background, values from various sources as compiled by Kabata-Pendias and Pendias (1999 and 2001), unless otherwise indicated.
[b] After Plant et al. (2004).
[c] Mean for World soils.
[d] World average for river waters (Gaillardet et al. 2003).
[e] After Szefer (2002a).
[f] After Reimann and Caritat (1998).

Selenium is allotropic and may occur in three forms: *(i)* "metallic" hexagonal, *(ii)* red, monoclinic crystals, and *(iii)* vitreous amorphous forms, black, red, and colloidal with its properties depending on the form. Commonly Se exists in four valence states, and the –2 state predominates in organic Se compounds. Due to the close crystallochemical behavior of Se and S, much of the geochemistry of Se is tied to the more abundant S. Elemental Se is the most stable form for most of the range of natural near-surface conditions. Selenium exhibits variable geochemical properties, chalcophilic and siderophilic, that affects its complex behavior in geochemical processes and results in relatively easy changes of its oxidation states (Table II-16.3).

Commonly occurring species, selenites (Se^{4+}) and selenates (Se^{6+}) in geochemical environments do not form stable compounds and are preferably absorbed by minerals, particularly clay minerals, and Fe and Mn oxides and hydroxides.

Approximately 50 Se minerals are known, of which the relatively common ones are: klockmanite, CuSe; berzelianite, $Cu_{2-x}Se$; clausthalite, PbSe; tiemannite, HgSe; and crookesite, $(Cu,Tl,Ag)_2Se$. Selenium is often associated with host minerals, such as pyrite, chalcopyrite, and sphalerite.

Table II-16.3. Impact of soil pH on the formation of soluble Se species (after Kabata-Pendias 1998)

Redox value (*Eh*, mV)	pH	Oxidation state of Se	Major species in soil solution
High, >400	7	+6	SeO_4^{2-} selenates
	<2	+6	SeO_4^-
Moderate, 200–400	>7	+4	SeO_3^{2-} selenites
	<7.3	+4	$HSeO_3^-$
Low, <200	>3.8	−2	HSe^- selenides
	<3.8	−2	H_2Se^0

The great affinity of Se to different organic compounds and substances results in a large number of organic compounds that are analogous to those of organic S compounds and are easily accumulated in some biolites, e.g., coals (up to 10.7 mg kg^{-1}), crude oils (up to 1.4 mg kg^{-1}), and bituminous shales. Extremely high Se concentrations, up to 6 500 mg kg^{-1}, occur in Chinese stone-coal (Plant et al. 2004). Concentrations of as much as 0.1% Se are not uncommon in the immediate vicinity of some oxidized sandstone-type uranium deposits.

The easy methylation of Se, mainly due to biological processes causes formation of volatile Se compounds and has a significant role in the geochemical cycling of this element.

Selenium is one of the elements playing a most important role in human and animal health and is essential to all other organisms including bacteria and algae. In the environment, however, there are concerns with both deficiency and toxicity. Since 1983 there has been dynamic growth in both studies and understanding of Se cycling and importance in human and animal health. Since that time an explosion in scientific papers and review articles has been observed.

II-16.1.2
Production and Use

Global production of Se in 1995 was calculated as 2.31 kt (Reimann and Caritat 1998).

Compounds of Se are used in photoelectric cells of broad utility from photometers to photovoltaic-batteries. They are used as pigment (maroon and orange colors), mainly in glass and plastic production (approximately 50% of the total Se production) and serves as vulcanizing and galvanizing agents. Its addition to alloys increases the machinability of stainless steel. Selenide compounds (e.g., WSe_2) are used in lubricants for metals.

A relatively high proportion of Se (around 20% of the total production) is used as dietary supplements for humans and livestock. Selenium is a relatively common component of various cosmetics and medications as therapeutic agent and has been recently used in cardiology as a major antioxidant.

In agriculture, Se is used as an addition (mainly as sodium selenite, Na_2SeO_3), to insecticides, fertilizers and foliar sprays.

Se

II-16.1.3 Soils

The Se content in soils is inherited from parent material and its distribution strongly reflects soil-forming processes and atmospheric deposition. The lowest amounts of Se are in sandy soils developed under humid climate, particularly in podzols, with the highest contents occurring most often in organic and calcareous soils (Table II-16.2). Selenium contents of worldwide soils have been calculated at an average value of 0.33 mg kg^{-1}, but the range of its concentrations is very broad, from 0.005 to 3.5 mg kg^{-1}. Averages of Se in sandy soils are reported for some countries at values of (in mg kg^{-1}): 0.14 in Poland, 0.14 in Lithuania, 0.18 in Russia, 0.21 in Finland, and 0.27 in Canada (Antapaitis et al. 2004; Eurola et al. 2003; Kabata-Pendias and Pendias 2001). Swedish arable soils contain Se in the range of 0.11–0.53 mg kg^{-1}, at the mean value of 0.23 mg kg^{-1} (Eriksson 2001a).

Soils from the Se-deficient endemic areas (e.g., in China) contain very low water-soluble Se, in the range from 0.2 to 2 μg kg^{-1}, whereas soils from non-endemic regions contain easily soluble Se from 1 to 11 μg kg^{-1} (Xu and Jiand *vide* Kabata-Pendias and Pendias 2001). The total Se range in topsoils of New Zealand is between 0.1 and 4.0 mg kg^{-1} and the common range is 0.3–0.9 mg kg^{-1}. Selenium-responsive disease in sheep is observed in areas where soils contain <0.5 mg kg^{-1} (Oldfield 2002). In Se-deficiency areas of Yugoslavia, average levels of this element in soils range from 0.04 to 0.44 mg kg^{-1} (Maksimović and Djujić 1998).

Higher contents of Se are observed in surface layer of forest soils, organic rich soils, calcareous soils, and volcanic soils (Table II-16.2). Elevated Se levels in peat soils up to 230 mg kg^{-1} (mean 138) in Great Britain, and up to 360 mg kg^{-1} (mean 54) in Ireland are associated with Se toxicity in livestock (Fleming; Williams and Thornton *vide* Kabata-Pendias and Pendias 1999).

Behavior of Se in soil has been studied extensively and summarized in several publications (e.g., Vouri et al. 1994; Frankenberger and Endberg 1998; Kabata-Pendias and Pendias 2001). It can be generalized that the main factors controlling Se forms and behavior in soils are Eh and pH, however several other parameters like organic ligands, clays, and hydroxides also play very significant roles (Tables II-16.3 and II-16.4). Nakamaru et al. (2005) reported that the major factor for Se adsorption in Japanese agricultural soils were active-Al and active-Fe. The range of Se K_d values in these soils was very broad, from 12 to 1 060, and exhibited the best correlation with acid oxalate extractable Al (active-Al). Inorganic species of Se associated with defined soil parameters reveal variable properties:

- Selenates (Se^{6+}) are mobile in inorganic forms, especially in neutral and alkaline soils, and are not absorbed on hydrous sesquioxides (in particular $Fe_2O_3 \cdot H_2O$)
- Selenites (Se^{4+}) are slightly mobile in neutral and acid soils of humid temperate regions, and are easily absorbed on hydrous sesquioxides and organic matter
- Selenides (Se^{2-}) are rather immobile in acid soils due to the formation of stable mineral and organic compounds

The transformation between these species as well as the formation of elemental Se takes place in all soils, however, under most conditions these are very slow processes.

Several complex Se anions, organic compounds and chelates greatly modify the behavior of Se in each particular soils and affect its distribution along soil profiles. Most commonly, the Se accumulation in surface soil horizon resulted from its adsorption by organic matter. It has been calculated that absorbed forms of Se in soil range from 15 to 40% of its total content and that Se-organic compounds range from 4 to 22% of the total Se (Kabata-Pendias 1998). Nakamaru et al. (2005) found that 80–100% of the absorbed Se was in Al-bound (28–78%) and Fe-bound (14–53%) fractions depending on soil groups. Small proportions of adsorbed Se were loosely bound (0.2–1.7%).

Forms and concentrations of Se in soil solution are governed by various physical-chemical and biological factors, and common inorganic anions are: SeO_3^{2-}, SeO_4^{2-}, $HSeO_4^-$, $HSeO_3^-$, and $H_2SeO_4^-$ (Haygarth 1994; Kabata-Pendias and Sadurski 2004). Under oxidizing conditions, SeO_4^{2-} is the favored form, while in mild reducing conditions, SeO_3^{2-} is likely to dominate. This anion is strongly sorbed on oxides and precipitates such as $Fe(SeO_3)_3$, whereas SeO_4^{2-} is very weakly sorbed, especially at high pH. Hence, mobile and easily phytoavailable Se occurs in alkaline well aerated soils, most commonly in arid and semi arid regions. Organic matter has a strong tendency to form organometallic complexes which remove Se from soil solution.

High Se mobility can be expected in soils of high pH and Eh, and conversely low mobility in soils with high contents of hydroxides, organic matter and granulometric clay fractions (Table II-16.4). In acid soils Se is likely to occur as Se^{4+}, strongly adsorbed

Table II-16.4. Soils factors affecting the mobility of selenium (after Kabata-Pendias 1998)

Soil parameter	Se species	Mobility
pH		
High	Selenates	High
Moderate	Selenites	Moderate
Low	Selenides	Low
Eh		
High (high oxidation)	Selenites	High
Low (low oxidation)	Selenides	Low
Hydroxides (Fe, Mn)		
High content	Absorbed all Se forms	Low
Low content	Slight absorption	High
Organic matter[a]		
Undecayed	Absorbed all Se forms	Low
Decayed (e.g., peat)	Complexed	High
Enhanced biomethylation	Volatilized	High
Clays		
High content	Absorbed all Se forms	Low
Low content	Soluble all Se forms	High
Interacions with		
S, P, and N	Antagonistic effects	Rather low

[a] Variable impact of organic matter depends on its kind.

by Fe oxides to form ferric selenite, $Fe_2(OH)_4SeO_3$, and iron selenide, FeSe. Maximum Se adsorption occurs between pH 3 and pH 5 and decreases as the pH rises. In alkaline Se-rich soils, the predominated species is Se^{6+} which is very weakly adsorbed. Hence, selenates occur in soluble forms in soil of arid and semi-arid regions.

The stability of Se-Fe minerals and Se adsorption by metal hydroxides and clays control to a high degree Se behavior. Addition of PO_4 to soil may increase Se uptake by plants, because this ion displaces selenites from soil particles and thus increases Se mobility. The phytoavailability of different Se species in soils decreases in the following order: selenate > selenomethionine > selenocysteine > selenite > elemental selenium > selenide.

The hot-water soluble Se fraction is considered to be the fraction that is most easily available to plants. However, other extractants, such as K_2SO_4, NH_4OH, and DTPA solutions are reported to give a good correlation between Se extractability and Se availability to some plants. In seleniferous soils, hot-water-soluble Se is >0.05 mg kg^{-1} (www.ext.colostate.edu/pubs).

The Se contents of soils and its mobility have received much attention, especially in countries where its deficiency in human and animal has been recognized. However, elevated soil Se in some regions, due to both geochemical and anthropological factors, has also been of great concern. Soils developed on seleniferous parent material (e.g., Cretaceous limestone) and soils contaminated with Se from irrigation water and/or from coal power plant emission are now of real environmental concern.

A close relationship between Se and organic C has been observed in most soils. Microbial processes in both the formation and mineralization of organic Se (e.g., selenomethionine, selenocysteine) play a crucial role in Se cycling and especially in its volatilization from Se-contaminated soils (Martens and Suarez 1998). These processes are important, especially in the reduction of Se, principally through the reduction of Se^{4+} and Se^{6+}. The activity of microbiota reveals variable sensitivity to increased Se contents in soil. However, a 5 mM concentration of selenic acid inhibited the activity of all soil enzymes (Nowak et al. 2002). In poorly drained soils, an accumulation of insoluble Se^{2-} compounds is likely to occur. Due to methylation processes under anaerobic condition, Se may volatilize in the form of $(CH_3)_2Se$, as well as in forms of several other methane and sulfide Se compounds. A number of bacteria and fungi species are involved in Se volatilization processes (Frankenberger and Karlson 1994). As reported by these authors, organic amendments may significantly increase the rate of Se volatilization from soils.

The program of increasing the Se status of soils in Finland was started in 1984. It was based on Se (as sodium selenate) addition to inorganic fertilizers at levels from 6 to 10 mg kg^{-1}, with a permitted highest concentration of 15 mg kg^{-1} (Eurola et al. 2003). The mean hot-water extractable Se in soils before this program (in 1979) was 0.006 mg l^{-1}, and after the program (in 1998) was 0.01 mg l^{-1}. However, Se easily available to plants was present only in the first summer after fertilizing, and quickly turned into unavailable forms in the soil. These authors stated that there is not any long-term environmental risk since Se remains in soils as immobile species. Studies of Vuori et al. (1994) indicated that selenate was soluble for at least 3 months after the addition. The sorption of Se was correlated positively with clay content, Fe content, and surface area and negatively with sulfuric acid-extractable P. However, these authors have also stated that the behavior of selenate is not uniform in cultivated soils of Finland, and that added selenate sorbed during the first growing season can later be mobilized into soluble fractions.

II-16.1.4 Waters

Natural waters usually contain Se at a level of <1 µg l^{-1}. The Se concentrations in seawater commonly vary between 0.1 and 0.35 µg l^{-1}. The median Se concentration in oceans is 0.2 µg l^{-1} (Reimann and Caritat 1998). According to another sources of data, the average Se level in oceans is 0.09 µg l^{-1} (ATSDR 2002b), and in the North Pacific is 0.1 µg kg^{-1} (Nozaki 2005).

It has been estimated that 7.7–8.0 kt of Se is introduced into the sea annually (Schrauzer 2004b). Organic selenide, mainly dimethyl selenide, $(CH_3)_2Se$, makes up around 80% of total dissolved Se in ocean surface water and its outgassing may be an important removal mechanism for dissolved Se from aquatic systems (Steinnes 2003).

The global Se average in river-waters is given at a value of 0.07 µg l^{-1}, with a range of 0.02–0.5 µg l^{-1} (Gaillardet et al. 2003). However, some rivers, e.g., Colorado River, contain Se in the range of 1–4 µg l^{-1}, although much higher values, up to 400 µg l^{-1} are also reported (ATSDR 2002b). Much of this Se is thought to have been derived from industrial sources, e.g., oil refineries contribute up to 75% of the total Se load entering the San Francisco Bay (Plant et al. 2004). The world average riverine flux of Se is given as 2.6 kt yr^{-1} (Gaillardet 2003).

The common range of Se in rainwaters is 0.04–1.4 µg l^{-1} and in Polar ice its content averages 0.02 µg l^{-1} (Plant et al. 2004). The median Se concentration in rainwater from the remote region of the Kola Peninsula is 0.5 µg l^{-1} and increases up to 0.9 µg l^{-1} in the polluted area (Reimann and Caritat 1998). Rainwater collected during 1999 in Sweden contained Se at an average value of 0.15 µg l^{-1} and at a maximum value of 0.26 µg l^{-1} (Eriksson 2001a). Rainfall may have increased Se levels not only from fossil-fuel combustion, especially coal, but also from volcanic sources.

Ground waters usually contain more Se than surface waters. Especially high levels of Se, up to 1 000 µg l^{-1}, are found in ground water of some regions of China. Also waters of saline, seleniferous areas in some arid parts of various countries (e.g., USA, China, Pakistan, Venezuela) may contain Se > 2 000 µg l^{-1} (Plant et al. 2004). Usually, the labile Se in most soils and the Se deposited atmospherically onto soils are rapidly leached into groundwater (Haygarth 1994). Wang et al. (1994) reported that Se levels in stream and river waters of Finland increased up to an average of 180 µg l^{-1} and in bottom sediments up to around 4 mg kg^{-1}, after the Se-fertilizing program.

The threshold value for Se in drinking water has been established by the WHO as 10 µg l^{-1} (Table II-16.5) whereas in the USA it ranges between 10 and 45 µg l^{-1}. The MCL value for the Se concentration in waters of all states of the USA is 50 µg l^{-1}. The limit level for Se in water used for irrigation is 20 µg l^{-1} (ATSDR 2002b). In most countries the MPC for Se are (µg l^{-1}) for: drinking water for humans 10, drinking water for livestock 50, and irrigation water 20 (www.ext.colostate.edu/pub/natres).

The bioreactor channel systems (BCSs) based on channels filled with rice straw stimulate the bacterial reduction of Se^{6+} to elemental Se^0 are very effective in removing the excess of Se from water (Zhang and Frankenberger 2003). As Terry et al. (2000) stated, the problem of cleaning up Se from contaminated water might be resolved by the selection of wetland species that will volatilize Se from the ecosystem.

Selenium concentrations in sewage have been reported to be (in µg l^{-1}): 280 in raw, 45 in primary effluents, and 50 in secondary effluents (ATSDR 2002b).

Se

Table II-16.5.
Selenium in wheat grains from different countries[a] ($\mu g\ kg^{-1}$)

Country	Range	Average
Algeria[b]	910 – 1040	970
Australia	1 – 117	23
Canada (barley)	9 – 38	21
Czech Republic	11 – 49	25
Denmark	–	21
Egypt	140 – 430	340
Finland[c]	10 – 70	–
Finland[d]	76 – 190	142
Finland (rye)	66 – 130	90
France	30 – 53	36
Norway	1 – 169	33
Poland	–	42
Sweden[e]	2 – 71	14
USA	280 – 690	490
Yugoslavia[f]	<10 – 25	–

[a] As compiled by Kabata-Pendias and Pendias (1999 and 2001), unless otherwise indicated.
[b] Data for fresh weight samples (Klapinska and Saci 1994).
[c] Data before the Se supplementation program (Fordyce 2005).
[d] Data for spring wheat in the period 1998-2001 (Eurola et al. 2003).
[e] After Eriksson (2001a).
[f] After Maksimovic et al. (1992).

Aquatic biota. The guideline for Se in aquatic life of freshwater is 20 and 5 $\mu g\ l^{-1}$, for acute and chronic conditions, respectively. These values for marine water are higher and ranges between 71 and 300 $\mu g\ l^{-1}$, for chronic and acute hazards to aquatic organisms (ATSDR 2002b).

Increased levels of Se in drainage water from Se-contaminated soils have created serious hazards to aquatic organisms, especially birds.

II-16.1.5
Air

Concentrations of Se in the atmosphere are highly variable due to differentiated sources: *(i)* evaporation from ocean and sea surface, *(ii)* volcanic eruption, and *(iii)* industrial emissions. The Se concentration in air above the South Pole is 0.06 $ng\ m^{-3}$ and the average value for worldwide air from remote regions is 0.2 $ng\ m^{-3}$ (Table II-16.2) whereas the median for polluted areas is 4.0 $ng\ m^{-3}$ (Reimann and Caritat 1998).

There is evidence that the ocean is a significant source of Se to coastal areas. Significant enrichment of Se in marine aerosols results from the formation of volatile organoselenium compounds, mainly dimethyl selenide, $(CH_3)_2Se$. Increased Se levels in mosses (>1 mg kg^{-1}) and in peat (>2 mg kg^{-1}) in the marine regions clearly indicate the impact of Se volatilization from the surface of seawaters (Steinnes 1991, 2003). Also substantial amounts of Se may be removed from Se-contaminated soils through bacterial and phytovolatilization processes.

Selenium is released during fossil fuel combustion. Its annual emission from all sources in Europe (only) was 420 t in 1979 (Schrauzer 2004b). Its global emission is >6 kt yr^1, in both small particles and volatile compounds which make around 40% of the total aerial abundance.

Inhalation exposure limits have been established by the US EPA as follows (in µg m^{-3}): 12 700 for hydrogen selenide, 400 for Se-hexafluoride, and 200 for other Se-compounds (Fordyce 2005). According to the guidelines presented by ATSDR (2002b), the concentration of Se in air may vary from 160 to 1 000 µg m^{-3}. The recommended threshold limit value for Se in workplace is 200 µg m^{-3}, whereas in Germany, the MAK value at the workplace has been established at 50 µg m^{-3}, and in Russia at 100 µg m^{-3} (Schrauzer 2004b).

II-16.1.6 Plants

Most plants contain rather low foliar Se, around 25 µg kg^{-1} and rarely exceed 100 µg kg^{-1}. However, some plants exhibit a great capability to accumulate Se and they may concentrate Se to extremely high levels (over 1 000 mg kg^{-1}).

The uptake of Se by plants depends on several factors, such as climate, soil parameters, and plant capacity to accumulate. When present in soluble forms, Se is readily absorbed by plants, although differences between plant species are very pronounced. Primarily, Se is taken up from the soil as selenate, SeO_4^{2-} or selenite, SeO_3^{2-} (Ellis and Salt 2003). In most cases there is a positive linear correlation between Se in plant tissues and Se contents of soils. However, the complex impact of variable factors on Se uptake by plants can significantly alter the relation between Se in plants and soils.

The Se content of crops recently received much attention because of its importance in the food chain. Thus, most available data are for food and forage plants. Cereal grains, as the most common source of Se in diets, have received special attention and have been broadly analyzed. In general, mean concentrations of Se in grains are higher in countries from arid climates than in countries from humid climates. Finland is an exception, having implemented the program of Se addition to fertilizers.

Although Se is not an essential element for plants, with some exceptions, is being added to soil to ensure that both food and feed products contain adequate amounts for the dietary needs. However, plants may also accumulate high amounts of Se that may be toxic to humans and animals.

The range of Se in cereals at the worldwide level is estimated as 100–800 µg kg^{-1} FW (Fordyce 2005). The range of mean Se varies from 142 to 970 µg kg^{-1} for countries with high Se levels in grains, and from 14 to 90 µg kg^{-1} for countries with low Se levels in grains (Table II-16.5). These variations do not indicate a significant impact of climatic conditions, because several other factors also control the Se absorption by plants. The addition of Se to soil (at 10 g ha^{-1}) affected its contents in grains of barley and oats,

Se

from 19 to 260 μg kg^{-1} and from 32 to 440 μg kg^{-1}, respectively (Gupta and Gupta 2000). Johnson et al. (2000) concluded that Se-grain status is the best means of defining its phytoavailability in the soil.

Food plants of the USA contain fairly similar amounts of Se that does not exceed 100 μg kg^{-1} FW (Table II-16.6). Its mean concentration is higher in roots and tubers (13 μg kg^{-1} in potato and 17 μg kg^{-1} in carrot) than in fruits (from 1 μg kg^{-1} in oranges

Table II-16.6.
Selenium in food plants and selected food in the USA (μg kg^{-1} FW) (adapted from ATSDR 2002b)

Plant and food item	Range[a]	Average
Food plants		
Carrot	6 – 29	17
Potatoes	4 – 23	13
Cauliflower	0 – 103	9
Tomato	0 – 19	2
Onion, bulbs	0 – 39	6
Green beans	0 – 13	1
Apple	3 – 6	4
Orange	0 – 12	1
Strawberries	0 – 12	3
Food items		
Bread, whole wheat	280 – 670	40
Bread, white	230 – 540	320
Rye bread	155 – 400	260
Corn flakes	26 – 120	63
Potato chips	0 – 27	26
Egg noodles, cooked	140 – 420	190
Cottage cheese	52 – 68	60
Chicken, cooked	170 – 260	210
Beef, cooked	190 – 510	260
Pork, ham	190 – 510	330
Salmon, canned	310 – 1490	750
Shrimp, canned	210 – 1610	640
Brazil nuts	200 – 253000	14700
Brazil nuts[b]	70710 – 87320	–
Coconut[b]	405 – 678	–

[a] Due to lack of the detection limit, the minimum value is given as 0 (zero).
[b] After Kápolna et al. (2004), data for Brazil coconut are on defatted base.

to 4 μg kg^{-1} in apples). Se-enriched garlic has often been reported as an important dietary supplement (Ip and Lisk 1993; Ip et al. 2000; Maksimović et al. 1995). Commonly reported increased levels of Se in Brazil nuts (*Bertholletia excelsa*) may vary highly depending on the growth conditions. For instance, nuts that were grown on soils derived from Cretaceous sediments enriched in Se by volcano emissions contain more Se than those grown on lower Se soils.

Mean Se contents in some food plants in Finland, after the Se supplementation program, varied in the period from 1989 to 2001 (in μg kg^{-1}) as follows: potatoes, 110 and 31; carrots, 40 and 20; white cabbage, 580 and 160; and strawberries, 10 and 11 (Eurola et al. 2003). All food plants grown on seleniferous or Se-contaminated soils contain much more Se than plants grown on low Se soils (Table II-16.7). Ge and Yang (1993) analyzed a number of food and forage samples from China and found that samples contain Se at the levels (in μg kg^{-1}) of: <20, from severely Se-deficient areas; 30–50, from deficient areas; 60–90, from moderate areas; and >100, from adequate areas.

The Se content of forage plants has also been of great concern. Leguminous plants (clover and alfalfa) usually contain more Se than grasses (Table II-16.8). Soil properties, particularly pH and SOM, have a significant impact on the Se uptake by grasses (Fig. II-16.1). Temperature is especially known to affect Se uptake by plants, with much higher uptake at temperatures >20 °C than during cooler seasons at <15 °C. Among soil factors, the pH seems to be most important and promotes higher Se availability at higher pH values.

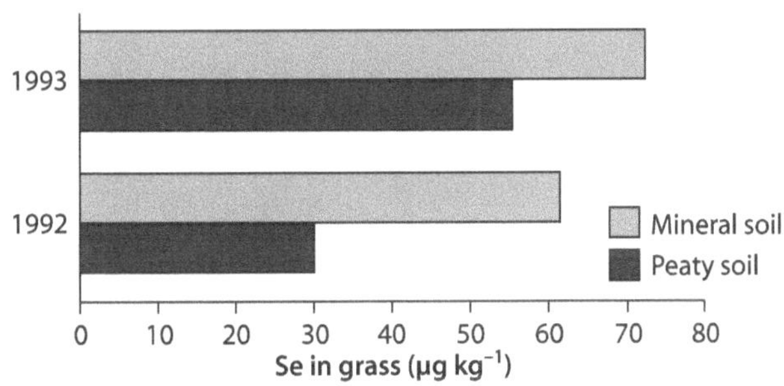

Fig. II-16.1. Impact of soil properties on Se status of grasses grown during 1992–1993 (Borowska and Łyszcz 1996, with the authors permission)

Table II-16.7. Selenium in food plants from areas of different Se levels in soils (μg kg^{-1}) (as compiled by Kabata-Pendias and Pendias 1999 and 2001)

Plant, country	Se levels of soil		
	Low	Moderate	High
Corn, grains, China	–	50	6900
Oat meals, USA	440	110	280
Sweet corn, China	0.5 – 3	75 – 85	2800
Wheat, whole grains, China	6	40	1100
Wheat, whole grains, USA	200	450	1800
Cabbage, China	–	2 – 38	2900
Potatoes, China	2	6	400

Se

Table II-16.8. Selenium in forage plants from different countries (μg kg^{-1}) (after Kabata-Pendias and Pendias 1999, 2001, unless otherwise stated)

Country	Grass		Clover or alfalfa	
	Range	Average	Range	Average
Canada	8 – 3	13	5 – 31	15
Czech Republic[a]	75 – 85	–	–	–
Denmark	–	32	–	35
Germany	20 – 210	110	5 – 130	90
Finland[b]	120 – 240	–	–	–
France	19 – 134	47	36 – 39	38
India	200 – 250	352	440 – 870	672
Japan	4 – 174	43	6 – 287	33
Sweden	11 – 64	30	18 – 40	–
USA	10 – 40	32	30 – 880	320

[a] Range of mean values (Wagner et al. 2004).
[b] Range of mean Se contents in silage, after the Se supplementation program (Eurola et al. 2003).

The highest Se averages of leguminous plants are reported for India and the USA, 672 and 320 μg kg^{-1}, respectively, while for other countries it varies from 15 (Canada) to 90 μg kg^{-1} (Germany). Grasses most often contain smaller amounts of Se than legumes. Average values of Se in grasses vary from 13 (in Canada) to 352 μg kg^{-1} (in India), however it is difficult to observe any regular variation. As cited by Oldfield (2002), some times a grass (*Agrostis tenuis*) may accumulate 2 to 7 times as much Se as white clover (*Trifolium repens*).

There is a great variation in plants' capability to absorb Se from soils, especially from seleniferous ones. Thus, plants have been divided into three categories: *(i)* plants that are accumulators and contain high amounts of Se and presumably require this element, *(ii)* plants that absorb medium quantities, up to around 100 μg Se kg^{-1}, and *(iii)* plants non-accumulators, containing usually below 30 μg Se kg^{-1}, under field conditions. Plants that are able to hyperaccumulate Se are species mainly of *Astragalus* and several other herbaceous species, e.g., *Oonopsis*, *Stanleya*, *Xylorhiza*, and *Machaeranthera*. These plants can accumulate well above 100 μg Se kg^{-1} when grown on Se rich soils, but even when grown on soils of low-to-medium Se content can accumulate up to 100 μg Se kg^{-1}. A Se content >5 000 μg kg^{-1} was reported in certain *Astragalus* species (Streit and Stumm 1993). Not all plant species grown on seleniferous soils are Se accumulators, and may contain only a few μg Se kg^{-1}. The plants variable capability to either concentrate or exclude Se from the growth media is closely related to the complex mechanism of Se compounds transformation and differences in Se pathways (Terry et al. 2000).

Selenium is easily absorbed directly from atmospheric deposition, which is clearly shown by its high accumulation in both mosses and mushrooms. Mosses from the

Table II-16.9. Mean selenium content of mushrooms from some European countries (mg kg^{-1}) (as compiled by Kabata-Pendias and Pendias 1999 and 2001, unless otherwise indicated)

Mushroom species	Finland	Norway	Poland[a]	Switzerland[a]
Boletus edulis	17	–	8.7 – 23[b]	14; 13
Macrolepiota procera	4.8	–	4.9	2.4; 0.7
Agaricus compestris	2.7	2.9	5.9	5.7; 3.6
Lecciunum testaceoscabrum	0.6	1.3	1.1; 6.9	1.2
Lactorius deliciosus	0.9	–	2.0	1.2; 0.6
Cantharellus cibarius	0.07	0.17	0.23; 6.23	0.05; 0.17
Sarcodon imbricatus	–	0.18	3.06; 0.36	1.87; 1.66

[a] Two values given for mushrooms of two different forest ecosystems.
[b] Range of mean values (Falandysz et al. 2003).

Scandinavian countries contain Se at an average of 390 µg kg^{-1}, up to 2 900 µg kg^{-1} (Berg and Steinnes 1997). Mosses sampled in Germany during the period 1995/1996 contained Se in the range from <140 to >560 µg kg^{-1} and increased Se concentrations were observed in western and eastern heavily industrialized regions (Oldfield 2002). High Se accumulation in rootless mushrooms also indicates its ready absorption from atmospheric sources (Table II-16.9). Large variation in the contents of Se in mushroom fruiting bodies seems to indicate that neither species nor sites have a significant impact on its absorption. Also grasses and other monocotyledonous and dicotyledonous plants absorb volatile Se from the atmosphere *via* the leaf surface. This Se is transported and accumulated in roots in various forms of inorganic selenite, and organic compounds such as SeGSH (selenoglutathione), SeMet (selenomethionine), and protein-bound SeMet (Terry et al. 2000). Haygarth (1994) calculated that the absorption of Se by plants from atmospheric deposition contributes from 33 to 82% of its total amounts in plants.

Plants absorbing Se from the atmosphere can also volatilize this element from their tissues. The complex pathways of assimilation and volatilization of Se for both non-accumulators and accumulators are described by Terry et al. (2000). The final products of Se compound transformation are dimethylselenide (DMSe) and dimethyldiselenide (DMdSe) which are easily volatilized from both roots and above ground plant parts as well as from bacteria and fungi (Haygarth 1994). The type of Se compounds added to soil (contamination or supplementation) as well as several plant factors strongly influence these processes.

Plants can assimilate soluble inorganic Se, mainly selenates, in association with sulfate anions via a sulfate transporter in the root plasma membrane (Terry et al. 2000). Unlike for selenate, there is no evidence that the uptake of selenite is mediated by membrane transporters. Plants can also actively absorb organic forms of Se such as SeMet. The uptake of Se by roots is controlled by several plant factors as well as soil and climatic conditions. The role of *Rhizobium* bacteria is significant in Se uptake by plants. Plants inoculated with the bacteria have a fourfold higher rate of Se volatilization than non-inoculated plants, which is probably due to the microbial increase in Se uptake. Interactions among S, Se, and O-acetylserine at the root-soil surface can also control the Se phytoavailability (Terry et al. 2000). Zayed and Terry (1994) have described the volatilization of Se in roots and shoots under the impact of sulfate level.

The function of Se in plants has been investigated in many studies and there is still little evidence that Se is essential for all plants. As reviewed by Terry et al. (2000), there is some evidence that Se is required for the growth of algae, but its essentiality to higher plants is controversial and yet unresolved. However, there are some indications that this element may be required for Se-accumulating plants. Recent studies on some grasses and vegetables provide indications that at proper Se addition the growth rate of these plants may be enhanced (Hartikainen 2005; Turakainen et al. 2004; Xue and Hartikainen 2000).

Pyrzyńska (1995) cited that there are several naturally occurring organic Se species: *(i)* selenocysteine, *(ii)* methylselenocysteine, *(iii)* selenomethionine, *(iv)* selenotaurine, *(v)* seleniobetaine, *(vi)* seleniocholine, *(vii)* dimethylselenide, *(viii)* dimethyldiselenide, and *(ix)* trimethylselenonium. Although the essentiality of selenoproteins in higher plants has not been documented, syntheses of selenoproteins in some plants (e.g., sugar beet) have been reported (Terry et al. 2000). Several selenoamino acids, SeMet (selenomethionine), SeCys (selenocysteine), and SeMC (selenomethylocysteine) in association with glutathione peroxidases have been found in both bacteria and higher plants. Predominate forms of Se in plants, as reported by Djujić et al. (2001), are SeMet in cereal grains and legumes seeds, and SeMC in vegetables. Schaumlöffel et al. (2004) have cited that more than 20 Se compounds were present in Se-enriched yeast and 5 Se-species were identified in Se-enriched garlic, onion, and broccoli.

Results of recent investigations suggest that Se is an antioxidant that activates protective mechanisms, which alleviate oxidative stress in the chloroplasts and protect plants against UV induced stress (Xue and Hartikainen 2000; Seppänen et al. 2003). Positive effects of Se on carbohydrate metabolism in some plants (e.g., potatoes, lettuce) have also been observed (Turakainen et al. 2004). Djujić et al. (2004) reported that Se supplementation to plants (wheat and soybean) increases tolerance to oxidative stress and improve mineral nutrient balance and vitamin E level. These authors concluded that under Se supplementation, cereals and forage crops convert Se primarily into SeMet and incorporate it into proteins, although SeMet is not required for the growth of these plants. This form of Se dominates in most plants, and in wheat and soybean grown on Se rich soil, 80% of the total Se consists of SeMet.

The toxic concentrations of Se in non-accumulator plants, resulting in a 10% reduction of yield, without visible symptoms, range from Se contents of 2 mg kg^{-1} in rice to 330 mg kg^{-1} in white clover. In accumulators, Se concentration may reach 4 000 mg kg^{-1} without negative effects. Tolerance mechanisms involve processes of exclusion of active Se amino acids, thus preventing their incorporation into proteins and damaging effects on plant functions (Terry et al. 2000). Thus, the exclusion of Se from proteins in the accumulator plants is the basis for their tolerance to Se. Generally, food crops have a low Se tolerance, however, most other plants may accumulate amounts of Se that are toxic to humans and animals.

In non-tolerant plant species, an excess of Se may impair germination and growth and cause chlorosis and black spots on leaves. Increased Se levels in plants suppress their concentrations of N, P, and S as well as several amino acids. High concentrations of Se inhibit the absorption of metals, mainly Mn, Zn, Cu, and Cd. These relationships are dependent on the ratio between the elements, and therefore stimulating effects of high Se levels on uptake of some trace metals may sometimes be expected. The application of N, P, and S is known to help in detoxifying Se, which may be a result either of depressing the Se uptake by roots or of establishing a beneficial ratio of Se to these elements.

Plant capability to both hyperaccumulate and volatilize Se has been used for the removal of excess Se. The first example of the effective phytoremediation of Se-contaminated soils was the plant-based phytoextraction of Se, using mainly *Brassica*, *Astragalus*, and *Graminceae* species in California (San Joachim Valley), where drainage of Se-enriched water threatened human health by increased levels in food and forage plants (Mc Grath 1998). Some transgenic plants (e.g., Indian mustard) may develop a mechanism of both high absorption and immobilization of Se in their tissues. These plants, as well as other hyperaccumulators may serve in the phytoremediation of Se-enriched soils.

A notable problem with phytoextraction is that plants with elevated Se content may become available to wildlife potentially causing toxicity, therefore they should be harvested, removed and utilized elsewhere. Volatilization of Se, both directly from the soil or by plants, may have a practical application in specific ecosystems. Microorganisms (bacteria and fungi) and several higher plants can biomethylate inorganic Se species and exhibit a great capability to remove Se from soil (Frankenberger and Karlson 1994; Lin et al. 1999). The main Se species that volatilize from soils are those that also volatilize from plants (dimethylselenides) as well as other methane and sulfide-methane compounds. As Lin et al. (1999) calculated, the phytovolatilization of Se during the growing season of *Salicornia* species was 34.6 mg Se m^{-2} and exceeded about two times the amount of Se removed by phytoextraction. Phytovolatilization seems to be promising methods since it removes Se from soils directly into the atmosphere.

II-16.1.7
Humans and Animals

Humans. Selenium occurs in mammalian tissues in the range from 0.7 in heart tissue to 2.5 mg kg^{-1} in muscles. The average Se content in human soft tissues is estimated as 0.11 mg kg^{-1} (Li 2000). The distribution of Se in soft tissue of adults in Poland varies from 0.05 to 0.50 mg kg^{-1} FW, being the highest in kidneys. Concentrations of Se in kidneys of humans from other European countries are reported by Zduńska et al. (1994) as follows (in mg kg^{-1} FW): Bulgaria, 2.5; Germany 0.7; and Italy, 1.9. In human fluids, mean Se concentrations are (in µg l^{-1}): blood, 107; serum, 80; urine, 22; and milk, 13 (Li 2000; Reimann and Caritat 1998). The range of Se in milk of women in Poland varies from <9 to >11 µg l^{-1} (Zachara and Pileexi 2000).

The species of ingested and inhaled Se may influence the behavior and quantity of Se in indicator tissues. Nevertheless, Se contents of serum and whole-blood are good indicators of body Se levels. The level required to optimize glutathione peroxidase activity is estimated at about 100 µg l^{-1}. In the European population, mean blood Se concentrations vary from around 45 µg l^{-1} in Serbia to around 90 µg l^{-1} in France, Italy and Sweden (Rayman 2000). Combs (2001) presented a report on Se in blood and serum of worldwide healthy adults. Adults of China contain Se in serum (in µg l^{-1}) *(i)* in urban areas, 80–111; *(ii)* in rural non-Keshan disease areas, 42–98; *(iii)* in blood, from

Keshan disease areas, 9–29; and *(iv)* in blood from Kaschin-Beck diseases area, 23–3 480. Yang et al. (*vide* ATSDR 2002b) reported that blood Se in a population affected by Keshan disease averaged 18 $\mu g\,l^{-1}$ whereas in people from non affected areas this value was >60 $\mu g\,l^{-1}$. Relatively high and fairly uniform mean Se levels are in serum of the population from the whole territory of the USA, in the range between 80 and 167 $\mu g\,l^{-1}$. In Finland, the Se level in serum was between 0.63 and 0.76 $\mu mol\,l^{-1}$, and after the Se supplementation of fertilizers, it increased to the range of 1.2–1.4 $\mu mol\,l^{-1}$ (Hartikainen 2005).

Adequate Se status (e.g., 70–75 $\mu g\,l^{-1}$ of plasma) is prerequisite for proper iodine metabolism that is interrelated in the conversion of thyroxine to triiodothyronine (Thompson 2004). Effects of iodine deficiency are worsened if Se status is low (WHO/IPCS 2002).

A relatively good biomarker of Se status in whole blood seems to be its content in human hair. Selenium concentrations in animal hair have been used to diagnose both deficiency and toxicity. A very broad range of Se, from 0.3 to 13 $mg\,kg^{-1}$, is reported for hair (Jørgensen 2000). However, hair Se values have become unreliable indicators due to the use of Se-containing cosmetics and/or exposure to Se-containing impurities (Schrauzer 2004b). Johnson et al. (2000) emphasized that Se in hair is the most variable indicator reflecting the impact of various and complex factors.

Toenails also provide good information on the Se status of the human body. The Se level in toenails of the Finnish population was affected by Se fertilization (starting in 1985), and increased from around 0.4 $mg\,kg^{-1}$ in 1984 to >0.8 and >0.7 $mg\,kg^{-1}$ in 1992 and 1996, respectively (Aro et al. 1998).

Deficiency of Se in humans was first identified as Keshan disease (endemic cardiomyopathy) in central China and some parts of Russia and later defined as Kashin-Beck disease (atrophy, degeneration and necrosis of cartilage tissue and joint deterioration, probably resulting from deficiency of both Se and I). Over 20 structural selenoproteins and catalytic selenoenzymes have been identified in human metabolism. It is an agent in antioxidation and anti-inflammation processes and participates in thyroid hormone production. Its roles in chemoprevention against HIV and AIDS have been recently recognized (Clark et al. 1996; Baum et al. 2000). The correlation of the Se depletion with cardiovascular diseases, cirrhosis, and diabetes has also been reported (Navarro-Alarcón and Lopez-Martinez 2000).

Huge numbers of various publications (from Internet information to books) have presented physiological functions of Se and its compounds in humans and animals. Descriptions of symptoms of improper Se supply to humans can be summarized as follows:

- Deficiency:
  - Muscle weakness and pain
  - Inflammation of muscles
  - Fragile red blood cells
  - Degeneration of pancreas
  - Abnormal skin coloration
  - Heart muscle dysfunction (dilation, congestion and other failures)
  - Prolonged illness condition
  - Susceptibility to cancer
  - Keshan disease (KD), (cardiomyopathy)
  - Kashin-Beck disease (KBD), (osteoarthropahty, disorder of the bones and joints)

- Toxicity:
  - Liver and kidneys damage
  - Blood clotting
  - Necrosis of heart and liver
  - Skin lesions
  - Hair and nail loss
  - Nausea and vomiting

Etiology of KBD is still not well documented because, contrary to KD, the efficiency of Se supplementation in the prevention of KBD has never been satisfactory demonstrated (Ge and Yang 1993).

Environmental levels of Se are very often closely linked with local geology. This is confirmed in studies carried out by Fordyce et al. (2000) who observed that low Se in the food chain is associated with Jurassic siltstones and sandstones, whereas sources of high Se occur in areas underlain by Permian carbonaceous strata.

Metabolic interactions between inorganic and organic forms of Se have been relatively well recognized. Selenium is necessary for the function of several enzymes and proteins, particularly glutathione peroxidase, an enzyme that prevents oxidative damage to cells from a variety of peroxides. Rotruck et al. (1973) were first to demonstrate that Se is an obligatory component of this crucial antioxidant enzyme. The dependence of glutathione peroxidase on nutritional Se clarifies the essential role of this element.

More than 100 Se-proteins, enzymes and other compounds have been identified in biological samples. The most common organic compounds are:

- selenomethionine (SeMet)
- selenocysteine (SeCy)
- dimethylselenide (DMSe)
- dimethyldiselenide (DMDSe)
- selenomethyltranferase (SMT)
- Se-methylmethionine (SeMM)
- glutathione peroxidase (GSH-Px)

Because of the important role that Se plays in human and animal nutrition there is interest in measuring the "bioavailable" amounts especially in food-stuffs, using various bioassays. Most available are selenates and Se-aminoacids. The availability of Se increased in diets rich in small molecular proteins (e.g., methionine) and in vitamins E, A, and C. Diets enriched in metals and sulfur inhibit Se availability. The excessive ingestion of As (e.g., with drinking water in Bangladesh) accelerate the excretion of Se lowering the body's content of this element. It has also been observed that As added to animal fodder counteracts Se toxicity (Spallholz et al. 2004). Wuyi et al. (2003) described a possibility of mitigation of endemic arsenocosis with Se supplementation.

Adequate Se content of food and forage should range between 100 and 3 000 $\mu g\ kg^{-1}$ (Table II-16.10) and its common concentrations in food and fodder plants, including mushrooms, should cover the daily requirement. However, some Se-supplementation is advised due to variable Se availability from different kinds of food. The availability

Se

Table II-16.10. Thresholds values for Se in food and forage (adopted from Kabata-Pendias and Pendias 1999, ATSRD 2002b, and Fordyce 2005)

Material (unit)	Deficient	Adequate	Toxic
Animal fodder, chronic exposure ($\mu g\ kg^{-1}$)	<40	100 – 3000	>3000
Human food ($\mu g\ kg^{-1}$)	<50	100 – 3000	>3000
Human diet (WHO) ($\mu g\ day^{-1}$)	<40	40 – 200	>400
Drinking water (WHO) ($\mu g\ l^{-1}$)	–	10	>10 >50[a]

[a] Raised Se level by EPA in 1990 (Putman *vide* Schrauzer 2004b).

of ingested Se and its incorporation into various compounds in the body depends on the forms existing in plants. While selenomethionine (mainly in cereals) is easily incorporated into protein, Se-methyl-Se-cysteine (mainly in Se-fortified garlic and broccoli) is likely to be converted into methyl selenol that has a great anti-carcinogenic potential (Ip et al. 2000; Hartikainen 2005). The nutritional prevention of cancer with Se has been broadly investigated (Clarks et al. 1996). Also Se-responsive diseases have been studied (Fordyce et al. 2000).

There are several recommended dietary Se intakes, as well as estimates of TUIL for adults, set at 400 $\mu g\ d^{-1}$, and NOAEL, set at 800 $\mu g\ d^{-1}$ (Schrauzer 2004). Recommended dietary Se intakes depend highly on human life stage and were set at lower levels in the year 2000 as compared with levels proposed in 1980 (Table II-16.11). Recently recommended adults daily intakes range between 40 and 55 µg.

Data presented by Combs (2001) from all over the world has indicated very broad ranges in Se intakes (μg person^{-1} d^{-1}); between 11 and 95 in most European countries (at moderate Se-blood status), and between 60 and 350 in other countries, e.g., Canada, USA, Venezuela, Japan, and Greece. These values in China were 7–11 in Keshan disease areas, and 750–4990 in selenosis areas. The Finnish program to increase Se in cereal grains up to 100 $\mu g\ kg^{-1}$, by supplementation with Se of multinutrient fertilizers had the primary goal of elevating the Se intake to an adequate rate of 50–200 μg person^{-1} d^{-1} (Hartikainen 2005). Clinical signs of selenosis are observed at, and above intakes of 900 $\mu g\ d^{-1}$, but a maximal daily safe dietary Se intake of 400 µg has been suggested for adults (WHO/IPCS 2002).

Selenium and its compounds have been recently regarded as the most effective antioxidants in both prevention and cure of cancer (Spallholz 2001). Selenomethionine (SeMet) is involved in oxidation processes as both promotor and inhibitor agents. However, in association with glutathione, SeMet acts catalytically as a cellular antioxidant (Schrauzer 2003) and also enhances the antioxidant effect of vitamin E. There is ample evidence that Se may have a protective role in many diseases of humans. The most common ones have been mentioned above: cardiac arrhythmia, atherosclerosis, liver cirrhosis, and cancer (particularly of colon and prostate). Patients with acute myocardial infarction or with ischemic cardiomyopathy contain significantly lower serum Se than control groups (Navarro-Alarcón et al. 1999, 2000). It is also an effective

Table II-16.11. Recommended dietary selenium intakes[a] ($\mu g\ d^{-1}$) (adapted from Levander 2001, ATSDR 2002b, and WHO/IPCS 2002)

Life stage	1980, DDI[a]	1996, DRA[b]	2000, DRA[b]
Infants	10 – 60	6 – 12	15 – 20
Children			
<10 years	20 – 200	20 – 25	20 – 40
>10 years	50 – 200	30 – 36	40 – 55
Adults >14 years			
Males	50 – 200	40	40 – 55
Females	50 – 200	30	55
Women			
Pregnant	65	60	70
Lacting		–	70

[a] Daily dietary intake.
[b] Dietary recommended allowance.

detoxifier of some metals, e.g., due to its ability to bind Cd, Pb, Hg, and Ta, mainly as selenites, which decreases their toxicity. Its immune functions increase antibacterial and antiviral defenses of the organism. The biofortification of Se in food plants, and in particular of wheat and garlic has been studied recently (Ip and Lisk 1993; Djujić et al. 2001; Lyons et al. 2003). There is evidence that yeast, Brazil nuts and whole grains, when enriched in Se, provide an easily available source.

Animals. Deficiency, as well as excess of Se in farm animals seems to be more widespread than in humans. Most domestic animals need 3–6 times higher Se intake than humans.

The optimal Se concentration in serum of livestock, according to German regulation, ranges from 80 to 300 $\mu g\ l^{-1}$, whereas following the Czech Republic directives it is satisfactory between 40 to 150 $\mu g\ l^{-1}$ (Kvicala et al. 2002). Forage Se at a concentration above 100 $\mu g\ kg^{-1}$ satisfies the requirement of most animals (Table II-16.10). At lower Se content in forage its level in sheep blood decreases to below 50 $\mu g\ l^{-1}$ (Ermakov *vide* Kabata-Pendias and Pendias 1999).

The supplementation of feed by selenomethionine or Se yeast has become increasingly common in animal nutrition, especially in low-Se areas worldwide. For instance, most commercial animal feed has been supplemented with selenate in several countries since 1969. Observations have indicated that the supplementation of feed seems to have some prophylactic effect on humans who ingested animal products (Hartikainen 2005).

In general, selenites are more easily available and thus might be more toxic than selenates and selenides to animals. The toxicity of Se in animals, first reported in farm animals in China, is likely to occur in regions with a predomince of alkaline soils and/

or Se-enriched peats, with a high ratio of easily phytoavailable Se species. Alkali disease occurs when livestock is fed with grain containing 3–20 mg Se kg^{-1}. Schrauzer (2003) described toxicity symptoms in macaques, after 30 days of the administration of "maximum tolerable dose" (188 µg Se kg^{-1} BW per day) as evident in: mean body weight reduction, hypothermia, dermatisis, xerosis, cheilitis, and disturbance in menstruation. The toxic threshold values for Se in fodder has been set at above 3 mg kg^{-1} (Table II-16.10), and some have recommend 5 mg kg^{-1} in the diet as the dividing level between toxic and nontoxic feed (Schrauzer 2004b). There is no evidence of the toxicity of selenomethionine (SeMet) to humans but it has been observed in animals although not as severe as that from selenate.

Symptoms of improper Se supply to animals can be summarized as follows:

- Deficiency:
  - White muscle disease in sheep and cattle
  - Mulberry heart disease (myocardium dystrophy of the heart) in swines
  - Exudative diathesis in poultry
- Toxicity:
  - Dullness and lack of vitality "alkali disease"
  - Body weight reduction
  - Hair loss
  - Hoof deformation and soreness
  - Hypothermia
  - Dermatitis, xerosis, cheilitis
  - Stiffness and lameness

II-16.2
Tellurium (Te)

II-16.2.1
Introduction

The terrestrial abundance of Te is approximately 1 µg kg^{-1} and its content in rocks ranges from 1 to 5 µg kg^{-1}, being elevated in organic shales. Its increased amounts in coal and fly ash (20–2000 µg kg^{-1}) reveal an ability to be sorbed by organic matter. Volcanic gases and volcanic sulfur deposits are usually enriched in Te. It is also present in gold tellurite deposits.

Tellurium resembles Se and S in a number of its properties, but is more metallic in nature than either of these elements. The most common Te minerals are: tellurite, TeO_2; sylvanite, $(Ag,Au)Te_4$; hessite, Ag_2Te; calaverite, $AuTe_2$, tetradymite, Bi_2Te_2S, and nagayagite, $Au(Pb,Sb,Fe)_3(TeS)_{11}$. Minerals of Te are commonly found in pyritic ores. Its possible host minerals are: pyrite, chalcopyrite, molybdenite and pentlandite.

Tellurium shows variable valences from –2 to +6 and easily forms di- and trivalent oxides (Table II-16.1). During weathering, Te is oxidized to slightly mobile tellurite that is readily sorbed by Fe hydroxides. Its migration in hypergenic zones is rather limited, however, its complex anion, TeO_3H^-, is easily mobile.

II-16.2.2 Production and Use

The global Te production was estimated in 1999 at about 124 t (Kobayashi 2004) and in 2003 at 243 t (WMSY 2004). Tellurium is mainly a by-product of Cu smelting and/or electric refining of Cu. Also the refining of other metals, e.g., Pb, may be a source of Te.

Compounds of Te are used to improve metal alloys, e.g., add strength and machinability of Sn, Pb, Cu, and stainless steel. It is added as vulcanizing agent to rubber to increase resistance to heat and abrasion, and improve characteristics of the rubber. Due to its relatively high conductivity, Te is used in semiconductors and other electronic devices. It is also a catalyst and in vapor form is used in "daylight lamps". It is used in ceramics, but its addition to pottery glazes became recently limited. Also photographic and pharmaceutical industries utilize some quantities of this metal.

II-16.2.3 Soils

Limited data is available on Te occurrence in soils. In reference soils of China, Te contents vary from 0.03 to 4.0 mg kg^{-1} and in soils of the USA the range is between 0.02 and 0.69 mg kg^{-1} (Govindaraju 1994). Hashimoto et al. (*vide* Kobayashi 2004) reported soil Te to be 0.035 and 0.031 mg kg^{-1} in Japan and Korea, respectively. Swedish arable soils contain Te in the range of <0.08 to 0.12 mg kg^{-1} (Eriksson 2001a).

Since the biological cycling of Te resembles that of Se, the microbial metabolism of Te also seems to be similar. Thus, bacteria capable of methylating Se and As can also methylate Te, whereas the reduction of tellurite to elemental Te can readily occur under the influence of variety of microorganisms. Tellurite reductase was found in some bacteria and the reduction of selenate and tellurite by nitrate reductase has been reported (Sabaty et al. 2001).

II-16.2.4 Waters

The concentration of Te in natural waters ranges from 0.17 to 0.9 ng l^{-1}, the highest being in the Red Sea (Kobayashi 2004). In the North Pacific it averages 0.05 ng kg^{-1} (Nozaki 2005).

Rain waters sampled in 1984 in Florida contained 0.51–3.3 ng Te l^{-1} (Andreae *vide* Kobayashi 2004). Rainwater collected during 1999 in Sweden contained Te in the range from 0.1 to 1.1 ng l^{-1}, at the average value of 0.4 ng l^{-1}. Its wet deposition in Sweden is calculated at 22 mg ha^{-1} yr^{-1} (Eriksson 2001a).

II-16.2.5 Air

The concentration of Te in air averages 0.12 ng m^{-3} in Japan, and ranges from 0.35 to 50 ng m^{-3} in the USA (Kobayashi 2004).

II-16.2.6 Plants

Generally, the Te content in plants does not exceed 1 mg kg^{-1}, and ranges from <0.013 to 0.35 mg kg^{-1} FW (Kabata-Pendias and Pendias 1999). Barley and wheat grains from Sweden contain Te at the mean value of <0.001 (Eriksson 2001a). Plants uptake Te, similarly to Se, relatively easily and when its content in growth media is high, there is a corresponding increase in the Te content of plants.

Seleniferous plant species (e.g., *Astragalus*) also contain elevated quantities of Te (up to 6 mg kg^{-1}), particularly when grown on Te enriched soil. In vegetation growing in some mining districts, e.g., a copper mine in Nevada, an increased level of Te has been observed (Cowgill 1988).

Vegetables contain Te in the range from <0.013 to >0.35 mg kg^{-1} FW, the lowest being in apple fruits and the highest in onion and garlic. The garlicky odor of some plants is caused by vapors of dimethyl telluride.

Norwegian mosses sampled in the period 1990–1995 contained Te at average of 5.2 µg kg^{-1}, within the range from <2 to 89 µg kg^{-1} (Berg and Steinnes 1997).

II-16.2.7 Humans and Animals

Humans. The occurrence of Te in mammalian tissues was reported for muscle by Jørgensen (2000), at the level of 20 µg kg^{-1}. It is concentrated in kidneys, liver, bone, and brain. Apparently, bone contains around 90% of the total Te body burden. Bone-deposited Te slowly leaches out and its half-time is estimated to be 600 days (Hollins 1969). In human urine, the Te concentration is reported to be <1 µg l^{-1} (Reimann and Caritat 1998), whereas Kobayashi (2004) reported the range to be from <0.1 to 10 µg l^{-1}. This author found whole-blood Te concentrations to range from 0.15 to 0.3 µg l^{-1}.

Tellurium has no biological role and all its compounds are highly toxic. It is considered to be teratogenic agent, however acute poisoning is rare. Occupational exposure during production of the metal causes irritation of skin and eyes. Symptoms of Te intoxication are: *(i)* dreadful smelling breath and body odor, *(ii)* discoloration of skin, *(iii)* metallic taste in the mouth, and *(iv)* weight and hair loss. Workmen exposed to 0.5 to <10 µg Te m^{-3} have a garlic-like odor that is symptomatic of the Te intoxication.

The mechanism of Te toxicity is associated with impaired neurotransmission that affects saliva and sweat secretion in humans. Ingested and inhaled Te compounds, after reduction to tellurite (Te^{4+}), are methylated to dimethyl telluride which has a garlic-like aroma.

Content of Te in food varies from 1.1 to 4.8 mg kg^{-1} with the highest being in cereals and meats. The daily dietary intake of Te by adults from food has been estimated at around 0.1 mg kg^{-1} BW (Kobayashi 2004).

Permissible exposure limits for Te in inhaled air vary for different compounds, from 0.1 mg m^{-3} for elemental and dioxide Te, to 10 mg m^{-3} for telluride compounds (Kobayashi 2004).

Animals. An excess of tellurite (Te^{4+}) in animals damages the erythrocyte membrane and thus leads to hemolysis. Tellurate (Te^{6+}) ions do not have such an effect. Rats intoxicated with Te suffered from a segmental demyelination of the sciatic nerve and paralysis of the hind limbs (Kobayashi 2004).

II-16.3
Polonium (Po)

II-16.3.1
Introduction

Abundance of Po in the Earth's crust is on the order of 1 fg kg^{-1} (10^{-16}%). However, other sources estimated its possible occurrence in bulk continental crust as 2 pg kg^{-1} (10^{-13}%). Some accumulation may be expected in coal. Uranium ores contain Po up to about 0.1 μg kg^{-1}.

There are twenty-five known isotopes of Po with atomic masses ranging from 194 to 218 (Table II-16.1). Seven of those isotopes naturally occur. The only one occurring to any extent is ^{210}Po with a half-life of 138.4 days. This naturally occurring radioactive isotope is relatively ubiquitous in the environment but tends to be elevated in uraninite/pitchblende (at approximate concentration 10–100 ng kg^{-1}) and in phosphate-related materials.

II-16.3.2
Production and Use

Most commonly ^{210}Po is extracted from U-Po pitchblende deposits. Natural radioisotopes, including ^{210}Po and several other short-life radionuclides, are associated with the radioactive decay within the uranium chain U-Ra-Pb. More common sources of this radionuclide are Bi isotopes (209 and 210) that decay to ^{210}Po.

Uses of ^{210}Po include nuclear batteries, neutron sources and film cleaner. It is also used as a lightweight heat source for thermoelectric power in space satellites.

While using Po compounds, strict control is needed to minimize exposure hazards.

II-16.3.3
Soils and Plants

The natural abundance of ^{210}Po in soil is reported to range from 8 to 220 Bq kg^{-1}. It is elevated, in soils around radioactive and phosphate ore processing waste disposal sites. Arthur and Markhan (1984) reported that concentrations of ^{210}Po in soils and wild animals (deer and mice) collected near radioactive waste disposal sites in the vicinity of a phosphate processing plant in the state of Idaho, USA, were about three times greater than those from other sampling locations.

The radioactivity of ^{210}Po in terrestrial plants ranges from 8 to 12 Bq kg^{-1}. Elevated contents of this radionuclide is observed in plants, and especially in tobacco, grown on soil heavily fertilized with some phosphate-based materials. Increased levels of this radionuclide may also reflect an aerial source of pollution.

Relatively widespread use of radioactive phosphate fertilizers may increase levels of radioactive Po in food and feeds. Thus, diets composed of grains, vegetables, and meat in some regions may be the main source of this radionuclide. However, it is not clear yet how much radioactivity that fertilizer contributes to food and how much is naturally present.

II-16.3.4
Waters

The median of Po in ocean water is roughly estimated as 1 fg l^{-1} (Reimann and Caritat 1998). The mean ^{210}Po in public drinking water in Poland in 1998 was 0.48 mBq l^{-1}, and the annual intake, per person, of this isotope was calculated as 0.24 Bq (Skwarzec et al. 2001).

II-16.3.5
Humans

The alpha radiation of ^{210}Po has an effect on human and animal systems, similar to the impact of all other alpha-emitting radionuclides with carcinogenic and teratogenic changes observed. Due to this radiation, relatively common is lung cancer in both active and passive smokers.

The level of ^{210}Po in American tobacco tripled between 1938 and 1960 due to the increased use of chemical fertilizers. Thus, there is belief that tobacco is the largest source of radiation exposure among the American public, but research carried out with animals that inhaled ^{210}Po has not confirmed cancer indices (Smith's Health Newsletter, www.webspawner.com). However, ^{210}Po is considered to be a highly dangerous radioactive isotope that accounts for the very high level of cancer from smoking. There is evidence that Po from tobacco is carried by the blood and accumulates in the liver, kidneys, bone marrow, and blood vessel walls. The maximum allowable concentration for soluble Po compounds in air is about 2×10^{11} μCi cm^{-3} (www.webspawner.com).

Trace Elements of Group 17 (Previously Group VIIa)

The Group 17 of the Periodic Table is composed of three trace elements: fluorine (F), bromine (Br), and iodine (I). To this group of elements, called halogens, belongs also chlorine (Cl) which is relatively common in the biosphere, but some authors have treated it as a trace element. Iodine is the least reactive of the elements in this group. The last element of this group, astatine (At), occurs as unstable radioactive isotope of very short lives (the longest half-life of ^{209}At is 8.1 h).

In nature, halogens form simple anions, commonly with a 1 oxidation state (Table II-17.1). With an exception of F, halogens exhibit an affinity to combine with oxygen and play a significant role in several biochemical processes. These anions exhibit lithophilic properties and are highly mobile in the Earth's crust.

II-17.1 Fluorine (F)

II-17.1.1 Introduction

Fluorine, in the Earth's crust, is slightly concentrated in mafic rocks, up to 1 200 mg kg^{-1}, when compared with its content in acid igneous rocks (300–850 mg kg^{-1}) and to its

Table II-17.1. Selected properties of trace elements of Group 17

Element	Atomic number	Atomic mass	Atomic radius[a] (pm)	Density (20 °C, g cm^{-3})	Valence[b]	Melting point (°C)
F, fluorine	9	18.99	57	1.69[d]	**−1**[e]	−219.6
Cl, chlorine	17	35.42	97	3.21[d]	**−1**[e]	−100.9
Br, bromine	35	79.9	122	7.59[d]	**−1**[e]	−7.2
I, iodine	53	126.9	132	4.93	**−1**[e]	113.9
At, astatine	85	210[c]	143	–	**−1, +3, +5**	302.0

[a] Approximately average values.
[b] Valence values in bold are for the main oxidation states.
[c] Variable atomic mass of unstable isotopes from 196 to 219.
[d] Density in g l^{-1} (gases).
[e] Other oxidation states are from +1 to +7.

F

Table II-17.2. Abundance[a] of fluorine, chlorine, bromine and iodine in the environment

Environmental compartment	F	Cl	Br	I
Earth's crust ($mg\ kg^{-1}$)	525 – 650	145 – 640	1 – 2.4	0.15 – 1.4
Igneous rocks ($mg\ kg^{-1}$)				
Mafic	300 – 1200	40 – 300	0.2 – 3.0	0.007 – 0.5
Acid	300 – 850	130 – 300	0.2 – 4.5	0.01 – 0.5
Sedimentary rocks ($mg\ kg^{-1}$)				
Argillaceous	500 – 800	100 – 200	5 – 10	1 – 38
Sandstones	50 – 270	10 – 20	1 – 5	0.02 – 1.5
Calcareous	50 – 350	50 – 350	<1 – 6	0.5 – 3.0
Raw energy material ($mg\ kg^{-1}$)				
Coal, hard	75 – 100	210 – 1000	2.5 – 20	>1.1
Crude oil	0.0004	–	–	0.1 – 0.8
Petrol	<0.0003	–	–	–
Soils ($mg\ kg^{-1}$)[b]	300	300	10	2.8
Light sandy	83 – 205	115	<0.1 – 8	<0.1 – 10
Medium loamy and silty	175 – 462	–	–	0.3 – 10
Heavy loamy	470 – 1360	1000	50 – 100[c]	50 – 104[c]
Calcareous	470 – 680	–	68 – 130[c]	0.3 – 9.5
Organic	<0.1 – 350	–	–	1 – 10
Water ($\mu g\ l^{-1}$)	–	–	–	58[e]
Sea	30 – 1350	20000	65000	50 – 70
River	50 – 2700	8	2 – 4470	2 – 15
Rain	50[d]	–	–	–
Air ($\mu g\ m^{-3}$)[f]	1 – 7	0.23	–	–
Urban	–	4 – 10	50 – 500	100 – 350
Greenland	–	0.04 – 0.1	7 – 20	0.3 – 4
South Pole	–	0.003	0.4 – 1.4	0.08

[a] Presented are common, possible background, values from various sources as compiled by Kabata-Pendias and Pendias (1999 and 2001), unless otherwise indicated.
[b] Means for world soils. [c] Volcanic ash soils. [d] Rain of the Kola Peninsula (Reimann and Caritat 1998). [e] Mean content in the North Pacific (Nozaki 2005). [f] Data for Br and I are in $ng\ m^{-3}$.

abundance (650 mg kg^{-1}) in the upper continental crust (Table II-17.2). It is also likely to concentrate in black shales. High levels of F are known to be associated with some phosphate deposits over large areas of some continents (e.g., Africa, America). A strong association of F with phosphates is observed in both primary and secondary minerals. Concentrations of F in the form of fluorite also occur in some dolomites as fissure veins and other intrusions. Compounds of F are common constituents of magmatic and volcanic exhalations, and sometimes may occur in rocks in gaseous nebulae.

Fluorine is a very reactive element (most of the halogen group) and may combine with several other elements, including inert gases. Its most common mineral, fluorite (as a salt called also fluoride), CaF_2, is widely distributed in both the lithosphere and hydrosphere. Other commonly exploited minerals are cryolite, $Na_3(AlF_6)$; and fluorapatite, $CaF_2 \cdot 3Ca_3(PO_4)_2$. Topaz, $Al_2F_2SiO_4$, is a popular gemstone. In the hypergenic zones, cryolite is most readily soluble, and the mobile F is easily sorbed by clay minerals and phosphorites. Several minerals, like micas and amphiboles, may be associated with fluorapatite. Increased F contents in some combustion byproducts may be expected.

II-17.1.2 Production and Use

The global production of F in 2000 was estimated at 4.4 kt, given as fluorspars (WMSY 2004).

Fluorite is the most common commercial source of F. A great proportion of F goes to the Al-industry. In the steel industry, it is used for making hydrofluoric acid, and has been used as a flux for more than two centuries. It is also used in plastic, ceramic and glass production (glass etching). As a very oxidizing agent and exothermic in chemical reactions, F has been broadly used in various chemical processes. Fluorine is a common component of Freon gases, chlorofluorocarbons (CFCs), used for the refrigeration. Perfluorooctane sulfonate (PFOS), very persistent chemical, is applied as stain repellent.

It is used for the synthetic manufacture of cryolite and fluorapatite. Natural fluorapatite might be utilized for phosphate fertilizers. Fluorine easily reacts with other halogens to form interhalogen species: ClF, BrF, and IF.

Fluoride has long been added to municipal drinking water (at the level of 1.2–1.9 mg l^{-1}) for the prevention of dental caries. It is still added to tooth pastes.

II-17.1.3 Soils

The F content in soils is inherited from parent material, but its distribution reflects soil-forming processes and soil texture. A relatively close positive relationship between F and the clay fraction contents is evident. The lowest F amounts are found in sandy soils under a humid climate and the highest contents occur in heavy clay soils and in soils derived from mafic rocks (Table II-17.2).

The average F contents of worldwide soils have been calculated to be 329 and 360 mg kg^{-1} for European and the USA soils, respectively (Kabata-Pendias and Pendias

2001). Common concentrations for most soils seem to range from 150 to 400 mg kg^{-1}. Very high levels of F, above 1 000 mg kg^{-1}, are reported for uncontaminated soils derived from F-rich bedrocks that are associated with the province of endemic fluorosis (Fuge and Adrews 1988; Ermakov 2004).

The behavior of F in soils is controlled by several soil parameters, mainly by clay contents, pH, and concentrations of Ca, P, and hydroxides. Arnesen and Krogstad (1998) observed increased sorption of F in the B soil horizons, which usually have high levels of amorphous Al-oxides/hydroxides. The highest sorption occurred at pH 4.8 to 5.5.

Under natural soil conditions, F is slightly mobile, however, in acid soils its solubility usually increases due to the occurrence of readily soluble fluorides such as NaF, KF, and NH_4F, whereas AlF_3 and CaF_2 are known to be of low solubility. Fluorine mobility in soils is variable and highly controlled by pH; its higher solubility is at pH below 5 and above 6. In soil solution, F can occurs mainly as complex cations: AlF^{2+} and AlF_2^+, as well as anions: F^- and AlF_4^- (Kabata-Pendias and Sadurski 2004).

The most common form of F in soils is fluorapatite, however, several fluorides (e.g., AlF_3, CaF_2) and aluminosilicates (e.g., $Al_2(SiF_6)_2$) also are reported to occur in soils. In calcareous soils, the occurrence of slightly soluble fluorides and complexes with Fe, Al, and Si are responsible for the low migration of this element. In sodic soil, on the other hand, a high level of exchangeable Na affects increased solubility of F. The sorption of F by acid soils has been attributed to interactions with Al oxy/hydroxy species. In most soils, a low F affinity to SOM has been observed. However, in some tropical soils, organically bound F (monofluorinated compounds) may occur and affect its high availability to plants. The origin of these F compounds might be attributed to the synthesis by certain microorganisms.

Soil pollution by F in some regions has become of ecological importance. Several sources, but mainly Al smelters and P-fertilizer factories, are still significant emitters of F to the atmosphere.

Soils surrounding Al-processing industries, china clay industries, and P-fertilizer factories are reported to contain up to 3 200, 3 560, and 2 080 mg F kg^{-1}, respectively (Kabata-Pendias and Pendias 2001). In contaminated soils surrounding old mining regions in the UK, F concentrations range from 620 to 19 960 mg kg^{-1} (Fuge and Andrews 1988).

The input of F to arable soils with P fertilizers might be of great environmental concern. Easily soluble F-bearing fertilizers or sewage sludges may cause a remarkable increase in the phytoaccumulation of F from soils. However, the most important hazard of F contamination in soils concerns changes in soil properties due to the great chemical activity of hydrofluoric acid that may be formed from both solid and gaseous F pollutants. The decomposition of clays and other silica minerals in soils heavily polluted by F have been reported. Also the destruction of humic mineral complexes and loss of organic matter, as well as the reduction of enzymatic activity of some microorganisms has been observed in such soils. The sensitivity of microorganisms to increased levels of F varies, being higher in bacteria using organic nitrogen than of those utilizing inorganic nitrogen compounds (Grishko 1999).

Industrial emissions are sources of readily soluble fluorides and thus easily available to plants. The added F is either fixed by components of heavy soils (fine granulometric fraction, Ca, and P) or readily removed from light sandy soils by water. Effective amelioration of F-polluted soils is presently of great ecological concern.

II-17.1.4 Waters

The median F concentration in world ocean waters has been estimated as 1.3 mg l^{-1}, and the range between 0.03 and 1.35 mg l^{-1}. The North Pacific Ocean waters contain F at the average of 2.8 mg kg^{-1} (Nozaki 2005).

River waters contain F from 0.05 to 2.7 mg l^{-1}, whereas the median F value for stream waters of Finland is 0.008 mg l^{-1} (Reimann and Caritat 1998; Kabata-Pendias and Pendias 1999). Neal et al. (2003) have reported that in main rivers and tributaries of the UK the mean F concentrations varied from approximately 0.5 to over 2 mg l^{-1}.

Concentrations of F in surface waters vary highly within the order of magnitude 0.0X to X mg kg^{-1}, depending on geological regions and on pollution. Its levels are especially increased in some spa waters, and can reach up to 12 mg l^{-1}. Fluorine in groundwater is associated with both geological and anthropogenic factors. The excess F (up to 9000 mg kg^{-1}) is mainly reported for calcareous rocks and arid or semi-arid regions (e.g., India, Pakistan) where it is of great health concern (Jacks et al. 2005).

Fluorine in waters is either fixed, mainly by clay minerals, phosphates, and carbonates or evaporates to the atmosphere. Aerial deposition of F into water basins and surroundings is an important part of its natural cycling. Riverine flux of F from industrial pollution is in some regions of the environmental risk.

Rainwater and snow melt-water in both remote and polluted regions of the Kola Peninsula contain approximately 0.05 mg F l^{-1} (Reimann and Caritat 1998). Rainwater sampled around Italian volcanoes, during the period 1998–2001, contained F in the very broad range from 0.005 to 495.8 mg l^{-1} (Bellomo et al. 2003).

The level of F in drinking waters is of special concern since it is the main source of this element in nutrition. Its level in drinking waters of most countries varies between <0.1 and 3 mg l^{-1}. However, in some regions, mainly in dry and hot climates (i.e., Azerbejdzhan, Tanzania) its level can reach >5 mg l^{-1} (Kabata-Pendias and Pendias 1999). Enriched F concentrations in drinking water in Tanzania up to around 8 mg l^{-1} resulted in the endemicity of fluorosis (Aswathanarayana 1990).

The mandatory upper limit value for F in drinking water in the UK has been set at 1.5 mg l^{-1} (Neal et al. 2003). Fluorine has long been added to municipal drinking water at a level of 1.2 to 1.9 mg l^{-1}. Recently, in several countries, F addition to drinking water has ceased.

II-17.1.5 Air

Concentrations of F in the atmospheric air are highly variable due to differentiated sources: *(i)* evaporation from ocean and sea surface, *(ii)* volcanic eruption, and *(iii)* industrial emissions. In the gaseous phase, F occurs mainly as di-atomic particles (F_2) and hydrogen fluoride (HF). Various fluoride compounds, including cryolite, are emitted from industries.

The limit for F concentrations in air has been established in different countries from 1 to 7 µg m^{-3}. Longer influence of increased F levels in air has detrimental effect on plants, humans, and animals. Coniferous trees are especially sensitive to F injury. Thus, the admissible F concentration in the atmosphere for forested regions in Poland has been established as 0.02 µg m^{-3}, whereas for the whole country it is 2 µg m^{-3} (Gramowska and Siepak 2002).

II-17.1.6 Plants

Fluorine contents in food plants vary between 0.1 and 11 mg kg^{-1} and are significantly higher in above ground parts than in roots. Forage plants are likely to contain a bit more elevated amounts of F than other plant types (Table II-17.3). In general, tissue F concentrations in terrestrial plants from uncontaminated soils rather do not exceed 30 mg kg^{-1}. Plants can absorb F relatively easily from both soil and air and therefore its concentrations significantly increase in plants grown on contaminated sites (Table II-17.4) Plants exposed to volcano and/or industrial emissions may contain significantly elevated amounts of F. Some parts of plants, growing in acid soils can accumulate increased amounts of F, like in the case of older tea leaves, which can contain up to 300 times more of this elements than young leaves (Siemiński 2001).

Plants grown in contaminated sites may accumulate high amounts of F in above ground parts. However, a significant proportion of F remains on the surface of leaves and can be easily washed off with water (Table II-17.4). Nevertheless much F contamination can enter the food chain, and there is real concern on adequate F contents, particularly in forage crops. Opinions vary as to the toxic threshold values of F in fodder, but commonly reported ranges are fairly similar, 30–40 mg kg^{-1} and 20–50 mg kg^{-1} (Kabata-Pendias and Pendias 1999). Forage plants play the important role as a sink for emitted F and in its transfer to animals, especially to grazing animals.

Plants do not require F, and there is also no evidence of its toxicity when roots absorb this element. Several suggestions that F may stimulate some isoenzymes (e.g., acid phosphatase) have not been yet proven. The uptake of F by roots has been controlled to a great extent by soil pH. Mackowiak et al. (2003) have found that plant (rice) can easily uptake HF^0 while F^- uptake is restricted. According to these authors, F^- anions competed with HA for Ca, and soluble Ca resulted in the precipitation of CaF_2 on the root surface. Thus, increasing bioavailable Ca results in lower leaf F contents and increased root F contents.

The effects of atmospheric F on plants have been of great concern, because airborne F is considered to be one of the most phytotoxic trace pollutants. Its toxicity,

Table II-17.3. Fluorine in food and forage plants (mg kg^{-1}) (compiled from Kabata-Pendias and Pendias 1999, 2001)

Plant	Range	Mean
Wheat, grains	0.4 – 1.4	0.9
Barley, grains	0.5 – 5.5	–
Oats, grains	0.2 – 0.9	0.5
Lettuce, leaves	4.4 – 1.3	–
Potato, tubers	0.1 – 3.0	1.4
Apple, fruits	1.3 – 5.7	–
Pears, fruits	2.1 – 4.4	–
Forage legumes, tops	1.4 – 7.8	4.4
Grass, tops	3.0 – 6.8	–

Table II-17.4. Excessive levels of fluorine in plants grown in contaminated sites (mg kg^{-1}) (compiled from Kabata-Pendias and Pendias 1999, 2001, unless otherwise indicated)

Pollution source	Country	Plant, part	Maximum, range
Mine wastes	UK	Grass, tops	130 – 5450
Al-processing industry	Australia	Shrubs, leaves	150 – 500
	Czech Republik	Grass, tops	1330
	Poland	Grass, tops	75 – 340
	Norway	Birch, leaves	230
	UK	Lichens	27 – 241
Phosphate rock processing	Canada	Tree, foliage	71 – 900
	USA	Sagebrush, tops	100 – 360[a]
China clay industry	UK	Grass, tops	788; 1543[b]
Fiberglass plant	Canada	Vegetation, foliage	71 – 900

[a] AW basis.
[b] Means for grasses washed with distilled water and unwashed, respectively (Fuge and Andrews 1988).

however, is influenced by several ecological and biological factors. Plants exhibit a broad range of tolerance to foliar injury by F pollution. For example, around the Etna volcano symptoms of F toxicity were observed only in imported plants. This clearly suggests that the endemic plant species have developed resistance to increased F concentrations in growth media (Bellomo et al. 2003). A great variability in plant response to F accumulation in tissues is observed even between cultivars or genotypes of the same species. Susceptible plants can be injured by foliar F contents between 20 and 150 mg kg^{-1}, whereas highly tolerant plants do not exhibit injury at about 500 mg kg^{-1}. Among food plants, some F-tolerant species are: bean, cabbage, and carrot.

The most toxic effects of F on plant metabolism are related to:

- Decrease of oxygen uptake and respiratory disorder
- Assimilation decrease
- Reduction in chlorophyll content
- Inhibition of starch synthesis
- Inhibition of pyrophosphate function
- Altered cell membranes and metabolism of cell organelles
- Disturbance of DNA and RNA
- Synthesis of fluorooctane, a very hazardous F-compound

The reactions of plants exposed to F pollution, before any visible symptoms of F toxicity occur, are retarded growth, inhibited reproduction, and yield reduction. However, the greatest concern with increased F concentrations in plants is related to the toxicity to animals. Thus, the F content of forage crops is of a greater ecological concern than that of other plants.

Trees surrounding phosphate fertilizer factories are subjected to increased F deposition. Coniferous species differ highly in F accumulation. The total F concentrations in coniferous needles from the area around such a factory, were (in mg kg^{-1}): >200 in Scots pine, >100 in Norway spruce, and >50 in Douglas fir. These values exceeded several fold the F content of control trees that varied around 20 mg kg^{-1} (Karolewski et al. 2000).

II-17.1.7
Humans and Animals

Humans. Fluorine contents in soft mammalian tissue vary from 2 to 5 mg kg^{-1}, being a bit higher in liver and muscles (Jørgensen 2000). The total F contents in the "reference man" are established as 0.48 mg kg^{-1} in soft tissues and 250 mg kg^{-1} in the skeleton. Human fluids contain, on average, 1.2 mg l^{-1} and 0.017 mg l^{-1} in urine and milk, respectively (Li 2000).

Uptake of F by apatite of teeth and bones has been of interest for over a century. The first observations were centered on the findings that hydoxylapatite of teeth and bones incorporates F as a substitute for hydroxyl in the mineral structure. At low excess of F, mottling of tooth enamel occurs. At high levels of F excess, several symptoms of fluorosis occur, including such problems as hyperparathyroidism, calcification of soft tissues, interference with collagen formation, and severe skeletal deformity (Plumlee and Ziegler 2003).

Recently there has been a wide discussion on the health risk associated with increased intake of F. Excessive intake of F is mainly *via* ingestion, and most commonly from drinking water. The F intake of 20–70 mg d^{-1} by adults can cause heartburn symptoms due to displacement of Ca. Elevated F levels in drinking water can produce both mutagenic and carcinogenic changes in the kidneys. It has been observed that the mortality rate from cancer in the cities using fluorinated water increased significantly as compared with the cities that did not use fluorinated water (www.innvista.com/health/nutr). Increased intake of F has been noticed at the Tibetan population where fluorosis of teeth and bones due to drinking F-rich tea has been observed (Siemiński 2001). The endemity of fluorosis in Tanzania is associated with high F levels in drinking water that average 8 mg l^{-1} and caused the daily total F dose to be 33 mg (Aswathanarayana 1990).

Over the past 50 years there has been great interest in the fluoridation of water supplies and topical application of F to teeth as means of reducing dental caries (Binbin et al. 2005). In preventive dentistry, F is with no doubt, the most recommended element. Recently, in several countries, F addition to drinking water has ceased (see Sect. II-17.1.4).

Since the middle of the past century, there has been a great exposure of people to F from several sources, e.g., industrial, agricultural, domestic, and medical. Also domestic coal-burning may be a significant source of excess F. Ando et al. (*vide* Finkelman et al. 2003) estimated that 97% of exposure came from consumption of food dried over F-rich-coal stoves in some regions of China. Typical symptoms of fluorosis are mottling of tooth enamel and several skeletal and joint deformation including osteoporosis, spinal curvature and knock-knees problems. Binbin et al. (2005) have stated that effects of dental caries by high F content in drinking water (up to 3.7 mg l^{-1}) vary for different age groups, and the most dangerous is for 15- and 18-year-old groups.

Some individuals might be especially susceptible to F and its compounds. These include the elderly, persons with deficiencies of Ca, Mg, and vitamin C, as well as people with cardiovascular and kidney problems. Also in children with poor nutrition, dental and skeletal fluorosis can be enhanced.

There are several reports that excess intake of F coupled with low iodine absorption affects some nervous lesions, thyroid dysfunctions, and liver-parenchyma function (Mikhailest et al. 1996; Swarup et al. 1998; Wang et al. 2004).

Most commonly, the hyperfluoric status of populations has been observed in areas of volcanic activity, in some arid zones, and industrial regions where phosphorites and apatites are mined and processed, as well as in surroundings of the Al industry. There are several endemic fluorosis reported for China, India, Kenya, and Tanzania (Ermakov 2004).

Although no cases of F deficiency have been reported in humans, some symptoms of low F supply are suggested to be associated with dental decay, osteoporosis, and possible with growth retardation (Plumlee and Ziegler 2003). These symptoms are observed mainly in preschool children.

Animals. Alimentary F deficiency was provoked in goats by a semisynthetic diet and resulted in higher mortality than when adequate F was provided. F-deficient animals suffered from pre- and postnatal growth retardation, increased P levels in blood plasma, and reduced Ca concentrations (Anke et al. *vide* Ermakov 2004).

Symptoms of F toxicity to animals were observed centuries ago, after volcanic eruptions. Fluoride intoxication of dairy cattle resulted in a lower reproduction and lower milk production (Swarup et al. 1998). The maximum permissible levels of F in forage should be within the range between 20 and 50 mg kg^{-1} (Ermakov 2004).

II-17.2 Chlorine (Cl)

II-17.2.1 Introduction

Chlorine is a common halogen element in terrestrial and aquatic environments. Its concentrations are very variable, being the highest in mafic rocks, up 180 mg kg^{-1}, and the lowest in sandstone, up 20 mg kg^{-1}. In some calcareous rocks it can be concentrated up to around 350 mg kg^{-1} (Table II-17.2). Although Cl is not preferably accumulated by organic matter, its content, on average 41 mg kg^{-1}, can be elevated in some coals up to 8 800 mg kg^{-1} (Finkelman 1999).

In a small number of minerals Cl is present as a major constituent. Some of these are: halite, NaCl; carnallite, $KMg(H_2O)_6Cl_2$; chloroapatite, $Ca_5(PO_4)_3Cl$; and cerargyrite, AgCl. Most of these minerals are quantitatively unimportant, only halite is relatively abundant, as deposits of solid rock salt, and as a dissolved ingredient in seawater and subsurface brines.

Chloride is a lithophile, highly mobile in most conditions and may occur as a minor constituent of various minerals, like amphiboles, apatite, and micaceous and clay minerals.

This element is released during weathering, goes into solution as Cl^- ions, and accumulates in the oceans and seas.

II-17.2.2 Production and Use

Yearly global production of Cl has been estimated roughly as 40 Mt (Reimann and Caritat 1998).

Salt NaCl obtained either from solid sedimentary deposits or from seawater and brines, is the prime source of Cl. It is an important raw material and is prerequisite to the production of soda ash, which is utilized in glass and many chemical products. However, the most common use of NaCl is in the household. Also processing of several products for daily uses, especially associated with some fibers, like polyvinylchloride (PCV) involves Cl_2-based materials.

Chlorine is broadly used as ClO_2 (chlorine dioxide) utilized principally in paper pulp and cellulose bleaching. Additional uses are also in bleaching flour, fats, leather, and textiles (dye batches). In aqueous solution, ClO_2 is used as a sterilizing agent in hospitals, as bactericides, antiseptics, and deodorizers. It has been used as a strong disinfecting agent for supplied drinking water. It is also used to lower the risk of food contamination during food transport and food preparation. Chlorine effectively controls some bacteria, algae and fungi that can grow in vessels used in food processing and storage.

The proportion of Cl used in the USA, in 2003 (www.saltinstitute.org/16html), was as follows:

- Highway salt: 68%
- Water conditioning: 12%
- Chemical salt: 8%
- Agricultural salt: 6%
- Food grade salt: 6%

Chlorine is a component of several persistent pesticides (fungicides), and of other chlorinated organic compounds, like dioxins/furans and PCBs (polychlorinated biphenyls). Some of these compounds are industrial by-products and unintended by-products (e.g., waste material combustion, components of sewage sludge). The increased use of Cl-pesticides and growing release of polychlorinated hydrocarbons are presently a great threat to the environment and health. All these compounds seem to travel over long distances, concentrated in northern areas and are likely to be accumulated in the fat of mammals. PCBs and other polychlorinated compounds (dioxin-related compounds) have especially bioaccumulated in Arctic people and wildlife and are of environmental concern.

The observed dioxin load of the soil, water and foodstuffs indicates the capacity of polychlorinated organic compounds to penetrate the environment and to infiltrate the food chain (Coenen 2004).

II-17.2.3 Soils

There are not many data on Cl in non-saline soils, and its range has been estimated roughly at around 100 to 1 000 mg kg^{-1} (Table II-17.2). Like other halogens, contents of Cl in soils decrease with increasing distance from the sea coast. This is well illustrated by Yuita (1983) who found Cl in forest soils of the coastal plain in Japan to range from 91 to 486 mg kg^{-1}, with an average of 228, and in soils from upland fields, to range from 56 to 305 mg kg^{-1}, with an average of 114 mg kg^{-1}.

In soils of humid climate zones, Cl is easily leached down the profile and transported with drainage waters, whereas in soils of arid and semiarid climates Cl is concentrated in surface layers. Highly saline soils, solonetzs and solonchaks, are highly enriched in NaCl, as well as in other salts inherited from parent material. Contents of Cl in such soils usually vary from 0.0X to 0.X%, and in soil solution the Cl contents may reach the level of X.0%.

Excessive amounts of salt (NaCl) applied to soil are of environmental concern. Salt-affected soils due to anthropogenic activities result in most cases, from the applications of de-icing salt to roadways, field irrigation with highly mineralized waters, salt water spilled with the extraction of oil, natural gas deposits, and some coals. Wind erosion of salt evaporates also may affect the Cl enrichment in soils. High loadings of Cl in soils have deleterious effects on both soil properties and vegetation.

Elevated levels of Cl^- in soil solution influence the behavior of several cations. Increased solubility of Zn and Hg compounds has been already reported. However, the greatest environmental concern is associated with Cl affinity for forming soluble complexes with Cd (Kabata-Pendias and Pendias 2001).

II-17.2.4
Waters

Chlorine is the most abundant of the dissolved constituent of seawaters, making up around 55 wt% of the dissolved solids. Its median concentration in ocean waters is estimated as 19 400 mg l^{-1}, whereas in stream waters its mean concentrations vary from 1.4 to 5.3 mg l^{-1} (Reimann and Caritat 1998).

Easily soluble and mobile Cl is transported to oceans, seas, and closed water basins. Predominating in those waters, the Cl^- anion combined with Na^+ cation has resulted in the high salinity, which averages in seawaters around 2%, but in some salty lakes can exceed 20%.

The concentrations of Cl in water of the Odra River (below Wroclaw) have ranged from 35 to 40 mg l^{-1} (Kabata-Pendias and Pendias 1999). Its concentration in landfill leachate can reach 4 500 mg l^{-1} (Bjerg et al. 2003).

Precipitation collected over the Kola Peninsula contain Cl between <0.1 and 2.1 mg l^{-1} (Reimann and Caritat 1998), whereas Cl in precipitation over the UK are reported to contain on average 3 mg Cl l^{-1} (Fuge 1988).

Previously Cl has been used for the disinfection of drinking water and thus, its concentration could be significantly elevated. Chlorination is essential for safe drinking water, however, the Cl concentration in water should be controlled.

II-17.2.5
Air

Concentrations of Cl in the atmosphere vary highly, from 0.003 µg m^{-3} above the South Pole to around 10 µg m^{-3} in some urban areas (Table II-17.2). The highest Cl concentrations, up 24 µg m^{-3}, have been found in air of industrialized regions of the Kola

Penisula (Reimann and Caritat 1998). Above seas and oceans, Cl in air is increased due to evaporation. Volcanic exhalations are also sources of Cl, often in the form of HCl.

Reference limits for chronic exposure is proposed in the range of 0.2–15 µg m^{-3}, but also much higher concentrations are accepted for limited occupational exposure (Coenen 2004).

II-17.2.6 Plants

Contents of Cl in plants vary highly depending on plant types and growth media. In cereal grains Cl contents usually vary between 10 and 20 mg kg^{-1}. Sugar beet leaves contain this element within the range from 100 to 200 mg kg^{-1}, and potato tubers from 1 300 to 5 500 mg kg^{-1}. Plants grown close to sea coasts usually contain considerably higher amounts of Cl than plants from inland regions. For example, coconut palm leaves growing at the distance 5 km from the seacoast contain up to 7 000 mg Cl kg^{-1} (Bergmann *vide* Kabata-Pendias and Pendias 2001).

Chorine is passively absorbed by roots and is readily transported to above-ground parts of plants. It can also be absorbed directly by plant leaves. The availability of Cl to plants is related to the associated cations. The Cl^- ions are more easily absorbed in the presence of monovalent cations than di- and trivalent cations. Studies on radio-chlorine (^{36}Cl) uptake (which mimics the uptake of stable Cl) by plants have shown that more than half of this radionuclide passed from the arable layer of soil into radish, lettuce and aboveground parts of wheat during one vegetation period (Kashparov et al. 2005b).

Chlorine is considered a micronutrient for higher plants, but the requirement for metabolic processes is at relatively low levels. The significant concentration of Cl in chloroplasts suggests its role in photosynthesis. However, the Cl function in photosynthesis, as well as its suggested role in phosphorolysis and the activity of cytochromdase, remains in doubt (Wild and Jones 1988). This element plays an important role in the osmotic system of plants.

In some plants grown in soils poor in Cl (<2 mg kg^{-1}) symptoms of its deficiency may be observed. These are mainly chlorotic leaves and brown leaf edges. In most soils, amounts of Cl present meet plant requirements. Increased Cl levels in soils, up to 15 mg kg^{-1}, may be beneficial to some plants (e.g., tobacco).

Toxic Cl levels in plants vary greatly. Increased Cl in soil solution and in water culture may be especially harmful to plants. Urban trees are especially exposed to increased levels of Cl, and some, particularly oaks, have developed a resistance to its high contents in growth media. The accumulation of Cl in oak leaves collected along streets ranged from 490 to 790 mg kg^{-1}, whereas in plane-tree leaves it ranged up to 7 425 mg kg^{-1}. Symptoms of abiotic damages of leaves have occurred at Cl concentrations of 33 400 mg kg^{-1} (Dmuchowski 2003). Trees may secrete an excess of Cl, most commonly *via* leaves and their fall in autumn.

A common threshold concentration of Cl in soils has been estimated as 2 700 mg kg^{-1}. Less tolerant plants (e.g., beans, apple trees) are inhibited by Cl concentrations in soil or nutrient solution from 460 to 673 mg l^{-1}, whereas resistant plants (e.g., tobacco, tomato, spinach) can tolerate higher Cl concentrations, from 887 to 3 546 mg l^{-1} (Bergmann *vide* Kabata-Pendias and Pendias 2001). Increased levels of Cl may be associated with some fertilization and can adversely affect crops, especially potato tuber quality, partly due to easy formation of mobile complexes with Cd.

II-17.2.7 Humans and Animals

Humans. The Cl contents in soft mammalian tissues vary from 4 800 to 11 000 mg kg^{-1}, being the lowest in liver and the highest in skin (Jørgensen 2000). The total Cl content in the "reference man" is established as 1 400 mg kg^{-1} and in fluids as 3 100 and 400 mg l^{-1} in urine and milk, respectively (Li 2000).

Humans require Cl and uptake it in the form of salt. Thus, salt is used for nutrition. As Coenen (2004) has emphasized, there is a difference between the physiological role of Cl^- and Cl_2. Anionic Cl species plays a significant role in the cation-anion balance of organisms and is involved in the osmotic regulation at both cell and body levels.

Its intake by people is highly related to food and water quality. The Cl concentration in drinking water is regulated, whereas in food products it can differ widely. The lowest Cl contents are in cereal grains, 0.1–2 g kg^{-1}, and the highest are in seafoods, 14.6–28 g kg^{-1} (Table II-17.5).

Recently there has been great concern on eating too much salt that became to be sometimes called the "silent killer". Too much salt in food can raise blood pressure, cause stroke and other heart diseases, and sometimes cause overweight. There is the recommendation that adults should have less than 6 g d^{-1}, which is about a teaspoonful. However, it is necessary to consider that 75% of the salt that people eat is already in everyday foods. The daily limit of salt in the diet should be controlled through cooking without salt and eating healthy food (Food Standards Agency: www.salt.gov.uk/your_daily_limit.html).

The main routes of Cl secretion by humans are saliva, sweat, and hydrochloric acid in the stomach. The acid environment of the stomach is prerequisite for controlling digestive processes and the microbial community.

Table II-17.5. Chlorine in some foodstuffs (g kg^{-1}) (compiled from Coenen 2004)

Food	Mean, range
Cereals, grain	0.1 – 2
Fruits	0.07– 1.4
Milk	8 – 9
Meat	1.6 – 6.8
Bone meal	6.4
Fish	1.3 – 9.1
Shrimp	14.6
Mussel	28

The safety margins for the Cl_2 and its compounds are strictly regulated – the concentration of Cl_2 in drinking water is 1.2–4 mg l^{-1} and in swimming pool water 0.3–1 mg l^{-1} (Coenen 2004).

Chlorine dioxide (ClO_2) is toxic by inhalation and administration in solution by oral routes. However, when it was used for tap water treatment (between 1940 and 1955 in the USA) no significant effects on humans were observed (Dobson and Cary 2001).

The limited occupational exposure concentration of ClO_2 in air has been established in the UK as 0.28 mg m^{-3}.

Animals. Animals require Cl and uptake it in a form of salt. Thus, salt is used for the feeding. The intake of Cl by domestic animals differs greatly, from 78 to 290 mg kg^{-1} BW, being the lowest for hen and the highest for cattle (Coenen 2004).

II-17.3 Bromine (Br)

II-17.3.1 Introduction

The Br abundance in crustal rocks is about 2 mg kg^{-1} and it is relatively uniformly distributed among igneous rocks, whereas in argillaceous sediments it is likely to be concentrated up to 10 mg kg^{-1} (Table II-17.2). Minerals containing Br are not very common, due to its low abundance and high solubility. The common minerals are: bromyrite, AgBr; embolite, Ag(Cl,Br); and iodobromite, Ag(Cl,Br,I). The close correlations of Br and Cl in various rocks have often been observed. Bromine is a very volatile element and its salts are readily soluble and mobile in most environments. However, it is likely to be accumulated by coals and organic matter. Its concentrations in the US coal average 9.1 mg kg^{-1}, but can reach 160 mg kg^{-1} (Finkelman 1999).

Bromine is very reactive chemically and can have several valence states (Table II-17.1). However, as only one-electron atom it can form compounds mainly with oxygen (e.g., Br_2O, BrO_2, BrO_3^-).

II-17.3.2 Production and Use

Global annual production of Br in 1995 was estimated as 4 322 kt (Reimann and Caritat 1998). The major commercial source of Br is the sea, especially from K-salt water, although some amounts are also obtained from brine deposits.

The primary use of Br was as ethylene dibromide, a gasoline anti-knock additive, together with Pb. Due to recent environmental legislation, Br addition to gasoline has been significantly reduced.

Considerable amounts of Br (methyl bromide and ethylene dibromide) are used in agriculture as pesticides (fungicide, herbicide, and insecticide). It has been utilized in the pharmaceutical industry and also has been used in fire-retardant compounds and for dyestuff production. In photosensitive compounds, Br has been used for many years in the photographic industry.

New persistent chemicals, polybrominated diphenol ethers (PBDEs) are presently widely used as flame-retardants in a great variety of products.

All bromine-based compounds are very persistent and seem to travel over long distances in both the atmosphere and water.

II-17.3.3 Soils

There are only a few reports on the Br status of soils indicating its common range between 5 and 40 mg kg^{-1} (Table II-17.2). The lowest range of Br, from 7.9 to 8.2 mg kg^{-1}, was found in sandy podzols in Poland, and the highest, from 50–104, in volcanic ash soils in Japan. The reference soils from China contain Br in the range from 2.6 to 7.2 mg kg^{-1}, and soils from the USA, between 1.4 and 7.8 mg kg^{-1} (Govindaraju 1994). The average Br concentration in soils of Finland is 10 mg kg^{-1} at the 90th percentile range (Köljonen 1992). A somewhat similar value of the arithmetic mean, 5.7 mg kg^{-1}, has been estimated for surface soils of Austria (Gerzabek et al. 1999).

Observed increased Br contents in soils close to sea costs is an effect of Br evaporation from the sea. Yuita (1983) reported that soils at the Sea of Japan accumulated up to 495 mg Br kg^{-1}. Soils derived from volcanic ash contain increased amounts of Br due to both elevated Br contents in parent material and the impact of volcanic exhalation.

Anthropogenic sources of Br in soils have been associated with the impact of car exhaust, (when methylene dibromide was added to gasoline) and with a broad use of Br-pesticides. Due to these practices, soil Br could increase significantly, and particularly in soil solution since these Br compounds can be easily transformed to the very mobile Br^- anionic forms.

Soil fumigation with methyl bromine can result in almost complete eradication of the population of a wide variety of soil microflora and fauna and thus alter the trophic structure of the soils (Pavelka 2004).

II-17.3.4 Waters

The median Br concentration in world ocean water is estimated as 67.7 mg l^{-1}, whereas its contents in stream waters vary broadly between 0.002 and 4.47 mg l^{-1} (Reimann and Caritat 1998). In brine and salt lake waters, the Br concentration can be very high, e.g., in Dead Sea 0.4–0.6%. According to GERM data (www.earthref.org) the mean Br concentration in seawater is 65 mg l^{-1} and in the North Pacific it averages 67 mg kg^{-1} (Nozaki 2005).

Precipitation collected over the Kola Peninsula contain Br at <0.2 mg l^{-1} (Reimann and Caritat 1998), whereas its concentration in precipitation over the UK is reported to be 15 mg l^{-1} (Fuge 1988).

II-17.3.5 Air

The Br concentration in air above the South Pole is around 1.4 ng m^{-3}, and over Greenland and Shetland Islands its content ranges between 7 and 20 ng m^{-3} (Table II-17.2).

Bromine concentrations in air over urban, and particularly over industrialized regions might increase up to around 2 500 ng m^{-3}, but most commonly it ranges between 150 and 500 ng m^{-3} (Kabata-Pendias and Pendias 1999). Automobile exhaust and decomposition of Br-pesticides might also be significant sources of Br in the atmosphere.

Most common forms of Br in the atmospheric air are: Br_2O, BrCl, $BrONO_2$, and also methylated compounds (e.g., CH_3Br).

II-17.3.6 Plants

The natural Br contents of plants seem not to exceed about 50 mg kg^{-1}, and higher values are apparently related to pollution (Table II-17.6). It is likely to be more concentrated in leaves than in roots. There is evidence that Br in herbage does not correlate with any soil parameters, such as soil Br, texture, pH, and Eh (Wilkins *vide* Kabata-Pendias and Pendias 2001). Plants are known to contain more Br when grown in soils enriched in this element, but the Br-transfer mechanism from soil to plants has not yet been described.

Contents of Br in food plants vary from 2 to 26 mg kg^{-1}, being lower in cereal grains. Relatively high Br levels, up to 22, 26 and 36 mg kg^{-1} have been found in lettuce leaves, radish roots, and mushrooms, respectively. The highest amounts of Br are reported in grass tops (Table II-17.6). Recently reported by Pavelka (2004), the highest Br concentrations (in mg kg^{-1}) are as follows, in: wheat grain, up to 70; almonds, 100; cacao beans, 48; tobacco, 286; and mushrooms, 300.

Wyttenbach et al. (1997) found a high concentration of Br in Norway spruce needles at polluted and maritime sites, and very low at continental prealpine sites. The mean

Table II-17.6. Bromine in food and forage plants (mg kg^{-1}) (compiled from Kabata-Pendias and Pendias 1999, 2001)

Plant	Range	Mean
Barley, grains	2.1 – 6.4[a]	5.5
Oats, grains	–	3.1
Beans, seeds	–	15
Lettuce, leaves	20 – 22	–
Cucumber, fruits	10 – 20	–
Potato, tubers	4.2 – 14.2	–
Radish, roots	24 – 26	–
Selery, stalks	–	17
Forage legumes, tops	2.1 – 52	–
Grass, tops	3 – 119	–
Edible mushrooms[b]	2 – 36	–

[a] For inland and coastal regions, respectively.
[b] Kind not specified.

Br value in spruce needles from remote continental sites was 0.1 mg kg^{-1}. The highest Br content, up to 2 000 mg kg^{-1}, was found in plants grown in volcanic ash soil.

Bromine from Br-pesticide also has a significant impact on the elevated Br levels in both soils and plants. Increased Br contents as a result of using Br compounds in the plant cultivation, particularly in leaf vegetables, have often been observed. Lettuce leaves that were cultivated in a greenhouse, after fumigation with CH_3Br, contained up to 9 515 mg Br kg^{-1} (Kabata-Pendias and Pendias 2001).

Plants differ in their tolerance to soil Br. Some vegetables and flowers are known to be sensitive to high Br levels, they are: potato, spinach, sugar beet, onion, carnation, and chrysanthemum. Resistant to Br toxicity are: carrot, tobacco, tomato, celery, and melon. These plants can accumulate Br to over 2 000 mg kg^{-1} without showing any effects.

Bromine can substitute for Cl in plants. Symptoms of Br toxicity are variable and resemble excess salt effects, so these are most commonly chlorosis followed by leaf tip necrosis, and dark green color of young leaves. The reduction in growth of citrus seedlings is correlated positively with water-soluble soil Br and might be used as phytoindicator of Br toxicity.

Marine plants contain high levels of Br, as is reported for algae, on average, 266, 606, and 3 293 mg kg^{-1}, in green, brown, and red algae, respectively (Fuge and Johnson 1986). In some halophytic algae, Br can replace Cl in several processes. Marine sponges incorporate Br into organic compounds, however, in most plants only the Cl ion is involved in metabolism.

II-17.3.7
Humans

The Br contents in soft mammalian tissues vary from 4 800 to 11 000 mg kg^{-1}, being the lowest in liver and the highest in skin (Jørgensen 2000). The total Br content in the "reference man" is established as 2.9 mg kg^{-1} and in urine as 2.8 (Li 2000).

After oral ingestion, Br is rapidly absorbed and distributed in the extracellular fluids and no particular accumulation in any organ has been noted. It is relatively easily excreted from the body. Average Br intake by adults is estimated at around 8 μg d^{-1}. An antagonistic interaction between Br and I intake has been observed. With a deficiency of I, Br may concentrate in the thyroid gland.

There have been many reports of cases of methyl bromide (CH_3Br) poisoning in humans, involving skin and eye injuries, gastric irritation, kidney lesions, damage to the central nervous system and toxic effects to immune system. Symptoms of chronic Br intoxication (brominism) are relatively well known (Pavelka 2004). Excess inhalation of Br triggers allergic reactions, especially of mucous membranes.

The ADI value for Br intake has been established as 1 mg kg^{-1} BW, however, a non-effect level of 4 mg kg^{-1} BW has been recently proposed (Van Gelderen et al. *vide* Pavelka 2004).

Although Anke et al. (*vide* Pawelka 2004) reported reduced growth, lower conception rate, and decreased milk and fat production by goats feeding with low Br fodder, the essentiality of Br has not yet been proven. Deficiency symptoms of Br have not been observed, neither in humans nor in animals.

I

II-17.4
Iodine (I)

II-17.4.1
Introduction

The I content in crustal material is extremely variable, and it has been estimated between 0.15 and 1.4 mg kg^{-1} in the upper continental crust. Its highest content is in shales (mean 1.5 mg kg^{-1}) and limestones (mean 1 mg kg^{-1}), and the lowest in acid igneous rocks (mean 0.17 mg kg^{-1}). Also mafic rocks contain relatively small amounts of I, in the range of 0.007–0.5 mg kg^{-1} (Table II-17.1). High average I contents of 30 mg kg^{-1} in deep-sea carbonates and of 2.5 mg kg^{-1} in other sediments are reported (Muramatsu and Wedepohl 1998).

Iodine does not form many of its own minerals, but can occur as minor constituent of various minerals. Iodine association, like CuI and AgI, can be found in weathering zones of these (Ag and Cu) metal deposits. Iodates are rare minerals occurring only in very arid regions, and their structures are not completely understood. Anhydrous iodates are: lautarite, $Ca(IO_3)_2$ and bellingerit, $Cu_3(IO_3)_6 \cdot 2H_2O$; and iodate with hydroxyl, salesite, $Cu(IO_3OH)$. Other minerals such as iodoargirite, AgI; marshite, CuI; and dietzete, $Ca(IO_3)_2CrO_4$ are much less common.

Increased I contents in some phosphates, bituminous shales, and coal are associated with organic matter and/or organic carbon. Iodine is concentrated in some nitrate deposits, especially in Chile saltpeter, up to around 400 mg kg^{-1}. The suggestion of an atmospheric origin of this element seems to be most reasonable. Iodine, when heated, sublimates and goes from a solid directly to a vapor phase, without the liquid phase. During weathering, I is rapidly released and transported to ocean and sea basins. Most of I, nearly 70%, exists in ocean sediments.

Among many I isotopes, the only stable isotope in nature is ^{127}I. Two radioactive isotopes, ^{129}I and ^{131}I, are most commonly byproducts of atomic reactors and have been released in various proportion into the environment. The half-lives of these isotopes vary from 8 days (^{131}I) to 1.6×10^7 years (^{129}I), and both are of growing environmental and health concern. However, ^{129}I is also naturally occurring long- half-life radioisotope.

II-17.4.2
Production and Use

Annual global production of I in 1995 was around 14 kt and in 2000 was 18 kt (Reimann and Caritat 1998; WMSY 2004). It is primarily obtained from brines that are associated with gas and oil deposits. Also caliche (surface layers of some calcareous deposits) is a source of this element. Previously I was a byproduct during Chile's production from these deposits. Until 1959, ash of seaweeds was very important source of I. Now, the most economic I production is from caliche deposits.

Iodine is utilized in a number of chemicals and pharmaceutics. Iodine and its organic compounds are used both externally and internally in medicine. Radioactive ^{131}I is used in medical diagnosis.

KI-water solution and I-alcohol solution are strong disinfectants and used for external wounds. It is added (mainly as KI) to table salt to assure adequate iodine level in the human diet, and is used as supplement in animal feeds.

In chemistry, I compounds are used as catalyst. It is added to colorants and inks, and as AgI, a photosensitive compound, is used in photography.

The $^{129}I/^{127}I$ ratio is used as hydrological, geochemical and environmental tracers.

II-17.4.3 Soils

The grand mean of I in surface soils has been estimated as 2.8 mg kg^{-1}, whereas its concentrations range from <0.1 to 10 mg kg^{-1} (Table II-17.2). However, in certain soils of islands (Ireland, Japan, New Zealand) as well as in sea-coast areas, a higher I accumulation, up to around 150 mg kg^{-1} has been reported. Also volcanic ash soils usually contain elevated amounts of I, up to 104 mg kg^{-1}, as well as solonchak soils of arid and semiarid regions, up to 340 mg kg^{-1} (Kabata-Pendias and Pendias 2001).

Iodine ranges in the reference soils of China from 1.3 to 19.7 mg kg^{-1}, and in the USA soils, from 1 to 35 mg kg^{-1} (Govindaraju 1994). In Russian chernozems, I varies from 4 to 5.6 mg kg^{-1}, and is accumulated in the top layer (Protasova and Kopayeva 1985).

Most commonly, soils contain several times as much I as do the parent rocks. Its accumulation in surface and subsurface soil horizons is affected by aerial deposition and biogenic concentration. The I abundance in soils shows a relation with their texture, but as Gerzabek et al. (1999) have described, the I is positively correlated only with the clay content of noncalcareous soils, whereas in calcareous soils this relationship was observed for several soils parameters, e.g., clay, exchangeable calcium and organic carbon. Usually, light soils of humid climate regions are poor in I, whereas high humus and clayed soils are enriched in I. However, this may vary greatly because the I level in soils is highly dependent on the atmospheric precipitation. The distance from the sea and recent glaciations also has an influence on the soil I status. Soils close to coastal districts are enriched in I, whereas soils derived from recent glacial (Pleistocene) deposits are usually poor in this element.

Soils surrounding fossil fuel combustion plants and, previously common, kelp burning facilities, as well as in the close vicinity to high traffic roads have elevated I levels. Some sewage sludges applied on fields can add I to surface soil.

In general, the influence of soil parameters on I status is very diverse. Soil acidity favors I sorption by various soils components, including organic matter. On the other hand, in alkali and salt-affected soils, the I mobilization also is low. The amount and kind of organic matter, as well as fixation by microorganisms, play an important role in the increased sorption of this element. Dai et al. (2004) reported that iodate (IO_3^-) adsorption is negatively correlated with soil organic matter and positively with free Fe oxides, whereas no correlation exists between iodate adsorption and CEC and soil pH.

The alteration of I species, especially the oxidation of iodide to iodate occurs in soils. Also the exchange of volatile iodine compounds between soils and the atmosphere is relatively common process. The I species occurring in the aquatic phase of soil are mainly the following anions: I^-, I_3^-, IO_3^-, and $H_4IO_6^-$, of which the first two are most common (Kabata-Pendias and Sadurski 2004).

Most I occurs in fixed forms being sorbed by humic and fresh organic matter as well as fixed in or on clay and crystal lattice of minerals. A small fraction of the total I is mobile and thus easily available to plants. Cold and hot-water extractable I, estimated as phytoavailable ranges from <1 to 25% of its total content, depending on variable soil properties (e.g., moisture, pH, amount and kind of clay minerals and SOM). The absorption rate differs for I species and is higher for I^- than for IO_3^- (Yuita 1983). Although I is scarcely mobile under most soil conditions, in some extreme conditions, (e.g., submerged soil), it can be highly solubilized, and thus available to plants.

Recently, great attention has been focused on the distribution and fate of radioisotopes in soil. Iodine isotopes, ^{129}I and ^{131}I, released to the environment by nuclear bomb testing and nuclear accidents have been deposited on surface soils very irregularly, in a mosaic-like pattern. The occurrence in soils of these two radionuclides has become of great environmental concern, especially after the 1986 Chernobyl reactor accident. Soil processes that affect in immobilization of I nuclides are highly variable between soils and within the large body of soil. Although I isotopes added to soils are readily sorbed in the soil solid phase, they are desorbed into soil solution and migrate relatively fast down the soil profile (Muramatsu and Wedepohl 1998; Muramatsu and Yoshida 1998; Schimmack et al. 1989). The depletion of ^{125}I added to arable soil during one year after the contamination due to the volatilization was observed, whereas its vertical migration was rather slow (Kashparov et al. 2005a).

II-17.4.4 Waters

In the North Pacific Ocean, the mean concentration of I is 58 $\mu g\,kg^{-1}$ (Nozaki 2005). Other sources give fairly similar mean I concentrations in seawaters: 50 and 58 $\mu g\,l^{-1}$ (Fuge and Johnson 1986; Mineral Information Institute, www.mii.org/Minerals/photoiodine.html).

Iodate (IO_3^-) is dominate species of I in seawater, up to around 90% of its total concentration. Also iodide (I^-) and organically bound iodine are present in waters. However, some I species are very unstable and soon transfer to iodate in alkali media and to iodide in acid media. The photochemical oxidation of iodide ions to iodate in surface seawater is a significant source of this element in the atmosphere. Also the biological formation of methyl iodide that takes place at the sea-surface is the important source of I in the marine atmosphere (Fuge and Johnson 1986).

In river water, average I concentrations vary between 2.3 and 15.1 $\mu g\,l^{-1}$. Its maximum concentrations in stream water range from 110 to 198 $\mu g\,l^{-1}$ (Reimann and Caritat 1998). In some lake-water, e.g., in Central Kazakhstan, its concentration can reach 160 $\mu g\,l^{-1}$ (Fuge and Johnson 1986).

Waters of Swedish rivers show relatively high concentration of ^{129}I (up to 1.4×10^9 atoms l^{-1}) of which the main source is atmospheric deposition of this isotope, which apparently originated from two nuclear reprocessing facilities in France and England (Kekli et al. 2003).

II-17.4.5
Air

Median I concentrations in world air vary in remote regions from 0.08 to 4.4 ng m^{-3} (median 1.1) and in polluted regions from 100 to 350 ng m^{-3}. The I concentration in air above the South Pole is around 0.08 ng m^{-3}, and over Shetland Islands its content increases to 4 ng m^{-3} (Table II-17.2).

Iodine concentrations in air over urban, and particularly over industrialized regions, is highly variable, from 15 to 6 000 ng m^{-3} (Kabata-Pendias and Pendias 1999). According to data cited by these authors, air above seas and oceans contains increased amounts of I (mean values in ng m^{-3}): 1 090 at the Baltic Sea (Gdańsk); 200 at the Japanese Sea; and up to 48 000 above the Black Sea. Yoshida and Muramatsu (1995) reported much lower I concentrations, in the mean range from 14.1 to 17.5 ng m^{-3} in the coastal atmosphere of Japan. Organic I was the dominant species in air, in the range between 77 and 88% of gaseous phase. The most probable organic chemical species is methyl iodine, which is known to be decomposed to inorganic I within a few days.

Elemental I vapor in the atmosphere is readily sorbed onto most surfaces. However, the emission of I from the surface of seawater, as well as its volatilization from soils and from volcanoes gases control the balanced concentration of I in the atmosphere. Anthropogenic sources, especially from the combustion of fossil fuels, also contribute to I levels in air around industrial and urban regions. The wet and dry precipitation of I in areas under marine influence, can be relative high, and as Fuge and Jonson (1986) estimated can reach 7 mg I m^{-2} yr^{-1}.

Iodine radionuclides, ^{129}I and ^{131}I are present in the atmosphere principally as a result of nuclear weapons testing and nuclear reactor accidents. Both radionuclides can be transported by air for very long distances.

II-17.4.6
Plants

There is a great variation observed in the I contents of plants, and it seems to be a function of both plant species and growth conditions. Generally, vegetables and mushrooms contain less I than forage plants. It might be stated that dicotyledonous plants (herbs, legumes) contain more I than monocotyledonous grasses and cereals (Table II-17.7). The common range of I in cereal grains is between 60 and 100 µg kg^{-1}, whereas in legume fodder plants it varies from around 100 to above 200 µg kg^{-1}.

Besides the soil I status and plant characteristic, distance from the seacoast has crucial influence on the I content in plants. Anke (2004d) presented data for I contents in plants as a function of distance from the seaside; I concentration in field red clover decreased from 272 to 132 µg kg^{-1} with growing distance from 10–50 to 201–310 m, respectively. A similar relationship was observed for other crop plants. There is clear evidence that plants are capable of absorbing I directly from the atmosphere, both through the cuticle and as adhesive particles on the surface, particularly of hairy leaves. Thus, the atmospheric I can contribute significantly to its concentration in plants. Mosses from Scandinavian countries are reported to contain I in the range of 1.5–3.5 mg kg^{-1} (Rühling et al. 1987).

Table II-17.7.
Iodine in food and forage plants from Germany (μg kg^{-1}) (compiled from Anke 2004d)

Plant	Range
Barley, grains	62 – 95
Wheat, grains	69 – 98
Potato, tubers	85 – 110
Beet, tops	240 – 446
Grass, tops	59 – 104
Red clover	113 – 133
White clover	215 – 238
Lucerne	158 – 197
Fodder beets	221 – 319
Green maize	261 – 346

Table II-17.8.
Iodine concentrations in some marine plants and animals (mg kg^{-1}) (after Fuge and Johnson 1986, unless otherwise indicated)

Sample	Range	Mean
Algae		
Brown	55 – 8 800	2 489
Red	200 – 565	383
Green	21 – 130	58
Fish and molluscs		
Herring	–	2.2
Mackerel	1.11 – 1.41	1.28
Mackerel fillet[a]	–	2.07
Trout[a]	–	0.40
Oysters	1.00 – 4.48	2.4
Cockle shell	–	1
Turritella shell	–	6

[a] After Anke (2004d).

Marine plants contain much more I than terrestrial plants. Its high concentration in algae seems to be associated with the pigmentation; for example, algae contain, on average (in mg kg^{-1}): green 58, red 383, and brown 2 489 (Table II-17.8).

Iodine has not been shown to be essential to plants, and reports on stimulating effects on plant growth at low concentrations in nutrient solution have not been explained. Plants take up I relatively easily from the soil solution and as reported by Muramatsu et al. (1983) iodide (I^-) was much more available species than iodate (IO^-). Both I species were more concentrated in roots than in shoots which could be due to the absorption of the element onto the root surface. Organically bound I seems to be scarcely available to plants, but after the decomposition of organic matter becomes more phytoavailable. The dominant I species in the plant juice is apparently iodide, and about 65% of total iodine in plants is bound to proteins.

The I toxicity to plants, due to pollution, has not often been reported. However, there have been some toxicity symptoms observed, which were similar to those caused by excess Br (margin chlorosis in older leaves and a dark green color of young leaves) in plants cultivated on some coastal fields where an excess of kelp (calcined ashes of seaweed) was applied as fertilizer.

II-17.4.7 Humans and Animals

Humans. The I contents in soft mammalian tissues vary highly from 0.0015 mg kg^{-1} in liver to 4 mg kg^{-1} in brain (Jørgensen 2000). The total I content in the "reference man" is established as 0.19 mg kg^{-1} and in fluids as 0.12 and 0.056 mg l^{-1} in urine and milk, respectively (Li 2000).

Iodine is one of the very essential micronutrients and the human body could not do without this element. It is concentrated mainly in thyroxin and triiothyronine, the hormones produced by the thyroid. These hormones contain up to 80% of the total body store of I and are involved in most biological processes, from bone growth to reproduction. Thus, an adequate level of I in the body is very crucial.

The study of Iodine Deficiency Disorders (IDD) is included in the program of the International Council for the Control of Iodine Deficiency Disorders (ICCIDD). There is evidence that effects of I deficiency is worsened if Se status is low (WHO/IPCS 2002). Previously, I deficiency was known under the term "goiter", that comprises the spectrum of dysfunctions, such as:

- Fetus anomalies, including abortion and stillbirth
- Impaired mental function and retarded physical development, commonly associated with congenital anomalies of children and adolescents
- Hypothyroidims with various complications in adults, such as neurological cretinism, mongolism, impotence, and infertility
- Additional symptoms might be: body weakness, dry skin, and sensitivity to cold

Increase intake of I resulted in incidence of hyperthyroidism, which has been observed following iodized salt programs, as well as after nuclear accidents, especially that in Chernobyl in 1986, when a supplementation of KI (potassium iodide) to human diets was administrated in affected regions, in order to block ^{131}I uptake into the thyroid. Some symptoms of I excess are: sensitivity to touch, psychological instability, weight losses, tachycardia, and exophthalmos. Iodine allergies are relatively common.

The daily requirement of I by adults is around 150–200 µg. Intake below 100 µg d^{-1} has resulted in mild deficiency, and a dose below 20 µg d^{-1} has caused severe deficiency symptoms. The mean intake of I by German adults has been estimated to vary (in µg d^{-1}) between 47 and 83 for women and between 35 and 59 for man (Anke 2004d).

This author has reported that the normative I requirement of adults is 1 μg kg^{-1} BW, whereas the WHO/IPCS (2002) recommendation for the I intake is 2 μg kg^{-1} BW. Under certain circumstances, however, the I intake by adults at the level above 30 000 μg d^{-1} may not affect thyroid function states (Pais and Jones 1997). In non-iodine deficient regions, diets enriched in meat, and especially in seafood, meet human requirements for this element (Tables II-17.7 to II-17.9). Thus, the richest natural sources of I are seafood and seaweeds.

The largest areas of IDD are in Africa, and it has been calculated that around 150 million people out of the total population are at risk of I deficiency (Aswathanarayana 1990). It has been proposed iodinating of piped water as the easiest and cheapest prevention action.

Iodine is easily absorbed in the organism, quickly distributed and accumulated in the thyroid (65–90% of administrated I is stored in the thyroid) and rapidly excreted through urine (and milk). The level of the $^{129}I/^{127}I$ in urine is accurate index for recent exposure to the radioisotope (Hou et al. 2003).

The strong interaction between I and Se has resulted in the close positive relation between the availability and metabolism of both elements (Van der Heide et al. 1993). Other elements, such as As, F, Co, Cu, Ca, Hg, and Mn, involved in thyroid metabolic processes often inhibit the I bioavailability.

Animals. The requirement for I among animals is between 0.1 and 0.5 mg kg^{-1} BW, and its deficiency is much more common than of other elements. Sheep and cattle seem to be most sensitive to I deficiency and develop similar symptoms.

In young animals I deficiency results in:

- General weakness
- Hairless, early and/or dead born
- Retarded fetal brain development
- Swelling of the thyroid gland, goiter

Table II-17.9. Iodine in some foodstuffs and beverages not supplemented with iodine (compiled from Anke 2004d)

Plant foodstuff	Mean (μg kg^{-1})	Animal foodstuff	Mean (μg kg^{-1})	Beverages	Mean (μg l^{-1})
Wheat and rye bread	24	Cows' milk	195	White wine	11
Pearl barley	42	Beef	187	Red wine	12
Rice	21	Pork	159	Beer	10
Rolled oats	39	Mutton	102	Vermouth	17
Honey	35	Chicken	275	Sparking wine	6.6
Sugar	8	Hens' egg	177	Brandy	2.2
Coffee	3	Liver sausage	91	Coke	6.8
Cocoa	52	Pork sausage	100	Lemonade	5

In adult animals, the long-term I deficiency affects in:

- Impairment of female reproduction
- Depletion of male fertility (lower semen quality)
- Lower milk production

Various sings of hypothyroidism are observed in different animals, but goiter, cretinism, impaired reproduction and abnormal fetus development are the most common symptoms (Anke 2004d). The association of goiter symptoms with the excess of *Brassica* spp. plants in fodder has been investigated recently (Frøslie 1990).

Animals differ widely in their susceptibility to I toxicity. The iodism (I poisoning) does not occur under natural conditions because its toxic dose is very high. The maximum tolerable level of I in feed has been established (in mg kg^{-1}) as: 50 for sheep and cattle; 300 for poultry; and 400 for pigs (Frøslie 1990). Therefore, the most important problem associated with excess I in animal feeding is its transfer to human with animal foodstuff. The food of animal origin is, under normal conditions, a significant source of I in human diet (Table II-17.9).

References

Aaseth J, Norseth T (1986) Copper. In: Friberg L, Nordberg, GF, Vouk VB (eds) Handbook on the toxicology of metals. 2nd ed., vol 2: Specific metals. Elsevier, New York: 233–254

Abbasi SA (1987) Analysis of sub-microgram levels of palladium (II) in environmental samples by selective extraction and spectrophotometric determination with N-p-methaxyphenyl-2-furylacrylohydroxamic acid and 5-(diethyloamino-2-(2-pyridyzalo) phenol. Anal Lett 20:1013–1027

Abbaspour A, Baramakeh L (2005) Simultaneous determination of antimony and bismuth by beta correction spectrophotometry and an artificial n151al network algorithm.1st Intern Workshop on Antimony in Environ, Heidelberg, 52 p

Abdelouas A, Grambbow B (2004) Interaction between radionuclides and microorganisms. Goldschmidt, Copenhagen, A197

Abernathy CO, Cantilli R, Du JT (1993) Essentiality versus toxicity: some considerations in the risk assessment of essential trace elements. Hazadr Assess Chem 8: 81–113

Abrahams PW (2005) Geophagy and the involution ingestion of soil. In: Selinus O, Alloway BJ, Centento JA, Finkelman RB, Fuge R, Lindh U, Smeldey P (eds) Medical geology. Elsevier, Amsterdam, pp 435–458

Académie des Sciences (1998) Contamination des sols par les elements en traces: les risques et leur gestion. Rapport 42

Aceto M, Abollino O, Bruzzoniti MC, Mentasti E, Sarzanini C, Malandrin M (2002) Determination of metals in wine with atomic spectroscopy (flame–AAS, GF–AAS, ICP–AES); a review. Food Add Contam 19:126–133

ACGIH (1994) Threshold limit values and biological exposure indices for 1993–1994. Am Conf Governmental Industrial Hygienists, Cincinnati, OH

Achterberg EP, Ussher S, Gledhill M, Worsfold PJ, Bowie AR, Gehrels WR, Geider R, de Baar HJ (2004) Role of iron in marine biogeochemistry. Goldschmidt, Copenhagen, 327 p

Adamo P. Arienzo M, Pugliese M, Roca V, Violante P (2004) Accumulation history of radionuclides in the lichen *Stereocaulon vesuvianum* from Mt. Vesuvius (south Italy). Environ Pollut 127:455–461

Adams FS, Cole H Jr, Massie LB (1973) Element constitution of selected aquatic vascular plants from Pennsylvania: Submersed and floating leaved species and rooted emergent species. Environ Pollut 5: 117–147

Adams MD, Johns MW, Dew DW (1999) In: Schmidbaur H (ed) Gold progress in chemistry, biochemistry and technology. Wiley, Chichester, pp 65–104

Adriano DC (2001) Trace elements in terrestrial environments: biogeochemistry, bioavailability and risk assessments. 2nd ed., Springer-Verlag, Berlin Heidelberg

Adriano DC, Chlopecka A, Kaplan SI (1998) Role of soil chemistry in soil remediation and ecosystem conservation. In: Huang PM (ed) Soil chemistry and ecosystem health. SSSA 3 Special Publ 52, 261 p

Adriano DC, Wenzel WW, Vangronsveld J, Bolan NS (2004) Role of assisted natural remediation in environmental cleanup. Geoderma 122:121–142

Agapkina GI, Tikhomirov FA, Shcheglov AI, Kracke W, Bunzl K (1995) Association of Chernobyl-derived $^{239+240}$Pu, ^{241}Am, ^{90}Sr, and ^{137}Cs with organic matter in the soil solution. J Environ Radioactivity 29:257–269

Agapkina GI, Shcheglov AI, Tikhomirov FA, Meculova LN (1998) Dynamics of Chernobyl–fallout radionuclides in soil solutions of forest ecosystems. Chemosphere 36:1125–1130

Agbenin JO (2002) Lead in a Nigerian Savanna soil under long-term cultivation. Sci Total Environ 286:1–14

Ahl T, Jönsson E (1972) Boron in Swedish and Norwegian fresh waters. Ambio 1:66–70

Ahmed KM (1999) A review of available geochemical data and their implications to the origin of arsenic in Bangladesh groundwater. Seminar on groundwater arsenic contamination in the Bengal Delta Plains of Bangladesh. University of Dhaka, Bagladesh, 24 p

Ainsworth N, Cooke JA, Johnson MS (1990) Distribution of antimony in contaminated grassland: 1 – Vegetation and soils. Environ Pollut 65:65–77

Akter K, Naidu R (2006) Arsenic speciation in the environment. In: Naidu R, Smith E, Owens G, Bhattacharya P, Nadebaum P (eds) Managing arsenic in the environment: from soil to the human health. CSIRO Publishing, Collingwood, VIC, pp 59–71

Al Rmalli SW, Haris PI, Harrington CF, Ayub M (2005) A survey in foodstuffs on sale in the United Kingdom and imported from Bangladesh. Sci Total Environ 337:23–30

Alcântara MAK, Camargo OA (2004) Chromium movement in columns of two highly weathered soils. Comm Soil Sci Plant Anal 35:599–613

Aleksander-Kwaterczak U, Helios-Rybicka E (2004) Heavy metals distribution in the river sediment depth profiles of the Mała Panew River (southern Poland). Geologia 30:153–164 (in Polish)

Alibo DS, Nozaki Y (1999) Rare earth elements in seawater: Particle association, shale-normalization, and Ce oxidation. Geochim Cosmochim Acta 63:363–372

Alibo DS, Nozaki Y, Jeandel C (1999) Indium and yttrium in North Atlantic and Medetertanean waters: composition to the Pacific data. Geochim Cosmochim Acta 63:1991–1999

Allen HE (1993) The significance of trace metal speciation for water, sediment and soil quality criteria and standards. In: Sloof W, de Kruijf H (eds) Proc 2nd Intern European Conference on Ecotoxicoly, Sci Total Environ, Suppl 1, pp 23–45

Allen AG, Nemitz E, Shi JP, Harrison RM, Greenwood JC (2001) Size distribution of trace metals in atmosphere aerosols in the United Kingdom. Atmos Environ 35:4581–4591

Alloway BJ (1995) Heavy metals in soils. 2nd ed., Blackie Acad, London

Alloway BJ, Steinnes E (1999) Anthropogenic addition of cadmium in soils. In: McLaughlin MJ, Singh BR (eds) Cadmium in soils and plants. Kluwer Acad Publ, London, pp 97–123

Alloway BJ, Zhang P, Mott CJB, Nicholson F, Chambers B, Smith S, Carlton-Smith C (1999) Atmospheric deposition of heavy metals onto agricultural land in England and Wales. 5th Int. Conf. Biogeochem. Trace Elements, Vienna, 1, pp 414–415

Almås A, Singh BR, Sveitrup TE (1995) The impact of the nickel industry in Russia on concentrations of heavy metals in agricultural soils and grass in Sør-Varanger, Norway. Norwegian J Agric Sci 9:61–74

Alstad NEW, Kjelsberg BM, Asbjørn Vøllestad L, Lydersen E, Poléo ABS (2005) The significance of water ionic strength on aluminium toxicity in brown trout (*Salmo trutta* L.). Environ Pollut 133:333–342

Alumaa P, Kirso U, Petersell V, Steinnes E (2002) Sorption of toxic heavy metals to soil: short communication. Int J Hyg Environ Health 204:375–376

Álvarez E, Monterroso C, Fernández Marcos ML (2002) Aluminium fractionation in Galician (NW Spain) forest soils as related to vegetation and parent material. Forest Ecol Management 166:103–206

Álvarez E, Fernández-Marcos ML, Vaamonde C, Fernández-sanjurjo MJ (2003) Heavy metals in the dump of an abandoned mine in Galicia (NW Spain) and in the spontaneously occurring vegetation. Sci Total Environ 313(1–3):189–197

Alyoshin NE, Avakayan ERE, Lebedev EV, Alyoshin EP (1988) Influence of silicon and its antagonists on rice mitochondria. Inst Rice Res Newsl 13:10–13 (in Russian)

Alzieu C, Michel P, Tolosa I, Bacci E, Mee LD, Readman JW (1991) Organotin compounds in the mediterranean: a continuing cause of concern. Mar Environ Research 32:261–270

AMAP (Arctic Monitoring and Assessment Programme) (1998) Heavy metals. Assessment report: arctic pollution issues. Arctic Monitoring and Assessment Programme, pp 373–453

American Council on Science and Health (1997) Lead and human health. In: Juberg DR (ed) Lead and health: An update. Ph.D., New York

Amouroux D, Tessier E, Donard OFX (2000) Volatilization of organotin compounds from estuarine and coastal environments. Environ Sci Technol 34:988–995

Anbar AD (2004) Molybdenum stable isotopes: observations, interpretations and directions. Rev Miner Geochem 55:429–454
Anbar AD, Creaser RD, Papanastassiou DA, Wasserburg GJ (1992) Rhenium in seawater: confimation of generally conservative behavior. Geochim Cosmochim Acta 56:4099–4103
Anbar AD, Wasserburg GJ, Papanastossiou DA, Anderson PS (1996) Iridium in natural waters. Science 273:1524–1528
Anderson RA (1997) Chromium as an essential nutrient for humans. Regul Toxicol Pharmacol 26:S35–S41
Anderson MA, Bertsch PM, Miller WP (1988) The distribution of lithium in selected soils and surface waters of the southeastern U.S.A. Appl Geochem 3:205–212
Anderson DL, Kitto ME, McCarthy L, Zollar WH (1994) Sources of atmospheric distribution of particulate and gas phase boron. Atmos Environ 28:1401–1410
Anderson CWN, Brooks RR, Chiarucci A, LaCoste CJ, Leblanc M, Robinson BH, Simcock R, Stewart RB (1999a) Phytomining for nickel, thallium and gold. J Geochem Explor 67:407–415
Anderson CWN, Brooks RR, Stewart RB, Simcock R (1999b) Gold uptake by plants. Gold Bulletin 32:48–51
Andersson A (1979) Mercury in soils. In: Nriagu JO (ed) The biogeochemistry of mercury in the environment, 3. Elsevier/North Holland Biomedical Press, Amsterdam, pp 79–112
Andersson A (1987) Mercury content in the raw humus larer of forest soils. SNV Report 1987-03-23, Swedish Environmental Protection Agency, pp 1–26 (in Swedish)
Andrási E, Igaz S, Szoboszlai N, Farkas E, Ajtony Z (1999) Several methods to determine heavy metals in human brain. Spetrochim Acta, Part B 54:819–825
Andreae MO (1983) The determination of the chemical species of some of the Rauser WE (1990) Phytochelatins. A Review Biochem 59"hydride elements" (arsenic, antimony, tin and germanium) in seawater: methodology and results. In: Wong CS, Boyle E, Bruland KW, Burton JD, Goldberg ED (eds) Trace metals in sea water. NATO, Plenum Press, New York, pp 1–19
Andren AW, Bober TW (eds) (2002) Silver. Environment, transport, fate and effects. SETAC Press, Pensacola
Andren AW, Einger-Runge B, Sedlak D (1995) The occurrence and behavior of silver in natural waters. In: Andern AW, Bober TW (eds) The 3rd Int Conf Proceedings: Transport, Fate and Effects of Silver in the Environment. Washington D.C., pp 7–9
Angelone M, Bini C (1992) Trace elements concentration in soils and plants of Western Europe. In: Adriano DC (ed) Biogeochemistry of trace metals. Lewis Publ, Boca Raton, pp 19–60
Angelova V, Ivanova R, Ivanov K (2004) Heavy metal accumulation and distribution in oils crops. Comm Soil Sci Plant Anal 35:2551–2566
Anger JP (2004) Tin. In: Merian E, Anke M, Ihnat M, Stoepppler M (eds) Elements and their compounds in the environment. 2nd ed., Wiley-VCH, Weinheim, pp 1113–1124
Anke M (2004a) Essential and toxic effects of macro, trace and ultra elements in the nutrition of man. In: Merian E, Anke M, Ihnat M, Stoepppler M (eds) Elements and their compounds in the environment. 2nd ed., Wiley-VCH, Weinheim, pp 343–367
Anke M (2004b) Essential and toxic effects of macro, trace and ultra elements in the nutrition of animals. In: Merian E, Anke M, Ihnat M, Stoepppler M (eds) Elements and their compounds in the environment. 2nd ed., Wiley-VCH, Weinheim, pp 305–341
Anke M (2004c) Vanadium. In: Merian E, Anke M, Ihnat M, Stoeppler M (eds) Elements and their compounds in the environment. 2nd ed., Wiley-VCH, Weinheim, pp 1171–1191
Anke M (2004d) Iodine. In: Merian E, Anke M, Ihnat M, Stoeppler M (eds) Elements and their compounds in the environment. 2nd ed., Wiley-VCH, Weinheim, pp 1457–1495
Anke M, Angelov L (2004) Rubidium. In: Merian E, Anke M, Ihnat M, Stoepppler M (eds) Elements and their compounds in the environment.2nd ed., Wiley-VCH, Weinheim, pp 547–573
Anke M, Seifert M (2004) Titanium. In: Merian E, Anke M, Ihnat M, Stoeppler M (eds) Elements and their compounds in the environment. 2nd ed., Wiley-VCH, Weinheim, pp 1125–1140
Anke M, Arnhold W, Gleich M, Müller M, Illing H, Schäfer U, Jaritz M (1995a) Esseniality and toxicity of lithium. In: Kosla T (ed) Lithium in trophic chain, soil-plant-anila-man. Proc. Intern. Symp. Warszawa, pp 19–42
Anke M, Angelow L, Glei M, Müller M, Illing H (1995b) The biological importance of nickel in the food chain. Fresenius J Anal Chem 352:92–96

Anke M, Amhold W, Glei M, Müller M, Iling H, Schäfer U, Jaritz M (1995c) Esentiality and toxicity of lithium. Lithium in the trophic chain soil-plant-animal-man. In: Proc Intern Symp, 13–14 September, Warszawa, pp 19–42

Anke M, Seeber O, Glei M, Dorn W, Müller R (2000) Uranium in the food chain of humans in Central Europe – risk and problems. 4th Symp Roumanan Acad-Branch. Metal elements in Environment, medicine and biology. Publ House "Eurobit", Timişara, pp 7–22

Anke M, Jaritz M, Dorn W, Seifert M, Müller R, Christina H (2002) The biological importance of barium in the food chain of humans. In: Gârban Z, Drâgan P, Gârban G (eds) Metal elements in environment, medicine and biology. Publ House "Eurobit", Timişara, pp 5–28

Anke M, Illing-Günther H, Anke S, Müler R (2003) Vanadium – an essential and toxic element? 4th Intern Symp Trace Elements in Human, New Persp, Enthypossis, Athens, 2:1120–1153

Antal M (2006) Trace elements in animal and human nutrition and health, mechanism, interactions, food and feed safety. In: Szilàgyi M, Szentmihalayi K (eds) Intern Symp Trace Elements in the Food Chain, Budapest, pp 350–356

Ansari AA, Singh IB, Tobschall HJ (1998) Organotin compounds in surface and pore waters of Ganga Plain in the Kanpur-Unnano industrial region, India. Sci Total Environ 223:157–166

Antapaitis A, Lubite J, Antapaitis S, Staugaitis W (2004) Regularity in selenium distribution in arable soils of Lithuania. 3rd Inter Conf. Heavy metals radionuclides and elements-biofills in the environment, Semipalatinsk, Kazakhstan, 1, pp 159–164 (in Russian)

Arapis G. Petrayev E, Shagalova E, Zhukova O, Sokolik G, Ivanova T (1997) Effective migration velocity of ^{137}Cs and ^{90}Sr as a function of the type of soils in Belarus. J Environ Radioactivity 34:171–185

Aras NK, Kumpulainen J (1996) Trace element in Turkish wheat and human diet. In: Kumpulainen J (ed) Trace elements, natural antioxidants and contaminants in European food and diets. Food of Agricultural Organization of the United Nations, REU Technical Series 49, Rome, pp 103–118

Archer FC, Hodgson IH (1987) Total and extractable trace element content of soils in England and Wales. J Soil Sci 38:421–431

Armienta MA, Rodriquez R, Ceniceros N, Juárez F, Cruz O (1996) Distribution, origin and fate of chromium in soils in Guanajuato, Mexico. Environ Pollut 93:391–397

Arnesen AK, Krogstad T (1998) Sorption and desorption of fluoride in soil polluted from the aluminium smelter at Ardal in western Norway. Water Air Soil Pollut 103:357–361

Arnold GL, Anbar AD, Barling J, Lyons TW (2004) Molybdenum isotope evidence for widespread anoxia in Mid-Proterozoic oceans. Science 304:87–90

Aro A, Alfthan G, Ekholm P, Varo P (1998) Effects of selenium supplementation of fertilizers on human nutrition and selenium status. In: Frankenberger WT, Engberg RA (eds) Environmental chemistry of selenium. Marcel Dekker, New York, pp 81–97

Arpalahti J (1996) Platinum (II)-nucleobase interactions, a kinetic approach. In: Sigel A, Sigel H (eds) Metal ions in biological systems, vol. 32. Dekker, New York, pp 379–395

Arthur WJ, Markhan OD (1984) Po-210 in the environment around a radioactive waste disposal area and phosphate ore processing plant. Health Phys 46:793–799

Asami T (1988) Environmental pollution by beryllium in Japan. In: Wolf K, van den Brink W, Colon FJ (eds) Contaminated soil '88. Kluwer Ac Publ, Dordrecht, pp 261–272

Asami T, Yoshimo A, Kubota M, Gotoh S (1990) Background levels of indium and gallium in soils with special reference to pollution of the soils from zinc and lead smelters. Z Pflanzenernähr Bodenk 153:257–259

Asami T, Kubota M, Saito S (1992) Simultaneous determinaton of antimony and bismuth in soils by continous hydride generation-atomic absorption spectrometer. Water Air Soil Pollut 62:349–355

Asami T, Mizui C, Shimada T, Kubota M (1996) Determination of thallium in soils by flame atomic absorption spectrometry. Fresenius' J Anal Chem 356:348–351

Ashley PM, Craw D, Graham BP, Chappell DA (2003) Environmental mobility of antimony around mesothermal stibnite deposits, New South Wales, Australia and southern New Zealand. J Geochem Expl 77:1–14

Aswathanarayana U (1990) Geoenvironment-related disease endemicity in Africa, with special reference to Tanzania. In: Låg J (ed) Geomedicine, CRC Press, Boca Raton, pp 235–244

ATSDR (Agency for Toxic Substances and Disease Registry) (1988) The nature and extent of lead poisoning in children in the United States: a report to congress, July 1988

ATSDR (Agency for Toxic Substances and Disease Registry) (1990) Toxicological profile for copper. U.S. Public Health Service. Agency for Toxic Substances and Disease Registry, Atlanta, GA

ATSDR (Agency for Toxic Substances and Disease Registry) (1992a) Toxicological profile for boron and compounds. Agency for Toxic Substances and Disease Registry, U.S. Public Health Service, Atlanta, GA, pp 47–80

ATSDR (Agency for Toxic Substances and Disease Registry) (1992b) Toxicological profile for thallium. U.S. Public Health Service. Agency for Toxic Substances and Disease Registry, Atlanta, GA

ATSDR (Agency for Toxic Substances and Disease Registry) (1994) Toxicological profile for zinc. U.S. Dept of Health and Human Services. Agency for Toxic Substances and Disease Registry, Atlanta, GA

ATSDR (Agency for Toxic Substances and Disease Registry) (1999a) Toxicological profile for mercury. U.S. Dept of Health HumanServices, Agency for Toxic Substances and Disease Registry, Atlanta, GA

ATSDR (Agency for Toxic Substances and Disease Registry) (1999b) Toxicological profile for lead. U.S. Department of Health and Human Services, Agency for Toxic Substances and Disease Registry, Atlanta, GA

ATSDR (Agency for Toxic Substances and Disease Registry) (2000) Drinking water health advisory for manganese. U.S. Department of Health and Human Services, Agency for Toxic Substances and Disease Registry, Atlanta, GA

ATSDR (Agency for Toxic Substances and Disease Registry) (2002a) Draft toxicological profile for several trace elements. U.S. Dept Health Human Services, Agency for Toxic Substances and Disease Registry, Atlanta, GA

ATSDR (Agency for Toxic Substances and Disease Registry) (2002b) Draf toxicological profile for selenium. U.S. Department of Health and Human Services, Agency for Toxic Substances and Disease Registry, Atlanta, GA

Aubert D, Stille P, Probst A, Gather-Lafaye F, Pourcelot L, Del Nero M (2002) Characterization and migration of atmospheric REE in soils and surface waters. Geochim Cosmochim Acta 19:3339–3350

Augustyniak M, Migula P (2000) Body burden with metals and detoxifying abilities of the grasshopper *Chorthippus brunneus* (Thunberg) from industrial polluted areas. In: Markert B, Friese (eds) Trace elements – their distribution and effects in the environment. Ensevier Sci, Amsterdam, pp 423–454

Ayars JE, Hutmacher RB, Schoneman RA, Vail SS (1990) Long term use of saline water for irrigation. In: Proc 3rd Nation Irrig Symp, Phoenix, AZ, ASAE Pub 04-90, pp 368–373

Baekken T (1994) Effects of highway pollutants on a small Norwegian lake. Sci Total Environ 146/147:131–139

Baes C, Sharp R, Sjoreen A, Shor R (1984) A review and analysis of parameters for assessing transport of environmentally released radionuclides through agriculture. Prepared under contract No. DE-AC05-84OR21400. U.S. Department of Energy, Washington, DC

Bahnick D, Sauer C, Butterworth B, Kuehl D (1994) A national study of mercury contamination of fish. Chemosphere 29:537–546

Baize D, Mench M, Sappin-Didier V, Mocquot B, Gomez A, Proix N, Sterckeman T (2003) Phytodisponibilité des elements traces métalliques dans les grains de blue. Dossier l'environ I'INRA 25:45–62

Baker DE (1974) Copper: soil, water, plant relationship. Fed Proc Am Soc Exp Biol 33:1188–1193

Baker AJM, McGrath SP, Sidoli CMD, Reeves RD (1994) The possibility of in situ heavy metal decontamination of polluted soils using crops of metal-accumulating plants. Resources Conserv Recycling 11:41–49

Baldock IA, Nelson PN (2000) Soil organic matter. In: Somner ME (ed) Handbook of soil chemistry, CRC Press, Boca Raton, FL, pp B25–B84

Bankovitch V, Carrier G, Gagnon C, Normandin L, Kennedy G, Zayed J (2003) Total suspended particulate manganese in ambient air in Montreal 1981–2000. Sci Total Environ 308:185–193

Bañuelos GS, Ajwa HA, Caceres L, Dyer D (1999) Germination responses and boron accumulation in germplasm from Chile and the United States grown with boron-enriched water. Ecotoxi Environ Safety 43:62–67

Barber SA (1984) Soil nutrient bioavailability. Wiley, New York

Barcelo J, Poschenreider C (1990) Plant water relations as affected by heavy metals: a review. J Plant Nutr 13:1–37

Bargar JR, Webb SM, Fuller CC, Tebo BM (2004) Structural chemistry and formation mechanism of bacteriogenic manganese oxides. Goldschmidt, Copenhagen A157

Barnhart J (1997) Chromium chemistry and implications for environmental fate and toxicity. J Soil Contam, Spec Issue 6:561–568

Barrie LA, Hoff MA (1985) Five years of air chemistry observations in the Canadian Arctic. Atmos Environ 19:1995–2010

Barrows ME, Petrocelli SR, Macek KJ, Carnoll JJ (1978) Bioconcentration and elimination of selected water pollutants by bluegill sunfish. In: Haque R (ed) Dynamics, exposure and hazard assessment of toxic chemicals. Ann Arbor Sci Publ Inc, Ann Arbor, MI, pp 379–392

Bartlett RJ (1999) Characterizing soil redox behavior. In: Sparks DL (ed) Soil physical chemistry, 2nd ed., CRC Press, Boca Raton, FL, pp 371–397

Bartlett RJ, Kimble JM (1976) Behavior of chromium in soils. I. Trivalent forms, II. Hexavalent forms. J Environ Qual 5:379–386

Bartolf M, Brennan E, Price CA (1980) Partial characterization of a cadmium protein from the roots of cadmium-treated tomato. Plant Physiol 66:438–441

Bratoń HJ, Zachwieja Z, Folta M, Janusz-Grzybowska E, Stompór T, Bułowicz W, Brzezicka M (2003) A pilot study of aluminum level in plasma in healthy subjects in Poland. In: Proc 3rd Intern Conf. The problems of hygiene and ecology on access of Poland to the European Union, Kraków, pp 111–114 (in Polish)

Bassett RL (1980) A critical evaluation of the thermodynamic data from boron ions, ion pairs, complexes, and poly anions in aqueous solution at 298.15 K and 1 bar. Geochim Cosmochim Acta 44:1151–1160

Basta NT, Gradwohl R, Snethen KL, Schroder JL (2001) Chemical mobilization of lead, zinc and cadmium in smelter contaminated soils treated with exceptional quality biosolids. J Environ Qual 30:1222–1230

Batley GE (1996) The distribution and fate of tributyltin in the marine environment. In: De Mora SJ (ed) Tributyltin: a case study of an environmental contaminant. Cambridge Environ Chem Series, Cambridge Univ Press, Cambridge, pp 139–165

Baum MK, Migues-Burbano MJ, Campa A, Shor-Posner G (2000) Selenium and interleukins in persons infected with human immunodeficiency virus-type 1. Infect Disease 182(1):S69–S73

Beavington F, Cawse PA, Wakenshaw A (2004) Comparative studies of atmospheric trace elements: improvements in air quality near a copper smelter. Sci Total Environ 332:39–49

Becerril JM, Gonzalez-Murua C, Munoz-Rueda R, De Felipe MR (1989) The changes induced by cadmium and lead in gas exchange and water relations of clover and lucerne. Plant Physiol Biochem 27:913–918

Becker K, Kaus S, Krause C, Lepom P, Schultz C, Seifert M, Seifert B (2003) German environmental survey 1998 (GerEsIII): environmental pollutants in blood of the German population. Int J Hyg. Environ Health 2005:297–308

Becker-Heidmann P, Scharpensel WH (1990) Effects of radionuclide radiation caused by man. In: Låg J (ed) Geomedicine. CRC Press, Boca Raton, FL, pp 171–184

Becquer T, Quantin C, Sicot M, Boudot J-P (2003) Chromium availability in ultramafic soils from New Caledonia. Sci Total Environ 3001:367–373

Begerow J, Sensen U, Wiesmuller GA, Dunemann L (1999) Internal platinum, palladium and gold exposure in environmentally and occupationally exposed persons. Zbl Hyg Umweltmed 202:411–424

Bejarano-Sifuentes G, Nordberg E (2003) Mobilisation of arsenic in the Rio Dulce Alluvial Cone, Santiago del Estero Province, Argentina. M.Sc. Thesis. Kungl Tekniska Högskolan, Stockholm

Bellomo S, Allesandro WD, Longo M (2003) Volcanogenic fluorine in rainwater around active degassing volcanos: Mt. Etna and Stromboli Island, Italy. Sci Total Environ 301:175–185

Bencs L, Ravidra K, Van Grieken R (2005) Spatial and temporal variation of the level of anthropogenic platinum group elements in the environment. Abstract, Min Soc Winter Meeting, Bath Spa Univ

Bennett JP, Wetmore CM (2003) Elemental chemistry of four lichen species from Apostole Island, Wisconsin, 1987, 1995 and 2001. Sci Total Environ 305:77–86

Bercowy GM, Vo H, Rieders F (1994) Silicon analysis in biological specimens by direct current plasma-atomic emission spectroscopy. J Anal Toxicol 18:46–48

Berg T, Steinnes E (1997) Recent trends in atmospheric deposition of trace elements in Norway as evident from the 1995 moss survey. Sci Total Environ 208, 197–206

Berggren D, Fiskesjo G (1987) Aluminium toxicity and speciation in soil liquids – experiments with *Allium cepa* L. Environ Toxicol Chem 6:771–779

Berggren D, Mulder J (1995) The role of organic matter in controlling aluminium solubility in acidic mineral soil horizons. Geochim Cosmochim Acta 59:4167–4180

Bergkvist B (1987) Leaching of metals from forest soils as influenced by tree species and management. Forest Ecol Management 22:29–56

Berthet B, Amirad JC, Amird-Triquet C, Martoja M, Jeantet AY (1992) Bioaccumulation, toxicity, and physico-chemical speciation of silver in bivalve mollusks: Ecotoxicological and health consequences. Sci Total Environ 125:97–122

Bertine K, Goldberg ED (1971) Fossil fuel combustion and the major sedimentary cycle. Science 173: 233–235

Bertram HP, Bertram C (2004) Thallium. In: Merian E, Anke M, Ihnat M, Stoeppler M (eds) Elements and their compounds in the environment. 2nd ed., Wiley-VCH Verlag, Weinheim, pp 1099–1112

Besnard PM, heu C, Roberts M (1999) Distribution of copper in champagne vineyards soils, as influenced by organis amendments. 5th Int Conf Biogeochem Trace Elements, Vienna, pp 416–417

Bettinelli M, Baffi C, Beone GM, Spezia S (2000) Soil and sediment analysis by spectroscopic techniques. Part I: Determination of Cd, Co, Cr, Cu, Mn, Ni, Pb, and Zn. Atomic Spectroscopy 21:50–59

BGS (2001) Arsenic consumption of groundwater in Bangladesh, vol 2: Final report. British Geol Sur Techn Rep WC/00/19

Bhattacharya P, Mukherjee AR (2002) Management of arsenic contaminated groundwater in the Bengal Delta Plain. In: Chattewrji M, Arlosoroff S, Guha G (eds) Conflict management of water resources. Ashgate Publ Comp, Aldershot, Hampshire, pp 308–348

Bhattacharya P, Chatterjee D, Jacks G (1996) Options to safeguard groundwater from arseniferous aquifers in West Bengal, In: Pickford J, House S, Miles D, Ockelford J, Parr J, Saywell D, Skinner B, Smout I, Stear (eds) Proc 22nd WEDO Conf, New Delhi, pp 258–261

Bhattacharya P, Chatterjee D, Jacks G (1997) Occurrence of arsenic contaminated groundwater in alluvial aquifers from Delta Plains, Eastern India: Options for safe drinking supply. Intern J Water Res Manag 13:79–92

Bhattacharya P, Frisbie SH, Smith E, Naidu R, Jacks G, Sarkar B (2002a) Arsenic in the environment: A global perspective. In: Sarkar B (ed) Handbook of heavy metals in the environment. Marcel Dekker, New York, pp 147–215

Bhattacharya P, Jacks G, Ahmed KM, Khan AA, Routh J (2002b) Arsenic in groundwater of the Bengal Delta Plain aquifers in Bangladesh. Bull Environ Cont Toxicol 69:538–545

Biernacka M, Lipiński P, Sosińska A (2004) Radioactive contamination of soil and background gamma radiation in Poland, 1988-2001. Bibl Monit Środ, Warszawa (in Polish)

Biégo GH, Joyeux M, Hartemann P, et al. (1999) Determination of dietary tin intake in an adult French citizen. Arch Environ Contam Toxicol 36:227–232

Bills TP (1996) The influence of water quality parameters on acute silver toxicity to first instar *Daphina magna*. MS Thesis, Clemson University, Clemson, SC

Binbin W, Baoshan Z, Hongyin W, Yakun P, Yuehua T (2005) Dental caries in fluorine exposure areas in China. Environ Geochem Health 27:285–288

Birchall JD, Espie AW (1986) Biological implications of the interaction (via silanol groups) of silicon with metal ions. 121 Symp on Silicon biochemistry, Ciba Foundation, Wiley, New York, pp 140–153

Birke M, Rauch U (1994) Geochemical methods of investigation in the urban area of Berlin and its surroundings. 3rd Intern Symp Environ Geochem, Kraków, 49 p

Birke M, Rauch U (1997) Geochemical mapping in the Federal States of Germany. 4th Inter Symp Environ Geochem, Vail, CO, pp 11–12

Biró B, Füzy A, Kádár I, Posta K (2006) Sensititvity of mycorrhizal fungi inside the barley rhizosphere at long-term heavy metal stress. Intern Symp Trace Elements in the Food Chain, Budapest, pp 276–280

Bishop K, Lee Y-H (2004) Cycling of methylmercury and mercury in a boreal catchment: Input-output budget and Hg soil stores. 7th Intern Conf on Mercury as a Global Pollutant, Ljubljana, 2, pp 819–820

Bissé E, Epting T, Beil A, Lindinberg G, Lang H, Wieland H (2005) Reference values of serum silicon in adults. Analyt Biochem 337:130–135

Bissen M, Frimmel FH (2003) Arsenic – A review: Part I. Occurrences, toxicity, speciation, and mobility. Acta Hydrochim Hydrobiol 31:9–18

Bjerg PL, Albrechtsen H-J, Kjeldesn P, Christensen TH (2003) The groundwater geochemistry of waste disposal facilities. In: Holland HD, Turekian KK (eds) Treatise on geochemistry, vol. 9, Environmental geochemistry. Elsevier Pergamon, Oxford, pp 579–612

Bjernes E (1993) Alzheimer's disease and possible relation to aluminium. In: Låg J (ed) Geimedical problems, related to aluminium, iron and manganese. Nor Acad Sci Letters, Oslo, pp 31–40

Blancaflor EB, Jones DL, Gilory S (1998) Alterations in the cytoskeleton accompany aluminium-induced growth inhibition and morphological changes in primary roots of maize. Plant Physiol 118:159–172

Bloom NS (1992) On the chemical form of mercury in edible fish and marine invertebrate tissue. Can J Fisher Aquat Sci 49:1010–1017

Bloom NS, Fitzgerald WF (1988) Determination of volatile mercury species at the picogram level by low-temperature gas chromatography with cold-vapor atomic fluorescence detection. Anal Chim Acta 208:151–161

Bloom NS, Watras CJ, Hurley JP (1991) Impact of acidification on the methylmercury cycle of remote seepage lakes. Water Air Soil Pollut 56:477–491

Blunden SJ, Evans CJ (1990) Organotin compounds: In: Hutzinger O (ed) The handbook of environmental chemistry. vol 3, Part E, Anthropogenic compounds, Springer-Verlag, Berlin Heidelberg, pp 1–44

Blunden SJ, Hobbs LA, Smith PJ (1984) Chemistry of organotin compounds. Environ Chemistry 3:47–77

Boawn LC, Rasmussen PE (1971) Crop response to excessive fertilization of alkaline soil. Agron J 63:874–876

Boganegra MP, Lobartini JC, Orioli GA (2004) Iron-humate as a source of iron for plants. Comm Soil Sci Plant Anal 35:2567–2576

Bojakowska I. Sokołowska G, Sztyrak T (1992) Contamination of bottom sediments in Poland – result of monitoring in1991 year. PIOŚ, Warszawa (in Polish)

Bolt GM (ed) (1979) Soil chemistry, B. Physico-chemical models. Elsevier Sci Publ, Amsterdam

Borochoff RT (2001) Model for mercury deposition from smoke plumes. In: Veiga M, Le Clair K (eds) Intelligence in a materials world. CRC Press, Boca Raton, FL, pp 615–628

Borovička J, Řanda Z, Jelinek E (2005) Gold content of ectomycorrhizal and saprobic macrofungi from non-auriferous and unpolluted areas. Mycol Res 109:951–955

Borowska K, Łyszcz R (1996) The selenium concentration in some grass species from mineral and peaty soils. 16th Gen Meeting EU Grassland Federation, Grado, Italy, pp 707

Boudot J-P, Maitat O, Merlet D, Rouiller J (1996) Occurrence of non-monometric species of aluminium in undersaturated soil and surface waters: consequences for the determination of mineral saturation indices. J Hydrology 177:47–63

Bourtlès DI (1992) Beryllium isotopes in the Earth's environment. In: Nierenberg WA (ed) Encyclopedia of Earth system science. Acad Press, London,1, pp 337–352

Boust D, Saas A (1981) A selective chemical extraction procedure applied to trace metals compodition; comparison between several reagents on two types of sediments (Seine and Gironde estuaries). In: Proc Intern Heavy Metals in the Environment, CEP Const, Edinburgh, pp 709–711

Bowen HTM (1979) Environmental chemistry of the elements. Acad Press, London

Bowles JFW (2005) Platinum-group elements in the urban environment – should we worry? An introduction. Abst Min Soc Winter Meeting, Bath Spa, Univ Bath

Boyle EA, Sclater F, Edmond JM (1976) On the marine geochemistry of cadmium. Nature 263:42–44

Bradley RW, Morris JR (1986) Heavy metals in fish from a series of metal contaminated lakes near Sudbury, Ontario. Water Air Soil Polut 27:341–254

Brake SS, Jensen RR, Mattox JM (2004) Effects of coal fly and amended soils on trace element uptake in plants. Environ Geol 45:680–689

Braman RS, Tompkins MA (1978) Atomic emission spectrometric determination of antimony, germanium, and methylgermanium. Anal Chem 50:1088–1093

Bramley RGV (1990) Cadmium in New Zealand agriculture. NZJ Agric Res 33:505–519

Brandi K, Müller AS, Pallauf MWJ (2004) Influence of manganese deficiency on manganese enzyme activity and gene expression in growing rats. 22 Workshop, Macro and Trace Elements, Jena, 2, pp 1296–1302

Brännvall M-L, Bindler R, Renberg I, Emteryrd O, Bartnicki J, Billström K (1999) The medieval metal industry was the cradle of modern large-scale atmospheric lead pollution in northern Europe. Environ Sci Technol 33:4391–4395

Brasil DB, Almedia R, Nascimento EL, Oliveria RC, Myuyai RK, Bastos WR, da Silveita EG (2004) Mercury in intiofauna of Samuel Hydroelectric Reservoir, Amazon – Brazil. 7th Intern Conf on Mercury as a Global Pollutant, Ljubljana, 2, pp 865–868

Bratakos MS, Lazos ES, Bratakos SM (2002) Chromium content of selected Greek foods. Sci Total Environ 290:47–58

Brewer AK (1984) The high pH therapy for cancer test on mice and humans. Pharmacol Biochem Behav 21(1):1–5

Brigden K, Stringer R, Santillo D (2002) Heavy metal and radionuclide contamination of fertilizer products and phosphogypsum waste produced by the Lebanese Chemical Co, Lebanon. Greenpeace Research Laboratories, Tech Note 13/2002, University of Exeter: (www.greenpeace.to/publications_pdf/LCC_2002.pdf)

Brockhaus A, Dolgner R, Ewers U, Karmer U, Soddemann H, Wiegand H (1981) Intake and health effects of thallium among a population living in the vicinity of a cement plant emitting thallium containing dust. Int Arch Occup Environ Health 48:375–389

Brooks RR (1998) Phytochemistry of hyperaccumulators. In: Brooks RR (ed) Plants that hyperaccumulate heavy metals. CAB Intern, Wallinford, pp 15–53

Brooks SC, Herman JS (1998) Rate and extent of Co sorption in representative aquifer minerals in the presence of moderately strong organic ligands. Appl Geochem 13:77–88

Brooks K, Mertens J, Smolders E (2005) Toxicity of heavy metals in soil assessed with various soil microbial and plant growth assays: a comparative study. Environ Toxicol Chem 24:101–107

Brown PH, Hu H (1996) Phloem mobility of boron in species dependent. Evidence for phloem mobility in sorbitol rich species. Ann Bot 77:497–505

Brown PH, Shelp BJ (1997) Boron mobility in plants. In: Dell B, Brown PH, Bell RW (eds) Boron in soils and plants. Kluwer Acad. Publ., Boston, pp 85–101

Brüggemann J, Dörfner HH, Hecht H, Kumpulainen JT, Westermair Th (1996) Status of trace elements in stable foods from Germany 1990–1994. In: Kumpulainen JT (ed) Trace elements, natural andtioxidants and contaminants in European foods and diets. FAO/RE Techn Ser 49:5–8

Brügmann L (1979) Zum Spurenmetallgehalt des Ostseewassers. Concentration of trace metals in the water of the Baltic Sea. Chem Erde 38:292–297

Brügmann L, Gaul H, Rohde K-H, Ziebarth U (1991) Regional distribution and temporal trends of some contaminants in the water of the Baltic Sea. Dtsch Hydrgr Z 44:161–184

Brügmann L, Hallberg R, Larsson C, Löffler A (1998) Trace metal speciation in sea and pore water of the Gotland Deep, Baltic Sea, 1994. Appl Geochem 13:359–368

Brümmer GW (1986) Heavy metal species, mobility and availability in soils. In: Bernhard M, Brinckman FE (eds) The importance of chemical "speciation" inenvironmental processes. Springer-Verlag, Berlin Heidelberg, pp 37–52

Bryan GW, Langston WJ (1992) Bioavailability, accumulation and effects of heavy metals in sediments with special reference to United Kingdom estuaries: a review. Environ Polut 76:89–131

Brzezińska A, Trzosińska A, Żmijewska W, Wódkiewicz L (1984) Trace metals in some organisms from the southern Baltic. Oceanologia 18:79–94

Buat-Menard PE (1984) Fluxes of metal through the atmosphere and ocean. In: Nriagu JO (ed) Changing meal cycles and human health. Springer-Verlag, Berlin Heidelberg, pp 43–69

Buc J, Santini C-L, Giordani R, Czizek M, Wu L-F, Giordano G (1999) Enzymatic and physiological properties of the tungsten-substituted molybdenum TMAO reductase from *Escherichia coli*. Molec Microbiol 32

Buhl F, Połedniok J (2000) The comparison of gallium, vanadium and scandium content in soils from industrial and agricultural regions. Soil Sci Ann Poland 51:67–75

Bunzl K, Kracke W (1984) Distribution of 210 Pb, 210 Po, stable lead and fallout 137 Cs in soil, plants and moorland sheep of a heath. Sci Total Environ 39:143–159

Burd GI, Dixon DG, Glick BR (2000) Plant growth-promoting bacteria that decrease heavy metal toxicity in plants. Can J Microbiol 46:237–245

Burgess W, Ahmed KM (2006) Arsenic in aquifers of the Bengal Basin. In: Naidu R, Smith E, Owens G, Bhattacharya P, Nadebaum P (eds) Managing arsenic in the environment: from soil to the human health, CSIRO Publ, Collingwood, Vic, pp 31–56

Burns RG, Dick RP (eds) (2002) Enzymes in the environment. Marcel Dekker, New York

Burt R, Wilson MA, Mays MD, Lee CW (2003) Major and trace elements of selected pedons in the USA. J Environ Qual 32:2109–2121

Butterwick L, de Oude N, Raymind K (1989) Safety assessment of boron in aquatic and terrestrial environment. Ecotoxicol Environ Safety 17:339–371

Bylińska E (1993) Bioaccumulation of chromium and nickel in plants from areas of polymetallic deposits in Sudety Mts. In: Kabata-Pendias A (ed) Chromium, nickel and aluminum in the environment: ecological and analytical problems. Kom Nauk PAN MAB, Warszawa, pp 65–72 (in Polish)

Cabrera C, Teissedre PL, Cabanis MT, Cabanis JC (1996) Determination of platinum levels in wines samples. In: Sigel A, Sigel H (eds) Metal ions in biology and medicine, 4. Dekker, New York, pp 163–165

Cain DJ, Luoma SN (1990) Influence of seasonal growth, age, and environmental exposure on copper and silver in a bivalve indicator, *Macoma balthica*, in San Francisco Bay. Mar Ecol Prog, Ser 60: 45–55

Cain DJ, Luoma, SN, Carter JL, Fend SV (1992) Aquatic insects as bioindicators of trace element contamination in cobble-bottom rivers and streams. Canad J Fish Aquatic Sci 49:21441–2154

Call DJ, Polkinghorne CN, Markee, TP, Brooke LT, Geiger DL, Gorsuch JW, Robillard, KA (1999) Silver toxicity to Chironomus Tentans in two freshwater sediments. Environ Toxicol Chem 18:30–39

Callister SM., Winfrey HR (1986) Microbial methylation of mercury in upper Wisconsin River sediments. Water Air Soil Pollut 29:453–465

Camilleri C, Markich SJ, Noller BN, Turley CJ, Parker G, Dam RA (2003) Silica reduces the toxicity of aluminium to a tropical fish (*Mogurnda mogurnda*). Chemosphere 50:355–364

Campbell LC, Miller MH, Loneragan JF (1975) Translocation of boron to plant fruits. Aust J Plant Physiol 2:481–487

Campbell PGC, Paquin PR, Adams WJ, Brix KV, Juberg DR, Playle RC, Ruffing CJ, Wentsel RS (2002) Risk assessment. In: Andren AW, Bober TW (eds) Silver in the Environment: transport, fate and effects. SETAC, Pensacola, FL, pp 97–140

Canizares P, Castro L de, Valcarcel M (1994) Flow through fluorimetric sensor for determination of aluminum at the nanogram per milliliter level. Anal Lett 27:247–262

Cannon HL, Shacklette HT, Bystron H (1968) Metal absorption by *Equisetum* (Horsetail), U.S. Geol Surv Bull 1278a:21

Cao ZH, Hu ZY (2000) Copper contamination in paddy soils irrigated with waste water. Chemosphere 41:3–6

Cappel R (1997) Silver compounds. In: Howe-Grant M (ed) Kirk-Othmer Encyclopedia of Chemical Technology, 4th ed., vol 22, Wiley, New York, pp 179–195

Cappon CJ (1981) Mercury and selenium content and chemical form in vegetable crops grown on sludge-amended soil. Arch Environ Contam Toxicol 10:673–689

Cappon CJ (1987) Uptake and speciation of mercury and selenium in vegetable crops grown on compost-treated soils. Water Air Soil Pollut 34:353–361

Carlisle EM (1986) Silicon as an essential element in animal nutrition. Ciba Foundation Symp 121: 123–139

Carlton WW, Henderson W (1964) Studies on the copper requirement of growing chickens. Avian Diseases 8:227–234

Caroli S, Senofonte O, Violante N, D'Ilio S, Caimi S, Choodo F, Menditto A (1998) Diagnostic potential of hair analysis as applied to the goldsmith sector. Microchem J 59:32–44

Carvahlo FM, Silvany-Neto AM, Melo AM, et al. (1989) Cadmium in hair of children living near a lead smelter in Brazil. Sci Total Environ 84:119–128

Case CE, Hambidge KM (1980) Epidemiological aspects of human zinc deficiency. In: Nragu JO (ed) Zinc in the environment, Part II: health effects, Wiley, New York, pp 1–27

Castilhos ZC, Rodrigues-Filho S, Rodrigues AP, Villas-Bôas R, Veiga MM, Beinhoff C (2004a) Mercury contamination in fish from gold mining areas in the Amazon region and human health risk assessment. 7th Intern Conf on Mercury as a Global Pollutant, Ljubljana, 1, pp 26–30

Castilhos ZC, Rodrigues-Filho S, Rodrigues AP, Villas-Bôas R, Veiga MM, Beinhoff C (2004b) Bioagnification of Hg in fish from gold mining affected Amazonian aquatic systems. 7th Intern Conf on Mercury as a Global Pollutant, Ljubljana, 1, pp 31–34

Castilhos ZC, Rodrigues-Filho S, Rodrigues AP, Villas-Bôas R, Veiga MM, Beinhoff C (2004c) Mercury contamination in fish from gold mining areas in Indonesia. 7th Intern Conf on Mercury as a Global Pollutant, Ljubljana, 1, pp 35–38

Cavalcante de Oliveria R, Bastos WR, Glória da Silveira E, Fonseca M de F (2004) Human exposure to mercury in the riverine population of the Madeira River, Amazon, Brazil. 7th Intern Conf on Mercury as a Global Pollutant, Ljubljana, 1, pp 369–372

Cawse PA (1987) Trace and major elements in the atmosphere at rural locations in Great Britain, 1972–81. In: Coughtrey PJ, Martin MH, Unsworth MH (eds) Pollutant transport and fate in ecosystems. Blackwell Sci Publ, Oxford, pp 89–112

Celo V, Scott SL, Lean DRS (2004) Abiotic methylation of mercury in the aquatic environment. 7th Intern Conf on Mercury as a Global Pollutant, Ljubljana, 2, pp 915–918

CEPA (1993) Arsenic and its Compounds. Can Environ Protection Act, Quebec

Chaccour CJ, Gayral LF, Goldberger E (1977) Lithium and water metabolism. Encephale 3:211–225

Chamberlain AC (1983) Fallout of lead and uptake by crops. Atmos Environ 17:693–706

Chandra Sekhar K, Kamala CT, Chary NS, Balaram V, Garcia G (2005) Potential of *Hemidesmus indicus* for phytoextraction of lead from industrial contaminated soils. Chemosphere 58:507–514

Chaney RL, Malik M, Li YM, Brown SL, Brewer EP, Angle JS, Baker AJM (1997) Phytoremediation of soil metals. Current Opinion Biotechnol 8:279–284

Chaney RL, Angle JS, Li Y-M, Baker AJM, Reeves RD, Rosenberg RJ, Van Volk V, Kukier U, Peters C, Zdimal K, Nelkin JP (1999a) Phytoextraction of soil nickel using *Alyssum* species. 5th Inter Conf Biogeochem Trace Elements, Vienna, pp 14–15

Chaney RL, Brown SL, Stuczynski TI, Danies WL, Henry CL, Li Y-ML, Siebielec G, Malik M, Angle JS, Ryan JA, Compton H (1999b) In-situ remediation and phytoextraction of metals from hazardous contaminated soils. U.S.-EPAconference on conference on "Innovative clean-up approaches: Investments in Technology Development, Results and Outlook for the Future." Bloomingdale, IL, pp 1–24

Chaney RL, Kukier U, Siebielec G (2003) Risk assessment for soil Ni remediation of soil-Ni phytotoxicity *in situ* and by phytoextraction. In: Proc Mining Environ III. Sundbury, Laurentian Univ, Sundbury, Ontario

Chang AC, Granato TC, Page AL (1992) A methodology for establishing phytotoxicity criteria for chromium, copper, nickel and zinc in agricultural land application of municipal sewage sludges. J Environ Quality 21:521–535

Chang P-C, Yoshida S, Kim K-W (2003) The effect of citric acid on uranium accumulation in plants. 16th Inter Symp on Environ Biogeochem, Oirase, 2, 231 p

Chassaigne H, Vacchina V, Kutchan TM, Zenk MH (2001) Identification of phytochelatin-related peptides in maize seedlings exposed to cadmium and obtained enzymatically in vitro. Phytochemistry 56:657–668

Chatterjee J, Chatterjee C (2000) Phytotoxicity of cobalt, chromium and copper in cauliflower. Environ Pollut 109:69–74

Chaudry FMA, Wallace A, Mueller RT (1977) Barium toxicity in plants. Comm Soil Sci Plant Anal 8:795–797

Cheam V, Lechner J, Desrosiers R, Sekeka I, Lawson G, Mudroch A (1995) Dissolved and total thallium in Great Lakes waters. J Great Lakes Res 31:384–394

Chen Z-S (1999) Selecting indicators to evaluate soil quality (www.fftc.agnet.org)

Chen Q, Wang N (1990) Analysis of carcinogenic metals (As, Ni, Cd) in lung tissues of smokers with lung cancer. In: Tan Jian'an, Peterson PJ, Li Ribang, Wang W (eds) Environmental life elements and health. Sci Press, Beijing, pp 243–245

Chen Z-S, Mutoh M, Sumizawa T. Furukawa T, Haraguchi M, Tani A, Akiyama S (1997) Reversal of heavy metals resistance in multidrug-resistant human KB carcinoma cells. Biochem Biophysic Res Comm 236:586–590

Chen HM, Zheng CR, Tu C, She ZG (2000) Chemical methods and phytoremediation of soil contaminated with heavy metals. Chemosphere 41:229–234

Chen Y-W, Belzile N, Gunn JM (2001) Antagonostic effects of selenium on mercury assimilation by fish population ear Sunbery metal smelters. Limnol Oceanology 46 1814–1818

Chen B-D, Zhu Y-G, Jakobsen I (2005) Effects of abscular mycorrhizal fungi on plant uptake and accumulation of uranium from contaminated soil. ICOBTE, 8th Intern Conf, Adelaide, pp 490–491

Cheney MA, Ferruzzi G, Sposito G, Casey WH (2004) The energetics of and iodate adsorption onto goethite. Goldschmidt, Copenhagen, 130 p

Chimie et Ecologie (1998) Impact des metaux sur l'homme et l'environment influence de la speciation - le Cas du Plomb, Paris

Chisolom Jr JJ (2001) The road to primary prevention of lead toxicity in children. Pediatrics 107: 581–583

Chłopicka J, Nartoń H, Zachwieja Z, Schleiger-Zawadzka M, Fołta M, Krośniak M (1998) Lead content in hair of children near the three industrial centers at southern Poland. In: Kabata-Pendias A, Szteke B (eds) Lead in the environment: Ecological and analytical problems. Polish Acad Sci, Warszawa, pp 411–418 (in Polish)

Chojnicki J (2000) Rare earth elements in alluvial soils of central Vistula valley and Żuławy. Soil Sci Ann Poland 51:57–65

Chon HT, Lee JS (2004) Heavy metal contamination and human risk assessment at Au–Ag and base metal mine sites in Korea. 7th Intern Conf on Mercury as a Global Pollutant, Ljubljana 49 p

Chowdhury UK, Biswas BK, Roychowdhury T, et al. (2000) Groundwater arsenic contamination in Bangladesh and West Bengal, India. Environ Health Perspect 108:393–397

Christensen TH, Haung PM (1999) Solid phase cadmium and the reaction of aqueous cadmium with soil surfaces. In: McLaughlin MJ, Singh BR (eds) Cadmium in soils and plants. Kluwer Acad Publ, London, pp 65–96

Christensen JT, Stipp SLS, Baker.JA, Harstad AO (2004) Rare earth elements in natural calcite. Goldschmidt, Copenhagen, A512

Chuang MC, Shu GY, Liu JC (1996) Solubility of heavy metals in a contaminated soil: effects of redox potential and pH. Water Air Soil Pollut 90:543–556

Chukwuma SC (1995) Evaluating baseline data for copper, manganese, nickel and zinc in rice, yam, cassava and guinea grass from cultivated soils in Nigeria. Agric Ecosyst Environ 53:47–61

Chung J-B, Sa T-M (2001) Chromium oxidation potential and related soil characteristics in arable upland soils. Comm Soil Sci Plant Anal 32:1719–1733

Churey DJ, Gutenmann WH, Kabata-Pendias A., Lisk D (1979) Element concentrations in equilibrates of coal and lignite fly ashes. J Agric Food Chem 27:910–911

Cicchella D, De Vivo B, Lima A (2003) Palladium and platinum concentration in soils from the Napoli metropolitan area, Italy: possible effects of catalytic exhausts. Sci Total Environ 308:121–131

Cidu R, Fanfani I, Shand P, Edmunds WM, Vant Dack L, Gijbels R (1994) Determination of gold at the ultratrace level in natural waters. Anal Chim Acta 296:295–304

Ciesielski T, Mikhail V, Patukhov V, Fodor P, Bertenyi Z, Namieśnik J, Szefer P (2006) Relationships and bioaccumulation of chemical elements in the Baikal seal (*Phoca sibirica*) Environ Pollut 139:317–384

Clarke LB, Sloss LL (1992) Trace elements: emissions from coal combustion and gasification. IEA Coal Research, IEACR/49, London, pp 21–27

Clarks LC, Combs GF, Turnbull BW, et al. (1996) The nutritional prevention of cancer with selenium 1983–1993: a randomized clinical trial. J Am Med Assoc 276:1957–1963

Clarkson TW (1990) Human health risks from methylmercury in fish. Environ Toxicol Chem 9: 957–961

Clarkson TW, Marsh DO (1982) Mercury toxicity in man. In: Prasad AS (ed) Clinical, biochemical, and nutritional aspects of trace elements. Vol 6. AR Liss, New York, pp 549–568

Cleuvenbergen van RJA, Adams FC (1991) Organolead compounds. In: Hutzinger O (ed) Handbook of environmental chemistry. Vol 3E. Springer-Verlag, Berlin Heidelberg, pp 97–153

Coenen M (2004) Chlorine. In: Merian E, Anke M, Ihnat M, Stoeppler M (eds) Elements and their compounds in the environment. 2nd ed. Wiley-VCH Verlag, Weinheim, pp 1423–1443

Coleman SL, Bragg LJ (1990) Distribution and mode of occurrence of arsenic in coal. In: Chyi LL, Chou C-L (eds) Recent advances in coal geochemistry. Geol Soc Am Spec Pap 248:13–26

Coleman ME, Elder RS, Basu P (1992) Trace metals in edible tissues of livestock and poultry. JAOAC Int 75(4):615–625

Colle C, Masson M, Hurtevent P, Morello M, Hebert D (2002) Root uptake by cultivated vegetables from contaminated algae used as organic manure in soil. Radioprot-Colloques 37:C1-1329–C1-1333

Collery P, Shtemenko N, Bourleaud M, Etienne JC, Maymard L, Loriquet P (2004) Supplementation by rhenium compounds instead of iron compounds during the treatment by erythropoetin of anemia in cancer patients. In: Cser MA, László IS, Étienne J-C, Maymard Y, Centeno JA, Khassanova L, Collery P (eds) Metals in biology and medicine. John Libbey & Comp. Ltd., England, Paris, pp 534–537

Combs GF (2001) Selenium in global food systems. British J Nutr 85:517–547

Combs GF Jr (2005) Geological impacts on nutrition In: Selinus O, Alloway B, Centeno JA, Finkelman RB, Fuge R, Lindh U, Smedley P (eds) Essentials of Medical Geology, Impacts of the Natural Environment on Public Health. Elsevier Acad. Press, pp 162–177

Commission Regulation (EC) (2001) Setting maximum levels for certain contaminants in foodstuffs. No 466/2001, 16.3.2001, OJL 77, pp 1–12

Compeau G, Bartha R (1984) Methylation and demethylation of mercury under controlled redox, pH, and salinity conditions. Appl Environ Microbiol 48:1202–1207

Conley D (1997) Riverine contribution of biogenic silica to the pcean silica budget. Limon Oceanogr 42:774–777

Conley DJ (2002) Terrestrial ecosystems and global biogeochemical silica cycle. Global Biogeochem Cycles, 16(4), GB001894

Connor JJ, Shacklette HT (1975) Background geochemistry of some rocks, soils, plants and vegetables in the conterminous United States. USGS Prof. Pap. 574f

Copeland RA, Ayers JC (1972) Trace element distributions in water, sediment, phytoplankton, zooplankton and benthos of Lake Michigan: a baseline study with calculations of concentration factors and build up of radioisotopes in the food web. ERG Special Report

Cornelis R, De Kimpe J (1994) Elemental speciation in biological fluids. J Anal Atom Spectrometry 9:945–950

Cotton FA, Wilkinson GW (1989) Advanced inorganic chemistry. 5th ed., Wiley, New York

Coughlin JR, Nielsen FH (1999) Advances in boron essentiality research: In: Abdulla M, Bost M, Gamon S, Arnaud P, Chazot G (eds) New aspects of trace element research. Smith-Gordon Co, London, pp 33–41

Courchesne F, Séguin V, Dufresne A (2001) Solid phase fractionation of metals in the rhizosphere of forested soils. In: Gobran GR, Wenzel WW, Lombi E (eds) Trace elements in the rhizosphere. CRC Press, Boca Raton, FL, pp 189–206

Cowgill UM (1988) The tellurium content of vegetation. Biol Trace Elem Res 17:43–67

Cross HJ, Faux SP, Sadhra S, Sorahan T, Levy LS, Aw TC, Braithwaite R, McRoy C, Hamilton L, Calvert I (1997) Criteria document for hexavalent chromium. ICDA, Paris

Crössmann C (1994) Transfer und akkumulation von thallium bei Gemüse und obstarten. VdLUFT Kongress, Jena

Crowley JD, Traynor DA, Weatherburn DC (2000) Enzymes and proteins containing manganese: an overview. In: Sigel A, Sigel H (eds) Metal ions in biological systems. Manganese and its role in biological processes. Marcel Dekker, New York, pp 209–278

Cruz AC, Fomsgaard IS, Lacayo J (1994) Lead, arsenic, cadmium and copper in Lake Asosca, Nicaragua. Sci Total Environ 155:229–236

Cunningham SD, Berti WR (2000) Phytoextraction and phytoremediation: technical, economic, and regulatory consideration of the soil-lead issue. In Terry N, Banuelos G (eds) Phytoremediation of conminated soils and water. CRC Press, Boca Raton, FL, pp 359–376

Cunningham WC, Stroube WB Jr. (1987) Application of an instrumental neutron activation analysis procedure to analysis of food. Sci Total Environ 63:pp 29–43

Cunningham SD, Berti WR, Huang JM (1995) Phytoremediation of contaminated soils. TIBTECH 13: 393–397

Cunningham SD, Shann JR, Crowley DE, Anderson TA (1997) Phytoremediation of contaminated water and soil. In: Kruger EL, Anderson TA, Coats JR (eds) Phytoremediation of soil and water contaminants. ACS Symposium Series 664. Am Chem Soc, Washington DC, pp 2–19

Cuzba M, Moritimer DC (1980) Stability of methylmercury and inorganic mercury in aquatic plants (*Elodea densa*). Can J Botany 58:316–320

Czaban J, Wróblewska B (2005) Microbial transformation of cadmium in two soils differing in organic matter content and texture. Polish J Environ Studies 14:727–737

Czekała J (1997) Chromium in soil and plants – occurrence, sorption, and uptake related to its forms and rates, medium properties and fertilization. Rocz AR Rozp Nauk 274:3–90 (in Polish)

D'Haese PC, Verbeckmoes SC, Schrooten I, De Broe ME (2002) The effects of strontium on bone formation. Epidemiological experimental and cell biological aspects. Macro and trace elements, Workshop, Jena, pp 1223–1230

D'Haese PC, Lamberts LV, Broe ME (2004) Silicon. In: Merian E, Anke M, Ihnat M, Stoeppler M (eds) Elements and their compounds in the environment. 2nd ed. Wiley-VCH, Weinheim, pp 1273–1284

D'Itri FM (ed) (1972) The environmental mercury problem. CRC Press, Cleveland, OH

Da Silva EF, Zhang C, Pinto LS, Patinha C, Reis P (2004) Hazard assessment of arsenic and lead in soils of Castromil gold mining area, Portugal. Applied Geochem 19:887–898

Dalenberg JW, van Driel W (1990) Contribution of atmospheric deposition of heavy metal concentration in field crops. Netherlands J Agric Sci 38:365–379

Dąbrowski J (1992) Sources and evaluation of mercury pollution of cereals and feed. In: Kabata-Pendias A, Żmudzki J (eds) Mercury in the environment. Ecological and analytical problems. Polish Acad Sci, pp 27–40 (in Polish)

Dähn R, Schneidegger AM, Manceau A, Bayens B, Bradbury MH (2003) Uptake mechamisms of Ni(II) on montmorillonnnite as determined by X-ray absorption spectroscopy. Goldschmidt, Copenhagen, A164 p

Dai J-L, Zhang M, Zhu Y-G (2004) Adsorption and desorption of iodine by various Chinese soil. I. Iodate. Environ Intern 30:525–530

Dai X, Chai Ch, Mao H, Ouyang H (1998) Sorption and desorption of iridium in marine sediment–seawater system. 1st Intern conf. Trace metal speciation in biomedical, nutritional and environmental sciences, Neuherberg/München, 66 p

Dalenberg JW, van Driel W (1990) Contribution of atmospheric deposition to heavy-metal concentration in field crops. Netherland J Agric Sci 38:367–379

Danish EPA (2003) The elements in the second rank: an environmental problem now or in the future? Danish Environ Protection Agency, Environ Project, No 770

Dannecker W, Schöder B, Stechmann H (1990) Organic and inorganic substances in highway tunnel exhaust air. Sci Total Environ 93:293–300

Darling CTR, Thomas VG (2003) The distribution of outdoor shooting ranges in Ontario and the potential for lead pollution of soil water. Sci Total Environ 313:235–243

Das D, Samanta G, Mandal BK, Chowdhury RT, Chandra CR, Chowdhury PP, Basu GK, Chakraborti D (1996) Arsenic in groundwater in six districts of West Bengal, India. Environ Geochem Health 18:5–15

Das P, Samantaray S, Rout GR (1997) Studies of cadmium toxicity in plants: A review. Environ Pollut 98:29–36

Das AK, de la Guardia M, Cervera ML (2001) Literature survey on-line element speciation in aqueous solutions. Talanta 55:1–28

Dassani SD, McClellan E, Gordon M (1975) Submicrogram level determination of mercury in seeds, grains, and food products by cold-vapor atomic absorption spectrometry. J Agric Food Chem 23:671–674

Datta DK, Subramanian V (1998) Distribution and fractionation of heavy metals in the surface sediments of the Ganges-Brahmaputra-Meghna river system in the Bengal basin. Environ Geology 36:93–101

Dauchy X, Potin-Gautier M, Astruc A, Astruc M (1994) Analytical methods for the speciation of selenium compounds: a review. Fresenius J Anal Chem 348:792–805

Davies BE (1995) Lead. In Alloway BJ (ed), Heavy metals in soils. Blackie Academic, London, pp 206–223

Davies BE (1997) Deficiencies and toxicities of trace elements and micronutrients in tropical soils: Limitations of knowledge and future research needs. Environ Toxicol Chem 16:75–83

Davis S, Waller BA (1990) Quantitative estimation of soil ingestion in normal children between the ages of 2 and 7 years: population based estimates using aluminium, silicon and titanium as soil tracer elements. Arch Environ Health 45:112–122

De Cremer K, De Kimpe J, Cornelius R (1998) Biding of vanadium to different serum proteins. 1st Itern Conf Trace Metal Speciation in Biomedical Nutritional Environ Sci, Neuherberg/München, 40 p

De Miquel E, Llamas JF, Chacon E, et al. (1997) Origin and patterns of distribution of trace elements in street dust: unleaded petrol and urban lead. Atmospheric Environ 31:2733–2740

Debacker V. Jauniaux T, Coignoul F, Bouquegneau JM (2000) Heavy metals contamination and body condition of wintering guillemots (*Uria algae*) at the Belgian coast from 1993 to 1998. Environ Res Sec A 84:310–317

Dehn L-A, Follmann EH, Thomas DL, Sheffield GG, Rosa C, Duffy LK, O'Hara M (2005) Trophic relationships in an Arctic food web and implications for trace metal transfer. Sci Total Environ 362:103–123

Del Delumyea R, McKay T, Horowitz J (1997) Trends in air quality in northern Florida, 1945–1995 and comparison of air quality in Jacksonville, FL, with nine other cities in the United States 1972–1992. Fla Sci 60:175–192

Delhaize E, Ryan PR (1995) Aluminium toxicity and tolerance in plants. Plant Physiol 107:315–321

Dell B, Huang L (1997) Physiological response of plants to low boron. Plant Soil 193:103–110

De Miquel E, Llamas JF, Chacon E, et al. (1997) Origin and patterns of distribution of trace elements in street dust: unleaded petrol and urban lead. Atmospheric Environ 31:2733–2740

Denys S, Echevarria G, Leclerc-Cessac E, Morel JL (2005) Using stable elements and soil properties to assess the fate of radionuclides in the soil-plants system. ICOBTE, 8th Intern Conf, Adelaide, pp 450–451

Di Toro DM, Mahony JD, Carbonaro RF, Love JH, Morrissey JC (1996) The bioavailability of silver in sediments. 4th Argentum Intern Conf on Transport, Fate and Effects of Silver in the Environment, Madison, WI, pp 145–154

Dias GM, Edwards GC (2003) Differentiating natural and anthropogenic sources of metals to the environment. Human Ecol Risk Assess 9:699–721

Dickson EL, Stevens RJ (1983) Extractable copper, lead, zinc and cadmium in Northern Ireland Soils. J Sci Food Agric 34:1197–1205

Dienstbach F, Bächmann K, Thom C (1983) Untersuchungen des Plutoniumgehalts von Pflanzen. Z Pflanzenern Bodenkd 146:690–696

Dietrich D, Schlatter C, Blau N, Fisher M (1989) Aluminium and acid rain: mitigating effects of NaCl on aluminium toxicity to brown trout (*Salmo trutta* Fario) in acid water. Toxicol Environ Chem 16:267–283

Dinman BD (1983) Aluminium, alloys and compounds. In: Parmeggiani L (ed) Encyclopedia of occupational health and safety. ILO 1, Geneva, pp 131–135

Dirkx W, Lobinski R, Ceulemans M, Adams F (1993) Determination of methyl- and butyltin compounds in waters of the Antwerp harbour. Sci Total Environ 136:279–300

Dissanayake AB, Kritsotakis K (1984) The geochemistry and Au and Pt in peat and algal mats – A case study from Sri Lanka. Chem Geol 41:61–76

Divine KK, Goering PL (2004) Tantalum. In: Merian E, Anke M, Ihnat M, Stoeppler M (eds) Elements and their compounds in the environment, 2nd ed., Wiley-VCH, Weinheim, pp 1087–1097

Dixon HBF (1997) The biochemical action of arsenic acids especially as phosphate analogues. In: Sykes AD (ed) Advances in inorganic chemistry. Acad Press, San Diego, CA, pp 191–228

Dixon GR (1998) Modifying boron nutrition to minimize clubroot disease of *Brassicas*. Boron in Agriculture 18:1–4

Djingova R, Kovacheva P, Wagner G. Markert B (2003) Distribution of platinum group elements and other traffic related elements among different plants along some highways in Germany. Sci Total Environ 308:235–246

Djujić I, Jozanov-Stankov O, Milovac M, Bosnić O, Djermanović VT (2001) The impact of consuming wheat naturally enriched with selenium on trace elements and antioxidant defense in humans. 3rd Inter Symp Trace Elements in Human: New Perspectives, Athens, pp 281–304

Djujić I, Jozanov-Stankov O, Milovac M, Demajo A (2003) Manganese in the food chain of humans in Serbia. In: Ermakov VV (ed) Geochemical ecology and biochemical investigation on toxic substances in biosphere. Nauka, Moscow, pp 16–25

Djujić I, Milovac M, Jozanov-Stankov O (2004) Selenium protect plants from oxidative stress. 5th Intern Symp of Roumanian Academy-Branch, Timişara, 5, pp 65–70

Dmuchowski W (2003) Chlorine contents in the leaves of tree species characterized with relatively greater resistance to the adverse urban conditions. In: Gworek B, Misiaka J (eds) Obieg pierwiastków w przyrodzie, Inst Ochrony Środ, Warszawa, pp 309–312

Dmuchowski W (2005) Use of plant bioindicators in assessment of environmental contaminations with heavy metals. Rep Bot Garden of the Polish Academy of Sciences, Ser Monogr Treasises 7, 116 p
Dobbs AJ, French P, Gunn AM, Hunt DTE, Winnard DA (1990) Aluminium speciation and toxicity in upland waters. In: Lewis TE (ed) Environmental chemistry and toxicity of aluminium. Lewis Publ, Chelsea, MI, pp 209–228
Dobrowolski JW (2004) Yttrium. In: Merian E, Anke M, Ihnat M, Stoeppler M (eds) Elements and their compounds in the environment. 2nd ed., Wiley-VCH, Weinheim, pp 1193–1201
Dobson S, Cary R (2001) Chorine dioxide, WHO, Geneva, CICAD No 37
Dodd RT (1981) Meteorites a petrologic-chemical synthesis. Univ Press. Translated into Russian by Pietayeve Mi, Ulyanova AA (1986) Mir, Moscow
Doherty PJ (2004) Silver. In: Merian E, Anke M, Ihnat M, Stoeppler M (eds) Elements and their compounds in the environment. 2nd ed., Wiley-VCH, Weinheim, pp 751–764
Döhler H, Schultheiß U, Eckel H, Roth U (2002) Heavy metal concentrations in animal manure in Germany and the European Union. Wasser Boden 54:4–9
Dolgner R, Brockhaus A, Ewers U, Wiegand H, Majewsky F, Soddemann H (1983) Repeated surveillance of exposure to thallium in a population living in the vicinity of a cement plant emitting dust containing thallium. Intern Arch Occup Environ Health 52:79–94
Dolin VV, Bondarenko GN (2004) Geochemical conjugations between abiogenic speciation and biogenic migration of radionuclides in terrestrial ecosystems. Goldschmidt, Copenhagen, A505 p
Donard OFX, Shor FT, Webe JH (1987) Regulation of tin and methyltin compounds by the green alga Enteromorpha under simulated estuarine conditions. Can J Fish Aquat Sci 44:140–145
Drabek O, Boruvka L, Mladkova L, Kocarek M (2003) Possible method of aluminium speciation in forest soils. J Inorganic Biochem 97:8–15
Drasch G, Horvart M, Stoeppler M (2004) Mercury. In: Merian E, Anke M, Ihnat M, Stoeppler M (eds) Elements and their compounds in the environment. 2nd ed., Wiley-VCH, Weinheim, pp 931–1005
Drebickas V, Drulia B, Tamuklentait O, Maceika E (1996) The migration of radiocesium in pine forest biogeocenosis of the Curoniam Spit (Lithuania) after the Chernobyl accident. In: Anke M (ed) Mengen und Spuren Elements. 16. Arbeitstagung, Jena, pp 679–683
Driscoll CT, Wyskowski BJ, Destaffan P, Newton RM (1989) Chemistry and transfer of aluminium in a forested watershed in the Adirondack region of New York, USA, New Hampshire, and Jamieson Creek, British Colombia. Water Resour Res 24:195–200
Drobner U, Tyler G (1998) Conditions controlling relative uptake of potassium and rubidium by plants from soils. Plant Soil 201:285–293
Duda M, Pasternak K (2003) Lithium – its role and meaning in psychiatry, the mechanisms of action and the side effects. J Elementology 8:185–198 (in Polish)
Dudas MJ (1987) Accumulation of native arsenic in acid sulphate soils in Alberta. Can J Soil Sci 67:317–331
Dudka S (1992) The concentrations of some scandium subgroup trace metals in soils of Poland. Geoderma 52:279–289
Dudka S, Markert B (1992) Baseline concentrations of As, Ba, Be, Li Nb, Sr and V in surface soils of Poland. Sci Total Environ 122:279–290
Dudka S, Piotrowska M, Chlopecka A (1993) Forms of chromium and nickel in soils. In: Kabata-Pendias A (ed) Chromium, nickel and aluminum in the environment: ecological and analytical problems. Kom Nauk PAN, Warszawa, pp 15–22 (in Polish)
Duffus JH (2002) "Heavy metals" – A meaningless term? Pure Appl Chem 74:793–807
Dushenkov S, Sorochinsky B (1999) Alternatve agriculture and fate of ^{137}Cs and ^{90}Cs in the Chernobyl exclusion zone. 5th Intern Conf Biogeochem Trace Elements, Vienna, pp 60–61
Dybczyński R (1998) Reference materials and their role in quality assurance in inorganic trace analysis. In: Kabata-Pendias A, Szteke B (eds) Quality problems in trace analysis in environmental studies. 2nd ed. Wyd Edukacyjne Z. Dobkowska, Warszawa, pp 41–78 (in Polish)
Eaton FM (1944) Deficiency, toxicity, and accumulation of boron in plants. J Agric Res 69:237–277
Eaton RDP, Secord DC, Hewitt P (1980) An experimental assessment of the toxic potential of mercury in ringed-seal liver for adult laboratory cats. Toxicol Appl Pharmacol 55:514–521
Ebbs SD, Lasat MM, Brady DJ, Cornish J, Gordon R, Kochian LV (1997) Phytoextraction of cadmium and zinc from a contaminated soil. J Environ Qual 26:1424–1430

Ebinghaus R, Tripathi RM, Wallshiläger D, Lindberg SR (1999) Natural and anthropogenic mercury sources and their impact on the air-surface exchange of mercury on regional and global scales. In: Ebinghaus R, Turner RR, de Lacerda LD, Vasiliev O, Salomons W (eds) Mercury contaminated sites: Characterization, risk assessment and remediation. Springer-Verlag, Berlin Heidelberg, pp 3–50

EC (2003) Report from the EU Comm to the Council and the European Parliament on the Implementation of Community Waste Legislation – Directive 75/442/EEC on waste, Directive 91/689/EEC on hazardous waste, Directive 75/439/EEC on waste oils, Directive 86/278/EEC on sewage sludge and directive 94/62/EC on packaging and packing waste – for the period 1998–2000. COM/2003/0250 final/3

ECETOC (1997) Ecotoxicology of some inorganic borates – Interim report. Brussels, European Centre for Ecotoxicology and Toxicology of Chemicals. Special Report 11

Echevarria G, Morel JL, Fardeus JC, Leclerc-Cessas E (1998) Assessment of phytoavailabilit of nickel in soils. J Environ Qual 27:1064–170

Echevarria G, Vong PC, Leclerc-Cessac E, Morel JL (1997) Bioavailability of technetium-99 as affected by plant species and growth, application form, and soil incubation. J Environ Qual 26:947–956

Eckel WP, Jacob TA (1988) Ambient levels of 24 dissolved metals in U.S. surface and ground waters. Natl Meet Am Chem Soc Div Environ Chem 28:371–372

Eckel WP, Langley WD (1988) A background-based ranking technique for assessment of element enrichment in soils at hazardous waste sites. In: "Superfund 88", 9th Natl Conf, The Hazardous Materials Control Research Institute, Washington, DC, pp 282–285

Eckel H, Roth U, Döhler H, Nicholson F, Unwin R (eds) (2005) Assessment and reduction of heavy metal input into agro-ecosystems. Final rep. EU-Concerted Action AROMIS, KTBL-Schrift, Darmstadt

Edmunds E, Smeldey P (2005) Fluoride in natural waters. In: Selinus O, Alloway BJ, Centento JA, Finkelman RB, Fuge R, Lindh U, Smeldey P (eds) Medical geology. Elsevier, Amsterdam, pp 301–329

Edner H, Faris GW, Sunesson A, Svanberg S (1989) Atmospheric atomic mercury monitoring using differential absorption LIDAR technique. Appl Opt 28:921

Edwards SC, Macleod CL, Lester JN (1998) The bioavailability of copper and mercury to the common nettle (*Urtica dioica* L.) and the earthewoam (*Eisenia fetida*) from contaminated dredge soil. Water Air Soil Pollut 102:75–81

EEA (European Environmental Agency) (1995) Europe's Environment. Earthscan Publ, London

Efroymson RA, Will ME, Suter II GW, Wooten AC (1997) Toxicological bench marks for screening contaminants of potential concern for effects on terrestrial plants. East Tennessee Technology Park, Techn Inform Office, ES/ER/TM-85/R-3

Egan SK, Tao SS-H, Pennington JAT, Bolger PM (2002) U.S. food and drug administration's Total Diet Study: intake of nutritional and toxic elements, 1991–1996. Food Add Contam 19:103–125

Eichenberger E (1986) Concepts of metal ion toxicity: In: Siegel H (ed) Metal ions in biological systems 20, Marcel Dekker, New York, pp 67–100

Eisler R (1979) Copper accumulation in coastal and marine biota. In: Nriagu GO (ed) Copper in the Environment. Part I: Ecological cycling. Wiley, New York, pp 383–449

Eisler R (1981) Trace metal concentrations in marine organizms. Pergamon Press, New York

Eisler R (1985) Cadmium hazards to fish, wildlife and invertebrates: A synoptic review. U.S. Fish Wild Serv Biol Rep 85:1–46

Eisler R (1990) Boron hazards to fish, wildlife, and invertebrates: a synoptic review. U.S. Fish Wildl, Ser Biol Rep 85

Eisler R (1994) A review of arsenic hazards to plants and animals with emphasis on fishery and wildlife resources. In: Nriagu JO (ed) Arsenic in the Environment. Part II. Wiley, New York,

Eisler R (1997) Zinc hazards to plants and animals with emphasis on fishery and wildlife resources. In: Cheremisinoff PN (ed) Ecological issues and environmental impact assessment. Advances in environmental control technology series. Gulf Publ Comp, Houston, TX, pp 443–537

Eisler R (1998a) Silver hazards to fish, wildlife and invertebrates: A synoptic review. Contaminant Hazard Reviews Report 32, National Biological Service, U.S. Depart of Interior, Washington, DC

Eisler R (1998b) Copper hazards to fish, wildlife, and invertebrates: A synoptic review. Biological Science Report USGS/BRD/BSR-1997-0002, pp 5–146

Eisler R (2000) Mercury. Handbook of chemical risk assessment: Health hazards to humans, plants, and animals. vol 1, Metals, Lewis Publ, Boca Raton, FL, pp 313–409

Eisler R (2004) Gold concentration in abiotic materials, plants, and animals: A synoptic review. Environ Monitoring Assessment 90:73–88

El-Hassanin AS, Labib TM, Dobal AT (1993) Potential Pb, Cd, Zn and B contamination of sandy soils after different irrigation periods with sewage effluent. Water Air Soil Pollut 66:239–249

El Wardani SA (1958) Marine geochemistry of germanium. Geochim Cosmochim Acta 15:237–254

Elinder CG (1985) Normal values of cadmium in human tissue, blood and urine in different countries. In: Friberg L, Elinder CG, Kjellstrom T, et al. (eds) Cadmium and Health: A toxicological epidemiological appraisal. Vol 1. Exposure, dose, and metabolism. CRC Press, Boca Raton, FL, pp 81–102

Elinder CG (1992) Cadmium as an environmental hazard. IARC Sci Pub 1118:123–132

Ellis DR, Salt DE (2003) Plants, selenium and human health. Current Opinion Plant Biol 6:237–279

Elliott P, Arnold R, Barltrop D, Thronton I, House IM, Henry JA (1999) Clinical lead poisoning in England: an analysis of routine sources of data. Occup Environ Med 56:820–824

Elrashidi MA, O'Connor GA (1982) Boron sorption and desorption in soils. Soil Sci Soc Am J 46:27–31

Emelyanov EM (2001) Biogenic components and elements in sediments of the Central Baltic and their redistribution. Mar Geol 172:23–41

EMEP (2000) Heavy metal transboundary air pollution in Europe: Monitoring and results for 1997 and 1998. Meteorological synthesizing center-east, Moscow, EMEP Report 3/2000, 139 p

Engle M, Gustin M, Zhang H (2001) Quantifying natural source mercury emissions from the Ivanhoe Mining District, North-Central Nevada, USA. Atmosph Environ 35:3987–3997

Ensminger AH, Ensminger ME, Konlade JE, Roson JRK (1995) The concise encyclopedia of foods and nutrition, 2nd ed. CRC Press, Boca Raton, FL

Entry JA, Watrud LS (1998) Potential remediation of ^{137}Cs and ^{90}Sr contaminated soil by accumulation in Alamo switchgrass. Water Air Soil Pollut 104:339–345

Epstein E (1994) The anomaly of silicon in plant biology. Proc Natl Acad Sci USA 91:11–17

Epstein E (1999) Silicon. Ann Rev Plant Physiol Plant Mol Biol 50:641–664

Erämetsä O, Yliroukanen I (1971) The rare earths in lichens and mosses. Suom Kemistil 44b:121–128

Ericksen JA, Gustin MS, Schorran DE, Johnson DW, Lindberg SE, Coleman JE (2003) Accumulation of atmospheric mercury in forest foliage. Atmosph Environ 37:1613–1622

Erickson BE, Helz GR (2000) Molybdenum (VI) speciation in sulfidic waters: stability and lability of thiomolybdates. Geochim Cosmochim Acta 64:1149–1158

Ericson JE (2001) Enamel lead ionmarkers for prenatal exposure assessment. Environ Res Section A 87:136–140

Eriksson JE (1990) A field study on factors influencing Cd levels in soils and grains of oats and winter wheat. Water Air Soil Pollut 53:69–81

Eriksson JE (2001a) Concentrations of 61 trace elements in sewage sludge, farmyard manure, mineral fertilizers, precipitation and in oil and crops. Swedish EPA Rep 5159, Stockholm

Eriksson JE (2001b) Critical load set to "no further increase in Cd content of agricultural soils" consequences. In: Proc Intern Expert Group on Effect-based Critical Limits for Heavy Metals, Bratislava

Eriksson JE, Öborn I, Jansson G, Andersson A (1996) Factors influencing Cd content in crops – results from Swedish field investigations. Swed J Agric Res 26:125–133

Ermakov VV (2004) The halogens, Fluorine. In: Merian E, Anke M, Ihnat M, Stoeppler M (eds) Elements and their compounds in the environment. 2nd ed. Wiley-VCH Verlag, Weinheim, pp 1415–1421

Ernst E (2002) Toxic heavy metals and undeclared drugs in Asian herbal medicines. Trends Pharmacol Sci 23:136–139

Ernst WHO (1998a) Effects of heavy metals in plants at the cellular and organismic level. In: Schürmann G, Markert G (eds) Ecotoxicology. Wiley/Spektrum Acad, New York, pp 587–620

Ernst WHO (1998b) The origin and ecology of contaminated, stabilized and non-pristine soils. In: Vangronsveld J, Cunnungham SD (eds) Metal-contaminated soils. In situ inactivation and phytorestoration. Springer-Verlag, Berlin Heidelberg, pp 17–25

Eskew DL, Welch RM, Cary EE (1983) Nickel: an essential micronutrient for legumes and possible all higher plants. Science 222:621–623

Etat de l'environment Wallon (1995) Etat de'lenvironment Wallon-1994, vol 1. Ministère de la Région Wallonne – Direction Générale des Ressources Naturelles et de l'Environment (DGRNE), Belgium

Etris S (1997) Silver and silver alloys. In: Howe-Grant M (ed) Kirk-Othmer Encyclopedia of Chemical Technology, 4nd ed., Wiley, New York, pp 163–179

Eurola M, Alfthan G, Aro A, Ekholm P, Hietaniemi V, Rainio H, Rankanen R, Venäläinen E-R (2003) Results of the Finnish selenium monitoring program 2000–2001. MTT Agrif Res Finland Report 36, Jokioinen

Ewell WS, Gorsuch JW, Ritter M, Ruffing CJ (1993) Ecotoxicological effects of silver compounds. 1st Argentum Intern Conf on the Transport, Fate, and Effects of Silver in the Environment. Madison, WI, 9 p

Ewers U (1988) Environmental exposure to thallium. Sci Total Environ 71:285–292

Expert Panel on Mercury Atmospheric Processes (1994) Mercury atmospheric processes: a synthesis report, no TR-104214

Fagioli F, Locatelli C, Landi S (1994) Heavy metals in the Goro bay: sea water, sediments and mussels. Annali di Chimica 84:129–140

Falandysz J (1994) Some toxic and trace metals in big game hunted in the northern part of Poland in 1987–1991. Sci Total Environ 141:59–93

Falandysz J, Bona H, Danisiewicz D (1994a) Silver uptake by *Agaricus bisporus* from an artificially enriched substrate. Z Lebensm Unters Forsch 199:225–228

Falandysz J, Kotecka W, Kannan K (1994b) Mercury, lead, cadmium, manganese, copper, iron and zinc concentrations in poultry, rabbit and sheep from the northern part of Poland. Sci Total Environ 141:51–57

Falandysz J, Szymczyk K, Ichihashi H, Bielawski L, Gucia M, Frankowska A, Yamasaki S-I (2001) ICP/MS and ICP/AES elemental analysis (38 elements) of edible wild mushrooms growing in Poland. Food Add Contam 18:503–513

Falandysz J, Kubota R, Kunito T, Bielawski L, Brzozowski A, Gucia M, Jadrusiak A, Lipka K, Tanabe S (2003) Relationships between selenium and mercury in the fruiting bodied of some mushrooms growing in Poland. J Phys IV, France 107:443–446

FAO/WHO (1998) Committee on Food Additives and Contaminants. 31st Season, Poison Paper on Arsenic. CX/FAC 99/22, Alimentarius Commission, Food Standard Programme. Rome

FAO/WHO (2002) Codex Alimenatry Commission. CX/FAC 04/36/16

FAO/WHO (2003) Codex Alimenatry Commission. CX/FAC 04/36/16

Farago ME, Parsons PL (1994) The effects of various platinum metal species on the water plant *Eichhornia crassipes* (Mart.) Solms Chem Spec Bioavail 6:1–12

Farago ME, Kavanagh P, Blanks R, Kelly J, Kazatzis G, Thornton I, Simpson RS, Cook JM, Delves HT, Hall GEM (1998) Platinum concentrations in urban dust and soil, and in blood and urine in the United Kingdom. Analyst 123:451–454

Fairbridge RW (ed) (1972) The encyclopedia of geochemistry and environmental sciences. Van Nostrand Reinhold Company, New York

Farmer JG, Graham MC, Thomas RP, Licona-Manzur C, Paterson E, Campbell CD, Geelhoed JS, Lumsssdon DG, Meeussen JCL, Roe MJ, Conner A, Fallick AE, Bewley RJF (1999) Assessment and modeling of the environmental chemistry and potential for remediation treatment of chromium-contaminated land. Environ Geochem Health 21:331–337

Farmer JG, Thomas RP, Graham MC, Geelhoed JS, Lumsdon DG, Paterson E (2002) Chromium speciation and fractionation in ground and surface waters in the vicinity of chromite ore processing residue disposal sites. J Environ Monit 4:235–243

Farrell N (1996) Current status of structure-active relationships of platinum anticancer drugs: activation of the trans geometry. In: Sigel A, Sigel H (eds) Metal ions in biological systems, 32. pp 608–726

FDA (1998) Mercury in fish: causes for concern? U.S. Food and Drug Administration. FDA Consumer (http://vm.cfsan.fda.gov/~dms/mercury.html)

Feldmann J, Himer AV (1995) Occurrence of volatile metal and metalloid species in landfills and sewage gases. Intern J Environ Anal Chem 60:339–359

Felix HR, Kayser A, Schulin R (1999) Phytoremediation, field trails in the years 1993–1998. 5th ICOBTE Itern Conf, Vienna, pp 8–9

Fendorf S, Hansel CM, Wielinga B (2002) Operative pathways of chromate and uranyl reduction within soils and sediments. In: Zhang P-C, Brady PV (eds) Geochemistry of soil radionuclides. SSSA Spec Publ, Madison WI, pp 111–129

Fendorf S, La Force MJ, Li G (2004) Temporal changes in soil partitioning and bioaccessilbility of arsenic, chromium, and lead. J Environ Qual 33:2049–2055

Fent K (1996a) Ecotoxicology of organotin compounds. Crit Rev Toxicol 26:1–117

Fent K (1996b) Organotins in municipal wastewater and sewage sludge. In: Champ MA, Seligman PF (eds) Organotin – Environmental fate and effects. Chapman Hall, London, pp 1–25

Fergusson JE (ed) (1990) The heavy metals: Chemistry, environmental impact and health effects. Pergamon Press, New York

Fergusson JE, Kim ND (1991) Trace elements in street and house dusts: sources and speciation. Sci Total Environ 100:125–150

Ferrara R (1999) Mercury mines in Europe: Assessment of emissions and environmental contamination. In: Ebinghaus R, Turner RR, de Lacerda LD, Vasiliev O, Salomons W (eds) Mercury contaminated sites: Characterization, risk assessment and remediation. Springer-Verlag, Berlin Heidelberg, pp 51–72

Ferrara R, Maseti BE, Edner H, Ragnarson P, Svanberg S, Wallinder E (1991) Mercury in abiotic and biotic compartments of an area affected by a geological anomaly (Mt Amiata, Italy). Water Air Soil Pollut 56:219–233

Ferrari CP, Hong S, van de Velde K, Boutron CF, Rudniev SN, Bolshov M, Chisholm W, Rosman KJR (2000) Natural and anthropogenic bismuth in central Greenland. Atmos Environ 34:941–948

Ferretti M, Ghisi R, Merlo L, Vecchia FD, Passera C (1993) Effect of cadmium on photosynthesis and enzymes of photosynthetic sulphate and nitrate assimilation pathways in maize (*Zea mays* L.). Photosynthetica 29:49–54

Ferris FG, Hallberg RO, Lyvén B, Pedersen K (2000) Retention of strontium, cesium, lead and uranium by bacterial iron oxides from a subterranean environment. Appl Geochem 15:1035–1042

Filella M, Belzile N, Chen Y-W (2002) Antimony in the environment: a review focused on natural waters. Earth-Science Reviews 57:125–176

Filius A, Streck T, Richter J (1998) Cadmium sorption and desorption in limited topsoils as influenced by pH: Isotherms and simulated leaching. J Environ Quality 27:12–18

Finkelman RB (1999) Trace elements in coal. Environmental and health significance. Biol Trace Element Res 67:197–204

Finkelman RB, Belkin HE, Centeno JA, Baoshan Z (2003) Geological epidemiology: coal combustion in China. In: Skinner HC, Berger AR (eds) Geology and health: closing gap. Oxford Univ Press, Oxford, pp 45–50

Fishelson L, Yawetz A, Perry AS, Zuk-Rimon, Z, Manelis R, Dotan A (1994) The environmental Health Profile (EHP) for the Acre Valley (Isreal): xenobiotics in animals and physiological evidence of stress. Sci Total Environ 144:33–45

Fisher NS, Reinfelder JR (1995) The trophic transfer of metals in marine systems. In: Tessier A, Turner DR (eds) Metal speciation and bioavailability in aquatic systems. Wiley, London, pp 363–406

Fisher NS, Wang WX (1998) Trophic transfer of silver to marine herbivores: a review of recent studies. Environ Toxicol Chem 17:562–571

Fitz WJ, Wenzel WW (2002) Arsenic transformations in the soil-rhizosphere-plant system, fundamentals and potential application to phytoremediation. J Biotechnol 99:259–278

Fitzgerald WF (1994) Global biogeochemical cycling of mercury. DOE/FDA/EPA Workshop on Methylmercury and Human Health, Bethesda, MD

Flegal AR, Patterson C (1985) Thallium concentrations in seawater. Mar Chem 15:327–331

Flegal AR, Itoh K, Patterson CC, Wong CS (1986) Vertical profile of lead isotopic composition in the north-east Pacific. Nature 321:689–690

Fleming GA (1980) Essential micronutrients. Boron and molybdenum. In: Davies BE (ed) Applied soil trace elements. Wiley, New York, pp 155–197

Flora SJS, Bhadauria S, Pant SC, Dhaked RK (2005) Arsenic induced blood and brain oxidative stress and its response to some thiol chelators in rats. Life Sciences 77:2324–2337

Florence TM (1980) Speciation of zinc in natural waters. In: Nraigu JO (ed), Zinc in the environment. Part A, Chapter 7, Wiley, New York, pp 199–217

Floriańczyk B, Grzybowska L (2006) Concentraton of metallohionein and microelements in breast cancer. Polish J Environ Stud 15(2A):69–71

Flynn HC, Meharg AA, Bowyer PK, Paton GI (2003) Antimony bioavailability in mine soils. Environ Pollut 124:93–100

Font O, Querol X, López-Soler A, Chimenos JM, Fernández AI, Burgos S, Peña FG (2004) Ge extraction from gasification fly ash. Fuel, pp 1–9

Fordyce F (2005) Selenium deficiency and toxicity in the environment. In: Selinus O, Alloway BJ, Centeno JA, Finkelman RB, Fuge R, Lindh U, Smedley P (eds) Essentials of medical geology. Elsevier, Amsterdam, pp 373–415

Fordyce F, Guandi Z, Green K, Xinping L (2000) Soil, grain and water chemistry in relation to human selenium-responsive diseases in Enshi Disrict, China. Appl Geochem 15:117–132

Förstner U (1979) Sources and sediment association of heavy metals in polluted coastal regions. Ahrens LH, Protas G (eds) Origin and distribution of the elements. Pergamon Press, Oxford, pp 849–866

Forsyth DS, Sun WF, Dalglish K (1994) Survey of organotin compounds in blended wines. Food Add Contam 11:343–350

Forsythe BL II, La Point TW, Cobb GP, Klaine SJ (1996) Silver in an experimental freshwater ecosystem. 4th Argentum Intern Conf on Transport, Fate and Effects of Silver in the Environment, Madison, WI, pp 185–189

Frank A (2003) Molybdenosis leading to Type 2 Diabetes Mellitus in Swedish moose. In: Skinner HCW, Berger AR (eds) Geology and health-closing the gap. Oxford Univ Press, pp 79–81

Frank R, Stonefield KI, Suda P (1982) Impact of nickel contamination on the production of vegetables on an organic soil, Ontario, Canada. 1980–1981. Sci Total Environ 26:41–65

Frank A, Wibom R, Danielson R (2002) Myocardial cytochrome *c* oxydase activity in Swedish moose (*Alces alces* L.) affected by molybdenosis. Sci Total Environ 290:121–129

Frankel RB, Bazylinski DA (2003) Biologically induced mineralization by bacteria. In: Dove PM, De Yoreo JJ, Weiner S (eds) Reviews in mineralogy and geochemistry. Biomineralization. Min Soc Am Geochem Soc, Washington DC, pp 95–114

Frankenberger WT, Engberg RA (eds) (1998) Environmental chemistry of selenium. Marcel Dekker, New York

Frankenberger WT, Karlson U (1994) Microbial volatilization of selenium from soils and sediments. In: Frankerberger WT, Benson S (eds) Selenium in the environment. Marcel Dekker, New York, pp 369–387

Freda J (1991) The effects of aluminium and other metals on amphibians. Environ Pollut 71:305–328

Freedman B, Hutchinson TC (1980) Pollutant inputs from the atmosphere and accumulations in soils and vegetation near a nickel-copper smelter at Sudbury, Ontario, Canada. Can J Soil Bot 58:108–132

Freitas MC, Reis MA, Alves LC, Wolterbeek HTh (2000) Nuclear analytical techniques in atmospheric trace element studies in Portugal. In: Markert B, Friese K (eds) Trace elements – Their distribution and effects in the environment. Elsevier, Amsterdam, pp 187–213

Freitas H, Prasad MNV, Pratas J (2004) Plant community tolerant to trace elements growing on the degraded soils of São Domingos mine in the south east of Portugal: environmental implications. Environ Intern 30:65–72

Frescholtz T, Gustin M (2004) Soil and foliar mercury emission as a function of soil concentration. Water Air Soil Pollut 155:223–237

Friberg L, Vostal D (eds) (1974) Mercury in the environment. CRC Press, Cleveland, Ohio

Friberg L, Piscator M, Nordberg GF, Kjelström T (1976) Cadmium in the Environment. 2nd ed., CRC Press, Cleveland, Ohio

Friberg L, Elinder CG, Kjellström T, Nordberg GF (1986) Cadmium and Health: A Toxicological and Epidemiological Appraisal. vol 11, Effects and Responses. CRC Press, Boca Raton, FL

Frøslie A (1990) Problems on deficiency and excess of minerals in animal nutrition. In: Låg J (ed) Geomedicine. CRC Press, Boca Raton, FL, pp 37–60

Fuge R (1988) Sources of halogens in the environment, influences on humas and animal health. Environ Geochem Health 10:51–61

Fuge R (2005) Anthropogenic sources. In: Sellinus O (ed) Essential of medical geology-impact of the natural environment on public health. Elsevier, Amsterdam, pp 43–60

Fuge R, Adrews MJ (1988) Fluorine in the UK environments. Environ Geochem Health 10:96–104

Fuge R, Johnson CC (1986) The geochemistry of iodine – a review. Environ Geochem Health 8: 31–54

Gabrielides GP, Alzieu C, Readman JW, Bacci E, Aboul DO, Salihoglu I (1990) MED POL survey of organotins in the Mediterranean. Mar Pollut Bull 21:233–237

Gaillardet J, Viers J, Dupré B (2003) Trace elements in river waters. In: Drever JI (ed) Surface and ground water, weathering and soils In: Holland HD, Turekian KK (eds) Treatise on geochemistry. Elsevier, Oxford 5:225–227

Gainer JH (1974) Environ Health Prespect Exp 7:113–119, as cited in USEPA: Ambient Water Quality Criteria Doc: Lead p c–71, 1980 EPA 440/5–80–057

Galal-Gorchev H (1993) Dietary intake, levels in food and estimated intake of lead, cadmium, and mercury. Food Add Contam 10:115–128

Galani-Nikolakaki SM, Kallithrakas-Kontos NG (2007) Elemental content of wines. In: Szefer P, Nriagu JO (eds) Mineral components in foods. Taylor & Francis, CRC Press. Boca Raton, FL pp (in press)

Gallichio VS (2003) Lithium effects in virus, blood cell effects and signal transcuction pathways. 4th Intern Symp Trace Elements in Human. New Persp. Enthypossis, Athens, 1:499–502

Galster WA (1976) Mercury in Alaskan Eskimo mothers and infants. Environ Health Perspec 15:135–140

Garcia G, Batista RS, Parra AML, Faz Á, Cunha M (2002) Availability of heavy metals in nutrient solutions for plants growing in polluted soils. 12th Symp Intern Biodeterioration and Biodegradation, Prague, 203 p

Garrels JW, Christ CL (1965) Solutions, minerals and equilibria. Harper Row, New York

Garrett R (1996) Natural sources of cadmium. In: OECD Proceedings – Sources of cadmium in the environment. OECD, Paris, pp 149–161

Gartrell JW (1981) Ditribution and correction of copper deficiency in crop and pastures. In: Loneragan JF, Robson AD, Graham RD (eds) Copper in soils and plants. Acad Press, Sydney, pp 313–349

Gartrell MJ, Craun JC, Podrebarac DS, Gunderson EL (1986) Pesticides, selected elements, and other chemicals in adult total diet samples, October 1980–March 1982. J Assoc Off Anal Chem 69:146–161

Ge K, Yang GQ (1993) The epidemiology of selenium deficiency of endemic disease in China. Am J Clin Nutr suppl 57:259–263

Geelhoed JS, Meeussen CL, Lumsdon DG, Hiller S, Roe MJ Thomas RP, Bewley RJF, Farmer JG, Paterson E (2001) Modelling of chromium behaviour and transport at sited contaminated with chromite ore processing residue: implications for remediation methods. Environ Geochem Health 21:261–265

Geelhoed JS, Meeussen CL, Roe MJ, Hiller S, Thomas RP, Farmer JG, Paterson E (2003) Chromium remediation or release? Effect of iron (II) sulfate addition on chromium (VI) leaching from columns of chromite ore processing residue. Environ Sci Technol 32:3206–3213

Gehrke SG, Kulaksiz H, Herman T, Riedel H-D, Bents K, Veltkamp C, Stremmel W (2003) Expression of hepcidin in hereditary hemochromatosis: evidence for a regulation in response to the serum trasferrin saturation and to non-transferrin-bound iron. Blood 102:371–376

Genderbien A, Carlton-Smith C, Izzo M, Hall JE (1999) UK sewage sludge survey: National Presentation, Techn Report P165, Environ Agency, Swindon

Gensemer RW, Playle RC (1999) The bioavailability and toxicity of aluminium in aquatic environments. Critical Reviews Environ Sci Technol 29:315–450

Gerber GB, Léonard A (1997) Mutagenicity, carcinogenicity and tetratogenicity of germanium compounds. Mutation Research 387:141–146

Gerhardssom L (2004) Lead. In: Merian E, Anke M, Ihnat M, Stoeppler (eds), Elements and their compounds in the environment, 2nd ed., Wiley-VCH Verlag, Weinheim, pp 879–900

Gerth J, Brümmer GW, Tiller KG (1993) Retentiona Ni, Zn, and Cd by Si-associated goethite. Z Pflanzernähr Bodenk 156:123–129

Gerzabek MH, Muramatsu Y, Strebl F, Yoshida S (1999) Iodine and bromine contents of some Austrian soils and relation to soil characteristics. 5th Intern Conf Biogeochem Trace Elements, Vienna, pp 638–639

Gettier SW, Burkman WG, Adriano DC (1987) Factors affecting vanadium phytotoxicity. In: Lindberg SE, Hutchinson TC (eds) Heavy metals in the environment. CEP Consult, Edinburgh 1:469–472

Gettier SW, Adriano DC (1987) The availability of soil-added thallium and beryllium. In: Linberg SE, Hutchinson TC (eds) Heavy meals in the environment. CEP Consult, Edinburgh 1:472–473

Girling CA, Peterson PJ, Warren HV (1979) Plants as indicators of gold mineralization at Waston Bar, British Colombia, Canada. Econ Geol 74:902–907

Glasauer S, Beveridge TJ (2003) The dissimilatory reduction of Mn by *Shewanella putrefaciens*: mineralogical and biological constraints. 16th Itern Symp Environ Biogeochem, Oirase, Japan, p 112
Glei M (2004) German. In: Merian E, Anke M, Ihnat M, Stoeppler (eds), Elements and their compounds in the environment, 2nd ed., Wiley-VCH Verlag, Weinheim, pp 787–793
Gobran GR, Clegg S (1996) A conceptual model for nutrient availability in the mineral soil–root system. Can J Soil Sci 76:125–131
Göd R (1994) Geogene Arsengehalte außergewöhnlichen Ausmaßes in Böden, nördliche Saualpe – ein Beitrag zur Diskussion um Grenzwerte von Spurenelemente in Böden. Berg- und Hüttenmännische Monatshefte, Leoben 139:442–449
Göd R, Heiss G (1996) Die Arsenanomalie Feistriz am Wechsel (Niederösterreich). Jahrb Geol Bundesanstalt, Wien 139:437–444
Godal A, Langseth W, Sivertsen T, Lund W (1995) Determination of aluminium in liver from reindeer, moose and sheep by electrothermal atomic absorption spectrometry. Sci Tot Environ 168: 249–254
Goering PL (2004) The lanthanides. In: Merian E, Anke M, Ihnat M, Stoeppler M (eds) Elements and their compounds in the environment. 2nd ed., Wiley-VCH, Weinheim, pp 867–878
Goering PL, Ziegler TI (2004) Niobium (Nb) (Columbium). In: Merian E, Anke M, Ihnat M, Stoeppler M (eds) Elements and their compounds in the environment, 2nd vol. Wiley-VCH, Weinheim, 2nd ed., pp 1039–1046
Goldbach H, Amberger A (1986) Influence of B nutrition on cell wall polysaccharides in cell cultures of *Daucus carota* L. J Plant Physiol 123:263–269
Goldberg ED (1965) Minor elements in seawater. In: Riley JP, Skirrow G (eds) Chemical oceanography. Academy Press, London
Goldberg S, Glaubig RA (1986) Boron absorption on California soils. Soil Sci Soc Am J 50:1173–1176
Goldberg SP, Smith KA (1984) Soil manganese: E values, distribution of manganese-54 among soil fractions, and effects of drying. Soil Sci Soc Am J 48:559–564
Goldberg S, Forster HS, Lesch SM, Heick EL (1993a) Boron adsorption mechanisms on oxides, clay minerals, and soils inferred from ionic strength rffects. Soil Sci Soc Am J 57:704–708
Goldberg S, Forster HS, Lesch SM, Heick EL (1993b) Temperature effects on boron adsorption by reference minerals and soils. Soil Sci 156:316–321
Goldschmidt VM (1958) Geochemistry. Muir A (ed), Clarendon Press, Oxford
Gomez-Ariza JL, Beltran R, Morales E, Giraldez I, Ruiz-Benitez M (1995) Acid extraction treatment of sediment samples pf organotin speciation, occurrence of butyltin and phenyltin compounds on the Cadiz coast, south-west Spain. Appl Organomet Chem 9:51–64
Gommes R, Muntau H (1981) La composition chimique des limnophytes due Lac Majeur. Mem Ist Ital. Idrobiol 38:237–307
Gong H, Zhu X, Chen K, Wang S, Zhang C (2005) Silicon alleviates oxidative damage of wheat plants in pots under drought. Plant Sci 169:313–321
Gordeyev VV, Yegorov AS, Radayev VN, Zubow IV (1991) Gold as a possible tracer of hydrothermal influence on bottom waters. Oceanology 31:178–182
Gordeyev VV, Yegorov As, Lisitsyn AP, Letokhov VS, Pakhomov DY, Gulevich VM (1997) Dissolved gold in surface waters of the northeastern Atlantic. Geochem Int 35:1007–1015
Górski M, Uziak S (1991) ^{137}Cs in the selected plants and soils of the Tatra Mountain National Park. Ann UMCS Sect C. 46/1:1–6
Gorsuch JW, Klaine SJ (1998) Toxicity and fate of silver in the environment. Environ Toxicol Chem 17:537–538
Gortre J (1995) Zinc balance in the Netherland 1990. Deel 2: Reststoffen, afvallen en emissies. De totale balans. CBS, Kwartaaberichten Mileu 1995/1 (in Dutch)
Gough LP, Shacklette HT, Case AA (1979) Element concentrations toxic to plants, animals and man. U.S. Geol Surv Bull 1466:1–80
Gough LP, Severson RC, Shacklette HT (1988) Element concentration in soils and other surficial materials of Alaska. U.S. Geol Surv Prof Pap 1458
Govindarajan S, Rao VNR (1992) Accumulation of trace metals in the marine biota and their biological transport. Acta Botanica Indica 20:1–5

Govindaraju K (1994) Compilation of working values and sample description for 383 geostandards. Geostand Newsletters, Special Issue 18:1–158

Grabanov PG (1970) Contents of Ti in the soils of Kazakhstan, SSR and its accumulation by plants. Trudy Inst Pochvovedeniya Akad Nauk Kaz SSR, 18:68–75 (in Russian)

Grabowski D, Jakowska S, Muszyński W (1990) Impact of the Chernobyl reactor accident on radiological situation in Poland. Report CLOR, Warszawa (in Polish)

Grambrell RP (1994) Trace and toxic metals in wetlands: a review. J Environ Qual 123:883–891

Gramowska H, Siepak J (2002) The effect of the fluorine level on the state of leaves and needles of trees in Poznań city and its vicinities. Roczn Ochrona Środ 4:455–477 (in Polish)

Grankina VP, Chankina OV, Savchenko TI, Kovalskaya GA, Koutzenogii KP (2003) Multielement composition of *Glyryrrhiza uralensis* Fisch. (Fabaceae). 4th Intern Symp Trace Elements in Human, New Persp. Enthypossis, Athens, 2, pp 994–1000

Grant CA, Bailey LD (1997) Effects of phosphorus and zinc fertilizer management on cadmium accumulation in flax seed. J Sci Food Agric 73:307–314

Gray CW, McLarren RG, Roberts HC (2003) Atmospheric accession of heavy metals to some New Zealand pastoral soils. Sci Total Environ 305:105–115

Gray CW, McLaren RG, Günther D, Sinaj S (2004) As assessment of cadmium availability in cadmium-contaminated soils using Isotope Exchange Kinetics. Soil Sci Soc Am J 68:1210–1217

Greger M (1999) Metal availability and bioconcentration in plants. In: Prasad MN, Hagemeyer J (eds) Heavy metal stress in plants. Springer-Verlag, Berlin Heidelberg, pp 1–27

Greger M, Johansson M, Stihl A, Hamza K (1994) Foliar uptake of Cd by pea (*Pisum sativum*) and sugar beet (*Beta vulgaris*). Physiol Plant 88:563–570

Grew ES (2002) Mineralogy, petrology and geochemistry of beryllium: An introduction and list of beryllium minerals. In: Grew ES, Ribbe PH (eds) Review in Mineralogy and Geochemistry, vol 50, pp 1–76

Grin N, Stammose D, Guillou Ph, Genet M (1999) Mobility of ^{137}Cs and ^{90}Sr related to speciation studies in contaminated soils of the Chernobyl area. 5th Intern Conf Biogeochem Trace Elements, Vienna, July 11–15, pp 50–51

Grishko VN (1999) Role of soil microorganisms in biogeochemical fluorine cycle in ecosystem under contamination. 5th Intern Conf Biogeochem Trace Elements, Vienna, pp 175–176

Grive M, Duro L, de Pablo J, Bruno J (2004) Radionuclides sorption onto mineral surface. The role of carbonates as the linkage between uranium and iron cycling. Goldschmidt, Copenhagen, A503 p

Grobler SR, Maresky LS, Kotze TJ (1996) Lead reduction of petrol and blood lead concentrations of athletes. Arch Environ Health 47:139–142

Grodzińska K, Frintasyyeva M, Szrek-Łukaszewska G, Klich M, Kucharska-Fabiś A, Gundorina SF, Ostrovnaya TM (2003) Trace element contamination in industrial regions of Poland studied by moss monitoring. Environ Monit Assess 87:255–270

Grogan HW, Mitchell NG, Minski MJ, Bell JN (1987) Pathway of radionuclides from soils to wheat. In: Coughtrey PJ, Martin MH, Unsworth MH (eds) Pollutant transport and fate in ecosystem. Blackwell Sci Publ, Oxford, pp 353–70

Gu SH, Zhu JZ, Gu ZL (1989) Study on the critical lead content of red paddy soil. Agro environ Protect 8:17–22

Guha Mazumder DN, Gupta JD, Chakravorty AK, Chatterjee A, Das D, Chakraborti D (1992) Environmental pollution and chronic arsenicosis in South Calcutta. Bull WHO 70, pp 481–485

Gummow B, Botha CJ, Basson AT, Bastianello SA (1991) Copper toxicity in ruminants: air pollution as a possible cause. Onderstepoort J Veter Res 58:33–40

Gunderson EL (1988) FDA total diet study, April 1982–April 1984, dietary intakes pf pesticides, selected elements, and other chemicals. J Assoc Off Anal Chem 71:1200–1209

Gunther K, Henze W, Umland F (1987) Mobilisationverhalten von Thallium und Cadmium in einen Flusssediment. Fresenius' Z Anal Chem 327:301–303

Guo Y, George E, Marschner H (1996) Contribution of an arbuscular mycorrhizal fungus to the uptake of cadmium and nickel in bean and maize plants. Plant Soil 184:195–205

Gupta UC (1993) Deficiency, sufficiency, and toxicity levels of boron in crops. In: Gupta UC (ed) Boron and its role in crop production. CRC Press, Boca Raton, FL, pp 137–145

Gupta UC, Gupta SC (2000) Selenium in soils and crops, its deficiencies in livestock and humans: implications for management. Comm Soil Sci Plant Anal 31:1791–1807

Gzyl J (1995) Ecological impact and remediation of contaminated sites around smelters in Poland. J Geochem Explor 52:251–258

Gzyl J (1997) Assessment of Polish population exposure to lead and cadmium with special emphasis to the Katowice province on the basis of metal concentrations in environmental compartments. Centr Eu J Publ Health 5:93–96

Gzyl J, Marchwińska E (1995) State of heavy metal contamination in the allotments situated in the upper Silesia region of Poland. In: van den Brink WJ, Bosman R, Arendt F (eds) Contaminated soils '95, Kluwer Acad Publ, Dordrecht, pp 615–616

Haddadin MSY, Caretto SKD, Robison RK (2002) Potential intake of lithium by inhabitants of different regions in Jordan. Pakistan J Nutr 1:39–40

Hagner C (1999) Historical review of European Gasoline Lead Content Regulations and their impact on German Industrial Markets. GKSS Research Centre, Report No 99-E-30, Geesthacht

Hagner C (2000) European regulations to reduce lead emissions from automobiles – did they have an economic impact on the German gasoline and automobile markets? (http://www.springerlink.com/media/)

Hale M (1982) Pathfinder applications of arsenic, antimony and bismuth in geochemical exploration. In: Rose AW, Gundlach H (eds) Geochemical exploration 1980. J Geochem Explor 15:307–323

Hameed A, Vohora SB (1990) Neuro-psychiatric disorders. In: Vohora SB, Dobrowolski JW (eds) New horizons of health aspects of elements. Jamia Hamdard University, New Delhi, pp 115–142

Hamilton EI (1987) The periodic table of the elements: geochemical and biochemical associations. In: Coughtrey PJ, Martin MH, Unsworth MH (eds) Pollutant transport and fate in ecosystems. Blackwell Sci Publ, Oxford, pp 5–33

Hammel W, Debus R, Steubing L (2000) Mobility of antimony in soil and its availability to plants. Chemosphere 41:1791–1798

Häni H, Gupta S (1985) Reason to use neutral salt solution to assess the metal impact on plants and soils. In: Leschber R, Davis RD, l'Hermite R (eds) Chemical method for assessing bio-available metals in sludges and soils. Elsevier, New York, pp 42–48

Hansen JC (1996) Human health and diet in the Arctic. Sci Total Environ 186:135–142

Hanson EJ, Breen PJ (1985) Effects of fall boron sprays and environment factors on fruit set and boron accumulation in 'Italian' prune flowers. J Amer Soc Hort Sci 110:389–392

Hara S, Hayashi N, Hirano S, Zhong XN, Yasuda S, Komae H (1990) Determination of germanium in some plants and animals. Z Naturforsch [C] 45:1250–1251

Hartikainen H (2005) Biogeochemistry of selenium and its impact on food chain quality and human health. J Trace Element Med Biol 18:309–318

Hartikainen H, Pitkänen M, Kairesalo T, Tuominen L (1996) Co-occurrence and potential chemical competition of phosphorus and silicon in lake sediment. Water Res 30:2472–2478

Hasumi M, Suzuki K, Matsui H, Koike H, Ito K, Yamanaka H (2003) Regulation of metallothionein and zinc transporter expression in human prostate cancer cells and tissues. Cancer Letters 200:187–195

Hattori T, Lux A, Tanimoto E, Luxova M, Sugimoto S, Inanaga S (2001) The effect of silicon on the growth of sorghum under drought. 6th Symp Intern Soc of Root Research. Nagoya, Japan, pp 348–349

Haygarth PM (1994) Global importance and global cycling of selenium. In: Frankerberger WT, Benson S (eds) Selenium in the environment. Marcel Dekker, New York, pp 1–27

He M, Yang J (1999) Effects of different forms of antimony on rice during the period of germination and growth and antimony concentration in rice tissue. Sci Total Environ 243/244:149–155

He ZL, Yang XE, Stoffella PJ (2005) Trace elements in agroecosystems and impacts on the environment. J Trace Elem Med Biol 19:125–140

Hebert RFM (2004) Cadmium. In: Merian E, Anka M, Ihnat M, Stoeppler M. (eds) Elements and their compounds in the environment. 2nd ed., Wiley-VCH, Weinheim, pp 689–708

Hedrick JB (1995) The global rare-earth cycle. J Alloys Compouds 225:609–618

Hees van P, Lundström U, Danielsson R, Nyberg L (2001) Controlling mechanisms of aluminium in soil solution – an evaluation of 180 podzolic forest soils. Chemosphere 45:1091–1101

Heidam NZ (1986) Trace metals in the arctic aerosol. In: Nriagu JM, Davidson CI (eds) Toxic metals in the atmosphere. Wiley, New York, pp 267–293

Heim M, Wappelhorst O, Markert B (2002) Thallium in terrestrial environments – Occurrence and effects. Ecotoxycology 11:369–377

Heiserman DL (1997) Exploring chemical elements and their compounds. Prószyski S-ka, Warszawa (in Polish)

HELCOM (1987) First Baltic Sea pollution load compilation. Baltic Sea Environ Proc, Helsinki Commission, vol 20

HELCOM (1991) Airborne pollution load to the Baltic Sea 1986–1990. Baltic Sea Environ Proc, Helsinki Commission vol 39

HELCOM (1993) Second Baltic Sea pollution load compilation. Baltic Sea Environ Proc, Helsinki Commission vol 45

HELCOM (1998) The third Baltic Sea pollution l oad compilation (PLC-3) Baltic Sea Environ Proc, Helsinki Commission vol 70

HELCOM (2001) Environment of the Baltic Sea area, 1994–1998. Baltic Sea Environ Proc 82A:1–23

Helios-Rybicka E (1993) Phase-specific bonding of heavy metals in sediments of the Vistula River. Appl Geochem, Suppl Issue 2:45–48

Helios-Rybicka E, Strzebońska M, Protasowicki M, Poprawski L (1998) Distribution and fate of heavy metals in the flood 1997. Sediments of upper Odra River in Poland. Intern Odra Res Conf, Kraków, 1, pp 62–63

Heliövaara K, Väisänen R (1990) Concentrations of heavy metals in the food, faeces, adults, and empty cocoons of *Neodiprion sertifer* (Hymenoptera, Diprionidae). Bull Environ Contam Toxicol 45:13–18

Heliövaara K, Väisänen R (1991) Bark beetles and associated species with heavy metal tolerance. J Appl Entomology 111:397–405

Helmers E, Krümmerer K (1999) Anthropogenic platinum fluxes. Quantification of sources and sinks, an outlook. Environ Sci Pollut Res 6:29–36

Helmers E, Schwarzer M, Schuster M (1998) Comparison of palladium and platinum in environmental matrices polluted by automobile emission. Environ Sci Pollut Res 5:44–50

Henderson P (1982) Inorganic geochemistry. Pergamon, Oxford

Herber RFM (2004) Cadmium. In: Merian E, Anke M, Ihnat M, Stoeppler M. (eds) Elements and their compounds in the environment. 2nd ed., Wiley-VCH, Weinheim, pp 689–708

Herbert RB (1997) Partitioning of heavy metals in soils contaminated by mine drainage water, Dalarne, Sweden. Water Air Soil Pollut 96:39–59

Herrero TC, Martin LFL (1993) Evaluation of cadmium levels in fertilized soils. Bull Environ Contam Toxicol 50:61–68

Herter U, Kuelling (eds) (2001) Riskoanalyse zur Abfalldüngerverwertung in der Landwirtschaft. Forschungsanstalt für Agrarökologie und Landbau, FAL, Reckenholz, Zürich

Herzig PM, Hannington MD, Fouquet Y, Stacklberg UV, Petersen S (1993) Gold-rich polymetallic sulfides from the lau Back Arc and implications for the geochemistry of gold in sea-floor hydrothermal systems of the southwest Pacific. Econ Geol 88:2182–2209

Hildebrand HF (2004) Metals and ceramic implants. In: Merian E, Anke M, Ihnat M, Stoeppler M (eds) Elements and their compounds in the environment. 2nd, Wiley-VCH Verlag, Weinheim, pp 369–390

Hiller S, Roe MJ, Geelhoed JS, Fraser AR, Farmer JG, Paterson E (2003) Role of quantitative mineralogical analysis in the investigation of sites contaminated by chromite ore processing residue. Sci Total Environ 308:195–210

Hinton TG, Knox AS, Kaplan DI, Bell NC (2005) Phytoremediation potential of native trees in a uranium and thorium contaminated wetlands. J Radioanal Nucl Chem 264:417–422

Hirano S, Tsukamoto N, Higo S, et al. (1989) Toxicity of cadmium oxide instilled into the rat lung. II. Inflammatory in broncho-alveorar fluid technology. Toxicology 55:25–35

Hirsch MP (1998) Availability of sludge-borne silver to agricultural crops. Environ Toxicol Chem 17: 610–616

Hlawatsch S, Gabre-Schönbers CD, Lechtenberg F, Manceau A, Tamura N, Kulik DA, Kersten M (2002) Trace metal fluxes to ferromanagese nodules from western Baltic Sea as a record for long-term environmental changes. Chem Geol 182:697–709

Hlušek J (2000) Beryllium and its effefct on the quality of lettuce. PAN Z Probl Post Nauk Roln 472: 305–310

HMSO (1974) Lead in the environment and its significance to man. Her Majestry's Stationary Office, Pollution Paper 2, London

Hoch M (2001) Organotin compounds in the environment – an overview. Appl Geochem 16:719–743

Hodson PV, Borgmann U, Shear H (1979) Toxicity of copper to aquatic biota. In: Nriagu JO (ed) Copper in the environment. Part II. Health effect, Wiley, New York, pp 307–372

Hogstrand C, Wood CM (1998) Toward a better understanding of the bioavailability, physiology and toxicity of silver in fish: implications for water quality criteria. Environ Toxicol Chem 17:547–561

Hollins JG (1969) The metabolim of tellurium in rats. Health Phys 17:495–505

Holmgren GGS, Meyer MW, Chaney RL, Danies RB (1993) Cadmium, lead, zinc, copper and nickel in agricultural soils of the United States of America. J Environ Qual 22:335–348

Hood RD (1985) Cacodylic acid: agricultural uses, biologic effects and environmental fate. VA Monograph, U.S. Govt Printing Office, Washington, DC, 171 p

Hoppstock K, Sures B (2004) Platinum-group metals. In: Merian E, Anke M, Ihnat M, Stoeppler M (eds) Elements and their compounds in the environment, 2nd ed. Wiley-VCH Verlag, Weinheim, pp 1047–1086

Horril AD, Clint G (1994) Cesium cycling in heather moorland ecosystem. In: Ross SM (ed) Toxic metals in soil–plant system. Wiley, Chichester, pp 395–416

Horst WJ, Puschel A-K, Schmohl N (1997) Induction of callose formation is a sensitive marker for genotypic aluminium sensitivity in maize. Plant Soil 192:23–30

Horvat M, Bloom NS, Liang L (1993) Comparison of distillation with other current isolation methods for the determination of methyl mercury compounds in low level environmental samples. Analyt Chem Acta 281:135–152

Horvat M, Konti B, Kotnik J, Ogring N, Jereb V, Logar M, Faganeli J, Rajar R (2003) Remediation of mercury polluted sites due to mining activities. Crit Rev Anal Canad 33:291–296

Hou X, Malenchenko AF, Kucera J, Dahlgaard H, Nielsen SP (2003) Iodine-129 in thyroid in Ukraine and Denmark. Sci Total Environ 302:63–73

Houba VJ, Temminghof EJM, van Vark W (1999) Soil analysis procedure extraction with 0.01 M $CaCl_2$. Agric Univ. Wageningen

Howe PD, Malcolm HM, Dobson S (2004) Manganese and its compounds: environmental aspects. WHO, Geneva, CICAD No 63

Howell GO, Hill CH (1978) Biological interaction of selenium and other trace elements in chicks. Environ Health Persp 25:147–150

Howells G, Dalziel TRK, Reader JP, Solbe JF (1994) Aluminium and freshwater fish water and quality criteria. In: Howells G (ed) Water quality for fresh water fish. Gordon and Breach, Amsterdam, pp 55–115

HSDB (1995) Hazardous Substances Data Bank. National Library of Medicine, National Toxicology Program (via TOXNET), Bethesda, MD (http://www.inchem.org/documents/ehc/ehc.ehc015.htm)

Hu Q, Moran JE, Rose TP, Smith DK, Zhu Y (2005) Integrated approaches to studying speciation and behaviour of iodine and technetium. ICOBTE, 8th Intern Conf, Adelaide, pp 680–681

Huang JW, Cunningham SD (1996) Lead phytoextraction: species variation in lead uptake and translocation. New Phytologist 134:75–84

Huang J-H, Matzner E (2004) Biogeochemistry of organotin compounds and tin in a forested catchment in Germany. Sci Total Environ 332:231–241

Hultberg H, Munthe J, Iverfeldt A (1995) Cycling of methyl mercury and mercury – responses in the forest roof catchment to three years of decreased atmospheric deposition. Water Air Soil Pollut 80:415–424

Hung TC, Ling YC, Jeng WL, Huang CC, Han BC (1997) Marine environmental monitoring and QA/QC system in Taiwan. J Chinese Environ Prot Soc 20:60–90

Huq SMI, Naidu R (2005) Arsenic in groundwater and contamination of the food chain: Bangladesh scenario. In: Bundschuh J, Bhattacharya P, Chandrasekharam D (eds) Natural arsenic in groundwater: Occurrence, remediation and management. AA Balkema Publ, Lieded, pp 95–101

Hutchinson TC (1979) Copper contamination of ecosystem caused by smelter activities. In: Nriagu JO (ed) Copper in the environment. Part I: ecological cycling. Wiley, New York, pp 451–502

Hutchinson EJ, Pearson PJG (2005) An evaluation of the environmental and health effects of vehicle exhaust catalysts in the United Kingdom. Environ Health Persp 112:132–141

Hydes DJ, Kremling K (1993) Patchiness in dissolved metals (Al, Cd, Co, Cu, Mn, Ni) in North Sea surface waters: seasional differences and influence of suspended sediments. Cont Shelf Res 13: 1083–1101

IARC (1984) Polynuclear aromatic compounds, Part III, industrial exposures in aluminium production, coal gasification, coke production, and iron and steel founding. 34, WHO/IARC, Lyon, pp 89–91

IARC (1993) Cadmium and certain cadmium compounds. IARC monographs on the evaluation of the carcinogenic risk of chemicals to humans. Beryllium, cadmium, mercury exposures in the glass manufacturing industry. WHO/IARC, Lyon, pp 119–146; 210–236

Ibrahim D, Froberg B, Wolf A, Rusyniak DE (2006) Heavy metal poisoning: clinical presentations and pathophysiology. Clin Lab Med 26:67–97

ICRP (1993) Age-dependent doses to members of the public from intake of radionuclides. Intern Commission Radiol Protect. Pergamon Press, Oxford, part 2, publ. 67

Ilyin VB (1997) Monitoring of heavy metals in big industrial cities. Agrokhimiya 4:81–86 (in Russian)

Ilyin VB, Konarbajeva GA (1995) Arsenic in soils of western Siberia-regional environmental monitoring. Pochvoviedeniye 5:634–638 (in Russian)

Inadira M, Hoyano M, Noaki T (1984) Content of trace elements in tea leaves by neutron activation analysis. J Jpn Soc Nutr Food Sci 37:151–156 (in Japanese)

Indeka L, Karaczun ZM (1993) Aluminium, chromium and nickel in soils and cereals exposed to the emission of oil refinery plant in Płock. In: Kabata-Pendias A, Szteke B (eds) Aluminium, chromium and nickel in the environment. Ecological and analytical problems. Polish Acad Sci pp 85–91 (in Polish)

Injuk J, Otten P, Laane R, et al. (1992) Atmospheric concentrations and size distributions of aircraft sampled cadmium, copper, lead and zinc over the Bight of the Northern Sea. Atmos Environ 26A: 2499–2508

Ip C, Lisk DJ (1993) Bioavailability of selenium from selenium-enriched garlic. Nutr Cancer 20:129–137

Ip C, Birringer M, Block E, Kotreibai M, Tyson JF, Uden PC, Lisk DJ (2000) Chemical speciation influences comparative activity of selenium-enriched garlic and yeast in mammary cancer prevention. J Agic Food Chem 48:2062–2070

Ishida K (2004) Gold. In: Merian E, Anke M, Ihnat M, Stoeppler M (eds), Elements and their compounds in the environment, 2nd ed., Wiley-VCH, Weinheim, pp 765–773

Iskandar IK (2001) Environmental restoration of metal-contaminated soils. Lewis Publ, Boca Raton, FL

Islam S, Haque A, Islam R, Yusuf HKM (1999) Aluminium and calcium intoxication in young patients with writer's syndrome. In: Abdulla M, Bost M, Gamon S, Arnaud P, Chazot G (eds) New aspects of trace element research. 5th Intern Soc for Trace Element Research in Humans, London, pp 151–152

Isomura K, Higuchi M, Shibata H, Tsukada H, Iwashima K, Sugiyama H (1993) Distribution of radioactive cesium in moss and application of mosses fro monitoring of radioactive cesium. Radioisotopes 42:157–163 (in Japanese)

Isoyama N (2003) ^{137}Cs concentration in stingray (*Dasyatis akajei*). In: Proc Intern Symp Rokkasho, Radioecology and environmental dosimetry. Aomori, Japan, pp 301–305

ITDG Project (2002) Reducing indoor air pollution in rural households in Kenya. URL: (http://www.itdg.org/html/advocacy/docs/smoke_project_report.pdf)

Ivchenko VI (1988) Molybdenum in metabolism of plants. In: Pais I (ed) New results in the research of hardly known trace elements and their role in the food chain. Univ Horticulture Food Ind, Budapest, pp 51–65

Jacks G, Bhattacharya P, Chaudhary V, Singh KP (2005) Controls on the genesis of some high-fluoride groundwaters in India. Appl Geochem 20:221–228

Jackson AP, Alloway BJ (1992) The transfer of cadmium from agricultural soils to the human food chain. In: Adriano DC (ed) Biogeochemistry of trace metals. Lewis Publ, Boca Raton, FL, pp 109–158

Jagoe CH, Chesser RK, Smith MH, Lomakin MD, Lingerfelser SK, Dallas CE (1997) Levels of cesium, mercury and lead in fish, and cesium in pond sediments in an inhabited region of the Ukraine near Chernobyl. Environ Pollut 98:223–232

Jakubick AT (1979) Geochemistry and physics of plutonium migration. In: Ahrens LH (ed) Origin and distribution of the elements, Pergamon Press, Oxford, 2, pp 775–790

Jalkanen L, Häsänen E (1994) Heavy metals in aerosols on Finnish EMEP stations. EMEP Workshop on European Monitoring, Modelling, and Assessment of Heavy Metals and Persistent Organic Pollutants, Beekbergen

James BR (2001) Remediation-by-reduction strategies for chromate-contaminated soils. Environ Geochem Health 23:175–179

James BR, Peture JC, Vitale RJ, Mussoline GR (1997) Oxidation-reduction chemistry of chromium: relevance to the regulation and remediation of chromate-contaminated soils. J Soil Contam, Spec Issue 6:569–580

Jansson G (2002) Cadmium in arable crops: The influence of soil factors and liming. Doctoral Thesis, Swedish University of Agricultural Sciences, Uppsala

Jantzen E (1992) Analytische Untersuchungen zur Beurteilung der Bindungsformer (Speciation) des Zinns in Elbesedimenten unter Zuhilfenahme von GC-AAS, GC-MS, TRFA und GFAA. Dissertation, Universität Hamburg

Jaritz M (2004) Barium. In: Merian E, Anke M, Ihnat M, Stoeppler M (eds) Elements and their compounds in the environment. 2nd ed., Wiley-VCH, Weinheim, pp 627–634

Jarvis MD, Leung DWM (2002) Chelated lead transport in *Pinus radiate*: an ultrastructural study. Environmental and Experimental Botany 48:21–32

Jayasinghe R, Tsuji LJS, Gough WA, Karagatzides JD, Perera D, Nieboer E (2004) Determining the background levels of bismuth in tissues of wild game birds: a first step in addressing the environmental consequences of using bismuth shotshells. Environ Pollut 132:13–20

Jędrzejczak R (2004) Iron and manganese in food. Ann Nat Inst Hygiene, Suppl 55:13–20 (in Polish)

Jędrzejczak R, Ręczajska W (1996) Coper and molybdenum in food. In: Kabata-Pendias A, Szteke B (eds) Coper and molybdenum in the environment. Ecological and analytical problems. Polish Ac Sci, pp 174–180 (in Polish)

Jegadeeswari D, Raja AR, Krishnasamy R (2005) Plant uptake of radiocesium (^{137}Cs) in sorghum (*Sorghum bicolor* L.). ICOBTE, 8th Intern Conf, Adelaide, pp 460–461

Jeng MS, Jeng WL, Hung TC, Yeh CY, Tseng RJ, Meng PJ, Han BH (2000) Mussel watch: a review of Cu and other metals in various marine organisms in Taiwan, 1991–98. Environ Pollut 110:207–215

Jenkins DW (1980) Biological monitoring of toxic trace metals. Toxic trace metals in plants and animals of the world. U.S. EPA, Reps. 600/3-80-092, pp 1130–1148, and 600/3-80-091, pp 504–982

Jernelöv A (1969) Conversion of mercury compounds. In: Miller MW, Berg GG (eds) Chemical fallout, current research on persistent pesticides. Chas C Thomas, Springfield, Il, pp 68–74

Jernelöv A, Ramel C (1998) The Siberian mercury anomaly and a tentative answer to the missing mercury problem. Ambio 27:155

Jiménez B, Rodríguez-Estrella R, Merino R, Gómez G, Rivera L, González MJ, Abad E, Rivera J (2005) Results and evaluation of the first study of organochlorine contaminants (PCDDs, PCDFs, PCBs and DDTs), heavy metals and metalloids in birds from Baja California, México. Environ Pollut 133:139–146

Johansson K (1996) Cadmium and other heavy metals in Swedish forest soils. In: OECD: Sources of cadmium in the environment, pp 167–173

Johnson CC, Ge X, Green KA, Liu X (2000) Selenium distribution in the local environment of selected villages of the Keshan Disease belt, Zhangjiakou District, Hebei Province, People's Republic of China. Appl Geochem 15:385–401

Jones D L (1998) Organic acids in the rhizosphere – critical review. *Plant Soil* 205:25–44

Jones B (2005) Animals and medical geology. In: Selinus O, Alloway BJ, Centeno JA, Finkelman RB, Fuge R, Lindh U, Smedley P (eds) Essentials of medical geology. Elsevier, Amsterdam, pp 513–526

Jones DL, Kochian LV (1997) Aluminium interaction with plasma membrane lipids and enzyme metal binding sites and its potential role in Al cytotoxicity. FEBS Lett 400:51–57

Jones KC, Peterson PJ, Davies BE (1985) Silver and other metals in some aquatic bryophytes from streams in the lead mining districts of Mid-Wales, Great Britain. Water Air Soil Pollut 24:329–338

Jørgensen SE (1993) Removal of heavy metals from compost and soil by ecotechnological. Ecolog Engin 2:89–100

Jørgensen SE (2000) Principles of pollution abatement. Pollution abatement for the 21st century. Elsevier, Amsterdam

Jugdaohsingh R, Anderson SHC, Tucker KL et al. (2002) Dietary silicon intake and absorption. Am J Clin Nutr 75:887–893

Kabała C, Wilk J (2004) Sequential extraction of iron and zinc in soils irrigated with municipal wastewater. Ann Nat Inst Hygiene 55 Suppl:133–141 (in Polish)

Kabata-Pendias A (1973) Uptake of cobalt and copper by clover from minerals impregnated by these cations. Roczn Glebozn 24:273–287

Kabata-Pendias A (1994) Maintaining soil micronutrient status. In: Greenland DJ, Szabolcs I (eds) Soil resilience and sustainable land use. CAB Intern, Wallingford, pp 199–214

Kabata-Pendias A (1995) Biogeochemistry of lithium. Int Symp Lithium in the trophic chain soil-plant-animal-man, Warszawa, pp 9–14

Kabata-Pendias A (1998) Geochemistry of selenium. J Environ Pathol Toxic Oncol 17:173–177

Kabata-Pendias A (2004) Soil-plant transfer of trace elements-an environmental issue. Geoderma 122:143–149

Kabata-Pendias A, Brümmer GW (1991) Ecological consequences of As, Cd, Hg and Pb enrichment in European soils. In: Hutchinson TC, Meema K, Gordon C (eds) Lead, mercury, cadmium and arsenic in the environment. SCOPE 32, Wiley, New Delhi, pp 107–114

Kabata-Pendias A, Dudka S (1991) Baseline data for cadmium and lead in soils and some cereals of Poland. Water Air Soil Pollut 57–58:723–731

Kabata-Pendias A, Krakowiak A (1995) Soil parameters as a base for the calculation of background heavy metal status. In: Wilkens R-D, Förstner U, Knöchel A (eds) Heavy metal in the environment. CEP, Edinburgh, 1, pp 398–401

Kabata-Pendias A, Krakowiak A (1998) Useful phytoindicator (dandelion) for trace metal pollution. 5th Intern Conf. Transport Fate and Effects of Silver in the Environment, Hamilton, Ontario, pp 145–149

Kabata-Pendias A, Motowicka-Terelak T (1996) Heavy metals and sulfur in indicator plants as a base for the land use program in Poland. Rep KBN, Warszawa (in Polish)

Kabata-Pendias A, Pendias H (1999) Biogeochemistry of trace elements, 2nd ed., Wyd Nauk PWN, Warszawa (in Polish)

Kabata-Pendias A, Pendias H (2001) Trace elements in soils and plants, 3rd ed., CRC Press, Boca Raton, FL

Kabata-Pendias A, Piotrowska M (1999) Impact of Zn and Pb smelter flue-dust on Cd, Zn and Pb speciation in soil and their availability to spring barley. Bull l'Acad Serbe Sci Arts 119:77–82

Kabata-Pendias A, Sadurski W (2004) Trace elements and compounds in soil. In: Merian E, Anke M, Ihnat M, Stoepppler M (eds) Elements and their compounds in the environment, Wiley-VCH, Weinheim, 2nd ed., pp 79–99

Kabata-Pendias A, Terelak H (2004) Regional variation of trace elements (Cd, Cu, Ni, Pb, Zn) contents of native grassland grasses in Poland. 3rd Intern Conf Heavy Metals Radionucludes and Elements-Biofills in the Environment, Semipalatinsk, Kazakhstan, 1, pp 28–33

Kabata-PendiasA, Wiącek K (1985) Excessive uptake of heavy metals by plants from contaminated soil. Soil Sci Ann 36:33–42

Kabata-Pendias A, Dudka S, Chlopecka A, Gawinowska T (1992) Background levels and environmental influences on trace metals in soils of the temperate humid zone of Europe. In: Adriano DC (ed) Biogeochemistry of trace metals. Lewis Publ, Boca Raton, pp 19–60

Kabata-Pendias A, Piotrowska M, Witek T (1993) Preliminary standards for the assessment of soil and crop pollution with heavy metals as a guideline for agricultural land use. Rep IUNG, Pulawy (in Polish)

Kachur AN, Arzhanova VS, Yelpatyevsky PV, von Braun MC, von Lindern IH (2003) Environmental conditions in the Rudnaya River watershed – a compilation of Soviet and pos-Soviet era sampling around a lead smelter in the Russian far east. Sci Total Environ 303:171–185

Kadunas V, Budavičius R, Gregorauskienė V, Katinas, Kliugiene E, Radzevicius A, Taraskievicius R (1999) Geochemical Atlas of Lithuania. Geol Inst, Vilnius

Kagan LM, Kadatsky VB (1996) Depth migration of Chernobyl originated ^{137}Cs and ^{90}Sr in soils of Belarus. J Environ Rdioactivity 33:27–39

Kalac P, Burda J, Stascova I (1991) Concentrations of lead, cadmium, mercury, and copper in mushrooms in the vicinity of a lead smelter. Sci Total Environ 105:109–119

Kanerva T, Sarin O, Nuorteva P (1988) Aluminium, iron, zinc, cadmium and mercury in some indicator plants growing in south Finnish forest areas with different degree of damage. Ann Bot Fenn 25:275–279

Kaneta M, Hikichi H, Endo S, Sugiyama N (1983) Isolation of a cadmium-binding protein from cadmium treated rice. Agric Biol Chem 47:417–418

Kannan K, Tanabe S, Iwata H, Tatsukawa R (1995) Butyltins in muscle and liver of fish collected from certain Asian and Oceanian countries. Environmental Pollution 90:279–290

Kannan K, Corsolini S, Focardi F, Tanabe S, Tatsukawa R (1996) Accumulation pattern of butyltin compounds in dolphin, tuna and shark collected from Italian coastal waters. Arch Environ Contam Toxicol 31:19–23

Kannan K, Senthilkumar K, Elliott JE, Feyk LA, Giesy JP (1998) Occurrence of butyltin compounds in tissues of water birds, and seaducks from the United States and Canada. Arch Environ Contam Toxicol 35:64–69

Kaplan DI, Knox AS (2004) Enhanced contaminant desorption induced by phosphate mineral additions to sediment. Environ Sci Technol 38:3153–3160

Kaplan DI, Knox SA, Coffey C (2002) Reduction of contaminant mobility at the TNX outfall delta through the use of apatite and zero-valent ion as soil amendments (U). WSRC-TR-2002-00370, Rev. 0, Westinghouse Savannah River Company, Aiken, SC

Kaplan DI, Hinton TG, Knox AS (2005) Cesium-137 partitioning to wetland sediments and uptake by plants. J Radioanal Nucl Chem 264:393–399

Kápolna E, Gergely V, Dernovics M, Fodor P (2004) Selenium speciation of some raw food materials. In: Cser MA, László IS, Étienne J-C, Maymard Y, Centeno JA, Khassanova L, Collery P (eds) Metal ions in biology and medicine, 8. John Libbey & Comp. Ltd., England, Paris, pp 174–177

Kapu MM, Basak B, Job A,Umaru IO, Kalla SM, Mohammad BY, Harun BA,Schaeffer DJ (1989) Studies on human exposure to environmental lead in Zaria, Nigeria. Trace Elements Med 6:178–181

Karamushka VI, Gadd GM (1999) Interaction of *Saccharomyces cerevisiae* with gold: and accumulation. Biometals 12:289–294

Karczewska A (2002) Heavy metals in soils polluted by emissions from copper smelters – forms and solubility. Z Nauk AR Wroclaw 432:1–159 (in Polish)

Karczewska A (2004) Function of iron and manganese oxides in the sorption of heavy metals in polluted soils as related to the sequential extraction. Ann Nat Inst Hygiene, Suppl 55:133–141 (in Polish)

Karen D, Ownby DR, Day BS, Brown RW, Shupack DP, Cobb GP, La Point TW (1998) Influence of varying water quality parameters on toxicity and bioavailability of Ag^+ to rainbow trout (*Oncorhynchus mykiss*). 5th Argentum, Intern Conf on Transport, Fate and Effects of Silver in the Environment, Hamilton, pp 275–285

Karim Md. M (2000) Arsenic in groundwater and health problem in Bangladesh. Water Res 34:304–310

Karolewski P, Siepak J, Gramowska H (2000) Response of Scots pine (*Pinus sylvestris*), Norway spruce (*Picea abies*) and Douglas fir (*Pseudotsuga menziesii*) needles to environment pollution with fluorine compounds. Dendrology 45:41–46

Kashparov V, Colle C, Zvarich S, Yoshenk V, Levchuk S, Lundin S (2005a) Soil to plant halogens transfer studies-1. Root uptake of radioiodine by plant. J Environ Radioactivity 79:187–204

Kashparov V, Colle C, Zvarich S, Yoshenk V, Levchuk S, Lundin S (2005b) Soil to plant halogens transfer studies-2. Root uptake of radiochlorine by plant. J Environ Radioactivity 79:233–253

Kasimov NS (ed) (1995) Ecogeochemistry of urban areas. Izd Moscow University, Moscow (in Russian)

Kavas-Ogly A, Luftllayev FE, Ratiev AN Onishchenko VA, Fidirko EV, Rish MA (1996) Cadmium and lead self-selected diets of Samarkan residents. In: Anke M, Arnhold W, Bergmann H, Bitsch R (eds) Mengen und Spuren-Elemente. Arbeitstagung, Friedr.-Schiller-Univ., Jena, pp 651–654

Kayser A, Wenger K, Keller A, Attinger W, Felix H, Gupta SK, Schulin E (2000) Enhancement of phytoextraction of Zn, Cd, and Cu from calcareous soil: the use of NTA and sulfur amendments. Environ Sci Technol 34:1778–1783

Kazantzis G (2000) Thallium in the environment and health effects. Environ Geochem Health 22:275–280

Kekli A, Aldahan A, Meili M, Possner G, Buragio N, Stepanauskas R (2003) ^{129}I in Swedish rivers: distribution and sources. Sci Total Environ 309:161–172

Keller C (1997) Some microscopic and mineral particles of biological origin and soil solution. Eur J Soil Sci 48:193–199

Keller C (2006) Efficiency and limitations of phytoextraction by high biomass plants: the example of willows. In: Prasad MNV, Sajwan KS, Naidu R (eds) Trace elements in the environment: Biogeochemistry, biotechnology, and bioremediation. CRC Press, Boca Raton, FL, pp 611–630

Keller C, Kayser A, keller A, Schulin R (2001) Heavy-metal uptake by agricultural crops from sewage sludge treated soils of the upper Swiss Rhine Valley and the effect of time. In: Iskandar IK (ed) Environmental restoration of metals-contaminated soils. Lewis Publ, Boca Raton, FL, pp 273–291

Kennish MJ (1998) Trace metal-sediment dynamics in estuaries: Pollution assessment. Res Environ Contam Toxicol 155:69–110

Keren R, Mezuman U (1981) Boron adsorption by clay minerals using a phenomenological equation. Clays Clay Miner 29:198–204

Keren R, Gast RG, Bar-Yosef B (1981) pH-dependent boron adsorption by Na-montmorillonite. Soil Sci Soc Am J 45:45–48

Khan MA, Islam MR, Panaullah GM, Duxbury JM, Jahiruddin M, Loeppert RH, Meisner CA (2005) Movement of arsenic in irrigated rice soil. 8th Intern Conf on Biogeochemistry of Trace Elements, Adelaide (CD-ROM)

Kikkilä O (2002) Remediation through mulching with organic matter of soil polluted by a copper-nickel smelter. Ph.D Thesis, University of Helsinki, Helsinki

Kim EY, Goto R, Iwata H, Masuda Y, Tanabe S, Fujita S (1998) Preliminary survey of lead poisoning of Steller's eaglee (*Haliaeetus pelagicus*) and white-tailed sea eagle (*Haliaeetus albicilla*) in Hokkaido. Japan EnvironToxicol Chem 18:448–451

Kimmel G, Vu V (2001) Framework for human health risk assessment. Human Ecol Risk Assess 7: 153–156

Kinraide TB (1991) Identity of the rhizotoxic aluminum species. In: Wright et al. (eds) Plant-soil interactions at low pH. Kluwer Publ, The Netherlands, pp 717–728

Kist (1994) Investigation of element speciation in atmosphere. Biol Trace Elem Res 43–54:259–266

Kitagishi K, Yamane I (eds) (1981) Heavy metal pollution in soils of Japan. Japan Sci Soc Press, Tokyo

Kitano Y (1992) Water Chemistry. In: Nierenberg WA (ed) Encyclopedia of Earth system science, Acad Press, San Diego, 4, pp 449–470

Kjellstrom T, Nordberg GF (1978) A kinetic model of cadmium metabolism in the human being. Environ Res 16:246–269

Kjølholt J (1997) Sources of agricultural soil contamination with organic micropollutants and heavy metals. Spec Conf on Management and Fate of Toxic Organics in Sludge Applied to Land. Copenhagen

Klaassen CD (1985) Heavy metals and heavy metal antagonists. In: Gilman AG, Goodman LS, Rall TW, Mural F (eds) The pharmacological basis of therapeutics. MacMillan, New York, pp 1605–1627

Kleczkowski M, Kluciński W, Sikora J, Zdanowicz M (2004) Role of antioxidants in the protection against oxidative stress in cattle – trace elements and enzymatic mechanisms, Polish J Veterin Sci 7:233–240

Klein CB (1996) Carcinogencity and genotoxicity of chromium. In: Chang LW (ed) Toxicology of metals. CRC Lewis Publ, pp 205–219

Kloke A, Suerbeck DR, Vetter H (1984) The contamination of plants and soils with heavy metals and the transport of metals in terrestrial food chains. In: Nriagu JO (ed) Changing metal cycles and human health. Dahlem Conf., Springer-Verlag, Berlin Heidelberg, pp 113–114

Kłapcińska B, Saci A (1994) Selenium in foods produced and consumed in North-West Algeria. In: Kabata-Pendias A, Szteke B (eds) Arsenic and selenium in the environment-ecological and analytical problems, Polish Acad Sci ZN 8:94–99 (in Polish)

Knatko VA, Aggets VU, Shmigelskaya IV, Ivashkevich II (2000) Soil-to-potato transfer of ^{137}Cs in an area of Belarus: regression analysis of the transfer factor against ^{137}Cs deposition and soil characteristic. J Environ Radioactivity 48:171–181

Knox AS, Erdinger AP, Adriano DC, Kolka RK, Kaplan DI (1999) Source and practices contributing to soil contamination. In: Adriano DC, Bollag J-M, Frenkerberger WT, Sims RC (eds) Bioremediation of contaminated soils. Am Soc Agron, Madison, WI, pp 53–87

Knox AS, Seaman JC, Mench MJ, Vangronsveld J (2000a) Remediation of metal- and radionuclides-contaminated soils by in situ stabilization techniques. In: Iskandar IK (ed) Environmental restoration of metals-contaminated soils. Lewis Publ., New York, pp 21–60

Knox AS, Seaman J, Adriano DC (2000b) Chemo-phytostabilization of metals in contaminated soils. In: Wise DL, Trantolo DJ, Cichon EJ, Inyang HI, Stottmeister (eds) Bioremediation of contaminated soils. Marcel Dekker, New York, pp 811–836

Knox AS, Hinton TG, Kaplan DI (2001) Bioavailability of radioactive cesium in Old R Discharge Canal, R-Canal, Pond A, and the adjacent flood plain. WSRC-TR-2001-00455, Rev.0. West Savannah River Com, Aiken, SC

Kobayashi R (2004) Tellurium. In: Merian E, Anke M, Ihnat M, Stoeppler M (eds) Elements and their compounds in the environment. 2nd ed., Wiley-VCH, Weinheim, pp 1407–1414

Kodak Publication (1983) Silver recovery with the Kodak Chemical Recovery Cartridge, Type P. Kodak Publication J-9, Eastman Kodak Co., Rochester, New York

Kohl HC Jr, Oertli J (1961) Distribution of boron in leaves. Plant Physiol 36:420–424

Koivisto S (2003) Finnish exposure scenarios for antifouling products. Finnish Environment Institute (SYKE), Helsinki, Finland

Köljonen T (1992) The geochemical atlas of Finland. Geol Survey of Finland, Espoo, J Environ Qual 32:2230–2237

Konek C, Al-Abadleh H, Mifflin A, Geiger F (2004) Chromate binding to organic adlayers at silica/water interfaces studied by second harmonic generation. Goldschmidt, Copenhagen, A370 p

Korobova E, Ermakov A, Linnik V (1998) ^{137}Cs and ^{90}Sr mobility in soils and transfer in soil-plant systems in the Novozybkov district affected by the Chernobyl accident. Appl. Geochem 13:803–814

Koroleva M, Manngin A (2004) Uranium (VI) uptake by calcium carbonates and lake sediments. Goldschmidt, Copenhagen, A496 p

Korzh VD (1991) Geochemistry of elemental composition of the hydrosphere. Nauka, Moscow (in Russian)

Kośla T, Skibniewska E, Debski B, Urbańska-Slomka G (2001) Rubidium in the trophic chain soil-plants-animals. Trace Elements Eletrolytes 19:171–176

Kośla T, Cieślik L, Skibniewski M, Wrzesień R, Skibniewska EM, Michalik P (2004) Hepcidin – a hormone in iron metabolism in man and animal. Part I. The role of hepcidin in absorption of iron of digestive track and iron transport. Ann Nat Inst Hygiene, 55 Suppl:1–6 (in Polish)

Kosta I, Zelenko V, Ravnik V (1974) Trace elements in human thyroid with special reference to the observation of mercury following long-term exposure. Comp Studies Food Environ Contam, IAEA, Vienna, 541 p

Kostyuk O, Bunnerberg (1999) The role of microbiota in the behavior of radionuclides in seminatural ecosystems. 5th Intern Conf Biogeochemistry of Trace Elements, Vienna, 1, pp 44–45

Kotaś J, Stasicka Z (2000) Chromium occurrence in the environment and methods of its speciation. Environ Pollut 107:263–283

Kotliński R (1999) Metallogenesis of the World's Ocean against background of oceanic crust evolution. Polish Geol Inst, Special Papers 4:5–59

Kovalevskiy AL (1991) Biogeochemistry of plants. Nauka, Novosibirsk (in Russian)

Kovalskiy VV (1974) Geochemical ecology. Nauka, Moscow (in Russian)

Kovalskiy VV, Letunova SV, Aleksieyeva SA (1980) Accumulation of nickel and other elements in the microbiota of a soil in South Ural Subregion of the biosphere. In: Proc Nickel Symp, Friedrich-Schiller Univ. Jena, pp 163–167

Krám P, Hruška J, Driscoll JA (1998) Beryllium chemistry in the Lysina catchment, Czech Republic. Water Air Soil Pollut 105:409–415

Krämer U, Cotter-Howells JD, Charnock JM. Baker AJM, Smith AD (1996) Free histidine as a metal chelator in plants that accumulate nickel. Nature 379:635–637

Krämer U, Golldack D, Kiriazidou G, Dietz KJ (1999) Induction of plants pathogenesis-related proteins by heavy metals. 5th Conf Biogeochem of Trace Elements, Vienna, pp 1158–1159

Krampitz G (1980) Die biologische Bedeutung von Beryllium-Verbindungen. In: Proc Arsenic Symp, Friedrich-Schiller Univ, Jena, pp 245–252

Kraus T (2004) Tungsten. In: Merian E, Anke M, Ihnat M, Stoeppler M (eds) Elements and their compounds in the environment, 2nd ed., Wiley-VCH, Weinheim, pp 1141–1145

Kravtsov VA, Emelyanov EM (1995) Trace metal behaviour in the waters of the Gulf of Finland. 4th Mar Feol Conference – the Baltic, Uppsala, Ser Ca 86:99–109

Kreij de C, van Elderen CW, Basar H (1996) Iron availability in peat substrate assessed by water, calcium chloride, ammonium acetate, DTPA, and calcium chloride/DTPA. Commun Soil Sci Plant Anal 27:1813–1819

Krüger J, Winkler P, Lüderitz E, Lück U, Wolf HU (1985) Ullmann's Encyclopedia of Industrial Chemistry. 5th ed., A4, VCH Verlag, Weinheim, pp 171–189

Krümmerer K, Helmers E (1997) Hospital effluents as a source for platinum in the environment. Sci Total Environ 193:179–184

Kuballa J, Jantzen E, Wilken RD (1996) Organotin compounds in sediments of the rivers Elbe and Mulde. In: Calmano W, Förstner U (eds) Sediments and toxic substances. Springer-Verlag, Berlin Heidelberg, pp 245–270

Kucharzewska K, Borkowski J, Kucharzewski A (1991) Contents of heavy metals in sediments and alluvial soils of river Odra. Conf Geological Aspects of Environ Protect, AGH, Kraków, pp 172–176 (in Polish)

Kuehnelt D, Lintschiger J, Goessler W (2000) Arsenic compounds in terrestrial organisms. VI. Green plants and lichens from an old arsenic smelter site in Austria. Appl Organomet Chem 14:411–420

Kumpiene J, Ore S, Mench M, Maurice C (2005) Remediation of CCA contaminated soil with zerovalent iron. ICOBTE, 8th Intern Conf, Adelaide, pp 264–265

Kumpulainen JT (1996) (ed) Trace elements, natural antioxidants and contaminants in European food and diets. FAO/UN, REU Techn Series 49, Rome

Kunachowicz H, Nadolna I, Przygoda B, Iwanow K (1998) Food composition tables. National Food and Nutrition Institute, 85, Warszawa (in Polish)

Kurtz AC, Derry LA, Chadwick OA (2002) Germanium-silicon fractionation in the weather environment. Geochim Cosmochim Acta 66:1525–1537

Kuzniecow WA, Generalova WA (1996) Trace elements, radionuclides, and their forms in buried Holocene soils. Wiesti Ak Nauk Bielarusi, ser Chim Nauk. 2:87–92 (in Belarussian)

Kvicala J, Jirànek V, Ther R, Kroupová V (2002) Selenium status of the cattle in some regions of the Czech Republic. 21st Workshop on macro and trace elements, Friedrich-Schiller Univ, Jena, pp 460–467

Kwapuliński J, Wiechuła D, Malara P, Brodziak B (2004) A concept of cationic balance in human ecology. Problemy Ekologii 1:27–30 (in Polish)

Kyzioł J (2002) Sorption and binding force of organic substance (peat) for selected cations of heavy metals. PhD Thesis, Polish Acad Sci, Zabrze (in Polish)

Lacerda LD, Salomons W (1991) Mercury in the Amazon: A chemical time bomb. A report sponsored by the Dutch Ministry Housing, Physical Planning Environ

Lagerweff JV (1972) Lead, mercury and cadmium in environmental contaminants. In: Mortvedt JJ, Giordano PP, Lindsay WL (eds) Micronutrients in agriculture. Soil Sci Soc of Am, Madison, WI, pp 593–636

Lai MS, Hsueh YM, Chen CJ, Shyp MP, Chen SY, Wu MM, Kuo TL, Tai TY (1994) Ingested inorganic arsenic and prevalence of diabetes mellitus. Am J Epidemiol 139:484–492

Lamy I, Bourgeois S, Bermond A (1993) Soil cadmium mobility as a consequence of sewage disposal. J Environ Quality 22:731–737

Landa ER (2003) Uranium mill tailing: nuclear waste and natural laboratory for geochemical and radiological investigation. 16th Intern Symp Environ Biogeochem, Oirase, Abstr 1:81–83

Landner L, Lindeström L (1998) Zinc in society and in the environment, Swedish Environmental Research Group, Stockholm, pp 16–30

Langston WJ, Burt GR (1994) Bioindicators of Ag availability in UK estuaries. 4th Argentum Intern Conf on Transport, Fate and Effects of Silver in the Environment, Madison, WI, pp 125–130

Laperdina TG, Melnikova MV, Khvostova TE (1999) Gold mining in Siberia as a source of mercury contamination of the environment. In: Ebinghaus R, Turner RR, de Lacerda LD, Vasiliev O, Salomons W (eds) Mercury contaminated sites: Characterization, risk assessment and remediation. Springer-Verlag, Berlin Heidelberg, pp 357–374

Larsen EH, Andresen NL, Møller A, Petersen A, Mortensen GK, Petersen J (2002) Monitoring the content and intake of trace elements from food in Denmark. Food Add Contam 19:33–46

Laukel M, Chistoserdova L, Lidstrom ME, Vorholt JA (2003) The tungsten-containing formate dehydrogenase from *Methylbacterium extorquens* AM1: purification and properties. Eur J Biochem 270:325–333

Laul JC, Weimer WC, Rancitelli LA (1979) Biochemical distribution of rare earths and other trace elements in plants and soils. In: Ahrens LH (ed) Origin and distribution of the elements, Pergamon Press, Oxford, 2, pp 819–830

Lawrence JF, Michalik P, Tam G, Conacher HBS (1986) Identification of arsenobetaine and arsenocholine in Canadian fish and shellfish by high-performance liquid chromatography with atomic absorption detection and conformation by fast atom bombardment mass spectrometry. J Agric Food Chem 34:315–319

Laxen DPH, Harrison RM (1983) The physio-chemical speciation of selected metals in the treated effluent of a lead-acid battery manufacturer and its effect on metal speciation in the receiving water. Water Res 17:71–80

Lazo O, Cullaj A, Deda T (2007) Arsenic in soil environments in Albania. In: Bhattacharya P, Mukherjee AB, Zevenhoven R, Loeppert R (eds) Arsenic in soil and groundwater: Biogeochemical interactions. Elsevier Book Series, Amsterdam

LDAI (2002) Lead Development Association International, London. (http:/www.ldaint.org/default.htm)

Lebeau T, Bagot D, Jézéquel K, Fabre B (2002) Cadmium biosorption by free and immobilized microorganisms cultivated in a liquid soil extract medium: effects of Cd, pH and techniques of culture. Sci Total Environ 291:73–83

Leblanc J-Ch, GuérinT, Noë L, Calamassi-Tran G, Volatier J-L, Verger P (2005) Dietary exposure estimates of 18 elements from the 1st French Total Diet Study. Food Add Contam 22:624–641

Ledin M, Krantz-Rülcker C, Allard B (1999) Microorganisms as metal sorbent: comparison with other soil constituents in multi-compartment systems. Soil Biol Biochem 31:1639–1648

Lei G, Boström K (1995) Mineralogical control on transition metal distribution in marine manganese nodules. Marine Geology 123:253–261

Leinonen KP (1989) The influence of soil preparation on the levels of aluminium, manganese, iron, copper, zinc, cadmium, and mercury in *Vacccinium myrtillus*. Chemosphere 18:1581–1587

Leiterer M, Einax JW, Streibert M (2004) Bestimmung von Platinumgruppenelemten in Pflanzen mit Quadrupol-ICP-MS. In: Anke M, Flachowsky G, Kisters K, Müller R, Schäfer U, Schenkel H, Seifert M, Stoeppler M (eds) Workshop on macro and trace elements. Jena, pp 470–479

Lepp NW, Dickinson NM (1994) Fungicide-derived copper in tropical plantation crops. In: Ross SM (ed) Toxic metals in soil-plant systems. Wiley, Cichester, pp 367–396

Lepp NW, Dickinson NM (2003) Natural bioremediation of metal polluted soils – A case history from the United Kingdom. 7th Intern Conf on the Biogeochem of Trace Elements, Uppsala 2:320–321

Letunova SV, Kovalskiy VV (1978) Geochemical ecology of microorganisms. Izdat Nauka, Moscow (in Russian)

Levander OA (2001) Evolution of human dietary standards for selenium. In: Hatfield DL (ed) Selenium: Its molecular biology and role in human health. Kluwer Publ, Boston, pp 299–311

Lewandowski I, Schmidt U, Londo M, Faaij A (2005) The economic value of phytoremediation function – Assessed by the example of cadmium remediation by willow (*Salix* ssp.). Agricul Systems 89:68–89

Lewis TE (1990) Introduction. In: Lewis TE (ed) Environmental chemistry and toxicology of aluminium. Lewis Publ, Chelsea, MI, pp 1–2

Lewis BL, Andreae MO, Froelich PN, Mortlock RA (1988) A review of the biogeochemistry of germanium in natural waters. Sci Total Environ 73:107–120

Li Y-H (ed) (2000) A compendium of geochemistry: From solar nebula to the human brain, Princeton Univ Press, Princeton, Oxford

Li Z, Shuman LM (1996) Heavy metal movement in metal contaminated soil profiles. Soil Sci 161:656–666

Li X, Thornton I (1993) Arsenic, antimony and bismuth in soil and pasture herbage in some old metalliferous mining areas in England. Environ Geochem Health 15:135–144

Li F, Shan X, Zhang S (2001) Evaluation of single extractants for assessing plant availabily of rare earth elements in soils. Comm Soil Sci Plant Anal 32:2577–2587

Liang YC, Ding RX (2002) Influence of silicon on micro-distribution of mineral ions in roots of rice and wheat to silicon in calcareous soils. Commun Soil Sci Plant Anal 25:2285–2297

Liang YC, Yang CG, Shi HH (2001) Effects of silicon on growth and mineral composition of barley grown under toxic levels of aluminium. J Plant Nutr 24:229–243

Liang YC, Chen Q, Liu Q, Zhang WH, Ding RX (2003) Exogenous silicon (Si) increases antioxidant enzyme activity and reduces lipid peroxidation in roots of salt-stressed barley (*Hordeum vulgare* L.). J Plant Physiol 160:1157–1164

Liang T, Zhang S, Wang L, Kung H-T, Wang Y, HU A, Ding S (2005) Environmental biochemical behavior of rare earth elements in soil-plant systems. Environ Geochem Health 27:301–311

Licht OAB (ed) (2005) Geoquimica de solo do Estado do Parana, vol 1/2, Mineropar, Curitiba, Parana (in Spanish)

Lide DR (ed) (1996) CRC Handbook of Chemistry and Physics, 77th ed. CRC Press, Boca Raton, FL

Lieth H, Markert B (1985) Concentration cadaster of chemical elements in contrasting ecosystems. Naturwisseschaften 72:322–324

Lim J-M, Salido AL, Butcher DJ (2004) Phytoremediation of lead using Indian mustard (*Brassica juncea*) with EDTA and electrodes. Microchem J 76:3–9

Limbong D, Kumampung J, Rimper J, Arai T, Miyazaki N (2003) Emissions and environmental implications of mercury from artisal gold mining in north Sulawesi, Indonesia. Sc Total Environ 302:227–236

Lin T-S, Nriagu JO (1998) Speciation of thallium in natural waters. In: Nriagu JO (ed) Thallium in the environment. Advances in Environmental Science and Technology, vol 29, Wiley, New York, pp:31–43

Lin T-S, Nriagu JO (1999) Thallium speciation in river waters with Chelex-100 resin. Analyt Chim Acta 395:301–307

Lin Z, Zayed A, Terry N (1999) Role of selenium volatilization in the management of selenium-laden agricultural drainage water. 5th Intern Conf Biogeochem Trace Elements, Vienna, pp 878–879

Lindberg S, Zhang H, Gustin M, Casimir A, Ebinghaus R, Edwards G, et al. (1999) The role of rainfall and soil moisture on mercury emissions from mercuriferous desert soils. J Geophysical Res-Atmosph 104(D17):21879–21888

Lindberg SE, Brooks S, Lin C-J, Scott KJ, Landis MS, Stevens RK, Gooste M, Richter A (2002) Dynamic oxidation of gaseous mercury the the Arctic troposphere at polar sunrise. Environ Sci Technol 36:1245–1256

Lindqvist O, Johansson K, Aastrup M, Andersson A, Bringmark L, Hovsenius G, Håkanson L, Iverfeldt A, Meili M, Timm B (1991) Mercury in the Swedish Environment-Recent research on causes, consequences and corrective methods. Water Air Soil Pollut 55, (all chapters of the volume)

Lis J, Pasieczna A (1995) Geochemical atlas of Poland. PIG, Warszawa

Lisk DJ, Bache CA, Essick LA, Reid CH, Rutzke H, Crown K (1988) Absorption and extretion of selenium and barium in humans from consumption of Brazil nuts. Nutr Rep Intern 38:183–191

Little CA, Whicker FW, Winsor TF (1980) Plutonium in a grassland ecosystem at Rocky Flats. J Environ Qual 9:350–359

Liu Z (1988) The effect of rare elements on growth of crops. In: Pais I (ed) New results in the research of hardly known trace elements and their role in food chain. Univ Hort Food Industry, Budapest, pp 23–35

Liu Z, Zhu Q-Q, Li-Hua T (1980) Boron deficient soils and their distribution in China. Acta Ped Sin 17:228–239

Liu XJ, Liu QQ, Cheng J, Zhang LP, Mu BF, Xue ZN et al. (2000) Comparison of trace elements in drinking water between high and low incidence districts of gastric carcinoma. Nuclear Instruments and Methods in Physics, Res Sec B, Beam Interactions with Materials and Atoms 160:167–171

Liu J, Li K, Xu J, Zhang Z, Ma T, Lu X, Yang J, Zhu Q (2003) Lead toxicity, uptake, and translocation in different rice cultivars. Plant Sci 165:793–802

Liu H, Probst A, Liao B (2005) Metal contamination of soils and crops affected by the Chenzhou lead/zinc mine spill (Hunan, China). Sci Total Environ 339:153–166

Livett EA (1992) Heavy metal pollution of the atmosphere. In: Nierenberg WA (ed) Encyclopedia of Earth system science. Acad Press, San Diego, 2, pp 507–518

LlorensJF, Fernández JL, Querol X (2000) The fate of trace elements in a large caol-fired power plant. Envriron Geology 40:409–416

Lockhart HB (1983) Silver compounds. In: Grayso M (ed) Krik-othmer Encyclopedia of Chemical Technology, 2nd ed., New York, 21, pp 16–32

Lodenius M (1981) Regional distribution of mercury in *Hypogymnia physodes* in Finland. Ambio 10:44–45

Lodenius M, Herranen M (1981) Influence of chloralkali plant on the mercury contents of fungi. Chemosphere 10:313–319

Lodenius M, Laaksovirta K (1979) Mercury content of *Hypogymnia physodes* and pine needles affected by a chloralkali works at Kuusankoski, SE Finland. Ann Bot Fennici 16:7–10

Loebenstein JR (1994) The materials flow of arsenic in the United States. U.S. Bureau of Mines Information

Loomis WD, Durst RW (1992) Chemistry and biology of boron. Bio Factors 4:229–239

López-Arce P, Garcia-Guinea J, Ferro JLG (2003) Manganese micro-nodules on ancient brick walls. Sci Total Environ 302:267–274

Loppi S, Riccobono F, Zhang ZH, Savic S, Ivanov D, Pirintsos SA (2003) Lichens as biomonitors of uranium in the Balkan area. Environ Pollut 125:277–280

Lottemoser BG (1995) Noble metals in municipal sewage sludges of Southern Australia. Ambio 24:354–357

Louekari K, Mäkelä-Kurtto R, Virtanen JPV, Sippola J, Malm J (2000) Cadmium in fertilizers. Risk to human health and the environment. Ministry Agric Forestry in Finland, Helsinki

Lowe TP, May TW, Brumbaugh WG, Kane DA (1985) National contaminant biomonitoring program: Concentrations of seven elements in fresh-water fish, 1978–1981. Arch Environ Contamin Toxicol 14:363–388

Lu JY, Shroeder WH, Barrie LA, Steffen A, Welch HE, Martin K, Lockhart L, Hunt RV, Boila G, Richter A (2001) Magnification of atmospheric mercury deposition to polar regions in spring time: the link to troposphere ozone depletion chemistry. Geophys Res Letters 28:3219–3222

Ludwicki JK, Wiadrowska B (1992) Mercury in food. In: Kabata-Pendias A, Żmudzki J (eds) Mercury in the enviroment. Ecological and analytical problems. Polish Ac Sci, Warszawa, pp 19–25 (in Polish)

Lukin A, Dauvalter V, Kashulin N, Yakovlev, V, Sharov A, Vandysh O (2003) Assessment of copper-nickel industry impact on a subarctic ecosystem. Sci Total Environ 306:73–83

Lukšienė B, Druteikienė R (2004) Evaluation of ultratrace elements ($^{239,240}Pu$) in Lithuanian forests using mosses as bioindicators. In: Anke M, Flachowsky G, Kisters K et al. (eds) Workshop Macro and Trace Elements, Jena 1, pp 824F–829F

Lund U, Fobian A (1991) Pollution of two soils by arsenic, chromium and copper, Denmark. Geoderma 49:83–103

Lundström US, Giesler R (1995) Use of aluminium species composition in soil solution as an indicator of acidification. Ecol Bull 44:114–122

Luoma SN (1983) Bioavailability of trace metals to aquatic organisims – a review. Sci Total Environ 28:1–22

Luoma SN, Ho YB, Bryan GW (1995) Fate, bioavailability and toxicity of silver in estuarine environments. Mar Pollut Bull 31:44–54

Lux A, Luxová T, Hattori T, Inanaga S, Sugimoto Y (2002) Silicification in sorghum (*Sorghum bicolour*) cultivars with different drought tolerance. Physiol Plant 115:87–92

Lyday P (1993) Boron. U.S. Department of the Interior, Washington, DC

Lydersen E, Rukke NWA, Jensen JGB, Kjelsberg BM, Tornsjø B, Vogt RD, Vøllestad LA, Poléo ABS (2002) Seasonal variation in mortality of brown trout (*Salmo trutta*) in an acid aluminium-rich lake. Water Air Soil Pollut 51:203–215

Lyons G, Stangoulis J, Graham R (2003) High-selenium wheat: biofortification for better health. Nutr Res Rev 16:45–60

Lytle PE (1984) Fate and speciation of silver in public owned treatment works. Environ Toxicol Chem 3:21–30

Łobiński R, Potin-Gautier M (1998) Metals and biomolecules – bioinorganic analystical chemistry. Analusis 26:21–24

Łukaszewicz Z, Zembrzuski W, Piela A (1996) Direct determination of ultratraces of thallium by flow-injection-differential-pulse anodic stripping voltammetry. Anal Chem 318:159–165

Ma W (1987) Heavy metal accumulation in the mole, *Talpa europea*, and eathwarms as an indicator of metal bioavailability in terrestrial environments. Bull Environ Contam Toxicol 39:933–940

Ma JF (2000) Role of organic acids in detoxification of aluminium in higher plants. Plant Cell Physiol 41:383–390

Ma YB, Uren NC (1998) Transformations of heavy metals added to soil – application of a new sequential extraction procedure. Geoderma 84:157–168

Ma LQ, Komar KM, Zhang TuC, Cai WH, Kennelley ED (2001) A fern that hyperaccumulates arsenic: A hardy, versatile, fast-growing plant helps to remove arsenic from contaminated soils. Nature 409:p 579

Ma YB, Lombi E, Nolan AL, McLaughlin M (2006) Short-term natural attenuation of copper in soils: effect of time, temperature, and soil characteristics. Environ Toxic Chem 25:652–658

Maas EV (1986) Salt tolerance of plants. Appl Agr Res 1:12–26

McBride M, Sauvé S, Hendershot W (1997) Solubility control of Cu, Zn, Cd and Pb in contaminated soils. Eur J Soil Sci 48:337–346

Mackowiak CL, Grossl PR, Bugbee BG (2003) Biogeochemistry of fluoride in a plant-solution system. J Environ Qual 32:2230–2237

Madden EF, Anderson CJ, Goering PL (2004) Indium. In: Merian E, Anke M, Ihnat M, Stoeppler M (eds) Elements and their compounds in the environment. 2nd ed., Wiley-VCH, Weinheim, pp 801–809

Maenhut W, Cornille P, Pacyna JM, Vitols V (1989) Trace element composition and origin of the atmospheric aerosol in the Norwegian Arctic. Atmos Environ 23:2551–2569

Maessen O, Freedman B, McCurdy R (1985) Metal mobilization in home well water systems in Nova Scotia. J Am Water Works Assoc 77:73–80

MAF (1997) Rep. of Ministry of Agricultural and Forestry in Finland

Magiera T, Lis J, Nawrocki J, Strzyszcz Z (2004) Magnetic susceptibility of soils in Poland. Intern Symp on Iron and Manganese in the Environment. Ecol Analyt Poblems. Warszawa, 26 p (in Polish)

Magos L (1997) Physiology and toxicology of mercury. In: Sigel A, Sigel H (eds) Metal ions in biological systems 34. Mercury and its effects on environment and biology. Marcel Dekker, New York, pp 321–370

Maján GY, Kozák M, Püspöki Z, McIntosch R, Mikó L (2001) Environmental geological examination of chromium-contamination in Eastern Hungar. Environ Geochem Health 23:229–233

Makdessi K, Andreesen JR, Pich A (2001) Tungstate uptake by a highly specific ABC transporter in *Eubacterium acidaminophilum*. J Biol Chem 276:24557–24564

Maksimović ZJ, Djujić I (1998) Selenium research in Serbia, Yugoslavia. J Environ Pathol Toxic Oncol 17:165–171

Maksimović ZJ, Djujić I, Jović V, Ršumović M (1992) Selenium deficiency in Yugoslavia. Biol Trace Res 33:187–195

Maksimović ZJ, Nikolić M, Jorga J, Ršumović M, Radošević P (1995) Selenium status of soil, grain, garlic and human serum and indices of cancer in the community Barajevo (Belgrade). Conf of Selenium, Serbian Acad Sci Art 68, pp 165–169

Malek MA, Nakahara M, Nakamura R (2003) Retention and removal of ^{137}Cs in Japanese catfish (*Silurus asotus* Linnaceus) in dressing and cooking method. Intern Symp, Radioecology and Environmental Dosimetry. Rokkasho, Aomori, pp 306–313

Malinowska E, Szefer P, Falandysz J (2004) Metals bioaccumlation by bay bolete, *Xerocomus badius*, from selected sites in Poland. Food Chem 84:405–416

Malle KG (1992) Zink in der Umwelt. Acta Hydrochim Hydrobiol 20:196–204

Mandal B, Adhikari TK, De DK (1993) Effect of lime and organic matter application on the availability of added boron in acid alluvial soils. Comm Soil Sci Plant Anal 24:1925–1935

Mandal B, Roychowdhury T, Samanta G, Basu Gk, Chowdhury PP, Chandra CR et al. (1996) Arsenic in groundwater in seven districts of West Bengal, India: the biggest arsenic calamity in the world. Cur Sci 70:976–986

Manecki A (1984) Transport and input of air pollutants in the Niepołomice Forest area. In: Grodziński W, Weiner J, Maycock PF (eds) Forest ecosystem in industrial regions. Springer-Verlag, Berlin Heidelberg, pp 36–68

Manecki A, Shejbal-Chwastek M, Tarkowski J (1988) Mineral and chemical characteristic of dust air pollutants from area affected by short-and long-range industrial emissions. Prace Min PAN, Oddz Krakow 78:27–42

Mankovska B, Steinnes E (1995) Effects of pollutants from an aluminium reduction plant on forest ecosystems. Sci Tot Environ 163:11–23

Markert B (1987) The pattern of distribution of lanthanide elements in soils and plants. Phytochemistry 26:3167–3170

Markert B (1992) Multi-element analysis in plant materials – analytical tools and biological questions. In: Adriano DC (ed) Biogeochemistry of trace elements Lewis Publ, Boca Raton, FL, pp 401–428

Markert B, Lieth H (1987) Element concentrarion cadasters in Swedish biotopes. Fresenius' Z Anal Chem 327:716–718

Markert B, Vtorova VN (1995) Concentration cadaster of chemical elements in plants of eastern European forest ecosystems. Biol Bull 22:453–460

Marmiroli M, Robinson B, Bolan N, Maewstri E, Marmiroli N, Clothier B (2006) Phytotechnology for the treatment of dairy effluent. Abs. Meeting, Phytotechnologies lessons from pilot and filed scale, Lisbon, 58 p

Marschner H, Treeby M, Römheld V (1989) Role of root-induced changes in the rihizosphere for iron acquisition in higher plants. Z Pflanzenernähr Bodenk 52:197–204

Marschner P, Rengel Z (2005) Interactions between rhizosphere microorganisms and higher plants for iron and manganese – an overview. ICOBTE, 8th Intern Conf, Adelaide, pp 488–489

Mart L, Nürnberg HW (1986) Cd, Pb, Cu, N, and Co distribution in the German Bight. Mar Chem 18:197–213

Martens DA, Suarez DL (1998) Sequential extracion of selenium oxidation states. In: Frankenberger WT, Engberg RA (eds) Environmental chemistry of selenium. Marcel Dekker, New York, pp 61–79

Martin F, Ildefonse P, Hazeman J-L, Petit S, Grauby O, Decarreau A (1996) Random distribution of Ge and Si in synthetic talc and EXAFS and FTIR study. Eur J Mineral 8:289–299

Martin-Bouyer G, Barin C, Beugnet A, Cordier J, Guerbois H (1978) Epideminological aspects of the encephalopathies resuling from oral administration of bismuth. Therapie 35:307–313

Marzec Z, Schlegel-Zawadzka M (2004) Exposure to cadmium, lead and mercury in the adult population from Eastern Poland, 1990–2002. Food Add Contam 21:963–970

Marzec A, Marzec Z, Zaręba S (2004a) Iron and manganese in food products for infants and children. Ann Nat Inst Hygiene, Suppl 55:41–45 (in Polish)

Marzec Z, Bertrandt J, Kłos A, Ciagadlak T (2004b) Daily food rations as a source of iron for men nourished in the arrest and at home. In: Cser MA, László IS, Étienne J-C, Maymard Y, Centeno JA, Khassanova L, Collery P (eds) Metal ions in biology and medicine, 8. John Libbey & Comp. Ltd., England, Paris, pp 418–420

Marzec Z, Marzec A, Zaręba S (2006) Study of metal release from cookware. Polish J Environ Stud 15(2A):139–142

Mason RP, Fitzgerald WF, Morel FMM (1994) The biogeochemistry cycling of elemental mercury: Anthropogenic influences. Geochim Cosmochim Acta 58:3191–3198

Massoura ST, Echevarria G, Becquer T, Ghanbaja J, Leclerc-Cessac E, Morel L (2005a) Nickel bearing phases and availability in natural and anthropogenic soils. ICOBTE, 8th Intern Conf, Adelaide, pp 180–811

Massoura ST, Norini M-P, Echevarria G, Shaw G, Leclerc-Cessac E, Morel J-L (2005b) Determination of nickel-63 availability on organic matter rich soils. ICOBTE, 8th Intern Conf, Adelaide, pp 468–469

Matczak-Jon E (1995) Aluminium in environment. Wiadomości Chemiczne 49:483–499 (in Polish)

Matschullat J (1997) Trace elements fluxes to the Baltic Sea: problems of input budgests. Ambio 26: 363–368

Matschullat J (2000) Arsenic in the geosphere – a review. Sci Total Environ 249:297–312

Matschullat J, Tobschall HJ, Voigt H-J (1997) Geochemie und Umwelt. Springer-Verlag, Berlin Heidelberg

Matsi T, Keramidas VZ (2001) Alkaline fly ash effects on boron sorption and desorption in soils. Soil Sci Soc Am J 65:1101–1108

Matsumoto A, Hinkley TK (2001) Trace metals in Antarctic pre-industrial ice are consistent with emissions from quiescent degassing of volcanoes worldwide. Earth and Planet Sci Lett 186:33–43

Matsumoto H, Senoo Y, Kasai M, Maeshima M (1996) Response of the plant root to aluminium stress: analysis of the inhibition of the root elongation and changes in membrane function. J Plant Res 109:99–105

Matsumoto H, Yamamoto Y, Devi SR (2001) Aluminium toxicity in acid soils: plant responses to aluminium. In: Prasad MNV (ed) Metals in the Environment: Analysis by Biodiversity. Marcel Dekker, New York, pp 289–320

Maturana H, Lagos U, Flores V, Gaete M, Cornejo L, Wiertz JV (1993) Integrated biological process for the treatment of a Chilean complex gold ore. FEMS Microbiol Rev, pp 215–220

Mazur A, Feilletr-Coundry Ch, Romier B, Bayle D, Gueux E, Ruivard M, Coundy Ch, Rayssiquier Y (2003) Dietary iron regulates hepatic hepcidin 1 and 2 mRNAs in mice. Metabolims 52:1229–1231

McBride MB (1994) Environmental chemistry of soils. Oxford University Press, New York
McBride MB, Sauvé S, Hendershot, W (1997) Heavy metal uptake and chemical changes in the rhizosphere of *Thlaspi caerulescens* and *Thlaspi ochroleucum* grown in contaminated soils. Plant Soil 188:153–159
McDonald DJ, Duncan HJ (1979) Atmospheric levels of trace elements in Glasgow. Atm Environ 13:413–417
McGeer JC, Eddy FB (1998) Ionic regulation and nitrogenous excretion in Rainbow Trout exposed to buffered and unbuffered freshwater of pH 10.5. Physiol Zool 71:179–190
McGrath SP (1992) Soil quality in relation to agricultural uses. Integrated soil and sediment research: A basic for proper protection. Kluwer Publ, London/Boston, pp 187–198
McGrath SP (1998) Phytoextraction for soil remediation. In: Brooks RR (ed) Plants that hyperaccumlate heavy metals. CAB Intern,Wallingford, pp 261–385
McGrath SP, Loveland PJ (1992) The geochemical survey of topsoils in England and Wales. In: Beck BD (ed) Trace Subst Environ Health 25:39–51
McKenzie RM (1975) The mineralogy and chemistry of soil cobalt. In: Nicholas DJD, Agan AR (eds) Trace elements tin soil-plant-animal systems. Acad Press, New York, pp 83–93
McKnigtht DM, Duren SM (2004) Biogeochemical processes controlling midday ferrous iron maxima in stream waters affected by acid rock drainage. Appl Geochem 19:1075–1084
McLaughlin MJ (2001) Bioavailability of metals to terrestrial plants. In: Allen HE (ed) Bioavailability of metals in terrestial ecosystems. Importance of partitioning for bioavailability to invertebrates, microbes and plants, SETAC Press, Pensacola, FL, pp 39–68
McLaughlin MJ, Tiller KG, Beech TA, Smart MK (1994) Soil salinity causes elevated cadmium concentrations in field-grown potato tubers. J Environ Qual 23:1013–1018
McLaughin MJ, Tillee KG, Hamblin A. (1995) Managing Cd concentration of agricultural land. OECD Workshop, Saltsjöbaden, pp 16–20
McShehy S, Szpunar J (2000) Speciation of arsenic in edible algae bibidmensional HPLC with dual ICP MS and electrospray Ms/MS detection. J Anal Atom Spectrom 15:79–83
McShehy S, Pohl P, Łobiński R, Szpunar J (2001) Investigation of arsenic speciation in oyster test reference material by multidimensional HPCL-ICP-MS and electrospray tandem mass spectrometry (ES-MS-MS). Analyst 126:1055–1062
Meador JP (1991) The interaction of pH, dissolved organic carbon, and total copper in the determination of ionic copper and toxicity. Aquatic Toxicology 19:13–32
Megonigal JP, Hines ME, Vischer PT (2003) Anaerobic metabolism: Linkages to trace gases and aerobic processes. In: Holland HD and Turekian KK (eds) Treatise on geochemistry, Vol 8 Biogeochemistry, Elsevier/Pergamon, Amsterdam, pp 317–442
Meharg AA (1994) Integrated tolerance mechanisms-constructive and adaptive plant responses to elevated metal concentrations in the environment. Plant Cell Environ 17:989–993
Meharg AA, Macnair MR (1991) The mechanisms of arsenate tolerance in *Deschampsia cespitosa* (L.) Beauv and *Agrostis capillaries* L. New Phytologist 119:291–297
Meharg AA, Mazibur MD (2003) Arsenic contamination of Bangladesh paddy field soils: implications for rice contribution to arsenic consumption. Environ Sci Technol 37:229–234
Meij R, Winkel H Te (2004) The emissions and environmental impact of PM_{10} and trace elements from a modern coal-fired power plant equipped with ESP and wet FGD. Fuel Processing Technology 85:641–656
Meili M, Iverfeldt A, Hakanson L (1991) Mercury in the surface water of Swedish Forest Lakes – concentrations, speciation and controlling factors. Water Air Soil Pollut 56:439–453
Melo D, Burkart W (2004) Actinum, proactinum and transuranic actinides. In: Merian E, Anke M, Ihnat M, Stoeppler M (eds) Elements and their compounds in the environment. 2nd ed., Wiley-VCH, Weinheim, pp 1165–1169
Menard O, Adocat T, Ambrosi JP, Michera A (1998) Behavior of actinides (Th, U, Np and Pu) and rare earths (La, Ce and Nd) during aqueous leaching of nuclear glass under geological disposal condition. Appl Geochem 13:103–126
Mench MJ, Vangronsveld NW, Lepp NW, Edwards R (1998) Physico-chemical aspects and efficiency of trace element immobilization by soil amendments. In: Vangronsveld J, Cunningham SD (eds) In situ inactivation and phytorestoration of metal-contaminated soils. Landes Biosciences, Georgetown, TX, pp 151–182

Mendoza J, Borie F (1998) Effect of *Glomus etunicatum* inoculation on aluminium, phosphorus, calcium, and magnesium uptake of two barley genotypes with different aluminium tolerance. Commun Soil Sci Plant Anal 29:681–695

Meranger JC, Subramanian KS, Chalifoux C (1979) A national survey for cadmium, chromium, copper, lead, zinc, calcium, and magnesium in Canadian drinking water supplies. Environ Sci Technol 13:707–711

Mercier A, Pelletier É, Hamel J-F(1994) Metabolism and subtle toxic effects of butyltin compounds in starfish. Aquat Toxicol 28:259–273

Merian E (1991) Metals and their compounds in the environment: Occurrence, analysis, and biological relevance. VCH Publ, Weinheim

Meriwether IR, Beck JN, Keeley DF, Langley MP, Thompson RH, Young JC (1988) Radionuclides in Luisiana soils. J Environ Qual 17:562–569

Merry RH, Tiller KG, Alston AM (1983) Accumulation of copper, lead and arsenic in some Australian orchard soils. Austr J Soil Res 21:549–561

Messerschmidt J, von Bohlen A, Alt FR, Klockenkamper R (2000) Separation and enrichment of palladium and gold in biological and environmental samples, adapted to the determination by total reflection X-ray fluorescence. Analyst 125:397–399

Middelburg JJ, Hoede D, van der Sloot HA, van der Weijden CH, Wijkstra J (1988) Arsenic, antimony and vanadium in the North Atlantic Ocean. Geochim Cosmochim Acta 52:2871–2878

Mielke HW, Gonzales CR, Powell E, Shah A, Mielke PW (2002) Natural and anthropogenic processes that concentrate Mn in rural and urban environments of the Lowe Mississippi River Delta. Environ Res Sec A 90:157–168

Miettinen V, Verta M, Erkomaa K, Järvinen O (1985) Chlorinated hydrocarbons and heavy metals in fish in the Finnish coastal areas of the Gulf of Finland. Finn Fish Res 6:77–80

Mikhailest ND, Balabolkin MI Rakitin VA Danilov IP (1996) Thyroid function during prolonged exposure to fluorides. Problemy Endokrinologii 42:6–9

Milhaud GE, Mehannaoui S (1998) Indicators of lead, zinc and cadmium exposure in cattle. I. Results in pollution area. Vet Human Toxicol 30:513–517

Miller PA, Munkittrick KR, Dixon DG (1992) Relationship between concentrations of copper and zinc in water, sediment, benthic invertebrates, and tissues of white sucker (*Catostomus commersoni*) at metal-contaminated sites. Can J Fish Aquatic Sci 49:978–984

Mishra VK, Kim K-H, Kang C-H, Choi KC (2004) Wintertime sources and distribution of airborne lead in Korea. Atmos Environ 38:2653–2664

Mitra S (ed) (1986) Mercury in the Ecosystem.Trans Tech Publ, Switzerland, pp 1–190

Mitrovic B, Milacic R, Pihlar B, Simoncic P (1998) Speciation of trace amount of aluminium in environmental samples by cation-exchange FPLC-ETAAS. Analusis 26:381–388

Miyasaka SC, Buta JG, Howell RK, Foy (1991) Mechanism of aluminium tolerance in snapbeans: Root exudation of citric acid. Plant Physiol 96:737–743

Mniszek W (1996) Emission of mercury to the atmosphere from industrial sources in Poland. Environ Monit Assess 41:291–299

Moiseenko T, Dauvalter V, Kagan L, Sharov A, Vandysh O, Kashulin N, Yakovlev V (1995) Air pollution effects on the mountain lakes of the Kola, Northern Russia. Acidification of mountain lakes. Paleolimnology and ecology. Report 2/1995, INEP, Apatity, Kola Science Centre, Russian Acad Sci, pp 1–53

Molas A (1997a) Range of tolerance, limits and toxicity symptoms of ionic and chelatic nickel forms in cabbage plants. Z Probl Post Nauk Roln 448b:203–209 (in Polish)

Molas A (1997b) Changes in morphological and anatomical structure of cabbage (*Brassica oleacea* L.) outer leaves and in ultrastructure of their chloroplasts caused by an *in vitro* excess of nickel. Phoytosynthetica 34:513–522

Moldovan M, Rauch S, Gomez M, Palacios AA, Morrison GM (2001) Bioaccumulation of palludium, platinum and rhodium from urban particulates and sediments by freshwater isopod Asellus aquaticus. Water Res 35:4175–4183

Momčilović B (2004a) ^{54}Manganese whole body retention and the gastrointestinal transit time of the women in their reproductive age are intensively related to the ferritin status. 22 Workshop, Macro and Trace Elements, Jena, 2:1715–1722

Momčilović B (2004b) The copper group. In: Merian E, Anke A, Inhat M, Stoeppler M (eds) Elements and their compounds in the enviroment. 2nd ed., VCH, Weinheim, pp 731–750

Momčilović B, Ivičić N (2003) The analysis of human blood molybdenum (Mo^{6+}) with differential pulse anodic stripping voltammetry. 4th Intern Symp trace elements in human: New perspectives. Entypossis, Athens, pp 1339–1349

Momot O, Synzynys B (2005) Toxic aluminium and heavy metals in groundwater of Middle Russia: health risk assessment. Intern J Environ Res Public Health 2:214–218

Monbeshora C, Osibanjo O, Ajayi SO (1983) Pollution studies on Nigerian Rivers: The onset of lead pollution of surface waters in Ibadan. Environ Intern 9:81–84

Montgomery JM (1985) Water treatment principles and design. Wiley, New York

Moore WS (1972) Radium: element and geochemistry. In: Fairbridge RW (ed) The encyclopedia of geochemistry and environmental sciences. Van Nostrand Reinhold Comp, New York, pp 1006–1008

Moore JW (1991) Inorganic contaminants of surface water – research and monitoring priorities (Beryllium). Springer Verlag, New York, pp 50–56

Moreira JC (1996) Threats by heavy metals: human and environmental contamination in Brazil. Sci Total Environ 188:S61–S71

Morel FMM, Prince NM (2003) The biochemical cycles of trace metals in the oceans. Science 300:944–949

Morgado P, Pereira V, Pinto MS (2001) Chromium in Portuguese soils surrounding electroplating facilities. Environ Geochem Health 23:225–228

Morrell BC, Lepp NW, Phipps DA (1986) Vanadium uptake by higher plants: some recent developments. Environ Geochem Health 8:14–18

Morrissey JR, Edds DR (1994) Metal pollution associated with a landfill: concentration in water, sediment, crayfish, and fish. Transactions on the Kansas Acad of Sci 97:18–25

Mortvedt JJ, Cox FR, Shuman LM, Welch RM (eds) (1991) Mincronutrients in agriculture, 2nd ed., Chemical methods. Soil Sci Agr Am Inc, Madison, WI

Mosbaek H, Tjell JC, Sevel T (1988) Plant uptake of mercury in background areas. Chemosphere 17:1227–1236

Moseman RF (1994) Chemical deposition of boron in animals and humans. Environ Health Persp 102(7):113–117

Moskalyk RR (2004) Review of germanium processing worldwide. Mineral Engineering 17:393–402

Motowicka-Terelak T, Terelak H, Pietruch C (1997) Content of lithium in soils of some regions of Poland. Z Probl Post Nauk Roln 448b:217–224 (in Polish)

Mounicou S, Szpunar J, Andrey D, Blake C, Lobinski R (2003) Concentrations and bioavailability of cadmium and lead in cocoa powders and related products. Food Addit Contamin 20:343–352

Mukherjee AB (1994) An initial assessment of the production, consumption and emission of silver in the Nordic countries. 2nd Intern Conf on Transport, Fate and Effects of Silver in the Environment. Madison, WI, pp 43–48

Mukherjee AB (1997) The use and release of silver in Finland. Finnish Environment Institute, Report No 33, Helsinki

Mukherjee AB (1999) Nickel: a review of occurrence, uses, emission and concentration in the environment in Finland. Environ Rev 6:173–187

Mukherjee AB (2001) Behavior of heavy metals and their remediation in metalliferous soils. In: Prasad MNV (ed) Metals in the Environment: Analysis and Biodiversity. Marcel Dekker, New York, pp 433–471

Mukherjee AB, Bhattacharya P (2001) Arsenic in groundwater in the Bengal Delta Plain: slow poisoning in Bangladesh. Environ Rev 9:189–220

Mukherjee AB, Hartikainen H (2002) Emissions and occurrence of zinc in Europe with special reference to its behavior in soil, water and plants. In: Kabata-Pendias A, Szteke B (eds) Zinc in the environment. Ecological and analytical problems, Polish Acad Sci, Warsawa, pp 19–28

Mukherjee AB, Lahermo PW (1995) Silver loadings in sewage treatment plants and in the natural aquatic environment of Finland. 3rd Intern Conf Proceedings: Transport on Fate and Effects of Silver in the Environment, Washington, DC, pp 185–193

Mukherjee AB, Bhattacharya P, Jacks G, Banerji DM, Ramanathan AL, Mahanta C, Chandrashekharam D, Chatterjee D, Naidu R (2006) Groundwater arsenic contamination in India: Extent and severity. In: Naidu R, Smith E, Owens G, Bhattacharya P, Nadebaum P (eds) Managing arsenic in the environment: From soil to human health. CSIRO Publ, Collingwood VIC, pp 553–593

Müller R (2006) The biological and toxicological importance of copper – the copper content of invertebrates, mammals and humans. Inżynieria Ekologiczna 16:44–45

Munno De G, Geday MA, Medaglia M, Anastassopoulou J, Theophanides T (1996) Manganese and cobalt cytosine (Cyt) and 1-methycytosine (1-Mecyt) complexes. In: Collery Ph, Corbella J, Dominnngo JL, Etienne J-C, Llobet JM (eds) Metal ions in biology and medicine 4, Libbey Eurotext, Paris, pp 3–5

Muramatsu Y, Wedepohl KH (1998) The distribution of iodine in the Earth's crust. Chem Geol 147:201–216

Muramatsu Y, Yoshida S (1998) On the environmental behavior of radioactive and stable iodine. Intern Worshop Comp Evaluation Health Effects Environ Toxic Derived from Adv Technol, Suppl NIRS-M-130:45–58

Muramatsu Y, Christoffers D, Ohmomo Y (1983) Influence of chemical forms on iodine uptake by plants. J Radiation Res 24:326–338

Murray FJ (1995) A human health risk assessment of boron (boric acid and borax) in drinking water. Regul Toxicol Pharmacol 22:221–230

Nabrzyski M (1998) Toxicological evaluation of selected tarce metals in food. In: Kabata-Pendias A, Szteke B (eds) Quality problems of trace analysis in environmental studies. Wyd Ed. "Żak", Warszawa, pp 13–40 (in Polish)

Naidu R, Nadebaum PR (2003) Geogenic arsenic and associated toxicity problems in the groundwater-soil-plantanimal-human continuum. In: Skinner HCW, Berger AR (eds) Geology and health. Closing Gap. Oxfort Univ Press, New York, pp 37–44

Naidu R, Krishnamurti GSR, Bolan NS, Wenzel W, Megharaj M (2001) Heavy metal interactions in soils and implications to soil microbial biodiversity. In: Prasad MN (ed) Metals in the environment: analysis by biodiversity, Marcel Dekker, New York, pp 401–431

Naidu R, Smith E, Owens G, Bhattacharya P, Nadebaum P (eds) (2006) Managing arsenic in the environment. CSIRO Publ, Melbourne

Nakamaru Y, Tagami K, Uchida S (2005a) Distribution coefficient of selenium in Japanese agricultural soils. Chemosphere 58:1347–1354

Nakamaru Y, Tagami K, Uchida S (2005b) Sorption behaviour of antimony in Japanese agricultural soils. 8th Intern Conf Biogeochemistry of Trace Elements, Adelaide (CD-ROM)

Nakao A, Funakawa S. Kosaki T (2003) Kinetics of Cs adsorption on soils with different mineralogical composition. In: Proc Intern Symp Radioecology Environmental Dosimetry. Rokkasho, Aomori, pp 191–195

Nakashima S (1992) Complexation and reduction of uranium by lignite. Sci Total Environ 117/118:425–437

NAS (1977) Medial and biological effects of environmental pollutants: arsenic. Nat Acad Sci, Washington, DC

NAS (1978) An assessment of mercury in the environment. Nat Acad Sci, Washington, DC

Nasseem MG, Abdalla YH (2003) Cobalt status in the North Western Coast Soils of Egypt in relation to cobalt content of barley for ruminants. 6th Itern Symp Environ Biogeochem, Edynburgh

Navarro-Alarcón M, López-Marinez MC (2000) Essentiality of seleniumin the human body: relationship with different diseases. Sci Total Environ 249:347–371

Navarro-Alarcón M, López-Garcia MC, de la Serrana H, Pérez-Valero V, López-Martinez MC (1999) Serum and urine selenium concentrations in patients with cardiovascular diseseases and relationship to other nutritional indexes. Ann Nutr Metabolism 43:30–36

Neagoe AD (2004) Bismuth. In: Merian E, Anke M, Ihnat M, Stoeppler M (eds) Elements and their compounds in the environment. 2nd ed., Wiley-VCH, Weinheim, pp 671–688

Neal C, Fox KK, Harrow ML, Neal M (1998) Boron in the major UK rivers entering the North Sea. Sci Tot Environ 210/211:41–52

Neal C, Neal M, Davies H, Smith J (2003) Fluoride in UK rivers. Sci Total Environ 314/316:209–231

Neal C, Neal M, Reynolds B, Maberly SC, May, L, Ferrier RC, Smith J, Parker JE (2005) Silicon concentration in UK surface waters. J Hydrology 304:75–93

Neff JM (1997) Ecotoxicological of arsenic in the marine environment. Environ Toxicol Chem 16:917–927

Neilsen T (1982) Atmospheric occurrence of organolead compounds. CRC, Atlanta, GA

Nelson YM, Lion LW (2003) Formation of biogenic manganese oxides and their influence on the scavenging of toxic trace metals. In: Selim HM, Kindery WL (eds) Geochemical and hydrochemical reactivity of heavy metals in soils. Lewis Publ, pp 169–186

Newman DR, Dolhopf DJ (1993) Lead levels in blood from cattle residing near a lead smelter. J Environ Qual 21:181–184

Nicholson K, Wood J (1997) Environmnetal geochemistry of boron. 4th Intern Symp Environ Geochem, Vail, CO, pp 66–67

Nicholson FA, Smith SR, Alloway BJ, Carlton-Smith C, Chambers BJ (2003) An inventory of heavy metals inputs to agricultural soils in England and Wales. Sci Total Environ 311:205–219

Nicks LJ, Chambers MF (1998) A pioneering study of the potential of phytomining for nickel. In: Brooks RR (ed) Plants that hyperaccumulate heavy metals. CAB Intern, Wallingford, pp 313–325

Niçolas G, Bennoun M, Porteu A, Mativet S, Beamont C, Grandchamp B, Sirito M, Sewadogo M, Kahn A, Vaulont S (2002) Severe iron deficiency anaemia in transgenic mice expressing liver hepcidin. Proc Nat Acad Sci 99:4596–4601

Niechayeva EG (2002) Heavy metals in taiga zone of Western Siberia. 2nd Intern Conf Heavy Metals, Radionuclides and Elements-Biofills in the Environemnt, Semipalatynsk, Kazakhstan 1:115–121 (in Russian)

Nielsen FH (1991) The saga of boron in food: from a banished food preservative to a beneficial nutrient for humans. Curr Top Plant Biochem Physiol 10:274–286

Nielsen FH (1994) Biochemical and physiological consequences of boron deprivation in humans. Environ Health Perspectives 102(7):59–62

Nielsen FH (2004) Boron. In: Merian E, Anke M, Ihnat M, Stoeppler M (eds) Elements and their compounds in the environment. 2nd ed., Wiley-VCH, Weinheim, pp 1251–1260

Nieminen TM, Derome J, Saarsalmi A (2004) Te applicability of needle chemistry for diagnosing heavy metal toxicity to trees. Water Air Soil Pollut 157:269–279

Niklińska M, Chodak M, Stefanowicz A (2004) Community level physiological profiles of microbial communities from forest humus polluted with different amounts of Zn, Pb and Cd – preliminary study with BIOLOG ecoplates. Soil Sci Plant Nutr 50:941–944

Nilsson SI, Bergkvist B (1983) Aluminium chemistry and acidification processes in a shallow podzol on the Swedish west coast. Water Air Soil Pollut 20:311–329

Niu H, Volesky B (1999) Characteristics of gold biosorption from cyanide solution. J Chem Technol Biotechnol 74:778–784

Niu H, Volesky B (2000) Gold-cyanide biosorption with L-cysteine. J Chem Technol Biotechnol 75:436–442

Nkwenkeu S-F, Kennedy G, Philippe S, Zayed J (2002) Oral manganese intake estimated with dietary records and with direct chemical analysis. Sci Total Environ 287:147–153

Noël F (1999) Indium and Indium compounds. In: Elvers B, Hawkins S, Ravenscroft M, Rounsaville J, Schulz G (eds) Ullmann's Encyclopedia of Industrial Chemistry, 5th ed. A14:157–166

Noël L, Leblanc J-C, Guerin T (2003) Determination of several elements in duplicate meals from catering establishments using closed vessel microwave digestion with inductively coupled plasma mass spectrometry detection: estimation of daily dietary intake. Food Add Contam 20:44–56

Norderhaug M, Brun E, Mollen GU (1977) Barentshavets sjøfugleressurser. Norska Polarinst Meddel No 104 (in Norwegian)

Nordstrom DK (2000) An overview of arsenic mass-poisoning in Bangladesh, and West Bengal, India. In: Young C (ed) Minor elements 2000: Arsenic, Antimony, Selenium, Tellurium and Bismuth. Soc Mining Metallurgy and Exploration, pp 21–30

Nowak J, Kaklewski K, Klódka D (2002) Influence of various concentrations of selenic acid (IV) on the activity of soil enzymes. Sci Total Environ 291:105–110

Nowak J, Niedźwiecki E, Dziel M (1999) Influence of heavy metals on the soil enzyme activity. Sol Sci Ann (Poland) 50:61–69 (in Polish)

Nozaki Y (2005) A fresh look at element distribution in the North Pacific, AGU (http://www.agu.org/eos.elec/97025e.html.)

NRC (1999) Arsenic in drinking water. National Research Council, Natl Acad Press, Washington, DC

NRCC (1978) Effects of arsenic in the Canadian environment. Natl Res Council of Canada, NRCC No 1539

Nriagu JO (1979) The Biogeochemistry of Mercury on the Environment. Elsevier/North Holad, Biomedical Press, New York

Nriagu JO (1980) Human influence on the global cadmium cycle. In: Nriagu JO (ed), Cadmium in the environment, Wiley, New York, pp 81–89

Nriagu JO (1998) History, production and uses of thallium. In: Nraigu JO (ed) Thallium in the environment. Advances in Environmental Science and Technology, vol 29, Wiley, New York, pp 1–14
Nriagu JO, Pacyna JM (1988) Quantiative assessment of worldwide contmiantion of air, water and soils by trace metals. Nature 1669:14–139
NTP (1991) Cadmium and certain cadmium compounds. 7th Annual Report on Carcinogens, Summary, U.S. Public Health Service, Dept of Health and Human Services, pp 114–121
Nuorteva P (1990a) Metal distribution patterns and forest decline. Seeking Achilles' Heels for metals in Finnish forest biocoenoses. Depart Environ Conservation, Helsinki University, Helsinki
Nuorteva P (1990b) Factors affecting the forest decline 2: Collection of publications from the Department of Environmental Conservation Publications of the Department of Environmental Conservation at the University of Helsinki, No 12
OECD (1993) Lead: Risk reduction monograph No. 1. Background and National Experience with Reducing Risk. Organisation for Economic Co-operation and Development, OCDE/GD (93) 67, Paris
OECD (1994) Risk reduction monograph No 5. Cadmium Environmental Directorate OECD, Paris
Oertli JJ (1993) The mobility of boron in plants. Plant Soil 155/156:301–304
Oertli JJ (1994) Non-homogenity of boron distribution in plants and consequence for foliar diagnosis. Comm Soil Sci Plant Anal 25:1133–1147
Oertli JJ, Richardson WF (1970) The mechanism of boron immobility in plants. Physiol Plant 23:108–116
Offelder A, Wiessner U, Kuschk P, Wobus A, Münch Ch, Köser H (2006) Removal of arsenic and dynamics of sulphur compounds in a laboratory-scale constructed wetland. Meeting, phytotechnologies lessons from pilot and filed scale, Lisbon, 54 p
Ohlendorf HM, Hoffman DJ, Saiki MK, Aldrich TW (1986) Embryonic mortality and abnormalities of aquatic birds: apparent impacts of selenium from irrigation drainwater. Sci Total Environ 52:49–63
Ojo JO, Hasan A, Ogbonna O, Ogunba O, Owolawase H, Olaniyi B (2006) Daily dietary intake sof somne toxic and essential elements in two tin mining communities on the Jos Plateau. Intern Symp Trace Elements in the Food Chain, Budapest, pp 502–506
Oldfield JE (2002) Selenium World Atlas. Updated Edn 2002. STDA, Grimbergen
Oldiges H, Glaser U (1986) The inhalative toxicity of different cadmium compounds in rats. Trace Elem Med 3:72–75
Oliver DP, Hannam R, Tiller KG, Wilhelm NS, Merry RH, Cozens GD (1994) Heavy metals in the environment. The effects of zinc fertilization on cadmium concentrations in wheat grain. J Environ Qual 23:705–711
Olson GL (1986) Microbial intervention in trace element-containing industrial processes streams and waste products. In: Berhard M, Brinckman FE Sandler PJ (eds) Dahlem Konferenzen, Springer-Verlag, Berlin Heidelberg, pp 493–512
Olsson I-M, Eriksson J, Öborn I, Skerfving S, Oskarsson A (2005) Cadmium in food production systems: A health risk for sensitive population groups. Ambio 34:344–351
Oluwole AF, Ojo JO, Durosinmi MA, Asubioja OI, Akanle OA, Spyrou NM, Filby RH (1994) Elemental composition of head hair and fingernails of some Nigerian subjects. Biol Trace Elem Res 43/45:443–452
Orians KJ, Bruland KW (1988) The marine geochemistry of gallium: a comparison with dissolved aluminium. Geochim Cosmochim Acta 52:2955–2962
Ottosen LM, Hansen HK, Laursen S, Villumsen A (1997) Electrodialytic remediation of soil polluted with copper from wood preservation industry. Environ Sci Technol 31:1711–1715
Ou SH (1985) Rice diseases. 2nd ed., Commonwealth Mycological Institute, Kew
Ou L-T, Jing W, Thomas JE (1995) Biological and chemical degradation of ionic ethyllead compounds in soil. Environ Toxicol Chem 14:545–551
Outola L. Pehrman R, Jaakola T (2003) Effect of industrial pollution on the distribution of ^{137}Cs in soil and the soil-to-plant transfer in a pine forest in SW Finland. Sci Total Environ 303:221–230
Ozaki H, Watanabe I, Kuno K (2004) Investigation of the heavy metal sources in relation to automobiles. Water Air Soil Pollut 157:209–223
Ozoliniya GR, Kiunke LM (1978) Content of little investigated elements in parts of flax, barley and lettuce. In: Fizyologo-biokhimicheskiye issledovniya rasteniy. Zinante, Riga, pp 111–119 (in Russian)
Paasikallio A (1999) Effects of biotite, zeolite, heavy clay, bentonite and apatite on the uptake of radiocesium by grass from peat soil. Plant Soil 206:213–222

Pacha J (1989) Impact of chromium on biological and physical-chemical soils properties. Uniw Śląski, Katowice (in Polish)

Pacha J, Galimska-Stypa R (1988) Mutagenic properties of selected tri- and hexavalent chromium compounds. Acta Biol Sil, Katowice 9:30–36

Pacyna JM (1990) Estimation of the atmospheric emission of trace elements from anthropogenic sources in Europe. Atmos Environ 18:41–50

Pacyna JM (1995a) Global fluxes of heavy metals to the air. Intern Conf Heavy Metals in the Environment, Hamburg, 1, pp 77–80

Pacyna ZM (1995b) The origin of air pollutants: lessons learned and future research. Sci Total Environ 160/161:39–53

Pacyna JM (1998) Source inventories for atmospheric trace metals. In: Harrison RM, van Grieken R (eds) Atmospheric particles. Wiley, New York, pp 385–423

Pacyna JM (2002) Estimation of mercury emission in Europe. European Workshop: Mercury Abatement in Industry. Forschungszentrum Karlsruhe Technik und Umwelt, Karlsruhe

Pacyna JM, Oehme M (1988) Long range transport of some organic compounds to the Norwegian Arctic. Atmos Environ 22:243–257

Pacyna EG, Pacyna JM (2001a) Global emission of mercury from anthropogenic sources in 1995. Water Air Soil Pollut 137:149–165

Pacyna JM, Pacyna EG (2001b) An assessment of global and regional emissions of trace metals to the atmosphere from anthropogenic sources. Environ Rev 9:269–298

Page WG (1981) Comparison of groundwater and surface water for patterns and levels of contamination by toxic substances. Environ Sci Technol 15(12):1475–1480

Page AL, Chang AC, El-Amamy M (1987) Cadmium levels in soils and crops in the United States. In: Hutchinson TC, Meena KM (eds) Lead, mercury, cadmium and arsenic in the environment, SCOPE 31, Wiley, New York, pp 119–146

Pain DJ, Donald PF (2002) Outside the reserve: pandemic threats to bird diversity. In: Norris K, Pain DJ (eds) Conserving bird diversity: general principles and their application. Cambridge Univ Press, Cambridge, pp 157–179

Pais I, Jones JB (1997) The handbook of trace elements. St. Lucie Press, Boca Raton, FL

Pakkanen TA, Loukkola K, Korhonen CH, Aurela M, Mäkelä T et al. (2002) Sources and chemical composition of atmospheric fine and coarse particles in the Helsinki area. Atmos Environ 35:5381–5391

Pałasz A, Czekaj P (2000) Toxicological and cytophysiological aspects of lanthanides action. Acta Biochim Polonica 47:1107–1114

Paliouris G, Hutchinson TC (1991) Arsenic, cobalt, nickel tolerances in two populations of *Silence vulgaris* (Moench) Garcke from Ontario, Canada. New Phytologist 117:449–459

Pamphlett R, Danscher G, Rungby J, Stoltenberg M (2000) Tissue uptake of bismuth from shotgun pellets. Environ Research 82:258–262

Panin MS (2004) Main sources of pollution in Kazakhstan territory with radionuclides. 3rd Intern Conf Heavy Metals, Radionuclide and Elements-Biofills in Environ. Semipalatynsk, Kazahstan, pp 14–27 (in Russian)

Panova JG (1988) Biogeochemical indication of technogenic anomaly of tungsten in conditions of semidesert landscape. In: Biogeochemical indication of surrounding environment. Nauka, Leningrad, pp 47–52 (in Russian)

Parker H, Holm H (1990) Patterns of nutrient and energy expenditure in female. Common eiders nesting in the High Arctic. A global assessment of natural sources of atmospheric trace elements. Auk 107:660–668

Parker DR, Kinraide TB, Zelazny LW (1989) On the phytotoxicity of mononuclear hydroxy aluminium complexes. Soil Sci Soc Am J 53:779–789

Pasternak K (1997) Magnesium and zinc in stomach cancer. Biul Magnezol 2:222–225 (in Polish)

Patric K, Robinson TN, Alemi F, Eng TR (1999) Policy issues relevant to evaluation of interactive health communication applications. Am J Prev Med 16:35–42

Pattenden NJ, Bransow JR, Fisher EMR (1982) Trace element measurements in wet and dry deposition and airborne particulate at an urban site. In: Georgii HW, Pankrath J (eds) Deposition of atmospheric pollutants. Reidel Publ, Boston, pp 173–184

Paul EA, Clark FE (2000) Soil microbiology and biochemistry. Wyd UMCS, Lublin (Polish translation)
Pavelka S (2004) Bromine. In: Merian E, Anke M, Ihnat M, Stoeppler M (eds) Elements and their compounds in the environment. 2nd ed. Wiley-VCH, Weinheim, pp 1445–1455
Pavlikova D, Pavlik M, Tlustos P, Balik J, Vašičkova S, Száková J (2005) Separation of organic binding risk elements in plant biomass. Intern Sci Conf Innovation and Utility in the Visegrad Fours, Nyiregyhaza, 1, pp 277–282
Pédro G (1979) Caractérisation générale des processus de l'alterétion hydrolytique. Base des anaden géochimique et thermodynamique. Science Du Sol 2:93–105
Peganova S, Edlet K (2004) Zinc. In: Merian E, Anke M, Ihnat M, Stoeppler M (eds) Elements and their compounds in the environment. 2nd ed., Wiley-VCH, Weinheim, pp 1203–1239
Pekka L, Ingri J. Widerlund A, Mokrotovarova O, Riabstseva M, Öhlander B (2004) Geochemistry of the Kola River, northwestern Russia. Appl Geochem 19:1975–1995
Pempkowiak J (1991) Enrichment factors of heavy metals in the Southern Baltic surface sediments dated with ^{210}Pb and ^{137}Cs. Environ Intern 17:421–428
Penland JG (1994) Dietary boron, brain function, and cognitive performance. Environ Health Perspectives 102(7):65–71
Percival SS (1995) Neutropenia caused by copper deficiency: possible mechanisms of action. Nutrition Reviews 53:59–66
Pereira R, Ribeiro R, Gonçalves F (2004) Scalp hair analysis as a toll in assessing human exposureto heavy metals (S. Dominos mine, Portugal). Sci Total Environ 327:81–92
Peters SG (2002) Geology, geochemistry and geophysics of sedimentary rock-hosted Au deposits in P.R. China. U.S. Geological Survey, Open-File Report 02–131 (http://geopubs.wr.usgs.gov/open–file/of02–131/)
Petersen G, Munthe J, Bloxam R (1996) Numerical modeling of regional transport, chemical transformations and deposition fluxes airborne mercury species. In: Baeyens W, Ebinghaus R, Vasiliev O (eds) Global and regional mercury cycles: sources, fluxes and mass balances. Kluwer Publ, Dortrecht, pp 191–217
Petersen G, Munthe J, Pleijel K, Bloxam R, Vinod Kumar A (1998) A comprehensive eulerian modeling framework for airborne mercury species: development and testing of the troposphere chemistry module (TCM). Atmos Environ 32:829–843
Peterson PJ, Burton MAS, Gregosn M, Nye SM, Porter EK (1976) Tin in plants and surface waters in Malaysian ecosystems. Trace Subs Environ Health 10:123–135
Phillips DJH (1995) The chemistries and environmental fates of trace metal and organochlorides in aquatic ecosystem. Marine Pollut Bull 31:193–200
Pieper T, Keppler BK (1998) Tumor-inhibiting ruthenium complexes – formulation and analytical characterization. Analusis Mag 26:84–87
Pieri De LA, Buckley WT, Kowalenko CG (1996) Micronutrient concentrations of commercially grown vegetables and of soils in the Lower Fraser Valley of British Columbia. Can J Soil Sci 67:173–182
Pierzynski GM, Jacobs LW (1986) Extractability and plant availability of molybdenum from inorganic and sewage sludge sources. J Environ Qual 15:322–326
Pierzynski GM, Sims JT, Vance GF (eds) (2000) Soils and environmental quality. 2nd ed., CRC Press, Boca Raton, FL
Pietrzak-Flis Z, Rosiak L, Suplińska MM, Chrzanowski E, Dembińska S (2001) Daily intakes of ^{238}U, ^{234}U, ^{232}Th, ^{230}Th, ^{228}Th and ^{226}Ra in the adult population of central Poland. Sci Total Environ 273:163–169
Pietukhova NN, Fedenya VM, Matveyeve VI (1996) Assessment of soil heavy metal contamination of soil of Belarus. Prirodnye Resursy 1:20–23 (in Belarussian)
Piispanen R, Lähdesmäki P (1983) Effects of vanadium on some water plants. Geol Fören Stockholm Förh 105:23–27
Piltz K, Anke M, Klinger G (2004) Nickel in der Nahrungskette des Menschen – gesundheitliche Probleme und Risiken. In: Anke M, Flachowsky G, Kisters K et al. (eds) 22nd Workshop on Macro and Trace Elements, Jena, pp 1563–1568
Plankey BJ, Patterson HH (1987) Kinetics of aluminium – fulvic acid complexation in acid waters. Env Sci Technol 21:595–601
Plant JA, Kinniburgh DG, Smedley PL, Fordyce FM, Klinck BA (2004) Arsenic and selenium. In: Dollar BS (ed) Environmental geochemistry 9:17–66

Platts Metals Week (2003a) Lead-acid batteries top recycling, 74(30), July 28, p 10

Platts Metals Week (2003b) EC proposal targets collection, recycling of batteries, 74(48), December 1, p 12

Playle RC, Goss GG, Wood CM (1989) Physiological disturbances in rainbow trout (*Salmo gairdneri*) during acid and aluminium exposures in soft water of two calcium concentrations. Can J Zoology 67:314–324

Plumlee GS, Ziegler TL (2003) The medical geochemistry of dusts, soils, and other earth materials. Treatise on geochemistry 9:263–310

Pluta I (2001) Barium and radium discharged from coal mines in the Upper Silesia, Poland. Environ Geol 40:345–348

Pluta I (2005) Mine waters of the Upper Silesian Coal Basin-origin, pollutants and purification. Prace Nauk Głównego Instytutu Górnictwa, 865 (in Polish)

Podlešáková E, Němeček J (1994) Contamination of soils in the north-Bohemian region by hazardous elements. Rostl Výrob 40:123–130 (in Czech)

Pohl C, Hennings U (1999) The effect of redox processes on the partitioning of Cd, Pb, Cu and Mn between dissolved and particulate phases in the Baltic Sea. Mar Chem 65:41–53

Pohlmeier A (1999) Metal chelation and complexing ligands in plants. In: Prasad MNV, Hagemeyer J (eds) Heavy metal stress in plants. Springer-Verlag, Berlin Heidelberg, pp 29–50

Pokrovski G, Gout R, Schott J, Zotov A, Harrichoury JC (1996) Thermodynamic properties and stoichiometry of As (III) hydroxide complexes at hydrothermal conditions. Geochim Cosmochim Acta 60:737–749

Poléo ABS, Hytterød S (2003) The effect of aluminium in Atlantic salmon (*Saloma salar*) with special emphasis on alkaline water. J Inorganic Biochem 97:89–96

Polizzi S, Pira E, Ferrara M, Bugiani M, Papaleo A, Albera R, Palmi S (2002) Neuro toxic effects of aluminium among foundry workers and Alzheimer's disease. Neuro Toxicol 23:761–774

Ponizovsky AA, Thakali S, Allen HE, Di Toro DM, Ackerman AJ (2006) Effect of soil properties on copper release in soil solution at low moisture content. Environ Toxic Chem 25:671–682

Porvari P (1995) Mercury levels of fish in Tucurui hydroelectric reservoir and in River Moju in Amazonia, in the state of Para, Brasil. Sci Tot Environ 175:109–117

Poschenrieder Ch, Barceló J (2004) Water relations in heavy metal stressed plants. In: Prasad MNV (Ed) Heavy metal stress in plants. Springer-Verlag, Berlin Heidelberg, pp 249–270

Power PP, Woods WG (1997) The chemistry of boron and its speciation in plants. In: Dell B, Brown PH, Bell RW (eds) Boron in soils and plants: Reviews. Kluwer Publ, London, vol 77, pp 1–13

Prasad MNV (1995) Cadmium toxicity and tolerance in vascular plants. Environ Exp Bot 35:525–545

Prasad MNV (2004a) "Phytoferritins" for improving human health – an integrated approach of "metallomics". Intern Symp Iron and Manganese in Environ – Ecological Analytical Problems. Warszawa

Prasad MNV (2004b) Phytoremediation of metals in the environment for sustainable development. Plant Indian Nat Sci Acad B70, 1:73–100

Prasad MNV (2006) Plants that accumulate and/or exclude toxic trace elements play an important role in phytoremediation. In: Prasad MNV, Sajwan KS, Naidu R (eds) Trace elements in the environment: Biogeochemistry, biotechnology and bioremediation. CRC Press, Boca Raton, FL, pp 523–547

Prasad MNV, Hagemeyer J (1999) Heavy metal stress in plants. Springer-Verlag, Berlin Heidelberg

Pratt PF (1972) Quality criteria for trace elements in irrigation waters. California Agricultural Experiment Station

Proctor DM, Shay EC, Scott PK (1997) Health based soil levels for trivalent and hexavalent chromium: a comparison to state and federal standards. J Soi Contam 6:595–648

Prospero JM, Barrett K, Church T, Dentener F, Duce RA, Galloway JN, Levy H, Moody J, Quinn P (1996) Atmospheric deposition of nutrients to the North Atlantic Basin. Biogeochemistry 35:27–37

Protasova NA, Kopayeva MT (1985) Trace and dispersed elements in soils of Russian Plateau. Pochvovedeniye 1:29–37 (in Russian)

Prüeß A (1994) Einstufung mobiler Spurenelemente in Böden. In: Rosenkranz D, Bachmann G, Einsele G, Harreß H-M, (eds) Bodenschutz. Ergänzbares Handbuch der Maßnahmen und Empfehlungen für Schutz, Pflege und Sanierung von Böden, Landschaft und Grundwasser – 1. Band, 15 Lfg. 1/94, Erich Schmidt, Berlin

Przysławski J, Duda G (2000) Trace elements: chromium, nickel and cobalt in food. In: Karaś Z (ed) Chromium, nickel and cobalt in food-ecosystem, friends or enemies. PTTŻ, Poznań, pp 83–97

Punshon T, Adriano DC (1999) Nickel uptake by hybrid poplar: assessing phytoremediation potential. 5th Intern Conf Biogeochem Trace Metals, Vienna, pp 920–921

Purcell TW, Peters JJ (1998) Sources of silver in the environment. Environ Toxicol Chem 17:539–546

Puwastien P, Burlingarne B, Raroengwicht M, Sungpung P (2000) ASEAN Food Composition Tables, 1st ed., Institute of Nutrition, Mahidol University, ASEAN Regional Database Center

Puxbaum H, Limbeck A (2004) Chemical compounds in the atmosphere. In: Merian E, Anke M, Ihnat M, Stoepppler M (eds) Elements and their compounds in the environment, 2nd ed., Wiley-VCH, Weinheim, pp 17–46

Pyatt FB (1999) Comparison of foliar and stem bioaccumulation of heavy metals by Corsian pines in the Mount Plympus area of Cyprus. Ecotoxicol Environ Safety 42:57–161

Pyatt EB, Pyatt AJ, Walker C, Sheen T, Grattan JP (2005) The heavy metal content of skeletons from an ancient metalliferous polluted area in southern Jordan with particular reference to bioaccumulation and human health. Ecotoxycol Environ Safety 60:295–300

Pyrzyńska K (1995) Solid phase extraction for preconcentration and separation of selenium. Solvent Ext Ion Exchange 13:369–389

Pyrzyńska K (2000) Monitoring of platinum in the environment. J Environ Monit 2:99N–103N

Queirolo F, Stegen S, Restovic M, Paz M, Ostapczuk P, Schwuger MJ, Munoz L (2000) Total arsenic, lead and cadmium levels in vegetables cultivated at the Andean villages of northern Chile. Sci Total Environ 255:75–84

Quevauviller P, F OFX (1990) Variability of butyltin determination in water and sediment samples from European coastal environments. Appl Organometal Chem 4:353–367

Quinn KA, Byrne RH, Schuf J (2004) Ytrium and rare element adsorption onto freshly preacipitaded hydroxides of Fe (III), Al, Ga, and In. Goldschmidt, Copenhagen, A328 p

Quintela MJ, Gallego M, Valcarcel M (1993) Flow injection spectrophotometric method for the speciation of aluminium in river and tap waters. Analyst 118:1199–1203

Rakovič M, Latýnová E, Foltýnová V, Výborný S, Kučera J, Pilecká N, Glagoličová A (1996) Feasibility of using INAA for biomonitoring of heavy metals and other elements in human ectoderm derivates. J Radioanal Nucl Chem Letters 212:281–186

Ramsden D (2000) Thallium. In: Waring RH, Steventon B, Mitchell SC (eds) Molecules of death. World Scientific, London, pp 304–311

Ranau R, Oehlenschäger J, Steinhart H (2001) Aluminium levels of fish fillets backed and grilled in aluminium foil. Food Chem 73:1–6

Randall P, Hedrick E, Grimmet P, Engle M, Ilyushchenko M (2004) Observations and analysis of mercury in the topsoil with a 100m radius of a chlor-alkali plant in northern Kazakhstan using EPA method 7473. 7th Intern Conf Mercury as a Global Pollutant, Ljubljana 1:207–211

Rashed MN, Awadallah RM (1998) Trace elements in faba bean (*Vicia faba* L.) plant and soil as determined by atomic absorption spectroscopy and ion selective electrodes. J Sci Food Agric 77:18–24

Rasmussen PE (2004) Elements and their compounds in indoor environments. In: Merian E, Anke M, Ihnat M, Stoepppler M (eds) Elements and their compounds in the environment, 2nd ed., Wiley-VCH, Weinheim, pp 215–234

Rasmussen PE, Subramanian KS, Jessiman BJ (2001) A multi-element profile of housedust in relation to external dust and soils in the city of Ottawa, Canada. Sci Total Environ 267:125–140

Raspor B (2004) Elements and elemental compounds in waters and the aquatic food chain. In: Merian E, Anke M, Ihnat M, Stoeppler M (eds) Elements and their compounds in the environment. Wiley-VCH, Weinheim, 1:127–147

Ratie HT (1999) Bioaccumulation and toxicity of silver compounds: A review. Environ Toxicol Chem 18:89–108

Rau T, van Eldik R (1996) Mechanistic insight from kinetic studies on the interaction of model palladium (II) complexes with nucleic acid components. Metal ions in Biological Systems 32:339–378

Rauser WE (1990) Phytochelatins. A Review Biochem 59:61–86

Ravindra K, Bencs L, Van Grieken R (2004) Platinum group elements in the environment and their health risk. Sci Total Environ 318:1–43

Rayman MP (2000) The importance of selenium to human health. Lancet 356:233–241

Read HJ, Martin MH (1993) The effect of heavy metals on populations of small mammals from woodlands in Avon (England) with particular emphasis on metal concentrations in *Sorex araneus* L. and *Sorex anaden* L. Chemosphere 27:2197–2211

Reed JS, Henningson JC (1984) Acid precipitation and drinking water supplies. J Am Water Works Assoc 76:60–65

Reichman SM, Parker DR (2005) Probing the diural patterns of phytosidrophore release in wheat (*Triticum aestivum*). ICOBTE, 8th Intern Conf, Adelaide, pp 476–477

Reif J, Průša J, Mejstřík V Prochazka L, Liska A, Blaha P, Duras R et al. (2004) Survey of land contaminated with mercury at PVC–SODA factory in Vlora, Albania. 7th Intern Conf Mercury Global Pollutant, Ljubljana, 1, pp 220–228

Reimann C, de Caritat P (1998) Chemical elements in the environment. Springer-Verlag, Berlin Heidelberg

Reimann C, Halleraker JH, Kashulina G, Bogatyrev I (1999) Comparison of plant and precipitation chemistry in catchments with different level of pollution on the Kola Peninsula, Russia. Sci Total Environ 243/244:169–191

Reisdorf RP, D'Orlando DC (1984) Survey of health hazards control systems for mercury use and processing. NTIS PB85-107241. National Technical Information Service, Springfield, VA

Renaud P, Pourcelot L. Métivier J-M, Morello M (2003) Mapping of ^{137}Cs deposition over eastern France 16 years after the Chernobyl accident. Sci Total Environ 309:257–264

Rengel Z (2004) Biogeochemistry of manganese in agricultural, native and disturbed ecosystems. Symp Iron and Manganese in Environ-Ecologic Analyt Problems. Warszawa

Rengel Z, Römheld V, Marschner H (1998) Uptake of zinc and iron by wheat genotype differing in zinc deficiency. J Plant Physiol 52:433–438

Richmond KE, Sussman M (2003) Got silicon? The non-essential beneficial plant nutrient. Plant Biology 6:268–272

Rikhanov LP (1997) General and regional problems of radioecology. Publ House Tomsk Polytech Univ, Tomsk (in Russian)

Ringelman JK, Miller, MW, Andelt WF (1993) Effects of ingested tungsten-bismuth-tin shot on captive mallards. J Wildlife Manag 57:725–732

Riog-Navarro AF, López FJ, Serrano R, Hernández F (1997) An assessment of heavymetals and boron contamination in workplace atmospheres from ceramic factories. Sci Total Environ 201:225–234

Rish MA (2004) Antimony. In: Merian E, Anke M, Ihnat M, Stoeppler M (eds) Elements and their compounds in the environment. 2nd ed., Wiley-VCH, Weinheim, pp 659–670

Risher JF, Amler SN (2005) Mercury exposure: Evaluation and intervention; the inappropriate use of chelating agents in the diagnosis and treatment of putative mercury poisoning. Neuro Toxicology 26:691–699

RIZA (1989) Opportunities for water organisms: ecological support for quality goals for water and water-soil) DBW/RITZ-note 89.016, Lelystad (in Dutch)

Roberts AHC, Longhurts RD, Brown MW (1994) Cadmium status of soils, plants, and grazing animals in New Zealand. NZ J Agric Res 37:119–129

Robinson WWO, Bastron H, Murata KJ (1958) Biogeochemistry of rare-earth elements with particular reference to hickory trees. Geochim Cosmochim Acta 14:55–67

Robinson BH, Green S, Mills T (2003) Assessment of phytoremediation as best management practice for degraded environment. In: Currie LD, Stewart R, Anderson CWN (eds) Environmental management using soil-plant-system. Fertilizer and Lime Research Center, Massey University, Palmerston North, Occasional Report 16:39–49

Rocher V, Azimi S, Gasperi J, Beuvin L, Muller M, Moilleson R, Chebbo G (2004) Hydrocarbons and metals in atmospheric deposition and roof runoff in central Paris. Water Air Soil Pollut 159:67–86

Röderer G (1984) Toxic effects in plant organisms. In: Grandjean Ph (ed) Biological effects of organolead compounds. CRC Press, Boca Raton, FL, pp 63–95

Rodrigues FÁ, Vale FXR, Korndörfer GH, Prabhu AS, Datnoff LE, Oliveria AMA, Zambolim L (2003) Influence of silicon on sheath blight of rice in Brazil. Crop Protection 22:23–39

Rogall H, Römheld V (2002) Role of leaf apoplast in silicon-mediated manganese tolerance of *Cucumis sativus* L. Plant Cell Environ 25:549–555

Rosiak L, Pietrzak-Flis Z (1998) Uptake of ^{226}Ra from soil by Graminae and Umbelliferae. Report CLOR, Warszawa (in Polish)

Ross SM (1994) Sources and forms of potentially toxic metals in soil-plant systems. In: Ross SM (ed) Toxic metals in soil-plant systems. Wiley, Chichester, pp 4–25

Rosseland BO, Skogheim OK, Sevaldrud IH (1986) Acid deposition and effects in Nordic Europe. Damage to fish populations in Scandinavia continue apace. Water Air Soil Pollut 30:65–71

Rossmann R (1984) Trace metal concentrations in the offshore waters of lakes Erie and Michigan. Special Rep No. 108, Great Lakes Research Division, University of Michigan, Madison, MI

Rossmann MD (2004) Beryllium. In: Merian E, Anke M, Ihnat M, Stoeppler M (eds) Elements and their compounds in the environment. 2nd ed., Wiley-VCH, Weinheim, pp 575–586

Roszak W (1991) Chemical composition of shallow groundwaters in Odra valley near Wrocław. University of Wrocław, Wrocław (in Polish)

Rotruck JT, Pope AI, Ganther HE, Swanson AB, Hafeman DG, Hoestra WG (1973) Selenium: biochemical role as a component of glutathione peroxidase. Science 179:588–590

Routh J, Bhattacharya P, Jacks G, Ahmed KM, Khan AA, Rahman MM (2000) Arsenic geochemistry of Tala groundwater and sediments from Satkhira District, Bangladesh. Eos Trans Amer Geophy Union 81:550

Roy AK (1996) Innovative methods to manage sheath blight of rice. J Mycopathol Res 34:13–19

Rüdel H, Wenzel A, Terytze K (2001) Quantification of soluble chromium (VI) in soils and evaluation of ecotoxicological effects. Environ Geochem Health 23:219–224

Ruf M (1981) Kongreβvortäge Wasser Berlin '81, Colloquium Verlag, Otto H. Hess, Berlin, pp 415–428

Rufyikiri G, Dufey JE, Nootens D, Devaux B (2000) Effect of aluminium on bananas (Musa spp.) cultivated in acid solutions. II. Water and nutrient uptake. Fruits 56:3–14

Rugh CL, Wilde HD, Stack NM, Thomson DM, Summers AO, Meagher RB (1996) Mercuric ion reduction and resistance in transgenic *Arabidopsis thaliana* plants expressing a modified bacterial merA gene. Proc Natl Acad Sci USA 93:3182–3187

Rühling Å (ed) (1994) Atmospheric heavy metal deposition in Europe – estimation based on moss analysis. Nord 9:53

Rühling Å, Tyler G (2004) Changes in the atmospheric deposition of minor and rare elements between 1975 and 2000 in south Sweden, as measured by moss analysis. Environ Pollut 131:417–423

Rühling Å, Rasmussen L, Pilegaard K, Mäkinen A, Steinnes (1987) Survey of atmospheric heavy metal deposition in Nordic Countries in 1985. Report for Nordic Council of Ministers, Kopenhagen

Ruiz JM, Bachelet G, Caumetter P, Donard OFX (1996) Three decades of tributyltin in the coastal environment with emphasis on Arcachon Bay, France. Environ Poll 93:195–203

Rule JH (1994) Use of small plants as trace element phytomonitoring, with emphasis on the common dandelion, *Taraxacum officinale*. In Adriano DC, Chen Z-S, Yang S-S (eds) Biogeochemistry of trace elements. Sci Technol Letters, Bukhurst Hill, Essex, pp 627–654

Rule JH (1999) Trace metal cation adsorption in soils: selective chemical extractions and biological availability. In: Dabrowski (ed) Adsorption and its applications in industry and environmental protection. Elsevier, Amsterdam, pp 319–349

Rulkens WH, Honders A (1996) Clean-up of contaminated sites: Experiences in the Netherlands. Water Sci Tech 34:293–301

Rulkens WH, Grotenhuis JTC, Soczó A (1993) Remediation of contaminated soil: State of the art and desirable future developments. In: Arendt F, Annokkée GJ, van den Brink WJ (eds) Contaminated soils '93, Kluwer Acad Publ, pp 1007–1018

Runde W (2002) Geochemical interactions of actinides in the environment. In: Zhang P-C, Brady PV (eds) Geochemistry of soil radionuclides, SSSA Spec Publ, Madison, WI, pp 21–44

Rush MC, Lee FN (1992) Sheath blight. In: Webster RK, Gunnell PS (eds) Compendium of rice diseases. Am Phytopath Soc, St Paul, MN, pp 22–23

Ruskin I, Kumar PBAN, Dushenkov S, Salt DE (1994) Bioconcentration of heavy metals by plants. Current Option in Biotechnology 5:285–290

Sabadell JE, Axtmann RC (1975) Heavy metal contamination from geothermal sources. Environ Health Perspect 12:1–7

Sabaty M, Avazeri C, Pignol D, Vermeglio A (2001) Characterization of the reduction of selenate and tellurite by nitrate reductases. Appl Environ Microbiol 67:5122–5126

Sabbioni E, Ceotz L, Bignoli G (1984) Health and environmental implications of trace metals released from coal-fired power plants: an assessment study of the situation in the European Community. Sci Total Environ 40:141–154

Sabti H, Hossain MM, Brooks RR, Stewart RB (2000) The current environmental impact of base-metal mining at the Tui Mine, Te Aroha, New Zealand. J Roy Soc NZ 30:197–208

Sager M (1998) Thallium in agricultutal practice. In: Nriagu JO (ed) Thallium in the environment. Advances in Environmental Science and Technology, vol 29, Wiley, New York, pp 59–87

Sakata M (1998) Estimation method for contribution from emission sources to trace metal concentrations in urban atmosphere based on lead isotope ratios. CRIEPI Rep. No T97050 (in Japasene)

Salifoglou A (2003) Vanadium ions in pursuit of insulin mimetic action in humans. 4th Intern Symp Trace Elements in Human: New Persp. Enthypossis, Athens 1, pp 538–542

Salomons W, Förstner U (1984) Metals in the hydrocycle. Springer-Verlag, Berlin Heidelberg

Salomons W, Kerdijk HN (1986) Cadmium in fresh and estuarine waters. In: Mislin H, Ravera O (eds) Cadmium in the environment. Birkhäuser Verlag, Experientia Suppl 50:24–28

Salt DE, Smith RD, Raskin I (1998) Phytoremediation. Ann Rev Physiol Plant Mol Biol 48:643–668

Samanta G, Mandal BK, Chowdhury TR, Biswas BK, Chowdhury et al. (1999) Arsenic in groundwater and sufferings of people in eight districts of West Bengal, India. 5th Intern Conf Biogeochem Trace Elements. Vienna, pp:256–257

Samecka-Cymerman A (1988) Tolerance of *Scapania undulata* to barium present in the environment, Pol Arch Hydrobiol 35:33–44

Samecka-Cymerman A, Kempers AJ (2003) Biomonitoring of water pollution with *Elodea anadensis.* A case study of three small Polish rivers with different levels of pollution. Water Air Soil Pollut 145:139–153

Samecka-Cymerman A, Kempers AJ (2004) Toxic metals in aquatic plants surviving in surface water polluted by copper mining industry. Ecol Environ Safety 59:64–69

Samecka-Cymerman A, Stepien D, Kempers AJ (2004) Efficiency in removing pollutants by constructed wetland purification system in Poland. J Toxicol Environ 67:265–275

Sanchez-Cabeza J-A, Merino J, Masqué O, Mitchell PI, Vitró LL, Schell WS, Cross L, Cakbert A (2003) Concentration of plutonium and americium in plankton from western Mediterranean Sea. Sci Total Environ 311:233–245

Santa-Maria GE, Cogliatti DH (1998) The regulation of zinc uptake in wheat plants. Plant Sci 137:1–12

Sañudo-Wilhelmy SA, Flegal AR (1992) Anthropogenic silver in the Southern California Bight: A new trace of sewage in coastal waters. Environ Sci Technol 26:2147–2151

Sarong LQ, Bouldin DR, Reid WS (1989) Total and labile zinc concentrations in water extracts of rhizosphere and bulk soils of oats and rice. Comm Soil Sci Plant Anal 20:271–289

Sarosiek J, Letachowicz B, Wozakowska-Natkaniec H (1998) Beryllium and germanium bioindication in the environment pollution from Karkonsze Mountain with the use of *Salix caprea* L. Symp Geoekol Problem Karkonoszy, Aracus, Poznan, pp 253–256

Sata F, Araki S, Murata K, Aono H (1998) Behavior of heavy metals in human urine and blood following calcium disodium ethylenediamine tetraacetate injection. Observations in metal workers. J Toxicol Environ Health, Part A 54:167–178

Savant NK, Korndörfer GH, Datnoff LE, Snyder GH (1999) Silicon nutrition and sugarcane production: a review. J Plant Nutr 22:1853–1903

Savinov VM, Gabrielsen GW, Savinova GW (2003) Cadmium, zinc, copper, arsenic, selenium and mercury in seabirds from the Barents Sea: levels, inter-specific and geographical differences. Sci Total Environ 306:133–158

Scansetti G (1992) Exposure to metals that have recently come into use. Sci Total Environ 120:85–91

Schafer SG, Fermfert V (1984) Tin – a toxic heavy metal? A review of the literature. Regul Toxicol Pharmacol 4:57–69

Schäfer U (1997) Essentiality and toxicity of lithium. J Trace Microb Techiques 15:341–149

Schäfer U (2004a) Alkali metals. Lithium. In: Merian E, Anke M, Ihnat M, Stoepppler M (eds) Elements and their compounds in the environment, Wiley-VCH, Weinheim, pp 479–496

Schäfer U (2004b) Manganese. In: Merian E, Anke M, Ihnat M, Stoeppler M (eds) Elements and their compounds in the environment, 2nd ed., Wiley-VCH, Weinheim, pp 901–930

Schäfer U, Anke M. Seifer M, Fischer AB (2002) Manganese intake, extretion and balance of adults. I. Different influences on the manganese concentration of the consumed food. 21th Workshop Macro Trace Elements, Jena, pp 1074–1080
Schaller K-H (2004) Zirconium In: Merian E, Anke M, Ihnat M, Stoeppler M (eds) Elements and their compounds in the environment, 2nd ed., Wiley-VCH, Weinheim, pp 1241–1247
Schaumlöffel D, Ourdane L, Bonyssiere B, Łobiński R (2003) Speciation analysis of nickel in the latex of a hyperaccumulating tree *Sebertia acumulata* by HPLC and CZE with ICP MS and electrospray MS–MS detector. J Anal At Spectrom 18:120–127
Schaumlöffel D, Szpunar J, Łobiński R (2004) Species-selective analysis for metals and metalloids in plants. In: Prasad MNV (ed) Heavy metal stress in plants: From biomolicules to ecosystem. 2nd ed., Springer-Verlag, Berlin Heidelberg, pp:409–433
Schauss AG (1991) Nephrotoxicity and neurotoxicity in humans from organogermanium compounds and germanium dioxide. Biol Trace Elem Res 29:267–280
Scheinost AC, Bernhard G, Selenka-Pobell S (2004) Fate of uranium in the environment. Goldschmidt, Copenhagen, A526
Schenck RU, Bjorksten J, Yeager L (1990) Composition and consequences of aluminium water, beverages and other ingestibles. In: Lewis TE (ed) Environmental chemistry and toxicology of aluminium. Lewis Publ, Chelsea, MI, pp 247–269
Schimmack W, Bunzl K, Zelles L (1989) Initial rates of migration of radionuclides from the Chernobyl fallout in undisturbed soils. Geoderma 44:211–218
Schlegel-Zawadzka M, Nolwak G (2000) Alterations in serum and brain trace element levels after antidepressant treatment. Part II. Copper. Biol Trace Element Res 73:37–45
Schlegel-Zawadzka M, Zięba A, Dudek D, Krośniak M, Szymanek M, Nowak G (1999) Serum trace elements in animal model human depression. Part II, Copper. Human Psychopharmacol Clin Exp 14:447–451
Schlegel-Zawadzka M, Zięba A, Dudek D, Krosniak M, Szymczek M, Nowak G (2000) Effect of depression and of antidepressant therapy on serum zinc levels. In: Roussel AM, Anderson RA, Favrier AE (eds) Trace elements in man and animals 10. Kluwer Publ, Dortrecht, pp 607–610
Schmitt CJ, Brumbaugh WG (1990) National contamination biomonitoring program: Contamination of As, Cd, Cu, Pb, Hg, Se and Zn in the U.S. freshwater fish 1976–1984. Arch Environ Contam Toxicol 19:731–747
Schoer J (1984) Thallium. In: Hutzinger O (ed) Handbook of environmental chemistry. Springer-Verlag, Berlin Heidelberg, pp 143–214
Scholl W (1980) Bestimmung von Thallium in verschiedenen anorganischen und organischen Matrices. Ein einfaches photometrisches Routineverfahren mit Brillantgrün. Landwirtsch Forsch 37:275–286
Schrauzer GN (2003) The nutritional significance, metabolism and toxicology of selenomethionine. In: Taylor SL (ed) Advances in food and nutrition research. Elsevier, Amsterdam 47:91–112
Schrauzer GN (2004a) Cobalt. In: Merian E, Anke M, Ihnat M, Stoeppler M (eds) Elements and their compounds in the environment. 2nd ed., Wiley-VCH Verlag, Weinheim, pp 825–839
Schrauzer GN (2004b) Selenium. In: Merian E, Anke M, Ihnat M, Stoeppler M (eds) Elements and their compounds in the environment. 2nd ed., Wiley-VCH Verlag, Weinheim, pp 1365–1406
Schroeder WH, Munthe J (1998) Atmospheric mercury – An overview. Atmosph Environ 29:809–822
Schroeder HA, Kanisawa M, Frost DV, Mitchener M (1968) Germanium, tin and arsenic in rats: Effects on growth, survival, pathological lesions and life span. J Nutr 96:37–45
Schroeder WH, Dobson M, Kane DM, et al. (1987) Toxic trace elements associated with airborne particulate matter: A review. J Air Pollut Control Assoc 37:1267–1285
Schuhmacher MS, Domingo JL, Corbella J (1994) Zinc and copper levels in serum and urine. Relationship to biological, habitual and environmental factors. Sci Total Environ 148:67–72
Schultz R (1987) Vergleichende Betrachtung des Schwermetallhaushalts verschiedener Waldökosysteme Norddeutschlands. Berichte des Forschungszentrums Waldökosysteme der Universität Göttingen Reigh A 32
Schümann K, Elsenhans B (2004) Iron. In: Merian E, Anke M, Ihnat M, Stoeppler M (eds) Elements and their compounds in the environment. 2nd ed., Wiley-VCH Verlag, Weinheim, pp 811–824

Schuster E (1991) The behavior of mercury in the soil with special emphasis on complexation and adsorption processes: A review of the literature. Water Air Soil Pollut 56:667–680

Ścibior A (2003) Total antioxidant status in plasma of rats in vanadium-chromium interaction. 4th Intern Symp Trace Elements in Human: New Persp, Enthypossis, Athens 2, pp 1120–1153

Sedlak W (1967) The role of silicon in biochemical evolution of life. PWN, Warszawa (in Polish)

Seebold KW, Datnoff LE, Correa-Victoria FJ, Kucharek TA, Snyder GH (2000) Effect of silicon rate and host resistance on blast, scald and yield of upland rice. Plant Dis 84:871–876

Seebold KW, Kucharek TA, Datnoff LE, Correa-Victoria FJ, Marchetti MA (2001) The influence of silicon on components of resistance to blast in susceptible, partially resistance, and resistant cultivars of rice. Phytopathology 91:63–69

Séguin V, Gagnon C, Courchesne F (2004) Changes in water extractable metals, pH and organic carbon concentrations at the soil-root interface of forested soils. Plant Soil 260:1–17

Seifert M (2004) Strontium. In: Merian E, Anke M, Ihnat M, Stoeppler M (eds) Elements and their compounds in the environment, 2nd ed., Wiley-VCH, Weinheim, pp 619–626

Seifert M, Arnhold W, Jaritz M, Holzinger S, Anke M (2002) Strontiumverzehr, -ausscheidung und -bilanz Erwachsener. Workshop Macro Trace Elements, Jena, pp 1217–1221

Seiler HG, Sigel H (eds) (1987) Handbook on toxicity of inorganic compounds. Marcel Dekker, New York

Senwo ZN, Tazisong IA (2004) Metal contents in soils of Alabama. Commun Soil Sci Plant Anal 35: 2837–2848

Seppänen M, Turakainen M, Hartikainen H (2003) Selenium effects on oxidative stress in potato. Plant Sci 165:311–319

Shacklette HAT, Boerngen JG (1984) Element concentration sin soils and other surficial materials of the conterminous United States. U.S. Geol Survey Prof Papers 1270:1–105

Shafer MM (1995) Sampling and analytical techniques for silver in natural waters. 3rd Intern Conf Transport, Fate and Effects of Silver in the Environment. Washington, D.C. pp 99–108

Schäfer U (2004a) Lithium. In: Merian E, Anke M, Ihnat M, Stoeppler M (eds) Elements and their compounds in the environment. 2nd ed., Wiley-VCH Verlag, Weinheim, pp 479–496

Schäfer U (2004b) Manganese. In: MerianE, Anke M, Ihnat M, Stoeppler M (eds) Elements and their compounds in the environment. 2nd ed., Wiley-VCH, Weinheim, pp 901–930

Schäfer U, Anke M. Seifer M, Fischer AB (2002) Manganese intake, extretion and balance of adults. I. Different influences on the manganese concentration of the consumed food.In: Anke M, Müller R, Schäfer R, Stoeppler M (eds) Workshop Macro and Trace Elements, Jena, pp 1074–1080

Schmidt U (2003) Festlegung von Schwermetallen in Böden. Endbericht an das Regierungspräsidium Freiburg, Innoterr, Freiburg

Shagalova E, Zkukova O, Germenchuk M, Matveenko I, Bakarikova Zh (2000) Dynamics of radiation situation of the territory of Belarus and migration of radionuclides in different types of soils after Chernobyl catastrophe. J Radioanal Nuclear Chem 246:521–525

Sharma M, Wasserburg GJ (1997) Osmium in the rivers. Geochim Cosmochim Acta 61:5411–5416

Sharma SS, Schat H, Vooijs R, van Heerwaadens LM (1999) Combination toxicology of copper, zinc, and cadmium I binary mixtures: concentration dependent antagonistic, nonadditive, and synergistic effects on root growth in *Silene vulgris*. Environ Toxicol Chem 18:348–353

Shaw CF (1991) Gold. In: Merian E (ed) Metals and their compounds in the environment. VCH, New York, pp 931–938

Shelp BJ, Shattuck VI (1987) Boron nutrition and mobility, and its relation to elemental composition of greenhouse grown root crops. I Rutabega. Comm Soil Sci Plant Anal 18:187–201

Shennan C, Grattan SR, May DM, Hillhouse CJ, Schachtman DP, Wander M, Raberts B, Tafoya S, Burau RS, McNeish C, Lelinshi C (1995) Feasibility of cyclic reuse of saline drainage in a tomato cotton rotation. J Environ Qual 24:476–486

Sheppard SC (2003) Transfer parameters – when do we know enough? In: Inaba J, Tsukada H, Takeda A (eds) Radioecology and environmental dosimetry. Inst Environ Sci, Japan, pp 3–12

Shibita Y, Morita M, Fuwa K (1992) Selenium and arsenic in biology: their chemical forms and biological functions. Adv Biophys 18:31–80

Shiller AM (1988) Enrichment of dissolved gallium relative to aluminium in natural water. Geochim Cosmochim Acta 52:1879–1882

Shiller AM (2003) Dissolved gallium in the Atlantic Ocean. Marine Chem 61:87–99
Shiller AM, Frilot DM (1996) The geochemistry of gallium relative to aluminium in California streams. Geochim Cosmochim Acta 60:1323–1328
Shorrocks VM (1997) The occurrence and correction of boron deficiency. In: Dell B, Brown PH, Bell RW (eds) Boron in soils and plants: Reviews. Kluwer Publ, London, vol 77, pp 121–148
Shotyk W (1996) Natural and anthropogenic enrichments of As, Cu, Pb, Sb, and Zn in ombrotrophic versus minetrophic peat bog profiles, Jura Mountain Switzerland. Water Air Soil Pollut 90:375–405
Shtangeeva I (2006) Phytoremediation of trace element contaminated soil with cereal crops: Role of fertilizers and bacteria on bioavailability. In: Prasad MNV, Sajwan KS, Naidu R (eds) Trace elements in the environment: Biogeochemistry, biotechnology, and bioremediation. CRC Press, Boca Raton, FL, pp 549–581
Shtangeeva I, Ayrault S (2004) Phytoextraction of thorium from soil and water media. Water Air Soil Pollut 154:19–35
Shtangeeva I, Ayrault S (2005) About behaviour of europium in plants. ICOBTE, 8th Intern Conf, Adelaide, pp 470–471
Shtangeeva I, Ayrault S (2006) Impact of Eu and Ca on plant yield and mineral nutrition. Environ Exp Botany (in press)
Shtangeeva I, Vouorinen A, Riets B, Christiansen G, Carlsn L (2001) Decontamination of polluted urban soils by plants. Our possibilities to enhance the uptake of heavy metals. J Radioanal Nuclear Chem 249:369–374
Shtangeeva I, Ayrault S, Jain J (2004) Scandium bioaccumulation and its effect on uptake of macro- and trace elements during initial phase of plant growth. Soil Sci Plant Nutr 50:877–883
Shtangeeva I, Ayrault S, Jaind J (2005a) Thorium uptake by wheat at different stage of plant growth. J Environ Radioactivity 81:283–293
Shtangeeva I, Kehelin H, Laiho H (2005b) Effects of soil fertilization on uptake of macro- and trace elements. Intern Workshop Fate and Impact of Persistent Pollutants in Agroecosystems. IUNG, Pulawy, pp 154–155
Siebielec G, Chaney RL (2006) Manganese fertilizer requirement to prevent manganese deficiency when liming to remediate Ni-phytotoxic soils. Comm Soil Sci Plant Anal 37:1–17
Siegel SM, Siegel BZ (1988) Temperature determinants of plant-soil-air mercury relationships. Water Air Soil Pollut 40:443–448
Siegert E, Anke M, Baumann W, Bräunlich H, Brückner C, Groppel B. (1986) 5th Symp Zinc supply of plants and animals in middle Europe, Karl-Marx-Univ, Leipzig, Friedrich-Schiller Univ, Jena, pp 487–493
Siemiński M (2001) Environmental health hazard. Wyd Nauk PWN, Warszawa (in Polish)
Sikora FJ, Stevenson FJ (1988) Silver complexation by humic substances: conditional stability constants and nature of reactive sites. Geoderma 42:353–359
Sillanpää M (1982) Micronutrient and nutrient status of soils – a global study. FAO Soil Bull 48, Rome
Sillanpää M, Jansson H (1992) Status of cadmium, lead, cobalt and selenium in soils and plants of thirty countries. FAO Soils Bull 65, Rome
Simon L, Balazsy S. Balogh A, Pais I (1990) Effects of titanium on the growth of *Bradyrhizobium japonicum* and *Bradyrhizobium lupuli* strains. Acta Microbiol Pol 39:51–59
Simon L, Szegvári I, Prokisch J (2001) Enhacement of chromium phytoextraction capacity of fodder radish with picolinic acid. Environ Geochem Health 23:313–316
Singh BR, McLaughlin (eds) (1999) Cadmium in soils and plants. Kluwer Publ, Dordrecht, vol 85, pp 257–265
Sinha R, Saxena R, Singh S (2002) Comparative studies on accumulation of Cr from metal solution and tannery effluent under repeated metal exposure by aquatic plants: toxic effects. Environ Monit Assess 80:7–31
Sivaguru M, Horst WJ (1998) The distal part of the transition zone is the most aluminium sensitive apical root zone of maize. Plant Physiol 116:155–163
Sjögren B, Elinder C-G, Iregren A, McLachlan DRC, Rithimäki V (1997) Occupational aluminium exposure and its health effects. In: Yokel RA, Golub MS (eds) Research issues in aluminium toxicity. Taylor & Francis, Washington DC, pp 165–183

Skibniewska KA, Markiewicz K, Radzymińska M, Mozolewski W (2004) Daily intake of iron and manganese by children and adolescents. Ann Nat Inst Hygiene, Suppl 55:47–50 (in Polish)

Skinner WM, Martin RR, Naftel SJ, Macfie S, Courchesne F, Séquin V (2005) Multi-technique studies of the distribution of metals between the soil, rhizosphere and roots of *Populus tremuloides* growing in forest soil. ICOBTE, 8th Intern Conf Adelaide, pp 488–489

Skjelkvåle BL, Anderssen T, Fjeld E, Mannio J, Wilander A, Jahansson K, Jensen JP, Moiseenko T (2001) Heavy metal survey in Nordic lakes; concentrations, geographic patterns and relation to critical limits. Ambio 30:2–10

Skotnicki AB (1990) Effect of lithium ions on immune reactivity. In: Vohora SB, Dobrowolski JW (eds) New horizons of health aspects of elements. Jamia Hamdard University, New Delhi, pp 23–35

Skwarzec B (1997) Polonium, uranium and plutonium in selected components of the Southern Baltic Sea. Ambio 26:113–117

Skwarzec B (2006) Studies on contamination of lake Rożnowskie with trace metals and numerological modeling of contaminant transport using surface-water model system. Ph. D. Thesis, Univ of Mining & Metallurgy, Kraków (in Polish)

Skwarzec B, Bojanowski R (1992) Distribution of plutonium in selected components of the Baltic ecosystem within the Polish economic zone. J Environ Radioactiv 15:249–263

Skwarzec B, Strumińska DI, Boryło A (2001a) Bioaccumulation and distribution of plutonium in fish from Gdansk Bay. J Environ Radioactivity 55:167–178

Skwarzec B, Strumińska DI, Boryło A (2001b) The radionuclides ^{234}U, ^{238}U and ^{210}Po in drinking water in Gdańsk agglomeration (Poland). J Radioanal Nuclear Chem 250:315–318

Slikkerveer A, de Wolff FA (1989) Pharmacokinetics and toxicity of bismuth compounds. Med Roxicol Adverse Drug Exp 4:303–323

Slikkerveer A, Jong HB, Helmich RB, de Wolff FA (1992) Development of a therapeutic procedure of bismuth intoxication with chelating agents. J Lab Clinical Medicine 119:259–537

Smedley PL, Kinniburgh DG (2002) A review of the source, behaviour and distribution of arsenic in natural waters. Appl Geochem 60:517–568

Smichowski P, Gómez D, Giné MF, Sánchez Bellato AC, Rosa S (2005) Antimony in airborne particulate matter emitted by natural and anthropogenic sources: A case study in Argentina. 1st Intern Workshop on Antimony in the Environment, Heidelberg, 43 p

Smidley P, Kinniburgh DG (2005) Arsenic in groundwater and the environment. In: Selinus O, Alloway BJ, Centento JA, Finkelman RB, Fuge R, Lindh U, Smeldey P (eds) Medical geology. Elsevier, Amsterdam, pp 263–299

Śmiechowska M, Dmowski P, Stasiuk E (2004) The content of bioactive elements (iron, manganese) in tea imported to Poland. 22nd Workshop Macro and Trace Elements, Jena, 2, pp 1201–1205

Smith SR (1996) Agricultural recycling of sewage sludge and the environment. CAB Intern, Oxford

Smith GR (2004) Lead. U.S. Geological Survey Mineral Commodity Summaries, pp 94–95

Smith I (2005) Land uses of coal fly ash, benefits and barriers. IEA Coal Research, London, UK, CCC/96

Smith IC, Carson BL (1977a) Trace metals in the environment. 2 – Silver. Ann Arbor Sci Publ, Ann Arbor, MI

Smith IC, Carson BL (1977b) Trace metals in the environment. 1 – Thallium, Ann Arbor Sci Publ, Ann Arbor, MI, pp 138–139

Smith IC, Carson BL (1981) Trace metals in the environment. 6 – Cobalt. Ann Arbor Sci Publ Inc, Michigan, WI

Smith GJ, Flegal AR (1993) Silver in San Francisco Bay estuarine waters. Estuaries 16:547–558

Smith KS, Huyck HLO (1999) An overview of the abundance, relative mobility, bioavailability, and human toxicity of metals. In: Plumlee GS, Logsdon JJ (eds) The environmental geochemistry of mineral deposits. Part A. Processes, techniques, and health issues. Review Econ Geol 6A:29–70

Smith IC, Carson BL, Ferguson F (1978a) Trace metals in the environment. Ann Arbor Sci Publ, Ann Arbor, MI

Smith PJ, Luitjen JGA, Klimmer OR (1978b) Toxicological data on organontin compounds. ITRI Publ No. 538, Intern Tin Research Institute, London

Smith E, Naidu R, Alston AM (1998) Arsenic in the soil environment: a review. Advances Agronomy 64:149–195

Smith DS, Adams NWH, Kramer JR (1999) Resolving uncertainty in chemical speciation determinations. Geochim Cosmochim Acta 63:3337–3347
Smith AH, Lingas EO, Rahman M (2000) Contamination of drinking water by arsenic in Bangladesh: a public health emergence. Bull WHO 78:1093–1103
Smith JVS, Jankowski J, Sammut J (2003) Vertical distribution of As(III) and As(V) in a coastal sandy aquifer: factors controlling the concentration and speciation of arsenic in the Stuarts Point groundwater system, Northern New South Wales, Australia. Appl Geochem 18:1479–1496
Smolders E (2005) Risk assessment of metals – cadmium as a case study. In: Eckel H, Roth U, Döhler H, Nicholson F, Inwin R (eds) Assessment and regulation of heavy metal input into agro-ecosystem. KTBL, Darmstadt, pp 229–232
Smolders E, Buekers J, Oliver I, McLaughlin MJ (2004) Soil properties affecting microbial toxicity of Zn in laboratory-spiked and field-contaminated soils. Environ Toxicol Chem 23:41–48
Songchitsomboom S, Komindr S (1996) Serum zinc and copper in healthy adults living in Bangkok and surrounding districts, J Med Assoc Thai 79:550–557
Spallholz JE (2001) Selenium and prevention of cancer. Part I: Envidence of carcinostatic activity of selenium compounds. The Biul Selenium-Tellurium (www.stda.be)G
Spallholz JE, Boylan LM, Rhaman MM (2004) Environmental hypothesis: Is poor dietary selenium intake an underlying factor for arsenicosis and cancer in Bangladesh and West Bengal, India. Sci Total Environ 323:21–32
Sparks DL (1995) Environmental soil chemistry. Acad Press, San Diego
Spokes L, Jickells T, Jarvis K (2001) Atmospheric inputs of trace metals to the northeast Atlantic Ocean: the importance of south-easterly flow. Marine Chem 76:319–330
Sprague JB (1986) Toxicity and tissue concentrations of lead, zinc, and cadmium for marine mollusks and crustaceans. Intern Lead Zinc Res Org pp 1–74
Sracek A, Bhattacharya P, Jacks G, Gustafsson JP (2001) Mobility of arsenic and geochemical modelling in groundwater environment. In: Jacks G, Bhattacharya P, Khan AA (eds) Groundwater arsenic-contamination in the Bengal Delta Plain of Bangladesh. KTH-Dhaka Seminar, University of Dhaka, Bangladesh, KTH Special Publ, TRITA-AMI Report 3084, pp 9–20
Srikanth R, Reddy SR-J (1991) Lead, cadmium and Cr levels in vegetables grown in urban sewage sludge – Hyderabad, India. Food Chemistry 40:229–234
Stäb JA, Traas TP, Stroomberg G, van Kesteren J, Leonards P, van Hatttum B, Brinkman UAT, Cofino WP (1996) Determination of organotin compounds in the food-web of a shallow freshwater lake in the Netherlands. Arch Environ Contam Toxicol 31:319–328
St Louis VL, Sharp MJ, Steffen A, May AI, Barker J, Kirk JL, Kelly DJA, Arnott SE, Keatley B, John P, Smol JP (2005) Some sources and sinks of monomethyl and inorganic mercury on Ellesmere Island in the Canadian High Arctic. Environ Sci Technol 39:2686–1701
Stadnichenko TM, Zubovic P, Sheffey NB (1961) Beryllium content of Americal coals. U.S. Geol Survey Bull 1084–K:253–295
Stanek EJ III, Calabrese EJ, Zorn M (2001) Soil ingestion distributions for Monte Carlo Risk Assessment in children. Human Ecol Risk Assess 7:357–368
Stauton S, Bonafos B, Leclerc-Cessas E (1999) Possible effects of root action in the rhizosphere on the adsorption of trace amounts of nickel by various soils. 5th Intern Conf Biogeochem Trace Elements, Vienna, pp 46–47
Stehouwer RC, Macneal KE (2004) Effect of alkalline-stabilize biosolids on alfalfa molydnenum and copper content. J Environ Qual 33:133–140
Steinnes E (2003) Biogeochemiacl cycling of iodine and selenium and potential geomedicval relevance. In: Skinner HC, Berger AR (eds) Geology and health: closing gap. Oxford Univ Press, Oxford, pp 57–60
Steinnes E, Rühling Å (2002) Heavy metals in surface soils in Fennoscandia. Nordic Council Ministry, Copenhagen
Steinnes E, Allen RO, Petersen HM, Rambaek JP, Varskog P (1997) Evidence of large scale heavy-metal contamination of natural surface soils in Norway from long range atmospheric transport. Sci Tot Environ 205:255–266
Stemme K, Meyer U, Flachowsky G, Scholtz H (2002) Kobalt-Bedarf von Wiederkäuern – sind die derzeitigen Bedarfsempfehlungen für laktierende Milchkühe ausreichend? In: Anke M, Müller R, Schäfer U, Stoeppler M (eds) Macro and Trace Elements 21:403–420

Stilianakis N (2005) Environmental entry and health effects of platinum group elemen particles. An assessment. Min Soc Winter Meeting Bath Spa Univ, Bath

Stoecker B (2004) Chromium. In: Merian E, Anke M, Ihnat M, Stoeppler M (eds) Elements and their compounds in the environment. 2nd ed., Wiley-VCH Verlag, Weinheim, 2, pp 709–729

Stoeppler M (2004) Arsenic. In: Merian E, Anke M, Ihnat M, Stoeppler M (eds) Elements and their compounds in the environment. 2nd ed., Wiley-VCH, Weinheim, pp 1321–1364

Stoewsand GS, Anderson JL, Rutzke M, Lisk DJ (1988) Deposition of barium in the skeleton of rats fed Brazil nuts. Nutr Rep Itern 38:259–262

Storm GL, Fosmire GJ, Bellis ED (1994) Persistence of metals in soil and selected vertebrates in the vicinity of the Palmerton zinc smelters. J Environ Quality 23:508–514

Strączyk D, Somerlik U (2003) Barium in rivers and streams of Podkarpacie region. J Elementology 8:107–110

Streit B, Stumm W (1993) Chemical properties of metals and the process of bioaccumulation in terrestrial plants. In: Markert B (ed) Plants as biomonitors. Wiley-VCH, Weinheim, pp 31–62

Strzelecki R, Wołkowicz S, Szewczyk J, Lewandowski P (1993) Radioecological maps of Poland. Part 1. PIG Warszawa

Stumm W, Brauner PA (1975) Chemical Speciation. In: Riley JP, Skirrow G (eds) Chemical oceanography, 2nd ed., Acad Press, London, pp 173–239

Sudaryanto A, Takahashi S, Iwata H, Tanabe S, Ismail A (2004) Contamination of butyltin compounds in Malaysian marine environment. Environ Pollut 130:347–358

Sudaryanto A, Takahashi S, Iwata H, Tanabe S, Muchtar M, Razak H (2005) Organotin residues and the role of anthropogenic tin sources in the coastal marine environment of Indonesia. Marine Poll Bull 50:226–235

Sugiyama H, Terada H, Isomura K (1993) Radiocesium uptake mechanisms in wild and culture mushrooms. Radioisotpes 42:683–690 (in Japanese)

Sundby B, Philippe M, Gobeil C (2004) Comparative geochemistry of cadmium, rhenium, uranium, and molybdenum in continental margin sediments. Geochim Cosmochim Acta 68:2485–2493

Sunderman Jr. FW (2004) Nickel. In: Merian E, Anke M, Ihnat M, Stoeppler M (eds) Elements and their compounds in the environment. 2nd ed., Wiley-VCH, Weinheim, pp 841–865

Suter II GW, Vermeire T, Munns, Jr WR, Sekizawa J (2003) Framework for the integration of health and ecological risk assessment. Human Ecol Risk Assess 9:281–301

Swarup D, Dwivedi SK, Dey S, Ray SK (1998) Fluoride intoxication in bovine due to industrial pollution. Ind J Animal Sci 68:605–608

Szefer P (1998) Distribution and behavior of selected heavy metals and other elements in various components of the southern Baltic ecosystem. Appl Geochem 13:287–292

Szefer P (2002a) Metals, metalloids and radionuclides in the Baltic Sea Ecosystem. Elsevier, Amsterdam

Szefer P (2002b) Metal pollutants and radionuclides in the Baltic Sea – an overview. Oceanologia 44:129–178

Szefer P, Nriagu JO (eds) (2006) Mineral components in food. Taylor & Francis – CRC Press, Boca Raton FL

Szefer P, Pempkowiak J, Skwarzec B, Bojanowiski R (1994) Distribution and coassociations of selected metals in seals of the Antarctic. Environ Pollut 83:341–349

Szentmihalayi SM, Anke M, Regius A (1985) The importance of lithium for plants and animals. Intern Symp New results in the research of hardly known trace elements. University of Hort Food Indust, Budapest, pp 136–140

Szentmihalayi K, Hajdu M, Then M (2006) Trace elements in medicinal plants and extracts and their potential, beneficial and toxic effects. Intern Symp Trace Elements in the Food Chain, Budapest, pp 130–134

Szilàgyi M (2004) Hafnium. In: Merian E, Anke M, Ihnat M, Stoeppler M (eds) Elements and their compounds in the environment. 2nd ed., Wiley-VCH, Weinheim, pp 795–800

Szkoda J, Żmudzki J (2000) Alive test for evaluation of cadmium contamination in horses. In: Kabata-Pendias A, Szteke B (eds) Cadmium in the environment. Ecological and analytical problems, Polish Acad Sci, Warszawa, pp 403–409 (in Polish)

Szpunar J, Lobinski R, Prange A (2003) Hyphenated techniques for elemental speciation in biological systems. Appl Spectroscopy 57:102A–112A

Szteke B (2006) Monitoring of trace elements in edible plants. Polish J Environ Studies 15(2A):189–194
Szteke B, Jędrzejczak R, Ręczajska W (2004) Iron and manganese in selected edible plants. Ann Nat Inst Hygiene, Suppl 55:21–27 (in Polish)
Tagami K, Uchida S (1998) Some consideration on the fate of ^{99}Tc in paddy fields. In: Inaba J, Nakamura Y (eds) Int Worshop Comp Evaluation Health Effects Environ, Toxic. Suppl NIRS-M-130, pp 113–119
Tagami K, Uchida S (2005) Technetium-99 transfer from soil to plants collected in Chernobyl area. ICOBTE, 8th Intern Conf, Adelaide, pp 108–109
Takahashi S, Tanabe S, Kubodera T (1997) Butyltin residies in deep-sea organisms collected from Suruga bay, Japan. Environ Sci Technol 31:3103–3109
Takahashi S, Mukai H, Tanabe S, Sakayama K, Miyazaki T, Masuno H (1999) Butyltin residues in livers of humans and wild terrestrial mammals and in plastic products. Environ Pollut 106:213–218
Takayanagi K, Cossa D, Martin J-M (1996) Antimony cycling in the western Mediterranean. Mar Chem 54:303–312
Takeda A, Kimura K, Yamasaki S-I (2004) Analysis of 57 elements in Japanese soils, with special reference to soil group and agricultural use. Geoderma 119:291–307
Takeda A, Tsukada H, Nanzyo M, Takaku Y, Uemura T, Hisamatsu S, Inaba J (2005a) Effects of long-term fertilizer application on the concentration and solubility of major and major trace elements in cultivated Andisol. Soil Sci Plant Nutr 51:251–260
Takeda A, Tsukada H, Takaku, Y, Hisamatsu S, Inaba J, Nanzyo M (2005b) Assessment of Sr phytoavailability in soils for prediction of Sr concentration in komatsu (*Brassica rapa* L. var. *perviridis*) ICOBTE 8th Intern Conf, Adelaide, pp 426–427
Takeuchi T, Eto K, Tokunaga H (1989) Mercury level and histochemcal distrbution in a human brain with Minamata disease following a long-term clinical course of twenty-six years. Nerotoxicology 10:651–657
Tan KH (1998) Principles of soil chemistry. 3rd ed., Marcel Dekker, New York
Tanabe S, Prudente M, Mizuno T, Hasegawa J, Iwata H, Miyazaki N (1998) Butyltin contamination in marine mammals from North Pacific and asian coastal waters. Environ Sci Technol 32:193–198
Tani Y, Miyata N, Ohashi M, Iwahori K, Soma M, Seyma H (2003) Interaction of Co(II), Zn(II) and As(III/V) with manganese oxides formed by Mn-oxidizing fugus. 16th Itern Symp Environ Biogeochem, Oirase, 113 p
Tasdemir Y, Kural C, Cindoruk SS, Vardar N (2005) Assessment of trace element concentrations and their estimated deposition flux in an urban atmosphere. Atmos Research 81:17–35
Taylor MD, Percival HJ (2001) Cadmium in soil solutions from a transect of soils away from a fertilizer bin. Environ Pollut 113:35–40
Taylor DM, Lehmann M, Planas-Bohne F, Seidel A (1983) The metabolism of radiohafnium in rats and hamsters: a possible analog of plutonium for metabolic studies. Radiat Res 95:339–358
Taylor HE, Garbarino JR, Brinton TI (1990) The occurrence and distribution of trace metals in the Mississippi River and its tributaries. Sci Total Environ 97/98:369–384
Tebo BM, Webb SM, Bargar JR, Toyoda K, Johson HA, Parker DL, Verity RB, Dick GJ, McCarthy JK (2003) Unraveling the molecular mechanisms of Mn oxide biomineralization. 16th Itern Symp Environ Biogeochem, Oirase, Japan, pp 7–8
Teissèdre PL, Krosnika M, Portet K, Gase F, Waterhouse AL, Serrano JJ, Cabanis JC, Cros G (1998) Vanadium levels in Franch and Californiam wines: Influence on vanadium dieary intake. Food Add Contam 15:585–591
Temmerman LR, Vandeputte R, Guns M (1986) Biological monitoring and accumulation of airborne mercury in vegetables. Environ Pollut 41:139–151
Templeton D, Ariese F, Cornelis R, Danielsson L-G, Muntau H, van Leeuwen HP, Łobiński R (2000) Guideline for terms related to chemical speciation and fractionation of elements. Definitions, structural aspects, and methodological approaches. Pure Appl Chem 72:1453–1470
Terashima S, Katayama H, Itoh S (1991) Geochemical behaviour of gold in coastal marine sediments from the southeastern margin of Japan Sea. Mar Mining 10:247–257
Terelak H, Piotrowska M (1998) Contents of lead in agricultural land of Poland and some selected voivodeships. In: Kabata-Pendias A, Szteke B (eds) Lead in the environment: Ecological and analytical problems. Polish Acad Sci, Warszawa, pp 19–24 (in Polish)

Terelak H, Motowicka-Terelak T, Stuczyński T, Pietruch Cz (2000) Trace elements (Cd, Cu, Ni, Pb, Zn) in agricultural soils of Poland. IOŚ, Warszawa (in Polish)

Terelak H, Motowicka-Terelak T, Maliszewska-Kordybach B, Pietruch Cz (2002) Monitoring of chemical properties of arable soils in Poland. Bibl Monit Środowiska, Warszawa (in Polish)

Terry N, Zayed AM, de Souza MP, Tarun AS (2000) Selenium in higher plants. Ann Rev Plant Physiol Plant Mol Biol 51:401–432

Tessier A, Cambell PGC, Bisson M (1979) Sequential extraction procedure for the speciation of particulate metals. Anal Chem 51:844–851

Tessier E, Amouroux D, Donard OFX (2002) Volatile organotin compounds (butylmethyltin) in three European estuaries. Biogeochemistry 59:161–181

The Silver Institute (1994) World silver survey. Washington DC, pp 370–377

Thewys T (2006) Phytoremediation: socio-economic aspects. Meeting Phytotechnologies Lessons from Pilot and Filed Scale, Lisbon, pp 1

Thomas DW (1991) Bismuth. In: Merian E (ed) Metals and their compounds in the environment. Wiley-VCH, Weinheim, pp 789–801

Thompson CD (2004) Interaction between selenium and iodine in thyroid status. In: Cser MA, László IS, Étienne J-C, Maymard Y, Centeno JA, Khassanova L, Collery P (eds) Metal ions in biology and medicine, 8. John Libbey & Comp. Ltd., England, Paris, pp 552–554

Thompson DR, Stewart FM, Furness RW (1990) Using seabirds to monitor mercury in marine environments – the validity of conversion ratios for tissue comparison. Mar Pollut Bull 21:339–342

Thorat DD, Mahadevan TN, Ghosh DK et al. (2001) Beryllium concentrations in ambient air and its source identification. Environ Monit Assess 69:49–61

Thornton I (1990) A survey of lead in the British urban enrironment: an example of research in urban geochemistry. In: Lieth H, Markert B (eds) Element concentration cadaster in ecosystems. Wiley-VCH, Weinheim, pp 221–240

Tiller KG (1992) Urban soil contamination in Australia. Austr J Soil Res 30:937–957

Ting YP, Mittal AK (1999) An evaluation of equilibrium and kinetic models for gold biosorption. Resour Environ Biotechnol 2:311

Ting YP, Teo WK, Soh CY (1995) Gold uptake by *Chlorella vulgaris*. J Appl Phycol 7:97

Tokunaga TK, Wan J, Firestone MK, Hazen TC, Olson KR, Herman DJ Sutton SR, Lanzirotti A (2003) In situ reduction of chromium (VI) in heavily contaminated soils through organic carbon amendment. J Environ Qual 32:1641–1649

Toner CV, Sparks DL (1995) Chemical relaxation and double layer model analysis of boron adsorption on alumina. Soil Sci Soc Am Proc 36:746–752

Tonkin JW, Balistrieri LS, Murray JW (2004) Modeling sorption of divalent cations on hydrous manganese oxide. Appl Geochem 19:29–53

Toro C, Arunachalam J, Martin C, Vingare F, Ostapczuk P (1995) Poplar leaves and pine needles as biomonitors for air pollution. In: Wilken R-D, Förstner U, Knöchel A (eds) Heavy metals in the environment, 1. pp 180–183

Treguer P, Nelson DM, van Bennekom AJ, DeMaster DJ, Leynaert A, Quéguiner B (1995) The silica balance in the world ocean: a reestimate. Science 268:375–379

Tremel A, Masson P, Streckeman T, Baized D, Mench M (1997a) Thallium in French agrosystems: I. Thallium contents in arable soils. Environ Pollut 95:293–302

Tremel A, Masson P, Garraud H, Donard OFX, Baized D, Mench M (1997b) Thallium in French agrosystems: II. Concentration of thallium in the field-grown rape and some other plant species. Environ Pollut 97:161–168

TRI96 (1998) Toxic chemical release inventory. National Library of Medicine, National Toxicological Information Program, Bethesda, MD

Trojanowska M (2003) Characteristics of the occurrence of chromium in the Radomka River Basin by tannery wastes. Phytoremediation of chromium. In: Gworek B, Misiak J (eds) Cycling of elements in nature, 2. pp 301–304 (in Polish)

Trueman ER, Black S, Read D, Hodgson ME (2004) Secondary phase formation on depleted uranium metal in soil. Goldschmidt, Copenhagen, A498 p

Tsadilas CD, Yassoglou N, Kosmas CS, Kallianou CH (1994) The availability of soil boron fractions to olive trees and barley and their relationships to soil properties. Plant Soil 162:211–217

Tschenko D, Kandeler E (1997) Ecotoxicological effects of fluorine deposits on microbial biomass and enzyme activity in grassland. Eur J Soil Sci 48:329–335

Tsubaki T, Takahashi H (eds) (1986) Recent advances in Minamata disease studies. Methylmercury poisoning in Minamata and Niigata, Japan. Kodansha, Tokyo

Tsukada H, Hasegawa H (2002) Soil-to-plant transfer of ^{137}Cs and other essential and trace elements in cabbage plants. J Radioanal Nuclear Chem 252:219–224

Tsukada H, Nakamura Y (1999) Transfer of ^{137}Cs and stable Cs from soil to potato in agricultural fields. Sci Total Environ 228:111–120

Tsukada H, Hasegawa H, Hisamatu S (2002) Distribution of alkali and alkaline earth metals in several agricultural plants. Radioprotection Colloq 37:C1535–C1540

Tsukada H, Hisamatu S, Inaba (2003) Transfer of ^{137}Cs and stable Cs in soil-grass-milk pathway in Aomori, Japan. J Radioanal Nuclear Chem 255:455–458

Tsukada H, Shibata H, Sugiyama (1998) Transfer of radiocesium and stable cesium from substrata to mushrooms in a pine forest in Rokkasho-mura, Saomoriu, Japan. J Environ Radioactivity 39:149–160

Tsuruta T (2003) Bioaccumulation of thorium using *Streptomyces albus*. 16th Intern Symp Environ Biogeochem, Oirase, 2, pp 115–116

Tsuruta T (2004) Cell-associated adsortion of thorium and uranium from aqueous system using various microorganisms. Water Air Soil Pollut 159:35–47

Tu C, Ma LQ (2003) Effects of arsenate and phosphate on their accumulation by an arsenic-hyperaccumulator *Pteris vittata* L. Plant Soil 249:373–382

Tu C, Ma LQ, Bondada B (2002) Arsenic accumulation in the hyperaccumulator Chinese brake and its utilization potential for phytoremediation. J Environ Qual 31:1671–1675

Tukker A, Buijst H, van Oers L, van der Voet L (2001) Risks to human and the environment related to the use of lead products. TNO Report STB-01-39

Turakainen M, Hartikainen H, Seppänen M (2004) Effects of selenium treatment on potato (*Solanum tuberosum* L.). Growth and concentrations of soluble sugars and starch. J Agric Food Chem 52:5378–5382

Turnau K, Ślusarczyk E, Mleczko P (1996) Localization of copper in ectomycorrhizas. In: Kabata-Pendias A, Szteke B (eds) Copper and molybdenum in the environment – ecologic analytic problems. Polish Acad Sci, Warszawa, pp 111–116 (in Polish)

Turnau K, Ryszka P, Jurkiewicz A, Orłowska E (2005) Use of mycorrhizal fungi in restoration of heavy metal rich industrial wastes. Intern Workshop Current Development in Remediation of Contaminated Sites, IUNG, Puławy, 24 p

Turner MA (1997) Effect of cadmium treatment on cadmium and zinc uptake by selected vegetable species. J Environ Qual 2:118–119

Turnlung JR, Keyes WR, Pfeiffer GL (1993) A stable isotope study of the dietary molybdenum requirement of young men. In: Anke A, Meissner D, Mills CF (eds) Trace elements in man and animals, Verlag Media Touristik, Gersdorf, pp 189–193

Turski R, Wójcikowska-Kapusta A, Kukier U (1989) Barium balance in a lysimeter experiment. Prace Kom PTG, Warszawa, pp 74–79 (in Polish)

Twidwell LG, Beam CW (2002) Potential technologies for removing thallium from mine and process wastewater: an abbreviated annotation of literature. Eur J Miner Process Environ Prot 2:1–10

Tyler G (1983) Rubidium – availability and uptake in natural soils. Comm Soil Sci Plant Anal 14: 1075–1089

Tyler G (2004) Ionic charge, radius, and poyential controlroo/soil concentration ratios of fifty cationic elements in the organic horizon of a beech (*Fagus sylvatica*) forest podzol. Sci Total Environ 329:231–239

Tyler G (2005) Changes in the concentrations of major, minor and rear-earth elements during leaf senescence and decomposition in a *Fagus sylvatica* forest. For Ecol Manag 206:167–177

Tyler G, Olsson T (2002) Conditions related to solubility of rare and minor elements in forest soils. J Plant Nutr Soil Sci 165:594–601

Tzou YM, Loeppert RH, Wang MK (2003) Light-catalyzed chromium (VI) reduction by organic compounds and soil minerals. J Environ Qual 32:2076–2084

UBA (2001) Grundsätze und Maßnahmen für eine vorsorgeorientierte Begrenzung von Schadstoffeinträgen in landbaulich genutzten Böden. UBA Texte 59/01 Ministry of Environment, Berlin

Uchvatov VP, Bashkin VN (1997) Hydrochemistry of heavy metals in surface waters of Central-European Russia. Symp Heavy Metals in Surrounding Environment. Pushchino, pp 130–145 (in Russian)
UK DETR (Dept. Environ., Transport and Regions) (1998) Expert panel on air quality standards: lead. Dept Environ, Transport and Regions, UK
Umemura T, Asaka K, Seizawa K, Odake T, Tsunoda K, Satake K, Wang Q, Huang B (2001) Preliminary investigation of the application of electrospray ionization mass spectrometry to the study of aluminium speciation. Anal Sci 17:149–152
Umemura T, Usimi Y, Aizawa S, Tsunoda K, Satake K (2003) Seasional change in the level and the chemical forms of aluminium in soil solution under a Japanes ceder forest. Sci Tot Environ 317:149–157
Umfitseva MD, Terekhina NV (2005) Phytoindicatin of ecological state of urban geosystems in Saint Petesburg, Nauka, Saint Petersburg (in Russian)
Uren NC (1992) Forms, rections, and availability of nickel in soils. Advance in agronomy 48:141–203
Ursinova M, Masanova V (2005) Cadmium, lead, and mercury in human milk from Slovakia. Food Add Contam 22:579–589
UNEP/CHW.7/1 (2004) Draft technical guidelines on the environmentally sound recycling/reclamation of metals and metal compounds (R4)
UNEP/WHO (2002) Environmental health criteria 226. Palladium, Geneva
United Nations (1996) Report of the UN Conference on Human Settlements (Habitat II), Istanbul, A/CONF.165/14, UN, New York
WHO (n.d.) URL:http://www.who.int/water_sanitation_health/dwq/guidelines/enindex.html
Ursinova M, Masanova V (2005) Cadmium, lead, and mercury in human milk from Slovakia. Food Add Contam 22:579–589
U.S. Bureau of Mines (1989) Mineral commodity summaries. Washington, DC
U.S. Dept of Health & Human Services (2004) Mercury levels in fish: FDA monitoring pogram (1999–2003)
U.S. EPA (1973) National emission inventory of sources and emissions of silver. U.S. Environmental Protection Agency, Office of Air and Water Programs, Research Triangle Park, NC, EPA–450/3/74–011, pp 1–35
U.S. EPA (1980a) Ambient water quality criteria for silver. EPA-440/5-80-071. Office of water regulations and standards. Criteria and Standard Division. Washington, DC
U.S. EPA (1980b) Exposure and risk assessment for zinc. Office of Water Regulations and Standards (WH-553) EPA 440/4-81-016. PB85-212009
U.S. EPA (1981) Treatability manual. Vol 1. U.S. Environmental Protection Agency. EPA 600/2–82–001a, [14.7–1 to 1.4.7–5]
U.S. EPA (1982) An expose and risk assessment for lead. Washington DC. Office of Water Regulation Standard. Monitoring and data, Support Division, USEPA 440/4-85/010, NTIS PB85 – 220 606
U.S. EPA (1984) Review of national emissions standards for mercury. EPA-450/3-84-014. Emissions Standards and Engineering Division, Research Triangle Park, NC
U.S. EPA (1986) Quality criteria for water. Environmental Protection Agency, Office of water planning and standards, Washington DC, EPA 440/5-86-001
U.S. EPA (1987) Toxic air pollutant/source crosswalk: A screening tool for locating possible sources emitting toxic air pollutants. U.S. Environmental Protection Agency, Research Triangle Park, NC, USA, Office of Air Quality Standard. EPA-450/4-87-023a
U.S. EPA (1991) Ambient water quality criteria for copper. U.S. Environmental Protection Agency Report 440/5-80-036
U.S. EPA (1993) Summary review of health effects associated with mercuric chloride. Office of health and environment assessment, Washington, DC, EPA/600/R–92/1993
U.S. EPA (1997a) Mercury study report to congress. 3. Fate and transport of mercury in the environment. EPA–452/R–97–005, Office of Air Quality Planning and Standards, Research Triangle Park, NC
U.S. EPA (1997b) Mercury study report to congress. 2. An inventory of anthropogenic mercury emissions in the United States. EPA–452/R–97–004, Office of Air Quality Planning and Standards, Research Triangle Park, NC
U.S. EPA (1998) Listing of fish and wildlife advisories – 1997 (for Pb), U.S. Environmental Protection Agency, Office of Water, Washington, DC

U.S. EPA (1999) National recommended water quality criteria correction. EPA-822-Z-99-001, Washington, DC
U.S. EPA (2000) Aquatic toxicity information retrieval database (AQUIRE)
U.S. EPA (2002) National primary drinking water regulations. Consumer fact sheet on: copper. Wysiwyg: (//196/http://epa.gov/safewater/dwh/c-ioc/copper.html)
U.S. EPA (2003) United States Environmental Protection Agency. National primary drinking water regulations (http://www.epa.gov/safewater/dwh/t-ioe/thallium.html)
U.S. Public Health Service (PHS) (1990) Toxicological profile for silver. Agency for Toxic Substances and Disease Registry TP-90-24
UNEP/CHW.7/1 (2004) Draft technical guidelines on the environmentally sound recycling/reclamation of metals and metal compounds (R4)
USGS (2004) Mineral commodity summary. USGS, Reston VA
Vacchina V, Lobinski R, Oven M, Zenk MH (2000) Signal identification in size-exclusion HPLC-ICP-MS chromatograms of plant extracts by electrospraytandem mass spectrometry (ES MS/MS). J Anal At Spectrom 15:529–534
Vaessen HAM, Szteke B (2000) Beryllium in food and drinking water – a summary of available knowledge. Food Add Contam 17:149–159
Vaganov PA, Nizharadze TN (1981) On microelements in the loesslike and Cretaceous sediments. Geokhimiya 1:149–153 (in Russian)
Vahter ME (1988) Arsenic. In: Clarkson TW, Friberg L, Nordberg Gf, Sager PR (eds) Biological montoring of toxic metals, Plenum Press, New York, pp 303–321
Van Assche F, Van Tilborg W, Waetershoot H (1996) Environmental risk assessment for essential elements – Case study – Zinc. Intern Workshop Risk Assessment of Metals and their Inorganic Compounds, ICME, Ottawa, pp 171–180
Van der Heide D, Schröder-van der Elst JP, Binnerts WT (1993) Selenium and iodine status of animals and man: Is there a relation to persistent enlargement of thyroids in adolescents. In: Anke M. Messner D, Mills CF (eds) Trace elements in man and animals – TEMA 8. Verlag Media Touristik, Gersdorf, pp 1054–1057
Van der Weijden CH, Middleburg JJ, De Lange GJ, van der Sloot HA, Hoede D, Woittiez JRW (1990) Profiles of the redox-sensitive trace elements As, Sb, V, Mo, U in the Tyro and Bannock Basins (eastern Mediterranean). Mar Chem 31:171–186
Veiga MM (1997) Mercury in artisanal gold mining in Latin America: facts, fantasis and solutions. UNIDO – Expert Group Meeting – Introducing new technologies for abatement of global mercury pollution deriving from artisanal gold mining, Vienna, 23 p
Veiga MM, Baker RF (2004) Protocols for environmental and health assessment of mercury released by artisanal and small scale gold miners. Global Mercury Project, UNIDO, Vienna
Veiga MM, Meech JA (1995) HgEX – A heuristic system on mercury pollution in the Amazon. Water Air Soil Pollution 80:123–132
Vera Tome F, Blanko Rodriguez MP, Lozano JC (2003) Soil-to-plant transfer factors for natural radionuclides and stable elements in a Mediterranean area. J Environ Radioactiv 65:161–175
Vermes L (1989) Heavy metals in waters, soils and plants of a sewage irrigated lysimetric experiment at Keckemet. 6th Intern Trace Element Symp 1:199–207
Vernadsky VI (1998) The biosphere. Copernicus. An imprint of Springer-Verlag, New York
Veronese I, Giussani A, Cantone MC, Birattari C, Bonardi M, Groppi F, Höllriegl V, Roth P, Werner E (2004) Influence of the chemical form on the plasma clearance of ruthenium in humans. Appl Radiat Isot 60:7–13
Veronese I, Giussani A, Cantone MC, de Bartolo D, Roth P, Werner E (2001) Kinetics of systemic ruthenium in human blood using a stable tracer. J Radiol Prot 21:31–38
Veselý J, Norton SA, Skřivan P, Majer V, Krám P, Navrátil T, Kaste JM (2002) Environmental chemistry of beryllium. In: Grew ES, Ribbe PH (eds) Review in mineralogy and geochemistry, 50, pp 291–317
Vietrov VA (2002) Eko-geochemical desctription of Bajkal at the end of XX century as base for assessment of technogenic pollution of the lake. 2nd Intern Conf Heavy Metals, Radionuclides and Elements – Biofills in the Environemnt, Semipalatynsk, Kazakhstan, 1, pp 85–94 (in Russian)

Vignati D (2004) Trace metal partitioning in freshwater as a function of environmental variables and its implications for metal bioavailability. Ph. D. Thesis, Université de Genève
Vink BW (1993) The behavior of thallium in the (sub)surface environment in terms of Eh and pH. Chem Geol 19:119–123
Vitosh ML, Warncke DD, Lucas RE (1994) Secondary and micronutrients for vegetables and field crops. Ext Bull E-486, Michigan State Univ, East Lansing, MI
Vlasov KA (1966a) Gallium. In: Vlasov KA (ed) Geochemistry of rare elements, vol 1, Israel Program for Scientific Translations, Jerusalem, pp 437–458
Vlasov KA (1966b) Germanium. In: Vlasov KA (ed) Geochemistry of rare elements, vol 9. Acad Sci USSR, State Geological Committee USSR, pp 523–552
Voigt HR (2003) On heavy metal concentrations in certain bottom sediments and benthic animals of the Tvärminne area. Ympäristö ja Terveys-Lehti 34:39–41 (in Finnish)
Voigt HR (2004) Concentrations of mercury (Hg) and cadmium (Cd) and the condition of some coastal Baltic fishes. Ph.D. Thesis, Environ Fennica, No. 21, University of Helsinki
Von Wiren N, Marschner H, Romheld V (1996) Roots of iron-efficient maize also adsorb phytosidophore-chelated zinc. Plant Physiol 111:1119–1125
Vorholt JA, Thauer RK (2002) Molybdenum and tungsten enzymes in C1 metabolism. In: Sigel A, Sigel H (eds) Metal ions in biological systems, 39, Marcel Dekker, New York, pp 571–619
Vouk V (1986) Germanium. In: Friberg L, Nordberg GF, Vouk V (eds), Handbook on toxicology of metals. Elsevier, Amsterdam, pp 255–266
Vuori E, Vääriskoski J, Hartikainen H, Kumpulainen J, Aarnio T, Niinivaara K (1994) A log-term study of selenate sorption in Finnish cultivated soils. Agric Ecosyst Environ 48:91–98
Wagner H, Petrova I, Angelow L (2004) Distribution of selenium along the food chain of sheep reared in mountain pasture areas. In: Anke M, Flachowsky G, Kisters K et al. (eds) Workshop Macro and Trace Elements, Jena, pp 424–429
Walker F (1971) Experimental argyria: A model for basement membrane studies. Br J Exp Pathol 52:589–593
Walker WJ, Cronan CS, Bloom PR (1990) Aluminium solubility in organic soil horizons from northern and southern forested watersheds. Soil Sci Soc Am J 54:369–374
Wallace A, Romney EM (1971) Some interaction of Ca, Sr and Ba in plants. Agr J 63:245–247
Wang D, Alfthan G, Aro A (1994a) The impact of selenium fertilization on the distribution of selenium in rivers in Finland. Agric Ecosyt Environ 50:133–149
Wang L, Reddy KJ, Munn LC (1994b) Comparison of ammonium bicarbonate-DTPA, ammonium carbonate, and ammonium oxalate to assess availability of molybdenum in mine spoils and soils. Comm Soi Sci Plant Anal 10:1319–1330
Wang XC, Yan WD, An Z, Lu Q, Shi WM, Cao ZH, Wong MH (2003) Status of trace elements in paddy soil and sediment in Taihu Lake region. Chemosphere 50:707–710
Wang J, Yaming G, Ning H, Wand S (2004) Effects of high fluoride and low iodine on biochemical indexes of the brain and learning-memory of offspring rats. Fluoride 37:201–208
Ward NI, Abou-Shakra FR (1993) Rubidium levels in human tissues and fluids. In: Trace elements in man and animals. TEMA 8, Verlag Media Touristik, Gersdorf, pp 81–85
Wąsowicz W, Reszka E, Gromadzka J (2003) Trace elements (Se, Zn, Cu) and lung cancer risk. 4th Intern Symp Trace Elements in Human: a New Perspective, Part I, Enthypossis, Athens, pp 54–65
Wastras CJ, Bloom NS (1998) Bioaccumulation of mercury in pelagic freshwater food webs. Sci Total Environ 219:183–208
Watanabe ME (1997) Phytoremediation on the brick of commercialization. Environ Sci Technol 31:182A–198A
Watmough SA, Dikinson NM (1995) Dispersal and mobility of heavy metals in relation to tree survival in an aerially contaminated woodland soil. Environ Pollut 90:135–142
Wegenke M, Junge M, Diemer J, Nittka J (2005) Monitoring of antimony in the environment of Bavaria. 1st Intern Workshop Antimony in the Environment, Heidelberg, 45 p
Weidenhaupt A, Arnold C, Müller SR, Haderlein SB, Schwarzenbach RP (1997) Sorption of organotin biocides to mineral surfaces. Environ Sci Technol 31:2603–2609
Weis P, Weis J (1991) The developments toxicity of metals and metalloids in fish. In: Newman MC and McIntosh AW (eds) Metal ecotoxicology: concepts and applications. Lewis Publ, Chelsea, MI, pp 145–169

Welch AH, Lico MS (1998) Factors controlling As and U in shallow groundwater, southern Carson Desert, Nevada. Appl Geochem 13:521–539

Welch RM, Norvell WA (1999) Mechanisms of cadmium uptake, translocation and deposition in plants. In: McLaughlin MJ, Singh BR (eds) Cadmium in soils and plants, 85. Kluwer Publ, pp 125–150

Wells DE, Gee AS, Battarbee RW (1986) Sensitive surface waters – a UK perspective. Water Air Soil Pollut 31:631–668

Wenzel WW, Bunkowski M, Puschenreiter M, Horak O (2004) Rhizosphere characteristics of indigenously growing nickel hyperaccumulator and excluder plants on serpentine soil. Environ Pollut 123:131–138

White P, Bowen H, Broadley M, Hammond J, Hamptom C, Payne K (2003) The mechanism of cesium uptake by plants. Intern Symp Radioecology and Environmental Dosimetry. Rokkasho, Aomori, pp 255–262

WHO (1980) Tin and organotin compounds: A preliminary review. Environ Health Criteria 15, Geneva

WHO (1984) Guidelines for drinking water quality. Recommendations, vol 1. Geneva

WHO (1987) Air Quality Guidelines for Europe, Europe Series 23

WHO (1989) Toxicological evaluation of certain food additives and contaminants. Food Additives, ser 24, Univer Press, Cambridge, pp 116–117

WHO (1992) Cadmium. Environmental Health Criteria, Geneva 134

WHO (1995) Inorganic lead. Environmental Health Criteria 165, Geneva

WHO (1997) Aluminium. Environmental Health Criteria 194, Geneva

WHO (1998a) Boron. Environmental Health Criteria 204. Geneva

WHO (1998b) Copper. Environmental Health Criteria 2000. Geneva

WHO (1999) Summary of evaluations performed by the Joint FAO/WHO/JECFA. Geneva

WHO (2001) Global Environmental Monitoring System, Food, Europe. Report of the 1st Advisory Committee Meeting, Rome

WHO (2004) Guidelines for drinking-water quality. 3rd ed., vol 1, Geneva, pp 488–496

WHO (2005) Revised Air Quality Guidelines for Europe, 2nd ed., URL(http://www.euro.who.int/air/activities/20050223_4)

WHO/IPCS (1991) Platinum. Environmental Health Criteria 125. Geneva

WHO/IPCS (2002) Principles and methods for the assessment of risk from essential trace elements. Environmental Health Criteria 228. Geneva

Wieland E, Stumpf T. Tits J, Fanghänel T (2004) Wet chemistry and TRLFS studies of Eu(III) and Cm(III) uptake by cementious materials. Goldschmidt, Copenhagen, A169 p

Wiener JG, Krabbenhoft DP, Heina GH, Schuhammer AM (2003) Ecotoxicology of mercury. In: Hoffman DJ, Rattner BA, Burton GA, Cairns J (eds) Handbook of ecotoxicology. 2nd ed., Lewis Publ, CRC Press, Boca Raton, FL, pp 409–463

Wiersma D, van Goor BJ, van der Veen NG (1986) Cadmium, lead, mercury, and arsenic concentration in crops and corresponding soils in the Netherlands. J Agric Food Chem 34:1067–1074

Wierzbicka M, Szarek-Łukaszewska G, Grodzińska K (2004) Highly toxic thallium in plants from the vicinity of Olkusz (Poland). Ecotoxicol Environ Safety 59:84–88

Wilber GG, Smith L, Malanchuk JL (1992) Emissions inventory of heavy metals and hydrophobic organics in the Great Lakes basin. In: Schnoor JL (ed) Fate of pesticides and chemicals in the environment. Wiley, New York, pp 27–50

Wild A, Jone LHP (1988) Mineral nutrition of crop plants. In: Wild A (ed) Russell's soil condition and plant growth. Longman Sci. Tech Publ, Harlow, Essex, pp 69–112

Willén E (1991) Planktonic diatoms – an ecological review. Algological Studies 62:69–106

Williams CA, Ward NI, Brodie K (1995) The analysis of gold in plant materials from an operating F by ICP-MS. In: Wilken R-D, Förstner U, Knöchel A (eds) Heavy metal in the environment, vol 2, CEP Consultants Ltd Edinburgh, UK, pp 395–398

Wilson SA, Briggs PK, Mee JS, Siems DF (1994) Determination of thirty-two major and trace elements in three NIST soil SRMs using ICP-AES and WDXRF. Geostandards Newsletter 18:85–89

Wilson AR, Lion LW, Nelson YM, Shuler ML, Ghiorse WC (2001) The effects of pH and surface composition on Pb adsorption to natural freshwater biofilms. Environ Sci Technol 35:3182–3189

Wilson NJ, Craw D, Hunter K (2004) Antimony distribution and environmental mobility at an historic antimony smelter site, New Zealand. Environ Pollut 129:257–266

Winfrey MR, Rudd JWM (1990) Environmental factors affecting the formation of methylation in low pH lakes. Environ Toxicol and Chem 9:853–869

Wingert-Runge B, Andren AW (1994) Desorption behavior of silver from natural sediments under fresh water and marine conditions. 2nd Intern Conf Transport, Fate and Effects of Silver in the Environment. University of Wisconsin, Madison, WI, pp 89–93

Winter-Sorkina de R, Bakker MI, van Donkersgoed G, van Klaveren JD (2003) Dietary intake of heavy metals (cadmium, lead and mercury) by the Dutch population. RIVM report 32010001

Wiśniowska-Kielan B, Klima K (2005) Trace elements in liquid phase of surface run-off as a potential source of the watercourse pollution in a mountain region. Intern Sci Conf Innovation and Utility in the Visegrad fours, Nyiregyhaza, 1, pp 219–226

Witczak S, Adamczyk A (1995) Catalog of selected physical and chemical indicies for ground water contamination, and analytical methods. Bibl Monitor Środowiska, Warszawa (in Polish)

Wittbrodt PR, Palmer CD (1996) Reduction of Cr (VI) by soil humic acids. Eur J Soil Sci 47:151–162

Witting R (1993) General aspects of biomonitoring heavy metals by plants. In: Markert B (ed) Plants as biomonitors. VCH, Weinheim, pp 3–27

WMSY (2004) World Metal Statistics Yearbook, World Bureau of Metal Statistics

Wojciechowska-Mazurek M, Karlowski K, Starska K (1996) Contents of lead, cadmium, copper and zinc in Polish cereal grain, flour and powdered milk. In: Kumpulainen JT (ed) Trace elements, natural antioxidants and contaminants in European foods an diets. FAO Technical workshop, pp 93–101

Wolt JD (1994) Soil solution chemistry. Applications to environmental science and agriculture. Wiley, New York

Wolz S, Fenske RA, Simcox NJ, Palcisko G, Kissel JC (2003) Residential arsenic and lead levels in an agricultural community with a history of lead arsenate use. Environ Research 93:293–300

Wong MH, Zhang ZQ, Wong JWC, Lan CY (1998) Trace metal contents (Al, Cu, and Zn) of tea: tea soil from tea plantation, tea products from different provinces of China. Environ Geochem Health 20:87–94

Wood CM, McDonald DG (1987) The physiology of acid/aluminium stress in trout. Ann de la Societe Royale Zoologique de Belgique 117(1):339–410

Wood CM, Playle RC, Hogstrand C (1999) Physiology and modeling of mechanisms of silver uptake and toxicity in fish. Environ Toxic Chem 18:71–83

Working Group on Mercury (2001) Ambient air pollution by mercury (Hg): Position Paper, pp 1–29

Wren CD (1986) A review of metal accumulation and toxicity in wild mammals. I. Mercury. Environ Res 40:210–244

Wu YJ, Chen YH, Liu RF, Zou ZH, Li CF (2000) Thallium in slag of sulphuric acid is leached under acid rain. Environ Chem 19:447–454 (in Chinese)

Wuyi W. Linsheng Y, Shaofan H, Jian'an T (2003) Mitigation of endemic arsenocosis with selenium: an example from China. In: Skinner HC, Berger AR (eds) Geology and health: closing gap. Oxford Univ Press, Oxford, pp 51–55

Wyness AJ, Parkman RH, Neal C (2003) A summary of boron surface water quality data throughout the European Union. Sci Total Environ 314–316:255–269

Wyszkowska J, Kucharski J (2004) Effects of chromium on multiplication of bacteria in artificial media. J Elemenotology 9:165–174 (in Polish)

Wyttenbach A, Bajo S, Furrer M, Langenauer M, Tobler L (1997) The accumulation of arsenic, bromine and iodine in needles of Norway spruce (*Picea abiens* [L.] Karst.) at sites with low population. Water Air Soil Pollut 94:417–425

Wyttenbach A, Furrer V, Schleppi P, Tobler L (1998) Rare earth elements in soils and in soil-grown plants. Plant Soil 199:267–273

Xianmao L, Guogang H, Huijan W (1990) Effects of molybdenum on etiology, pathogenesis and prevention of esophageal cancer. In: Jian'an T (ed) Environmental life elements and health. Beijing, pp 309–317

Xiao T, Guha J, Boyle D, Liu C-Q, Chen J (2004a) Environmental concerns related to high thallium levels in soils and thallium uptake by plants in southwest Guizhou, China. Sci Total Environ 318:223–244

Xiao T, Guha J, Boyle D, Liu C-Q, Zheng B, Wilson GC et al. (2004b) Naturally occurring thallium: a hidden geoenvironmental health hazard? Environ Int 30:501–507

Xie ZM, Huang CY (1994) Relationships between lead zinc and arsenic contents and rice tillering in soil-rice system. J Zhejiang Agric Univ 20:67–71

Xie WB, Chen YH, Chen SL, Bi XZ, Fan XY (2000) Study on thallium distribuion in pyrite cinders and its environmental effect. Bull Mineral Petrol Geochem 19:204–206 (in Chinese)

Xijie L, Zian F, Jinxing D, Quancheng L (1986) Studies on the application of rare earth elements on the narcissus. J Fuijan Agric College 15:246–251

Xu JM, Wang K, Bell RW, Yang YA, Huang LB (2001) Soil boron fraction and their relationship to soil properties. Soil Sci Soc Am 65:133–138

Xu X, Zhu W, Wang Z, Witkamp G-J (2002) Distribution of rare earth and heavy metals in field-grown maize after application of rare earth-containing fertilizer. Sci Tot Environ 293:97–105

Xue T, Hartikainen H (2000) Association of antioxidative enzymes with the synergic effect of selenium and UV irradiation in enhancing plant growth. Agric Food Sci Finland 9:177–186

Yamada M, Zheng J, Aono T (2004) Temporal variation of $^{239+240}$Pu profiles in the water columns of the North Pacific. Goldschmidt, Copenhagen, A490 p

Yamamoto T, Tamaoka T, Fujita T, Isoda C (1973) Boron content in marine plankton. Res Oceanogr Works Jpn 12:13–21

Yan R, Gauthier D, Flamant G (2001) Partitioning of trace elements in the flue gas from coal combustion. Combust Flame 125:1660–1661

Yanagisiva K, Muramatsu Y (1993) Transfer factor of technetium from soil to vegetable. Radio-chimica Acta 63:83–86

Yanagisava K, Muramatsu Y (1995) Transfer of technetium from soil to paddy and upland rice. J Radiation Res 36:171–178

Yang C, Chen Y, Peng P, Li C, Chag X, Xie C (2005) Distribution of natural and anthropogenic thallium in the soils in an industrial pyrite slag disposing area. Sci Total Environ 341:159–172

Yeo AR, Flowers SA, Rao G, Welfare K, Senanayake N, Flowers TJ (1999) Silicon reduces sodium uptake in rice (*Oryza sativa* L.) in saline conditions and this is accounted for a reduction in the transpirational bypass flow. Plant Cell Environ 22:559–565

Yi W, Halliday AN, Lee DC, Christensen JN (1995) Indium and tin in basalts, sulfides and the mantle. Geochim Cosmochim Acta 59:5081–5090

Yin Y, Allen E, Huang CP, Sanders PF (1997) Adsorption/desorption isotherms of Hg (II) by soil. Soil Sci 162:35–45

Yokel RA (2004) Aluminum. In: Merian E, Anke M, Ihnat M, Stoeppler M (eds) Elements and their compounds in the environment. 2 ed., Wiley-VCH, Weinheim, pp 635–658

Yoshida S, Muramatsu Y (1994) Radiocesium concentrations in mushrooms collected in Japan. J Environ Radioactivity 22:141–154

Yoshida S, Muramatsu Y (1995) Determination of organic, inorganic and particulate iodine in the coastal atmosphere of Japan. J Radioanal Nuclear Chem 1996:295–302

Yoshida S, Muramatsu Y (1997) Determination of major and trace elements in mushroom, plant, and soil samples collected from Japanese forest. Intern J Environ Anal Chem 6:49–58

Yoshida S, Muramatsu Y, Tagami K, Uchida S (1998) Concentrations of lanthanide elements, Th and U in 77 Japanese surface soils. Environ Intern 24:275–286

Yoshida S, Muramatsu Y, Linkov I, Dvornik A, Zhuchenko T (2003) Equilibrium of radiocesium with stable cesium within the biological cycle of contaminated forest ecosystem. Intern Symp. Radioecology and environmental dosimetry. Rokkasho. Aomori, pp 228–235

Yost KJ (1986) Cadmium, the environment and human health: an overview. In: Mislin H, Ravera O (eds) Cadmium in the environment. Birkhäuser Verlag, Stuttgart, pp 137–144

Ysart G, Miller P, Crews H, Robb P, Baxter M, De L'Argy Ch, Sargen Ch, Harrison N (1999) Dietary exposure estimates of 30 elements from UK Total Diet Study. Food Add Contam 16:391–403

Yudovich YE, Ketris MP (2005) Arsenic in coal: A review. Intern J Coal Geol 61:141–196

Yuita K (1983) Iodine, bromine and chlorine in soils and plants of Japan. Soil Sci Plant Nutr 9:403–407

Zachara BA, Pileexi A (2000) Selenium concentration in the milk of breast-feeding mothers and its geographic distribution. Environ Health Persp 108:1043–1046

Zaman K, Pardini RS (1996) An overview of the relationship between oxidative stress and mercury and arsenic. Toxic Subst Mechanisms 15:151–181

Zayed AM, Terry N (1994) Selenium volatilization in roots and shoots: effects of shoot removal and sulfate level. J Plant Physiol 143:8–14

Zayed AM, Terry N (2003) Chromium in the environment: factors affecting biological remediation. Plant Soil 249:139–156

Zayed AM, Lytle CM, Qian J-H, Terry N (1998) Chromium accumulation, translocation and chemical speciation in vegetable crops. Planta 206:239–299

Zduńska H, Bloch-Bogusławska E, Przygońska J, Śliwka K, Zachara BA (1994) Selenium concentration and glutathione peroxidase activity in the organs of people died a violent death, In: Kabata-Pendias A, Szteke B (eds) Arsenic and selenium in the environment. Ecologic analyt problem. Polish Ac Sci, pp 190–193 (in Polish)

Zereini F, Alt F (eds) (2006) Palladium emissions in the environment: analytical methods, environmnetal assessment and health effects. Springer-Verlag, Berlin Heidelberg

Zereini F, Alt F, Rankenbur K, Beyer J-M, Altelt S (1997) The distribution of the platinum group elements (PGE) in the environmental compartments of soil, mud, roadside dust, road sweepings and water. Emission of the platinum group elements (PGE) from motor vehicle catalytic converters. Umweltwiss Schadstoff Forsch 9:193–200 (in German)

Zereini F, Skerstupp B, Rankenburg K, Dirksen F, Beyer JM, Torsten C, Urban H (2001) Anthropogenic emission of platinum-group elements into the environment. Concentration, distribution and geochemical behavior in soils. J Soil Sedim 1:444–449

Zevenhoven R, Mukherjee AB, Bhattacharya P (2006) Arsenic flows in the environment of the European Union: A synoptic review In: Bhattacharya P, Mukherjee AB., Zevenhoven R, Loepper RH (eds) Arsenic in soil and groundwater: biogeochemical interactions. Elsevier Book Series (in press)

Zevenhoven-Onderwater M (2001) Ash-forming matter in biomass fuels. Ph.D. Thesis, Faculty of Chemical Engineering, Åbo Akademi, Turku, Finland

Zhang Y, Frankenberger, Jr WT (2003) Removal of selenate in simulated agricultural drainage water by rice straw bioreactor channel system. J Environ Qual 32:1650–1657

Zhang Z, Zhang B, Long J, Zhang X, Chen G (1998) Thallium pollution associated with mining of thallium deposits. Sci China, Ser D 41:75–81

Zhang P-C, Krumhansi JL, Brady PC (2002) Introduction to properties, sources and characteristics of soil radionuclides. In: Zhang P-C, Brady PV (eds) Geochemistry of soil radionuclides, SSSA Spec Publ, Madison WI, pp 1–20

Zhang GY, Dou Z, Toth JD, Ferguson J (2004a) Use of fly ash as environmental and agronomic amendments. Environ Geochem Health 26:129–134

Zhang MK, He ZL, Calvert DV, Stoffella PJ (2004b) Leaching of minerals and heavy metals from muck-amended sandy soil columns. Soil Sci 169:528–540

Zhao FJ, Hamoin RE, McLaughlin MJ (2001) Root exudates of the hyperaccumulator *Thlaspi caerulescens* do not enhance metal mobilization. New Phytologist 151:613–621

Zhou LX, Wong JWC (2003) Behavior of heavy metals in soils: effect of dissolved organic matter. In: Selim HM, Kingery WL (eds) Geochemical and hydrological reactivity of heavy meals in soils. Lewis Publ, Chelse, MI, pp 245–269

Zhu J, Zhong G, Kennedy M, de Leer EWB, Bosman R, Kong L, Alaerta GJFR (1997) The distribution of rare earth elements (REEs) in Chinese soils. 4th Intern Conf Biogeochem Trace Elements, Berkeley, CA, pp 31–302

Zhu Y-G, Liu W-J, Chen Z, Geng C-N (2005) The role of iron in controlling the dynamics of arsenic in the rhizosphere of rice plants. ICOBTE, 8th Intern Conf, Adelaide, pp 208–209

Zhuk LI, Mikholskaya IN, Danilova EA, Kist AE (1994) Mapping using human blood composition data. Biol Trace Elem Res 43–45:371–381

Ziegler TI, Divine KK, Goering PL (2004) Galium. In: Merian E, Anke M, Ihnat M, Stoeppler M (eds) Elements and their compounds in the environment. 2nd ed., Wiley-VCH, Weinheim, pp 775–793

Zielecka M (1996) Silicon-universal element. Part I & II.Wiadomości Zielarskie 6:1–2; 9; 6–7 (in Polish)

Zimdahl RL, Koeppe DE (1979) Uptake by plants. In: Boggess WR, Wixson BG (eds) Lead in the environment. Castle House Publ, Washington, DC, pp 99–104

Zonn SV (1982) Iron in soils. Publ House, Nauka, Moscow (in Russian)

Żmudzki J, Szkoda J (2000) Cadmium in animal tissues and food of animal origin. In: Kabata-Pendias A, Szteke B (eds) Cadmium in the environment. Ecological and Analytical Problems, Polish Acad Sci, Warszawa, pp 381–389 (in Polish)

Index

A

B

D

E

H

J

K

M

Q

R

U

FSC
www.fsc.org
MIX
Papier | Fördert
gute Waldnutzung
FSC® C083411

Kolibri 360 GmbH
Ferdinand-Jühlke-Straße 7
99095 Erfurt, Deutschland
produktsicherheit@kolibri360.de